ADHESION SCIENCE AND ENGINEERING – 1

Series Editor

A.V. Pocius

ADHESION SCIENCE AND ENGINEERING – 1

THE MECHANICS OF ADHESION

Edited by

D.A. Dillard

Virginia Tech
Department of Engineering Science and Mechanics
120 Patton Hall
Blackburg, VA 24061
U.S.A.

A.V. Pocius

3M Adhesive Technologies Center
3M Center, Building 201-4N-01
St. Paul, MN 55144-1000
U.S.A.

2002

ELSEVIER

Amsterdam – Boston – London – New York – Oxford – Paris – San Diego
San Francisco – Singapore – Sydney – Tokyo

ELSEVIER SCIENCE B.V.
Sara Burgerhartstraat 25
P.O. Box 211, 1000 AE Amsterdam, The Netherlands

First edition 2002

Library of Congress Cataloging in Publication Data
A catalog record from the Library of Congress has been applied for.

British Library Cataloguing in Publication Data
A catalogue record from the British Library has been applied for.

ISBN: 0 444 51140 7 (set)

⊗ The paper used in this publication meets the requirements of ANSI/NISO
Printed in The Netherlands.

Preface

Adhesion science and technology is inherently an interdisciplinary field, requiring fundamental understanding of mechanics, surfaces, and materials, the topics emphasized in this Adhesion Science and Engineering series. This volume focuses attention on the contribution of mechanics principles and solutions to understanding the fabrication, design, analysis, and testing of adhesive bonds. Building on the fundamentals laid by such noted mechanicians as Winkler, Timoshenko, Volkersen, Goland, Reissner, Williams, Gent, and Johnson, this volume offers a comprehensive overview of the current understanding of stresses, deformation, and fracture parameters associated with a range of adhesive bonds.

Starting with a background and introduction to stress transfer principles (Chapter 1), fracture mechanics and singularities (Chapter 2), and an energy approach to debonding (Chapter 3), the volume continues with analysis of structural lap (Chapter 4) and butt (Chapter 5) joint configurations. The volume continues with discussions of test methods for strength and constitutive properties (Chapter 6), fracture (Chapter 7), and peel (Chapter 8). Chapter 9 covers coatings, the case of adhesion to a single substrate, and Chapter 10 addresses elastomeric adhesives such as sealants. The role of mechanics in determining the locus of failure in bonded joints is discussed (Chapter 11), followed by a chapter on rheology relevant to adhesives and sealants (Chapter 12). Pressure sensitive adhesive performance (Chapter 13), the principles of tack and tack measurements (Chapter 14), and contact mechanics relevant to wetting and surface energy measurements (Chapter 15) are then covered. The volume concludes with sections on fiber-matrix bonding and reinforcement (Chapter 16), durability considerations for adhesive bonds (Chapter 17), ultrasonic non-destructive evaluation of adhesive bonds (Chapter 18), and design of adhesive bonds from a strength perspective (Chapter 19).

The references cited in the chapters herein represent only a portion of the intense effort and contributions made by the mechanics community working in adhesion science and technology. Through compiling these important topics covered by the respective chapters, along with the surfaces and materials issues addressed in the companion volumes, the editors hope to address the need for a current and comprehensive series of books that provides an overview of this broad and interdisciplinary field. As the scientific and engineering community comes to a better understanding of the mechanics, surface, and materials issues, improved models of representing complex bonded systems will develop. Technical challenges remain for understanding how properties of the adhesive and adherends affect the performance and durability of bonded joints and structures, how designers can reliably utilize adhesive bonds in a variety of service conditions, and how mechanics' insights can efficiently be translated into improving surfaces and adhesives. One can only wonder what new science will be developed as we move into the nanomechanics world, where the applicability of continuum mechanics becomes questionable as atomic scales are approached.

The editors of this volume would like to express sincere appreciation to each of the authors for their invaluable contributions to this volume. Their collective expertise represents many years of industrial and academic experience in the field of adhesion science and technology, and we are very grateful for the time and effort they have devoted to the task. We would also like to thank the employers of each of the contributors for allowing them to take on the added responsibilities associated with this volume. We would like to thank Mrs. Shelia L. Collins for her diligence in contacting authors, maintaining files, and assembling materials prior to publication. Assistance from the Engineering Science and Mechanics Department and Center for Adhesive and Sealant Science at Virginia Tech is specifically acknowledged. Finally, we would like to thank our wives for their patience as we compiled and edited the volume.

ALPHONSUS V. POCIUS   DAVID A. DILLARD
Editor   *Associate Editor*

Corporate Scientist   *Professor of Engineering Science and Mechanics*
3M Company   *Director, Center for Adhesive and Sealant Science*
St. Paul, MN, USA   Virginia Polytechnic Institute and State University
                             Blacksburg, VA, USA

Contents

Chapter 1

Fundamentals of stress transfer in bonded systems

DAVID A. DILLARD [*]

Engineering Science and Mechanics Department, Virginia Polytechnic Institute and State University, Blacksburg, VA 24061, USA

1. Introduction

Since before recorded history, mankind has been joining materials together to produce useful items. To increase effectiveness and efficiency, many prehistoric as well as modern devices required the assembly of several components, often involving dissimilar materials. Stone points retained their sharpness and provided mass for arrows, whose wooden shafts provided lightweight strength and stiffness, which in turn were outfitted with feathers mounted at the tail to maintain stability in flight. Woven baskets could hold water when sealed with tar, pitch, or other naturally occurring resins. Whether lashing with natural fibers, or sealing with resins or gums, mankind has, from the earliest times, been involved in joining various materials. Over time, the sophistication of our joining methods has increased to include a wide variety of mechanical fasteners, numerous welding methods, and the use of adhesives, sealants, mortars, cements, and other binders to hold components together. Joining offers us the ability to have structures much larger than could be made or transported as a single entity. By combining multiple materials, the resulting structure acquires useful features of each constituent, often making the whole greater than the sum of the parts. Joining allows us to fabricate efficient, lightweight, open structures with tailored properties and performance matched to the intended use. Indeed, even life in the Stone Age would have been very different without the use of a variety of joining methods.

The use of functional joining, however, predates human contributions, having been widely employed throughout the created world. Nature is replete with a variety of applications of adhesion mechanisms and, in some cases, specific

---

[*] E-mail: dillard@vt.edu

adhesives. Certain types of rocks and minerals have taken on their form through adhesion forces acting among the individual grains, e.g. the sandstones and limestones that are widely used to make buildings and other structures. Life itself would be impossible without the preferential adsorption on organic surfaces. The shells of crustaceans involve intricate assemblies of weak inorganic materials with organic binders to produce amazingly strong and effective shells. Organic adhesives are of particular interest in this series of volumes and several biological systems are worth noting where organic adhesives are specifically used. One of the most obvious and most widely studied are the secretions of crustaceans such as barnacles and mussels, which allow attachment to almost any substrate in harsh marine environments. Scientists continue to be fascinated by the excellent adhesive characteristics of this substance, and the tenacity and durability of these bonds formed under very unfavorable conditions. In addition to tests conducted to measure the structural performance of such materials [1], a considerable amount of study has been devoted to isolating and synthesizing these materials for possible commercialization. Other examples of natural adhesives include: the tortoise beetle (*Hemisphaerota cyanea*), which apparently secretes a substance from its feet when disturbed, making it very difficult to remove from the object on which it stands; lignin which serves as an adhesive binding cellulose/hemicellulose fibrils into one of the most widely used structural materials, wood; and the list goes on and on.

Although mankind's early joining methods employed the use of natural materials, the actual methods of joining were quite distinct from the manner in which natural creatures are held together. Lashing, stitching, pinning, and nailing all continue to be used, as do the use of a wide variety of screws, bolts, and rivets that are more recent inventions. These discrete mechanical fastening systems, although relatively simple to implement, have few counterparts in the natural world. Instead, natural creatures are much more likely to be held together by continuous adhesive layers. These amazing natural structures are able to withstand tremendous forces, harsh environments, and at the same time allow for growth and repair. Their unique properties come from both the optimal molecular [2] and micromechanics [3] designs. We continue to marvel at the complexity and efficiency of joining in the natural world, and are beginning to mimic these bonded systems. Increasingly complex and sophisticated structures are being fabricated as we learn from this growing field of biomimetics [4].

Mechanical fasteners such as bolts, screws, rivets, and nails have been widely and successfully employed in building the manmade world around us. A number of advantages continue to make these appropriate joining techniques in certain instances. Using them often requires no surface preparation, although drilling is needed in most cases. Unlike many adhesives, mechanical fasteners have a very long shelf life, generally have less environmental concerns, and may facilitate repair because they can often be removed and reinstalled with little or no damage to the joined components. Mechanical fasteners facilitate inspection; a loose or missing rivet may be easily seen and repaired. However, mechanical fasteners and welds are not practical in many situations. One of the key factors is simply that drilling a hole induces stress concentrations that weaken the components to be joined. In fact, the components may need to be made thicker simply to withstand the higher stresses imposed by holes, especially loaded holes associated with load bearing mechanical fasteners [5]. Adhesive bonding is becoming an increasingly viable alternative for joining materials for structural, non-structural, and semi-structural applications.

In designing and fabricating modern structures, the decisions whether to use adhesives, mechanical fasteners, some type of welding, or some combination of these methods often fall to the engineers involved in the design process. A number of factors should be considered when making such decisions. From a feasibility standpoint, adhesives are often the joining method of choice where thin, flexible, or dissimilar adherends are involved. Certain brittle or damage-prone adherends are difficult to drill for traditional mechanical fasteners. From a performance standpoint, adhesives may offer certain advantages in that they may eliminate stress concentrations associated with mechanical fasteners in loaded holes, prevent loss in temper associated with welding processes, and provide weight savings that can prove to be quite significant due to the snowballing effect weight has on lightweight structures. Continuous beads of adhesives can significantly stiffen structures when compared to discrete mechanical fasteners or welds, and viscoelastic polymeric adhesives also offer damping capabilities. This increase in stiffness and damping reduces noise, vibration, and harshness (NVH) and leads to quieter, better performing automobiles.

Although they may be used alone, adhesives in such applications are also used to augment mechanical fasteners or spotwelds. Weldbonding involves using adhesives and spotwelding in conjunction, and offers a number of significant advantages in the automotive industry. By eliminating the stress concentrations around loaded holes, many adhesively bonded joints may exhibit improved fatigue performance over other joining methods, although this is dependent on the particular system and service environment. From a fabrication standpoint, adhesives can be cost-effective in many situations by eliminating drilling and reducing handwork. However, the need for proper surface preparation, jigs during curing, careful curing procedures, and other factors can offset the benefits. Furthermore, adhesives may be less robust than mechanical fasteners because of the need to store and use them properly. From a durability standpoint, adhesives can effectively seal joints, keeping water out of the bondline. In other situations, however, adhesive durability in the presence of excessive heat or cold, water, organic solvents, or other media can be a significant limitation. Such exposure can degrade the adhesive properties, deteriorate the interface, or both. Non-destructive evaluation as discussed in Chapter 18 can play an important role in identifying weak, degraded, or missing bonds for some applications, offering better reliability in certain critical structures. Solvent-free and waterborne adhesives are more environmentally friendly than some of their predecessors. However, the difficulty in disassembling adhesively bonded joints may contribute to waste in our throw-away society. Overall, we see that many factors may influence the decisions for how to join various materials to build modern structures and assemblies.

This chapter will review some of the key concepts relevant for understanding how loads are transferred from one adherend to another through an adhesive layer. The resulting stress fields are often quite complex, being highly non-uniform within the adhesive layer and often involving both normal and shear stresses acting in several directions. These complex, non-uniform, triaxial stress states serve to remind us that tests of an adhesive joint are really tests of a *structure*, requiring a more involved analysis. This is in sharp contrast with tests of a material using, for example, a uniaxial dogbone specimen, where the stress field is quite uniform except near the gripped ends. Thus even the simplest adhesive joints are found to be relatively complex structures involving highly non-uniform stress states. An understanding of these complexities is important in selecting, conducting, and interpreting adhesion tests, in developing meaningful design criteria, and in designing and analyzing bonded structures. In addition to the chapters of this volume, a number of books are recommended for their treatment of the mechanics of bonded joints, providing additional details and insights into many of the concepts introduced herein [1,6–12].

2. Applications of mechanics to bonded systems

A fundamental understanding of adhesively bonded joints requires an understanding of the substrates and surfaces being bonded, the behavior of the adhesive, and also the mechanics principles that govern the stress and strain states within the bonded joint. Before beginning our discussion of how mechanics can be utilized to understand adhesive joints, we will first briefly review the concepts of stress and strain that will be used throughout this volume. Gordon has written an interesting book for the layperson that provides a nice introduction to many of these mechanics concepts and applications [13].

2.1. Definitions of stress and strain

Stress is defined as the force acting over a given area divided by that area, and represents the force carried per unit area by the material. Stresses are normally broken into normal and shear components acting perpendicular and parallel to the surface of interest, respectively. The average normal stress acting over an area is thus given by the normal component of the force divided by the area. Although average stress is of importance in certain cases, one is often interested in the local stress state and how it may vary over the surface. This localized stress or stress at a point is defined as one takes a vanishingly small area

$$\sigma = \lim_{\Delta A \to 0} \frac{\Delta F_n}{\Delta A}, \qquad \tau = \lim_{\Delta A \to 0} \frac{\Delta F_t}{\Delta A}, \tag{1}$$

where σ and τ represent the normal and shear stresses, respectively, ΔF_n is the component of force acting normal or perpendicular to the surface, ΔF_t is the tangential component of the force, and ΔA is the increment of area as shown in Fig. 1a.

A more general notation recognizes that stresses are second-order tensors, meaning that they have magnitude and two directions associated with them. Stresses are often listed with two subscripts representing these two directions, viz. σ_{ij}; the first subscript refers to the outward normal of the plane on which the stress acts, and the second subscript indicates the direction in which the increment of force is acting. Here, i and j are indices that may denote any of the three mutually orthogonal directions of a coordinate system. When the increment of force acts perpendicular to the plane, the indices i and j are the same and denote a normal stress. For components of force acting tangential or parallel to the plane, the indices refer to perpendicular directions, resulting in a shear stress. From moment equilibrium, one can easily show that $\sigma_{ij} = \sigma_{ji}$, implying that the stress tensor is symmetric. Although the tensorial notation uses the σ_{ij} form for both normal and shear stresses, common engineering usage often denotes normal stresses with σ and shear stresses with τ. When using these forms, a single subscript is sufficient for the normal stress, since it is understood that the outward normal of the plane and the direction of the stress must coincide for normal stresses. Shear stresses must continue to have two subscripts to avoid ambiguity about the face and direction in which the stress acts. Both notations may be encountered in different chapters within this volume. Fig. 1b illustrates the stresses on a plane element, using both the tensorial and engineering notations that are commonly encountered. Fig. 1c illustrates the general three-dimensional stress state that exists within many loaded structures, including adhesive bonds.

Strain is defined as the deformation divided by the distance over which this deformation takes place, and is a dimensionless quantity. Normal strains, often denoted by ε, represent deformations that are measured in the same direction as the original length, as shown in Fig. 2.

$$\varepsilon = \frac{\Delta L_n}{L}. \tag{2}$$

Taking the limit of this, as the original length, L, goes to zero, provides the engineering normal strain at a point. Engineering shear strains are given as angles

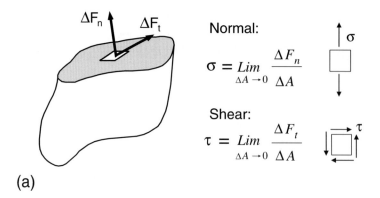

Normal:

$$\sigma = \underset{\Delta A \to 0}{Lim} \frac{\Delta F_n}{\Delta A}$$

Shear:

$$\tau = \underset{\Delta A \to 0}{Lim} \frac{\Delta F_t}{\Delta A}$$

(a)

σ_{ij} ⎡ i denotes plane on which stress acts
⎣ j denotes direction stress acts in

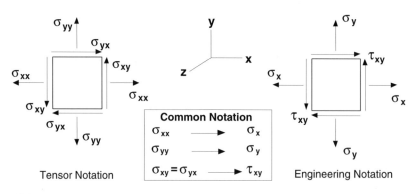

Common Notation

σ_{xx}	→	σ_x
σ_{yy}	→	σ_y
$\sigma_{xy} = \sigma_{yx}$	→	τ_{xy}

Tensor Notation

Engineering Notation

(b)

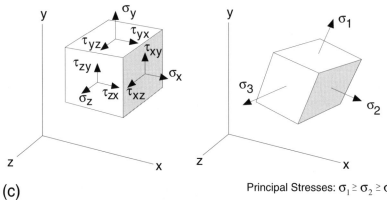

(c)

Principal Stresses: $\sigma_1 \geq \sigma_2 \geq \sigma_3$

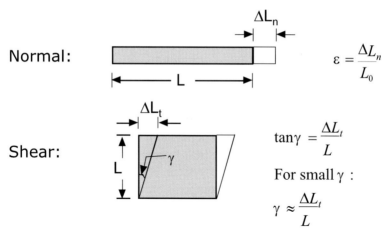

Fig. 2. Definitions of (engineering) normal and shear strains.

measured in (dimensionless) radians, where

$$\tan \gamma = \frac{\Delta L_t}{L}$$

and ΔL_t is the component of deformation perpendicular to the original length. Thus γ is a measure of the angular distortion from a right angle. For small shear strains,

$$\gamma \approx \frac{\Delta L_t}{L},$$

taking on the same form as the normal strain. A limit as the length, L, becomes infinitesimally small is used to obtain shear strain at a point.

Like stresses, strains are also second-order tensors, so may be given with two subscripts indicating the plane over which the strain occurs, and the direction in which the strain occurs, respectively. Provided the deformations are small, tensorial strains are given by

$$\varepsilon_{ij} = \frac{1}{2} \left(\frac{\partial u_i}{\partial x_j} + \frac{\partial u_j}{\partial x_i} \right), \tag{3}$$

where x_i is the ith coordinate axis and u_i is the displacement in the ith direction. Other strain definitions are often used for large deformations and for other situations, as are given in standard continuum mechanics texts [14].

Fig. 1. (a) Definitions for normal and shear stresses at a point in terms of the normal and tangential components of the force increment acting over the infinitesimal area, ΔA. (b) Two-dimensional representation of tensorial and engineering notations for stresses acting on a plane stress element. (c) General three-dimensional stress state and corresponding stress state in principal planes.

Although normal and shear stresses are the same in both tensorial and engineering stress definitions, for strains, the tensorial shear strain components ε_{ij}, where $i \neq j$, represent half of the corresponding engineering strains ($\varepsilon_{ij} = \gamma_{ij}/2$). Normal strains, however, are the same in both systems.

Although the definitions of normal and shear stresses are closely related, as are normal and shear strains, the distinction between normal and shear might suggest that these are separate and distinct quantities. In fact, the magnitude of these normal and shear quantities is dependent on the reference coordinate system chosen. Thus, a pure uniaxial normal stress becomes a combination of normal and shear stresses in a rotated coordinate system; a pure shear stress state becomes pure normal stresses in tension and compression when one rotates the coordinate system by 45°. Fig. 3a illustrates these two examples. Second-order tensors such as stress and strain can all be transformed by certain coordinate transformation equations, or by a graphical procedure known as Mohr's circle [15], also illustrated in Fig. 3a. A simple application of this stress transformation concept is illustrated in Fig. 3b, where two members of cross-sectional area, A_0, are bonded together across a bond plane that may be oriented at different angles with respect to the axial load axis. This plot shows how the average normal stress (perpendicular to the bond plane) and average shear stress (with respect to the bond plane) vary as the bevel angle is changed. A nominal uniaxial stress state is resolved into normal and shear stresses that vary in magnitude as the orientation of the coordinate system is rotated, and the bond area changes accordingly.

One property of second-order tensors such as stress and strain is that one can identify certain principal planes on which extreme values of the magnitudes occur. In general, complex, three-dimensional stress states such as shown in Fig. 1c can be resolved into principal stresses as shown in the same figure. Maximum and minimum normal stresses occur on these principal planes where shear stresses vanish. These principal stresses (or strains) are of significant importance in several failure criteria for homogeneous, isotropic materials [16], and may be important in the failure of adhesives as well [17]. Because of the natural planes associated with the bond plane, however, normal and shear stresses acting on these bond planes are often examined and reported in adhesion-related literature.

The stress and strain measures listed above are based on the original area and original length, respectively. These simple quantities are known as engineering stress and engineering strain. Several other forms of stress and strain are often encountered, especially when deformations are large. For such situations, the final area or length may be significantly different than the original area or length, and more sophisticated treatments are warranted. The interested reader may consult texts on continuum mechanics for alternate definitions [14].

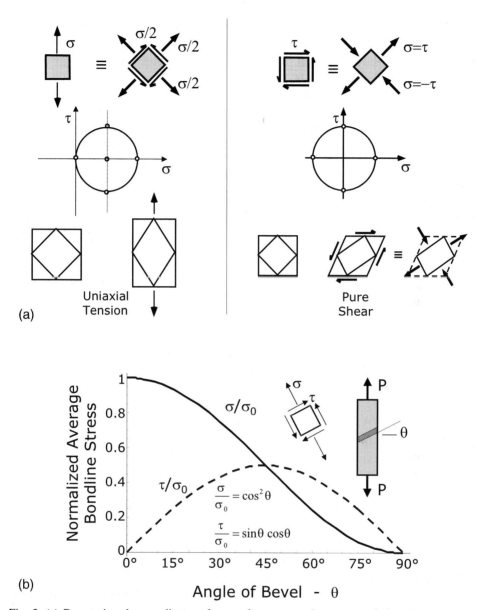

Fig. 3. (a) By rotating the coordinate reference frame, normal stresses and shear stresses vary, obeying Mohr's circle, a graphical transformation technique for analyzing second-order tensors. Here are illustrations for uniaxial tension and pure shear. (b) Illustration of how average normal and shear stresses across a bondline vary with angle in a beveled joint. Results are normalized with respect to the nominal stress, $\sigma_0 = P/A_0$, the load divided by the cross-sectional area of the adherends.

2.2. Constitutive relations between stress and strain

Stresses and strains are related through constitutive equations; for linear elastic, homogeneous, isotropic materials, these relations include

$$\varepsilon_x = \frac{1}{E}\left[\sigma_x - v\left(\sigma_y + \sigma_z\right)\right],$$

$$\varepsilon_y = \frac{1}{E}\left[\sigma_y - v(\sigma_x + \sigma_z)\right],$$

$$\varepsilon_z = \frac{1}{E}\left[\sigma_z - v\left(\sigma_x + \sigma_y\right)\right],$$

$$\gamma_{xy} = \frac{\tau_{xy}}{G}, \qquad \gamma_{yz} = \frac{\tau_{yz}}{G}, \qquad \gamma_{xz} = \frac{\tau_{xz}}{G},$$

$$\varepsilon_x + \varepsilon_y + \varepsilon_z = \frac{\sigma_x + \sigma_y + \sigma_z}{K}, \tag{4}$$

where E is Young's modulus or modulus of elasticity, G is the shear modulus or modulus of rigidity, K is the bulk modulus, and v is Poisson's ratio. The simple Hooke's law relationship, $\sigma = E\varepsilon$, is only valid for uniaxial loading; more general stress states require the use of the generalized Hooke's law relationships above. For linear elastic, isotropic materials, only two constitutive properties are independent. For example, the shear and bulk moduli can be given in terms of E and v as

$$G = \frac{E}{2(1+v)}, \qquad K = \frac{E}{3(1-2v)}. \tag{5}$$

Much of the treatment contained within this volume is limited by the assumptions that the adhesives, and usually the adherends, are linear elastic, homogenous, and isotropic. For bulk adhesives, the assumption of isotropy is usually justified, although instances do arise where preferred orientation of filler particles or crystalline regions can lead to anisotropic behavior. Common adherends such as fiber-reinforced composites, wood, and cold-drawn metals often exhibit anisotropic behavior that can significantly affect joint behavior.

Homogeneity may be appropriate in a global sense, although local inhomogeneities may exist for several reasons. Most practical adhesives and sealants are filled with a variety of polymeric, metallic, or inorganic fillers. Phase-separated rubber or thermoplastic regions are often used to toughen thermosetting resins. Metallic fillers may be added to achieve electrical or thermal conductivity, or provide desired magnetic behavior. A wide variety of metal oxides, clays, and other inorganic fillers are often added to improve strength, stiffness, toughness, or abrasion resistance, or to reduce cost of the material. Filler particles may be well dispersed, but often aggregate into complex clusters and dispersions. If the filler particles are small compared to other geometric features of the bond, and are well dispersed, the global behavior of the adhesive is often assumed to be homogeneous. Locally, however, filler particles induce a complex state of stress that can significantly alter the performance.

Local anisotropy and inhomogeneity may exist in the interphase [18], that complex region between the bulk adhesive and substrate in a bonded joint. Here molecular alignment, variations in stoichiometry, molecular weight gradients due to entropic effects, transcrystalline growth, and other factors may alter the chemical composition and affected mechanical properties. Although such variations can be very important in the performance of bonded systems, they are seldom considered in classical mechanics treatments of adhesive joints. Intense recent efforts in graded interfaces, however, are leading to significant progress in understanding some of these challenging areas.

The constitutive relationships given above were based on linear elastic behavior. Although certain brittle thermosets or highly filled systems may exhibit very limited ductility, most fully formulated, commercially available adhesive products have been designed to provide at least a modest amount of plastic flow when loaded sufficiently. Yielding and plastic deformation are important in improving the toughness of a material. This inelastic behavior is important in relieving stress concentrations at corners and debond tips, and improving fracture toughness, impact performance, and usually fatigue resistance. Such materials are more forgiving, and tend to produce more durable bonded structures for systems that fail by debond or crack propagation. Although this ductility is often very beneficial, nonlinear behavior can be quite complex to analyze. Finite element techniques are often needed to address such material complexities in a meaningful way. In some limited situations, however, closed form solutions are possible. For example, Hart-Smith's classical analysis of adhesive plasticity in lap joints was based on elastic–perfectly plastic material behavior [19]; Williams has used bilinear models to predict bending behavior of peeled adherends [20]; and Thouless and coworkers have used power law hardening models to predict adherend bending of impact specimens [21]. Liechti et al. [22] have made extensive use of Ramburg–Osgood models for adhesive behavior in their finite element representations of a variety of adhesive bonds to accurately represent adhesive behavior.

When dealing with polymeric materials, in particular, time-dependence is often of considerable importance. Viscoelastic behavior of the polymer gives rise to a variety of observed behaviors of adhesively bonded joints. Changes in adhesive modulus with time cause load sharing changes [23,24] that exhibit a delay time followed by creep and rupture for lap joints [25,26]. Time-dependent deformation at a debond tip gives rise to the common stick-slip phenomenon in bonded joint failure, and results in significant rate-dependent fracture energies. Viscoelastic flow can alter the propensity of a pressure-sensitive tape to flag [27]. Chapter 12 will provide a mathematical description of various viscoelastic representations, along with applications to flow of viscoelastic fluids relevant to fabricating bonded joints. The reader may refer to books that cover viscoelastic analysis methods

[28–31] for details on solving viscoelastic boundary value problems involving the correspondence principle and other approaches. Chapters 13 and 14 will cover some of the effects of time-dependence in pressure-sensitive adhesives, and Chapter 17 will examine joint durability.

2.3. Determining stress and strain fields in boundary value problems

Boundary value problems refer here to objects subjected to loads (tractions) and displacements prescribed along the boundary. Mechanics solutions for boundary value problems can be obtained through several approaches. The simplest involves solutions for stresses, strains, displacements, and energies based on the assumptions made in the elementary mechanics of materials classes taught at the undergraduate level in engineering curricula. Although lacking in some detail, these so-called mechanics of materials solutions have served as the basis for most engineering designs until fairly recent times. These closed-form solutions may not exactly satisfy all equilibrium or kinematic requirements, but in many situations have been sufficiently accurate for practical design procedures. The solutions presented within this chapter all fall within this classification of mechanics of materials solutions. Exact solutions, based on elasticity theory [32], began flourishing in the 19th century. As the name implies, the solutions exactly satisfy all equilibrium and kinematic equations, at least in some sense. For example, using the principles of elasticity theory may allow one to construct an 'exact' solution valid in two dimensions, although the use of this solution for structures of finite thickness again leads to some deviation from satisfying all equilibrium and kinematic relations. While these principles provide powerful tools for solving a variety of problems, they are often difficult to apply to the complex geometries, loading scenarios, and multiple material structures that are commonly encountered in modern engineering design.

The third major category of mechanics solutions involves numerical approximations of the stress and strain states. Perhaps the most powerful of these tools is the finite element method, which involves discretizing the structure of interest into a series of small elements. By approximating the relevant mechanics quantities over each finite element, the entire structure can be analyzed. By refining the discretization process, solutions of the desired accuracy can often be obtained. The finite element method is an extremely powerful tool to analyze stresses, strains and other quantities in any structure. Complex geometries, multiple materials, nonlinear behavior, and a host of other complexities can often be accurately represented with this approach. A number of commercial software packages are available for use, and many of these have become quite user-friendly, greatly facilitating their use. These codes are often able to analyze other physical phenomena, such as heat transfer, electromagnetic fields, fluid flow, etc. that can sometimes be important in bonded assemblies as well.

Real adhesive joints are typically three-dimensional structures. Three-dimensional solutions are notoriously difficult when using analytical approaches and often tedious and computationally intensive when using numerical solution methods. Although some adhesive bond problems are inherently 3-D, many can be reasonably approximated by 2-D or even 1-D models. For 2-D models, axisymmetric solutions are appropriate for circular geometries. The plane stress approximation assumes that there are no stresses perpendicular to the plane being analyzed, as is often appropriate for narrow beam geometries. When the strains perpendicular to the plane of analysis are zero, plane strain conditions are assumed. These are often appropriate for wide structures where strains in the width direction are zero or a constant (generalized plane strain). These conditions might be appropriate for a long sealant joint in pavement, or for an adhesive constrained between two stiff adherends. In some cases, even 1-D approximations can provide useful qualitative, if not quantitative insights. The shear lag and beam on elastic foundation models described below fall into this category; the former considers the relative deformation parallel to the bond plane, and the latter addresses the out-of-plane deformation.

3. Structural elements in bonded joints

Before examining the stress states within bonded joints, it is instructive to review several structural elements involved in typical bonded joints. An understanding of these elements will prove useful in considering the stresses, deformations, and energy stored in the members that make up a bonded joint or structure. The interested reader will find more details on these and other elements in elementary mechanics of materials texts [15].

One of the basic structural elements encountered in bonded joints is the straight, axially loaded bar as illustrated in Fig. 4. Except near the ends or near holes or other discontinuities that might exist in the member, the normal stress is considered to be uniform across the cross-sectional area. If the material behaves in a linear elastic fashion, the overall deformation in the rod may easily be determined, as may the stored elastic energy within the rod. Expressions for these quantities are also given in Fig. 4. In addition to the axial deformation, which occurs due to the applied axial load, deformations in the transverse directions are also observed, and are related to the axial strains through the Poisson's ratio of the material. Specifically, for uniaxial loading, Poisson's ratio is defined as the negative of the ratio of the transverse strain to the axial strain. Theoretically ranging from -1 to 0.5, Poisson's ratio for most engineering materials ranges only from around 0.2 to nearly 0.5. Poisson's ratio can have significant effects on the stress states present in bonded joints, giving rise to complex three-dimensional stress states. As Poisson's ratio approaches a value of 0.5, as occurs

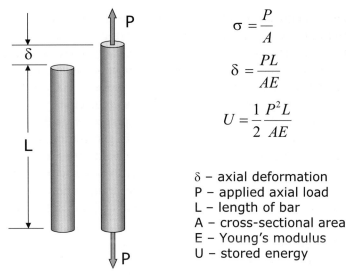

$$\sigma = \frac{P}{A}$$

$$\delta = \frac{PL}{AE}$$

$$U = \frac{1}{2}\frac{P^2 L}{AE}$$

δ – axial deformation
P – applied axial load
L – length of bar
A – cross-sectional area
E – Young's modulus
U – stored energy

Fig. 4. Illustration of an axially loaded member, including formulae for stress, deformation, and stored energy (in a linear elastic material).

with elastomeric systems, unique stress profiles can present themselves [33], as will be seen in Chapter 10.

Although perhaps less common in published treatments of adhesive joints, torsional elements are also encountered in many bonded joints and structures. For arbitrarily shaped cross-sections, formulae for stresses and deformations may be obtained using the principles of elasticity [32]. Two special cases are easily analyzed and are especially relevant to bonded joints; their solutions are given here. For circular cross-sections (either solid or hollow), cross-section planes remain plane even when torsional loading is applied. Shear strains, $\gamma_{z\theta}$ [1], may be easily shown to vary linearly from zero at the center to a maximum value at the outermost radius of the cross-section. For linear elastic materials, the shear stresses, $\tau_{z\theta}$, must also vary linearly, remaining proportional to the shear strains through the shear modulus, G. The total torque resisted by the resulting shear stresses is the integral of the shear stresses, weighted by the moment arm, ρ, over the cross-sectional area. Since the shear stress increases linearly with radius and the moment arm also increases linearly with radius, torque and moment loadings naturally give rise to a term that represents the second moment of area [2]. For torque

---

[1] Here we use the cylindrical coordinate system where r is the radial coordinate axis, θ is in the circumferential direction, and z is the axial coordinate.

[2] Often referred to as the moment of inertia, this latter term is sometimes reserved for the mass moment of inertia used in flywheels and other rotating objects, where linear velocity and moment arm both vary linearly with radius, again giving rise to a squared radial dependence.

of circular cross-sections, the second polar moment of area about the center is the appropriate term. Using this quantity, simple expressions for shear stress, angle of twist, and stored energy may easily be written [15], as given in Fig. 5a. The second torsional geometry discussed here is the thin-walled element of arbitrary shape, a classic example of which would be an aircraft wing or other airfoil. Torsional stiffness arises primarily from shear stresses carried within the airfoil skin. These shear stresses can be obtained from the equations [15] given in Fig. 5b. When thin-walled tubular sections are joined together, these latter formulae are useful in estimating the magnitude of the shear loading that must be carried by the bond.

Of significant importance in many bonded structures is the beam or bending element. Before examining the deformation of beams, we first examine the concepts of shear and bending moment that exist within beam-like structures. Consider a beam element as shown in Fig. 6a. The beam is subjected to a lateral, distributed loading, $q(x)$. The internal shear force, $V(x)$, and bending moment, $M(x)$, within the beam are related to $q(x)$ through

$$\frac{\mathrm{d}V}{\mathrm{d}x} = -q(x), \qquad \frac{\mathrm{d}M}{\mathrm{d}x} = V(x). \tag{6}$$

These quantities are often plotted in shear and bending moment diagrams [15], such as those illustrated in Fig. 6b for a simply supported beam subjected to a central load, and a cantilever beam subjected to a uniform distributed load.

Whereas torques (moments applied parallel to the shaft axis) tend to twist structural elements, moments applied perpendicular to the axis of the beams tend to bend them. The resulting radius of curvature is used to characterize the deformation of beams. In a manner analogous to torsional loading of a circular shaft, the resulting normal strains vary linearly from the neutral axis, an unstrained plane passing through the centroid of the cross-section. For linear elastic materials, the normal stresses also vary linearly, giving rise to a second moment of area term about the centroidal axis. The equation for stress (as given in Fig. 7) is similar to that in torsion of circular shafts, but the resulting relationship for deformation is quite different. The equations given in Fig. 7 are for the case of a moment applied about an axis of symmetry or principal axis of the beam cross-section; for more complex cross-sections or combined loading situations, consult an elementary mechanics of materials text [15].

The relation for beam bending relates the radius of curvature, ρ, to the applied moment, M, the Young's modulus of the material, E, and the second moment of area, I, as shown in Fig. 7. For linear elastic materials, we may write the curvature (defined as the reciprocal of the radius of curvature) as

$$\frac{1}{\rho} = \frac{M}{EI}, \tag{7}$$

where ρ and M are understood to be functions of position, x. The curvature is

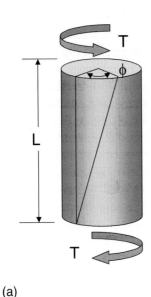

$$\tau = \frac{T\rho}{J} \qquad J = \int \rho^2 dA = \frac{\pi}{2}\left(r_o^4 - r_i^4\right)$$

$$\phi = \frac{TL}{JG}$$

$$U = \frac{1}{2}\frac{T^2 L}{JG}$$

τ – shear stress
φ – angle of twist
ρ – distance from center
T – applied torque
L – length of shaft
J – 2nd polar moment of area
G – shear modulus
U – stored energy

(a)

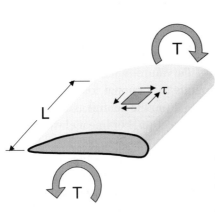

$$\tau = \frac{T}{2t\mathcal{A}}$$

$$\phi = \frac{TL}{4\mathcal{A}^2 G} \oint \frac{ds}{t}$$

$$U = \frac{T^2 L}{8\mathcal{A}^2 G} \oint \frac{ds}{t}$$

τ – shear stress
φ – angle of twist
T – applied torque
𝒜 – enclosed area
L – length of tube
t – wall thickness
G – shear modulus
U – stored energy

(b)

Fig. 5. (a) Illustration of a circular member loaded in torsion, including formulae for stresses, deformation, and stored energy (in a linear elastic material). (b) Illustration of a thin-walled member loaded in torsion, including formulae for stress, deformation, and stored energy (in a linear elastic material).

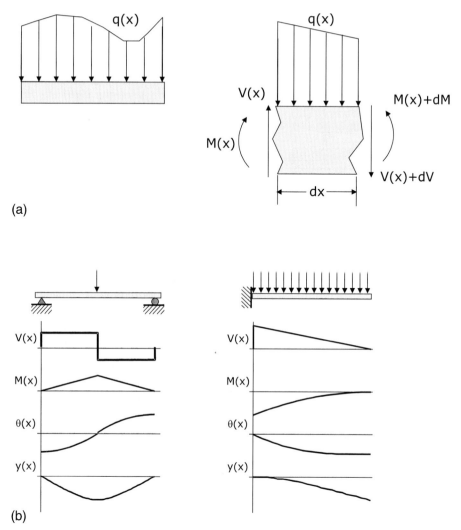

(a)

(b)

Fig. 6. (a) A differential beam element subjected to a distributed lateral load and internal shear forces and bending moments. (b) Illustration of shear, bending moment, slope, and deflection diagrams for two beam configurations, a simply supported beam subjected to a centered concentrated force, and a cantilever beam subjected to a uniform load distribution.

given in terms of the slope $\frac{dy}{dx}$ and second derivative of the beam displacement $\frac{d^2y}{dx^2}$:

$$\frac{1}{\rho} = \frac{\dfrac{d^2y}{dx^2}}{\left[1+\left(\dfrac{dy}{dx}\right)^2\right]^{3/2}}. \tag{8}$$

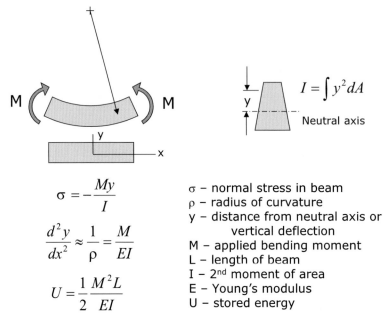

$$\sigma = -\frac{My}{I}$$

$$\frac{d^2 y}{dx^2} \approx \frac{1}{\rho} = \frac{M}{EI}$$

$$U = \frac{1}{2} \frac{M^2 L}{EI}$$

σ – normal stress in beam
ρ – radius of curvature
y – distance from neutral axis or
         vertical deflection
M – applied bending moment
L – length of beam
I – 2nd moment of area
E – Young's modulus
U – stored energy

Fig. 7. Bending of a beam due to applied moments, including formulae for stresses, deformation, and stored energy (in a linear elastic material).

If the slopes are sufficiently small, the curvature, $1/\rho$, can be approximated as the second derivative of the deflection, y, with respect to x. Subject to this assumption, we have the widely used equation for beam deflections

$$\frac{d^2 y}{dx^2} = \frac{M(x)}{EI}. \tag{9}$$

Noting that the second derivative of moment with respect to x is $-q(x)$, the lateral loading on the beam, one can also write an alternate form of the deflection relationship

$$\frac{d^4 y}{dx^4} = -\frac{q(x)}{EI}. \tag{10}$$

Applications of this relationship to beams with various boundary conditions and loadings are relatively straightforward, and are tabulated in elementary engineering texts [15] as well as in more detailed treatments [34]. Fig. 8 shows several elementary loading cases for cantilever beams, along with the resulting deflection equations. These linear solutions may readily be combined through superposition to obtain relationships for more complex combinations of loading. The first derivative of the beam deflection is the slope. Slope and deflection curves for two beam configurations are also shown in Fig. 6b.

One additional aspect of beam bending is the shear stresses that occur within beams subjected to transverse loads as illustrated in Fig. 9. When subjected to

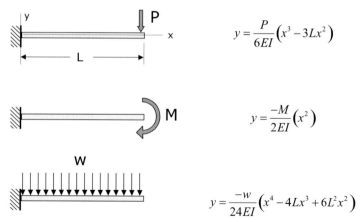

Fig. 8. Superposition formulae for deflections, $y(x)$, of representative cantilever beams (subject to assumptions listed in text).

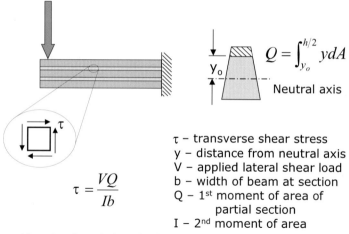

$$Q = \int_{y_o}^{h/2} y\,dA$$

$$\tau = \frac{VQ}{Ib}$$

τ – transverse shear stress
y – distance from neutral axis
V – applied lateral shear load
b – width of beam at section
Q – 1st moment of area of
        partial section
I – 2nd moment of area

Fig. 9. When subjected to lateral shear loads, transverse shear stresses are required within the beam to allow buildup of axial stresses associated with changing bending moment.

such lateral loads, the axial stresses vary not only from top to bottom of the beam, but also along the length due the change in bending moment that occurs. This variation in stress along the length requires shear stresses to transfer forces within the beam. These stresses may easily be calculated. For built-up or laminated beams, illustrated in Fig. 10, these stresses may be carried through mechanical fasteners such as nails, bolts, and rivets, through discrete or continuous welds, or through adhesives joining the components. These load-transferring components allow the individual beams to act together as a single beam rather than as a combination of flexible, independent beams. This composite action is of critical importance in obtaining lightweight, efficient structural elements. For example,

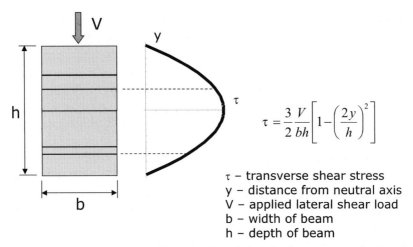

$$\tau = \frac{3}{2}\frac{V}{bh}\left[1 - \left(\frac{2y}{h}\right)^2\right]$$

τ – transverse shear stress
y – distance from neutral axis
V – applied lateral shear load
b – width of beam
h – depth of beam

Fig. 10. Within a rectangular beam subjected to lateral shear loads, transverse shear stresses are distributed in a parabolic fashion throughout the depth of the beam. If this beam is laminated, adhesive stresses at any given bond plane are easily determined, as illustrated by the dashed lines.

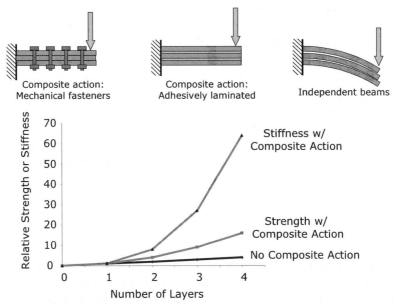

Fig. 11. Beams joined together by discrete mechanical fasteners or by continuous adhesive layers function as a single entity through composite action, significantly increasing the strength and stiffness over an assembly of independent beams.

as shown in Fig. 11, the strength and stiffness of a series of independent beams go up linearly with the number of layers; through composite action, however, the bending strength of bonded layers goes up as the square of the number of layers, and the stiffness goes up as the cube of the number of bonded layers. Adhesives thus can play a significant role in this important strengthening and stiffening mechanism.

Beam solutions are applicable to relatively narrow members that are free to deform along the length as well as across the width. Beam theory assumes that no stresses are applied perpendicular to the plane of curvature. This plane stress solution results in curvatures along the length as well as an anticlastic curvature perpendicular to the length, whose curvature is equal to the negative of Poisson's ratio times the longitudinal curvature. This tendency can have effects on bonded joints [35]. As the beam becomes wider, the anticlastic bending can no longer develop freely, and the beam bends as a plate. Although considerably more complicated to analyze, the analysis of wide, plate-like structures including bonded joints with various boundary conditions are related to the beams covered above. The reader is referred to books [36] on plates and shells for a discussion of this topic, which is beyond the scope of this chapter.

These three simple structural elements (axial member, torsional member, and bending beam) along with extensions of the latter to plate and shell geometries, can be used to represent the adherends used in many commonly used joints. The simple treatments considered herein ignore complications that occur near the ends where loads may be introduced, and also ignore the possibilities of holes, changes in size or shape, and other discontinuities that often occur in real structures. Nonetheless, their analysis serves as an important component in understanding stresses and strains within many bonded joints. Although addressed individually in the above treatment, many bonded joints involve combinations of the above loading modes. For example, and as will be shown in Chapter 4, the axial loads applied to single lap joints actually produce moments within the adherends, resulting in stresses in the adherends that can be as much as four times those predicted by the axial formula alone. Clearly these combined loading cases are also of significant importance for bonded joints and structures. For cases where the solutions are linear and uncoupled, one can simply superimpose the relevant solutions to obtain the answers to combined loading problems.

Solutions for stresses and strains within even simple structures discussed above often involve statically indeterminate solutions requiring information beyond that available through the equations of equilibrium alone. In addition to equilibrium equations, kinematic relations linking displacements, and constitutive relations relating stresses and strains are normally required for these solutions. Such solutions may be obtained at the mechanics of materials level, involving simple solutions for basic structural elements as given above, or they may require more sophisticated analytical or numerical methods.

Finally, it should be noted that although the simplistic treatment of these important structural elements has provided the basis of much of the analysis and design that has gone into developing the engineering structures we are familiar with, many complications can arise that require more detailed analysis. Chapter 2 will introduce the stresses and failure processes that may occur in bonded structures that contain cracks and debonds, a topic of importance in many of the chapters of this book. Chapters 2 and 5 describe the singular stress fields that may arise around other details associated with bond terminations.

4. The shear lag model

The shear lag model, first published by O. Volkersen in 1938 [37], is one of the most fundamental concepts in the transfer of load between two members joined by either discrete connections, such as mechanical fasteners, or by a continuous layer such as an adhesive. For any type of bonded joint involving adherends laid side by side and loaded axially in tension or compression, as shown in Fig. 12a, the adhesive layer serves to transfer load from one adherend to the other through shear stresses distributed along the length of the bond. If the adherends are relatively rigid, the transfer occurs in a more uniform fashion, with load being gradually transferred across the bond plane at a fairly uniform rate. If the adherends are noticeably deformable in the axial direction, however, significant peaks in the shear stresses will occur at the ends of the joints. The following analysis is meant to illustrate how this determination of stress distributions is made for this widely applicable shear lag concept.

The basic shear lag model is based on several assumptions as follows.

(1) The adhesive does not carry any significant axial force, because it is more compliant in the axial direction than the adherends, and because it is relatively thin compared to the adherends.

(2) The adherends do not deform in shear, implying that the shear modulus of the adherends is much greater than that of the adhesive. This assumption becomes especially suspect with anisotropic materials such as wood- or fiber-reinforced composites.

(3) Out-of-plane normal stresses are ignored in both the adhesive and adherends.

(4) The effect of the load eccentricity or couple is ignored, and bending of the adherends is specifically ignored.

(5) Adhesive and adherends are assumed to behave in a linear elastic manner.

(6) Bonding is assumed to be perfect along both bond planes.

(7) The effects of the bond terminus are ignored.

(8) Plane stress conditions are assumed, ignoring complications arising from different Poisson contractions in the bonded region and single adherend regions.

In spite of the assumptions, the shear lag analysis does provide a great deal of useful insight. Perhaps the most questionable assumption involves the failure to consider the couple, thus the analysis is conducted on an object that is not

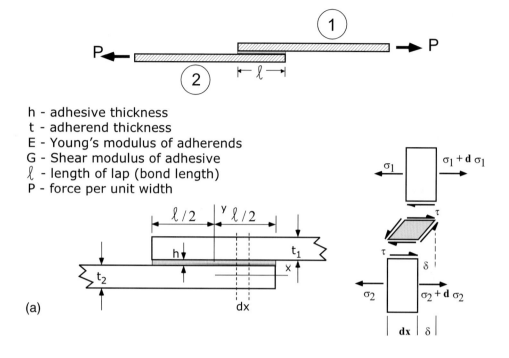

h - adhesive thickness
t - adherend thickness
E - Young's modulus of adherends
G - Shear modulus of adhesive
ℓ - length of lap (bond length)
P - force per unit width

(a)

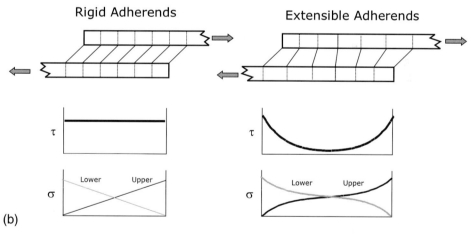

(b)

Fig. 12. (a) The shear lag model for a single lap joint is formulated using the indicated terms and the differential elements. Bending in the adherends is specifically ignored. (b) Exaggerated deformations of loaded lap joints illustrating the effect of adherend extensibility on adhesive shear stress, τ, and adherend axial stresses, σ. After Adams and Wake [11].

in moment equilibrium. This prevents the model from recognizing the adherend bending and severe peel stresses that often account for failure in single lap joints. Because of this, the model is not particularly appropriate for single lap joints, an important geometry that will be covered in more detail in Chapter 4. If one applies the shear lag model to half of a double lap joint, the results are more meaningful since bending is less pronounced in this symmetric geometry. Ignoring bending, the derivation proceeds as follows.

The upper and lower adherends are denoted by 1 and 2, respectively. Each adherend has a Young's modulus E_i and a thickness t_i. The adhesive has a shear modulus of G and a thickness of h. The joint length is ℓ, as shown in Fig. 12a. Solving this statically indeterminate problem involves the equilibrium equations based on a differential element as shown in Fig. 12a:

$$\frac{d\sigma_2}{dx} = -\frac{1}{t_2}\tau, \qquad \frac{d\sigma_1}{dx} = \frac{1}{t_2}\tau, \tag{11}$$

the kinematic expressions

$$\frac{d\delta}{dx} = \varepsilon_1 - \varepsilon_2, \qquad \gamma = \frac{\delta}{h}, \tag{12}$$

and the constitutive relationships

$$\varepsilon_1 = \sigma_1/E_1, \qquad \varepsilon_2 = \sigma_2/E_2, \qquad \gamma = \tau/G, \tag{13}$$

where σ_1 and σ_2 are the axial stresses in the upper and lower adherends, respectively, ε_1 and ε_2 are the corresponding axial strains, and δ represents the relative horizontal displacement between the upper and lower adherend across the bondline; each is a function of position, x, along the length of the bond.

Combining these equations, one obtains the governing differential equation

$$\frac{d^2\tau}{dx^2} - \omega^2\tau = 0, \quad \text{where} \quad \omega = \sqrt{\frac{G}{h}\left(\frac{E_1t_1 + E_2t_2}{E_1t_1E_2t_2}\right)}. \tag{14}$$

The solution to this second-order differential equation is

$$\tau(x) = A\cosh\omega x + B\sinh\omega x. \tag{15}$$

This governing differential equation will be useful for several different geometries and loading scenarios. The boundary conditions needed to solve for the unknown coefficients, A and B, depend on the configuration being considered. The simplest and yet most general approach seems to be based on establishing first derivatives of the shear stress at the bond ends by combining the kinematic equations to obtain

$$\frac{d\tau}{dx} = \frac{G}{h}(\varepsilon_1 - \varepsilon_2).$$

Since ε_1 and ε_2 vanish at the right and left ends of adherends 1 and 2, respectively, because these are free ends, we obtain

$$\left.\frac{d\tau}{dx}\right|_{-\ell/2} = -\frac{GP}{hE_2t_2}, \qquad \left.\frac{d\tau}{dx}\right|_{\ell/2} = \frac{GP}{hE_1t_1},$$

which are the two boundary conditions that allow us to determine the final solution for the shear lag model of the lap joint

$$\tau(x) = \frac{P\omega}{2\sinh(\omega\ell/2)}\cosh\omega x + \frac{P\omega}{2\cosh(\omega\ell/2)}\left(\frac{E_2t_2 - E_1t_1}{E_1t_1 + E_2t_2}\right)\sinh\omega x, \qquad (16)$$

where P is the axial load per unit width of the joint.

For the balanced adherend case where the E_it_i products for the upper and lower adherends are the same, the coefficient for the hyperbolic sine term becomes zero, and the shear stress distribution is symmetric about the center of the joint. Adams and Wake [11] have provided insightful figures illustrating the shear stress distributions within joints consisting of either rigid or extensible adherends. As seen in Fig. 12b, if the adherends are relatively rigid (ω very small), the adherends translate relative to one another, and the shear stress is uniform (based on the Volkersen assumptions). If the adherends are extensible, those sections of the adherend with more axial load will deform more, resulting in greater elongation. The shear strains within the adhesive are seen to vary significantly along the length of the bond, obeying the characteristic hyperbolic cosine form. Average axial stresses within the adherends are related to the integrals of the shear stress, as indicated in Eq. 11, so their form may easily be determined as well.

Fig. 13 illustrates the shear stress distributions in balanced joints of various lengths. When non-dimensionalized by the average shear stress, as shown in Fig. 13a, the shear lag model predicts that the maximum stress becomes larger for longer joints sustaining the same average shear stress. Fig. 13b shows the same data, but non-dimensionalized by $P\omega$, effectively comparing different length joints supporting the same applied load. Beyond a point, increasing the joint length does not reduce the maximum shear stress predicted by the shear lag model. These figures also provide convincing evidence about the dangers of reporting average shear stress at break when testing lap joints, especially if the properties of the adherends or adhesives are changing, or if the geometry of the joint is not the same. Many sources, including ASTM, strongly caution against the use of the apparent shear strength (used in many lap shear test methods [38]) for design purposes [39].

A key feature to be gained from the Volkersen shear lag result is that there is a relatively uniform shear stress distribution only for the case of 'short joints'. For longer joints, there are peaks in the shear stress anywhere there are relative changes in the stiffness of the adherends. Thus near joint ends, large shear stress

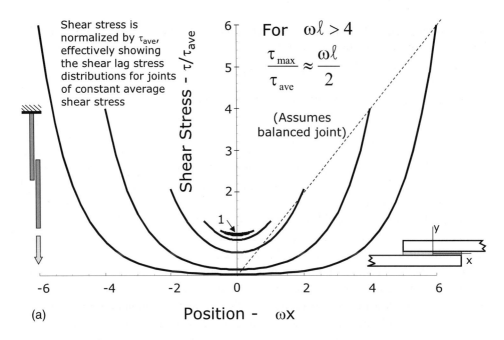

(a)

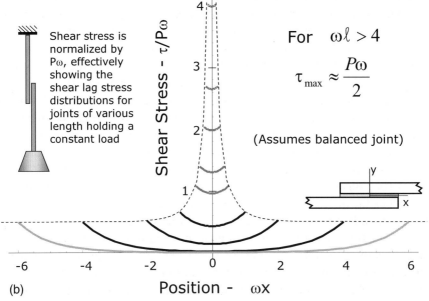

(b)

Fig. 13. (a) Shear lag predictions of shear stresses within lap joints of various lengths, normalized by the average shear stress in the joint. Note that, relative to the average shear stress, the maximum predicted stress increases as the bond length becomes longer. (b) Shear lag predictions of shear stresses within lap joints of various lengths, normalized by the applied load. Note that according to the shear lag model, the maximum predicted shear stress under a given load does not continue to decrease as overlap length is increased.

peaks are expected. The words 'long' and 'short' refer to the actual length of the joint compared to the characteristic length, ω^{-1}, the reciprocal of the eigenvalue. The characteristic length is a measure of the distance necessary to transfer the load, equilibrating the strains in the adherends, giving rise to the spatial lag required to reach compatible strains in the adherends.

For situations where the adherends are not balanced, the hyperbolic sine term does not vanish, and the shear stress distribution is skewed to result in higher shear stresses at the end of the stiffest adherend. Since stress trajectories[3] always tend to follow the stiffest path, we expect the less stiff adherend to shed the majority of the load to equilibrate the axial strains in the adherends. Because the shear stresses are reduced near the end of the softer (thinner) adherend, one might recognize that tapering the adherends could result in a more uniform shear stress distribution. Indeed, the stresses are much more uniform for scarf or bevel joints. It has been shown, however, that unless the adherends are tapered to less than 10% of the thickness, there may be little benefit from tapering [40]. Because of the difficulty in machining feather-edges on complex adherends, there is a tendency to use the discrete version of a bevel joint, the step lap joint. This geometry is easy to machine, and especially well suited to manufacture laminated composite joints. The key feature to note is that there will be shear stresses anywhere there is a relative change in adherend stiffness. For tapered adherends, there is a distributed change along the entire length of the joint, resulting in a uniform shear stress distribution over this region. For the case of the step lap joint, there are discrete changes in relative stiffness, resulting in discrete regions over which there are high shear stresses [41]. While not as efficient as tapered joints, step lap joints are able to carry considerably higher loads than lap joints in which the adherends are of constant stiffness. Step lap joints can be optimized to minimize the likelihood of adherend and adhesive failure. In designing joints for aircraft, the author has found that computer codes developed by Hart-Smith [42] were able to accurately predict strength for step lap joints with composite and titanium adherends. Additional details on design aspects can be found in Chapter 19.

The original Volkersen shear lag analysis is inadequate for single lap joints because it ignores the moments, shear forces, peel stresses, and other complications. It is more applicable to the double lap joint in which these terms are significantly reduced, although not eliminated. In spite of the limitations, however, the shear lag model is useful in estimating the manner in which load is transferred between adherends. The underlying concepts of the shear lag model can be applied to a wide variety of related geometries. For the case of thin-walled torsional members, one finds that the basic equations translate directly. However, the shear modulus

---

[3] Stress trajectories are the paths stresses tend to follow, and are akin to the more familiar concept of streamlines in fluid flow.

of the adherends rather than the Young's modulus appears in the expression for ω. Because the shear moduli are smaller than the Young's moduli, ω is larger and the characteristic shear lag distance is shorter, resulting in more prominent peaks in shear stress [43]. For the case of torsion of circular tubes bonded to tubes or shafts, similar relationships are found involving the second polar moments of area of the individual tubes [44]. Extensions to materials reinforced with fibers or even to steel-reinforced concrete are possible. The load is transferred from the reinforcing fiber to the surrounding material through a shear lag process that has been modeling in a variety of ways depending on the boundary conditions and the assumptions made [45,46]. Chapter 16 will provide additional details on these applications to fiber-reinforced systems. Such models are useful in determining how load is transferred around a broken fiber or a broken ply in laminated systems [47,48]. Indeed, the shear lag model is a widely used concept in understanding load transfer in bonded systems, having wide applicability to many different configurations when appropriately modified.

5. The beam on elastic foundation model

Another fundamental mechanics solution that has many applications in bonded joints is that of a beam on an elastic foundation. Emil Winkler first reported this analysis in 1867 [49]. The method has been widely applied to a variety of problems, perhaps most obviously that of trains passing over rails supported by the earth, and has been included in most texts on advanced mechanics of materials [16]. Since many bonded joints have beam-like adherends supported by a more flexible adhesive layer, this model of a beam on an elastic foundation is also of great importance for a variety of joints ranging from the lap shear specimen to fracture specimens, from peel specimens to the loop tack test.

The beam on elastic foundation analysis begins by considering a beam supported by a continuous foundation, both of which are assumed to be linearly elastic in their behavior. The flexural rigidity of the beam (EI) and the foundation stiffness (k) determine the form of the resulting stress distributions and deflections. Here, k represents the force per unit length required to produce a unit deflection of the foundation, and has units of force/length^2. Although often illustrated as a series of discrete springs supporting the beam, as shown in Fig. 14, the model actually involves a continuously distributed support of independent springs. Deflections of the beam then result in distributed forces along the length of the beam that are given by $q(x) = ky(x)$, where y is the vertical deflection of the beam. Recalling and allowing

$$\lambda = \sqrt[4]{\frac{k}{4EI}}, \tag{17}$$

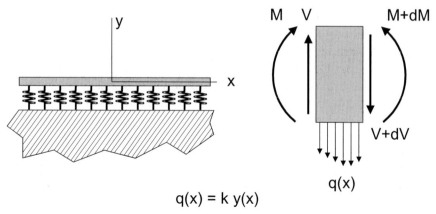

$$q(x) = k\,y(x)$$

Fig. 14. A beam on an elastic foundation represented as a series of discrete, independent springs. At the right is a differential element of the beam showing moment, shear, and distributed loading.

we obtain the governing differential equation

$$\frac{d^4 y}{dx^4} + 4\lambda^4 y = 0, \tag{18}$$

which has solutions of the form

$$y(x) = e^{-\lambda x}(A\cos\lambda x + B\sin\lambda x) + e^{\lambda x}(C\cos\lambda x + D\sin\lambda x). \tag{19}$$

The coefficients are determined by the relevant boundary conditions.

To cast this solution in terms of adhesive bonds, the adhesive, of thickness h, is assumed to be linearly elastic with a modulus of E_a, unaffected by deformations in the surrounding material, and acting over a bond width w on the beam. Under these conditions, the foundation stiffness becomes

$$k = \frac{E_a}{h} w. \tag{20}$$

This extension of Winkler's model is strictly valid only when each infinitesimal column of adhesive is unaffected by the deflections of the adjacent columns (i.e. if Poisson's ratio of the adhesive is zero and if shear stresses within the adhesive are negligible; alternate forms are obtained when the supporting layer is elastomeric [50]). Furthermore, only the adhesive layer is assumed to strain in the vertical or out-of-plane direction, an assumption that becomes particularly inaccurate when anisotropic materials such as wood or composites are used as adherends. Nonetheless, the approach has been widely used for adhesive joints since the pioneering single lap joint analysis of Goland and Reissner [51], and provides useful insights into the deflections and stresses present within a wide range of bonded joints.

One of the more useful sets of boundary conditions we can apply to bonded layers is the application of a concentrated lateral force or a concentrated moment

D.A. Dillard

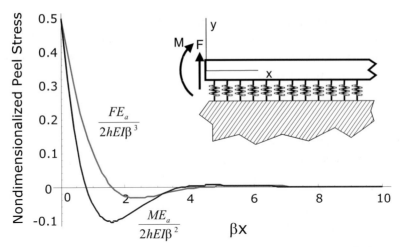

Fig. 15. A semi-infinite beam on an elastic foundation is subjected to an applied moment and force at the end. The foundation is illustrated by a discrete set of springs, although the analysis and applications to adhesive layers are for continuous support.

at a free end, as shown in the inset of Fig. 15. For a relatively long beam supported on such a foundation, the resulting peel stress distribution is given by

$$\sigma(x) = \frac{E_a}{2hEI\beta^3}e^{-\beta x}[F\cos(\beta x) + M\beta(\cos(\beta x) - \sin(\beta x))], \tag{21}$$

where F is the applied force, M is the applied moment, and

$$\beta = \sqrt[4]{\frac{E_a w}{4EIh}}. \tag{22}$$

If the beam is shorter than $5/\beta$, the above expression for the peel stress should be corrected for end effects [52]. The reciprocal of β has units of length, and is a measure of a characteristic distance over which the stresses are distributed. As β becomes smaller, the stresses are distributed over wider areas, effectively reducing the peak stresses.

The above results are plotted in Fig. 15 individually for an applied lateral force and a moment. The trigonometric terms in Eq. 21 suggest that the solution will alternate between tensile and compressive stresses. The exponential decay is so rapid, however, that oscillations beyond the first tension and compression zones are barely evident in graphs of the stresses. For the case of the applied load, the integral of the stresses (over the area) within the adhesive must equal the applied load. For the case of the applied moment, the area under the stress curve must equal zero, and the first moment of the area must equate to the applied couple. The compressive and tensile zones counteract one another so that no net force is present, although they do constitute a couple. Although the areas under these respective portions of the curve are equal, the peak of the region at the end of the bond is about five times larger in magnitude than the inner peak. Reversing the direction of the applied moment will reverse this comparison, and has useful implications for design purposes [53]. Since adhesives are considerably weaker in tension than in compression, having the outer (and larger magnitude) peak be tensile is of greater concern. This is particularly the case since the environment has access to the outermost region, and may increase the likelihood for debonding, especially over time.

The beam on elastic foundation solution arises in numerous solutions for stresses within adhesive layers. Goland and Reissner's seminal work examining peel stresses within the single lap joint, for example, result in these classical beam on elastic foundation profiles for the case where the adhesive is considered to be flexible in comparison with the adherends [51]. Solutions for residual stresses induced by moisture gradients in bonded joints also exhibit these characteristic profiles [54]. The case of tubular joints loaded in tension results in closely related solutions that recognize the increased bending stiffness of the tubular adherends associated with stretching of the walls of the tubes [55]. The beam on elastic foundation model is directly applicable to the case where adherends with different curvatures are brought together and bonded with an adhesive [53,56]. Any bond geometry related to peeling will typically include the beam on elastic foundation approximation for the bonded region. This beam on elastic foundation formulation was a key feature in the early analysis of peel geometries by Kaelble [57,58] and as corrected by Dahlquist [59]. Kaelble also demonstrated the magnitudes of these stress profiles experimentally [60]. One can easily demonstrate the compressive zone by lightly draping a typical pressure-sensitive adhesive tape on a sheet of glass or other smooth surface, and then lifting the free end of the tape. Even though external pressure was never applied to cause the adhesive to wet the glass, a significant removal force is required. In fact, the adhesive will wet the adherend due to the compressive region that precedes any tensile zone that may develop. One can easily see the compressive front as it moves along the tape just ahead of the debond region. This phenomenon is the basis for several tack tests designed to measure the aggressiveness of the adhesive. The loop tack test and the quick stick test quantify the energy release rate required to debond tapes wetting the surface based solely on this compressive front. The stresses within double cantilever beam specimens bonded with flexible adhesives also obey the beam on elastic foundation form [61–64]. Anytime one deals with flexible adherends on an adhesive foundation, these combined tension and compression regions of the peel stress associated with the beam on elastic foundation are manifest. Those bonded joints that may be classified as 'cleavage' tests will often exhibit these characteristic stress profiles. Even when the assumptions associated with this model no longer apply, the same qualitative trends are often observed.

6. Residual stresses in adhesively bonded joints

Residual stresses arise in materials due to several phenomena related either to a misfit in dimensions, or a gradient of some type. For homogeneous materials, residual stresses often develop during cooling from their processing temperatures because of non-uniform temperatures within the material. For bimaterial or multimaterial systems such as composites or bonded joints, different material properties, most notably the coefficient of thermal expansion, lead to residual stresses that can prove to be very detrimental to the performance of these material systems. In fact, for many bonded systems, the residual stresses formed during processing are often as significant as the mechanically applied stresses experienced in service. Although this chapter will not go into extensive details regarding residual stress fields, it is instructive to see the relevance of the shear lag and beam on elastic foundation models for several common scenarios.

Residual stresses can result from shrinkage associated with the cure mechanism, drying, or crystallization, and from changes in temperature. Both of these can contribute significantly to the residual stress state present in adhesives used above their glass transition temperature. For adhesives used well below their glass transition temperature, the residual stresses typically begin building significantly when the temperature drops below the glass transition temperature of the adhesive. Below the glass transition temperature, the CTE of a polymeric adhesive is only about a third of the value above the T_g. The modulus, however, is significantly higher, perhaps by three orders of magnitude. Thus significant stresses begin building at the stress free temperature, which can vary slightly from the T_g, depending on the processing conditions [65].

For the case of similar adherends bonded with an adhesive possessing a different CTE (or for the case of a coating), the adhesive will often have an equal biaxial normal stress present within the plane of the bond. For relatively stiff adherends made of the same material, the biaxial in-plane stress within the bondline may be estimated by

$$\sigma_0 = -\frac{E_a}{1 - \nu_a}(\alpha_a - \alpha)\Delta T, \tag{23}$$

where E_a is the modulus of the adhesive, ν_a is the Poisson ratio of the adhesive, α_a is the CTE of the adhesive, α is the CTE for the substrates, and ΔT is the temperature change measured from the stress-free temperature. If we assume that the CTE of the adhesive is larger than that of the adherends, and that the bond is cooled down from a stress-free temperature where stresses begin to increase rapidly, the in-plane residual stress state within the adhesive layer is tensile, as is commonly the case with polymeric adhesives and coatings.

Except near the edges of the bond, there are no significant stresses (shear or peel) across the interface. Along the edges, interfacial shear stresses are present,

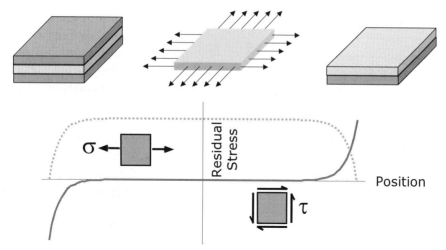

Fig. 16. In-plane tensile stresses present within adhesive layers and coatings drop to zero at free edges; interfacial shear stresses are non-zero only near free edges, as predicted by shear lag model.

as illustrated in Fig. 16, obeying relationships that are qualitatively related to the shear lag formulation presented earlier in this chapter. Due to the coupling between shear and peel stresses, peel stresses also arise along the edges. This key understanding that interfacial stresses are limited only to the edges (external edges as well as internal edges at cracks, holes, holidays, or other flaws) is an important concept for adhesive layers and for coatings. These peel and shear stresses decay rapidly as one moves away from the free edge. Solutions for these coating problems have been given by Suhir [66] and others. Further discussions of this important class of adhesion problems will be covered in Chapter 9.

Where dissimilar adherends are joined with an adhesive, the Volkersen shear lag model is directly applicable. Using the same geometry as defined in Fig. 12a, one finds that the shear stresses within the adhesive layer are given by

$$\tau(x) = \left[\frac{(\alpha_2 - \alpha_1)\Delta T \omega}{\left(\dfrac{1}{E_1 t_1} + \dfrac{1}{E_2 t_2}\right)\cosh(\omega \ell / 2)} \right] \sinh(\omega x), \qquad (24)$$

if bending is prevented. Here, the terms are as defined earlier, with α_1 and α_2 representing the coefficients of thermal expansion of the upper and lower adherends, respectively. The similarity with Eq. 16 is obvious; predictions are shown in Fig. 17.

The above solution is based on the shear lag model and ignores bending in the adherends. Another situation of considerable importance is the bending of beams and plates made with two or more materials. For example, consider the case of

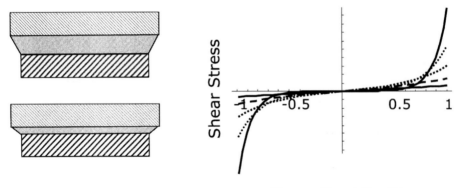

$$\tau(x) = \frac{G(\alpha_1 - \alpha_2)\Delta T}{h\omega \cosh\left(\frac{\omega \ell}{2}\right)} \sinh(\omega x)$$

Fig. 17. Shear lag predictions for shear stresses within adhesive layer joining two dissimilar adherends having different coefficients of thermal expansion. Thinner bond lines result in more significant peaks in shear stress. Bending is ignored for this shear lag approximation.

a bimaterial strip (plane stress) subjected to a change in temperature from its flat (stress-free) condition. This could be applicable to two dissimilar adherends joined by a thin adhesive layer, or to the case of a coating on a substrate. Timoshenko [67] showed that the resulting curvature for the geometry illustrated in Fig. 17 could be given by

$$\frac{1}{\rho} = \frac{6(\Delta\alpha)(\Delta T)(1+m)^2}{h\left[3(1+m)^2 + (1+mn)\left(m^2 + (mn)^{-1}\right)\right]}, \tag{25}$$

where $\Delta\alpha = \alpha_2 - \alpha_1$, $m = t_1/t_2$, and $n = E_1/E_2$. This behavior can produce large stresses within the adherends as well as within the adhesive layer, and can cause significant problems with dimensional stability. Alternatively, the approach has proven to be quite useful in characterizing the coefficients of thermal expansion, residual stresses, and stress-free temperature of adhesives and coatings [68].

Significant residual stresses within the adhesive layer can result when two adherends with dissimilar curvatures are bonded together. Based directly on the beam on elastic foundation solution, the moment required to bend the adherends to conform must be reacted by stresses within the adhesive layer [55]. These stresses take on the characteristic beam on elastic foundation solution as illustrated in Fig. 18, the magnitude of which is affected by geometry and modulus parameters. One can imagine gradients of temperature or moisture within the adherends over time, and how these gradients induce bending and residual peel stresses within

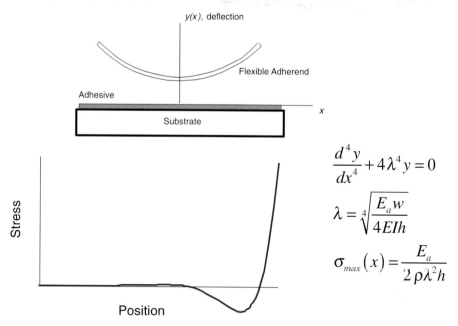

Fig. 18. Residual peel stresses predicted within the adhesive layer due to mismatch in curvature of the two adherends.

the adhesive layer and eventually lead to flagging. As with many residual stress solutions, the resulting geometry can be used as constant strain energy release rate fracture specimens, and has been successfully used to study time-dependent debonding and durability [69].

7. Applications of fundamental stress transfer solutions

The solutions covered in this introductory chapter all fall into a class of mechanics solutions known as mechanics of materials solutions because they involve assumptions that are typical of those made in the undergraduate level mechanics of materials courses. These closed form solutions are easy to apply, and can provide fundamental insights into the stress fields present within many idealized bonded joints. The shear lag concept is of fundamental importance to any bonded configuration where load is transferred from one adherend to another, primarily through shear stresses within the adhesive layer. The beam on elastic foundation solution provides the basis for explaining the nature of bonded beams or plates subjected to lateral loads or applied moments. The material on residual stresses and curvature are important in understanding the significant stresses that can result from mismatches in properties such as the coefficients of thermal expansion.

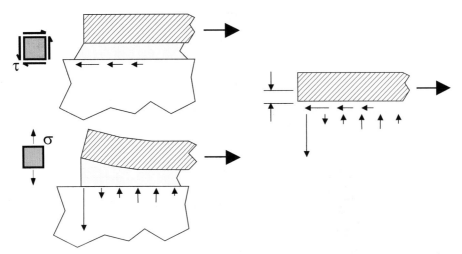

Fig. 19. Schematic illustration of the coupling between shear and peel stresses within the bondline. The eccentric loading in the adherend produces a couple, which tends to bend the adherend. Peel stresses must be present in the form of a couple in order to resist the edge lifting phenomenon.

These fundamental solutions are applicable, at least qualitatively, to a wide range of practical adhesive joints. Although the basic trends predicted by these methods are of critical importance in understanding the resulting stress profiles, more sophisticated solutions may be required in many practical geometries in order to accurately establish the resulting stress field.

The above treatments have considered peel and shear stresses in bonds as independent quantities. In fact, coupling between shear and normal stresses occurs in many bonded joints. Consider for example a section of a bonded joint in which an axial force is present within the top adherend. This axial force will pass near the centroid of the adherend. Resisting this horizontal force is an equal and opposite horizontal force composed of the shear stresses along the adhesive interface, as shown in the free body diagram of the adherend section, as shown in Fig. 19. Since these forces are not collinear, they compose a couple or moment that tends to overturn the section of the adherend. This overturning moment must be reacted by either peel stresses present within the adhesive layer, or by a resisting moment within the adherend. Either of these involves an out-of-plane deformation of the adherend and the associated peel stresses. These peel stresses take on the characteristic shape predicted by the beam on elastic foundation solution. This simple illustration provides convincing evidence that the shear stresses cannot, in general, occur in isolation in bonded joints. Shear stresses will often require corresponding peel stresses in order to maintain equilibrium. Shear is often also associated with peeling forces. For example, when a flexible adherend is peeled away from a substrate, as is illustrated in Fig. 20, bending of

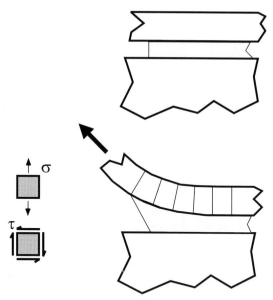

Fig. 20. For peel loading, bending of the flexible adherend results in relative shear deformations within the bondline, again demonstrating coupling often seen between shear and peel stresses in bonded joints.

the adherend results in a relative horizontal translation of the adherend surface with respect to substrate. Thus a peeling action produces not only peel stresses, but also shear stresses. Clearly, the stresses present within even the most simple joint configuration can be quite complex and highly non-uniform.

In applying the simple models described herein, errors can arise based on the assumed partitioning of deformations and stresses among the adherends and adhesive. One common assumption is that since adhesives are often more flexible than the adherends they join, certain deformations within the adherends can be assumed to be negligible compared with the corresponding deflections within the adhesive layer. This approach was taken for shear deformations in the shear lag model and in out-of-plane deformations for the beam on elastic foundation model. Although a reasonable assumption in some situations, one should note that because the thickness of the adhesive is often significantly less than the adherends, the total deformation within the adherend can actually be as large or larger than in the adhesive [11,51]. A second approach is to assume that because the adhesive is so thin, it contributes little to the total deformation. Such assumptions are useful in modeling the common double cantilever beam for fracture studies of adhesives [70], for example. Here, the displacement of the applied loads is due primarily to the deformations of the beam-like adherends. Although the above two assumptions are very useful in many elementary solutions for bonded joints, either approach can lead to significant errors when assumptions about moduli and thicknesses are inappropriate. Caution is recommended when using solutions based on either of these assumptions.

One issue of importance in many bonded joints is the nonlinear nature of certain solutions. Where adherend or adhesive plasticity occurs, for example, the resulting stress distributions will obviously not depend linearly on the applied load. Geometric nonlinearities may also arise with many bonded joints where the adherend configuration changes appreciably as load is increased. The eccentric nature of the single lap joint, for example, results in a significant geometric nonlinearity as was first analyzed by Goland and Reissner [51]. Because the load axis does not pass through the centroid of the adherends for much of their length, a moment results in the adherends, producing bending. As the load increases, the moment arm decreases, so the increment of moment decreases with a given increment of load as the load is increased. Some disagreement has appeared in the literature over these stresses, although careful numerical analysis using the geometric nonlinear analysis has resolved some of the discrepancies [71]. The original analysis of Brussat et al. [72] of the crack lap shear geometry neglects the significant geometric nonlinearity that can occur for certain configurations of this specimen. Either geometrically nonlinear analytical [73] or numerical [74] methods are needed to capture these details on these and related geometries exhibiting eccentric load paths.

Finally, the distinction between mechanics of materials solutions and elasticity solutions is illustrated in a very simple example. From the shear lag model, the maximum shear stresses are found to occur at the ends of the bonds. By considering the details of the idealized bond termination, however, one recognizes that the shear stress at the very edge of the adhesive layer must be identically zero since this is a free edge. The 'exact' solution recognizes that the shear stress must be zero at the edge, and includes modifications of the basic shear lag predictions as one approaches discontinuities such as bond terminations. Indeed, elasticity theory predicts that the peel stresses become infinite as one approaches the bond termination, as will be seen in Chapters 2 and 5. Here, the 'exact' solution mathematically predicts stresses that no real material can bear. In reality, ductile adhesives will locally yield near the bond terminations, and the resulting stress state may actually closely resemble the mechanics of materials solutions. Thus although solutions for stresses may be approached from several levels, real joints made with real materials often defy any of these idealizations. Although qualitative agreement with these solutions is common, quantitative discrepancies may arise as the joints being modeled deviate from the assumptions made. Numerical analysis, when properly applied, can become a powerful tool for incorporating these and other complexities.

8. Strength and fracture approaches to adhesive bonds

Throughout this volume we will encounter two basic approaches to examining adhesive bonds, whether test specimens for obtaining material properties, or designing actual bonded structures. One criterion is based on the idea that failure will occur when the maximum stress (or strain) within a bonded joint reaches a critical value. Slightly more sophisticated approaches recognize that the failure criterion is met when some combination of stresses (for example, the distortional energy [4]) reaches a critical value. Such strength-based criteria are widely used to interpret failures within test specimens, and also to design adhesively bonded joints. Further coverage of this approach will be given for several test methods described in Chapter 6.

The second major approach to interpret failures and on which to base design is the fracture mechanics approach. Within this method, several different criteria can be used. The stress intensity approach scales linearly with the applied stress level, and provides a measure of the severity of the stresses in front of a crack or debond tip. The strain energy release rate approach scales as the square of the applied stress (for linear systems), and provides an intuitively meaningful measure of the energy required to debond a unit area of the propagating crack. Fracture specimens are made with intentional flaws that can be monitored during testing. By determining the energy required to propagate debonds, one can characterize the adhesive bond's resistance to failure by fracture.

Within this volume, the reader will find several approaches within this general framework. Chapter 2 introduces the concept of fracture mechanics, which is treated in more detail in Chapters 7 and 8. Chapter 3 provides an elegant overview of the energy approach to adhesion. Stresses and driving energies for contact problems relevant to adhesion are given in Chapter 11, and Chapter 15 uses fracture mechanics concepts to help interpret the failure modes occurring in bonded joints.

Although both the strength and fracture approaches have strong proponents and detractors, both approaches can prove useful in the design of engineering structures and components, including adhesively bonded joints. Rather than viewing them as contradictory approaches, they should instead be seen as complementary approaches, the former ensuring that a design is strong enough to withstand the design loads, and the latter ensuring that if a debond is present, it will not propagate to a point where catastrophic failure can occur. Indeed adhesive joints must be both strong enough and tough enough to sustain the multitude of loads to which they may be exposed. Mohammed and Liechti [75] have shown that the use of the

---

[4] Known as the von Mises criteria, this approach has been applied accurately to yield phenomena in a variety of ductile materials.

cohesive zone model can be used to predict both the initiation of debonding from a singular corner as well as predict debond propagation. One recent approach for modeling the strength of adhesive joints through the use of Weibull statistics is intriguing in offering possible paths to link strength and toughness criteria [76].

9. Summary and conclusions

Unlike tests on monolithic test coupons involving stress states that are relatively uniform within the test section, stresses present within adhesive joints are, with few exceptions, highly complex and non-uniform. When interpreting the results from tests on adhesive bonds, one must recognize that adhesive bonds are actually *structures* in the sense that they involve inherent complexities in stress distributions. Understanding the stresses within test specimens and bonded components requires detailed analysis, the sophistication of which depends on the complexities of the joint and the desired accuracy of the solution. A failure to recognize these complexities can lead to erroneous design procedures and incorrect interpretation of joint failures.

This introductory discussion of stresses within adhesive joints began by reviewing the concepts of stress and strain, and several of the common structural elements involved in bonded joints. Axial members, torsional members, beams, plates, and shells are all frequently encountered components in adhesively bonded structures. Knowing the stress distributions within these members, along with the deformations and energy stored within loaded components, provides a useful basis upon which to build solutions for such components that are then joined with adhesives. Furthermore, several relevant analyses are commonly encountered in the analysis of bonded joints, and provide insights into even relatively complex bonded structures. Although they may not be quantitatively accurate in certain situations, they nonetheless provide a valuable qualitative understanding for many of the bonded joints observed in practice.

The shear lag analysis is useful in understanding how load is transferred from one adherend to another in lap-type adhesive joints. Numerous versions of this fundamental concept are available for many lap configurations, including single and double-lap joints, embedded fibers, torsion of tubular joints, etc. The analysis is applicable to both mechanical loading, and to thermal loads related to differences in coefficients of thermal expansion. The basic concept is that a finite distance is required to transfer stresses from one adherend to another across an adhesive layer in lap joints. This characteristic lag distance depends on the geometric and constitutive properties of the adherends and adhesive layer. The resulting shear stress solution is characterized by hyperbolic sine and cosine functions along the length of the bond. For non-balanced adherends, the largest shear stresses are predicted to occur at the end of the stiffest adherend.

The concept of the beam on elastic foundation analysis finds wide applicability when lateral loads or bending moments are applied to a flexible adherend supported by an adhesive layer. The resulting peel stress solution is an oscillatory function that decays rapidly with an exponential decay. The resulting solution involves characteristic tensile and compressive regions whose magnitude and spatial size depend on the geometric and constitutive properties of the adherends and the adhesive layer. The solution has relevance for many geometries including the single lap joint, fracture and peel specimens of various types, moisture ingression and swelling of the adhesive, adherend curvature mismatches, etc.

Residual stresses are also very important in many adhesive bonds. When adherends with similar coefficients of thermal expansion are bonded with an adhesive, a biaxial tensile stress often results within the adhesive layer. Interfacial stresses, obeying shear lag distributions, are limited to the edges and around holes and defects where free edges are present. When dissimilar adherends are bonded together, significant stresses can result in both the adherends and the adhesive, as can curvatures of the bonded system. These residual stresses can often be very significant when compared with mechanically induced stresses.

Although these simplistic models presented in this chapter have certain assumptions and limitations, they nonetheless provide important insights into the stress states that will develop within a wide range of adhesive joints. For simple joint configurations, minor changes to the solutions presented herein can provide reasonably accurate estimates of the stresses present. Although more sophisticated numerical analyses are required for more complex cases, the characteristic stress distributions predicted by these models are still qualitatively present in many cases. Clearly, these solutions form the basis for understanding stress distributions in many adhesively bonded joints. The following chapters will now proceed to more accurately analyze a variety of related bonded joint geometries.

References

1. Anderson, G.P., Bennett, S.J. and DeVries, K.L., *Analysis and Testing of Adhesive Bonds*. Academic Press, New York, 1977.
2. Smith et al., B.L., Molecular mechanistic origin of the toughness of natural adhesives. fibers, and composites. *Nature*, **399**, 761–763 (1999).
3. Nicholson, P.S., Nature's ceramic laminates as models for strong, tough laminates. *Can. Ceram. Q.*, 24–30 (Feb. 1995).
4. Almqvist, N., et al. Methods for fabricating and characterizing a new generation of biomimetic materials. *Mater. Sci. Eng.*, **C7**, 37–43 (1999).
5. Matthews, F.L. (Ed.), *Joining Fibre-Reinforced Plastics*. Elsevier, Amsterdam, 1987.
6. Pocius, A.V., *Adhesion and Adhesives Technology: An Introduction*. Hanser, Munich, 1977.
7. Kinloch, A.J., *Adhesion and Adhesives: Science and Technology*. Chapman and Hall, London, 1987.
8. Lees, W.A., *Adhesives in Engineering Design*. Springer, Berlin, 1984.

9. Tong, L. and Steven, G.P., *Analysis and Design of Structural Bonded Joints*. Kluwer, Boston, MA, 1999.

10. Adams, R.D., Comyn, J. and Wake, W.C., *Structural Adhesive Joints in Engineering*, 2nd ed. Chapman and Hall, London, 1997.

11. Adams, R.D. and Wake, W.C., *Structural Adhesive Joints in Engineering*. Elsevier, Amsterdam, 1984.

12. Brinson, H.F. (Ed.), *Engineered Materials Handbook, Vol. 3. Adhesives and Sealants*. ASM International, Metals Park, 1990.

13. Gordon, J.E., *Structures: or Why Things Don't Fall Down*. Da Capo, New York, NY, 1978.

14. Frederick, D. and Chang, T.S., *Continuum Mechanics*. Scientific Publishers, Boston, MA, 1972.

15. Beer, F.P. and Johnson, E.R., *Mechanics of Materials*, 2nd ed. McGraw-Hill, New York, NY, 1992.

16. Seely, F.B. and Smith, J.O., *Advanced Mechanics of Materials*, 2nd ed. Wiley, New York, NY, 1952.

17. Adams, R.D. and Harris, J.A., The influence of local geometry on the strength of adhesive joints. *Int. J. Adhes. Adhes.*, **7**, 69–80 (1987).

18. Sharpe, L.H., *J. Adhes.*, **4**, 51–64 (1972).

19. Hart-Smith, L.J., *Adhesive-Bonded Double-Lap Joints*. Tech. Rep. NASA CR-112235, NASA Langley Research Center, Hampton, VA, 1973.

20. Williams, J.G., Root rotation and plastic work effects in the peel test. *J. Adhes.*, **41**, 225–239 (1993).

21. Thouless, M.D., Adams, J.L., Kafkalidis, M.S., Ward, S.M., Dickie, R.A. and Westerbeek, G.L., Determining the toughness of plastically deforming joints. *J. Mater. Sci.*, **33**, 189–197 (1998).

22. Liechti, K.M., Ginsburg, D. and Hanson, E.C., A comparison between measurements and finite element predictions of crack opening displacements near the front of an interface crack. *J. Adhes.*, **23**, 123–146 (1987).

23. Delale, F. and Erdogan, F., Viscoelastic analysis of adhesively bonded joints. *J. Appl. Mech.*, **48**, 331–338 (1981).

24. Groth, H.L., Viscoelastic and viscoplastic stress analysis of adhesive joints. *Int. J. Adhes. Adhes.*, **10**, 207–213 (1990).

25. Allen, K.W. and Shanahan, M.E.R., The creep behaviour of structural adhesive joints, I. *J. Adhes.*, **7**, 161–174 (1975).

26. Allen, K.W. and Shanahan, M.E.R., The creep behaviour of structural adhesive joints, II. *J. Adhes.*, **8**, 43–56 (1976).

27. Sancaktar, E., An analysis of the curling phenomenon in viscoelastic bimaterial strips. *J. Adhes.*, **40**, 175–187 (1993).

28. Christensen, R.M., *Theory of Viscoelasticity: An Introduction*, 2nd ed. Academic Press, New York, NY, 1982.

29. Haddad, Y.M., *Viscoelasticity of Engineering Materials*. Chapman and Hall, London, 1995.

30. Lakes, R.S., *Viscoelastic Solids*. CRC Press, New York, NY, 1999.

31. Flugge, W., *Viscoelasticity*, 2nd rev. ed. Springer, New York, NY, 1975.

32. Timoshenko, S.P. and Goodier, J.N., *Theory of Elasticity*. McGraw-Hill, New York, NY, 1934.

33. Gent, A.N. and Lindley, P.B., Internal rupture of bonded rubber cylinders in tension. *Proc. R. Soc. London, Ser. A, Math. Phys. Sci.*, **249**, 195–205 (1959).

34. Young, W.C., *Roark's Formulas for Stress and Strain*, 6th ed. McGraw-Hill, New York, NY, 1989.

35. Andruet, R.H., Dillard, D.A. and Holzer, S.M., Two- and three-dimensional geometrical nonlinear finite elements for analysis of adhesive joints. *Int. J. Adhes. Adhes.*, **21**, 17–34 (2001).

36. Timoshenko, S. and Woinowsky-Krieger, S., *Theory of Plates and Shells*. McGraw-Hill, New York, NY, 1959.

37. Volkersen, O., Die Nietkraft Verteilung in zugbeanspruchten Nietverbindungen mit konstanten Laschenquerschnitten. *Luftfahrtforschung*, **15**, 41–47 (1938).

38. ASTM D1002, Standard test method for apparent shear strength of single-lap-joint adhesively bonded metal specimens by tension loading (metal-to-metal). *Annual Book of ASTM Standards*, **15.06**, 42–45 (1994).

39. ASTM D 4896-95, Standard guide for use of adhesive-bonded single lap-joint specimen test results. *Annual Book of ASTM Standards*, **15.06**, 419–424 (2001).

40. Thamm, F., Stress distribution in lap joints with partially thinned adherends. *J. Adhes.*, **7**, 301–309 (1976).

41. Hart-Smith, L.J., Design of adhesively bonded joints. In: Mathews, F.L. (Ed.), *Joining Fibre-Reinforced Plastics*. Elsevier, Amsterdam, 1987, pp. 271–311.

42. Hart-Smith, L.J., *Adhesive-Bonded Scarf and Stepped-Lap Joints*, Tech. Rep. NASA CR-112237, NASA Langley Research Center, Hampton, VA, 1973.

43. Krieger, R.B., *Engineered Materials Handbook, Vol. 3. Adhesives and Sealants*. ASM International, Metals Park, 1990.

44. Adams, R.D. and Peppiatt, N.A., Stress analysis of adhesive bonded tubular lap joints. *J. Adhes.*, **9**, 1–18 (1977).

45. Chua, P.S. and Piggott, M.R., The glass fibre-polymer interface, I. Theoretical consideration for single fibre pull-out tests. *Compos. Sci. Technol.*, **22**, 33–42 (1985).

46. Penn, L.S. and Lee, S.M., Interpretation of experimental results in the single pull-out filament test. *J. Compos. Technol. Res., JCTRER*, **11**, 23–30 (1989).

47. Reifsnider, K.L., Some fundamental aspects of the fatigue and fracture response of composite materials. In: *Proceedings of the 14th Annual Meeting of the Society of Engineering Science*. Lehigh University, 14–16 November, 1977.

48. Moore, R.H. and Dillard, D.A., Time-dependent matrix cracking in cross-ply laminates. *Compos. Sci. Technol.*, **39**, 1–12 (1990).

49. Winkler, E., *Die Lehre von der Elasticitaet und Festigkeit*, Teil 1, 2. H. Dominicus, Prague, 1867. (As noted in L. Fryba, History of Winkler Foundation, *Vehicle System Dynamics Supplement*, **24**, 7–12 (1995).)

50. Dillard, D.A., Bending of plates on thin elastomeric foundations. *J. Appl. Mech.*, **56**, 382–386 (1989).

51. Goland, M. and Reissner, E., The stresses in cemented joints. *J. Appl. Mech.*, **77**, A17–A27 (1944).

52. Hetényi, H., *Beam on Elastic Foundation*, 9th ed. University of Michigan Press, Ann Arbor, MI, 1971.

53. Dillard, D.A., Stresses between adherends with different curvatures. *J. Adhes.*, **26**, 59–69 (1988).

54. Weitsman, Y., Stresses in adhesive joints due to moisture and temperature. *J. Compos. Mater.*, **11**, 378–394 (1977).

55. Lubkin, J.L. and Reissner, E., Stress distribution and design data for adhesive lap joints between circular tubes. *J. Appl. Mech.*, **78**, 1213–1221 (1956).

56. Corson, T., Lai, Y.-L. and Dillard, D.A., Peel stress distributions between adherends with varying curvature mismatch. *J. Adhes.*, **33**, 107–122 (1990).

57. Kaelble, D.H., *Trans. Soc. Reol.*, **3**, 161 (1959).

58. Kaelble, D.H., *Trans. Soc. Reol.*, **4**, 45 (1960).
59. Dahlquist, C.A., In: *Technical Seminar Proceedings, Pressure Sensitive Tape Council,* **XI**. Pressure Sensitive Tape Council, Deerfield, IL, 1988, pp. 19–46.
60. Kaelble, D.H., *Trans. Soc. Reol.*, **9**, 135 (1961).
61. Lefebvre, D.R., Dillard, D.A. and Brinson, H.F., The development of a modified double cantilever beam specimen for measuring the fracture energy of rubber to metal bonds. *Exp. Mech.*, **28**, 38–44 (1988).
62. Ouezdou, M.B., Chudnovsky, A. and Moet, A., Re-evaluation of adhesive fracture energy. *J. Adhes.*, **25**, 169–183 (1988).
63. Ouezdou, M.B. and Chudnovsky, A., Stress and energy analysis of toughness measurement for adhesive bonds. *Eng. Fract. Mech.*, **29**, 253–261 (1988).
64. Williams, J.G., The fracture mechanics of delamination tests. *J. Strain Anal.*, **24**, 207–214 (1989).
65. Dillard, D.A., Park, T., Chen, B., Yu, J.-H., Guo, S., Cao, Y., Williams, S. and Xu, S., Using DMA/TMA equipment for bimaterial curvature technique. *NATAS Notes*, **32**, 10–14 (2001).
66. Suhir, E., Interfacial stresses in bimetal thermostats. *J. Appl. Mech.*, **56**, 595–600 (1989).
67. Timoshenko, S., *J. Opt. Soc. Am.*, **11**, 233–255 (1925).
68. Dillard, D.A. and Yu, J.-H., Revisiting bimaterial curvature measurements for CTE of adhesives. *Adhes. Meas. Films Coatings*, **2**, 329–340 (2001).
69. Randow, C.L., *Mechanisms and Mechanics of Non-Structural Adhesion*. MS Thesis, Virginia Tech, Blacksburg, 1996.
70. ASTM D3433, Standard test method for fracture strength in cleavage of adhesives in bonded metal joints. *Annual Book of ASTM Standards*, **15.06**, 218–224 (1994).
71. Tsai, M.Y. and Morton, J., An evaluation of analytical and numerical solutions to the single-lap joint. *Int. J. Solids Struct.*, **31**, 2537–2563 (1994).
72. Brussat, T.R., Chiu, S.T. and Mostovoy, S., *Fracture Mechanics for Structural Adhesive Bonds*. AFNL-TR-77-163, Air Force Materials Lab, Wright-Patterson AFB, OH.
73. Lai, Y.H., Rakestraw, M.D. and Dillard, D.A., The cracked lap shear specimen revisited — a closed form solution. *Int. J. Solids Struct.*, **33**, 1725–1743 (1996).
74. Johnson, W.S., Stress analysis of the cracked-lap-shear specimen: an ASTM round-robin. *J. Test. Eval.*, **15**, 303–324 (1987).
75. Mohammed, I. and Liechti, K.M., Cohesive zone modeling of crack nucleation at bimaterial corners. *J. Mech. Phys. Solids*, **48**, 735–764 (2000).
76. Towse, A., Potter, K.D., Wisnom, M.R. and Adams, R.D., The sensitivity of a Weibull failure criterion to singularity strength and local geometry variations. *Int. J. Adhes. Adhes.*, **19**, 71–82 (1999).

Chapter 2

Fracture mechanics and singularities in bonded systems

KENNETH M. LIECHTI [*]

Center for the Mechanics of Solids, Structures and Materials, The University of Texas at Austin, Austin, TX 78712, USA

1. Introduction

Structures that are designed for stress levels below the yield or fatigue strengths of their constituent materials have long been observed to experience material failure by cracking. In many cases, the failure could be traced to stress concentrations, preexisting flaws or some previously generated localized damage. These observations led to the conclusion that structures could fail at relatively low levels of applied load due to the presence of cracks. Cracks and sharp corners give rise to singular stress and strain fields in the material surrounding them. A singular stress has the form $\sigma \sim r^{\lambda-1}$, where r is the distance from the crack tip or corner and λ is known as the order of the singularity. If $0 < \lambda < 1$, the stress $\sigma \to \infty$ as $r \to 0$. This infinite stress is known as a singular stress and is more severe than the stress concentration at, for example, a hole. It is not physically realistic to expect infinite stresses in a material because breakdown mechanisms (not accounted for in analyses that yield singular stresses) will dominate at high stress levels. Nonetheless, as we will see, singular stress fields can still be useful for describing crack nucleation and propagation.

Fracture mechanics is perhaps the best example of how the fiction of singular stresses can be acknowledged, while still providing a framework for a quantitative measure of the severity of cracks. In addition, fracture mechanics has been developed for determining the resistance of materials to the growth of cracks, as well as for designing damage-tolerant structures. Bonded systems are quite rich in their array of potential fracture mechanisms. There can be cohesive cracks that grow entirely within the adhesive layer. Another possibility is cracks that grow

---

[*] Corresponding author. E-mail: kml@mail.utexas.edu

along an interface to create adhesive fracture. Cracks may branch into substrates or oscillate within the adhesive layer. All these can be dealt with by fracture mechanics. Cracks are not the only sources of high stresses or singularities in bonded systems: there are generally a multitude of so-called bimaterial corners that also excite singular stresses. They are admittedly of a slightly different nature from those generated by cracks, but nonetheless can often be handled in a similar manner.

The main emphasis of this chapter will be on the basic fracture mechanics concepts for cohesive and adhesive fracture with some extension to crack branching and crack nucleation from bimaterial corners. Most of the current fracture mechanics practice in testing adhesives and designing of adhesively bonded joints is limited to linear elastic fracture mechanics concepts. The development of the background material presented here will therefore be similarly constrained, except for the last section. Historically, fracture mechanics developed from energy balance concepts and examinations of stresses around crack tips. The adhesive fracture community has tended to favor the former, but both have useful features and will be carried forward in the discussions that follow.

Fracture in bonded systems can be viewed at several scales. In many cases, the adhesive layer itself may be ignored. Thus, if the adherends are the same, the crack will appear to be cohesive, or one in a homogeneous material. If the adherend materials differ and a crack is growing in the adhesive layer, but it is being ignored, then the crack appears to be an adhesive one, growing along the interface between the two different adherends. When the adhesive layer is accounted for, then cohesive and adhesive cracking are again possible, albeit from a slightly different perspective. As a result, the chapter will be divided into sections that deal with adhesive and cohesive fracture on the macroscopic and microscopic scales. The final section gives a brief overview of nonlinear effects.

2. Cohesive cracking

Cracks are considered to be cohesive when they grow entirely within the adhesive layer. The chemists and surface scientists are usually happy when this happens because it means that the bond between the adhesive and adherend is performing well. Cracks may also be considered to be cohesive if the adherends are the same and the adhesive layer is being ignored in the analysis.

2.1. Crack-tip stress analysis

If a linear elastic stress analysis is conducted to determine the stress distribution in the region surrounding a crack tip, it can be shown [21] that the stress state is given, with reference to Fig. 1, as:

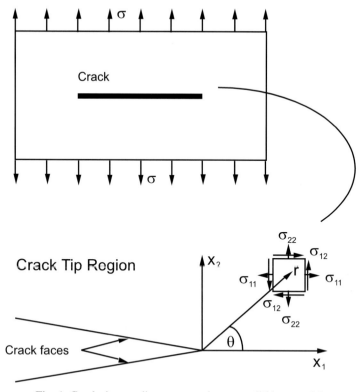

Fig. 1. Crack-tip coordinate system in a monolithic material.

$$\left\{\begin{array}{c} \sigma_{11} \\ \sigma_{22} \\ \sigma_{12} \end{array}\right\} = \frac{K_{\mathrm{I}}}{\sqrt{2\pi r}} \left\{\begin{array}{c} \sigma_{11}^{\mathrm{I}} \\ \sigma_{22}^{\mathrm{I}} \\ \sigma_{12}^{\mathrm{I}} \end{array}\right\} + \frac{K_{\mathrm{II}}}{\sqrt{2\pi r}} \left\{\begin{array}{c} \sigma_{11}^{\mathrm{II}} \\ \sigma_{22}^{\mathrm{II}} \\ \sigma_{12}^{\mathrm{II}} \end{array}\right\} + \left\{\begin{array}{c} T \\ 0 \\ 0 \end{array}\right\} + O\left(r^{1/2}\right). \tag{1}$$

The two-dimensional stress state with components σ_{11}, σ_{22}, and σ_{12} is represented in Eq. 1. The trigonometric functions are

$$\left\{\begin{array}{c} \sigma_{11}^{\mathrm{I}} \\ \sigma_{22}^{\mathrm{I}} \\ \sigma_{12}^{\mathrm{I}} \end{array}\right\} = \cos\theta/2 \left\{\begin{array}{c} 1 - \sin\theta/2 \sin 3\theta/2 \\ \sin\theta/2 \cos 3\theta/2 \\ 1 + \sin\theta/2 \sin 3\theta/2 \end{array}\right\} \tag{2a}$$

and

$$\left\{\begin{array}{c} \sigma_{11}^{\mathrm{II}} \\ \sigma_{22}^{\mathrm{II}} \\ \sigma_{12}^{\mathrm{II}} \end{array}\right\} = \left\{\begin{array}{c} -\sin\theta/2(2 + \cos\theta/2 \cos 3\theta/2) \\ \cos\theta/2(1 - \sin\theta/2 \sin 3\theta/2) \\ \sin\theta/2 \cos\theta/2 \cos 3\theta/2 \end{array}\right\} \tag{2b}$$

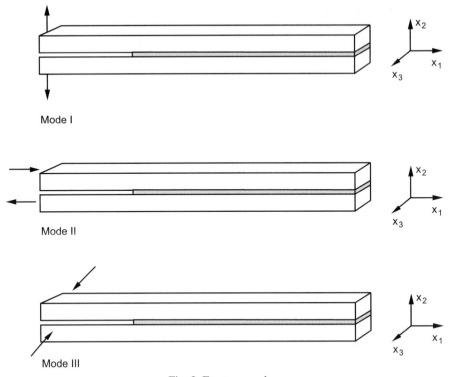

Fig. 2. Fracture modes.

The quantity T is the 'T-stress', a non-singular normal stress in the x_1-direction. The solution was obtained for a planar body under the action of in-plane loads and by enforcing the stress-free condition of the crack faces ($\theta = \pm\pi$). The solution given is the first two terms in a series and is valid in a region quite close to the crack tip and well removed from any external boundaries of the component. The stress-free condition and the sharp crack lead to the square root singular and trigonometric terms in the solution. These are the same for all cracks. The coefficients K_I and K_{II} are known as the mode I and II stress intensity factors, respectively, and reflect symmetric tensile opening and asymmetric in-plane shear sliding components due to the nature of the applied loads in Fig. 2, upper and middle panel, respectively. The stress intensity factors in Eq. 1 are undetermined from the asymptotic analysis and depend upon the globally applied loads and the complete geometry of the configuration, thus requiring a separate stress analysis. However, the important consequence of the asymptotic analysis is that the stress intensity factor distinguishes the local crack-tip stress distribution from one cracked configuration to another. Alternatively, one can say that, when two cracked configurations have the same stress intensity factor, the stress distributions are the same in the vicinity of the crack tips.

The mode I and II stress intensity factors are defined as

$$K_I = \lim \left\{ \sqrt{2\pi r}\,(\sigma_{22})_{\theta=0} \right\} \quad \text{and} \quad K_{II} = \lim \left\{ \sqrt{2\pi r}\,(\sigma_{12})_{\theta=0} \right\}, \tag{3}$$

and are generally a function of the applied load or stress, σ and geometry, represented here by the crack length a. Thus

$$K = K(\sigma, a). \tag{4}$$

The in-plane displacement components around the crack tip are given by

$$\begin{Bmatrix} u_1 \\ u_2 \end{Bmatrix} = \frac{K_I}{2\mu}\sqrt{\frac{r}{2\pi}} \begin{Bmatrix} \cos\theta/2(\kappa - 1 + 2\sin^2\theta/2) \\ \sin\theta/2(\kappa + 1 - 2\cos^2\theta/2) \end{Bmatrix}$$

$$+ \frac{K_{II}}{2\mu}\sqrt{\frac{r}{2\pi}} \begin{Bmatrix} \sin\theta/2(\kappa + 1 + 2\cos^2\theta/2) \\ -\cos\theta/2(\kappa - 1 - 2\sin^2\theta/2) \end{Bmatrix}, \tag{5}$$

where μ is the shear modulus of the material, $\kappa = 3 - 4\nu$ is for plane strain, $\kappa = (3 - \nu)/(1 + \nu)$ is for plane stress, and ν is the Poisson ratio of the material.

For antiplane shear (Fig. 2, lower panel), the mode III stress intensity factor is introduced to obtain near-tip stresses and displacements:

$$\begin{Bmatrix} \sigma_{31} \\ \sigma_{32} \end{Bmatrix} = \frac{K_{III}}{\sqrt{2\pi r}} \begin{Bmatrix} -\sin\theta/2 \\ \cos\theta/2 \end{Bmatrix}, \tag{6}$$

$$u_3 = \frac{2K_{III}}{\mu}\sqrt{\frac{r}{2\pi}}\sin\theta/2, \tag{7}$$

$$K_{III} = \lim_{r \to 0} \left\{ (2\pi r)^{1/2}\sigma_{32}|_{\theta=0} \right\}. \tag{8}$$

Again, it is the stress intensity factor that distinguishes the crack-tip stress distribution from one loading and crack geometry to another.

A large variety of methods exist for determining the stress intensity factor associated with a particular configuration as can be seen from the compilation by Tada et al. [55]. When finite element methods are used for the stress analysis of cracked components, stress intensity factors may be extracted by examining the displacement solution near the crack and making use of Eqs. 5 and 7. Some finite element codes make use of so-called hybrid elements that contain fracture parameters as degrees of freedom of the elements and hence yield them directly as part of the solution without further post-solution processing. Many other codes use energy principles and the relationship between the stress intensity factor and the energy-release-rate parameter. Quite often, the stress intensity solution for a particular loading and geometry can be cast in the form

$$K = \sigma\sqrt{\pi a}\,Q(a/W), \tag{9}$$

where σ represents the applied stress, a the crack length and $Q(a/W)$ is known as

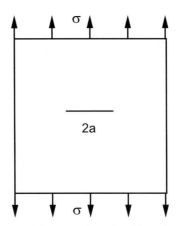

Fig. 3. Infinite plate with a central crack subjected to tensile loading.

the configuration factor and W is another characteristic dimension of the cracked component. When $Q = 1$, we have the stress intensity factor for an infinite plate with a central crack of length $2a$ under a remote tensile stress σ (Fig. 3). In that case, the stress intensity factor is K_I, the mode I stress intensity factor. If the remote loading had been a shear stress τ, then $K_\mathrm{II} = \tau \sqrt{\pi a}$. The infinite plate case is essentially the baseline and the configuration factor is what is determined from handbook solutions or finite element analyses.

2.2. Crack growth criteria

A natural consequence of the asymptotic solutions given above is that the stress intensity factor may be used as a crack initiation criterion. Suppose that a crack initiation experiment had been conducted on a given material with recordings of the applied load and crack length at the onset of crack growth. These could be substituted into the appropriate expression for the stress intensity factor in Eq. 9, thereby yielding the value associated with crack initiation. A component made of the same material, but having a different geometry and/or loading, would be expected to crack at the same critical value of the stress intensity factor that was noted in the first experiment. The criterion for a crack to become what is known as a fast crack is expressed as

$$K(\sigma, a) = K_\mathrm{c}, \tag{10}$$

where K_c is the critical value of the total stress intensity factor and serves as a measure of the resistance of the material to crack initiation or fast crack growth. The quantity K_c is also known as the fracture toughness of the material. For the purposes of this discussion K_c is a 'generic' fracture toughness under any mode or combination of fracture modes.

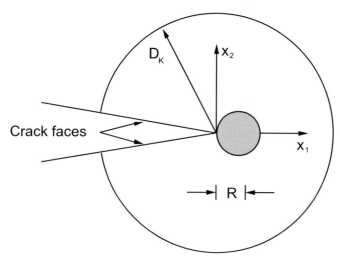

Fig. 4. Domain of validity of the singular elastic solution.

The elasticity solutions for the stresses in Eqs. 1 and 6 suggest that the stresses at the crack tip are infinite. In practice, the stresses exceed the yield strength of the material and zones of damaged and inelastically deformed material surround the crack tip. The damage may take the form of voids, crazes, etc., whereas the inelastic deformations could be due to plasticity, viscoelasticity or viscoplasticity. The extent of the plastic zone, R, under mode I conditions may be obtained by considering σ_{22} ahead of the crack ($\theta = 0$) and equating it to the tensile yield strength of the material, (σ_y).

Eqs. 1 and 2 indicate that

$$R = \frac{1}{2\pi} \left(\frac{K_I}{\sigma_y} \right)^2. \tag{11}$$

Within this region (Fig. 4), the linear elastic solution is invalid. However, if the region $r < R$ is much smaller than the region over which the elastic asymptotic solution dominates ($r < D_K$), then the stress intensity factor can still be used to characterize the stress distribution that controls crack initiation.

There are a number of situations where crack growth may occur at stress intensity factor values that are below (K_c), the toughness of the material, which marks the onset of fast growth. This subcritical growth arises under cyclic fatigue loadings, when the material's time-dependent behavior is important and also from environmental effects. Under these conditions, resistance to crack growth is characterized by correlation of crack growth rates with the stress intensity factor as shown schematically in Fig. 5. Alternatively, one can say that for fatigue growth

$$\frac{da}{dN} = f(\Delta K) \tag{12}$$

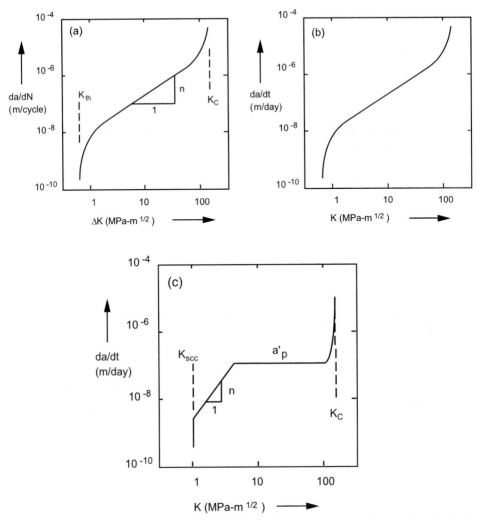

Fig. 5. Representations of the resistance to subcritical crack growth: (a) fatigue crack growth; (b) viscoelastic crack growth; (c) environmentally assisted crack growth.

represents the resistance to fatigue crack propagation, where N is the number of cycles and ΔK is the change in stress intensity factor over one cycle. If σ_{max} and σ_{min} are the maximum and minimum stresses in a constant-amplitude loading, then

$$\Delta K = (\sigma_{max} - \sigma_{min}) \sqrt{\pi a} \, Q(a/W). \tag{13}$$

The linear portion of the double logarithmic plot in Fig. 5 suggests a power law of the form

$$\frac{da}{dN} = A(\Delta K)^n, \tag{14}$$

where A and n are material properties that represent the resistance of the material to fatigue crack growth. The power law is also known as the Paris Law after a pioneering fracture mechanics researcher.

Once A and n have been determined, we are in a position to predict fatigue crack growth in a structural component. This is accomplished by determining the stress intensity factor solution for the most likely crack path in the structure being designed and substituting it into the integral of Eq. 14. In this way, the number of cycles N that it will take to reach a crack length a, starting from an initial crack length a_0 is given by

$$N = \int_{a_0}^{a} \frac{da}{A(\Delta K)^n}. \tag{15}$$

If the stress intensity factor solution for the cracked component has the form of Eq. 9 and it is subjected to constant amplitude loading, then Eq. 15 becomes

$$N = \frac{1}{A(\sigma_{max} - \sigma_{min})^n} \int_{a_0}^{a} \frac{da}{\left(\sqrt{\pi a}\, Q(a/W)\right)^n}. \tag{16}$$

The number of cycles it will take for the fatigue crack to become a fast one is obtained by making the upper limit of the integral a_c, which can be determined from

$$\sqrt{\pi a_c}\, Q(a_c/W) = \frac{K_c}{\sigma_{max}}. \tag{17}$$

The initial crack length a_0 may correspond to minimum detectable flaw sizes or some other convenient scale.

In the case of time-dependent growth due to viscoelastic effects [22,32], the following correlation can often (Fig. 5b) be made

$$\frac{da}{dt} = C(K)^m, \tag{18}$$

which can then be integrated in a similar manner to the fatigue case in order to predict the crack growth history so that inspection frequencies and the probable lifetime of the component can be established.

Time-dependent crack growth may also arise when solvents are being absorbed into the adhesive layer [10,46,49]. In this case, the correlation between crack growth rates and stress intensity follows the behavior shown schematically in Fig. 5c. The initial rising portion of the curve is also a power law. Diffusion effects are dominant in this regime. When crack growth rates outrun diffusion, the crack speeds become independent of the stress intensity factor in what is known as the plateau region. Thereafter, fast crack growth mechanisms begin to dominate and the crack speeds rise sharply with stress intensity factor.

Combinations of cyclic loading, viscoelastic effects and solvent diffusion may also occur. Their synergistic effects remain an open question at this time with no experimental or analytical/numerical works that address all three simultaneously. Some that perhaps come the closest to addressing all three issues include Liechti and Arzoumanidis [30], Birksham and Smith [3], Jethwa and Kinloch [20], Sancaktar [50], and Wylde and Spelt [63].

The procedures outlined above can be classified as a finite life design approach. However, crack growth rates, particularly for fatigue crack growth, may be so high that it is better to preclude altogether the further growth of any preexisting flaw. This can be done by noting the threshold values shown in Fig. 5, where K_{th} and K_{SCC} are the fatigue and environmentally assisted thresholds, respectively. For stress intensity factors below these values no crack growth will occur. A more conservative design approach is therefore to consider the smallest preexisting flow that can be detected in a structure and make sure that the stress intensity factor associated with the maximum load is lower than K_{th} or K_{SCC}.

2.3. Energy concepts

The analysis of crack initiation in brittle materials was initially approached from an energy balance viewpoint [15]. It was postulated that during an increment of crack extension, da, there can be no change in the total energy, E, of the cracked body. The total energy E was viewed as being composed of the potential energy of deformation, Π, and the surface energy, S. Therefore, during crack extension:

$$dE = d\Pi + dS = 0. \tag{19}$$

The rate of change of potential energy with respect to crack extension, da, in a planar component of thickness, b is defined as the energy release rate G

$$G = \frac{-d\Pi}{bda}. \tag{20}$$

If the surface energy density is denoted by γ then $dS = 2\gamma b\,da$ for the two increments of fracture surfaces formed during crack extension. Eqs. 19 and 20 can then be combined to yield

$$G = 2\gamma \tag{21}$$

as the criterion for crack extension in a brittle solid. Because G is derivable from a potential function it is often referred to as a crack driving force. Thus Eq. 21 represents the balance that is achieved at the point of crack initiation between the

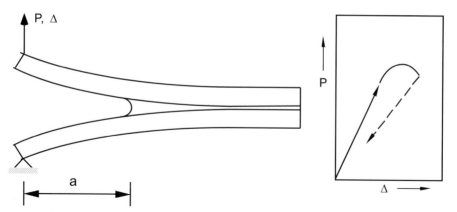

Fig. 6. A cracked adhesive joint and its load-displacement response during loading, crack growth and unloading.

energy provided by the loaded component and the energy required for the creation of new surface or the fracture resistance. The fracture resistance is a characteristic of the material, whereas the energy release rate depends upon the loading and geometry of the crack component.

Perhaps the simplest and most common method of determining the energy release rate is to consider the change in component compliance as a crack grows in it. With reference to Fig. 6, we can see that a cracked adhesive joint is being subjected to a force P and Δ is the associated displacement through which P does work. The potential energy of the component is the difference between the strain energy, U, and the work done by the force:

$$\Pi = U - P\Delta. \tag{22}$$

For a linearly elastic material, and small displacements, the strain energy $U = 1/2 P\Delta$ and Eqs. 20 and 22 yield

$$G = \frac{P}{2b}\left(\frac{d\Delta}{da}\right)_P. \tag{23}$$

The compliance, C, of a linearly elastic component is given by $C = \Delta/P$, and the energy release rate in Eq. 23 becomes

$$G = \frac{P^2}{2b}\frac{dC}{da}. \tag{24}$$

The same expression can be obtained for a fixed grip displacement loading and a compliant loading device. This simple result is quite powerful because it can be used to determine energy release rates directly from load-displacement records of fracture toughness tests without the need for any further stress analysis. Furthermore, in the event that laminated beams are used for fracture tests, the energy release rate solutions can be obtained from relatively simple beam theory analyses. Eq. 24 does not hold for blistering thin films or peeling where there may be geometrical and/or material nonlinearities.

Because the stress intensity factor and energy release rate approaches both apply to linearly elastic components, they can, in fact, be related [21] through

$$G = \frac{1}{\bar{E}} \left(K_{\mathrm{I}}^2 + K_{\mathrm{II}}^2 \right) + \frac{1}{2\mu} K_{\mathrm{III}}^2, \tag{25}$$

where $\bar{E} = E/(1 - v^2)$ for plane strain and $\bar{E} = E$ for plane stress. Therefore, for a mode I crack at initiation

$$G_{\mathrm{c}} = K_{\mathrm{c}}^2/\bar{E}. \tag{26}$$

When G_{c} is computed from K_{c} as indicated in Eq. 26, it is generally several orders of magnitude higher than the surface energy 2γ. Only for very brittle materials is the equality $G_{\mathrm{c}} = 2\gamma$ preserved. In the tougher materials most of the energy released at crack initiation is dissipated in inelastic deformation near the crack tip. Because of the equivalence between K and G noted above, the energy release rate can also be used to characterize fatigue and environmentally assisted crack growth in a manner similar to that depicted in Fig. 5.

3. Adhesive cracking

In this case, cracks run along the interface between two materials due to interactions between the stress field in the adhesive layer and spatial variations in fracture properties. The cracks are not generally free to evolve as mode I cracks, as was the case for cohesive cracks, and mixed-mode fracture concepts (combinations of tension and shear) have to be considered. Mode II or shear components are induced, even in what appear to be nominally mode I loadings, due to differences in moduli about the interface. Again, if the presence of the adhesive layer is being ignored and the adherends are dissimilar, then a crack appears to be adhesive (i.e. an adhesion failure) on the macroscopic scale.

3.1. Crack-tip stresses

The dominant stresses near the tip of an interface crack with material 1 above material 2 (Fig. 7) are given by

$$\sigma_{\alpha\beta} = \frac{\mathrm{Re}[K r^{i\varepsilon}]}{(2\pi r)^{1/2}} \sigma_{\alpha\beta}^{\mathrm{I}}(\theta, \varepsilon) + \frac{\mathrm{Im}[K r^{i\varepsilon}]}{(2\pi r)^{1/2}} \sigma_{\alpha\beta}^{\mathrm{II}}(\theta, \varepsilon). \tag{27}$$

The bimaterial constant $\varepsilon = \frac{1}{2\pi} \ln \left(\frac{1-\beta}{1+\beta} \right)$, where $\beta = \frac{\mu_1(\kappa_2-1)-\mu_2(\kappa_1-1)}{\mu_1(\kappa_2+1)+\mu_2(\kappa_1+1)}$, is one of the Dundurs [11] parameters for elastic bimaterials. The quantity $\kappa_i = 3 - 4v_i$

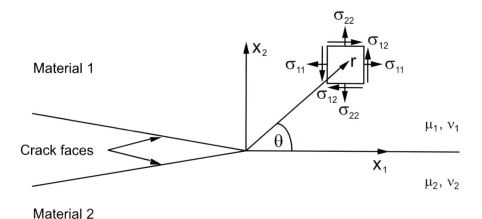

Fig. 7. Crack-tip coordinate system for a crack at a bimaterial interface.

for plane strain, $\kappa_i = (3 - \nu_i)/(1 + \nu_i)$ for plane stress and the μ_i and ν_i are the shear moduli and Poisson's ratios of the upper and lower materials. The quantity $K = K_1 + iK_2$ is the complex stress intensity factor. Its real and imaginary parts are similar to the mode I and II stress intensity factors for monolithic materials. The functions $\sigma_{\alpha\beta}^{I}(\theta, \varepsilon)$ and $\sigma_{\alpha\beta}^{II}(\theta, \varepsilon)$ are given in polar coordinates by Rice et al. [45]. The transformation to Cartesian coordinates is routine but tedious, so the results are not given here.

The stresses are normalized so that the stresses ahead of the crack tip are given by

$$\sigma_{22} + i\sigma_{12} = \frac{(K_1 + iK_2)}{(2\pi r)^{1/2}} r^{i\varepsilon}, \tag{28}$$

where $r^{i\varepsilon} = \cos(\varepsilon \ln r) + i \sin(\varepsilon \ln r)$. This is an oscillating singularity, which leads to interpenetration of the crack faces as can be seen by examining the crack flank displacements $\delta_i = u_i(r, \pi) - u_i(r, -\pi)$, which are given by

$$\delta_1 + i\delta_2 = \frac{8}{(1 + 2i\varepsilon)\cosh(\pi\varepsilon)} \frac{K_1 + iK_2}{E^*} \left(\frac{r}{2\pi}\right)^{1/2} r^{i\varepsilon}, \tag{29}$$

where

$$\frac{1}{E^*} = \frac{1}{2}\left(\frac{1}{\bar{E}_1} + \frac{1}{\bar{E}_2}\right) \tag{30}$$

and $\bar{E}_i = E_i/(1 - \nu_i^2)$ for plane strain and $\bar{E}_i = E_i$ for plane stress.

The energy release rate for crack advance along the interface is

$$G = \frac{(1 - \beta^2)}{E^*}\left(K_1^2 + K_2^2\right). \tag{31}$$

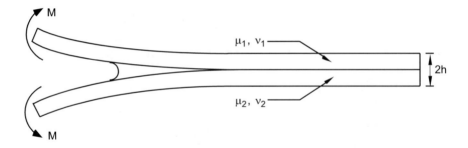

Fig. 8. Double cantilever beam specimen under uniform bending.

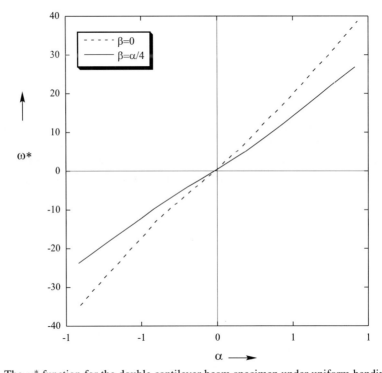

Fig. 9. The ω^* function for the double cantilever beam specimen under uniform bending [19].

A number of stress intensity factor solutions have been developed over the years. Several solutions are given in the review article by Hutchinson and Suo [19]. One example is the stress intensity factors for a bimaterial double cantilever beam subjected to uniform bending (Fig. 8). In this case

$$K_1 + i K_2 = 2\sqrt{3} M h^{-3/2 - i\varepsilon} \left(1 - \beta^2\right)^{-1/2} e^{i\omega^*(\alpha,\beta)}. \tag{32}$$

The function ω^* is given in Fig. 9. The quantity $\alpha = \left(\bar{E}_1 - \bar{E}_2\right)/\left(\bar{E}_1 + \bar{E}_2\right)$ is the other Dundurs [11] parameter for bimaterial systems. An examination of Eq.

32 reveals that, even though the loading is symmetric, the mode II component is nonzero, due to the difference in material properties across the interface. This so-called mode-mix is defined as the ratio of the stress intensity factors through

$$\psi = \tan^{-1}\left[\frac{\mathrm{Im}\left(Kl^{i\varepsilon}\right)}{\mathrm{Re}\left(Kl^{i\varepsilon}\right)}\right], \tag{33}$$

where l is a suitable reference length. If l is chosen so that it lies within the zone of K dominance ($r < D_K$ in Fig. 4), then an equivalent expression for the mode-mix is

$$\psi = \tan^{-1}\left[\frac{\sigma_{12}}{\sigma_{22}}\right]_{r=l}. \tag{34}$$

The choice of l is arbitrary, but is often based on some material length scale such as the plastic zone size. If ψ_1 is the mode-mix associated with a length l_1, then the mode-mix ψ_2 associated with a different length scale l_2 is

$$\psi_2 = \psi_1 + \varepsilon \ln(l_2/l_1). \tag{35}$$

This transformation is very useful when comparing toughness data. The toughness of many interfaces is a function of mode-mix and it is important to note what length scale is used in the measure of mode-mix. Two sets of toughness data reported on different length scales can be brought into registration using Eq. 35.

3.2. Crack growth criteria

Just as we did for cohesive cracks, we distinguish between fast and slow adhesive cracks. However, whereas cohesive cracks tend to follow a path where $K_{\mathrm{II}} = 0$, adhesive cracks must, by definition follow the interface. This usually means that a mixed-mode fracture criterion that involves tensile and shear components must be developed. For fast crack growth, the most common approach has been to plot the critical value of the energy release rate as a function of the fracture mode-mix ψ. We can think of this as a two-parameter criterion that involves energy and stress intensity parameters. One example for a glass–epoxy interface [28,31] is shown in Fig. 10. There are two sets of data, one for 6 mm (1992) and the other for 2 mm (1995) thick specimens. The toughness rises sharply for positive and negative shear, but not in the same way. There is an asymmetry to the shear-induced toughening, which, in this case was caused by differences in the amount of plastic deformation that are induced by positive and negative shear [52]. Although a by-product of the explanation just given is that the toughness envelope for a bimaterial interface can now be predicted via cohesive zone modeling once the intrinsic (minimum) toughness and inelastic deformation characteristics are known, many designers of bimaterial interfaces will make use of measured envelopes. Predictions of fast cracking at bimaterial interfaces will involve a

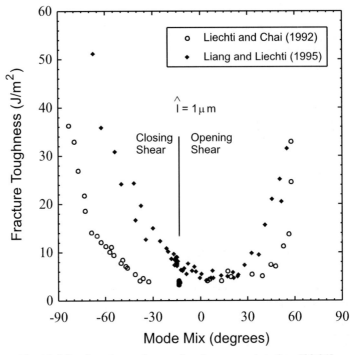

Fig. 10. Mixed-mode toughness of a glass epoxy interface [28,31].

determination of energy release rate and mode-mix in the cracked component followed by a comparison with the toughness envelope at the appropriate value of mode-mix.

Durability analyses of slow crack growth under fatigue or static loadings are usually conducted in the same way as was described earlier for cohesive cracking. The fracture parameter that is generally used for correlating with crack growth rates is the energy release rate. This generally seems to be sufficient for accounting for mode-mix effects [47,48].

4. Crack growth in sandwiched layers

The analyses described in the previous sections can be applied to adhesive joints with, respectively, similar and dissimilar adherends, but ignoring the adhesive layer itself. However, it is possible to move down one scale level and account for various types of crack growth (Fig. 11) within the adhesive layer by making use of the results of Fleck et al. [14]. These are restricted to situations where the adherends are the same. Nonetheless they are rather powerful and simple, because they can be used to extend existing analyses at the macroscopic level to the microscopic one without extensive numerical computations.

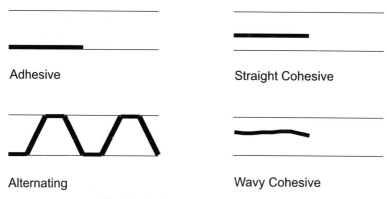

Fig. 11. Crack paths in adhesive layers.

We first consider the case where a straight crack is growing at some distance c from the lower adherend in an adhesive layer of thickness h (Fig. 11, right). At the scale level of the adherends, we can suppose that we have the stress intensity factors K_I^∞ and K_{II}^∞ for a cracked homogeneous joint. Then the macroscopic or global energy release rate is

$$G^\infty = \frac{1}{\bar{E}_1}\left(K_I^{\infty 2} + K_{II}^{\infty 2}\right). \tag{36}$$

At the same time, the energy release rate associated with the crack in the adhesive layer itself is

$$G = \frac{1}{\bar{E}_2}\left(K_1^2 + K_2^2\right). \tag{37}$$

From the path independence of a quantity known as the J integral [43], which is also the energy release rate when linear elastic fracture mechanics holds, we have that $G = G^\infty$. This, together with the fact that stresses are linearly related, allows the local stress intensity factors to be related to the global ones through

$$(K_1 + i K_2) = \left(\frac{1-\alpha}{1+\alpha}\right)^{1/2}\left(K_I^\infty + i K_{II}^\infty\right)e^{i\varphi(c/h,\alpha,\beta)}, \tag{38}$$

where $\varphi = \psi - \psi^\infty$ can be thought of as the shift in phase angle between the global and local stress intensity factors. Fleck et al. [14] have shown that

$$\varphi = \varepsilon \ln\left(\frac{h}{c} - 1\right) + 2\left(\frac{c}{h} - \frac{1}{2}\right)\tilde{\varphi}(\alpha,\beta), \tag{39}$$

where $\tilde{\varphi}$ is given in Fig. 12. If a joint is loaded under globally mode I conditions and the crack runs along the mid thickness of the layer, then Eqs. 38 and 39 indicate that

$$K_1 = \left(\frac{1-\alpha}{1+\alpha}\right)^{1/2} K_I^\infty \quad \text{and} \quad K_2 = 0. \tag{40}$$

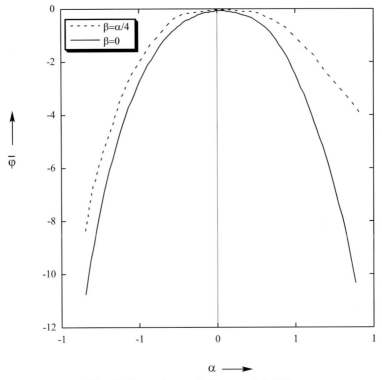

Fig. 12. Dependence of $\tilde{\varphi}$ on α and β [19].

Thus, if a polymeric adhesive layer is being used to join two stiffer metallic, composite or ceramic adherends $\alpha > 0$ and $K_1 < K_I^\infty$. This means that the crack in the adhesive layer is shielded from the global loading.

For an interface crack we have

$$K_1 + iK_2 = \left(\frac{1-\alpha}{1-\beta^2}\right)^{1/2} \left(K_I^\infty + K_{II}^\infty\right) h^{-i\varepsilon} e^{i\omega}, \tag{41}$$

where the ω function is plotted in Fig. 13.

One might think that an initially straight cohesive crack (Fig. 14) would turn (kink) upon the slightest application of K_{II}^∞. However, it turns out that the elastic mismatch that is contained in Eq. 38 allows a straight crack to continue as such for approximately $K_{II}^\infty \leq 0.1 K_I^\infty$. To see this, we take the common assumption that cracks in homogeneous materials grow in such a way that the local mode II component $K_{II} = 0$. This condition, when substituted into Eq. 38 yields a relationship between the location of the crack (c/h) and the global mode-mix ψ^∞ as

$$\varphi(c/h, \alpha, \beta) = -\psi^\infty. \tag{42}$$

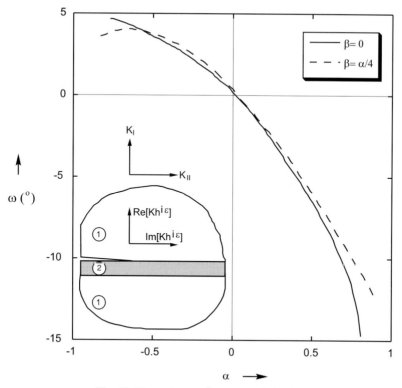

Fig. 13. Dependence of ω on α and β [19].

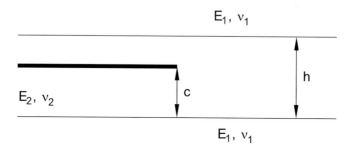

Fig. 14. The location of a straight crack trapped in a sandwich layer.

The solution is plotted in Fig. 15 for several values of α and $\beta = \alpha/4$. There can be no straight paths for zero mismatch ($\alpha = \beta = 0$). The 10% level of K_{II}^{∞} that was referred to earlier can lead to straight cracks when the mismatch is relatively large.

The next question that arises is whether or not an initially straight cohesive

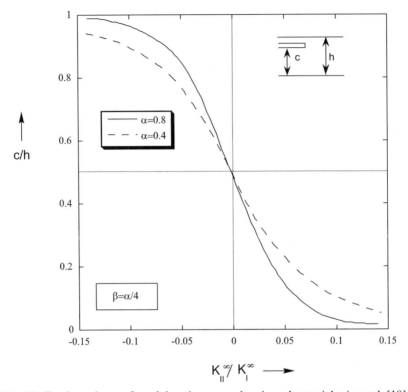

Fig. 15. The dependence of crack location on mode-mix and material mismatch [19].

crack can remain so in the presence of slight perturbations. This was first addressed for cracks in monolithic materials by Cottrell and Rice [9]. They postulated that the T-stress controls the directional stability of cracks. Chai [4–6] made the extension to adhesively bonded joints and identified an oscillating crack path, where the crack periodically touched the top and bottom adherends (composite or aluminum) under nominally mode I loading. The period was quite consistent at 3–4 times the bond thickness. Fleck et al. [14] and Akinsanya and Fleck [1,2] conducted an analytical investigation of the problem. Crack paths were categorized into those that settled on the centerline no matter what the original elevation, those that oscillated gently about the centerline, those that approached the interface gradually and those that approached it at a large angle. Just which pattern dominates in any particular situation was driven by the signs of two parameters: the local T-stress and $\frac{\partial K_{\mathrm{II}}}{\partial c}$. For nominally mode I loading ($K_{\mathrm{II}}^\infty = 0$), the local T-stress is given by

$$T = \frac{1-\alpha}{1+\alpha} T^\infty + \sigma^R + \left(\frac{1-\alpha}{1+\alpha}\right)^{1/2} c_{\mathrm{I}} \frac{K_{Ic}}{\sqrt{h}}, \tag{43}$$

which shows that it is controlled by the global T-stress T^∞, thermal and intrinsic residual stresses and the value of the global mode I stress intensity. The latter is related to the local toughness K_{Ic} via Eq. 40. The coefficient c_I (and its mode II partner for cases where $K_{II}^{\infty} \neq 0$) was tabulated for several values of $c/h, \alpha, \beta$. The global T-stress T^{∞} had previously been determined by Larsson and Carlsson [24] for several common bonded joints. Chen and Dillard [7] controlled the residual stress levels in the adhesive layer by pre-stretching double cantilever beam specimens prior to fracture testing. This gave them sufficient control over the local T-stress that a number of the crack patterns that had been predicted were indeed observed. Interestingly, the toughness of the joints was quite similar, irrespective of crack pattern. A subsequent paper [8] dealt with the effects globally mixed-mode loading and load rates. It was found that crack paths were stabilized at the interface for all T-stress levels when the mode II component was greater than 3%. Higher crack propagation rates led to more cohesive cracking and waviness. The toughness of specimens decreased with increasing mode II component. While seemingly at odds with previous results on mode-mix effects, this latter result can explained by the different crack paths that were taken as the mode-mix changed.

5. Crack nucleation from bimaterial corners

Adhesively bonded joints abound with bimaterial corners. These can be sources of crack nucleation due to the stress concentrations that can be associated with them. The simplest situation arises when one of the materials is comparatively rigid. In that case, we have a plate, which is clamped along one boundary and free on the other (Fig. 16). This is one of several combinations of boundary conditions at a corner that Williams [61] considered. With a polar coordinate system originating at the corner, it can be shown that the stresses have the form

$$\sigma \propto r^{\lambda-1}. \tag{44}$$

The value of λ depends upon the corner angle θ_1 of the plate and the Poisson ratio of the material (Fig. 17 with $\nu = 0.3$). For $\theta_1 > 60°$, $\lambda < 1$, which leads to singular

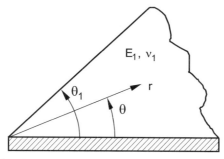

Fig. 16. The corner of a plate which is clamped along one boundary and free on the other.

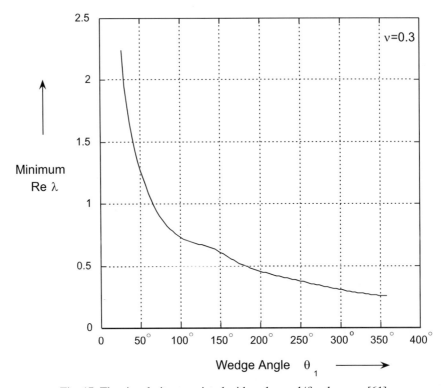

Fig. 17. The singularity associated with a clamped/fixed corner [61].

stresses that tend to infinity as the corner is approached. At the same time it can be seen that, for $\theta_1 < 60°$, the stresses are non-singular and the stress concentration does not exist. Thus, where possible, corner angles should be kept below 60°. Sometimes this can happen naturally when adhesive spews from the joint during processing. Notice that, when $\theta_1 = \pi$, the crack situation is recovered and $\lambda = 0.5$.

When both materials are compliant, singular stresses can also occur. The situation is more complicated because now there may be multiple singularities and they depend on the elastic properties of each material in addition to the corner angles in each material (Fig. 18). There are several ways to present the state of stress near a bimaterial corner. One common approach is given below.

$$\sigma_{\alpha\beta} = \sum_{i=1}^{N} K_{ai} r^{\lambda_i - 1} \bar{\sigma}_{\alpha\beta i}(\theta) + K_{a0} \bar{\sigma}_{\alpha\beta 0}(\theta) \quad (\alpha,\beta = r,\theta). \tag{45}$$

This indicates that there can be N singularities with strength $(\lambda_i - 1)$. The angular variation in the stresses is given by the $\bar{\sigma}_{\alpha\beta i}(\theta)$ terms and depends on the elastic constants of the materials, the corner angles (θ_1, θ_2) in each material and the boundary conditions on the sides that are not joined. The singularities are determined on the basis of asymptotic analyses that satisfy the local boundary

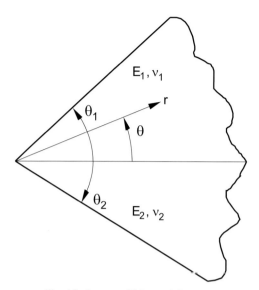

Fig. 18. A general bimaterial corner.

conditions near the corner. The singularities can be real, as was the case for the crack in a monolithic material, or complex [17], as in the case of the interface crack. The stress intensity factors K_{ai} have to be determined from the overall or global boundary conditions. When there is only one singular term, the interpretation of Eq. 45 is analogous to the crack problem, where $\lambda = 0.5$. It turns out that 90° corners give rise to just one singular term. Quite a lot of work has been done with them both on the stress analysis front and in determining the corner toughness K_{ac} for various combinations of adhesives and adherends ([38–42] and Chapter 5). Other corner angles and material combinations that give rise to single eigenvalues have been studied by Qian and Akinsanya [37] and Dunn et al. [12]. As a result, it has been possible to predict when cracks in one configuration will nucleate, based on experiments to determine K_{ac} on another. Needless to say, the corner angle and material combination have to be the same, only the global conditions differ. If one has a choice of corner angles in a particular design then a consequence of this approach is that the corner toughness K_{ac} must be found for each corner angle [16]. Mohammed and Liechti [33], recently remedied this situation for cracks that nucleate along the interface by making use of an energy approach instead. A cohesive zone model was used to represent the behavior of the interface. The cohesive zone model parameters, which can be thought of as representing the toughness and strength of the interface, were determined from experiments on and analysis of an interface crack ($\alpha = 0$) between the materials of interest. Since the interface was still the same for all corner angles and the crack nucleated along the interface, the same cohesive zone parameters were used with other corners and found to predict (Fig. 19) the nucleation load and near-corner

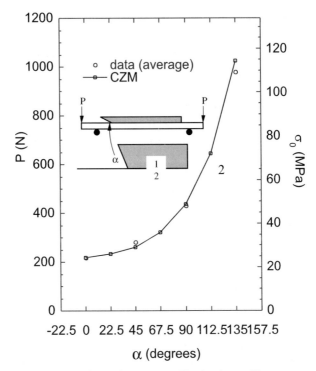

Fig. 19. Crack nucleation loads for various epoxy (1)–aluminum (2) corner angles ($\alpha = \pi - \theta_1$ and $\theta_2 = \pi$) [33].

displacements very well. Thus in this approach, only one corner was needed for calibration purposes. Another advantage is that it can be used for corners that give rise to multiple singularities. Such an energy-based approach could presumably be used even if the crack did not grow along the interface, although a suitable calibration specimen would have to be employed.

Corner cracks may also initiate under fatigue loading. In fact this may be the most common form of nucleation. Nonetheless, this problem seems to have received relatively little attention in the open literature. Lefebvre and Dillard [26,27] considered an epoxy wedge on an aluminum beam under cyclic loading. They chose corner angles (55°, 70° and 90°) that resulted in one singularity. A stress intensity factor based fatigue initiation envelope was then developed.

6. Nonlinear effects

All of the analyses of cracks and corners that were described above were based on the linearly elastic responses of the materials. It was recognized that this would lead to some physically unreasonable results very close to the crack front or corner where yielding would more likely occur. Nonetheless, as long as this yielding zone was small in scale, the elastic analysis was sufficient for characterizing the stress state. Nonlinear effects enter when the yielding or plastic zone becomes a dominant feature.

For materials that exhibit a strain hardening or a rising stress–strain curve response, the stresses are still singular. Hutchinson [18] and Rice and Rosengren [44] demonstrated this for cohesive cracks using a J_2-deformation plasticity approach. Under these conditions, the stress field became known as the HRR stresses with

$$\sigma_{\alpha\beta} = \left(\frac{J}{\alpha\varepsilon_y\sigma_y I_n r}\right)^{1/(n+1)} \bar{\sigma}_{\alpha\beta}(\theta) \quad (\alpha,\beta = 1,2), \tag{46}$$

where the material behavior followed the Ramberg–Osgood relation

$$\frac{\varepsilon}{\varepsilon_y} = \frac{\sigma}{\sigma_y} + \alpha\left(\frac{\sigma}{\sigma_y}\right)^n \tag{47}$$

and ε_y and σ_y are its yield strain and strength, respectively. The quantity I_n depends on the exponent n in Eq. 47 and whether plane stress or plane strain is dominant. The quantity J in Eq. 46 is the path-independent J-integral introduced by Rice [43]. It reduces to the energy release rate when the material behaves in a linearly elastic manner. In Eq. 46, J appears as the intensity of the HRR stress field and forms the basis for plastic fracture mechanics. With reference to Fig. 20, if the fracture process zone (region where material breakdown occurs) is small compared to the region where the asymptotic solution Eq. 46 dominates (i.e. $R \ll D_J$), then J controls fracture. As a result, fracture occurs when $J(\sigma,a) = J_c$. There are many instances when J must be greater than J_c for further crack growth to occur. This is known as resistance curve behavior (Fig. 21). The J integral can still control crack growth if the crack extension at any time is such that $\Delta a \ll D_J$ and $\frac{b}{J}\frac{dJ}{da} \gg 1$, where b is the thickness of the specimen.

In the approach just described, the actual behavior of the material in the fracture process zone is ignored. However, with recent advances in computer power and numerical methods, cohesive or fracture process zone modeling has become practical [35,36,59]. It has been applied to the fracture of interfaces [52,53,57,58], bonded joints [54,56,64], thin film debonding [29,51,60] and rate-dependent cohesive crack growth [23] and rubber/metal debonding [62]. In all these cases, the constitutive behavior or the traction-separation law of the failing region is an explicit component of the analysis. Calibrations of traction-separation laws have been conducted on the basis of measurements local quantities such as crack-tip displacements or crack length. As a result, it has become possible to reproduce resistance curve behavior numerically.

Elastoplastic analyses of bimaterial corners have also been conducted, albeit more recently. A singularity analysis based on J_2-deformation theory was devel-

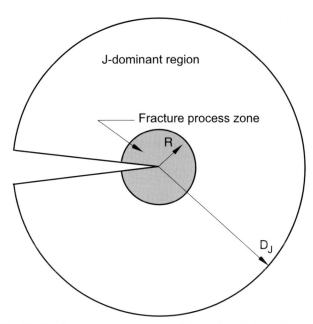

Fig. 20. The fracture process zone and the region of J dominance.

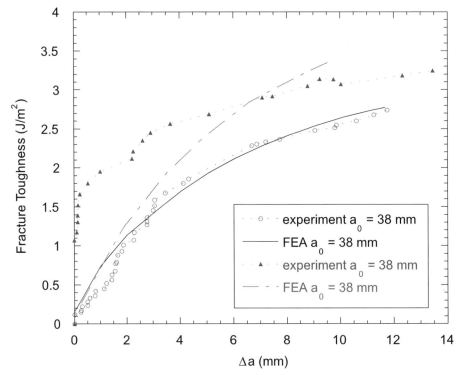

Fig. 21. Resistance curve behavior in a composite–epoxy–composite adhesively bonded joint under globally mode I loading [54].

oped by Lau and Delale [25] for corners and by Duva [13] for joints. Reedy [39] obtained the stress intensity factors for butt joints using the above theory. Mohammed and Liechti [34] extended Reedy's work to consider all possible angles in scarf joints.

7. Crack opening interferometry

One way of providing an overview of much of the foregoing material is to consider measurements of displacements near a crack front. It will also provide some points to ponder. A method that the author has used successfully is crack opening interferometry. This requires that at least one of the components of the bonded joint under consideration is transparent to the wavelength of the radiation being used. At first, especially considering the visible spectrum, this may seem to be rather restrictive. However, infrared opens up many practical microelectronics applications due to silicon's transparency to it. The technique also requires that the crack surfaces be quite planar. This condition is not always met by cracks in monolithic materials, but interfacial or sub-interfacial cracks often meet the planarity requirements.

A schematic of the apparatus used to measure the displacements of the crack faces (or crack opening displacements) is shown in Fig. 22. A beam of monochromatic light is introduced at zero angle of incidence to the interface via a beam splitter and a transparent adherend. The upper crack face reflects some of the incident beam back, but some light is also transmitted across the air gap separating the crack faces to the second crack face, where it is also reflected back. The amount being reflected back depends on the degree of transparency of the second material. The two reflected beams are out of phase with one another due to the

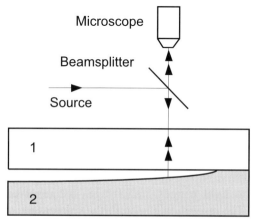

Fig. 22. Schematic of crack opening interferometry.

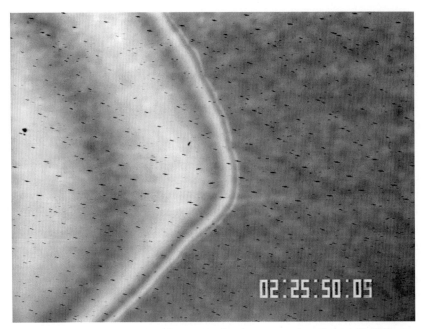

Fig. 23. Crack opening interferogram of a glass–epoxy interface crack [52].

extra distance traveled by the second beam. As a result, an interference pattern is set up that is very similar to the fringes (colored bands) that can often be seen when oil is floating on water. The fringe pattern is viewed by a video camera and recorded on a high-resolution recorder for subsequent digital image processing.

A typical fringe pattern caused by a crack is shown in Fig. 23. The dark region to the right is where the materials are still bonded. The darkness is due to the fact that the indices of refraction of the bonded materials are close enough that most of the incident beam is transmitted. The crack front is clearly outlined by the boundary between the dark region and the first, quite thin, bright fringe. It is certainly not as straight as we often assume! Higher-order fringes can be seen to the left, appearing much like contours on a map and in fact are contours of constant crack opening displacements. The crack opening displacements are given by

$$\Delta u_2 = \frac{m\lambda}{2}, \tag{48}$$

where m is the order (number) of the dark fringes. For example, in Fig. 23, the highest-order dark fringe is the third one. Counting dark fringes, the resolution is $\lambda/2$. For a wavelength $\lambda = 546$ nm, this corresponds to 0.273 µm. If the count is made from dark to light fringes, half fringes are resolved and the resolution becomes 0.137 µm. The resolution can be further increased by measuring the light intensity between fringes using image analysis equipment. This can be seen from

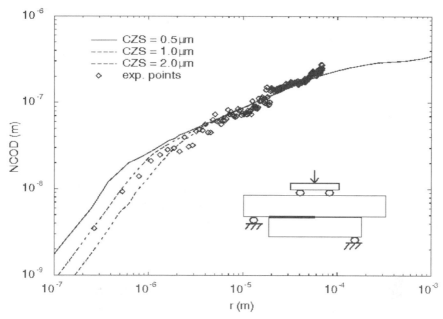

Fig. 24. Normal crack opening displacements of a glass–epoxy interface crack [52].

the relation

$$\frac{I}{I_{pp}} = \frac{1}{2}\left[1 - \cos\left(\frac{4\pi\,[\Delta u_2]}{\lambda}\right)\right],\tag{49}$$

where I is the measured light intensity and I_{pp} is the difference in the intensity between bright and dark fringes. In this way, Swadener and Liechti [52] were able to achieve an accuracy of 10 nm in NCOD.

A plot of NCOD versus distance from the crack front is shown in Fig. 24 [52]. The crack geometry that was considered here was a four-point bend sandwich specimen, made of glass–epoxy–aluminum with a crack at the glass–epoxy interface. The NCOD data are compared with finite element solutions that made use of three different traction-separation laws that resulted in cohesive zone sizes of 0.5, 1.0 and 2.0 μm. The data and solutions are all in good agreement for distances greater than 2 μm from the crack front. This is the K-dominant region where the slope of the data and solutions is 0.5. This is expected from the $r^{1/2}$ term in the linear elastic asymptotics (Eq. 29). Note that 2 μm is not close enough to the crack front for the oscillatory term ($r^{i\varepsilon}$) to have any effect. Closest to the crack front, it can be seen that the traction-separation law that gave rise to the best agreement with measurements was the one that resulted in a 1-μm cohesive zone size. The region that is intermediate to the K-dominant and cohesive zone likely exhibits inelastic effects. However, this transition region was quite small here, making it difficult to detect HRR zones.

8. Conclusions

The singular nature of stress fields near cracks and corners in adhesively bonded joints has been examined. The main emphasis has been on the elastic behavior of the adherends and adhesives. For cracks, this led to a presentation of linear elastic fracture mechanics concepts and how they are applied to bonded joints. Included in the fracture mechanics concepts were the criteria for fast and slow growth corresponding to overload and either static or repeated loading situations in wet or dry environments, respectively. We also saw that, in certain cases, crack nucleation from bimaterial corners could be treated in a similar manner. Although the emphasis of the chapter was on stress states, energy concepts were also introduced for the analysis of cracks or corners. Finally, the possibility of larger-scale inelastic effects was acknowledged and stress- and energy-based approaches were introduced for handling this class of problems.

References

1. Akinsanya, A.R. and Fleck, N.A., *Int. J. Fracture*, **55**, 29–45 (1992).
2. Akinsanya, A.R. and Fleck, N.A., *Int. J. Fracture*, **58**, 93–114 (1992).
3. Birksham, P. and Smith, G., *Int. J. Adhes. Adhes.*, **20**, 33–38 (2000).
4. Chai, H., *Composites*, **14**, 277 (1984).
5. Chai, H., *Eng. Fract. Mech.*, **24**, 413 (1986).
6. Chai, H., *Int. J. Fract.*, **32**, 211 (1987).
7. Chen, B. and Dillard, D.A., *Int. J. Adhes. Adhes.*, **21**, 357–368 (2001).
8. Chen, B., Dillard, D.A. and Clark Jr., R.L., *Int. J. Fract.*, **114**, 167–190 (2002).
9. Cottrell, B. and Rice, J.R., *Int. J. Fract.*, **16**, 155 (1980).
10. Dillard, D.A., Liechti, K.M., LeFebvre, D.R., Lin, C. and Thornton, J.S., In: Johnson, W.S. (Ed.), *Adhesively Bonded Joints: Testing, Analysis, and Design*. American Society for Testing and Materials Special Technical Publication #981, American Society for Testing and Materials, West Conshohocken, PA, 1988, pp. 83–97.
11. Dundurs, J., *J. Appl. Mech.*, **36**, 650–652 (1969).
12. Dunn, M.L., Cunningham, S.J. and Labossiere, P.E.W., *Acta Materialia*, **48**, 735–744 (2000).
13. Duva, J.M., *J. Appl. Mech.*, **55**, 361–364 (1988).
14. Fleck, N.A., Hutchinson, J.W. and Suo, Z., *Int. J. Solids Struct.*, **27**, 1683–1703 (1991).
15. Griffith, A.A., *Philos. Trans. R. Soc.*, **A221**, 163–197 (1920).
16. Hattori, T., Sakata, S. and Murakami, G., *J. Electron. Packaging*, **111**, 243–248 (1989).
17. Hein, V.L. and Erdogan, F., *Int. J. Fract.*, **7**, 317–330 (1971).
18. Hutchinson, J.W., *J. Mech. Phys. Solids*, **16**, 337–347 (1968).
19. Hutchinson, J.W. and Suo, Z., *Adv. Appl. Mech.*, **29**, 63–199 (1992).
20. Jethwa, J.K. and Kinloch, A.J., *J. Adhes.*, **61**, 71–95 (1997).
21. Kanninen, M.F. and Popelar, C.H., *Advanced Fracture Mechanics*. Oxford University Press, New York, 1986.
22. Knauss, W.G., *J. Compos. Mater.*, **5**, 176–192 (1971).
23. Landis, C.M., Pardoen, T. and Hutchinson, J.W., *Mech. Mater.*, **32** 663–678 (2000).
24. Larsson, S.G. and Carlsson, A.J., *J. Mech. Phys. Solids*, **21**, 263 (1973).

25. Lau, C.W. and Delale, F., *J. Eng. Mater. Technol.*, **110**, 41–47 (1988).
26. Lefebvre, D.R. and Dillard, D.A., *J. Adhes.*, **70**, 119–138 (1999).
27. Lefebvre, D.R. and Dillard, D.A., *J. Adhes.*, **70**, 139–154 (1999).
28. Liechti, K.M. and Chai, Y.-S., *J. Appl. Mech.*, **59**, 295–304 (1992).
29. Liechti, K.M., Shirani, A., Dillingham, R.G., Boerio, F.J. and Weaver, S.M., *J. Adhes.*, **73**, 259–297 (2000).
30. Liechti, K.M., Arzoumanidis, G.A. and Park, S.J., *J. Adhes.*, **78**, 383–412 (2002).
31. Liang, Y.-M. and Liechti, K.M., *Int. J. Solids Struct.*, **32**, 957–978 (1995).
32. Mignery, L.A. and Schapery, R.A., *J. Adhes.*, **34**, 17–40 (1991).
33. Mohammed, I. and Liechti, K.M., *J. Mech. Phys. Solids*, **48**, 735–764 (2000).
34. Mohammed, I. and Liechti, K.M., *Int. J. Solids Struct.*, to appear (2001).
35. Needleman, A., *J. Appl. Mech.*, **54**, 525–531 (1987).
36. Needleman, A., *Int. J. Fract.*, **42**, 21–40 (1990).
37. Qian, Z. and Akinsanya, A.R., *Int. J. Solids Struct.*, **46**, 4895–4904 (1998).
38. Reedy Jr., E.D. and Guess, T.R., *Int. J. Solids Struct.*, **30**, 2929–2936 (1993).
39. Reedy Jr., E.D., *J. Appl. Mech.*, **60**, 715–720 (1993).
40. Reedy Jr., E.D. and Guess, T.R., *J. Adhes. Sci. Technol.*, **9**, 237–251 (1995).
41. Reedy Jr., E.D. and Guess, T.R., *J. Adhes. Sci. Technol.*, **10**, 33–45 (1996)
42. Reedy Jr., E.D., *Int. J. Solids Struct.*, **37**, 2429–2442 (2000).
43. Rice, J.R., *J. Appl. Mech.*, **35**, 379–386 (1968).
44. Rice, J.R. and Rosengren, G.F., *J. Mech. Phys. Solids*, **16**, 1–12 (1968).
45. Rice, J.R., Suo, Z. and Wang, J.S., In: Ruhle, M., Evans, A.G., Ashby, M.F. and Hirth, J.P. (Eds.), *Metal–Ceramic Interfaces*. Pergamon Press, New York, NY, 1990, pp. 269–294.
46. Ripling, E.J., Mostovoy, S. and Corten, H.T., *J. Adhes.*, **3**, 107–123 (1971).
47. Ritter, J.E., Lardner, A.J., Stewart, A.J. and Prakash, G.C., *J. Adhes.*, **49**, 97–112 (1995).
48. Ritter, J.E., Lardner, T.J., Grayeski, W., Prakash, G.C. and Lawrence, J., *J. Adhes.*, **63**, 265–284 (1997).
49. Ritter, J.E., Fox, J.R., Hutko, D.J. and Lardner, T.J., *J. Mater. Sci.*, **33**, 1–8 (1998).
50. Sancaktar, E., In: Panontin, T.L. and Sheppard, S.D. (Eds.), *Fatigue and Fracture Mechanics, Vol. 29*, American Society for Testing and Materials Special Technical Publication #1333, American Society for Testing and Materials, West Conshohocken, PA, 1999, pp. 764–785.
51. Shirani, A. and Liechti, K.M., *Int. J. Fract.*, **93**, 281–314 (1998).
52. Swadener, J.G. and Liechti, K.M., *J. Appl. Mech.*, **65**, 25–29 (1998).
53. Swadener, J.G., Liechti, K.M. and de Lozanne, A.L., *J. Mech. Phys. Solids*, **47**, 223–245 (1999).
54. Swadener, J.G., Liechti, K.M. and Liang, Y.-M., *Int. J. Fract.*, to appear (2001).
55. Tada, H., Paris, P.C. and Irwin, G.R., *The Stress Analysis of Cracks Handbook*, 3rd ed. American Society of Mechanical Engineers, New York, 2000.
56. Thouless, M.D., Adams, J.L., Kafkalidis, M.S., Ward, S.M., Dickie, R.A. and Westerbeek, G.L., *J. Mater. Sci.*, **33**, 189–197 (1998).
57. Tvergaard, V. and Hutchinson, J.W., *J. Mech. Phys. Solids*, **40**, 1377–1397 (1992).
58. Tvergaard, V. and Hutchinson, J.W., *J. Mech. Phys. Solids*, **41**, 1119–1135 (1993).
59. Ungsuwarungsri, T. and Knauss, W.G., *J. Appl. Mech.*, **55**, 44–58 (1988).
60. Wei, Y. and Hutchinson, J.W., *Int. J. Fract.*, **93**, 315–333 (1998).
61. Williams, M.L., *J. Appl. Mech.*, **19**, 526–528 (1952).
62. Wu, J.-D. and Liechti, K.M., *J. Mech. Phys. Solids*, **49**, 1039–1072 (2001).
63. Wylde, J.W. and Spelt, J.K., *Int. J. Adhes. Adhes.*, **18**, 237–246 (1998).
64. Yang, Q.D., Thouless, M.D. and Ward, S.M., *J. Mech. Phys. Solids*, **47**, 1337–1353 (1999).

Chapter 3

Energy analysis of adhesion

KEVIN KENDALL [*]

Chemical Engineering, University of Birmingham, Birmingham, UK

1. Introduction

The energy balance method for analysing adhesion failure goes back at least to Galileo who described the basic principle in his book "Two Sciences" published almost 400 years ago [1]. Fig. 1 shows part of his diagram which depicts adhesive failure of the cellulose fibres in a wooden beam loaded by a force applied to its free end.

Galileo knew about the effect of defects on failure and mentioned "the imperfections of matter which is subject to many variations and defects". But he also knew that cracking did not normally take place at knot holes and other minor imperfections, but occurred along the top line of the beam where it was embedded in the wall, because there was a sharp corner and elastic mismatch in that locality.

Galileo took the imaginative step of considering the beam to be pivoted around the axis shown in the diagram. By taking moments about this pivot, knowing that force times distance (i.e. energy) must balance, he equated fd with FL, so that he could work out the local tensile force f of failure where the beam cracked. This was his famous energy balance. He then introduced the idea of a stress criterion of failure. In other words, he proposed that a certain stress was necessary to overcome the adhesion between the cellulose fibres, and he used the average stress across the rectangular section $\sigma = f/bd$. This was not a very good approximation, but he still managed to come up with a useful failure criterion by combining the two equations to give the breaking stress

$$\sigma = FL/bd^2. \tag{1}$$

Although Galileo's stress equation was later shown to have a slight numerical error of 6, because the tensile stress across the outer fibres of a cantilevered elastic beam is $6FL/bd^2$, the overall basis of deflection and strength of beams was established

---

[*] Corresponding author. E-mail: k.kendall@bham.ac.uk

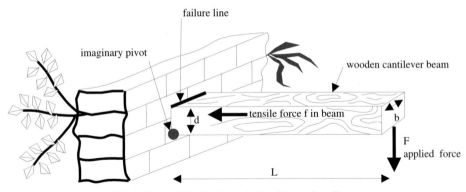

Fig. 1. Part of Galileo's analysis of beam bending.

by the above argument, showing why long beams fail at low loads. This idea led Galileo to the general principle that "the larger the structure is, the weaker . . . it will be".

Several lessons emerge from this analysis.

- The failure path needs to be known to apply the energy balance; thus the method explains failure rather than predicting it.
- Some mechanism has to be introduced with the energy balance; the cracking mechanism tends to be more important than Galileo's stress mechanism.
- Defects can be important but corners and modulus mismatch often dominate.
- Size effects readily emerge because the energy balance usually scales with length and not area.

In hindsight, we can see that Galileo's argument was somewhat over-simplistic because he presumed that only one variable in the equation of state, i.e. pressure (stress) was important. We now know that failure of fibrous composites like wood is dictated by the separation of the micron-scale fibres shown in Fig. 2a. This is an exquisitely complex structure whose failure depends on several variables, including stress. Also, at the molecular scale, the cellulose polymer chains must separate to allow failure, and this must surely depend on the interaction parameters at the molecular level, as suggested in Fig. 2b.

Thus, a global energy balance can lead to useful results, but closer inspection of the detailed mechanisms is rather important, at both micrometre and nanometre scales.

2. Key advances in the energy argument

Although Newton studied adhesion of glass lenses shortly after Galileo's death, noting "ye apparition of a black spot at ye contact of two convex glasses" [2], and measuring the interference fringes which indicated very close contact of the

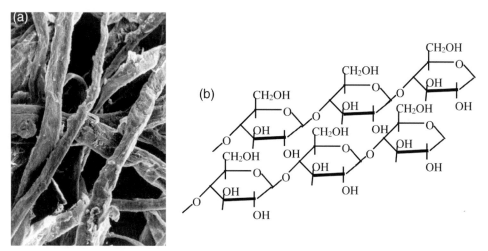

Fig. 2. (a) Cellulose fibres in wood at the micrometre level. (b) Cellulose molecules in contact at the nano-scale.

surfaces, three hundred years were to elapse before the energy balance ideas advanced significantly through the work of Obreimoff [3]. He worked in the Physics Institute of Leningrad, where large sheets of a perfect type of muscovite mica were available from the White Sea area near Chupa. Obreimoff had observed that freshly split mica foils could be put back together to adhere with considerable force and set out to investigate this unique effect. His paper was most significant because it identified for the first time the three processes involved in adhesion: the jumping into contact of the smooth surfaces, the equilibration of the black spot in molecular contact, and the pulling apart of the mica sheets by a cracking mechanism. In addition, Obreimoff saw that evacuating the apparatus (Fig. 3) improved the adhesion, and also found electrical discharges which proved that adhesion was essentially an electromagnetic phenomenon. But most importantly, he discovered that the energy balance theory fitted his results.

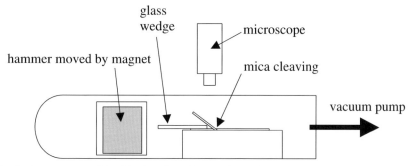

Fig. 3. Apparatus used by Obreimoff to cleave mica and observe its subsequent adhesion.

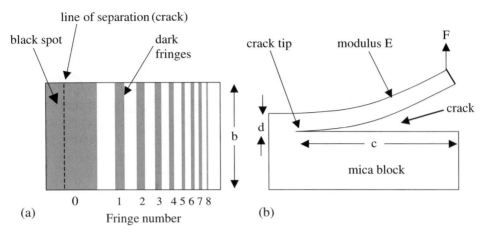

Fig. 4. (a) Interference fringes seen in gap between mica foil and block. (b) Interpretation of shape of bent foil in terms of a reversible crack.

The mica was cut using a razor blade into 2 by 20 by 50 mm blocks, and a glass wedge with a smooth and rounded end was used to split a foil 0.1 mm thick from the top surface. The experiment was placed in an evacuated tube and a glass hammer containing an iron mass was moved with a magnet to press the wedge into the mica to promote splitting. The region at which separation occurred was viewed under a microscope and Newton's black spot and interference bands were seen in the narrow gap between the split surfaces. By measuring the positions of the interference fringes, Obreimoff was able to determine the shape of the mica strip which was being wedged from the block, as shown in Fig. 4.

The point of separation of the mica foil from the block, that is the crack line, could not be seen directly because this kind of interference experiment only detected gaps down to about 50 nm. But the shape of the bent mica foil could be accurately measured and was shown to be cubic. In other words, the strip was behaving as a simple cantilever, just like Galileo's beam, and its shape was not affected by the molecular adhesion forces. This was an important observation because it proved that the molecular forces were only acting across the very small gap near the line of separation. Thus the molecular forces could be neglected in terms of the large-scale behaviour of the system. A one-parameter model of adhesion can be made to work in such circumstances.

Obreimoff's energy balance was very neat. He presumed that a reversible crack was operating, just like a Griffith [4] crack, and that no energy was lost as adhesion or fracture occurred. However, he did not recognise that a much simpler energy balance arises if the beam is very long. This was the case studied by Rivlin [5] in 1944, peeling adhering films from surfaces by hanging a load on the long film (Fig. 5). In this case, for non-stretchable material, the elastic deformation of the

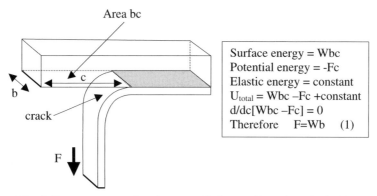

Surface energy = Wbc
Potential energy = -Fc
Elastic energy = constant
U_{total} = Wbc –Fc +constant
d/dc[Wbc –Fc] = 0
Therefore   F=Wb   (1)

Fig. 5. Energy conservation theory applied to peeling of an elastic film.

film remained constant as the crack progressed so that the elastic energy term disappeared from the conservation equation.

During peeling, the crack could be observed moving at steady speed along the interface by looking through the glass with reflected light. After a while, the crack moved a distance c. The area of interface broken by this crack movement is bc where b is the width of the peeling film. Therefore, the energy expended to create new surfaces by breaking the molecular bonds is Wbc where W is the thermodynamic work of adhesion (i.e. the reversible energy required to break one square metre of molecular bonds at the interface). The work done by the force is force times distance, i.e. Fc, which is all presumed to go into the surface energy Wbc, because energy must be conserved. Therefore, the peel equation is $F = Wb$. More formally, the equilibrium is calculated mathematically by writing down the sum of all the energy terms, then differentiating with respect to crack length to determine the minimum energy condition, as shown in the box of Fig. 5.

Of course there is elastic deformation energy in the bent elastic film, from the time when the force was first hung on the film. But this remains constant during peeling and so does not supply any energy to the surfaces. It is merely a constant energy term which moves along with the crack. Consequently, it does not change during the energy balance. Rivlin also assumed that there were no stretching or dissipation terms as the film detached.

3. Wedging

The same idea can be applied to the wedging situation as shown in Fig. 6. Perhaps the easiest way to detach a film from a surface is to scrape it with a sharp blade, driving a wedge along the interface. As Fig. 6 shows, this process opens a crack ahead of the wedge, and this crack progressively detaches the film as the wedge

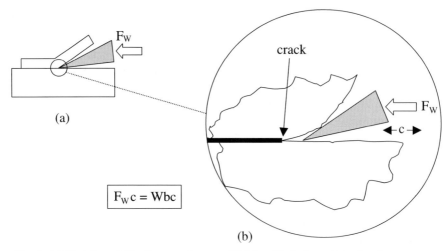

Fig. 6. (a) Wedging of film from a substrate. (b) Magnified view of the detachment zone.

is pushed in further. The simple connection between the wedge force and the adhesion in the case of zero friction [6] is shown in Fig. 6b. As the wedging force F_w drives the knife a distance c, the crack also moves the same distance. The work done by the knife is the force times the distance, $F_w c$, if there is no frictional resistance. At equilibrium, this work is converted completely into creating new surfaces which requires work of adhesion times the area of broken interface, Wbc, where b is the width of the film. Since the elastic deformation remains constant as the film detaches uniformly, the elastic energy in the system can be ignored. It does not change with crack length and so cannot drive the crack. Energy is conserved if there is zero friction, so the final equation for equilibrium fracture is

$$F_w = Wb, \tag{2}$$

the same as Rivlin's peeling equation.

There are several lessons to be learned from this theoretical argument:
- the mechanism is known to be cracking along the interface;
- the assumption of a smooth energy minimum is made;
- no recourse to a stress criterion is necessary in these examples.

This theory demonstrates that, under reversible cracking conditions, a very small force is needed to wedge a film from a surface, even if the bonding is the strongest chemical bonding available, because wedge cracking is a direct mechanism for converting mechanical energy into surface energy. A strong bond would give a work of adhesion around 10 J m^{-2}, leading to a wedging adhesion force of 10 N m^{-1} of film width, a feeble resistance to failure. In practical applications, such as polymer-coated steel sheets, the adhesion energy needed industrially is at least 100 times this value, or better yet 10,000 times, so it is important to consider mechanisms which can produce such amplification, as shown later.

This theory presumes that the crack can also heal at the same force. In practice, the force has to be slightly reduced for healing to be seen. For the most perfect elastic system, there is a force which can be suspended on the film whereby the crack does not know whether to peel or heal. The crack is essentially in thermodynamic equilibrium in which a slight increase in force will cause separation, and a slight decrease will cause healing.

4. Elasticity in the energy balance

Obreimoff was the first person to consider adhesive fracture of smooth mica joints driven only by elastic deformation of the film around the crack. Essentially, he used the Griffith [4] crack theory applied to wedging of an adhesive film. He showed experimentally that the film bends like a thin beam fixed at the tip of the crack, so that the deflection w at the end of the beam is $4Fc^3/Ebd^3$, from simple beam theory. Here, F is the force, c the crack length, E the Young modulus, b the width and d the thickness of the beam. Actually, the deflection will be slightly more than this because of extra deflection around the crack tip, which tends to increase the angle of the bent film at the tip from zero to around $5°$ [7]. However, crack-tip energy can largely be ignored because it is much smaller than the other terms. The stored elastic energy in the beam is then given by $Fw/2$, that is $2F^2c^3/Ebd^3$. This can also be expressed in terms of deflection w rather than force F as $Ebd^3w^2/8c^3$.

There are three energy terms which contribute to the cracking: the potential energy in the load F, the elastic energy stored in the bent film, and the surface energy in the cracked interface. Writing down these terms:
– potential energy $= 0$ because the force cannot move;
– elastic energy $= Ebd^3w^2/8c^3$;
– surface energy $= Wbc$.
Thus the total energy in the system is

$$U = Wbc + Ebd^3w^2/8c^3.$$

For equilibrium, $dU/dc = 0$. Therefore

$$W = 3Ed^3w^2/8c^4. \tag{3}$$

Alternatively the equilibrium can be expressed in terms of the vertical force applied by the wedge

$$F = b\left(WEd^3/6c^2\right)^{1/2}. \tag{4}$$

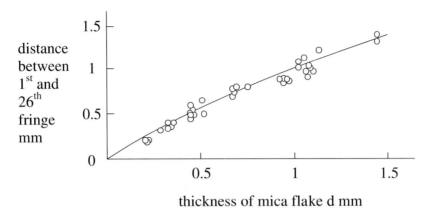

Fig. 7. Obreimoff's results for wedging of mica films from a mica block.

Eq. 3 is useful because it shows that the work of adhesion can be measured by looking at the crack length ahead of the wedge. Obreimoff's results are shown in Fig. 7 for various wedge and film thicknesses. They fit the theory remarkably well.

The value of this method, as Obreimoff demonstrated, is that it can be used to follow environmental changes in the work of adhesion as the crack is exposed to water, acid, etc. He showed, for example, that the adhesion energy in vacuum was 40 J m^{-2}, whereas in normal air conditions, the crack progressed further, so the adhesion energy dropped to about 3 J m^{-2}.

Eq. 4 is interesting because it demonstrates that the force of adhesion varies grossly as the geometry changes. The force needed to lift off the film when the force is applied vertically is much different than when the force is applied horizontally as in Eq. 2. Putting in some reasonable numbers for a low-adhesion polymer surface coating, $W = 0.1$ J m^{-2}, $E = 10^7$ Pa, $d = 10^{-3}$ m, $c = 10^{-3}$ m, the vertical force is 12 N per metre of film width whereas the horizontal force is only 0.1 N. This extra force is needed because the elastic linkage has to take energy from the elastic field to drive the crack, whereas the direct linkage converts mechanical energy directly into molecular crack energy.

5. Change in elastic linkage as the crack progresses

The cracking of an adhesive film interface by a normal force is complex because the elastic mechanism alters as the crack extends, as shown in Fig. 8. A very short crack exists at the start (Fig. 8a), where c is much less than d, and this is like a Griffith crack, requiring a high force given by

$$F = b(W E d \cdot 2d/\pi c)^{1/2}. \tag{5}$$

This is written in a somewhat unfamiliar form, but may be seen to depend on

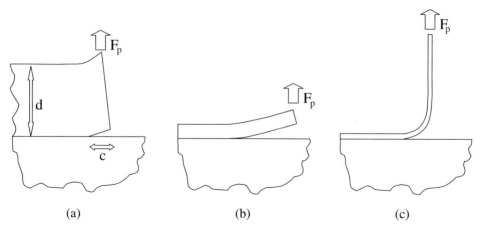

Fig. 8. Changes in the elastic linkage as the crack extends: (a) short crack; (b) intermediate-length crack; (c) long crack.

the parameter $(WEd)^{1/2}$ which appears in all the elastic linkage equations for cracking, but scaled by a d/c term, in this case to the power $1/2$.

After the crack has extended such that the crack length c is comparable to d, then Eq. 4 applies and the scaling is with d/c to the power 1. This obviously requires a lower force. When the crack extends still further, the elastic energy term becomes constant with crack length and then the force is independent of elasticity, giving a direct linkage between force and molecular adhesion, corresponding to the Rivlin peel equation $F = Wb$, which leads to a still smaller detachment force.

Thus, the adhesion force for an elastic film can range over many different values depending on the elastic linkage, which is governed by the specific geometry and force application in the test method. All this range of adhesion forces can be explained by the mechanics of cracking in terms of the single molecular parameter W, the work of adhesion, which remains constant. In conclusion, the mechanism dictates the adhesion force, which can vary substantially even though the molecular adhesion remains the same. In the above examples the elastic linkage makes detachment more difficult in terms of adhesion force. Now let us consider two examples where peeling is made easier by elastic linkage.

6. Elastic linkages easing failure of film adhesion

There are some situations in which the elastic energy helps to propagate the crack and thus decrease the adhesion force. The first, shown in Fig. 9a, occurs when the peeling angle of the film is reduced from 90° to lower values. As Rivlin showed, the potential energy in the load is now changed to $-Fc(1 - \cos\theta)$, so that the force must be raised to continue peeling. When the force is raised, the elastic film

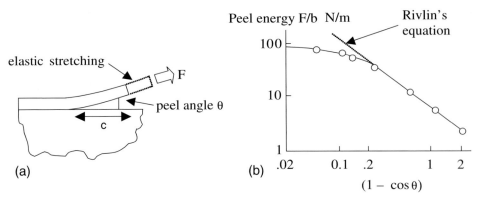

Fig. 9. (a) Reduction in peel angle causes elastic stretching of the film. (b) Results showing how the peel force is reduced by the stretching mechanism [8].

begins to stretch significantly, storing elastic energy F^2c/bEd in the uniformly extended elastic material. The condition for equilibrium of the crack is then

$$(F/b)^2/2Ed + F/b(1 - \cos\theta) - W = 0. \tag{6}$$

The results for peeling of an elastomer from glass at various angles are shown in Fig. 9b. As the angle was reduced, the peel force rose, but eventually levelled out at

$$F = b(2WEd)^{1/2}, \tag{7}$$

which is the equation for lap failure of a flexible film in contact with a rigid surface, which applies also to shrinkage of films, to lap joints and to testing of composite materials. This equation applies especially to the situation shown in Fig. 10b, in which a stress is applied to compress a film which then detaches by cracking along the interface. Eq. 7 was originally postulated and proved in 1973

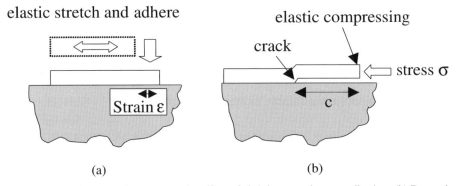

Fig. 10. (a) Experiment to demonstrate the effect of shrinkage strain ε on adhesion. (b) Removing an adhering film by applying a compressive stress.

[9] and was rather similar to the equation for fibre debonding of Gurney and Hunt [10] and Outwater and Murphy [11].

Shrinkage is another phenomenon which demonstrates how elastic energy can reduce peeling force, or even cause spontaneous detachment of an adhering coating. The shrunk film was formed by taking a strip of smooth elastomer, stretching it and adhering it to the glass surface in the stretched condition as in Fig. 10a. It was demonstrated [9] that the elastic energy stored in the film for a crack length c was $cbEd\varepsilon^2/2$, where ε was the residual strain in the film, b the width, E the Young modulus and d the thickness. On cracking, this energy was converted into surface energy Wbc, so the criterion for cracking under a peel force was

$$F = Wb - bEd\varepsilon^2/2 \tag{8}$$

and thus for spontaneous cracking under zero peel force

$$W - Eds^2/2 \quad \text{or} \quad W - d\sigma^2/2E. \tag{9}$$

In this equation, essentially similar to the condition for cracking of a lap joint or removal of a film by applying a compression (Fig. 10b), W is the work of adhesion, E is the Young modulus of the film, d its thickness and ε the residual elastic strain (or σ the elastic stress) in the coating. Experimental results confirmed this theory for elastomers adhering to glass. For biaxial tension, σ^2 is replaced by $\sigma^2(1 - \nu^2)$, where ν is the Poisson ratio.

This is an interesting result because it is evident that there is a strong size effect, with thinner films adhering better at the same strain (or stress) condition. The fracture mechanics analysis for elastic linkage must give a size effect since the crack is driven by stored energy. Thin films cannot store as much shrinkage energy because of their low volume and so adhere better, even when the shrinkage strain is very large. Thus, the stress in the film at failure can vary enormously. In other words adhesion strength is not a constant.

7. Calculation of the elastic energy terms

It may be seen from the above examples that there is always an elastic energy term which must be inserted in the energy balance, though sometimes this can be ignored, as in the case of peeling or frictionless wedging, because it remains constant as the crack moves. Only energy change can drive the crack by providing the surface energy needed for the new interfaces as the crack extends. In other words the elastic energy term must be a function of crack length to influence the energy balance, i.e. $U = f(c)$. The differentiation of this energy with respect to crack length dU/dc then provides the elastic crack driving force or 'strain energy release rate'.

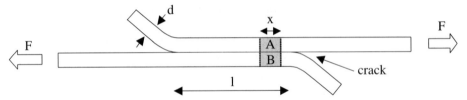

Fig. 11. A long lap joint peeling under the force F. As the crack penetrates further, the shaded elements A and B change their energy state.

This change in energy with crack length can be calculated in several ways:
- by estimating the largest energy terms;
- by examining an exact stress solution;
- by the Irwin fracture mechanics approach;
- by computer estimation of the elastic energy change with crack length.

An example of the first method is the derivation of the criterion for fracture of a long lap joint shown in Fig. 11 which shows a long joint which has already cracked substantially [12]. The equation for failure can be obtained by applying the energy balance theory of adhesion to the elements A and B in Fig. 11. Consider what happens to each element as the crack penetrates along the interface. Element A stretches as the crack goes through, whereas element B relaxes because it ends up with no force on it after the crack has passed. There is no energy change around the crack tip because this remains constant as the crack moves a short distance through a long joint. Thus the region around the crack tip can be ignored in the calculation because only changes in energy can drive the crack.

Consider the elastic energy stored in the shaded regions A and B before the crack passes. Elastic energy is calculated from the work done in stretching the element as the force is applied and this is shown in Fig. 12.

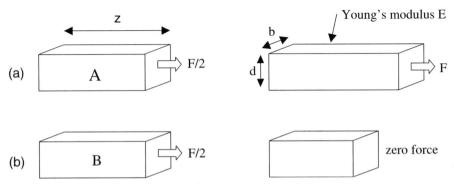

Fig. 12. The changes in the elements A and B in the lap joint as the joint peels. Element A expands as the force increases from $F/2$ to F. Element B shrinks as the force is relaxed from $F/2$ to 0.

The original elastic energy is equal in both A and B and is

$$(\text{half} \cdot \text{stress} \cdot \text{stress} \cdot \text{volume}/\text{modulus}) = \tfrac{1}{2}(F/2bd)(F/2bd)bd z/E$$
$$= F^2 z/8bd E, \tag{10}$$

giving a total elastic energy of $F^2 z/4bd E$.

The final energy in element B is zero while that in A is $F^2 z/2bd E$. Thus the overall elastic energy increases during the peeling by $F^2 z/4bd E$.

However, the force F also moves when it stretches the element A and the work done in this movement is

$$(\text{force} \cdot \text{distance}) = F^2 z/2bd E = 2F^2 z/4bd E. \tag{11}$$

So the total work released as the crack moves through the lap joint is $F^2 z/4bd E$. In equilibrium, because energy is conserved, this excess work must equal the surface energy created by revealing the new open surfaces, i.e. bzW. The equation for joint failure is therefore

$$F^2 z/4bd E = bzW.$$

Hence

$$F = b(4WEd)^{1/2}. \tag{12}$$

This equation for lap joint failure is surprising in a number of ways. It is equivalent to Griffith's brittle fracture theory for glass [4]. Moreover, it fits the puzzling historic results for lap joint failure which showed that the overlap length was not important for long joints, and the strength increased with sheet thickness d and stiffness E. Additionally, it is now clear why chemical environment can weaken the joint because the failure depends on work of adhesion W, which decreases markedly with surface contamination.

The most intriguing questions raised by this argument are related to the well-established idea of lap shear strength which is much used by engineers in designing bridges and other load-bearing structures. In the first place, it is obvious from Eq. 12 that the idea of strength cannot easily be applied to this joint, because the failure force is not proportional to area. The so-called 'shear strength' F/bl can be made into any number you desire merely by adjusting the values of l, d and E. A lap joint can have any strength you want! These ideas have been fully explored in a recent book [13].

A second issue is the use of the word shear to describe the fracture. It is evident from the calculation used to obtain Eq. 12 that shear is not mentioned. The joint peels but does not slide or shear (Fig. 11). Only tension forces and displacements are needed to explain the failure of the joint. In fact it would be far more logical to describe this failure as a tension failure, just as the Griffith equation describes tension failure. Of course, shear stresses exist around the crack tip, as in every

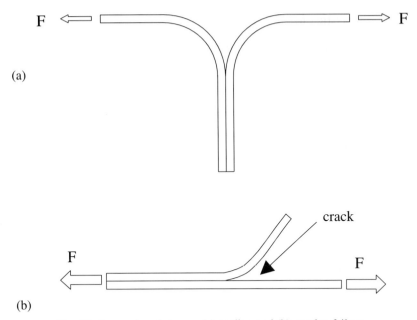

(a)

(b)

Fig. 13. Comparison between (a) peeling and (b) overlap failure.

crack geometry known, but the energy associated with these stresses remains constant as the crack moves and therefore cannot drive the crack.

The most interesting comparison is between the T peel test and the overlap test, as illustrated in Fig. 13. Both of these joints fail by a cracking mechanism. Both cracks travel at constant speed under steady load. This distinguishes such cracks from Griffith cracks which accelerate. But the overlap joint requires considerably more force because the energy is injected into the crack by elastic stretching and not by direct movement as in the peel joint. This was first demonstrated with rubber strips in 1975 [12]. The strips were smooth after casting on glass surfaces, and showed reversible adhesion which increased with peeling speed as shown by the full line in Fig. 14.

The theoretical prediction of overlap failure force was calculated from Eq. 12 and plotted as the broken line in Fig. 14. Experimental measurements of lap joint cracking over the same speed range confirmed good agreement between theory and practice.

8. Exact stress analysis

The exact stress analysis method is limited because only a few such analyses exist. Just one finite cracked specimen has been properly studied, that of the radially

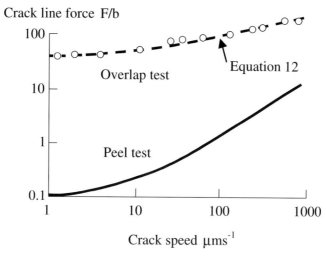

Fig. 14. Test results on same rubber strips for peeling and overlap failure.

edge cracked disc [14–16]; all the others are infinite or semi-infinite geometries including the rigid punch and adhering sphere analyses.

The radially edge cracked disc is a neat geometry for adhesive testing of a very thin interface as shown in Fig. 15. Half-discs of poly (methyl methacrylate) (pmma) are prepared and cemented together with a small amount of acetone to produce a complete disc which can be cracked along the interface. A slot is cut in the disc to allow steel pull-rods to be inserted, and the sample was pre-cracked by wedging with a razor blade, then stretched on a tensile testing machine to drive a crack along a radial direction towards the centre of the disc.

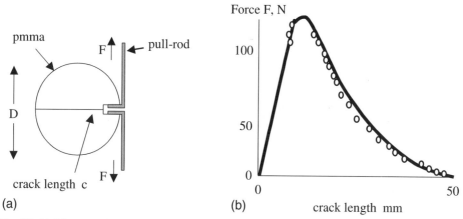

Fig. 15. (a) Disc sample for studying adhesion [17]. (b) Results for fracture of interface compared to exact theory.

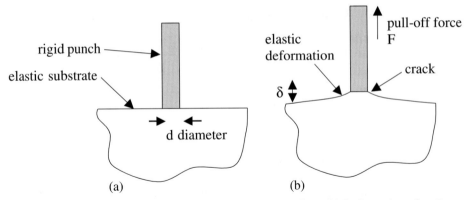

Fig. 16. (a) A rigid punch making contact with an elastic surface. (b) Deformation of surface as pull-off force is applied.

A force of around 100 N was applied by means of the loading machine and the pre-crack was seen to extend slightly. To prevent catastrophic propagation the machine was controlled by hand, obtaining a crack speed of 0.1 mm s^{-1} by gradually reducing the force as the crack progressed. The force is plotted in Fig. 15b for comparison with the theory below.

$$F/b = (EW)^{1/2}c/2D^{1/2}\left\{\left[c/0.3557(D-c)^{3/2}\right]+\left[2/0.9665(D-c)^{3/2}\right]\right\}^{-1},$$
(13)

where F was the tensile force, b the width of the disc, D its diameter, E its modulus, W the work of adhesion and c the crack length. The theory fitted the results quite well when $(EW)^{1/2}$ was taken to be 1.12 MPa m$^{1/2}$.

An exact solution is known for a rigid cylindrical punch adhering to a semi-infinite block of elastic material, as shown in Fig. 16. If the punch has a sharp edge, then the tensile stress at the corner is theoretically infinite, but the crack cannot propagate until sufficient energy is available [18]. This condition can be worked out knowing the displacement δ under a force F

$$\delta = \left(1 - v^2\right)F/Ed.$$
(14)

As the pull-off force is applied, the elastic substrate deforms into the shape shown in Fig. 16b and a crack starts at the edge of the contact and moves through the interface to cause rapid and unstable fracture. The energy balance analysis can be carried out for this geometry by considering the three energy terms involved in the cracking.

(1) Surface energy:

$$U_s = -W\pi d^2/4.$$
(15)

(2) Potential energy. The deflection δ of a rigid punch diameter d in contact with an elastic substrate under load F was given by Boussinesq in 1885 [19,20]

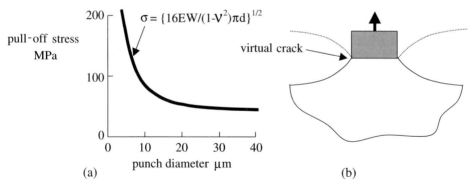

Fig. 17. (a) Pull-off strength for smaller punches. (b) Virtual crack at edge of rigid contact.

(Eq. 14):

$$\delta = (1 - v^2)F/Ed.$$

Therefore the potential energy is

$$-F\delta = -(1 - v^2)F^2/Ed = U_{\rm p}. \tag{16}$$

(3) The elastic energy. This is half the potential energy and of opposite sign, that is

$$U_{\rm e} = (1 - v^2)F^2/2Ed. \tag{17}$$

Adding these three terms and applying the condition of energy conservation as the contact diameter d decreases [18],

$$\mathrm{d}\left(U_{\rm s} + U_{\rm p} + U_{\rm e}\right)/\mathrm{d}d = 0,$$

gives:

$$F = \left\{\pi d^3 EW/(1 - v^2)\right\}^{1/2}. \tag{18}$$

The conclusion from this argument, which was verified by experiment, was that the adhesion force decreased as the punch diameter was reduced. But instead of the force going down with contact area, that is d^2, it went with $d^{3/2}$. In other words, the stress required for adhesion failure increased for finer punches. The adhesion seemed to get stronger with $d^{-1/2}$ as shown in Fig. 17a.

The reason for this strengthening of the smaller joints is the cracking mechanism. Although there does not seem to be a crack at the edge of the punch, there is a virtual crack because the rigid material can be replaced by an elastic half-space as shown in Fig. 17b. The stress in the elastic material rises to infinity at the edge according to the Boussinesq analysis, because of the $(1 - r^2/a^2)^{-1/2}$ pressure distribution. This infinite stress is similar to that causing cracking in the original Griffith theory [4]. However, as Maugis and Barquins [20] have discussed, this infinite stress was a problem for 100 years until 1971 [18] when it was realised that molecular adhesion could resist it. The upper limit of the curve in Fig. 17a is the stress to debond a single atom from the surface.

In his book, Maugis [20] describes the method which Irwin [21] used in 1957 for satisfying the energy balance when the stresses are known exactly around the crack tip. From this analysis, widely used in fracture mechanics, the limit of stress approaching the stress singularity is obtained to give the same answer as Eq. 18.

The same arguments apply to adhesion of two spheres where the contact spot is much smaller than the sphere diameter. In this case the stress inside the contact is the sum of the Hertzian hemispherical distribution and the flat punch Boussinesq distribution. This problem was solved by Johnson, Kendall and Roberts in 1971 [22] to give the diameter d of the equilibrium contact spot for equal spheres diameter D

$$d^3 = 3(1 - v^2)D \left\{ F + 3\pi W D/4 + \left[3\pi W D F/2 + (3\pi W D/4)^2 \right]^{1/2} \right\} \Big/ E \quad (19)$$

Finite element methods have also been employed to calculate the energy balance. Gent and his colleagues have computed the energy changes as a crack passed through an adhesive joint and worked out the cracking force from this [23].

9. The nature of the cracking equilibrium

The fracture mechanics theory described above is a global continuum model which satisfies the conservation of energy principle and the particular equation of state of the materials at large scales. However, we must realise that adhesive failure occurs at an atomistic level. Here the polymer molecules are fluctuating in rapid thermal motion, which can form adhesive bonds and also break them in a dynamic equilibrium of Brownian movement.

Clearly the macroscopic fracture model and the molecular Brownian model are consistent with each other and merge when the crack is looked at over a wide range of scales as in Fig. 18. At macroscopic resolution, the crack seems to be in equilibrium at a particular loading (Fig. 18a) and we can treat it by the method of continuum mechanics above. There does not seem to be any motion at the crack tip. However, when viewed at the molecular level (Fig. 18b) the crack tip is seen to be in rapid thermal Brownian motion. The reacted molecules form the adhered region to the left of the crack tip, whereas the unreacted molecules lie to the right at the open crack surface. The crack tip is not a static point in this model. It is wandering kinetically from right to left as the molecules spontaneously break and then rebond. Cracking is thus viewed as a chemical reaction between molecules at the crack tip. The force applied to open or close the crack is not the cause of reaction, i.e. peeling or healing, at the crack tip. The reaction is happening spontaneously and equally in both directions, causing the crack to open

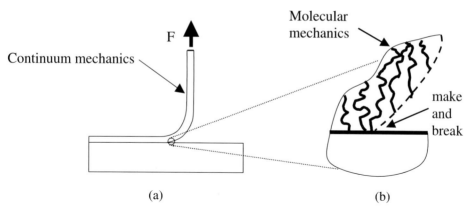

Fig. 18. (a) Schematic picture showing how continuum mechanics applies at large scales. (b) Molecular modelling taking over at nanometre scales.

and close spontaneously at the molecular scale. Applying the crack driving force merely shifts the chemical equilibrium in one particular direction, either opening or closing the crack. This argument presumes that adhesion is reversible and not more than kT for each bond, which can therefore make and break many times at room temperature.

This model allows us to define the work of adhesion W as the sum of all the molecular adhesion energies ε over one square metre of interface. Therefore, if n is the number of molecular bonds per square metre, then the work of adhesion W is given by $W = n\varepsilon$. In this model, it is assumed that the molecular bonds are independent such that the energies are simply additive. This assumption is true for elastomers which are cross-linked networks of long-chain molecules, largely moving independently. The individual molecules are above their glass transition temperature, and so can move like independent fluid chains at the rubber surface while restrained globally by the elastic network.

At the nanometre level, it is no longer a good approximation to assume that the range of the molecular forces is small [24,25]. In order to define adhesion of fine particles sticking together by one bond, the bond must be described by two parameters, ε the energy of the bond and λ its range.

Consider a dilute dispersion of uniform spherical polymer particles as shown in Fig. 19. These spheres experience Brownian motion and therefore diffuse in all directions, causing collisions between the particles. If an adhesion bond forms between the surface molecules, then a collision has a chance of creating a doublet, that is, two particles adhering together at the single molecular bond which forms at the point of contact. If the adhesive bond is weaker than kT, then thermal collisions can break this bond in a period of time. The spheres will then separate and move apart. Thus there is a dynamic equilibrium between joining and separation, giving a certain number of doublets in the suspension at equilibrium,

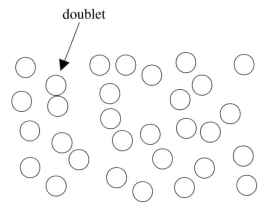

Fig. 19. Dispersion of spherical particles.

after a suitable time has elapsed for diffusion to take place. High adhesion should give a larger number of doublets and lower adhesion a smaller number. Hence there is a definite connection between molecular adhesion and the equilibrium number of doublets observed in a dilute suspension.

The most interesting consequence of the above idea, that molecular adhesion may be measured by observing the number of doublets at equilibrium in a dilute particle suspension, is that an exact mathematical solution can be found under certain circumstances, depending on the interaction between particles when they collide. The simplest situation is that shown in Fig. 20 where a particle approaches its neighbour at constant speed until, at a certain separation λa, the particles are

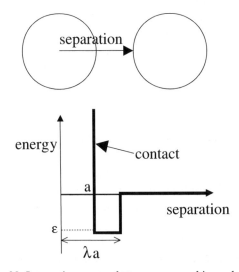

Fig. 20. Interaction energy between approaching spheres.

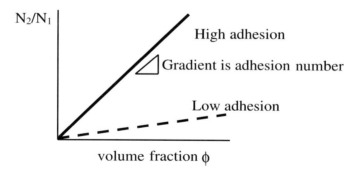

Fig. 21. Defining the adhesion number for single molecular bonds.

attracted to each other with an energy ε. If this energy remains constant until the spheres meet a hard repulsion at the point of contact, then the square well potential is revealed. The approaching sphere travels at constant speed, is accelerated into the potential well, reflects rigidly on contact, and then is decelerated as the particles move apart. This 'hard sphere square well' which was first used by Alder and Wainwright [26] in 1961 can be solved exactly to predict the number of doublets in a suspension.

The mathematical result is that the ratio of doublets to singlets N_2/N_1 is proportional to the volume fraction ϕ of the cells and depends on the range λ and the energy ε of the molecular bond according to the equation below.

$$N N_2/N_1^2 = 4\phi(\lambda^3 - 1)\exp(\varepsilon/kT) \approx N_2/N_1 \tag{20}$$

The conclusion of this argument is that a plot of doublet to singlet ratio versus particle volume fraction should yield a straight line passing through the origin. The gradient of the line is a measure of the adhesion which depends on range and energy of the interactions. Thus a high gradient signifies high adhesion and a low gradient low adhesion as shown in Fig. 21. Therefore, an adhesion number can be defined as the gradient of this plot, to give a measure of the bonding of the particles.

10. Measure of molecular bonding energy and range

The important conclusion from these arguments is that adhesion must depend on particle concentration. Particles will appear to stick more in proportion to their volume fraction. Of course this is a general law which applies to all reversibly adhering Brownian particles.

Red blood cells, erythrocytes, were used experimentally to verify this theory because of their low and reversible adhesion [24]. Cells were prepared from three species, human blood from North Staffordshire Hospital, fresh horse blood

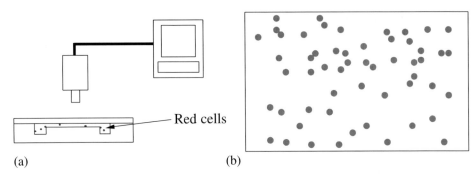

Fig. 22. (a) Video camera apparatus for observing red cells. (b) Field showing red cell doublets.

in EDTA and fresh rat blood from Central Animal Pathology Ltd. Each blood sample was washed six to seven times in phosphate-buffered saline to remove the non-red cell components, before suspending in physiological saline solution, then examined by both optical and Coulter tests. Each species of cell was treated in three ways to judge the effect of surface adhesion molecules, i.e. by adding glutaraldehyde, fibronectin and papain.

The optical apparatus is shown in Fig. 22. The cells were placed in an accurately defined 10-μm space within a glass apparatus, imaged using a video microscope at 40× magnification. Each erythrocyte could then be clearly seen moving around with Brownian movement, while not overheating as occurred at 100× magnification. Pictures of the particles were taken at random locations in the cell and the numbers of doublets and singlets were counted by image analysis software. Taking the ratio of doublets to singlets, the adhesion number was obtained.

The collision and adhesion events could be observed in experiments as shown in Fig. 22b which shows one field of view. There were several doublets which could be counted.

The second set of experiments to measure the doublet numbers used the Coulter counter, which was set up in standard mode to count the individual red cells, as shown by the results of Fig. 23a. The strong peak showed a symmetrical distribution of single cells at a volume fraction near 10^{-5}.

At higher concentration, a shoulder appeared at a 13% higher diameter, Fig. 23b, and this was interpreted as a doublet peak. At still higher concentration of the red cells, the shoulder increased in size, Fig. 23c, indicating that more doublets formed as the blood cells became more numerous. The number of doublets was measured and divided by the singlet peak to obtain the ratio N_2/N_1. This was then plotted as a function of cell volume fraction to give the curve shown in Fig. 24. The results showed the doublets increasing in proportion to concentration and allowing the adhesion number to be found by determining the gradient. For human cells this was 420.

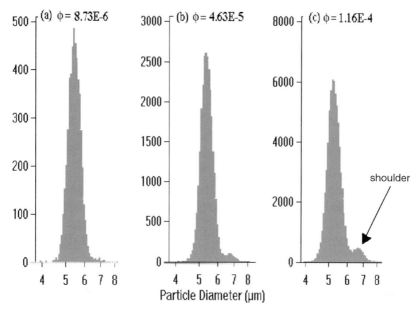

Fig. 23. (a) Coulter counter result for human cells. (b) Result at higher concentration showing shoulder. (c) Larger shoulder at higher concentration.

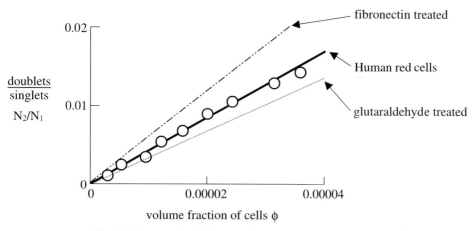

Fig. 24. Increase in doublets with concentration of red cells.

Horse and rat erythrocytes were then tested in the same way and shown to give significantly higher adhesion. Baskurt et al. [27] have shown that the aggregation of such cells is increased over human cells, but volume fraction effects were not taken into account. Popel et al. [28] recognised that horse cells stick better and this was attributed to the athletic nature of the animal. Table 1 quantifies the difference of adhesion in terms of the adhesion number $N_2/N_1\phi$ [29].

Table 1

Comparison between adhesion of various red cells

Animal	Adhesion number $N_2/N_1\phi$
Horse	1488 ± 200
Rat	750 ± 4
Human	420 ± 5

The interesting thing about this molecular model of adhesion is that energy and force of the molecular bonds do not decide the adhesion. Instead it is evident that molecular adhesion depends on the product of molecular adhesion energy and range.

11. Non-equilibrium adhesion

The problem with the above equilibrium energy balance arguments is that they cannot explain the kinetic effects observed in adhesion phenomena. For example, adhesion is often much larger than expected from the molecular bond energies. Adhesion usually depends also on time of contact, on rate of detachment and on hysteresis effects. In order to understand these influences, it is necessary to consider the barriers that exist between molecules, and how these energy barriers can be overcome.

The fundamental problem of adhesion measurement is that the measured energies required to break interfaces apart are too wide-ranging when compared to the equilibrium work of adhesion W. If adhesion energy R is defined as the experimental energy measured to break 1 square metre of interface, then measurements show that this practical adhesion energy can range from negative values, where joints fail spontaneously when immersed in liquid, to very small values when colloidal particles remain separate and stable for long periods, to the very large values of 10^5 J m^{-2} needed in engineering adhesive joints which hold aircraft together. By contrast, the values of the theoretical molecular bond energies, i.e. equilibrium work of adhesion W, occupy only a small range from 0.1 to 10 J m^{-2}, and cannot possibly explain the full scale of measured adhesion. These values are plotted in Fig. 25, and we can see immediately that we need two logarithmic scales to describe the results, one for attractions and another for repulsion. How can this extraordinary range of values be explained [13]? The fact is that simple, clean chemical bonding can only explain the range of adhesion energies from around 0.1 to 10 J m^{-2}.

Surface roughness or contamination by foreign molecules is necessary to explain reductions in adhesion. Further contamination can also be the source of

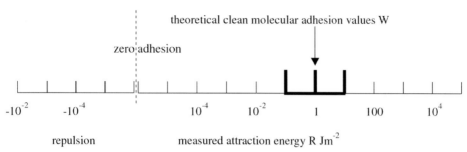

Fig. 25. Range of measured adhesion energies compared to theoretical clean molecular values.

negative, in other words repulsive, adhesion which pushes joints apart spontaneously as a result of wetting along the interface.

Obreimoff [3] was the first person to study these effects systematically using smooth mica surfaces. After splitting the mica, Obreimoff found that the surfaces would spontaneously jump back together when he removed the glass wedge. He measured the energy of this spontaneous adhesion and it was about 0.8 J m^{-2}, substantially less than the original adhesion energy. Then he pushed the wedge back in to measure the adhesion formed between the foil and the block by the jumping process and found that the adhesion energy was around 1.2 J m^{-2}, so it was taking more energy to split the adhering mica than was recovered on the jumping together. Thus he found some energy loss or adhesive hysteresis in this process. He also noticed that it took some time for the splitting to reach equilibrium; the fringes moved for quite a time after the wedge was fixed, around 15 s. This was the first observation of the rate effect on adhesion.

Perhaps of most significance was the result obtained when the air was evacuated from the vessel around the mica. Adhesion was increased as the air was pumped out. This observation showed that adhesion was reduced by air molecules contaminating the mica surfaces. As these contaminant molecules were removed by the evacuation, the energy of adhesion was then increased to 40 J m^{-2}, and impressive electrical discharges were seen around the mica samples at 1 nanobar pressure. This proved that the adhesion was connected with electromagnetic forces between the atoms in the mica crystal. However, it did not seem possible for the bonds themselves to have such a high energy, so the conclusion was that huge energy dissipation was occurring during this high vacuum adhesion test.

12. Causes of the non-equilibrium adhesion

Obviously, roughness is a barrier which inhibits molecular contact and strong adhesion between surfaces. But more significant, the roughness does not usually deform elastically as the surfaces make contact, as shown in Fig. 26. Thus, when

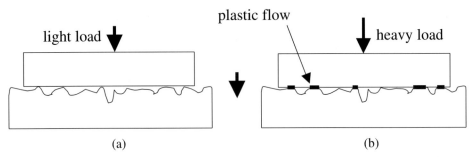

Fig. 26. (a) Light load gives elastic asperity deformation. (b) Heavy load gives plastic flow.

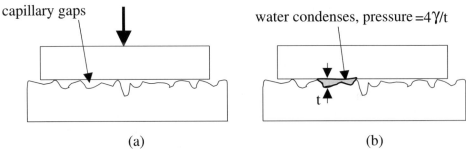

Fig. 27. (a) Capillary gaps formed by roughness. (b) Water condenses in gaps.

the surfaces are pushed together, plastic deformation of the asperities occurs and this dissipates energy, thereby providing a hysteresis mechanism. In the ultimate development of this process, the surfaces are deliberately sheared together during the contacting step, thus causing large local heating and softening at the asperity contacts, which then become more squashed, to allow very strong adhesion. This is used in friction welding to join materials together.

Another roughness-dependent mechanism for adhesion hysteresis can be seen in humid atmospheres. As the rough surfaces are brought together, the asperities make contact, leaving capillary gaps between the surfaces, as shown in Fig. 27. Water condenses in these gaps.

Because the condensing water is in a thin-film form, it has two substantial effects on the adhesion. First it gives a capillary attraction resulting from the curvature of the meniscus and the surface tension γ of the liquid. For a wetting liquid in a narrow gap, this pressure can be very large, as shown in Fig. 27b. Thus, with time of contact, the adhesion between the surfaces can increase due to this capillary condensation. In addition, water can react to give hydration products at many surfaces, and these colloidal products can glue the surfaces together. An example is the rusting together of steel plates, in contact under humid conditions, to give extremely strong adhesion.

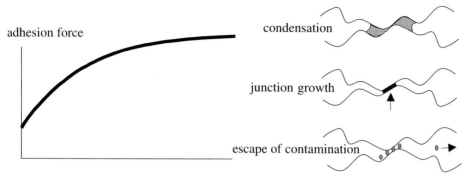

Fig. 28. The dwell-time effect, showing the increase in adhesion force with duration of contact, with three mechanisms: condensation, junction growth, and escape of contamination.

The second effect of the water in the surface gaps is the viscous resistance of the liquid inhibiting rapid separation. The more rapidly the surfaces are pulled apart, the more this viscous resistance is exhibited. This viscous resistance to separation was mentioned by Galileo and was quantified by Stefan 130 years ago. In summary, we can see that adhesive hysteresis can arise from roughness itself, and also from contamination condensing in the gaps caused by roughness.

A very general adhesive effect which stems from such mechanisms as those above is the 'dwell-time' phenomenon. Two surfaces are brought into contact and left for a time. The adhesion is then found to have increased. Further time of contact leads to further increase as shown in Fig. 28. This effect could for example be a result of the capillary condensation described above. When the surfaces are first in contact, the adhesion is low because roughness inhibits the short-range attractions. But as condensation occurs in the gaps, the adhesion rises with time.

Another possible cause of this effect is the creep of the interfacial contact caused by the gradual squashing of roughnesses. When two solids are placed in contact, the true atomic contact area tends to grow slowly with time because the material is not perfectly elastic. So, even when the atomic adhesion remains constant, hysteresis can occur from this junction growth. This has been measured particularly for polymers.

One of the best understood examples is that of mica surfaces, pressed into contact through water [30,31]. The plot of force versus separation is not smooth in this case, but shows sharp oscillations as layers of water molecules are squeezed out of the gap between the surfaces. When plotted as an energy diagram, this shows that the surfaces can exist in several metastable adhesive states, depending on how many water layers have been removed, as shown in Fig. 29. This type of diagram could be used to describe the growth of adhesion with time as mica sheets lie together over a period of time. The water molecules would gradually diffuse out from the gap and the mica would gradually move into closer contact. To

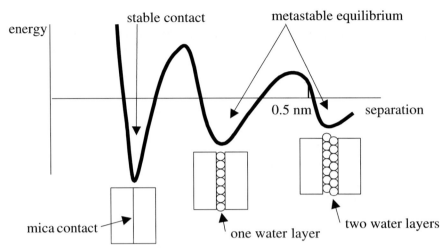

Fig. 29. Schematic energy diagram showing the three positions at which a mica contact stabilises.

achieve this, the mica has to overcome the energy barriers which become higher as true contact is approached. Thus we would expect the adhesion to depend both on temperature, which speeds up diffusion, and on the force applied to push the material closer. This is considered next.

13. Adhesive drag

Once we have recognised that the contaminant molecules introduce an oscillating interaction energy between surfaces, as illustrated in Fig. 29, then we see that more complex adhesion effects must follow [13]. For example, time effects must be observed because the contaminant molecules cannot get into position instantly. Molecules require time to diffuse into and out of the interface. Moreover, the contamination at the interface will depend on the force we apply to the joint and the Brownian energy kT of the molecules which drives the diffusion process as the contamination escapes.

Consider as an example the situation shown in Fig. 30. The interface between the two surfaces can exist in the two metastable states with adhesion energies W_1 and W_2. Imagine first that the surfaces are in state W_1 and we wish to pull them apart into state W_2. To do this we apply a peeling force and this must be sufficient to overcome the energy barrier, with the help of the Brownian energy kT. This problem of molecules coming apart across energy barriers was first solved by Eyring in the early 40s [32]. There are two forces required according to this theory: one to provide the reversible work of adhesion $W_2 - W_1$, and the second to overcome the energy barrier. Thus the total force can be expressed as the sum of

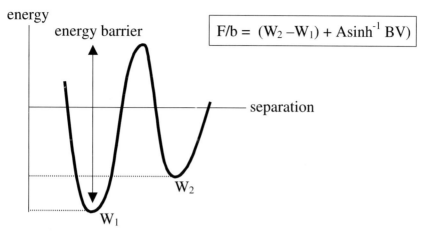

Fig. 30. Schematic of the energy barrier causing adhesive drag, in separating a joint from state W_1 to state W_2.

two terms in the peeling equation given in Fig. 30. The second term is an energy loss term which appears as heat. Clearly, this depends on the rate of peeling V, and also on the temperature variant constants A and B. The higher the rate of peeling, the greater the force required and hence the larger the energy dissipated. Also, the higher the temperature, the faster the peeling. Such behaviour is well-known for adhesive joints, such as those between silicone rubber and acrylic sheet as shown in Fig. 31 [33].

In this example, at low speeds, the adhesion levelled off at a low value, corresponding to an apparent reversible work of adhesion of 0.3 J m^{-1} at very

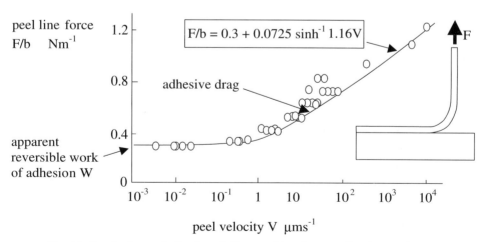

Fig. 31. Results for peel adhesion of silicone rubber from acrylic glassy polymer.

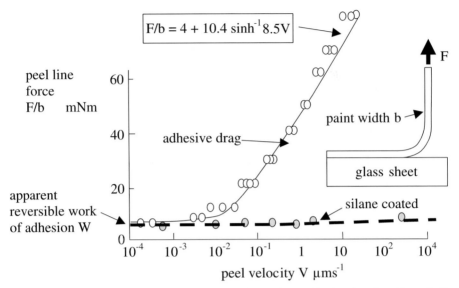

Fig. 32. Results for peel adhesion of paint from silicate glass [13], showing the effect of silane coating.

low velocities of peeling [13]. We will study the precise nature of this equilibrium value in the next section. But at high speeds, the adhesion increased very strongly. This is the adhesive drag effect. Very similar results were obtained by Russian experimentalists in the 50s [34].

The other significant aspect of adhesive drag is its relation to the surface contamination present on the surface. For example, an alkyd paint film was painted on a glass surface, cross-linked and then peeled off. For comparison, the same experiment was carried out on a glass surface coated with dimethyl dichloro silane, as shown in Fig. 32 [33].

These results gave similar behaviour to that of silicone on acrylic, with an apparent equilibrium work of adhesion at low speeds, plus a velocity-dependent peeling force at higher speed. However, there were two substantial differences: first the apparent work of adhesion was ten times too high at $4 \, J \, m^{-2}$; second, the presence of a silane coating had an enormous effect on the adhesive drag but not on the apparent equilibrium work of adhesion. This fall in adhesion due to one layer of molecules at the surface is akin to a catalytic effect: the monolayer is not changing the equilibrium, but is having a large effect on kinetics by reducing the energy barrier to peeling. Clearly, adhesive drag is not the whole story. There must be other energy losses in addition.

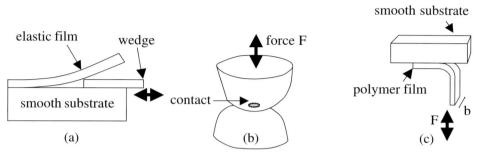

Fig. 33. Three experimental arrangements for studying equilibrium adhesion: (a) wedging; (b) sphere contact; (c) peeling.

14. Adhesive drag and hysteresis measurements

The problem of measuring adhesion, in general, is that the curves for peeling have a similar shape, with an apparent work of adhesion plus a large kinetic adhesion drag, but we are not sure exactly where the equilibrium is. So it is important to devise experiments to study both making and breaking the joint in order to define the precise equilibrium point. Three typical experiments are shown in Fig. 33 [13].

Fig. 33a shows a wedging experiment, rather like that used by Obreimoff on mica. The film is detached by wedging, then the wedge is withdrawn slightly to allow healing. Fig. 33b shows a sphere contact experiment, for example the JKR experiment, in which a smooth sphere is allowed to make contact with a surface, then detached with a force. Fig. 33c illustrates a peeling film experiment in which the peel force is raised to peel the film, then lowered to heal the strip back onto the smooth substrate. In each of these tests, the speed of movement of the crack front can be measured by observing the detachment line through the transparent materials, on both peeling and healing. The measured adhesion energy R, worked out from the force using the appropriate equation (e.g. $R = F/b$ for peeling), is then plotted against the crack velocity, on logarithmic scales as shown in Fig. 34. Both peeling and healing curves can be shown on the same logarithmic plot. This curve defines the adhesive drag on peeling and healing, and shows that equilibrium is not fully attained, but lies between the two asymptotes. Despite the fact that equilibrium is not reached, it has been found that the equilibrium equations can still be used by substituting R the measured adhesion energy at a certain crack speed v for W the equilibrium value at zero crack speed.

At very low speeds of crack propagation through the adhesive joint, to the left of Fig. 34, the peeling and healing curves should coincide. However, it was found experimentally that there was always a gap between the curves, which was small for silicone rubbers but larger for less elastic materials. This gap was defined as the adhesive hysteresis. The equilibrium work of adhesion was somewhere within this gap, around 70 mJ m^{-2}, but could not be found exactly in this experiment.

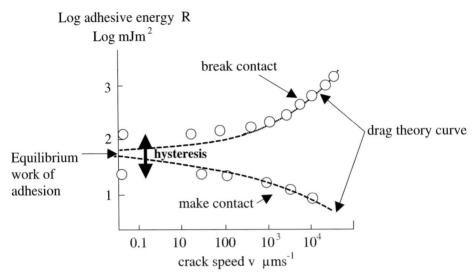

Fig. 34. Results showing the hysteresis for smooth cross-linked rubber on glass.

Only by removing all the energy losses in the experiment would it be possible to attain true adhesive equilibrium. Such energy losses could be caused by a number of mechanisms including roughness, impurities, inelastic deformation. etc.

However, one important energy loss which was explained was the effect of the visco-elastic behaviour of the polymer [35]. This was studied by varying the cross-link density of the rubber, to alter the loss of elastic energy as the material relaxed. As the viscoelastic loss increased, so did the adhesive hysteresis, as shown in Fig. 35.

These results demonstrated that the viscoelastic relaxation in the rubber could stop the peeling well away from the equilibrium point. This idea of crack stopping as a result of energy loss in the system is an interesting nonlinear mechanism which requires further investigation.

15. Conclusions

This paper has shown that energy analysis can allow transparent understanding of adhesion phenomena. In the first place, the adhesion equilibrium can be established under certain circumstances for elastic materials with smooth, clean surfaces and low adhesion. This equilibrium can be defined by a single parameter model based on the work of adhesion W. Adhesion forces can then be predicted for many different geometries and elasticities.

However, it is understood that adhesion of n molecules m^{-2} at the nanometre scale requires two parameters (both an energy $\varepsilon = W/n$ and a range λ) to define

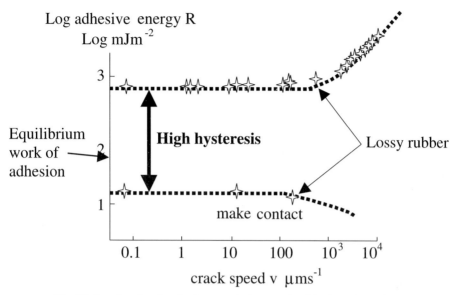

Fig. 35. Results showing the increase in hysteresis with viscoelastic loss.

its dynamic nature. In that case, the statistics of peeling and healing must be taken into account. Adhesion then varies with concentration of bonds and with the product of energy and range.

A more complex problem is to understand the kinetics of adhesion when the system departs from equilibrium. Various mechanisms can interpose a barrier between the molecules, to cause delayed adhesion, adhesive drag and hysteresis. Contemplating these mechanisms is a great challenge for the future.

References

1. Galileo 1638, *Two Sciences* (translated S. Drake). Wisconsin University Press, pp. 12–13.
2. Newton, I., In: Turnbull, H.W. (Ed.), *The Correspondence of Isaac Newton*, Vol. I. Cambridge University Press, Cambridge, 1959.
3. Obreimoff, J.W., *Proc. R. Soc. London*, **A127**, 290–297 (1930).
4. Griffith, A.A., *Philos. Trans. R. Soc. London*, **A221**, 163–198 (1920).
5. Rivlin, R.S., *Paint Technol.*, **9**, 215–218 (1944).
6. Kendall, K. and Fuller, K.N.G., *J. Phys. D*, **20**, 1596–1600 (1987).
7. Kendall, K., *J. Adhes.*, **5**, 105–117 (1973).
8. Kendall, K., *J. Phys. D*, **8**, 1449–1452 (1975).
9. Kendall, K., *J. Phys. D*, **6**, 1782–1787 (1973).
10. Gurney, C. and Hunt, J., *Proc. R. Soc. London*, **A299**, 508–524 (1967).
11. Outwater, J.O. and Murphy, M.C., *Mod. Plast.*, **47**, 160 (1970).
12. Kendall, K., *J. Phys. D*, **8**, 512–522 (1975).

13.   Kendall, K., *Molecular Adhesion and its Applications*. Kluwer Academic/Plenum, New York, 2001.
14.   Feddern, G.V. and Macherauch, E., *Z. Metallkd.*, **64**, 882 (1973).
15.   Gregory, R.D., *Math. Proc. Cambridge Philos. Soc.*, **81**, 497 (1977).
16.   Gregory, R.D., *Math. Proc. Cambridge Philos. Soc.*, **85**, 523 (1979).
17.   Kendall, K. and Gregory, R.D., *J. Mater. Sci.*, **22**, 4514–4517 (1987).
18.   Kendall, K., *J. Phys. D*, **4**, 1186–1195 (1971).
19.   Boussinesq, J., *Application des potentiels à l'étude de l'équilibre et du mouvement des solides élastiques*. Gauthier Villars, Paris 1885 (Blanchard, Paris 1969).
20.   Maugis, D., *Contact, Adhesion and Rupture of Elastic Solids*. Springer, Berlin, 1999.
21.   Irwin, G.R., *J. Appl. Mech.*, **24**, 361 (1957).
22.   Johnson, K.L., Kendall, K. and Roberts, A.D., *Proc. R. Soc. London*, **A324**, 301–313 (1971).
23.   Gent, A.N. and Wang, C., *Rubber Chem. Technol.*, **66**, 712–732 (1993).
24.   Kendall, K., Liang, W. and Stainton, C., *Proc. R. Soc. London*, **A454**, 2529–2533 (1998).
25.   Kendall, K. and Liang, W., *Colloids Surf.*, **131**, 193–201 (1998).
26.   Alder, B.J. and Wainwright, T.E., *J. Chem. Phys.*, **31**, 459–466 (1959).
27.   Baskurt, O.K., Farley, R.A. and Meiselman, M.J., *Am. J. Physiol.*, **42**, H2604–H2612 (1997).
28.   Popel, A.S., Johnson, P.C., Kavenesa, M.V. and Wild, M.A., *J. Appl. Physiol.*, **77**, 1790–1794 (1994).
29.   Attenborough, F.R. and Kendall, K., *J. Adhes.*, **74**, 42–51 (2000).
30.   Fisher, L.R. and Israelachvili, J.N., *Colloids Surf.*, **3**, 303–319 (1981).
31.   Pashley, R.M. and Israelachvili, J.N., *J. Colloid Interface Sci.*, **101**, 511–523 (1984).
32.   Glasstone, S., Laidler, K.J. and Eyring, H., *Theory of Rate Processes*. McGraw Hill, London, 1941, 339 pp.
33.   Kendall, K., *J. Adhes.*, **5**, 179–202 (1973).
34.   Krotova, N.A., Kirillova, Y.M. and Derjaguin, B.V., *Zhur. Fiz. Chim.*, **30**, 1921 (1956).
35.   Kendall, K., *J. Adhes.*, **7**, 55–72 (1974).

Chapter 4

Strength of lap shear joints

R.D. ADAMS * and R.G.H. DAVIES

Department of Mechanical Engineering, University of Bristol, Bristol BS8 1TR, UK

1. Introduction

Any engineering artefact that comprises of more than a single part will, almost inevitably, require some means of joining the constituent components. Structural adhesives, whilst being an ancient technology, have recently come to the fore in performing such a task efficiently and can impart a high degree of confidence in their ability to carry load safely.

The principal advantage of adhesives is in the ability to reduce stress concentration in joints by transferring load via the whole of the bonded area, as opposed to the discrete load points introduced by bolts or spot welds. In addition, the low density of adhesives imparted by their polymeric nature makes their use in joining technology attractive to the designers of weight-efficient structures and their uptake has therefore been widespread within the aerospace industry.

Another significant benefit of adhesives is their ability to join disparate materials and to accommodate complex joint geometries. Production benefits such as the absorption of manufacturing tolerances through the use of variable thickness bondlines also have obvious economic benefits. As a result, as adhesive technology has become more established then its application within other more commonplace industries, such as the automotive sector, has increased.

However, adhesives do not provide a panacea for the joining requirements of the designer or engineer. Inherent in their formulation is sensitivity to the environment in which they must operate and joints are often subject to premature failure brought on by moisture uptake and prolonged operation at elevated temperatures. Also, there remains a perceived lack of confidence in the ability to predict the behaviour of structural adhesive joints without investing significantly in programmes of extended mechanical testing. This reduces the economic attraction

---

* Corresponding author. E-mail: R.D.Adams@bris.ac.uk

of employing them, and one of the chief aims of this chapter is to improve this situation.

As the focus will be on the mechanical aspects of adhesives, no attempt will be made to explore the chemical and physical attributes of the adhesives and more information on these matters has been published in comprehensive reviews such as by Adams et al. [1].

To facilitate the inclusion of bonded joints within an engineering structure, several stages of suitability must be demonstrated. First the mechanical properties of the adhesive itself must be satisfactory and these are quantified by bulk materials tests. Once knowledge of the material itself is gained, then some idea of how well it performs in a bonded joint is required. This is most often achieved via standard test specimens such as the single or double lap joints.

Such tests are used to demonstrate that a particular adhesive and substrate combination will be able to carry adequate mechanical load. However, the mechanics of load transfer even within these joints is complex and in order to gain a detailed understanding of the joint behaviour, it is commonplace to perform some form of stress analysis.

With the recent exponential increase in the availability of computing power, it has become feasible to perform very complicated finite element analyses of these joint configurations and there is much work published to this effect. Such modelling has been shown to be capable of including the effects of material discontinuities, plasticity and complex joint geometry. However, there is a significant learning phase necessary even with commercial software. On the other hand, semi-closed form solutions have been proposed which, although being subject to various simplifying assumptions, have been shown to yield accurate, and often more easily implemented analyses of the detailed stress–strain distributions within bonded joints. In addition, such algebraic solutions are readily implemented in software tools that can be used by engineers without the need for specialist analysis knowledge.

The objective of this chapter is to review critically the available literature for predicting the strength of lap joints, and to indicate how the reader can carry out an appropriate analysis for a variety of adhesives and adherends. 'Appropriate' has implications of cost, ease of use, confidence, and accuracy. We will indicate where simple algebraic solutions can be used, and where it may be necessary to carry out a finite element analysis. And since the market place is full of salesmen whose job it is to sell extensive and comprehensive computer software, we indicate how far you need to go in order to get satisfactory results. Finally, we offer a solution methodology which needs little more than the clear space you may find on the reverse side of a cigarette package. It is remarkably accurate, but it has been extensively supported and checked out using finite element analysis!

2. Finite element analysis (FEA)

The finite element method has, within the past 30 years, provided increasingly sophisticated tools for the analysis of general engineering structures and in their book, Zienkiewicz and Taylor [2] give a very comprehensive review of the formulation and use of finite element models. As the speed and availability of computing power increases then the use of the finite element method for the analysis of adhesive joints has become commonplace. Each component of the adhesive joint is treated as a continuum and the geometry can be treated as either a simplified two-dimensional representation, or a full three-dimensional model of the joint. Complex material models are readily incorporated into the finite element method, and large displacements, such as those seen in the single lap joint, can also be simulated as well as thermal behaviour.

It was observed by Adams and Peppiatt [3] in one of the first papers applying FEA to joints that there was significant stress concentration at the ends of the adhesive layer, adjacent to the corner of the adherend and within the spew fillet. In fact, it was noted that the stresses at this point appeared to be singular in nature and the presence of these theoretical singularities has been the subject of many subsequent studies including Chapters 2 and 3 herein. Adams and Harris [4] employed a detailed model of the stresses at this embedded corner of the adherend and included material non-linearity in their formulation. It was concluded that for an elastic analysis that the stress and strain distributions were singular, and that the inclusion of plasticity in the model resulted in a singular strain field even if the stresses were no longer singular. The degree of rounding at these corners was found to have a significant influence on the predicted stress–strain distributions within the single lap joint. The effect of geometry on the behaviour of the single lap joint was also considered by Chai [5] who quantified the influence of bondline thickness using the finite element method.

Adams et al. [6] detailed axisymmetric analysis of a butt-tension joint and considered the influence of the detailed geometry at the edges of the joint on the stress distribution. The single lap joint was further analysed by Crocombe and Adams [7] using the finite element method and the effect of the spew fillet was included which was seen to significantly redistribute, and decrease, the stresses at the ends of the adhesive layer. In complementary work, Crocombe and Adams [8,9] analysed the mechanics of behaviour of the peel test and included the effects of non-linear deformations and also plasticity in their work. Harris and Adams [10] extended this work and accounted for the non-linear behaviour of the single lap joint. Crocombe et al. [11] quantified the influence of this non-linearity, both material and geometric.

It was noted in all of the aforementioned studies that the presence of the singularities makes the predictions of stress and strain highly dependent on the size of the finite elements used in the vicinity of the singularity. Theoretically, infinite stress or strain is predicted as the size of these elements approaches zero and evidently this cannot happen in practice. Adams et al. [1] state that, in practice, sharp corners do not exist and there is always some degree of rounding present at the embedded corner. Zhao [12] also showed that at a distance from the corner of the order of the degree of rounding present, then the stress–strain distributions reverted to those predicted by a model that did not include any rounding. This is of particular relevance to the application of any failure criteria that use values of maximum stress or strain. Richardson [13] used finite elements of the order of nanometres in dimension to produce a very detailed description of the stresses within an adhesively bonded cleavage joint. It was observed that the influence of any singularities present within the adhesive were highly localised. It was suggested, however, that it is imperative that the presence of these singularities be accounted for in any detailed analysis of adhesive joints.

Carpenter and Barsoum [14] formulated a specific finite element to simulate various closed form solutions to the stress and strain fields within a single lap joint. It was shown that the theoretical singularities within such a joint could be removed through use of incomplete strain-displacement equations. Beer [15] gave the formulation of a simplified finite element chiefly concerned with the correct representation of the mechanical properties of an adhesive within a structural model rather than the prediction of detailed stresses within the adhesive.

In order to validate the application of finite element methods to the analysis of single lap joints, Tsai et al. [16] compared predictions from a two-dimensional finite element model with the results from a photoelastic study of a single lap joint with quasi-isotropic composite adherends performed using Moiré interferometry. They noted a good general correlation between practical and theoretical results. However it was noted that at the joint edges there was some difference between prediction and practice at the edges of the joint. This was attributed to free-edge effects and because the coupling between bending, shearing and tension of a quasi-isotropic composite was not accounted for in their two-dimensional model.

Joints with mismatched and anisotropic adherends were considered by Adams et al. [17], including material and geometric non-linearity and with particular interest in the interlaminar stresses induced in the composite adherends. Good correlation was noted between their predictions of failure load and those seen in practice. This was attributed to the fact that the joints with composite adherends invariably failed via interlaminar failure which was remote from the influence of the singularities in the model and the values of stress predicted were therefore less sensitive to the details of the finite element mesh used. It was also noted in these papers that the addition of a spew fillet to a joint with composite adherends had the effect of considerably increasing the strength of the joint. This was shown to be due to the spew fillet smoothing the path of load transfer across the joint and therefore reducing the stress concentration within the adherend. As the interlaminar strength of such composite materials is relatively low then any means by which the through thickness stress can be reduced will evidently result in a higher joint strength.

Adams et al. [18] investigated the influence of temperature on the single lap joint using finite element methods and it was shown that significant stress could result from cure shrinkage and the use of mismatched adherends which can lead to large residual stress within the adhesive layer.

The three-dimensional nature of the stress distribution within the single lap joint was noted by Adams and Peppiatt [19], who used an approximate analytical method to predict the influence of Poisson's ratio effects on the stress across the width of the joint. Groth [20] used the method of sub-structuring to further investigate the detailed stress distribution across the joint width. Tsai and Morton [21] extended their earlier two-dimensional analyses into three-dimensions and again compared the results with practice using Moiré interferometry. It was demonstrated that the peel stresses within the single lap joint were highest at the middle of the joint and this is attributed to the influence of anticlastic bending. The remainder of the stress components are shown to be less sensitive to position across the width of the joint and compared well with predictions from a plane strain two-dimensional analysis. Analysis in three-dimensions was also undertaken by Zhao [12], who performed a simplified analysis in which the boundary conditions applied to the three-dimensional model of the joint overlap length were derived from a closed form solution. Zhao noted that the highest shear stresses were predicted towards the outer edges of the joint. Lyrner [22] used three-dimensional finite element analysis to quantify the stress distribution in the adhesive adjacent to small button-shaped voids and compared the results from three- and two-dimensional analyses. It was demonstrated that the plane strain condition imposed in the two-dimensional work was reproduced at the middle of the three-dimensional analysis.

Karachalios [23] considered the relationship between the three-dimensional stress distribution and the failure morphology observed in single lap joints with a spew fillet and included both material and geometric non-linearities. Of particular interest was the observation that the failure was seen to initiate at the centre of the joint where the highest peel stresses, and maximum principal stresses, were predicted to occur. Adams and Davies [24] also investigated the variation in stress distribution, using three-dimensional finite element analysis, within a single lap joint with composite adherends. Of particular interest was the variation in peel stress induced in the composite adherends across the width of the joint. It was also noted that the transverse shrinkage arising from Poisson's ratio effects was, in part, responsible for the increase in adhesive shear stresses towards the free edges of the adhesive layer. This was due to the nature of load transfer in the joint whereby there was differential transverse deformation between the upper and lower adherend at the joint ends, resulting in imposed shear strain across the adhesive layer.

Therefore, in order to gain a full understanding of the detailed behaviour of the single lap joint, the finite element method can be used to great effect. However, care must be taken to account for the niceties of application of finite elements to this geometry in that the presence of singularities must be accounted for and non-linear geometric and material effects must be included. In addition, it has been shown that the nature of the stress distribution within the single lap joint is three-dimensional and care must be exercised in the interpretation of two-dimensional results if an understanding of overall behaviour is to be achieved.

2.1. Some results from FEA

To illustrate the use of the finite element method, a series of single lap joints was analysed using the following combinations of adherend: titanium–titanium, titanium–composite, composite–composite, and aluminium–aluminium. For each adherend combination, both an epoxy and a bismaleimide adhesive were analysed and each joint was subjected to four different test temperatures: $-55°C$, $20°C$, $130°$ and $180°C$. In all, therefore, 32 different cases of single lap joints were considered. The thicknesses of the adherends were 2 mm for the composite, 1.6 mm for the aluminium and 1.2 mm for the titanium. The joint width was 25 mm, the length was 12.5 mm, and the bondline thickness was 0.1 mm.

A typical finite element mesh is shown in Fig. 1. Four elements were used through the adhesive thickness because of the very high stress gradients that were expected to occur at the ends of the adhesive layer. At these regions where high stress gradients were expected, the aspect ratio of the elements was kept close to unity to increase the accuracy of the stress predictions. Also, the sizes of the adhesive elements and adherend elements adjacent to the adhesive were kept constant for the different models. The reason for this was, as has been pointed out in the literature survey, that the solution for the stresses at the points of singularity was dependent on mesh size. By keeping the size of the elements at these points constant, it allowed direct comparisons to be drawn between the results from the different models without adding an extra variability in terms of element size affecting the solution. Further away from the adhesive layer a coarse mesh was used to minimise the size of the models.

Quadrilateral, reduced integration, generalised plane strain elements were used for both adherend and adhesive components. Generalised plane strain (or engineering plane strain) elements were used instead of the more commonly employed plane strain elements because they allow a certain amount of out-of-plane transverse deformation. If plane strain elements had been used then the out-of-plane strains would have been zero. These elements would consequently not allow any transverse thermal expansion to take place on the application of temperature loads, resulting in artificially high thermal stresses. The generalised plane strain elements, on the other hand, allow the cross-section of the model to

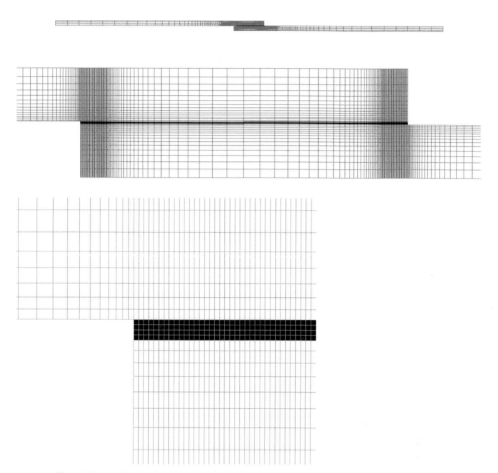

Fig. 1. Typical two-dimensional finite element mesh at increasing magnification.

deform uniformly and a certain degree of thermal expansion was therefore allowed.

In fact, in this case the transverse deformation was dominated by the properties of the relatively stiff adherends, much as would be observed in practice, and it was therefore a better approximation of the actual behaviour of the joint.

The geometry used in these analyses consisted of the single lap joint configuration with a relatively short overlap of 12.5 mm. In a parallel experimental programme, any spew fillet that had been formed on manufacture of the joints was machined off subsequent to the adhesive curing, so this was also removed in the model.

In similar joints analysed and tested by other authors [17] the lack of a spew fillet for single lap joints with composite adherends had been seen to lead to interlaminar failure of the composite. This was attributed to the high peel stress induced by the bending load within the laminate. However, it was demonstrated in the experimental programme that in all of the joints considered here, fracture was via cohesive failure of the adhesive layer and in none of the joints was interlaminar failure of the composite adherends witnessed.

The element sizes within the composite adherends were chosen such that two of the integration points of the second layer of elements adjacent to the adhesive were coincident with the interface between the first two plies of the laminate. Therefore, the stresses would be accurately predicted at the point where interlaminar failure was predicted by Adams et al. [17]. These Gauss point values were chosen as they are the only points within the finite element model at which the solution is fully integrated, and are therefore the most accurate predictions of stress and strain. The stresses and strains were monitored at these points in order to predict the interlaminar stresses so that the influence of temperature could be quantified.

2.2. Boundary conditions

The boundary conditions applied to each of the models were as follows. At the clamped ends of each adherend, restraining the vertical movement of the adherend at its centreline simulated the clamping condition. Although this models the restraint imposed by the clamp, it did not impose any through thickness stress as would be present in practice. However, such a load would have no influence on the stress distribution within the overlap regions and this was therefore felt to be an acceptable approximation. In addition, the horizontal displacement of the node at the outer edge of the upper adherend was also fixed in order to react the applied tensile load and to prevent rigid-body deformation when the temperature loads were applied.

Temperatures were imposed by specifying the initial temperature and subsequent changes of temperature at each node throughout the model. No temperature gradient within the joint was accounted for because when the actual joints were tested then they were allowed to dwell at the test temperature for a sufficient time that a uniform temperature was present throughout.

The temperature was first decreased from the cure temperature to 20°C in the first analysis step. If necessary, a second temperature change was then imposed in a second analysis step to bring the joint to its test temperature. Thus, the thermal loads induced by manufacture and subsequent test were represented. The actual cure temperature was dependent on the adhesive being modelled.

The tensile loads were applied to the faces of the lower adherend as a distributed pressure load. The load was applied incrementally in order that the stress–strain distribution could be monitored with increasing load and to enable the plasticity equations to be solved correctly. This incremental application of load also allowed for the application of failure criteria as the analyses progressed.

Table 1

Mean tensile loads (kN) at failure for single lap joints for a variety of adherends, for 2 adhesives, and at 4 temperatures

Adherends:	Ti–Ti		Ti–F655		Al–Al		F655–F655	
Adhesives:	FM350NA	HP655	FM350NA	HP655	FM350NA	HP655	FM350NA	HP655
−55°C	6.01	5.72	4.25	3.28	6.63	6.75	5.63	5.59
20°C	4.40	4.81	3.56	3.84	5.94	5.78	5.12	5.97
130°C	3.13	7.50	2.72	4.22	3.94	5.69	3.12	6.19
180°C	2.55	9.22	–	4.44	–	6.22	2.50	6.44

Table 1 gives the range of tensile loads that were applied to the models. The loads were the mean values of the measured failure loads of these joints.

2.3. Adhesive properties used in the FE model

Two commercial high-temperature adhesives were used in this set of results. This was because the most difficult case to model and to get agreement with test results is that where the adhesive is slightly non-linear and where the adherends are elasto-plastic. Also, we wanted to show the effects of a wide temperature range and to correlate these results with some which were available from an experimental programme.

The first adhesive was FM350NA which is an aluminium-filled epoxy resin manufactured by Cyanamid and formulated especially for use over a wide range of temperatures. The stress–strain data are shown in Fig. 2.

As can be seen in Fig. 2, the adhesive behaviour is non-linear and also varies with temperature. A Poisson's ratio of 0.38 was used, as given by the manufacturer and a ratio of yield stress in compression to yield stress in tension of 1.3 was assumed. Poisson's ratio, and the yield stress ratio were assumed to be independent of temperature. The measured elastic moduli at the four different temperatures are given in Table 2.

Table 2

Initial Young's modulus of adhesives at various temperatures

Temperature (°C)	Adhesive modulus (GPa)	
	FM350NA	HP655
−55	7.80	6.72
20	6.46	5.65
130	4.01	5.23
180	3.08	4.94

FM350NA at Various Temperatures

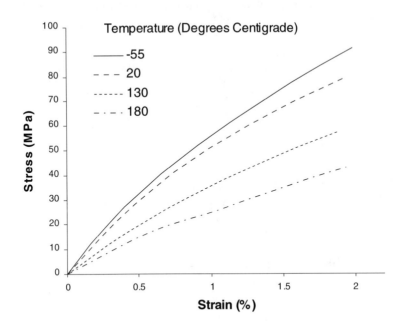

HP655 at Various Temperatures

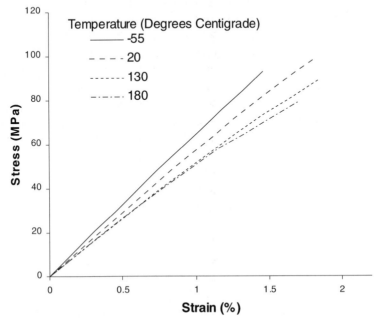

Fig. 2. FM350NA and HP655 tensile stress–strain curves to failure at different temperatures.

The coefficient of expansion for the FM350NA epoxy was 4.57×10^{-5} °C^{-1} and was assumed to be constant over the range of temperatures considered here.

The second adhesive was a bismaleimide adhesive, HP655, manufactured by Hexcel. As can be seen from Fig. 2, this is a relatively brittle material with little, if any, plastic deformation prior to failure. However, it maintains better mechanical properties over a wide range of temperatures and is specifically designed for high-temperature applications. The values of elastic moduli for the HP655 at the different test temperatures are given in Table 2.

The coefficient of expansion for the HP655 bismaleimide was 4.34×10^{-5} °C^{-1} and was assumed to be constant over the considered range of temperatures. A Poisson's ratio of 0.32 was used for the HP655.

Comparing the two adhesive systems, it was clear that they each had very different mechanical characteristics and, indeed, they were chosen for comparison because of this. The HP655 remains relatively brittle over the range of temperatures with little variation in its elastic modulus. On the other hand, the FM350NA shows a higher elastic modulus at −55°C, but this falls off rapidly at the elevated temperatures and significant plastic deformation is apparent. These are common characteristics of such bismaleimide and epoxy systems.

No account was taken of the chemical shrinkage which occurred on cure as no data were available quantifying its extent.

2.4. Adherend properties used in the FE model

Three different adherend materials were considered in this work. The first was 2024-T3, which is a high-strength aluminium alloy commonly used in the aerospace industry. The stress–strain curves shown in Fig. 3 were derived from data taken from the American Society of Materials (ASM) Handbook and, as can be seen, a bi-linear approximation to the material properties could be assumed. Clearly there is a severe degradation in properties shown at 204°C (400°F).

The onset of plasticity and subsequent hardening behaviour were modelled using the widely accepted von Mises criterion and isotropic hardening was assumed. The von Mises model within ABAQUS, the FE software used in this analysis, linearly interpolated between the curves at different temperatures in order to derive the mechanical properties at the required temperature. The elastic modulus of the 2024-T3 alloy was assumed to remain constant over the temperature range at 70 GPa and the Poisson's ratio was 0.33. As no experimental values were available for the strength of the aluminium adherend joints at 180°C, this joint configuration was not analysed. The coefficients of expansion over a suitable range of temperatures are given in Table 3, again from the ASM Handbook.

The titanium alloy used in this work was Ti–6Al–4V which is a commonly used aerospace grade. Bi-linear approximations to the stress–strain characteristics

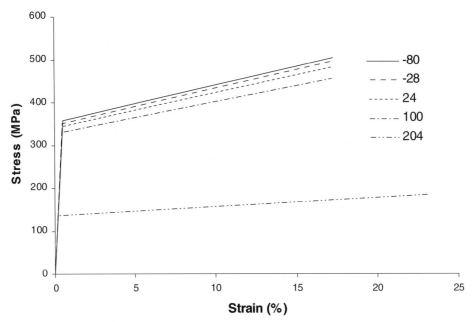

Fig. 3. 2024-T3 Aluminium adherend tensile stress–strain curves from ASM Metals Handbook Vol. 2, *Properties and Selection, Non-Ferrous Alloys and Special Purpose Materials*, 10th edn., ISBN 0871-703-785.

were assumed as shown in Fig. 4 and all data for this alloy were again taken from the ASM Handbook.

An elastic modulus of 110 GPa was assumed for the Ti–6Al–4V over the range of temperatures considered and a Poisson's ratio of 0.33 was used. The coefficients of thermal expansion are given in Table 3.

The final adherend material considered was a Hexcel composite, F655, which was an orthotropic laminate of high-strength T650 carbon fibres in a 5 harness satin weave impregnated with F655 bismaleimide resin. Ten of these plies were arranged to give a 2-mm-thick adherend with comparable properties to a balanced

Table 3

Aluminium and titanium thermal expansion coefficients

Temperature (°C)	2024-T3 expansion coefficient (°C^{-1})	Ti–6Al–4V expansion coefficient (°C^{-1})
−50		8.60×10^{-6}
−35	2.11×10^{-5}	
60	2.29×10^{-5}	
110	2.38×10^{-5}	
200		9.45×10^{-6}

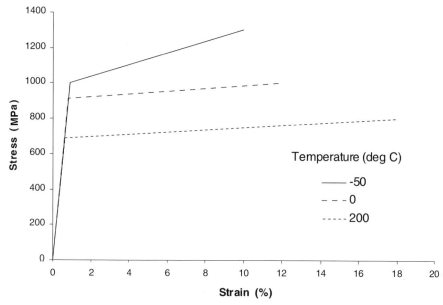

Fig. 4. Ti–6Al–4V titanium adherend tensile stress–strain curves.

laminate of 0/90° fibres. The mechanical behaviour of the laminate was assumed to remain linear elastic over the range of loads and temperatures considered and the properties are given in Table 4.

The adherend was assumed to behave in an orthotropic manner and no account was taken of the individual behaviour of the plies; instead, homogeneous orthotropic properties were applied to the model. All mechanical properties for the composite adherends were derived from data published by the manufacturer.

Table 4

F655 composite laminate properties

	Property	Value
Young's moduli (GPa)	E_{11}	86.72
	E_{22}	10.00
	E_{33}	86.72
Poisson's ratios	ν_{12}	0.297
	ν_{23}	0.031
	ν_{13}	0.031
Shear moduli (GPa)	G_{12}	7.00
	G_{23}	7.00
	G_{13}	7.00
Thermal expansion coefficient ($°C^{-1}$)	α_{11}	-0.9×10^{-6}
	α_{22}	27×10^{-6}
	α_{33}	-0.9×10^{-6}

2.5. Results

In all cases, non-linear static analyses were performed with the effects of non-linear geometry included. Both mechanical and temperature loads were applied incrementally if the adhesive or adherends were seen to yield and stress–strain distributions were reported at each load increment so that the joint behaviour could be monitored throughout the application of load.

In order to asses the influence of the thermal behaviour, Figs. 5–8 show the stress distributions at the failure load given in Table 1 for the composite–composite joint tested at 20°C with FM350NA epoxy resin. Two curves are shown on each diagram, one corresponding to an analysis in which the thermal shrinkage stresses from cure were not accounted for, and the second curve showing the influence of the cure shrinkage stress on the overall stress distribution. In each of the four figures, the stress at the mid-plane is given.

It can be seen that the effect of the residual stress arising from the thermal shrinkage gave rise to appreciable levels of longitudinal stress within the adhesive layer, with a large region of uniform longitudinal stress of 56 MPa. Without the cure shrinkage, the levels of longitudinal stress were significantly lower and, for the majority of the overlap length, were of the order of 8 MPa.

As can be seen from Fig. 6, the peel stress distribution is largely unaffected by the presence of cure shrinkage stresses with little discernible difference between the results from the two analyses. However, the peak values predicted by the analysis in which residual stress was included are slightly lower than those from

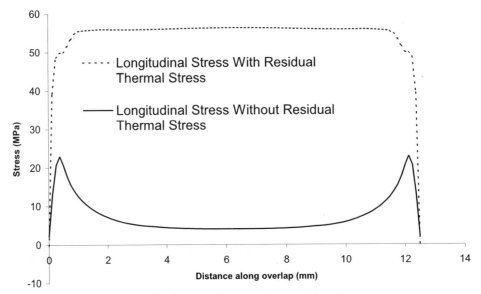

Fig. 5. Longitudinal stress with and without residual thermal stress.

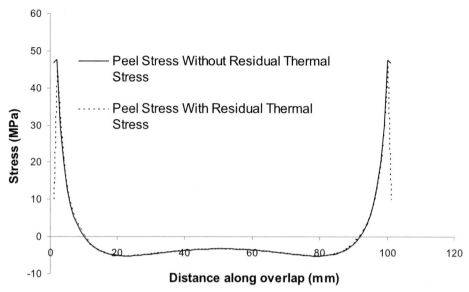

Fig. 6. Peel stress with and without residual thermal stress.

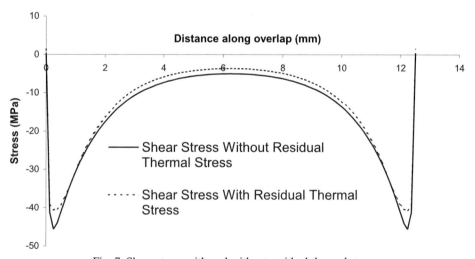

Fig. 7. Shear stress with and without residual thermal stress.

the analysis without residual stress. This was attributed to the fact that the state of initial tension induced by the cure shrinkage would have the effect of inhibiting the yield of the adhesive. This was again seen to be the case for the shear stress distribution shown in Fig. 7 and again, the values of stress for the two analyses were very similar, but the analysis including the effect of residual stress showed lower peak values.

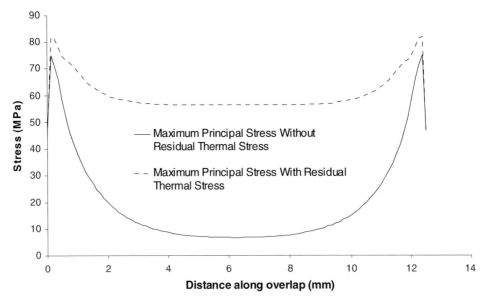

Fig. 8. Maximum principal stress with and without residual thermal stress.

The influence of the three aforementioned components of stress can be seen in Fig. 8 which shows the maximum principal stress within the adhesive. It is clear that the influence of the high induced longitudinal stresses is to increase slightly the prediction of peak maximum principal stress. The maximum predicted stress without thermal shrinkage stresses was 75 MPa whereas the maximum value predicted when cure shrinkage was accounted for was 82 MPa, which is almost a 10% increase.

What all of these figures show is the significant influence that the thermal shrinkage can have upon the prediction of the stress distribution within a lap joint. One conclusion which can be drawn from this is that if an accurate estimation of joint performance, be it a strength prediction or understanding of the joint behaviour, is to be achieved then it is imperative that the thermal effects be accounted for.

It was also instructive to consider the distribution of the stress within the adhesive layer. Fig. 9 shows the distribution of maximum principal stress at the end of the adhesive layer, and it can be seen that the effect of the residual thermal stress it to change the distribution of stress at the end of the overlap and to increase the area which is highly stressed. Whereas the majority of the high stress was focussed at the point of singularity for the analysis which neglected thermal stress, it is clear that including thermal effects increases the magnitude of imposed stress. This, again, has implications in terms of prediction of failure.

Similar trends were seen for the components of stress and strain within the other adhesive joints with the metallic adherends and the HP655 adhesive.

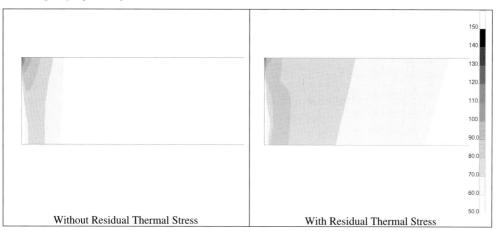

Fig. 9. Maximum principal stress distribution under load (MPa).

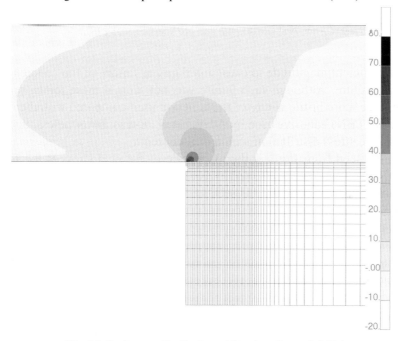

Fig. 10. Peel stress distribution within the adherend (MPa).

What was also of interest was to investigate the change in induced interlaminar stress in the composite induced by the thermal stress.

Fig. 10 shows the distribution of peel stress within the adherend and clearly there is a very high peel stress near the bi-material singularity at the adhesive corner. However, the region of high stress extends quite considerably into the adherend itself and Adams et al. [17] have shown that this component of stress

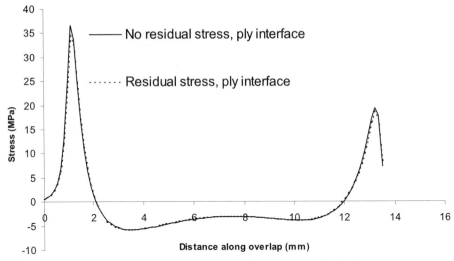

Fig. 11. Peel stress at the first ply interface under load.

can be of sufficient magnitude to cause interlaminar failure of the adherend prior to adhesive failure. Although such failure was not seen in these joints, probably because of the more brittle nature of the adhesive used compared with the rubber-toughened (CTBN) adhesive used by Adams et al., it was, nevertheless, important to consider the stress distribution within the adherend.

As can be seen from Fig. 11, the magnitude of the peel stress at the first ply interface is not significantly affected by the inclusion, or not, of the effects of the thermal shrinkage and, indeed, it is seen to drop slightly when the thermal effects were accounted for.

Let us also consider the titanium adherend joint with FM350NA adhesive, because it showed the widest variation of applied load. Fig. 12 shows the maximum principal stresses, averaged across the adhesive thickness, for joints loaded to the failure load normalised by dividing by the applied tensile stress.

What this shows is that the joint at 180°C is subject to the most severe state of stress relative to its intrinsic material properties when compared with the other temperature results. This is not surprising as it was shown in Fig. 2 that the FM350NA properties degrade quite severely at elevated temperatures. The curves were normalised in order to allow direct comparison to be made between them and it is clear that there is a significant influence of temperature on the stress distribution within the joint.

It has been shown by this two-dimensional modelling that if the stress distributions within the single lap joint were to be best represented and predicted then it was imperative that the thermal properties be accounted for. It has been shown that some stress components were more susceptible to the influence of thermal shrinkage than others and that the presence of thermal stress can inhibit yield.

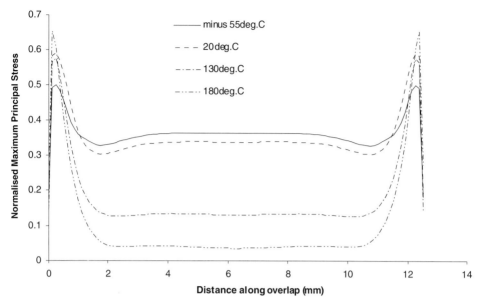

Fig. 12. Normalised maximum principal stresses at applied failure load.

3. Algebraic analyses

The finite element method requires considerable expertise in order to produce meaningful results and it is also subject to the requirement for relatively large amounts of computing power. The result of this is that the finite element method does not always hold favour over the often more straightforward implementations of simplified closed form algebraic solutions. The motivation for the derivation of closed form solutions was to provide an analysis tool which would enable solution of adhesive joint configurations and provide reasonably accurate behavioural and strength predictions. Comprehensive reviews of closed form methods have been presented [1,25–27].

Volkersen [28] presented a continuum mechanics approach to analyse the shear-lag configuration of the single lap joint. However, there were limitations to his analysis, most significant of which was its failure to account for the effect of the bending moments induced by the eccentricity of the applied load. Nor did it account for the adherend shearing, and it predicted the maximum shear stress to occur at the free surface at the end of the joint overlap. As this is a free surface, the shear stress here should, in practice, be zero.

Goland and Reissner [29] then improved upon Volkersen's analysis and proposed a two-stage methodology of analysis in which the bending moments were first calculated and then applied as boundary conditions to a model of the overlap layer. They considered two different configurations of joint with and without consideration being taken of the adhesive layer flexibility. The former was found to be more applicable to joints with metallic, or relatively stiff, adherends and was the most relevant analysis for the current study. In their theory, the shearing and normal stresses within the adherends are neglected but it was the first significant theory to include the influence of bending moments upon the stress distribution, especially the transverse or peel stresses.

Sneddon [26] and Adams and Peppiat [3] found an error in the initial formulation of Goland and Reissner's theory and presented a correction which derived an alternative bending moment factor. However, Carpenter [30] has since shown that Goland and Reissner's original solution was, in fact, correct. This was somewhat fortuitously achieved through cumulative typographical errors in the original paper.

Chen and Cheng [31] show a development of their earlier work [32] in which they extend their theory of lap-joint behaviour to include non-identical adherends. In order to allow the resulting system of equations to be solved using closed form methods, they assume a uniform shear stress distribution across the adhesive layer thickness. A complementary energy method was then employed to solve the final equations. This theory is also presented by Wu et al. [33], who also show that it decomposes to Goland and Reissner's solution if further simplifying assumptions are made.

Unbalanced joints with dissimilar adherends have also been analysed by Yang and Pang [34] who derived similar expressions to Chen and Cheng but solved them using a Fourier series method. A state of uniform shear stress through the adhesive thickness was again assumed. Good correlation was demonstrated between the results from a complementary finite element study and they also considered orthotropic (composite) adherends. Composite adherends are also dealt with in [35–38]. None of these approaches accounted for the coupling between bending, tension and shearing which can occur due to the anisotropic nature of composite adherends. Also, the stress-free boundary at the exposed faces remained unaccounted for, as did the variations in stress across the adhesive thickness. Delale et al. did, however, suggest a method for the inclusion of adhesive non-linearity.

Coppendale [39] and Weitsman [40] have studied the influence of the thermal expansion of the adhesive and have shown that it has a significant effect at the ends of the overlap.

Non-linearity in adhesives was accounted for by Hart-Smith [41] using a similar approach to Goland and Reissner but solving iteratively to account for the plasticity. However, the influence of the peel stress on the yield behaviour of the adhesive was not accounted for and it was the shear stress component alone that was used to control the plasticity. A bi-linear approximation to the adhesive stress–strain curve was employed in order to simplify the solution. An improvement to Goland and Reissner's solution was also presented in which better account was made of the flexibility of the adhesive layer.

A similar approach was presented by Adams and Mallick [42], who proposed the use of an effective modulus calculated using the strain energy from integrating the adhesive stress–strain curve that could then be input into a linear elastic solution. This was seen to give strength predictions which were reasonably close to those predicted by much lengthier calculations including full adhesive non-linearity. Adams and Mallick [42] also extended the work of Allman [43] and Chen and Cheng [32] by representing the behaviour of the single (and double) lap joint using four independent stress functions, resulting in a comprehensive description of the state of stress within the joint.

The main problem with closed form analyses is in accounting realistically for adhesive and adherend non-linearity. Also, even the elastic formulations produce complex equations which are difficult to solve. Fortunately, it is now relatively easy to set up these equations on desk top computers and to get almost instantaneous solutions which give a good indication of the stresses acting in a lap joint under tensile loading.

4. Prediction of failure for single lap joints

The application of failure criteria will be discussed here in the context of the finite element method, although they are also applicable to closed form predictions of stress/strain levels.

On considering how best to predict failure, the method which appears most obvious is simply to specify a stress or strain at which a particular material will fail. As adhesives perform well in shear it would appear best to specify a maximum shear stress limit to give some idea of joint strength. Greenwood et al. [44] performed such an analysis for single lap joints using closed form analysis, and found that the maximum shear stress occurred in the adhesive at 45° to the loading direction. The actual predictions of strength by this method were seen to underestimate performance, and this can be partly attributed to the fact that the use of principal stresses was not considered. The use of principal stresses is preferred because the tendency is for adhesives to fail through tensile mechanisms, even if loaded in shear, because this shear gives rise to tensile and compressive principal stresses.

Alternatively, the maximum peel stress can be used as a criterion for failure [45,41] and it was shown that reasonable predictions of strength could be achieved, albeit for a rather limited number of joint configurations. However, it was noted that the predictions were highly inaccurate for joints in which adherend yielding was present.

Harris and Adams [10] showed that using the maximum principal stress can give reasonable success when combined with non-linear finite element methods and their predictions were within 10% of experimental values. Various combinations of differing grades of aluminium adherends and epoxy adhesives were considered and the choice of whether a maximum principal stress or strain criterion was used was seen to depend on the specific joint configuration. This choice was not always that which would be expected intuitively, as a criterion of maximum principal stress was successfully applied to a highly ductile adhesive where one may have expected the application of a maximum strain criterion. It was shown that the choice of criterion depends on a comparison of the performance of the adhesive within the joint and in bulk form. It has also been subsequently noted by Adams and Harris [4] that the maximum principal stress is a function of the mesh density if a singularity is present and, unless this is taken into account, errors will be introduced. In this paper, a criterion of maximum stress at a distance was in fact specified, because the points at which the failure criterion was applied were not coincident with the singularity points, and instead were at element Gauss points (these are points at which the finite element method evaluates, strains and stresses with the greatest accuracy).

The maximum von Mises stress has also been employed as a measure of the stress to failure by Ikegami [46], but only with limited success. This was because the behaviour of adhesives is highly dependent on the state of hydrostatic stress within the joint, and this was not accounted for.

A criterion of critical stress or strain was employed by Lee and Lee [47] for tubular single lap joints. A combination of criteria were used, dependent on the adhesive thickness within the joint. A critical strain was applied to joints with a thick adhesive layer, and a maximum reduced stress for thin adhesives. Unfortunately, no means of choosing a particular criterion for an arbitrary thickness between the maximum and minimum thicknesses was given. The application of such criteria to experimental data has been investigated by Chai [5] through observing failure in notched flexure specimens and measuring the strain field in the adhesive. It was seen that critical shear strain decreased with increasing adhesive thickness. In addition, if the strain at the crack tip of the notch is compared to that seen in a napkin ring test, then the two values are seen to be similar. It was noted that failure can be expressed in terms of a critical fracture energy, but this too varies with adhesive thickness.

If plasticity is included in the adhesive characteristics, then an alternative failure criterion was proposed and implemented by Crocombe and Adams [9] using critical values of effective uniaxial plastic strain. The triaxial strain was expressed as an effective uniaxial strain, and then compared with the strain to failure of the bulk adhesive. Unfortunately, this too was seen to be dependent on the density of the finite element mesh, and in reality was a critical strain at a distance criterion. The choice of critical strain was also dependent on the adherends present as different plastic zone sizes will be given by different adherend combinations. The geometry considered in this paper was the peel joint.

An alternative criterion which includes adhesive plasticity is to use a value of critical plastic energy density. Harris and Adams [10] included a small degree of rounding on the adherend corners in order to remove the influence of the singularity. In fact this rounding should be of the order of twice the adhesive thickness to be effective, and this raises questions as to whether the degree of rounding will affect the maximum energy density predictions.

Crocombe et al. [11] used cleavage and compression specimens for the evaluation of failure criteria, and avoided the problem of singularities by using semi-closed form methods for analysis of the stress and strain distribution. They found that the maximum principal stress criterion gave reasonable success in the prediction of failure for untoughened epoxies, and for mode I loading in toughened epoxies. The maximum principal stress was seen to be a better criterion for mode II loading in toughened epoxies.

In all of the above criteria, there is an inherent problem of using stress or strain predictions at, or near, points of singularity which will not be accounted for correctly using numerical methods. If rounding is used to avoid the problem, then it is not clear what degree of rounding should be used. In essence, it is not straightforward to specify a general failure criterion from the above methods of failure prediction without careful consideration of the joint involved.

Zhao [12] used a criterion whereby if the average stresses over a certain distance within the single lap joint reached a critical value, then the joint was deemed to have failed. The distance picked was a line progressing into the adhesive from the singularity point. A criterion of critical average stress over the distance was applied to joints with a sharp adherend corner, or a small radius, whereas a criterion of maximum stress over the distance was used for larger radii. Unfortunately, no reasoning was given for the choice of critical distance.

This idea was extended by Clarke and McGregor [48] who predicted failure if the maximum principal stress exceeded the maximum uniaxial stress for a bulk adhesive over a certain zone normal to the direction of the maximum principal stresses. No justification was given for the choice of zone size. It was noted that the sensitivity to changes in local joint geometry, such as radius of the adherend corner, was reasonably low for this criterion.

Crocombe et al. [49] studied the failure of cracked and uncracked specimens subject to various modes of loading and used a critical peel stress at a distance from the singularity with some success. An alternative method was also proposed to use an effective stress, matched to the uniaxial bulk strength, at a particular distance. However, it was found for the latter criterion that the critical distance at which it should be applied varied with different modes of loading because of the change in plastic zone size which resulted. Again, no general criterion for a given adhesive was presented. Kinloch and Williams [50] also considered some cracked specimens, and applied failure criteria at critical distances with some success, but the work was not extended to consider uncracked continua. As for the earlier reviewed stress or strain criteria, there is no real physical justification for these criteria applied at a critical distance, and many of them are dependent on parameters such as adhesive thickness which means that no general criterion of failure is available within these methods.

Crocombe [51] applied a method of failure prediction where the adhesive was deemed to have failed if the whole of the adhesive layer was seen to yield under the applied load but this was only applicable for highly ductile adhesives. Schmit and Fraise [52] used a similar criterion for the prediction of the strength of stepped adhesive joints with some success, but again this was only really of use in predicting the behaviour of highly ductile adhesives.

Fracture mechanics is a study of the strength of structures which include flaws such as cracks and voids where stresses can be said to be singular. Fracture mechanics applies criteria to assess whether the conditions are such that failure will occur at these points, one such criterion being the critical energy release rate, G. Other criteria are also available, including crack opening displacement measurements, the J integral, and stress intensity measurements.

The energy release rate, G, has a strong physical meaning and is the easiest of the aforementioned parameters to obtain for adhesives, especially if the crack in the adhesive is near to the interface between the adherend and the adhesive. This view is supported by Toya [53]. It is possible to include the effects of rate of load application and temperature within these failure models.

In an adhesive joint, the adherends constrain the adhesive and this gives conditions of mixed mode loading. It can be shown that cracks will generally run perpendicular to the direction of the maximum principal tensile stress under straightforward loading [1]. Under mixed mode loading, as is present in the adhesive, the same is not true. Kinloch and Shaw [54] derived formulae which account for fracture under such loads. It was also noted that parameters such as stress intensity, K_{IC}, will vary with the geometry of the joint. Glueline thickness is seen to control G_C and, for a thin glueline, the induced tensile stresses will be increased and this will in turn increase the size of the plastic zone. For thicker gluelines the plastic zone size will be reduced, and G_C will be reduced. However, if the glueline is very thin, then the size which the plastic zone can achieve will be controlled by the thickness, and G_C will decrease. It was shown that G_{1C} and K_{1C} have their maximum values when the size of the plastic zone is equal to the thickness of the glueline, and that because of the constraining effect of the adherends these values can exceed those measured in the bulk adhesive.

Trantina [55] applied fracture mechanics to adhesive joints with some success and applied the failure criteria to a finite element model to find adhesive fracture energies. The influence of the glueline thickness was not accounted for. Hu [56] used a shear lag analysis and applied failure criteria in terms of J_C and it was shown that this gave good predictions of failure and was also able to account for the adhesive thickness. It was noted that this is consequently a good method of predicting failure for adhesive materials loaded in shear.

In order that fracture mechanics might be applied to continuous materials which do not contain a crack, various authors have studied the stress intensity at the bi-material interfaces which are present in adhesive joints. Gradin and Groth [57] applied this to a non-cracked specimen in cleavage tests and finite element methods were used to find a factor of stress intensity at the onset of failure, and was then used to predict failure with an accuracy of ±10%. Groth [58] then applied this to single lap joints without fillets and the stress intensity factor was shown to be independent of overlap length and glueline adherend thickness. Some dependence on the thickness of the adhesive was present though. This was then extended by Groth [59] to model joints including fillets, and it was noted that the criterion was only accurate in predicting the strength of joints with long overlap lengths. In this work, the analyses were elastic. However, no practical results were presented to give an idea of the accuracy of the strength predictions.

Crocombe et al. [11] used a combination of fracture mechanics and finite element analysis to predict failure in cleavage under mode I loading to within 7% of the actual failure strength. This was only applied to a very narrow range of joints, however, and further work is needed in order to assess whether a more generally applicable theory can be developed.

An alternative approach was taken by Fernlund et al. [60], whereby a fracture envelope is generated for a particular adhesive system of energy release rate versus the mode of loading. This method accounts for material and geometric parameters, such as adherend and adhesive thickness. The envelope was found by loading a double cantilever beam specimen with a specially designed loading jig. Various planar joint configurations were then analysed using closed form methods to give the joint's mode ratio, between mode I and mode II loading, and the energy release rate. This theoretical prediction is then compared with the fracture envelope to ascertain whether failure will take place via propagation of a crack in the bondline. It has been shown that this method gives very good predictions of bond strength, to within 9% of experimental values, for the joint geometries relevant to this project. The disadvantage of this method is that each fracture envelope is only valid for a single set of material and geometric parameters and further envelopes must therefore be generated if, for example, a study into the influence of adhesive thickness is to be undertaken. Further study is required to quantify the sensitivity of this method to variation of these parameters.

Damage modelling is a means of including the decrease in properties which occurs on failure of the adhesive and can be achieved by including softening of the material which initiates at some critical stress. This was done by Crocombe et al. [11] by modelling the softening with spring elements within the finite element mesh which are activated when the stress in the surrounding material is high enough. This gives predictions which can be related to the fracture mechanics method. To date this method of simulating damage has not been applied to an uncracked material, and plasticity has not been included.

Edlund [61] has applied continuum damage mechanics to adhesive joints, but only to limited joint geometries, as has Chow and Lu [62], but no strength predictions were made in the latter case. This is seen to be a promising area in which accurate methods of failure prediction may be developed, but to date little work has been published in its application to adhesive joints.

It may be concluded that there is an abundance of literature pertaining to the prediction of the strength of adhesive joints. Because stress or strain singularities exist in most modelled joints, and we know that adhesives (or adherends) cannot sustain infinite stresses or strains, it is tempting to use some means of calibration such as stress or strain at a distance. However, such calibrations can only give good approximations for a narrow range of joint materials and geometries. It is therefore necessary to present some guidance to the reader as to how to predict joint strength. For the difficult case of elasto-plastic adherends and relatively brittle (e.g. high strength) adhesives, it is recommended that a non-linear two-dimensional finite element analysis be carried out with four elements across the adhesive layer. The stresses at the 12 Gauss integration points are then arranged across the adhesive layer.

Using the average stress across the adhesive thickness makes the solution somewhat dependent on the mesh that is used to represent the adhesive. In order to calculate the average stresses, the values of stress were determined at the integration points within the finite element mesh as it was at these points that the solution was most accurate. In addition, it was shown that the closed form solution predicts very similar stresses to the finite element method and so it was decided just to report results from the finite element study to avoid repetition (Table 5).

The maximum principal stress in the adhesive was chosen as the component against which the ultimate strength would be compared as the analysis progressed and failure was deemed to occur once the two values of stress were equal, or the predicted maximum principal stress exceeded the failure stress. As there was prior knowledge of the failure loads for the test then the analyses were performed such that the majority of the known failure load was applied, and then as the analysis approached the failure load only small increments of load were applied. This allowed precise tracking of the state of stress within the adhesive layer. In addition, the load, which was applied to all of the joints, was set to exceed the

Table 5

Adhesive ultimate strength

Temperature (°C)	FM350NA ultimate tensile stress (MPa)	HP655 ultimate tensile stress (MPa)
−55	90	92
20	78	88
130	57	88
180	42	78

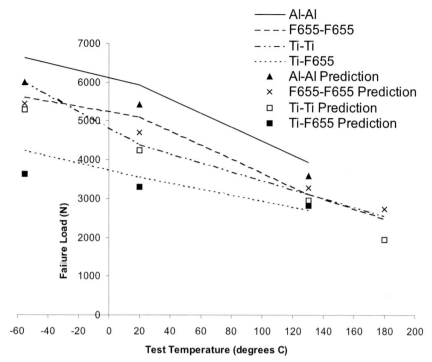

Fig. 13. FM350NA failure predictions compared with experimental results.

practical failure load in order to ascertain whether the failure criteria applied were overestimating the strength of the joint.

It can be seen from Fig. 13 that reasonable success was obtained in predicting the trends of the strength of the joints for the FM350NA joints in all of the different configurations. Bearing in mind that the failure criterion was somewhat simplistic, it was pleasing to note that the variation from the actual experimental values was quite small. The greatest error was for the aluminium joints and one possibility for this error was that a bi-linear approximation to the aluminium stress–strain curve had been used. This was derived from data which used the 0.2% proof stress as the initial yield. If a more comprehensive description of the adherend had been used then the prediction may have been closer to the experimental value. Nevertheless, the trend in joint strength was shown quite clearly.

No experimental data were available for the aluminium or the mixed joint at 180°C, and only two joints were analysed at this temperature for the FM350NA adhesive. It was thought that the errors in this case could be attributed to the fact that a maximum stress rather than maximum strain criterion had been applied. Given that the FM350NA undergoes reasonable plastic strain at the elevated temperature it may have been more appropriate to use a maximum principal strain criterion.

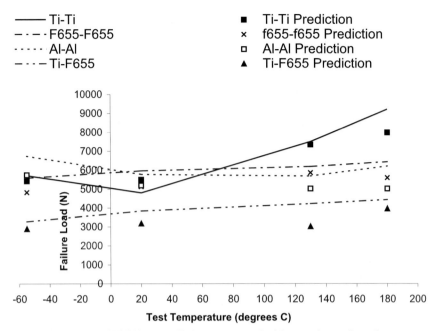

Fig. 14. HP655 failure predictions compared with experimental results.

In addition, the strength of the F655–F655 composite joint was predicted at 20°C using results from an analysis in which the residual stresses were not included. Using the maximum principal stress, a prediction of the failure load of 4328 N was predicted as compared to the prediction of 4700 N with the residual stresses accounted for. This compared to the actual failure strength of 5125 N.

The margin of error was approximately 10% for all of the joints, which is satisfactory given the direct and uncomplicated nature of the failure criterion.

Similar levels of accuracy were seen for the HP655 adhesive, although there was more scatter of the predictions, and the majority of the predictions underestimated the joint strength. Again, the key interest was in the predictions of the trends of behaviour with temperature and these were seen to agree reasonably well with the experimental values. Bismaleimide adhesives such as HP655 are expected to give increased joint strength with increasing temperature, as they develop a little plasticity but retain their strength, and this trend was observed both for the models and the experimental results (Fig. 14).

What was interesting in this study was that the short overlap composite joints did not fail via interlaminar failure and it was evident from the analysis that the stresses induced in the adhesive were sufficient to fail the adhesive before the transverse strength of the laminate was reached. The fact that a consistent failure mode was observed on test allowed the direct comparison of the analyses results for all of the different joint configurations. It has been shown that in all cases the influence of the thermal effects has been modelled and the trends of behaviour predicted with reasonable accuracy.

5. A simple predictive tool

Modern adhesives usually exhibit a degree of plasticity before failure. Typically, we might find 20% shear strain at a shear stress, τ, of 40 MPa. One simple predictive tool is to say that the absolute maximum strength for a lap joint is when the whole of the adhesive layer is at the shear yield strength. Thus, the maximum tensile load P which could be carried by a lap joint of width b and length l is given by

$$P = bl\tau.$$

However, we know that this simple rule is not sufficient when the adherends yield [1]. Observation of actual test failures with a wide variety of adhesives and adherends suggests that adherend yield has a very serious deleterious effect on joint strength. This is because a typical structural steel or aluminium alloy can extend under yield by a much higher strain than is possible for most structural adhesives. The adherend yield deformation can therefore rip the adhesive apart. We can use this knowledge to obtain a second prediction of joint strength which is particularly useful for ductile adherends.

Solid mechanics tells us that if we apply a bending moment M to a beam, the maximum stress is at the surface. For elastic deformation, this stress, σ_s, is given by

$$\sigma_s = 6M/bt^2,$$

where t is the beam thickness and b its width.

Using the theory of Goland and Reissner [29] we find that the bending moment at the edge of a lap joint is given by

$$M = kPt/2,$$

where k is the bending moment factor which reduces (from unity) as the lap rotates under load. The surface stress is therefore

$$\sigma_s = 3kP/bt.$$

In addition to the bending stress, there is a direct tensile stress in the adherends, σ_T, due to the applied load. This is given by

$$\sigma_T = P/bt.$$

Thus, the maximum surface stress, σ_m, is given by

$$\sigma_m = \sigma_s + \sigma_T = P(1+3k)/bt.$$

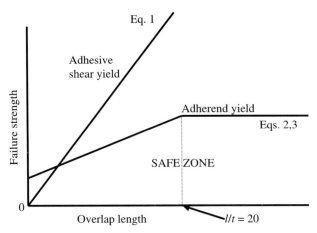

Fig. 15. Simple predictive methodology based on adhesive yield and/or adherend yield.

If σ_m is equal to the initial yield stress, σ_y, of the adherend, then the maximum load which can be carried which just creates adherend yield, is

$$P_{max} = \sigma_y bt/(1 + 3k). \tag{1}$$

For low loads and short overlaps, k is approximately 1. Therefore, for such a case,

$$P = \sigma_y bt/4. \tag{2}$$

However, for joints which are long compared to the adherend thickness, such that l/t is 20 or more, the value of k decreases and can be taken in the limit as zero. In this case, the whole of the cross-section yields and

$$P_{max} = \sigma_y bt. \tag{3}$$

Of course, this result takes no account of strain hardening which would increase σ_y. The result is, therefore, conservative. We can combine these three simple equations to give the strength of a lap joint. This is shown schematically in Fig. 15.

It is quite astonishing how well Fig. 15 predicts the strength of a wide variety of lap joints. Eq. 1 applies to very high-strength adherends which have a very high yield strength and in which the whole of the adhesive layer is yielding. No joint strength can exceed this line. On the other hand, when the adherends yield, then the ultimate load which can be carried is when the whole of the adherend is in yield and the edge of the joint exhibits a plastic hinge.

Fig. 16 shows these three equations applied to a single lap joint, with 1.6-mm-thick adherends, 25 mm wide, at various overlap lengths. The three adherends are hard steel, gauge steel and mild steel, which have initial yield points at 1800 MPa, 430 MPa, and 270 MPa, respectively. The adhesive is a modern epoxy AV119 by

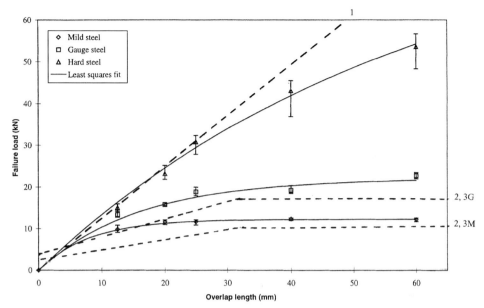

Fig. 16. Experimental joint strengths of three steel adherends with AV119 adhesive. The predicted lines are from Eqs. 1, 2 and 3.

Ciba, which has a measured maximum yield strength in shear of 49 MPa with about 30% shear strain to failure.

In Fig. 16, line 1 applies to Eq. 1. It predicts reasonably the strength of the high-strength steel joints, although it is clear that, at high loads, the measured strength is below that predicted. A more brittle adhesive used with these high-strength adherends gave results quite close to the AV119/gauge steel values. This is because the simple theory expects good ductility in the adhesive (at least 10% shear strain to failure) and will not work in the combination of a low-ductility adhesive (say 2% maximum shear strain) and high yield strength adherends. It is for this combination that the FE analysis must be used.

For the AV119 with the ductile gauge or mild steel, we have used a critical overlap/thickness ratio of 20 and assumed that beyond this, the adherend yields across the whole thickness. The predicted lines 2, 3G and 2, 3M fit the experimental points well, and validate the simple theory.

6. Conclusions

It has been shown that a very simple approach can be used for predicting the lap shear strength of adhesively bonded joints for ductile adherends and ductile adhesives. However, it cannot take into account such more complex situations as thermal stresses or changes in adherend thickness (stepped or tapered joints).

Nor is the simple approach satisfactory in all cases for adherends which do not yield, and especially for composite materials which have anisotropic strength and stiffness properties. It is therefore recommended that finite element analysis should always be used in these more complex cases. By using a two-dimensional FE analysis with four elements across the bondline and averaging the stresses at the Gauss points, a satisfactory predictive technology has been demonstrated for a variety of adherends, a wide range of temperatures, and the difficult case of a low-ductility adhesive. It is not necessary to worry about singularities, or to use three-dimensional finite element analysis in order to obtain satisfactory predictions of joint strength.

References

1. Adams, R.D., Comyn, J. and Wake, W.C., *Structural Adhesive Joints in Engineering*. Kluwer, Dordrecht, 1997.
2. Zienkiewicz, O.C. and Taylor, R.L., *The Finite Element Method*. 4th ed., V1 and V2, McGraw-Hill, New York, NY, 1989.
3. Adams, R.D. and Peppiatt, N.A., Stress analysis of adhesive-bonded lap joints. *J. Strain Anal.*, **9**, 185 (1974).
4. Adams, R.D. and Harris, J.A., The influence of local geometry on the strength of adhesive joints. *Int. J. Adhes. Adhes.*, **7**, 69 (1987).
5. Chai, H., Observation of deformation and damage at the tip of cracks in adhesive bonds loaded in shear and assessment of a criterion for fracture. *Int. J. Fract.*, **60**, 311 (1993).
6. Adams, R.D., Copperdale, J. and Peppiatt, N.A., Stress analysis of axisymmetric butt joints loaded in torsion and tension. *J. Strain Anal.*, **13**, 1 (1978).
7. Crocombe, A.D. and Adams, R.D., Influence of the spew fillet and other parameters on the stress distribution in the single lap joint. *J. Adhes.*, **13**, 181 (1981).
8. Crocombe, A.D. and Adams, R.D., Peel analysis using the finite element method. *J. Adhes.*, **12**, 127 (1981).
9. Crocombe, A.D. and Adams, R.D., An elasto-plastic investigation of the peel test. *J. Adhes.*, **13**, 241 (1982).
10. Harris, J.A. and Adams, R.D., Strength prediction of bonded single lap joints by non-linear finite element methods. *Int. J. Adhes. Adhes.*, **4**, 65 (1984).
11. Crocombe, A.D., Bigwood, D.A. and Richardson, G., Analysing structural adhesive joints for failure. *Int. J. Adhes. Adhes.*, **10**, 167 (1990).
12. Zhao, X., *Stress and Failure of Adhesively Bonded Lap Joints*. PhD Thesis, University of Bristol, 1991.
13. Richardson, G., *An Investigation of Interfacial Failure of Adhesive Joints*. PhD Thesis, Surrey University, 1993.
14. Carpenter, W. and Barsoum, R., Two finite elements for modelling the adhesive in bonded configurations. *J. Adhes.*, **30**, 25 (1989).
15. Beer, G., An isoparametric joint/interface element for finite element analysis. *Int. J. Numer. Methods Eng.*, **21**, 585 (1985).
16. Tsai, M.Y., Morton, J. and Matthews, F.L., Experimental and numerical studies of a laminated composite single-lap adhesive joint. *J. Compos. Mater.*, **29**, 1154 (1995).
17. Adams, R.D., Atkins, R.W., Harris, J.A. and Kinloch, A.J., Stress analysis and failure

properties of carbon-fibre reinforced plastic/steel double-lap joints. *J. Adhes.*, **20**, 29 (1986).

18. Adams, R.D., Coppendale, J., Mallick, V. and Al-Hamdan, H., The effect of temperature on the strength of adhesive joints. *Int. J. Adhes. Adhes.*, **12**, 185 (1992).

19. Adams, R.D. and Peppiatt, N.A., Effect of Poisson's ratio strain in adherends on stresses of an idealized lap joint. *J. Strain Anal.*, **8**, 134 (1973).

20. Groth, H.L., Calculation of stresses in bonded joints using the substructuring technique. *Int. J. Adhes. Adhes.*, **6**, 31 (1986).

21. Tsai, M.Y. and Morton, J., Mechanics of a laminated composite single-lap joint. *Mech. Compos. Rev.*, 1993.

22. Lyrner, T. and Marklund, P.E., *Finite Element Analysis of Adhesive Joints with Bondline Flaws*. Report No. 84-10, Department of Aeronautical Structures and Materials, Royal Institute of Technology, Stockholm, 1984.

23. Karachalios, V.F., *Stress and Failure Analysis of Adhesively Bonded Single Lap Joints*. PhD Thesis, University of Bristol, 1999.

24. Adams, R.D. and Davies, R.G.H., Strength of joints involving composites. *J. Adhes.*, **59**, 171 (1996).

25. Grant, P., Sanders, R.C. and Ward, A.P., *Developments in Bonded Joint Analyses*. BAC Report No. SOR(P) 137, 1983.

26. Sneddon, I., Adhesion. The distribution of stress in adhesive joints. In: Eley, D. (Ed.), Chapter 9. Oxford University Press, Oxford, 1961.

27. Tsai, M.Y. and Morton, J., An evaluation of analytical and numerical solutions to the single lap joint. *Int. J. Solids Struct.*, **33**, 2537 (1994).

28. Volkersen, O., Die Nietkraftverteilung in zugbeanspruchten Nietverbindungen mit konstanten Laschenquerschnitten. *Luftfahrtforschung*, **15**, 41 (1938).

29. Goland, M. and Reissner, E., The stresses in cemented joints. *J. Appl. Mech., Trans. Am. Soc. Mech. Eng.*, **66**, A17–A27 (1944).

30. Carpenter, W.C., Goland and Reissner were correct. Communication. *J. Strain Anal.*, **24**, 185 (1989).

31. Chen, D., Cheng, S. and Yupu, S., Analysis of adhesive bonded joints with non-identical adherends. *J. Eng. Mech.*, **117**, 605 (1991).

32. Chen, D. and Cheng, S., An analysis of adhesive bonded single lap joints. *J. Appl. Mech.*, **50**, 109 (1983).

33. Wu, Z.J., Romeijn, A. and Wardenier, J., Stress expressions of single-lap adhesive joints with dissimilar adherends. *Compos. Struct.*, **38**, 273 (1997).

34. Yang, C. and Pang, S.S., Stress–strain analysis of single lap composite joints under tension. *J. Eng. Mater. Technol.*, **118**, 247 (1996).

35. Delale, F., Erdogen, F. and Aydinoglu, M.N., Stresses in adhesively bonded joints — a closed form solution. *J. Compos. Mater.*, **15**, 249 (1981).

36. Renton, J.W. and Vinson, J.R., On the behaviour of bonded joints in composite material structures. *Eng. Fract. Mech.*, **7**, 241 (1995).

37. Srinivas, S., *Analysis of Bonded Joints*. NASA TN D7846/D7860, 1975.

38. Wah, T., Stress distribution in a bonded anisotropic lap joint. *ASME J. Eng. Mater. Technol.*, July, 1993.

39. Coppendale, J., *The Stress and Failure Analysis of Structural Adhesive Joints*. PhD Thesis, University of Bristol, 1997.

40. Weitsman, Y., An investigation of non-linear visco-elastic effects on load transfer in a symmetric double lap joint. *J. Adhes.*, **11**, 279 (1981).

41. Hart-Smith, L.J., *Adhesive Bonded Single Lap Joints*. Technical Report CR-112236, NASA, Langley Research Centre, 1973.
42. Adams, R.D. and Mallick, V., A method for the stress analysis of lap joints. *J. Adhes.*, **38**, 199 (1992).
43. Allmann, D.J., A theory for elastic stresses in bonded lap joints. *Q. J. Mech. Appl. Math.*, **30**, 415 (1977).
44. Greenwood, L., Boag, T.G. and McLaren, A.S., Stress distributions in lap joints. In: *Adhesion: Fundamentals and Practice*. McLaren, London, 1969.
45. Crocombe, A.D. and Tatarek, A., A unified approach to adhesive joint analysis. In: *Proceedings of Adhesives, Sealants and Encapsulants 85*. Plastics and Rubber Institute, London, 1985.
46. Ikegami, M., Kishimoto, W., Okita, K., Nakayama, H. and Shirato, M., Strength of adhesively bonded scarf joints between glass fibre reinforced plastics and metals. In: *Proceedings of Structural Adhesives in Engineering II*. Institute of Materials, London, 1989.
47. Lee, S.J. and Lee, G.L., Development of a failure model for the adhesively bonded tubular single lap joint. *J. Adhes.*, **40**, 1 (1992).
48. Clark, J.D. and McGregor, I.J., Ultimate tensile stress over a zone: a new failure criterion for adhesive joints. *Adhes.*, **42**, 227 (1993).
49. Crocombe, A.D., Richardson, G. and Smith, P.A., A unified approach for predicting the strength of cracked and non-cracked adhesive joints. *J. Adhes.*, **49**, 211 (1995).
50. Kinloch, A.J. and Williams, J.G., Crack blunting mechanisms in polymers. *J. Mater. Sci.*, **15**, 987 (1980).
51. Crocombe, A.D., Global yielding as a failure criteria for bonded joints. *Int. J. Adhes. Adhes.*, **9**, 145 (1989).
52. Schmit, F. and Fraisse, P., Fracture mechanics analysis of the strength of bonded joints. *Mater. Tech.*, **4–5** (1992).
53. Toya, M., Fracture mechanics of interfaces. *JSME Int. J. Ser. 1: Solid Mech., Strength Mater.*, **33**, 413 (1990).
54. Kinloch, A.J. and Shaw, S.J., The fracture resistance of a toughened epoxy adhesive. *J. Adhes.*, **12**, 59 (1981).
55. Trantina, G.C., Fracture mechanics approach to adhesive joints. *J. Compos. Mater.*, **6**, 192 (1972).
56. Hu, G.K., Non-linear fracture mechanics for adhesive joints. *J. Adhes.*, **37**, 261 (1992).
57. Gradin, P.A. and Groth, H.L., *A Fracture Criterion for Adhesive Joints in Terms of Material Induced Singularities*. Report No. 83-12, The Royal Institute of Technology, Stockholm, 1984.
58. Groth, H.L., *Prediction of Failure Loads for Adhesive Joints Using Singular Intensity Factors*. Report No. 85-2, The Royal Institute of Technology, Stockholm, 1985.
59. Groth, H.L., Stress singularities and fracture at interface corners in bonded joints. *Int. J. Adhes. Adhes.*, **8**, 107 (1988).
60. Fernlund, G., Papini, M., Mccammond, D. and Spelt, J.K., Fracture load predictions for adhesive joints. *Compos. Sci. Technol.*, **51**, 587 (1994).
61. Edlund, U., *Mechanical Analysis of Adhesive Joints: Models and Computational Methods*. Dissertation, University of Linkoping, 1992.
62. Chow, C.L. and Lu, T.J., Analysis of failure properties and strength of structural adhesive joints with damage mechanics. *Int. J. Damage Mech.*, **1**, 404 (1992).

Chapter 5

Strength of butt and sharp-cornered joints

E. DAVID REEDY Jr. [*]

Sandia National Laboratories, Albuquerque, NM, USA

1. Introduction

There are two common types of butt joints: a two-material butt joint (Fig. 1) and an adhesively bonded butt joint that joins two adherends together with a relatively thin adhesive layer (Fig. 2). Tensile-loaded butt joints are not often used in structurally demanding applications since load misalignment can introduce large bond-normal bending stresses. Perhaps the most common application of the butt joint is as a test geometry to characterize adhesive strength. Butt-joint test specimens usually bond two relatively rigid adherends together with a thin adhesive layer (e.g. ASTM D897-95a and D2095-96, American Society of Testing and Materials). The joint is loaded in tension, and the adhesive's apparent tensile strength is defined as the failure load divided by the bond area. This strength definition is based upon the premise that failure is associated with the nominal applied tensile stress at joint failure. In reality the stress state in an adhesively bonded butt joint is quite complex, and high stress concentrations are generated in the adhesive layer.

In the case of a glassy polymer (e.g. a high-strength epoxy) the relatively stiff adherends restrain the adhesive layer's tendency for Poisson's ratio-induced contraction, and as a consequence high shear and peel stresses are generated along the edges of the bond. Butt-joint adherends often have relatively sharp edges, and this further accentuates the magnitude of the edge stresses. The presence of high stress concentrations along the edges of the bond is consistent with the observed failure mode of epoxy-bonded, metal-adherend, cylindrical butt joints subjected to a tensile loading. Failure initiates adhesively (on the interface) in these joints, along a small segment of the specimen periphery [1–3]. Consequently, the nominal tensile strength is not a measure of the adhesive's cohesive strength, but reflects surface preparation and interfacial strength. Test results also show that

---

[*] Corresponding author. E-mail: edreedy@sandia.gov

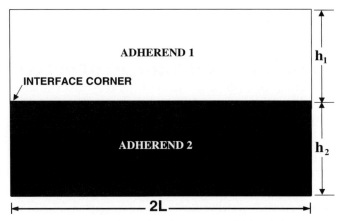

Fig. 1. Conventions used to define the two-material butt-joint geometry.

the nominal failure stress is not constant for a given adherend–adhesive material system, but can vary substantially with bond thickness. In one study the strength of adhesively bonded butt joints with steel adherends increased by a factor of 2 as bond thickness was decreased from 2.0 mm to 0.25 mm [2]. This chapter will focus on the failure of sharp-edged butt joints with failure initiating along the edges of the bond. It should be noted, however, that there are also instances when failure initiates away from the edges of the bond. This is particularly true for thin, highly constrained, elastomeric adhesive bonds [1]. Such failures are consistent

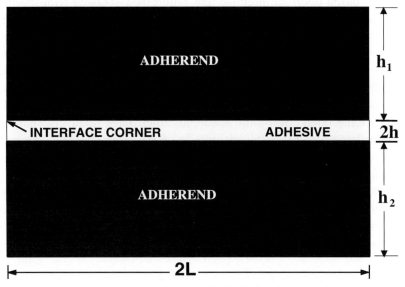

Fig. 2. Conventions used to define the adhesively bonded butt-joint geometry.

with the high hydrostatic tensile stress generated in the center of a sufficiently thin bond when the adhesive layer is a rubbery polymer (e.g. material with a Poisson ratio approaching 0.5).

There are three widely accepted methods for predicting the strength of bonded joints. One approach uses shear-lag-based, elastic–plastic stress analyses for bond stress and strain. Joint failure is predicted to occur at a critical adhesive shear strain [4]. In another approach, a detailed finite-element analysis of the joint is carried out using an elastic–plastic adhesive material model. Often geometric particulars like spew filets are included, but the joint is considered flaw-free (uncracked). It has been suggested that a maximum principal stress criterion works best for brittle adhesives, while a maximum principal strain condition should be used for toughened adhesives [5]. The third approach applies linear-elastic fracture mechanics concepts to bonded joints. A variety of adhesively bonded fracture specimens has been developed to measure the Mode I, Mode II, and mixed Mode I and II fracture toughness when the crack is within the adhesive layer [6]. Fracture mechanics techniques have also been developed to measure interfacial toughness [7]. One fracture mechanics-based approach uses a critical energy-release rate in conjunction with an inherent flaw size to predict joint strength [1].

In recent years a fourth approach for predicting the strength of bonded joints has shown promise when joints contain sharp corners. For example, a butt joint with sharp-cornered adherends can be thought of as two, edge-bonded, equal-width rectangles (Fig. 1). Failure in such joints often initiates at the interface corner, the point where the interface intersects the stress-free edge. Asymptotically, the interface-corner region looks like two edge-bonded 90° wedges (i.e. quarter planes). It is well known that, within the context of linear elasticity theory, a power-law singularity with a real exponent exists at the apex of bonded 90° wedges for certain material combinations [8–10]. That is,

$$\sigma \sim K_a r^{\lambda-1} \quad (-0.41 < \lambda - 1 < 0) \tag{1}$$

where the subscript 'a' on K denotes that this stress-intensity factor is associated with the apex of a wedge. The strongest singularity occurs in plane strain when one material is rigid and the other has a Poisson ratio of 0.5 (i.e. it is incompressible). The value of the stress-intensity factor K_a characterizes the magnitude of the stress state in the region of the interface corner. The calibration relation defining K_a is determined by the full solution and depends on loading, geometry, and elastic properties. It appears reasonable to hypothesize that failure occurs at a critical K_a value. Such an approach is analogous to linear-elastic fracture mechanics except, here the critical K_a value is associated with a discontinuity other than a crack. As is the case of linear-elastic fracture mechanics, the size of the yield zone must be small compared to the size of the region dominated by the stress singularity (i.e. much smaller than the bond thickness). This approach does not apply when there is large-scale yielding within the adhesive bond.

This chapter will describe methods for performing an interface-corner failure analysis, the limitations to such an approach, supporting experimental studies, and a discussion of unresolved issues in its application. Although the focus is on butt joints, the approach applies to sharp-cornered joints in general and results for a sharp-cornered, embedded inclusion are also presented.

2. Interface-corner stress state

The asymptotic stress state near the apex of dissimilar bonded wedges (i.e. interface corner) for plane stress or strain, when the wedge materials are isotropic and linear elastic, has the form

$$\sigma_{ij} = \sum_{n=1}^{N} K_{an} r^{\lambda_n - 1} \bar{\sigma}_{ijn}(\theta) + K_{a0} \bar{\sigma}_{ij0}(\theta) \quad (i, j = r, \theta) \tag{2}$$

where r, θ refer to a polar coordinate system defined at the interface corner. One or more power-law singularities of differing strength can exist, and the exponents can be real or complex [8–12]. The number of stress singularities N, the strength of these singularities $\lambda_n - 1$, and their angular variation $\bar{\sigma}_{ijn}(\theta)$ is determined by the asymptotic analysis and depends on nondimensional elastic properties (e.g. Dundurs' parameters), the local interface-corner geometry (i.e. wedge angle) and edge-boundary conditions (stress-free, fixed, etc.). The stress-intensity factors K_{an} determine the contribution of each singular term to the stress state in the region of the interface corner [13–18]. K_{an} depends on global geometry, applied loads and elastic properties. Although not explicitly shown in Eq. 2, there are certain special combinations of elastic properties, wedge angles, and edge loads that can also generate logarithmic singularities [11,19,20].

In some instances an asymptotic description that includes only the singular terms fails to accurately describe the stress state over a physically significant region about the interface corner. For example, when the r-independent regular term $K_{a0}\bar{\sigma}_{ij0}(\theta)$ in Eq. 2 exists, it must often be included in the asymptotic formula [21,22]. An r-independent regular term always exists for thermally induced strains and locally applied edge tractions but, except for certain special combinations of elastic properties and wedge angles [23], usually vanishes for remote applied loads.

The origin of the r-independent asymptotic term can be understood for the simple case of bonded rigid and elastic quarter planes subjected to a uniform temperature change (Fig. 3; the elastic quarter plane has Young's modulus E, Poisson's ratio v, and coefficient of thermal expansion α_T). If unbonded, a temperature change of ΔT would induce expansion relative to the rigid substrate. If plane stress is assumed and $v > 0$, a uniaxial stress parallel to the stress-free

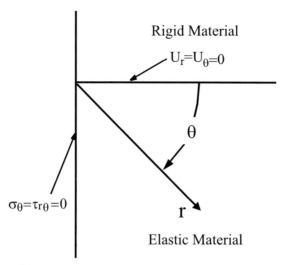

Fig. 3. Asymptotic problem for a butt joint with a rigid adherend. Note: the elastic material's stress-free edge is at $\theta = -\pi/2$.

edge (i.e. normal to the interface) of magnitude

$$\sigma_{\theta\theta}(0) = \frac{E\alpha_T \Delta T}{\nu} \tag{3}$$

negates the thermally induced strain that is parallel to the interface and makes the elastic quarter plane compatible with the rigid substrate. This stress is also consistent with the stress-free boundary condition. Consequently, $K_{a0}\bar{\sigma}_{ij0}(\theta)$ is equal to the uniform stress state defined by Eq. 3, since this stress state is the solution of the asymptotic problem [22]. Furthermore, it can be shown that Eq. 3 also holds for plane strain. The explicit form of the r-independent asymptotic term for the general case of two dissimilar bonded wedges subjected to a uniform temperature change, local edge tractions, or remote tractions has been determined [23–25].

3. Critical K_a fracture criterion

The choice of a failure criterion is obvious when interface-corner stresses are described by one real-valued, power-law singularity. In this case, a single interface-corner stress-intensity factor K_a characterizes the magnitude of the stress state in the region of the interface corner.

$$\sigma_{ij} = K_a r^{\lambda-1}\bar{\sigma}_{ij}(\theta) \quad (i,j=r,\theta) \tag{4}$$

where K_a is defined so that the stress component normal to the interface $\sigma_{\theta\theta}(r,0) = K_a r^{\lambda-1}$ (coordinate system defined in Fig. 3). Failure is assumed to occur at a critical value of K_a, referred to as the interface-corner toughness, K_{ac}. This approach is completely analogous to linear-elastic fracture mechanics (LEFM), except here the critical value of the stress-intensity factor is associated with a material and geometric discontinuity other than a crack. It is less apparent how to perform a failure analysis when the asymptotic stress state is described by multiple power-law singular terms (real or complex) or includes a significant r-independent term. When multiple singular terms exist, one cannot simply assume that the highest-order singularity dominates. Typically all singular terms must be included to get an accurate representation of the stress state over a reasonably large region [13,21,26]. Approaches for treating multiple singularities have yet to be developed. For this reason, the results presented in this chapter will focus almost exclusively on joints where the asymptotic interface-corner stress state is fully characterized by a single K_a. Fortunately, there are many joints of this type including butt joints, adhesively bonded butt joints, epoxy wedges on a stiff substrate, and encapsulated inclusions subjected to uniform cooling.

One notable feature of the interface-corner failure analysis is that it requires no detailed information about the failure process itself. Failure may be caused by a preexisting interfacial flaw or perhaps by a cavitation instability induced by the high levels of hydrostatic tension found at the interface. If a small interfacial crack is present, it could be sharp, bridged, or have a yield zone comparable to its length. To be applicable, it is only necessary that the failure process zone is deeply embedded within the region dominated by the interface-corner singularity at failure.

4. Calculating K_a

Several approaches have been developed for determining K_a for a specified joint geometry, material combination, and applied loading. Two commonly applied approaches post-process finite-element results for the joint of interest. In addition, broadly applicable numeric methods are being developed to solve complicated joint geometries that involve, for instance, multiple anisotropic materials.

4.1. Matching numeric solution with asymptotic form

One commonly used approach for determining K_a is to match detailed finite-element results with the known form of the asymptotic solution [14,17,18]. For example, displacements normal to the stress-free edge, in the region dominated by the stress singularity, were used to determine K_a values for an adhesively bonded butt joint with rigid adherends [18]. The asymptotic problem for this idealized joint is simply bonded rigid and elastic quarter planes, and the asymptotic solution is described by one real-valued, power-law singularity. The asymptotic solution for $U_\theta(-\pi/2)$, the normal displacement along the stress-free edge, requires that

$$U_\theta(-\pi/2) = \frac{K_a r^\lambda g_\theta(-\pi/2)}{E} \tag{5}$$

in the region dominated by the stress singularity. Note that the function $g_\theta(\theta)$ and the parameter λ ($\lambda - 1$ is the strength of the stress singularity) are fully determined by the asymptotic analysis. The value of K_a can be determined from the computed free-edge displacements by a linear least-squares fit

$$\frac{U_\theta(-\pi/2)}{r^\lambda g_\theta(-\pi/2)/E} = C_1 + C_2 r \quad \text{where} \quad C_1 = K_a \tag{6}$$

The finite-element mesh must be sufficiently refined in the region dominated by interface-corner singularity to accurately calculate the $U_\theta(-\pi/2)$ values used in Eq. 6. The adequacy of the mesh can be assessed by plotting $\log U_\theta(-\pi/2)$ vs. $\log(r)$. In the region dominated by the stress singularity, the log of the calculated edge displacement should vary linearly with the log of distance from the interface corner over a distance of several decades, and the line should have a slope λ that agrees with the asymptotic solution. Note that this approach is not limited to matching edge displacements; others often match interfacial stresses instead. Indeed when interfacial normal stress is matched, only the eigenvalue must be computed in the asymptotic analysis since $\sigma_{\theta\theta}(r,0) = K_a r^{\lambda-1}$ by definition (Eq. 4); the eigenvectors, which determine $\bar{\sigma}_{ij}(\theta)$, do not have to be computed.

The main advantage of using the matching approach to determine K_a values is that it can be easily applied. All that is needed to post-process the linear-elastic, finite-element results is a plotting program with a least-squares fit capability. The main disadvantage of the matching approach is the need to construct a finite-element mesh that is highly refined in the region of the interface corner. Otherwise, the extracted K_a values may lack accuracy. With due diligence in meshing, however, one can easily obtain highly accurate K_a results [27]. Finally note that, although the matching approach is most readily applied when the asymptotic solution is described by one real-valued, power-law singularity, it has been successfully adapted to cases where multiple power-law singularities exist [21,28].

4.2. Contour integral

Another widely used method for calculating interface-corner stress-intensity factors uses a path-independent integral. This integral is derived via an application of Betti's reciprocal work theorem [29–31]. The reciprocal work contour integral (RWCI) method was first used to determine stress intensities for crack and notch geometries in homogeneous bodies, but was later extended to notched bimaterials [32]. More recent enhancements include conversion of the line integral to an area integral for improved accuracy [27,33].

The path-independent RWCI used to calculate K_a is defined as follows:

$$K_a = \int_C (\tilde{\sigma}_{ij} u_i - \sigma_{ij} \tilde{u}_i) n_j \, \mathrm{d}s \tag{7}$$

where σ_{ij} and u_i are the numerically calculated stresses and displacements for the problem of interest, n_j is the unit outward normal to a counter-clockwise contour C that encircles the interface corner, and $\mathrm{d}s$ is an infinitesimal line segment of C. The auxiliary fields $\tilde{\sigma}_{ij}$ and $\tilde{u}_i$ are the asymptotic solution corresponding to $\lambda^* = -\lambda$, where λ is the strength of the singularity associated with K_a. The stress intensity of the auxiliary fields K_a^* is chosen so that Eq. 7 holds. The path-independent RWCI offers a convenient way for calculating accurate K_a values without constructing a highly refined finite-element mesh. The main disadvantage of the contour integral method is that it requires determining the full solution of the associated asymptotic problem (i.e. eigenvalues and eigenvectors) and then incorporating these fields in a post-processing program. Furthermore, one may still need to use a relatively refined mesh to determine the size of the region dominated by the K_a field and the applicability of small-scale yielding assumption.

4.3. Numeric solutions for complicated configurations

There are many corner configurations of practical importance for which the analytic solution of the asymptotic problem is not readily available. For example, an interface corner may be at the apex of three or more bonded wedges, and some if not all of the wedge materials may be anisotropic. In addition to complicated two-dimensional problems, there is a growing interest in fully three-dimensional, multimaterial geometries where singularities occur along edges and at three-dimensional corners. In complicated configurations such as these, it may be difficult to apply methods for calculating stress-intensity factors that require knowledge of the analytic form of singular stress fields (when using matching methods or contour integrals, for instance). Fortunately, broadly applicable numerical methods are being developed to solve such complicated asymptotic problems [34–37]. These methods typically determine the order of the stress singularities and the angular variation of the displacement fields numerically and then use the numerically determined fields in a solution scheme that calculates stress-intensity factors. For example, one approach extracts the stress-intensity factors in a post-processing step by using the wedge-tip displacements calculated from a full finite-element solution together with the asymptotic displacement fields determined by a modified Steklov method [36]. Another approach uses a finite-element eigenanalysis to obtain asymptotic displacements fields. These fields are then used to construct enriched finite elements that directly calculate stress-intensity factors [35].

5. K_a calibrations

The interface-corner stress-intensity factor, K_a, depends on the applied load, elastic properties, and the overall joint geometry. In recent years K_a calibrations have been published for a number of geometries of practical interest. These calibrations provide convenient formulas that can be used in a failure analysis without recourse to a detailed numerical analysis. They also provide insights regarding the dependencies of K_a on geometry and material properties. Note that all calibrations presented below are for joints whose asymptotic interface-corner solution is described by one real-valued, power-law singularity, and a single K_a, characterizes the magnitude of the stress state in the region of the corner.

Based upon dimensional considerations and linearity with applied load, K_a must have the form

$$K_a = \sigma^* h^{1-\lambda} A(\alpha, \beta, L/h, \ldots) \tag{8}$$

where σ^* is a characteristic stress, h is a characteristic length scale, and A is a function of nondimensional material and geometric parameters. As noted above, the strength of the stress singularity, $\lambda - 1$, is known from the asymptotic analysis and it depends on nondimensional elastic properties in addition to the local interface-corner geometry (e.g. wedge angle). Consequently, the K_a calibration is defined by specifying the choice of characteristic stress and length scale and prescribing the functional dependence of A on all relevant nondimensional geometric and elastic parameters.

Material-property dependence for traction-loaded, bimaterial-plane elasticity problems can be completely described in terms of two dimensionless parameters [38]. Consequently, Dundurs' elastic mismatch parameters, α and β, are commonly used to define material-property dependence in K_a calibrations. For a bimaterial with material 1 above the interface and material 2 below,

$$\alpha = \frac{\bar{E}_1 - \bar{E}_2}{\bar{E}_1 + \bar{E}_2}, \quad \beta = \frac{\mu_1(\kappa_2 - 1) - \mu_2(\kappa_1 - 1)}{\mu_1(\kappa_2 + 1) + \mu_2(\kappa_1 + 1)} \tag{9}$$

with $\bar{E}_i = E_i$ for plane stress and $\bar{E}_i = E_i/(1 - v_i^2)$ for plane strain, and where E_i, μ_i, and v_i (for $i = 1, 2$) are the Young modulus, shear modulus, and Poisson ratio of materials 1 and 2. Furthermore, $\kappa_i = (3 - v_i)/(l + v_i)$ for plane stress and $\kappa_i = (3 - 4v_i)$ for plane strain. Note that interchanging materials 1 and 2 simply changes the sign of α and β, and when material 1 is rigid,

$$\alpha = 1, \quad \beta = \begin{cases} \dfrac{1 - v_2}{2} & \text{in plane stress} \\[2mm] \dfrac{1 - 2v_2}{2(1 - v_2)} & \text{in plane strain} \end{cases} \tag{10}$$

In the following, K_a calibrations are presented for two classes of problems, butt joints and encapsulated inclusions. Note that one must be careful when comparing the K_a calibrations published by various researchers. Characteristic stress and length are often chosen differently, and even the definition of K_a can change (the convention used here is to include the characteristic length scale in the K_a definition as shown in Eq. 8). Also note that the characteristic stress for plane-stress and plane-strain conditions will in general differ.

5.1. Adhesively bonded butt joint with stiff adherends

A butt joint that bonds rigid adherends together with a thin adhesive layer is one of the simplest idealized joint geometries of practical importance (Fig. 2). The asymptotic problem is just a quarter plane with one edge fixed and the other edge stress-free (Fig. 3). This idealized joint is a reasonable representation when the adherends are much stiffer than the adhesive, such as the case of steel adherends and epoxy adhesive, and when the adhesive layer is thin relative to other joint dimensions. When the layer is relatively thin, it behaves as if it is semi-infinite and the stress in the adhesive layer is uniform and unaffected by the stress-free edge in regions sufficiently far from the stress-free edge. For typical epoxy properties (Young's modulus of 2.5 GPa, Poisson's ratio of 0.4), a bond-length to bond-thickness ratio of greater than 4 is sufficient for an elastic layer to be considered thin [22]. Restricting consideration to joints with rigid adherends and with a thin adhesive layer greatly simplifies the K_a calibration since layer thickness is the only length scale. As a result, the function A (Eq. 8) does not depend on geometric parameters but only on β or equivalently the adhesive layer's Poisson ratio (Eq. 10). The K_a calibration (i.e. function A) for the idealized, rigid-adherend, thin-bond butt joint has been determined for tension, uniform temperature change, and shear by matching finite-element solutions with their associated asymptotic form [18,39].

5.1.1. Tension

In this calibration, the characteristic length scale is one-half the bond thickness (h in Fig. 2). One particularly convenient choice for the characteristic stress σ^* is the uniform in-plane stress (i.e. stress directed parallel to the interface) found at the center of the layer in a region remote from the stress-free edge. With this definition for σ^*, the same K calibration relation applies to transverse tensile loading and to uniform adhesive shrinkage. This characteristic stress is related to the nominal, applied transverse (butt tensile) stress $\bar{\sigma}$ for an adhesive layer with Poisson's ratio ν by

$$\sigma^* = \begin{cases} \nu\bar{\sigma} & \text{plane stress} \\ \dfrac{\nu}{(1-\nu)}\bar{\sigma} & \text{plane strain} \end{cases} \tag{11}$$

Table 1

Nondimensional functions $A(1,\beta)$ used in the K_a calibration for a butt joint with a thin adhesive layer and rigid adherends, and also for a thin layer on a rigid substrate

β	ν [a]	$1-\lambda$	$A(1,\beta)$		
			Butt joint loaded in transverse tension or by uniform ΔT	Butt joint loaded in shear	Layer on substrate loaded by uniform ΔT
0.474	0.05	0.077	14.7	2.57	15.7
0.444	0.10	0.133	6.63	2.05	6.89
0.412	0.15	0.179	3.96	1.73	4.13
0.375	0.20	0.219	2.63	1.51	2.82
0.333	0.25	0.255	1.84	1.34	2.06
0.286	0.30	0.289	1.32	1.21	1.57
0.231	0.35	0.320	0.958	1.10	1.23
0.167	0.40	0.350	0.654	1.01	0.99
0.091	0.45	0.378	0.391	0.93	0.80

[a] ν value corresponding to the specified β when plane strain applies, see Eq. 10.

The corresponding $A(1,\beta)$ values [18] are listed in Table 1. These values are in good agreement with those reported by others [40,41]. Since the adherends are rigid and the adhesive layer is thin, the interface-corner stress field is the same for plane strain and axisymmetric geometries. Consequently, this K_a calibration also applies to a butt joint formed by bonding cylindrical rods together with a thin adhesive layer. The same K_a relation is also applicable to a transversely cracked elastic layer that is sandwiched between rigid layers (assuming a single, isolated crack). The stress-free edge can be considered a symmetry plane when the bounding layers are rigid.

5.1.2. Uniform temperature change

The K_a calibration defined for a tensile loading (i.e. same $A(1,\beta)$ values listed in Table 1 and with characteristic length $= h/2$) also applies when the adhesive undergoes a uniform temperature change, ΔT, if the characteristic stress is defined as

$$\sigma^* = \begin{cases} -E\alpha_T \Delta T & \text{plane stress} \\ -\dfrac{E\alpha_T \Delta T}{(1-\nu)} & \text{plane strain} \end{cases} \tag{12}$$

where α is the coefficient of thermal expansion. This is true because tensile and uniform adhesive-shrinkage load types are actually related [18]. The superposition of an appropriate uniform stress state with the solution for a uniform pressure applied to the layer's exterior edge yields solutions for transverse tension and also

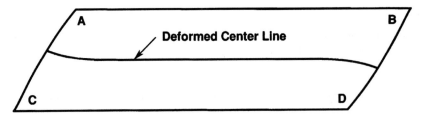

Fig. 4. Deformed shape of the elastic layer in an adhesively bonded butt joint with rigid adherends when subjected to a positive-shear loading.

for uniform adhesive shrinkage. The solution for the uniform edge pressure can be thought of as providing the fundamental singular solution, and the characteristic stress defined in Eqs. 11 and 12 is the stress superimposed with the edge pressure to obtain a stress-free edge.

5.1.3. Shear

The adhesively bonded butt joint's thin adhesive layer is sheared by tangentially displacing one rigid adherend relative to the other. The same asymptotic problem shown in Fig. 3 applies to both the tensile-loaded and shear-loaded rigid-adherend butt joint; the strength of the stress singularity is independent of the type of loading. In this calibration, the characteristic length scale is again taken to be one-half the bond thickness (h in Fig. 2), and the characteristic stress σ^* is the nominal shear stress $\bar{\tau}$ found at the center of the layer in a region remote from the stress-free edge. Values of $A(1, \beta)$ for shear [39] are listed in Table 1. These values are in good agreement with those reported by others [41]. Note that, for this antisymmetric loading, the magnitude of K_a is the same at the joint's four interface corners (A, B, C, and D in Fig. 4), but when the joint is subjected to positive shear, K_a is positive at B and C, and negative at A and D. This K_a calibration can be applied to an adhesively bonded, thick-adherend lap joint, either flat or tubular, when the adherends are much stiffer than the adhesive, provided that there is negligible bending induced by the loading.

5.1.4. Effect of adherend stiffness

Even though the rigid adherend idealization is reasonable for steel adherends joined by an epoxy adhesive, a noticeable deviation occurs when steel is replaced by aluminum [42]. Table 2 shows that the strength of the stress singularity decreases from −0.32 for rigid adherends to −0.30 for steel adherends to −0.27 for aluminum adherends (for an epoxy with $E = 3.5$ GPa, $\nu = 0.35$). Table 2 also lists $A(\alpha, \beta)$ values for these three adherend materials. Once again the characteristic length scale is the half-bond thickness (h in Fig. 2) and the characteristic stress

Table 2

Nondimensional function $A(\alpha,\beta)$ used in the K_a calibration for a butt joint with a thin adhesive layer and finite stiffness adherends

	α	β	$1-\lambda$	$A(\alpha,\beta)$
Rigid–epoxy	1.000	0.231	0.320	0.958
Steel–epoxy	0.966	0.222	0.302	0.989
Aluminum–epoxy	0.902	0.207	0.268	1.061

σ^* is the in-plane stress found at the center of the layer in a region remote from the stress-free edge. The relationship between σ^* and the nominal, applied tensile stress $\bar{\sigma}$ is

$$\sigma^* = \begin{cases} v_2\left[1 - \left(\dfrac{E_2}{E_1}\dfrac{v_1}{v_2}\right)\right]\bar{\sigma} & \text{plane stress} \\[2em] \left(\dfrac{v_2}{1-v_2}\right)\left[1 - \left(\dfrac{E_2}{E_1}\dfrac{v_1(1+v_1)}{v_2(1+v_2)}\right)\right]\bar{\sigma} & \text{plane strain} \end{cases} \qquad (13)$$

where E_i, v_i are Young's modulus and Poisson's ratio of the adherend (material 1) and the adhesive layer (material 2), respectively. Note that Eq. 13 reduces to Eq. 11 when the adherends are rigid ($E_1 \to \infty$). $A(\alpha,\beta)$ values for $-1 < \alpha < 1$ and $\beta = 0$, $\alpha/4$ are also available [43].

5.2. Thin layer on a thick substrate

5.2.1. Rigid substrate

A thin adhesive layer on a rigid substrate is another idealized geometry of practical importance. An example is a thin polymer film deposited on a silicon substrate. For convenience material 1 is taken to be rigid ($\alpha = 1$). The only length scale is layer thickness h_2 ($L \gg h_2$ in Fig. 1). The same asymptotic problem for an adhesively bonded butt joint with rigid adherends (Fig. 3) applies to a thin layer on a rigid substrate. Furthermore, a K_a calibration for a uniformly cooled layer can be defined using the same thermally induced in-plane stress (Eq. 12), but now the characteristic length scale is layer thickness h_2 (Fig. 1). Values of $A(1,\beta)$ are listed in Table 1 for a wide range of β.

5.2.2. Finite stiffness substrate

A K_a calibration for the case of a thin layer on a relatively thick, finite stiffness substrate ($h_2/h_1 \le 0.1$, $h_2/L \le 0.1$ in Fig. 1) has been published [44]. This calibration is presented as a polynomial expression that is a function of two variables: the strength of the singularity ($\lambda - 1$ in Eq. 4, where $\lambda - 1$ depends on Dundurs' parameters α, β) and the Poisson ratio of the thin layer. When specialized to the case of a rigid substrate, this calibration yields values that are within a couple of percent of those listed in Table 1.

5.3. Long bimaterial butt joint

The only length scale in a butt joint with h_1/L and h_2/L, both $\gg 1$ (Fig. 1), is joint width $2L$. A K_a calibration for the long-adherend idealization is available and is said to apply whenever h_1/L and h_2/L are greater than 2 [17]. For tensile loading, the characteristic stress σ^* is the nominal applied stress, and characteristic length is the half-joint width (L in Fig. 1). The function A is given as a polynomial fit

$$A(\omega) = 1 - 2.89\omega + 11.4\omega^2 - 51.9\omega^3 + 135.7\omega^4 - 135.8\omega^5 \tag{14}$$

where $\omega = 1 - \lambda$ is the negative of the strength of the stress singularity (Eq. 4). The calibration also applies to a uniform temperature change if the characteristic stress is defined by

$$\sigma^* = \frac{-E_1 E_2}{\nu_1(1+\nu_1)E_2 - \nu_2(1+\nu_2)E_1}[(1+\nu_1)\alpha_{T1} - (1+\nu_2)\alpha_{T2}]\Delta T \tag{15}$$

for plane strain.

Interestingly, for this particular geometry the nondimensional function can be seemingly correlated with a single parameter, i.e. the strength of the stress singularity rather than both of Dundurs' parameters. Although the strength of the singularity does itself depend on Dundurs' parameters, an unlimited number of α, β pairs can generate the same strength singularity. Graphs showing curves of constant $1 - \lambda$ for the range of all possible α and β pairs can be found in several references [9,17]. Furthermore the long-adherend butt joint exhibits a special symmetry that allows one to interchange the materials without changing the strength of the singularity (i.e. $\lambda(\alpha, \beta) = \lambda(-\alpha, -\beta)$) or the stress-intensity factor (i.e. $A(\alpha, \beta) = A(-\alpha, -\beta)$). It must be emphasized, however, that the dependence of function A on only $1 - \lambda$ is a special case, and, in general, this type of reduced dependence does not occur. Other recently reported values of $A(\alpha, \beta)$ for $\alpha = 0.2$, 0.5, 0.8, 0.99 and $\beta = 0$ or $\alpha/4$ [43] for the same joint geometry are within a few percent of those determined by Eq. 14. The accuracy of Eq. 14 is thought to be good for Poisson's ratios between 0.2 and 0.4, but there may be some divergence for Poisson's ratios outside of this range [44,45]. For this reason, another form for the function A has been proposed to improve the accuracy of the relationship for a wider range of Poisson's ratios [45]. Specifically,

$$A(\alpha, \omega) = e^{[-2.45\omega(1.1-\omega)]} - 0.0235\alpha^6 \tag{16}$$

where once again $\omega = 1 - \lambda$.

5.3.1. Long-adherend scarf joints

The long-adherend butt joint can be generalized to a scarf joint by allowing the interface to intersect the stress-free edge at a nonnormal inclination. A K_a calibration for the long-adherend scarf joint is available for both remote tension and bending [46]. Results are presented in tabular form for $\alpha = \pm 0.2, \pm 0.5, \pm 0.8,$ ± 0.99 with $\beta = 0$ or $\alpha/4$ for scarf angles of 15, 30, 45, 60, 75, and 90 degrees.

5.4. Bimaterial beam

K_a calibrations covering the full range of possible variations in butt-joint geometry $(h_1, h_2,$ and L in Fig. 1) are available [44]. These polynomial fits can be specialized to the important case of a bimaterial beam formed by two long strips $(h_1/L$ and $h_2/L \gg 1)$, where h_1 and h_2 are not necessarily equal. A second K_a calibration derived specifically for bimaterials beams is also available [47]. Calibration results are presented for $\alpha = \pm 0.3, \pm 0.5, \pm 0.7, \pm 0.90$ with $\beta = 0$ or $\alpha/4$ and h_1/h_2 varying from 1/32 to 32. The two calibrations are in good agreement.

5.5. Embedded inclusion

5.5.1. Rigid square inclusion in an epoxy disk

A simple, two-dimensional idealization of an encapsulated, sharp-cornered component or particle is a rigid, square inclusion embedded within a thin, isotropic material (Fig. 5). When viewed asymptotically, the inclusion tip is the apex of

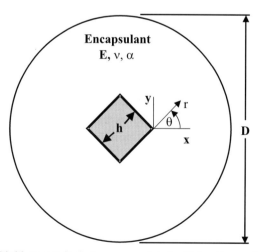

Fig. 5. Rigid, square inclusion encapsulated within a linear-elastic disk.

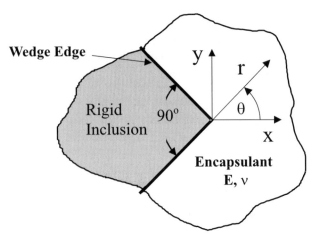

Fig. 6. Asymptotic problem for encapsulated, rigid wedge. Symmetry about x axis.

a wedge (Fig. 6). The solution of this asymptotic problem contains two singular terms with generally differing strengths. However, for a symmetric loading — such as that generated by uniform cooling — only one of these singularities is excited, and the magnitude of the asymptotic stress state is fully characterized by a single stress-intensity factor. The K_a calibration for a rigid, square inclusion embedded within a thin disk and subjected to uniform cooling for both fully bonded and unbonded (frictionless sliding) conditions has been determined for plane stress [48]. The K_a calibration uses inclusion edge length as the characteristic length (h in Fig. 5), and the characteristic stress is

$$\sigma^* = \frac{-E\alpha_T \Delta T}{1-\nu} \tag{17}$$

The function A depends on β ($\alpha = 1$) and the nondimensional geometric parameter D/h. Values of A for various β and D/h are listed in Table 3 for both bonded and unbonded inclusions.

The singular stress state generated by the bonded inclusion is very different from that generated by the unbonded inclusion. For an encapsulant with a Poisson ratio of 0.35, the strength of stress singularity for the bonded inclusion is -0.25, whereas the strength of the singularity for the unbonded inclusion is -0.67 (Fig. 7). The angular variation of the stress field also differs. When the inclusion is fully bonded, the magnitude of the radial stress in front of the inclusion tip ($\theta = 0°$, Fig. 8) is much larger than the hoop stress. When the inclusion is unbonded, hoop and radial stress have the same magnitude but are of opposite sign (Fig. 9). This suggests that an epoxy disk containing an unbonded inclusion is more likely to crack when cooled than a disk containing a fully bonded inclusion. When unbonded, the inclusion-tip stress field is fully characterized by a single, interface-corner stress-intensity factor, and the associated singular field

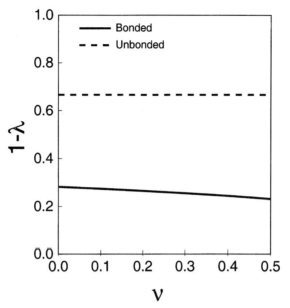

Fig. 7. Variation of the strength of the stress singularity, $-(1-\lambda)$, with Poisson's ratio ν for either a bonded or an unbonded rigid square inclusion.

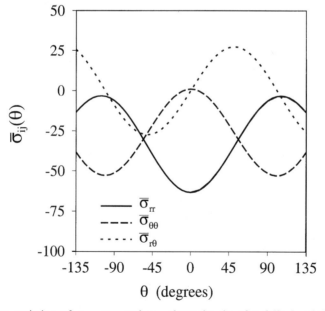

Fig. 8. Angular variation of σ_{rr}, $\sigma_{\theta\theta}$, and $\sigma_{r\theta}$ about the tip of a fully bonded, rigid square inclusion (encapsulant $\nu = 0.35$).

Table 3

Nondimensional $A(1,\beta)$ function used in the K_a calibration for a rigid square inclusion embedded in a thin, epoxy disk

Interface	β	ν [a]	D/h	$\lambda - 1$	$\bar{\sigma}_{rr}(0°)$	$\bar{\sigma}_{\theta\theta}(0°)$	$A(1,\beta)$
Bonded	0.325	0.35	7.07	−0.250	−63.25	1.00	0.0135
Unbonded	0.475	0.05	14.14	−0.667	−1.00	1.00	0.1779
Unbonded	0.425	0.15	14.14	−0.667	−1.00	1.00	0.1455
Unbonded	0.375	0.25	14.14	−0.667	−1.00	1.00	0.1182
Unbonded	0.325	0.35	14.14	−0.667	−1.00	1.00	0.0949
Unbonded	0.275	0.45	14.14	−0.667	−1.00	1.00	0.0748
Unbonded	0.325	0.35	7.07	−0.667	−1.00	1.00	0.0940
Unbonded	0.325	0.35	5.66	−0.667	−1.00	1.00	0.0935
Unbonded	0.325	0.35	4.71	−0.667	−1.00	1.00	0.0927
Unbonded	0.325	0.35	3.54	−0.667	−1.00	1.00	0.0910
Unbonded	0.325	0.35	2.83	−0.667	−1.00	1.00	0.0886
Unbonded	0.325	0.35	2.18	−0.667	−1.00	1.00	0.0831

[a] ν value corresponding to the specified β when plane stress applies, see Eq. 10.

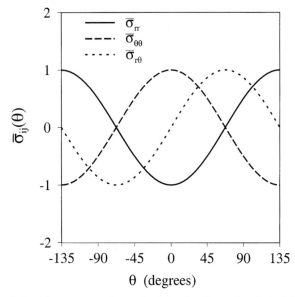

Fig. 9. Angular variation of σ_{rr}, $\sigma_{\theta\theta}$, and $\sigma_{r\theta}$ about the tip of an unbonded, rigid square inclusion (frictionless sliding).

dominates a relatively large region (roughly 15% of the inclusion edge length for an epoxy with $E = 3.5$ GPa and $\nu = 0.35$, Fig. 10). Elastic–plastic calculations for a thermally induced strain of -0.0004 show that, when the inclusion is unbonded, encapsulant yielding has a significant effect on the inclusion-tip stress

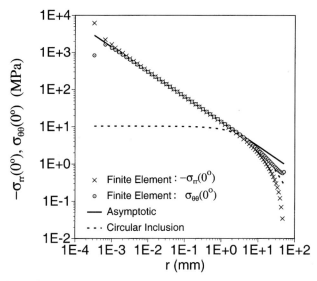

Fig. 10. Comparison of linear-elastic finite-element and asymptotic solutions for stress in front of an unbonded inclusion, embedded within an epoxy disk, with $h = 18$ mm and $\Delta T = -100°C$.

state ($\sigma_y = 74$ MPa, Fig. 11). Yielding relieves stress parallel to the interface and greatly reduces the radial compressive stress in front of the inclusion. As a result, the encapsulant is subjected to a nearly uniaxial tensile stress at the inclusion tip, and the calculated yield zone is embedded within the region dominated by the elastic hoop-stress singularity.

5.5.2. Inclusion in an infinite media

Chen and Nisitani [13] noted that for the case of bonded, dissimilar elastic wedges that together form a full plane, when $\beta(\alpha - \beta) > 0$, there are at most two power-law singular terms in the asymptotic expansion of the stress field with $-1 < \lambda - 1 < 0$. The exponent defining the strength of each of the singular terms is real, and the two exponents are generally different. They also found that one of the singular terms is associated with symmetric loading about the line bisecting the apex of the wedge, whereas the other is associated with an asymmetric loading about the bisecting line (i.e. x axis in Fig. 6). In another study Chen [49] reports K_a calibrations for a bonded inclusion embedded within an infinite plate subjected to uniaxial tension, biaxial tension, or shear at infinity. Results are presented for a variety of inclusion shapes, including rectangles, diamonds, and hexagons, and broad range of bimaterial combinations. Note that the K_a for a symmetric (asymmetric) loading will be associated with the stress singularity term corresponding to that mode of loading.

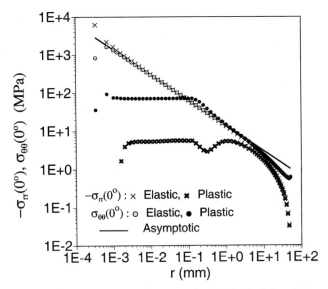

Fig. 11. Comparison of linear-elastic and elastic–perfectly plastic finite-element solutions with asymptotic solution for stress in front of an unbonded inclusion embedded, within an epoxy disk, with $h = 18$ mm and $\Delta T = -100°C$.

6. Limits on the applicability of a K_{ac} failure analysis

Two basic requirements must be met before the K_{ac} criterion can be applied. One obvious requirement is that failure must initiate at the interface corner. The other requirement is that the interface-corner singularity field (Eq. 4) must be a good approximation to the actual field in an annular region surrounding the corner. Consequently, the corner must appear sharp on a length scale commensurate with that defining the region dominated by the interface-corner stress singularity, and unstable crack growth must initiate from a fracture process zone that is deeply embedded within the region dominated by the interface-corner singularity. When these conditions are met, K_a is a unique measure of the intensity of the stress field at the corner, independent of geometry and details of the applied loading. This restriction is analogous to the familiar small-scale yielding requirement of LEFM [50,51].

6.1. Crack initiation away from the interface corner

Cracking in bonded or encapsulated bodies will often initiate at a sharp interface corner whenever such corners are present. For example, crack growth in epoxy-bonded, metal-adherend, cylindrical butt joints initiates adhesively (on the interface) along a small segment of the specimen periphery [1–3]. However, failure sometimes initiates away from the interface corner. This is particularly true for thin, highly constrained, elastomeric adhesive bonds when the adhesive's Poisson ratio has a value of nearly 0.5. For example, adhesively bonded butt joints with PMMA (polymethylmethacrylate) adherends and an elastomeric adhesive (Solithane 113) fail away from the edge when the bond is sufficiently thin [1]. The stress state in a thin elastomeric bond is quite different from that in a thin, glassy polymer bond, and this provides an explanation for the tendency for elastomeric bonds to sometimes fail away from a sharp interface corner.

Consider the case of a cylindrical butt joint with rigid adherends and a thin, elastic bond. In the center of the adhesive layer, far from the stress-free edge,

$$\sigma_{rr} = \sigma_{\theta\theta} = \frac{v}{1-v}\sigma_{zz} \qquad (18)$$

A typical glassy epoxy adhesive has a Poisson ratio of about 0.35; consequently, in-plane stresses are equal to roughly half the axial stress. Furthermore, a plane strain finite-element analysis of an adhesively bonded, rigid-adherend butt joint, when the adhesive layer has a Poisson ratio of 0.40, shows that stress within a thin bond is nearly uniform once more than four bond thicknesses from the edge, and $\sigma_z \approx \bar{\sigma}$, the nominal applied stress [22]. On the other hand, elastomeric adhesives often have a Poisson ratio approaching 0.5, and, as indicted by Eq. 18, a nearly hydrostatic tensile stress state is generated when the adhesive is highly constrained. Furthermore, an approximate analytic solution shows that stress varies continuously with radial distance in a rigid-adherend cylindrical butt joint when the bond's Poisson ratio approaches 0.5 [52,53]. In the limiting case of a bond with a large diameter-to-thickness ratio and with $v = 0.5$, this approximate solution for averages stress through the bond thickness yields

$$\sigma_{HT} = \sigma_{zz} = \sigma_{rr} = \sigma_{\theta\theta} = 2\bar{\sigma}\left[1 - \left(\frac{2r}{D}\right)^2\right] \qquad (19)$$

where r is radial distance, D is the bond diameter, and $\bar{\sigma}$ is the nominal applied stress [53]. The bond is subjected to a pure hydrostatic tension, σ_{HT}, and at the center of the bond $\sigma_{HT} = 2\bar{\sigma}$. Furthermore, a large region of the bond is subjected to elevated levels of hydrostatic tension ($\sigma_{HT} > \bar{\sigma}$ for $2r/D < 0.71$). These results suggest that, if interfacial flaws of sufficient size exist away from the bond edge, the relatively high hydrostatic stress state generated within a thin elastomeric bond may be sufficient to cause these flaws to propagate. A linear-elastic fracture mechanics analysis is generally applicable in those instances when failure initiates away from an interface corner. Note, however, that such an analysis should include compressibility effects, and model the actual bond-thickness to bond-diameter ratio. Calculations for a tensile-loaded, cylindrical butt joint with rigid adherends have shown that σ_{HT} is extremely sensitive to Poisson's ratio values and the bond-thickness to bond-diameter ratio [53,54]. For example, published results for an adhesively bonded cylindrical butt joint with rigid adherends indicate that for a bond diameter-to-thickness ratio of 50, $\sigma_{HT}/\bar{\sigma} \approx 2.00$, 1.15, and 1.00 for $\nu = 0.50$, 0.49, and 0.34, respectively, while for $\nu = 0.5$, $\sigma_{HT}/\bar{\sigma} \approx 1.9$, 1.7, and 1.1 for $D/h = 10$, 5, and 2, respectively [53].

6.2. Small-scale yielding

For K_{ac} to characterize the failure process, the fracture process zone must be deeply embedded within the region dominated by the interface-corner singularity. To check if this condition is met, one must determine the size of the region dominated by the interface-corner singularity and also estimate the size of the fracture process zone. An example of such a calculation is presented to illustrate the nature of the small-scale yielding requirement. This particular calculation is for an adhesively bonded butt joint with rigid adherends [55]. The failure load and epoxy properties used in the calculation are based on a previously reported series of steel–epoxy butt-joint tests [2]. The epoxy adhesive layer is 1.0 mm thick and has a Young modulus of 3.5 GPa, Poisson ratio of 0.35, a room temperature tensile yield strength of 70 MPa and a compressive yield strength of 100 MPa (at a strain rate of 0.0002 s^{-1}). If a linear dependence on pressure P is assumed, $\sigma_y = 82 + 0.53P$ (MPa). Failure of a joint with a 1-mm-thick bond occurs at a nominal applied stress $\bar{\sigma}$ of 30.7 MPa, a value consistent with an interface toughness of $K_{ac} = 12.7$ MPa mm$^{0.32}$.

The size of the region dominated by the stress singularity can be estimated by comparing finite-element results for the full-joint model with the singularity solution (Eq. 4). For a tensile-loaded butt joint with rigid adherends and a thin (essentially semi-infinite), epoxy bond, the calculated interfacial normal stress is in good agreement with the singular asymptotic solution for a distance of 15% of the bond thickness (Fig. 12). Consequently, not only must the corner appear sharp when viewed at this length scale, but also material yielding and subcritical crack growth must be limited to a few percent of the total bond thickness.

The size of the interface-corner fracture process zone is not known, but one can estimate the extent of yielding. Fig. 13 shows three different predictions for the interface-corner yield zone at joint failure. Epoxy yielding is rate- and temperature-dependent and is thought to be a manifestation of stress-dependent, nonlinear viscoelastic material response. A precise estimate of the size of the interface-corner yield zone is, of course, totally dependent on the accuracy of the epoxy constitutive model. This constitutive model must be valid at the extremely high strain and hydrostatic tension levels generated in the region of an interface corner. Unfortunately, accurate epoxy models of this type are not readily available. Nevertheless, simpler material models can be used to provide some insights. The crudest yield zone prediction shown in Fig. 13 uses a linear-elastic adhesive model to determine when the calculated effective stress exceeds the epoxy's yield

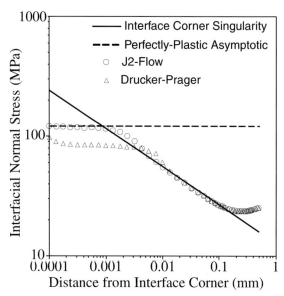

Fig. 12. Comparison of calculated interfacial normal stress when the adhesive layer yields with interface-corner singularity and perfectly plastic asymptotic solutions.

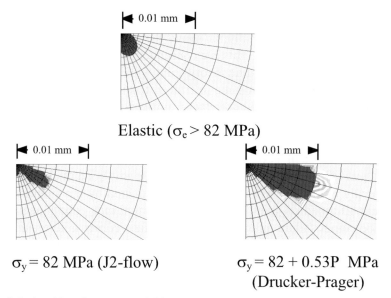

Elastic ($\sigma_e > 82$ MPa)

$\sigma_y = 82$ MPa (J2-flow)

$\sigma_y = 82 + 0.53P$ MPa (Drucker-Prager)

Fig. 13. Calculated interface-corner yield zones. Dark gray in the plasticity calculations corresponds to an equivalent plastic strain > 0.0001.

strength. Also shown are results using a standard, elastic–perfectly plastic $J2$-flow theory [56] and an elastic–perfectly plastic, Drucker–Prager model [57] that incorporates a pressure-dependent yield strength. A realistic adhesive model should

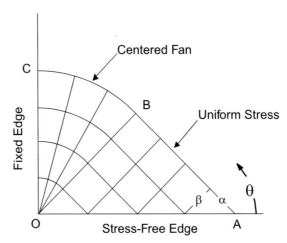

Fig. 14. Asymptotic, interface-corner slip-line solution for a perfectly plastic quarter plane bonded to a rigid substrate.

include pressure-dependent yielding. Fig. 13 shows that the size and shape of the calculated yield zone is a strong function of which adhesive constitutive model is used. The calculated yield zone determined when using elasticity and $J2$-flow theory is in poor agreement with that determined when pressure-dependent yielding is included. The Drucker–Prager solution shows the largest yield zone, and the zone is shifted towards the interface. This is a consequence of high levels of hydrostatic tension at the interface. A slip-line theory solution (Fig. 14) for a rigid–perfectly plastic adhesive predicts a hydrostatic interfacial tension of 1.5 σ_y [58].

Fig. 12 plots the calculated interfacial normal stress for the $J2$-flow and Drucker–Prager adhesive models along with interface-corner singularity and rigid–perfectly plastic slip-line solutions. As observed previously, the interfacial normal stress is in good agreement with the singularity solution over a distance equal to 15% of the 1-mm-thick bond. At this bond thickness, the yield zone is deeply embedded within the region dominated by the singularity. Tests have been performed on joints as thin as 0.25 mm with good agreement with an interface-corner failure analysis [2]. An interface-corner failure analysis cannot, however, be applied to arbitrarily thin bonds. The region dominated by the singularity solution scales with bond thickness and shrinks as bond thickness is reduced. Ultimately it reaches a size comparable to that of the yield zone at joint failure.

6.3. Small-scale cracking

In the case of joints composed of bonded brittle materials, the small-scale cracking problem provides insight into how much subcritical cracking can occur without

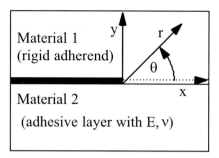

Fig. 15. Conventions used to define interfacial crack configuration.

invalidating the use of a K_{ac} criterion. The small-scale interfacial cracking problem for a crack originating at an interface corner is completely analogous to the small-scale yielding problem of traditional fracture mechanics [33,59]. Illustrative results for small-scale interfacial cracking will be presented below; however, first some commonly used interfacial fracture mechanics concepts will be defined. These definitions are specialized to plane strain and a rigid upper adherend (Fig. 15, Material 1).

The singular stress state at the tip of a crack lying on the interface between two dissimilar, linear-elastic, isotropic materials is well known [7,60]. The interfacial tractions directly ahead of the crack tip (Fig. 15) are given by

$$(\sigma_{yy} + i\sigma_{xy})_{\theta=0} = \frac{Kr^{i\varepsilon}}{\sqrt{2\pi r}} \tag{20}$$

where

$$K = K_1 + iK_2, \quad i \equiv \sqrt{-1}, \quad \text{and} \quad \varepsilon = \frac{-\ln(3-4\nu)}{2\pi} \tag{21}$$

Furthermore, the energy-release rate G for crack advance along the interface is related to the complex interfacial stress-intensity factor by

$$G = \frac{(1-\beta^2)}{E^*}|K|^2 \tag{22}$$

where

$$E^* = \frac{2E}{(1-\nu^2)} \quad \text{and} \quad |K|^2 = K_1^2 + K_2^2 \tag{23}$$

A generalized interpretation of mode measure has been suggested by Rice [60], and this definition is now widely used. Mode mixity $\hat{\psi}$ is defined as the ratio of interfacial shear stress to normal stress at a fixed distance $\hat{l}$ in front of the crack tip.

$$\hat{\psi}_{r=\hat{l}} = \tan^{-1}\left[\left(\frac{\sigma_{xy}}{\sigma_{yy}}\right)_{\theta=0, r=\hat{l}}\right] = \tan^{-1}\left[\frac{\mathrm{Im}(K\hat{l}^{i\varepsilon})}{\mathrm{Re}(K\hat{l}^{i\varepsilon})}\right] \tag{24}$$

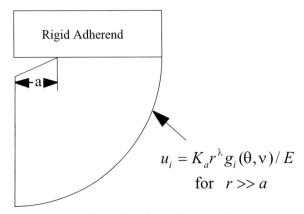

Fig. 16. Small-scale cracking problem.

The choice of reference length $\hat{l}$ is arbitrary and is sometimes based on a characteristic in-plane length of the body analyzed or on an intrinsic material length scale. In any case, $\hat{\psi}$ values corresponding to two different length scales, $\hat{l}_1$ and $\hat{l}_2$, are related by

$$\hat{\psi}_{r=\hat{l}_2} = \hat{\psi}_{r=\hat{l}_1} + \varepsilon \ln(\hat{l}_2/\hat{l}_1) \tag{25}$$

The solution of the small-scale cracking problem for a tensile-loaded, adhesively bonded butt joint with rigid adherends illustrates how much subcritical cracking can occur without invalidating the use of a K_{ac} criterion [55]. Consider a short, interfacial crack that is fully embedded within the region dominated by the interface-corner stress singularity (Fig. 16). The angular variation of displacements along the outer boundary ($r \gg a$) is known from the interface-corner solution, and K_a determines the magnitude of the loading. In accordance with linear elasticity, $|K|$, for the interfacial crack, must depend linearly on K_a. The only length scale in this asymptotic problem is the crack length a, and for the rigid-adherend case, the only nondimensional parameter that exists or can be formed is Dundurs' parameter β ($\alpha = 1$). Consequently, to be dimensionally correct, the energy-release rate must depend on crack length and K_a as

$$G = \frac{(1-\beta^2)}{E^*}|K|^2 = \frac{(1-\beta^2)}{E^*}K_a^2 a^{2\lambda-1} D(1,\beta) \tag{26}$$

Dimensional considerations require that mode mixity $\hat{\psi}_{r=a}$ is also only a function of β. The functions $D(1,\beta)$ and $\hat{\psi}_{r=a}$ are determined by solving the asymptotic problem for a range of β values. The G relation defined by Eq. 26 can be specialized to the case of a tensile-loaded, adhesively bonded butt joint with rigid adherends by substituting the K_a relationship for that geometry and loading (Eq. 8

Table 4

Parameters defining the small-scale cracking solution when the asymptotic solution is a quarter plane with one edge fixed and the other edge stress free (Fig. 3)

β	ν [a]	$D(1,\beta)$	$\hat{\psi}_{r=a}$ (°)
0.474	0.05	4.26	−12.3
0.444	0.10	4.51	−13.5
0.412	0.15	4.75	−14.4
0.375	0.20	5.00	−15.1
0.333	0.25	5.26	−15.7
0.286	0.30	5.55	−16.2
0.231	0.35	5.89	−16.5
0.167	0.40	6.30	−16.7
0.091	0.45	6.81	−16.7

[a] ν value corresponding to the specified β when plane strain applies, see Eq. 10.

in conjunction with Eq. 11 for plane strain and Table 1).

$$G = \frac{(1-\beta^2)}{E^*} \left(\frac{\nu}{1-\nu}\right)^2 A(1,\beta)^2 D(1,\beta) \left(\frac{h}{a}\right)^{1-2\lambda} \bar{\sigma}^2 h \tag{27}$$

When $\nu = 0.35$, $\lambda \sim 2/3$. Eq. 27 indicates that when small-scale cracking conditions exist, G, for an adhesively bonded butt joint with rigid adherends, varies as $h^{2/3}$ and as $a^{1/3}$. Table 4 lists $D(1,\beta)$ and mode mixity $\hat{\psi}_{r=a}$ values for a broad range of β. Note that the reported $D(1,\beta)$ apply not only to the adhesively bonded butt joint but also to any problem where the asymptotic problem is a quarter plane with one edge fixed and the other edge stress-free (e.g. debonding at the tip of a transverse crack in a thin layer on a rigid substrate). Eq. 26 relates G to any existing K_a relation.

Another useful result is the energy-release rate for a long, interfacial crack that is so far from the interface corner that it is no longer influenced by its presence. The steady-state asymptotic solution for plane strain is readily determined by a J-integral evaluation [61]

$$G_{ss} = \frac{(1+\nu)(1-2\nu)}{(1-\nu)E} \bar{\sigma}^2 h \tag{28}$$

A finite-element solution for a geometry approximating the long crack limit indicates that $\hat{\psi}_{r=2h} = -16.4°$.

A full, finite-element analysis of an adhesively bonded butt joint with rigid adherends has been performed to investigate the range of applicability of the asymptotic solutions. The failure load and epoxy properties used in these calculations are the same as those used in the small-scale yielding calculations for steel–epoxy adhesively bonded butt joints presented above (i.e. $E = 3.5$ GPa,

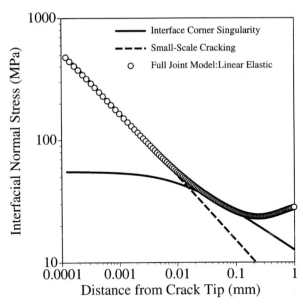

Fig. 17. Comparison of asymptotic interface-corner and small-scale cracking solutions for interfacial normal stress with finite-element results for the full joint model ($a = 0.01$ mm, $\bar{\sigma} = 30.7$ MPa).

$\nu = 0.35$, $K_{ac} = 12.7$ MPa mm$^{0.32}$, $\bar{\sigma} = 30.7$ MPa, $2h = 1.0$ mm). Finite-element calculations were carried out for a crack length that ranges from 0.001 to 10 times the bond thickness.

Fig. 17 illustrates the nature of the small-scale cracking problem when $a = 0.01$ mm and $2h = 1.0$ mm. Asymptotic interface-corner and small-scale cracking solutions for interfacial normal stress are compared to finite-element results for the full-joint model. The interface-corner stress solution is defined by Eq. 4 (K_a is defined by Eq. 8 in conjunction with Eq. 11, for plane strain, and Table 1). Note that the reason the plotted curve on this log–log plot is not a straight line is that stress is plotted as a function of distance from the crack tip, not distance from the interface corner. The small-scale cracking result is obtained by using Eq. 27 and the phase angle listed in Table 4 to determine the complex interfacial stress-intensity factor using Eqs. 22–25. The interfacial normal stress is then calculated using Eq. 20. The small-scale cracking solution merges with the full-joint solution at a distance of <0.01 mm. At a distance of 0.01 mm in front of the crack tip (a distance equal to the crack length), the small-scale cracking solution is within 10% of the full-joint model solution. The stress field associated with the interface crack is embedded within the field governed by the interface-corner singularity. Once beyond the region perturbed by the interface crack, the interface-corner singularity and the full-joint model solutions are within 10% out to a distance of 0.15 mm (15% of the total bond thickness).

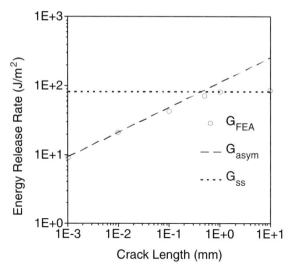

Fig. 18. Comparison of finite-element and asymptotic solutions for energy release rate ($\bar{\sigma} = 30.7$ MPa). G_{FEA}, the finite-element solution; G_{asym}, the small-scale cracking solution; G_{ss}, the long crack, steady-state solution.

Fig. 18 compares the calculated energy-release rate using the full-joint model (G_{FEA}) with the asymptotic solutions for small-scale cracking (G_{asym}) and steady-state cracking (G_{ss}) over a broad range of crack lengths. Together the asymptotic solutions form a fairly tight envelope of G_{FEA} over the full range of crack lengths. The small-scale cracking solution is within a few percent of G_{FEA} for $a/h < 0.02$ and differs by 20% at $a/h = 1$. The steady-state cracking solution is within a few percent of G_{FEA} for $a/h > 2$ and differs by 16% at $a/h = 1$. Fig. 19 plots a similar comparison for mode mixity. Here the phase angle is defined at a characteristic length scale of 0.01 mm. Again, the asymptotic solutions form a fairly tight envelope of the phase angle calculated using the full-joint model. These results suggest that the K_{ac} criterion can be applied even if a subcritical crack extends a distance equal to several percent of the bond thickness.

7. Experimental studies

7.1. Early studies (1980s)

It appears that the first study investigating the applicability of the K_{ac} criterion was published in 1982 [14]. Three different types of three-layer, steel–epoxy–steel model laminates were subjected to various loading conditions. The epoxy layer joining the steel adherends in these model laminates was relatively thick, 25 mm, and only eight samples were tested. The measured K_{ac} values were reasonably

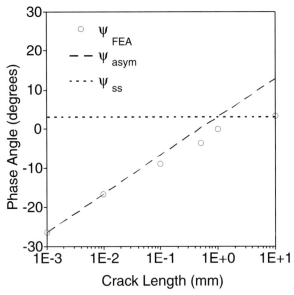

Fig. 19. Comparison of finite-element and asymptotic solutions for $\hat{\psi}_{r=0.01\ \text{mm}}$. $\hat{\psi}_{\text{FEA}}$, finite-element solution; $\hat{\psi}_{\text{asym}}$, small-scale cracking solution; $\hat{\psi}_{\text{ss}}$, long crack, steady-state solution.

consistent, with the exception of one outlying data point. In another study, single-lap joints with varying overlap lengths were tested [15]. These joints had a large spew fillet, and the asymptotic stress state at the sharp embedded corner, where failure presumably initiates, is described by two, real-valued power-law singularities. Neither of the two singularities dominated, so an 'equivalent-strength singularity' was used in the analysis. The agreement between measured strength and that predicted using a K_{ac} criterion was good for large overlap lengths, but was rather poor for smaller overlaps. The author indicates that the adhesive used in the tests is ductile, and that may be a factor affecting the accuracy of his predictions. Another study is notable for endeavoring to apply the K_{ac} failure criterion to a microelectronics packaging problem [16]. Three different epoxy/Fe–Ni alloy bimaterial configurations — each with a different asymptotic geometry (flush, or with either the epoxy or Fe–Ni alloy protruding) — were cooled from the cure temperature to induce delamination. The strength of the stress singularity $\lambda - 1$ differed in each case, and a K_{ac} vs. $\lambda - 1$ relation was constructed. This relationship was treated as a material property and used to predict the tendency of epoxy-encapsulated Fe–Ni inserts and Large Scale Integrated Circuit packages to delaminate. Test results were consistent with the predictions. Some of the configurations tested in this study had more than one singular term, and all had an r-independent term generated by cooling. The existence of a unique K_{ac} vs. $\lambda - 1$ relationship assumes that, even when the number of singularities and constant terms change with changes in the configuration, the highest-order term dominates. In general this is not true.

7.2. Adhesively bonded butt joints (1990s)

The need for a rigorous test of the applicability of the K_{ac} criterion motivated an extended experimental study that is reported in a series of five papers [2,3,42,62,63]. In this work a large number of adhesively bonded butt joints were tested. The adherends were solid metal cylinders (28.6 mm diameter by 38.1 mm long) that had been precision-machined to guarantee that the ends were flat and perpendicular to the cylinder axis and that the edges were left sharp. The epoxy bond was relatively thin. Joints with bonds as thin as 0.25 mm and as thick as 2.0 mm were tested. This test geometry was chosen because (1) it is widely used in the adhesives' community (e.g. ASTM D897 and D2095), (2) it is a relatively simple joint geometry to fabricate and test (7–10 joints were tested at each nominally identical condition), and (3) joint strength should vary with adhesive bond thickness in a definite and easily measurable way when the K_{ac} criterion applies.

7.2.1. Effect of bond thickness

The effect of bond thickness on joint strength was the focus of two test series [2,42]. One test series used steel adherends, while the other used aluminum adherends. Fabrication and test conditions for the two test series are summarized in Table 5. Thirty-seven joints with 303 stainless-steel adherends were tested, while 27 joints with 6061-T6 aluminum adherends were tested. The same triamine-cured

Table 5

Fabrication parameters for two sets of adhesively bonded butt joints, one with stainless steel adherends and one with aluminum adherends

	SS test series	AL test series
Adherends	303 stainless steel	6061 T6 aluminum
Surface	Sandblasted	Sandblasted
Preparation	Cleaned	Cleaned
	Passivated at RT using sodium dichromate/nitric acid	
Adhesive	Shell Epon 828 epoxy resin	Shell Epon 828 epoxy resin
	Huntsman T403 hardener	Huntsman T403 hardener
	100/36 weight ratio	100/36 weight ratio
Cure schedule	RT > 7 days	28°C for minimum of 1 day then stored at RT > 7 days before test

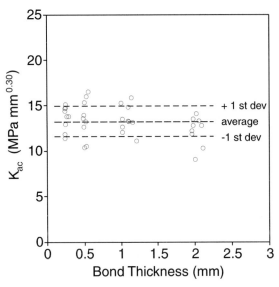

Fig. 20. Interface-corner fracture toughness for joints with steel adherends (st dev denotes standard deviation).

DGEBA epoxy (Shell's Epon 828 epoxy resin crosslinked with a Huntsman's Jeffamine T403 hardener using a 100/36 weight ratio) was used in both test series. This adhesive has a Young modulus of 3.5 GPa, a Poisson ratio of 0.35, and a compressive yield strength of 100 MPa (measured at room temperature and when loaded at a strain rate of 0.0002 s^{-1}). The adhesive bond thickness was varied, and had target bond thicknesses of 0.25, 0.50, 1.00, and 2.00 mm (actual thickness was determined after fabrication). The joints were cured near or at room temperature and tested at room temperature to minimize residual stress effects. The joints were tested in a conventional, screw-driven load frame. All specimens were loaded at a cross-head displacement rate of 0.2 mm s^{-1} using a load train that utilized a chain linkage to minimize misalignment effects. Time to failure ranged between 10 and 25 s, depending on joint strength. The calibration for an adhesively bonded butt joint with stiff adherends was used to calculate K_{ac} from measured joint strength and bond thickness (Eq. 8, in conjunction with plane strain σ^* from Eq. 13 and $A(\alpha, \beta)$ from Table 2). An examination of the failed joints showed that failure always initiated adhesively (on the interface) along a small segment of the specimen periphery.

Fig. 20 plots the K_{ac} data for the steel adherend joints [2]. The average value of K_{ac} is 13.3 MPa mm$^{0.30}$. Although there is moderate variability for each target bond thickness, K_{ac} values do not vary in any systematic way with bond thickness (standard deviation/average = 13%). This suggests that K_{ac} is indeed a material property. Fig. 21 plots the predicted variation in butt-joint strength with bond

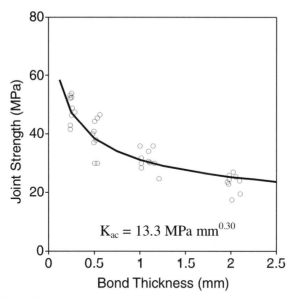

Fig. 21. Measured butt-joint strength vs. bond thickness for joints with steel adherends and prediction based upon an interface corner toughness of 13.3 MPa mm$^{0.30}$.

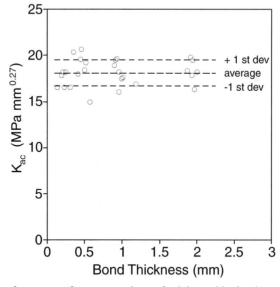

Fig. 22. Interface-corner fracture toughness for joints with aluminum adherends.

thickness when $K_{ac} = 13.3$ MPa mm$^{0.30}$ along with the underlying test data. The predicted relation is an excellent fit to the data.

Fig. 22 plots the K_{ac} data for the joints with aluminum adherends [42]. It is

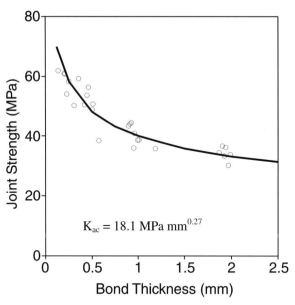

Fig. 23. Measured butt-joint strength vs. bond thickness for joints with aluminum adherends and prediction based upon an interface-corner toughness of 18.1 MPa $mm^{0.27}$.

again apparent that the measured K_{ac} values do not depend on bond thickness and can be considered a material property. The average value of K_{ac} is 18.1 MPa $mm^{0.27}$, with a standard deviation/average ratio of 8%. Fig. 23 shows that the predicted variation in joint strength with bond thickness for a K_{ac} value of 18.1 MPa $mm^{0.27}$ is a good fit to the underlying test data.

The applicability of the K_{ac} criterion to a bond thickness as small as 0.25 mm implies that the fracture process zone must be rather small. As discussed earlier, the interface-corner stress singularity dominates a region along the interface that equals 15% of the bond thickness in adhesively bonded butt joints. Consequently, when a bond is 0.25 mm thick, the interface singularity dominates a region extending about 40 μm along the interface. This suggests that the yield zone and fracture process zone extend no more than 10 μm along the interface before rapid crack propagation initiates. Subcritical cracking must be small with respect to this distance, and corner sharp must appear sharp on this scale.

7.2.2. Effect of adherend stiffness

If an adhesively bonded butt joint fails at a fixed K_{ac} value, then Eq. 8 (in conjunction with Eq. 13 for plane strain and Table 2) indicates that log(joint strength) vs. log(bond thickness) is a straight line with a slope equal to the order of the stress singularity for that joint. Because of the difference in Young's modulus, the order

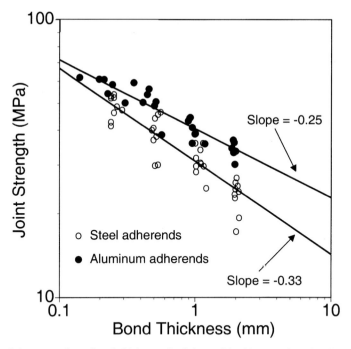

Fig. 24. Butt-joint strength vs. bond thickness for joints with either steel or aluminum adherends.

of the stress singularity for steel adherends is −0.30 and is −0.27 for aluminum adherends (Table 2). Fig. 24 compares test data plotted in Figs. 21 and 23, except log(joint strength) is plotted vs. log(bond thickness) [42]. A least-squares fit of the data for the joints with steel adherends has a slope of −0.33, while the fit of the data for joints with aluminum adherends has a slope of −0.25. Although variability in the strength data introduces some uncertainty in the measured strength–bond thickness relation, the difference in the measured joint strength–bond thickness relation for joints with aluminum and steel adherends seems to correlate with the difference in the order of the interface-corner stress singularity.

7.2.3. Variability in measured K_{ac} and the effect of surface preparation

Data drawn from five separate test series were used to assess the reproducibility of K_{ac} measurements and to examine the effect of surface preparation on K_{ac} values [3]. Table 6 lists fabrication parameters for the five sets of adhesively bonded butt joints. All had 6061-T6 aluminum adherends, and they were fabricated and tested over a period of 15 months. The aluminum bonding surfaces of most joints were lightly sandblasted (60 grit alumina oxide at 40 psi), although the surfaces of some joints in Sets 1 and 2 were left in the as-machined condition to assess the effect of surface roughness. The roughness of the as-machined and sandblasted

Table 6

Fabrication parameters for five different sets of adhesively bonded butt joints

Set	Surface roughness R_q (μm)	Cleaning method	Cure cycle [a] (°C)	Bond thickness (mm)	Number joints
1	1	Aqueous alkaline	50–40	0.2–0.5–1.0	27
	5			0.5	9
2	1	Aqueous alkaline	28–50–40	0.3–0.5–1.0	20
	5			0.5	7
3	5	Aqueous alkaline	28–50–40	0.6	10
4	5	Aqueous alkaline	28–50–40	0.3–0.5–1.0	41
5	5	Aqueous alkaline	28–50–40	0.3–0.5–1.0	12
		Solvent		0.3–0.5–1.0	12

[a] Held 24 h at each temperature.

surfaces was measured with a noncontact, optical probe. Table 6 lists root mean square roughness, R_q, as defined in Surface Texture (ASME B46.1-1995). A long wavelength cutoff of 0.8 mm was used for as-machined surfaces, whereas a cutoff of 2.5 mm was used for sandblasted surfaces. The adherends were cleaned just prior to bonding. An aqueous alkaline solution (Brulin 815 GD) was used in most cases, although some plugs in Set 5 were solvent-cleaned (trichloroethylene) to assess the effect of the method of cleaning. The same Epon 828/Jeffamine T-403 epoxy adhesive was used as in the studies discussed above, but here the adhesive was cured above room temperature, and the weight mix ratio of Epon 828 to Jeffamine T-403 was changed to 100:43. Set 1 was cured for 24 h at 50°C followed by an additional 24 h at 40°C. This cure schedule was chosen to minimize residual stress. The epoxy's glass transition temperature is 68°C (dielectric measurement technique). Note that Sets 2–5 added an initial 24 h at the 28°C step to the cure schedule to simplify handling. This modification of the cure schedule has minimal effect on the epoxy's properties. Compression tests of molded epoxy plugs cured with and without the 28°C step yielded nearly identical stress–strain relations. All except one set of joints included joints of several different bond thicknesses (e.g. 0.25, 0.5, and 1.0 mm).

Some variability in a toughness parameter like K_{ac} is expected when measured using nominally identical specimens and test procedures. Variability can be caused by variations in the flaw population and by variations in material and interface properties. Variations in fabrication and test procedures could also contribute. Many steps are required to make a butt-joint specimen. For example, the alkaline aqueous cleaning procedure involves ten separate steps, and many of these steps are carefully timed. Furthermore, some processes, such as sandblasting, are not fully controlled. The operator of the sandblaster manually directs the grit stream

Table 7

Variability in K_{ac} when measured using nominally identical specimens and test procedures

Set	Date fabricated	Nominal bond thickness (mm)	Joints tested	K_{ac} (MPa mm$^{0.27}$)	Standard deviation (MPa mm$^{0.27}$)
2	Apr. 98	0.5	7	14.9	0.9
3	Jun. 98	0.6	10	15.5	1.4
4	Oct. 98	0.3–0.5–1.0 [a]	41	14.4	0.9
5	Dec. 98	0.3–0.5–1.0 [a]	12	16.9	1.3

[a] Approximately equal number of samples of each thickness tested.

Table 8

Variability in K_{ac} with time interval between fabrication and test (Set 4)

Days between fabrication and test	Nominal bond thickness (mm)	Joints tested	K_{ac} (MPa mm$^{0.27}$)	Standard deviation (MPa mm$^{0.27}$)
1	0.3–0.5–1.0 [a]	20	14.4	1.0
14	0.3–0.5–1.0 [a]	21	14.5	0.9

[a] Approximately equal number of samples of each thickness tested.

across the adherend surface, and the grit can degrade or become contaminated with use. Table 7 shows K_{ac} data for four sets of nominally identical joints fabricated over a period of 8 months (all with sandblasted surfaces, cleaned using the alkaline aqueous method, and cured with the 28–50–40°C cure cycle). The coefficient of variation (standard deviation/average) of K_{ac} for the joints in Sets 2, 3, 4 and 5 is 6, 9, 6, and 8%, respectively. The variability of K_{ac} in the sets with multiple-bond thickness is no greater than sets with a single-bond thickness. Table 7 shows that, in spite of possible variations in processing during the 9-month period when Sets 2–5 were fabricated, the measured K_{ac} value for these sets is reasonably consistent. The average K_{ac} value is 15.4 MPa mm$^{0.27}$, with Set 5 showing the largest deviation from the mean (10%).

The time interval between the fabrication and testing of joints in Sets 2–5 varied from 1 to 17 days. Since physical aging can cause polymer properties to change with time, differences in the fabrication-to-test time interval are a potential contributor to the measured variability in K_{ac}. This issue was addressed in Set 4 tests. Forty-one joints were fabricated at one time. One-half of the joints was tested the day following fabrication, whereas the other half was tested 14 days later. Table 8 shows that there is no discernable difference in the measured K_{ac} for the two groups of joints. Physical aging of this epoxy adhesion does not appear to be a consideration for the cure cycle and time interval investigated.

Table 9

Sensitivity of K_{ac} to surface preparation

Set	Cleaning method	Surface roughness R_q (μm)	Joints tested	K_{ac} (MPa mm$^{0.27}$)	Standard deviation (MPa mm$^{0.27}$)
1	Aqueous alkaline	1	27	12.9	1.1
	Aqueous alkaline	5	9	17.6	1.7
2	Aqueous alkaline	1	20	8.0	0.7
	Aqueous alkaline	5	7	14.9	0.9
5	Aqueous alkaline	5	12	16.9	1.3
	Solvent	5	12	14.9	1.1

The dependence of K_{ac} on surface preparation was also studied. Both Set 1 and Set 2 contained joints with sandblasted ($R_q = 5$ μm) and as-machined ($R_q = 1$ μm) surfaces. Table 9 shows that K_{ac} is strongly dependent on surface roughness. The measured K_{ac} of Set 1 joints increased 36% when the as-machined surface was sandblasted. Set 2 joints showed an even greater increase in K_{ac} (86%). It should be noted that, although the as-machined surfaces of the Set 1 and Set 2 adherends have similar R_q values, the nature of the roughness is quite different. Set 1 plugs contain several sets of straight, parallel grooves. Each set of grooves covers the entire surface, but each set is rotated with respect to the others. The machining marks on Set 2 are concentric circles overlaid with short, arc-like scratches. Table 9 shows that the method of cleaning had only a modest effect on K_{ac}. Joints cleaned with the aqueous alkaline procedure have a 10% higher K_{ac} than the joints cleaned with a solvent.

7.3. Recent studies (1998–2000)

7.3.1. Epoxy wedge on aluminum substrate

An epoxy wedge cast onto an aluminum beam has been used to study the effect of wedge angle in two recent investigations. One study cast a 30-mm-long by 10-mm-high epoxy block at the clamped end of a 220-mm-long, 3.2-mm-thick cantilevered aluminum beam [64,65]. The side of the epoxy block opposite the clamped edge was wedge-shaped. The wedge-tip singularity for each of the three wedge angles tested — 55° (an acute epoxy angle), 70°, and 90° — is characterized by a single K_a. Tests were carried out to measure the number of cycles required to initiate an interfacial fatigue crack for a range of applied K_a. These data were used to define a K_a-based fatigue initiation envelope. It is also noted that since plastic zone size is a strong function of the applied K_a, a K_a-based fatigue criterion may be applicable in cases where static joint strength is so high that K_{ac} cannot be properly defined.

The aim of another study was to formulate a crack nucleation criterion that is independent of wedge angle [66]. In these tests a 76-mm-long by 12.7-mm-thick block of epoxy was cast on a 12.7-mm-thick aluminum beam. The aluminum beam was loaded in four-point bending. One side of the epoxy block was wedge-shaped, and the following four epoxy wedge angles were considered: 0° (interface crack), 45°, 90°, and 135°. Note that this range of epoxy wedge angles include cases with complex singularities and multiple K_a. Moiré interferometry was used to make high-resolution displacement measurements near the wedge tip. These measurements indicated that the interface was compliant and suggested the use of a cohesive zone model. Parameters for the cohesive zone model were determined by matching the measured interface crack-tip displacements (0° wedge). This fit reveals that the interface has a low toughness and a peak separation traction of only 3 MPa. Calculated results using this cohesive zone model reproduced the observed dependence of failure load on wedge angle.

7.3.2. Scarf joint

The K_{ac} criterion has been applied to adhesively bonded scarf joints with three different scarf angles [67]. Joints with scarf angles of 0° (i.e. adhesively bonded butt joint), 15°, and 30° and with bond thickness that ranged from 0.4 to 2 mm were tested. The adherends were 2014A-T4 aluminum with a 10-mm by 30-mm rectangular cross-section. Two types of adhesive were used. One adhesive was a brittle, high-temperature-cured epoxy (Ciba Geigy F922), while the other was a more ductile, room-temperature-cured epoxy (Ciba Geigy Araldite). Specimens bonded with the F922 epoxy were cured at either 160°C or 120°C to induce different levels of residual stress. It was assumed that the residual stress is associated with the temperature change from the cure temperature to room temperature (T_g data were not given). The $-100°C$ and $-140°C$ temperature changes generated a large fraction of the K_a value at joint failure. Indeed, it appears that the inferred residual stress contributed 85% or more of the K_a value at failure when the butt joints were cured at 160°C.

7.3.3. Glass–silicon anodic bonds

The K_a criterion has been applied to anodic bonding, a common process in the wafer-level packaging of microelectromechanical systems [68]. Anodic-bonded, glass–silicon test specimens with varying bond area were tested in bending. Fracture initiated at the silicon–glass interface corner on the tensile side of the specimen, and there was a significant variation in fracture stress with the bond area. The asymptotic interface-corner stress field for the sample geometry tested is characterized by two real power-law singularities: one has a strength of -0.497 (nearly as strong as that for a crack), and the other has a strength of -0.364.

The test results suggest that the observed variation in failure stress with bond area could be correlated using just the K_a associated with the stronger singularity. As an aside, Dunn and his colleagues have also used a K_a-based approach to successfully predict the fracture of homogeneous materials containing a sharp notch when the material is isotropic [69,70] and when the material is anisotropic [71,72].

8. Unresolved issues

Although the K_{ac} criterion has been applied with some notable successes, there are still numerous unresolved issues. These include the development of methods for treating time-dependent response, large-scale yielding, three-dimensional corners, and the development of a criterion that can be applied when the corner stress state is not characterized by a single K_a. Furthermore, a clearly desirable goal is the development of an approach for calculating K_{ac} from more fundamental quantities.

8.1. Time dependence and residual stress

Interface-corner stress-intensity factors are defined within the context rate-independent, linear elasticity theory. It is tacitly assumed that multiple loads can be superimposed, regardless of their duration, and stress does not change with time. For example, residual stress in an adhesively bonded butt joint is assumed to persist indefinitely, even when residual stress is developed and sustained over a much longer time scale than that associated with the applied loading. There are some test data that suggest that this assumption may not always be correct, especially when polymeric materials are involved [63]. Fig. 25 plots characteristic stress (Eq. 8) vs. bond thickness data for three series of butt-joint tests. The characteristic stress includes contributions from both the applied load at failure and the estimated residual stress (assuming linear-elastic response). Test-Series-1 specimens were cured and tested at room temperature to minimize residual stress. Test-Series-2 and Series-3 specimens were tested at temperatures well below their cure temperatures to introduce residual stress during cooling. The joint specimens were then loaded in roughly 10 s. If the K_{ac} criterion holds, the slope of each of the three lines in Fig. 25 should be roughly $-1/3$, the strength of the stress singularity for a steel–epoxy butt joint. This is clearly not so. Interestingly, the anticipated behavior holds when the residual stress term is neglected (Fig. 26). The reason why residual stress does not appear to contribute in the expected way is particularly puzzling for Series-3 test conditions since the magnitude of characteristic residual stress is comparable to the measured characteristic failure stress for a 1-mm-thick bond. These results seem to suggest that the residual stress generated during cooling from the cure temperature has little effect on joint strength.

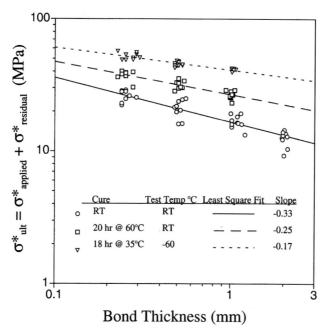

Fig. 25. Butt-joint strength vs. bond thickness. The characteristic failure stress includes both applied load and estimated residual stress.

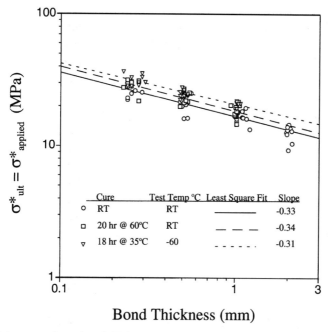

Fig. 26. Butt-joint strength vs. bond thickness. The characteristic failure stress includes only the applied load.

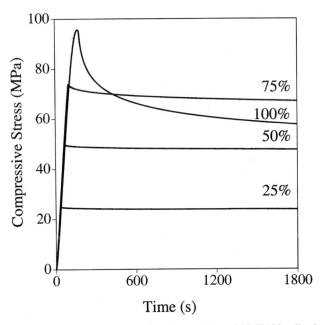

Fig. 27. Room-temperature stress-relaxation data for an Epon 828/T403 adhesive cured 20 h at 60°C. Specimens are loaded to 25%, 50%, 75%, and 100% of yield at 0.0002 s^{-1} prior to fixing the displacement.

Stress relaxation tests were carried out for the adhesive used in those butt-joint tests [63]. This adhesive was found to display a highly nonlinear, stress-level-dependent viscoelasticity at stress levels approaching the adhesive's yield strength (Fig. 27). Furthermore, significant stress relaxation was observed even at temperatures of more than 100°C below the adhesive's glass transition temperature (Fig. 28). These results indicate that the peak stress in an adhesive joint, in the yield zone at the interface corner where failure initiates, can decay significantly when given sufficient time. Note that it is not necessary that all residual stress in a bond relaxes out to affect bond strength; only the stress in the failure region is of importance. Consequently, it is possible that the first step (residual stress sustained over a long period of time) in this two-step loading process has little effect on the second loading step (mechanical load applied within 10 s). The influence of residual stress on joint strength might be much less than would be predicted by a linear analysis.

8.2. Large-scale yielding

Many modern adhesives are toughened, and yield zones can be large compared to the region dominated by an interface-corner singularity. Attempts to extend an interface-corner-type analysis to ductile materials are still in their initial

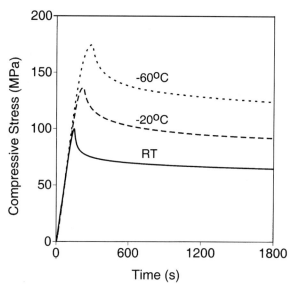

Fig. 28. Stress-relaxation data for an Epon 828/T403 adhesive cured 18 h at 35°C and tested at three temperatures. Each specimen is loaded to yield at 0.0002 s^{-1} prior to fixing the displacement.

phase. Several recent studies have investigated the nature of the asymptotic, interface-corner stress field in a power-law-hardening material. Most of this work has concentrated on determining the strength of the stress singularity [73–78], although there have been several studies where the associated stress-intensity factor has also been determined for a limited range of loadings [58,79,80]. This type of analysis has yet to lead to an experimentally verified fracture criterion for bonded materials undergoing large-scale yielding.

8.3. Three-dimensional interface corners

Work to develop an interface-corner-based failure analysis has for the most part focused on planar or axisymmetric geometries. However, many configurations have three-dimensional corners (e.g. points on the interface between bonded rectangular parallelepipeds where two edges intersect), and failure can initiate at three-dimensional corners. The strength of the singular stress field surrounding a three-dimensional interface corner differs from that along an edge where often two-dimensional analysis can be applied [34,81,82]. Only recently has a failure analysis based upon the stress intensity associated with three-dimensional corner singularity been subjected to experimental validation [82]. Aluminum–epoxy butt joints with a square cross-section and with edge lengths of 4, 6.2, 8.9, 12.5, and 17.8 mm were tested in four-point bending. The beam's square cross-section

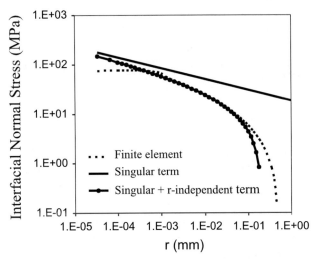

Fig. 29. Comparison of elastic–perfectly plastic finite-element and asymptotic solutions for normal stress for a thin layer on a rigid substrate subjected to uniform cooling.

was aligned to either subject a three-dimensional corner to the highest bending stress, or when rotated 45°, to subject an interface edge to the highest bending stress. The theoretical values for the strength of the stress singularity (−0.35 for three-dimensional, and −0.28 for two-dimensional) are in good agreement with those inferred from the tests (−0.33 for three-dimensional, and −0.27 for two-dimensional).

8.4. Fracture criterion when corner stresses are not characterized by a single K_a

Although there are many instances where the interface-corner stress state is fully characterized by a single K_a, there are others where it is not. For example, consider uniform cooling of a thin, elastic layer on a rigid substrate (Fig. 1, with adherend 1 rigid and $2L \gg h_2$). There is one power-law stress singularity with a real exponent in this case, and K_a is defined by Eq. 8, in conjunction with σ^* from Eq. 12 and $A(1, \beta)$ from Table 1. There is also an r-independent stress term for uniform cooling (Eq. 3). Fig. 29 compares plane-stress, finite-element results for interfacial normal stress with asymptotic results where either the r-independent term is or is not included ($h_2 = 1$ mm, $E = 4.0$ GPa, $\nu = 0.25$, $E\alpha_T \Delta T = -6.5$ MPa, $\sigma_y = 70$ MPa). There is a marked difference in the two asymptotic solutions at a physically significant distance of a micron or greater. Only when the r-independent term is included in the asymptotic solution is the interfacial normal stress in good agreement with the full, finite-element solution over any appreciable distance (~5% of the layer thickness). This clearly suggests that for such cases, an interface-corner fracture criterion must involve both K_a and the stress intensity K_{a0} associated with the r-independent term. One could envision an experimentally measured failure criterion that depends on both K_a and K_{a0}, but that would add complexity to the theory and increase the number of tests required to define the failure criterion.

8.5. Connection between K_{ac} and more fundamental properties

K_{ac} is a measured quantity. Its value will depend on the type of interface corner (e.g. butt joint vs. embedded inclusion), wedge angles, properties of the bonded materials, surface preparation (e.g. smooth or rough), loading rate, temperature, environment, and so on. Similar dependencies occur for other types of fracture toughness parameters. A clearly desirable goal would be the development of an approach for calculating K_{ac} from more fundamental quantities. In this way some of the stated dependencies could be determined without recourse to measurement. It is unlikely that such a goal will be fully achieved in the near future. An approach that uses a cohesive zone model does appear to hold some promise [66]. Nevertheless, to be successful, one must know much more about failure mechanisms and constitutive behavior at large strains than is currently known. In fact, one of the advantages of the K_{ac} criterion, when applicable, is that it can be applied without detailed information about the failure process itself.

9. Conclusions

There has been considerable progress in recent years towards developing a stress-intensity factor-based method for predicting crack initiation at a sharp, bimaterial corner. There is now a comprehensive understanding of the nature of multi-material, two-dimensional, linear-elastic, wedge-tip stress fields. In general, the asymptotic stress state at the apex of dissimilar bonded elastic wedges (i.e. at an interface corner) can have one or more power-law singularities of differing strength and with exponents that can be real or complex. There are, however, many configurations of practical importance (e.g. adhesively bonded butt joints, bi-material beams, etc.) where interface-corner stresses are described by one, real-valued power-law singularity. In such cases, one can reasonably hypothesize that failure occurs at a critical value of the stress-intensity factor: when $K_a = K_{ac}$. This approach is completely analogous to LEFM except that the critical stress-intensity factor is associated with a discontinuity other than a crack. To apply the K_{ac} criterion, one must be able to accurately calculate K_a for arbitrary geometries. There are several well-established methods for calculating K_a. These include matching asymptotic and detailed finite-element results, evaluation of a path-independent contour integral, and general finite-element methods for calculating K_a for complex geometries. A rapidly expanding catalog of K_a calibrations is now available for a number of geometries of practical interest. These calibrations provide convenient formulas that can be used in a failure analysis without recourse to a detailed numerical analysis.

The K_{ac} criterion has been applied with some notable successes. For example, the variation in strength of adhesively bonded butt joints with bond thickness and the dependence of this relationship on adherend stiffness is readily explained. No other one-parameter fracture criterion is able to make this sort of prediction. Nevertheless, the interface-corner fracture toughness approach is just in its initial states of development, and its strengths and limitations must be more clearly defined. There are still numerous issues yet to be resolved, including the development of methods for treating time-dependent response, large-scale yielding, three-dimensional corners, and the development of a criterion that can be applied when the corner stress state is not characterized by a single K_a.

Acknowledgements

This work was performed at Sandia National Laboratories. Sandia is a multi-program laboratory operated by Sandia Corporation, a Lockheed Martin Company, for the United States Department of Energy under Contract DE-AC04-94AL85000.

References

1. Anderson, G.P. and DeVries, K.L., *Int. J. Fract.*, **39**, 191–200 (1989).
2. Reedy Jr., E.D. and Guess, T.R., *Int. J. Solids Struct.*, **30**, 2929–2936 (1993).
3. Reedy Jr., E.D. and Guess, T.R., *Int. J. Fract.*, **98**, L3–L8 (1999).
4. Hart-Smith, L.J., In: Kedward, K.T. (Ed.), *Joining of Composite Materials, ASTM STP 749*. American Society for Testing and Materials, Philadelphia, PA, 1981, pp. 3–31.
5. Adams, R.D. and Wake, W.C., *Structural Adhesive Joints in Engineering*. Elsevier, London, 1984.
6. Liechti, K.M., In: Dostal, C.A. (Ed.), *Adhesives and Sealants*. ASM International, Materials Park, OH, 1990, Vol. 3, pp. 335–348.
7. Hutchinson, J.W. and Suo, Z., In: Hutchinson, J.W. and Wu, T.Y. (Eds.), *Advances in Applied Mechanics*. Academic Press, San Diego, CA, 1992, Vol. 29, pp. 63–191.
8. Bogy, D.B., *J. Appl. Mech.*, **35**, 460–466 (1968).
9. Bogy, D.B., *Int. J. Solids Struct.*, **6**, 1287–1313 (1970).
10. Williams, M.L., *J. Appl. Mech.*, **19**, 526–528 (1952).
11. Bogy, D.B., *J. Appl. Mech.*, **38**, 377–386 (1971).
12. Hein, V.L. and Erdogan, F., *Int. J. Fract.*, **7**, 317–330 (1971).
13. Chen, D.H. and Nisitani, H., *J. Appl. Mech.*, **60**, 607–613 (1993).

14. Gradin, P.A., *J. Compos. Mater.*, **16**, 448–456 (1982).
15. Groth, H.L., *Int. J. Adhes. Adhes.*, **8**, 107–113 (1988).
16. Hattori, T., Sakata, S. and Murakami, G., *J. Electron. Packaging*, **111**, 243–248 (1989).
17. Munz, D. and Yang, Y.Y., *J. Appl. Mech.*, **59**, 857–861 (1992).
18. Reedy Jr., E.D., *Eng. Fract. Mech.*, **36**, 575–583 (1990).
19. Chen, D.-H., *Int. J. Fract.*, **75**, 357–378 (1996).
20. Dempsey, J.P., *J. Adhes. Sci. Technol.*, **9**, 253–265 (1995).
21. Munz, D. and Yang, Y.Y., *Int. J. Fract.*, **60**, 169–177 (1993).
22. Reedy Jr., E.D., *Int. J. Solids Struct.*, **30**, 767–777 (1993).
23. Yang, Y.Y., *Arch. Appl. Mech.*, **69**, 364–378 (1999).
24. Munz, D., Fett, T. and Yang, Y.Y., *Eng. Fract. Mech.*, **44**, 185–194 (1993).
25. Yang, Y.Y. and Munz, D., *Int. J. Solids Struct.*, **34**, 1199–1216 (1997).
26. Liu, X.H., Suo, Z. and Ma, Q., *Acta Materialia*, **47**, 67–76 (1999).
27. Banks-Sills, L., *Int. J. Fract.*, **86**, 385–398 (1997).
28. Yang, Y.Y. and Munz, D., *Comput. Struct.*, **57**, 467–476 (1995).
29. Carpenter, W.C., *Int. J. Fract.*, **24**, 45–58 (1984).
30. Sinclair, G.B., Okajima, M. and Griffin, J.H., *Int. J. Numer. Methods Eng.*, **20**, 999–1008 (1984).
31. Stern, M., Becker, E.B. and Dunham, R.S., *Int. J. Fract.*, **12**, 359–368 (1976).
32. Carpenter, W.C. and Byers, C., *Int. J. Fract.*, **35**, 245–268 (1987).
33. Akisanya, A.R. and Fleck, N.A., *Int. J. Solids Struct.*, **34**, 1645–1665 (1997).
34. Pageau, S.S. and Biggers Jr., S.B., *Int. J. Numer. Methods Eng.*, **38**, 2225–2239 (1995).
35. Pageau, S.S. and Biggers Jr., S.B., *Int. J. Numer. Methods Eng.*, **40**, 2693–2713 (1997).
36. Szabó, B.A. and Yosibash, Z., *Int. J. Numer. Methods Eng.*, **39**, 409–434 (1996).
37. Yosibash, Z. and Szabó, B.A., *Int. J. Numer. Methods Eng.*, **38**, 2055–2082 (1995).
38. Dundurs, J., *J. Appl. Mech.*, **36**, 650–652 (1969).
39. Reedy Jr., E.D., *Eng. Fract. Mech.*, **38**, 273–281 (1991).
40. Ding, S., Meekisho, L. and Kumosa, M., *Eng. Fract. Mech.*, **49**, 569–585 (1994).
41. Wang, C.H. and Rose, L.R.F., *Int. J. Adhes. Adhes.*, **20**, 145–154 (2000).
42. Reedy Jr., E.D. and Guess, T.R., *Int. J. Fract.*, **88**, 305–314 (1997).
43. Akisanya, A.R., *J. Strain Anal. Eng. Des.*, **32**, 301–311 (1997).
44. Tilscher, M., Munz, D. and Yang, Y.Y., *J. Adhes.*, **49**, 1–21 (1995).
45. Fett, T., *Eng. Fract. Mech.*, **59**, 29–45 (1998).
46. Liu, D. and Fleck, N.A., *Int. J. Fract.*, **95**, 67–88 (1999).
47. Klingbeil, N.W. and Beuth, J.L., *Eng. Fract. Mech.*, **66**, 93–110 (2000).
48. Reedy, E.D. Jr. and Guess, T.R., *Int. J. Solids Struct.*, **38**, 1281–1293 (2001).
49. Chen, D.H., *Eng. Fract. Mech.*, **49**, 533–546 (1994).
50. Kanninen, M.F. and Popelar, C.H., *Advanced Fracture Mechanics*. Oxford University Press, New York, NY, 1985.
51. Kinloch, A.J. and Young, R.J., *Fracture Behavior of Polymers*. Applied Science Publishers, New York, NY, 1983.
52. Lindsay, G.H., *J. Appl. Phys.*, **38**, 4843–4852 (1967).
53. Anderson, G.P., DeVries, K.L. and Sharon, G., In: Johnson, W.S. (Ed.), *Delamination and Debonding of Materials, ASTM STP 876*. American Society for Testing and Materials, Philadelphia, PA, 1985, pp. 115–134.
54. Lai, Y.-H., Dillard, D.A. and Thornton, J.S., *J. Appl. Mech.*, **59**, 902–908 (1992).
55. Reedy Jr., E.D., *Int. J. Solids Struct.*, **37**, 2429–2442 (2000).
56. Simo, J.C. and Hughes, T.J.R., *Computational Inelasticity*. Springer, New York, NY, 1998, Vol. 7.

57. Desai, C.S. and Siriwardane, H.J., *Constitutive Laws for Engineering Materials, with Emphasis on Geologic Materials*. Prentice-Hall, Englewood Cliffs, NJ, 1984.
58. Reedy Jr., E.D. and Guess, T.R., *Int. J. Fract.*, **81**, 269–282 (1996).
59. Grenestedt, J.L. and Hallstrom, S., *J. Appl. Mech.*, **64**, 811–818 (1997).
60. Rice, J.R., *J. Appl. Mech.*, **55**, 98–103 (1988).
61. Rice, J.R., *J. Appl. Mech.*, **35**, 379–386 (1968).
62. Reedy Jr., E.D. and Guess, T.R., *J. Adhes. Sci. Technol.*, **9**, 237–251 (1995).
63. Reedy Jr., E.D. and Guess, T.R., *J. Adhes. Sci. Technol.*, **10**, 33–45 (1996).
64. Lefebvre, D.R. and Dillard, D.A., *J. Adhes.*, **70**, 119–138 (1999).
65. Lefebvre, D.R. and Dillard, D.A., *J. Adhes.*, **70**, 139–154 (1999).
66. Mohammed, I. and Liechti, K.M., *J. Mech. Phys. Solids*, **48**, 735–764 (2000).
67. Qian, Z. and Akisanya, A.R., *Acta Materialia*, **46**, 4895–4904 (1998).
68. Dunn, M.L., Cunningham, S.J. and Labossiere, P.E.W., *Acta Materialia*, **48**, 735–744 (2000).
69. Dunn, M.L., Suwito, W. and Cunningham, S., *Int. J. Solids Struct.*, **34**, 3873–3883 (1997).
70. Dunn, M.L., Suwito, W., Cunningham, S. and May, C.W., *Int. J. Fract.*, **84**, 367–381 (1997).
71. Suwito, W., Dunn, M.L. and Cunningham, S., *J. Appl. Phys.*, **83**, 3574–3582 (1998).
72. Suwito, W., Dunn, M.L., Cunningham, S. and Read, D.T., *J. Appl. Phys.*, **85**, 3519–3534 (1999).
73. Duva, J.M., *J. Appl. Mech.*, **55**, 361–364 (1988).
74. Duva, J.M., *J. Appl. Mech.*, **56**, 977–979 (1989).
75. Lau, C.W. and Delale, F., *J. Eng. Mater. Technol.*, **110**, 41–47 (1988).
76. Rudge, M.R.H., *Int. J. Fract.*, **63**, 21–26 (1993).
77. Zhang, N.S. and Joseph, P.F., *Int. J. Fract.*, **94**, 299–319 (1998).
78. Zhang, N.S. and Joseph, P.F., *Int. J. Fract.*, **90**, 175–207 (1998).
79. Reedy Jr., E.D., *J. Appl. Mech.*, **60**, 715–720 (1993).
80. Sckuhr, M.A., Brueckner-Foit, A., Munz, D. and Yang, Y.Y., *Int. J. Fract.*, **77**, 263–279 (1996).
81. Koguchi, H., *Int. J. Solids Struct.*, **34**, 461–480 (1997).
82. Labossiere, P.E.W. and Dunn, M.L., *J. Mech. Phys. Solids*, **49**, 609–634 (2001).

Chapter 6

Mechanical testing of adhesive joints

K.L. DEVRIES * and DANIEL O. ADAMS

Department of Mechanical Engineering, University of Utah, Salt Lake City, UT 84112, USA

1. Introduction

Adhesive joining is a common method for connecting two or more components together. In comparison with mechanical connections, adhesive joints reportedly allow for a more uniform distribution of load over a larger area, reducing localized stress concentrations. This does not imply, however, that the stresses are uniform or that the stress distributions are well understood in adhesive joints. Care must be taken to correctly design adhesive joints, just as it must be exercised in designing other mechanical joints. When designing adhesive joints, the two primary considerations are the adhesive to be used and the geometry of the joint. There are plenty of choices when selecting an adhesive; currently there are literally thousands of commercially available adhesives intended for a vast variety of applications. Similarly, there are numerous geometric parameters that can be varied in an adhesive joint design.

Mechanical testing of adhesives is an area that may at first appear straight-forward and, in principal, rather simple. Common perception among those with a technical background but with little or no experience with adhesives is that "adhesive strength" is "the stress required to cause the adhesive to fail" or "how tightly the material adheres to the substrate". While the stress analyses of joints are important and this 'adhering' feature is absolutely essential to the integrity of a joint, properties and events well removed from the so-called adhesive–substrate interface also dramatically affect joint strength. For the majority of practical ad-hesive joints, the actual path followed in the failure is somewhat removed from this 'interface'. Failure in which the adhesive appears to be pulled clean from the substrate is sometimes cited as evidence for environmental (moisture) degradation of the joint. Thus it is important for those involved in testing of adhesive joint strength to not only have an understanding of the mechanics of the joint, but

---

* Corresponding author. E-mail: kldevries@coe.utah.edu

also some awareness of the science and chemistry of adhesion. These factors provide an insight into the complexity of mechanical testing of adhesives and the importance on properly interpreting test results.

The strength of an adhesive joint is a *system property* dependent on the properties of the adhesive, the adherend(s), and the interphase and requires an understanding of mechanics, physics and chemistry. Accordingly, this chapter will provide a brief overview into proposed mechanisms (or molecular models) responsible for adhesion before embarking on a description of a number of test methods and discussing the meaning of some test results.

The fundamental mechanism that determines how one material adheres to another material has not been unambiguously identified. Indeed, it appears that different mechanisms might be active in different adhesive joints depending on a variety of factors. Despite extensive and careful research, no definitive, universally accepted relationship has been established between specific atomic or molecular parameters at or near an interface and the strength of an adhesive bond. While the purpose of this chapter is to explore the measurement of mechanical properties of joints, a brief description of proposed mechanisms, or theories, responsible for adhesion is presented to provide insight into the interpretation of physical test results. Details of these theories are available in references [1–4]. The following six theories are perhaps the most widely accepted mechanisms for one material adhering to another.

- *Mechanical interlocking* might be considered as being different from chemical adhesion. Nevertheless, this physical phenomenon influences the mechanical strength of many practical joints. Surface roughening and some surface modification treatments serve the purpose of improving mechanical interlocking or 'hooking'. The process of anodizing aluminum serves as an example of how mechanical interlocking is used to enhance adhesion. Venables et al. [5] have viewed this process as a model where in a common surface treatment of aluminum, a coherent, tightly attached, open mesh-like oxide layer is produced through which the adhesive may flow, thereby forming strong mechanical interlocks.
- The *diffusion theory* of adhesion is based on the hypothesis that one material inter-diffuses into and with another. This theory lends itself most readily to polymer bonding, in which the development of a boundary layer is envisioned, along which the polymer chains of the two materials are intertwined. For this to occur, at least one of the polymers must exhibit significant solubility in the other. Since adhesion occurs between materials without significant solubility, arguments exist against the general applicability of this mechanism. However, this theory is supported by the important role that viscosity, temperature, and polymer type play in determining joint strength. It appears reasonable that a diffusion mechanism is involved in solvent bonding, commonly used to bond two pieces of like materials such as PMMA, PVC, and ABS.

- *Absorption mechanisms* involve secondary molecular forces. Here it is hypothesized that molecules near the interface are attracted to each other by London dispersion forces, dipole–dipole interactions, hydrogen bonding, or other secondary molecular forces. The strength of these forces varies from 0.1 to 10 J/mol. Although most adhesives exhibit some dipole interactions, it is difficult to account for the relatively large strength of many practical joints purely on these secondary molecular forces.
- *Chemical reaction theories* propose that chemical reactions occur between the adhesive and the adherent forming primary chemical bonds. While it is unlikely that these theories are universally applicable to adhesives, chemical reactions may be present in some cases. Silanes, for example, are bifunctional molecules that are used as coupling agents [6]. One 'end' of the molecule is intended to interact with a polymeric adhesive. The other 'end' is intended to chemically react with atoms in the adherend's surface layer, such as oxygen in an oxide layer of a metal or the oxygen in a ceramic.
- *The electrostatic force model* of adhesion assumes that the electrons within the adhesive and the adherend occupy different energy levels and electron transfer occurs across the surface. The two surfaces are attracted to each other as a result of these opposite charges. It is generally accepted that these electrostatic forces are not primary contributors to the strength of most practical adhesive bonds.
- The *acid–base reaction* theory of adhesion has been proposed to explain a number of observed adhesive phenomena [7,8]. This theory is based on acid–base reactions at the surface. Initially, only the Bronsted concept of acid and bases were considered, but the more general current theory incorporates the Lewis acid (electron donor–acceptor) concept. The determination of the acidity or basicity of polymers is not as straightforward as might be accepted. Several approaches that have been used for this determination are described in [8].

The fact that it has not been possible to quantitatively identify the exact mechanisms responsible for adhesion has not prevented a phenomenal growth in commercially available adhesives. One catalog of adhesives, *The Desk Top Data Bank on Adhesives*, published by Cordura Publications, Inc., La Jolla, CA, lists thousands of different adhesives available from major U.S. manufacturers. Comparable numbers of adhesives are produced by foreign manufacturers and are becoming increasingly available in this country. While there is considerable overlap between many of the adhesives produced, there is also a great diversity in properties, characteristics, curing conditions, temperatures of application and use, materials and environmental conditions for which they are intended, etc. Adhesives also are available in a wide variety of forms and/or have widely differing application techniques and methods. When classified by their mode of processing or cure, some common adhesive groupings are: hot melt, anaerobic, cyanoacrylates, two-part curing, water-based, solvent-based, emulsion, contact, pressure-sensitive, and film adhesives.

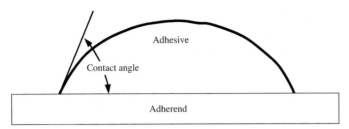

Fig. 1. Definition of contact angle for wetting tests of adhesives.

Because adhesion is a *system* phenomenon, the strength of practical adhesive joints involves many different factors including the chemical and physical properties of the materials involved, adhesive and adherend thicknesses, joint geometry, loading rate, and temperature. The growth of cracks in an adhesive may involve elements that are somewhat removed from the crack front itself. We often refer to an *interface* between an adhesive and adherend, but a simple plane of demarcation seldom, if ever, exists in practical joints. Adhesives generally form diffuse regions at the boundary between the adherends, which cannot realistically be represented by a simple plane. This observation has led to the use of the term 'interphase' rather than 'interface' to describe the region between an adhesive and adherend [9].

Other than the mechanical tests to be discussed in this chapter, one of the most commonly conducted adhesive tests are wetting experiments. The most common method for measuring the tendency of a liquid to 'wet' a surface is to measure the contact angle. In principle, such measurements of contact angle are simple, although the actual tests require great care and rather sophisticated lighting and microscope techniques. To measure how well a liquid wets a given surface, a small drop of the liquid is deposited on the surface and observed with the aid of a microscope; by 'sighting' parallel to the surface with back-lighting. As illustrated in Fig. 1, the contact angle that the liquid in the drop makes with the surface at the contact point (between liquid, solid, and air) is related to the 'surface energy of wetting' by the classical Young–Dupre equation. Thorough discussions of the Young–Dupre equation and means of measuring and interpreting the contact angle between a drop and a surface are found in many texts on surface chemistry and adhesion [1,10,11]. It is important to note that the work of adhesion, as determined by wetting experiments, represents only a small, but essential, portion of the total adhesive energy referred to in fracture mechanics, which will be described later. The work of adhesion involved in wetting involves only surface interactions while the adhesive fracture energy includes all other dissipative mechanisms involved during fracture. Some of these dissipative mechanisms may be somewhat removed from the locus of the fracture path. A one-to-one correlation has never been established between the contact angle of an adhesive in a wetting test and the subsequent strength of the adhesive bond of the cured adhesive. However, wetting is of critical importance, since during some phase of its application, the adhesive must thoroughly wet the adherend in order to form the intimate contact necessary for strong bonding. Depending on the adhesive system used, wetting may be 'in the melt' for hot melt adhesives, before cure for epoxies and other two part adhesives, before evaporation/absorption into the substrate of the solvent or carrier liquid for solvent or emulsion adhesives, etc.

The stresses in adhesive joints are complex and not always tractable analytically. This complex state of stress is produced by the geometric discontinuities in the joint configuration and by the drastically different material properties of the adhesive and adherends. The failure of an adhesive joint is also a complex phenomenon that involves many factors, including properties of the adhesive and adherends, surface preparation, the nature of the interface (or interphase), the cure cycle, the rate of loading, intricate details of geometry, temperature, and humidity. Therefore, it is difficult to specifically define the strength or quality of a given adhesive. For example, despite the fact that various handbooks and manufacturer's literature often list a shear strength as if it were a well defined property for the adhesive, it is important for the user to note that these values apply only for a given set of specific test conditions. The 'strength' of the adhesive in joints that differ only slightly from joints tested under ideal testing conditions may differ markedly from the 'published values'. Some of the reasons for these differences will be discussed later in this chapter.

Adhesive tests are performed for a number of reasons. Perhaps the most common reasons for testing adhesives are:
 (1) a qualitative comparison of two or more adhesives that are being considered for a given application;
 (2) to make certain that a given stock of adhesive has the same quality as previous shipments;
 (3) to ascertain if the properties of an adhesive are currently the same as when originally received;
 (4) to compare the effects of different surface treatments, coupling agents, anodizing, etc.;
 (5) to obtain parameters or 'properties' that might be used to design and/or predict the strength of, a practical joint to be used in a structure.

The first four of these reasons fall into a category of comparison or quality control testing. For these purposes, it might appear that almost any test could be used as long as care is taken to assure that the test conditions are held constant. However, an adhesive that tests superior to others may appear inferior if a different test is used. For example, some adhesives exhibit relatively high butt tensile strengths, but poor peel strengths, and vice versa.

The process of using results from standard laboratory tests to predict the strength of practical adhesive joints is generally more difficult than might be expected. The results of most adhesive tests are reported as the force at failure divided by the bonded area. In many practical joints, the maximum stress will be significantly higher than this average value. Since failure initiation in the joint is apt to be related to the maximum stress rather than the average stress values measured and reported, such tests are likely to be of limited use for design purposes. Further, the maximum stress is often difficult to determine, and is a function of the details of the joint geometry, or chemistry. There is considerable interaction between the mechanics, physics, and chemistry in adhesive joints. The value of the mechanical stress in a joint depends on the physical nature of the surface due to roughness, interlocking, extent of wetting, regions of poor adhesion or poor wetting, etc. Also, most adhesives are polymers whose properties are known to change with time, a process known as physical aging. Likewise, the chemistry of an adhesive has a large influence on its mechanical strength. Many adhesives are applied as liquids or pastes and through a chemical cure become viscoelastic solids. This cure can continue over significant periods of time. Furthermore, dimensional changes, applied stresses, etc. during the cure can dramatically effect the residual stresses present in the joint. These stresses can add to applied stresses leading to joint fracture. Therefore, test results are often only useful for predicting the strength of other joints that closely resemble the test geometry. Techniques currently being developed to lift this restriction are discussed later is this chapter.

2. Standard mechanical tests

A number of tests for evaluating adhesives have been formalized and standardized. The American Society for Testing and Materials (ASTM) and ISO (International Standards Organization) have compiled the most complete descriptions of these tests in this country and internationally, respectively. Most of the ASTM tests can be found in Volume 15.06 of the ASTM Book of Standards. Members of ASTM committees, and related agencies in other countries, perform a valuable service in designing and publishing details for standard methods for testing and reporting results, thereby facilitating comparisons of test results between laboratories. ASTM has developed standards to evaluate many different aspects of adhesives, but this chapter will focus on tests related to mechanical properties of adhesive joints.

Adhesive strength tests may be classified into three traditional categories and a fourth recent category of fracture mechanics. The three traditional categories are tensile tests, shear tests, and peel tests. All three categories will be described briefly followed by a discussion of tests specifically designed to yield fracture mechanics information. Before embarking on a discussion of these tests, simple screening tests referred to as qualitative tests will be described.

2.1. Qualitative tests

Most adhesive 'strength' tests are *quantitative* in nature, producing the average stress at failure (tensile and lap joints) or the force per unit width (peel tests). One drawback of many quantitative tests is the time and expense associated with specimen preparation and testing. *Qualitative* tests, or 'screening' tests, are useful for making preliminary adherence determinations to ascertain if the cost of conducting further quantitative tests is justified. Such quick screening of candidate adhesive/adherend pairs can result in significant time and cost savings by eliminating unlikely candidates and assisting in selecting those worthy of further study.

Along these lines, Committee D-14 of ASTM has formalized and adopted ASTM D3808, *Standard Practice for Qualitative Determination of Adhesion of Adhesives to Substrates*. The stated purpose of this practice is:

> *"a simple qualitative procedure for quickly screening whether an adhesive will, under recommended application conditions, bond to a given substrate without actually making bonded assemblies."*

The practice further claims:

> *"This is a quick, simple and inexpensive practice for qualitatively determining, without the need to prepare bonded test specimens whether the adhesive under consideration will bond to a particular substrate."*

In this test, small 'spots' of the adhesive are placed onto a substrate. Surface preparation, application procedures, and curing conditions are to be as similar as possible to those used in the quantitative test and/or the actual adhesively bonded joint. The adhesive spots are allowed to cure according to the manufacturers specifications. To test adhesion, this ASTM standard practice recommends the use of "a thin stainless steel spatula or similar probe as a prying lever" to assess the relative difficulty of removing the adhesive from the substrate. If the results are acceptable, standard quantitative adhesive test procedures can be used to obtain quantitative measurement of the adhesive's performance.

2.2. Tensile tests

Designers usually avoid using adhesives in a direct tensile loading mode. Overlapping, scarfing, or fingering the two pieces to be adhesively bonded can increase the bonding area significantly. Therefore, a significantly greater load carrying capacity may be obtained. Fig. 2 illustrates this point by comparing a butt joint with scarf and finger joint geometries. Not only is the adhesive area increased, but also the tensile loading of the adhesive is partially 'transformed' to shear. Additionally,

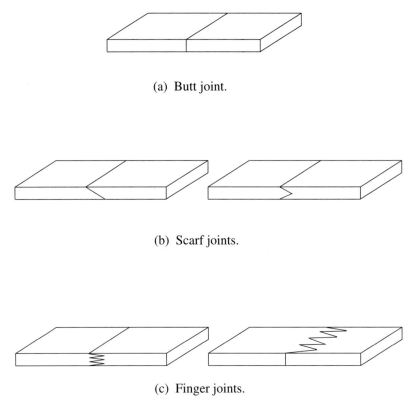

(a) Butt joint.

(b) Scarf joints.

(c) Finger joints.

Fig. 2. Comparison of bonding area produced in common adhesive joint configurations.

some adhesives demonstrate poor strength and high sensitivity to alignment when exposed to tensile loading as in the butt joint.

Although adhesives are rarely used in a direct tensile loading mode, several adhesive tensile tests are in common usage. One of the most commonly used adhesive tensile test method is described in ASTM D897. Fig. 3 shows two of the specimen configurations from ASTM D897 used to measure the strength of wood-to-wood and metal-to-metal bonds. To prepare the specimen for testing, the two halves of the spool configuration are bonded together with the adhesive to be tested. It is essential that surface preparation, adhesive application, and curing be performed according to specifications. The strength of the joint may depend on the thickness of the adhesive as well as the uniformity of the adhesive thickness. The enlarged ends of the spools are designed to mate with yokes of the load train, allowing the specimen to be pulled to failure in a universal testing machine. Detailed specifications for 'self-aligning' grips, intended to promote alignment of the test specimen upon loading, are described in detail in the standard. The authors' experience has been that even with such 'self-aligning' grips, it is often difficult to apply an axially centric load to the specimen. As a consequence, data

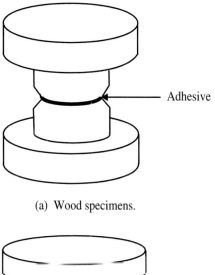

(a) Wood specimens.

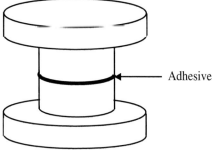

(b) Metal specimens.

Fig. 3. Specimen configurations for adhesive tensile testing (ASTM D897).

from these experiments often exhibit relatively large scatter unless extraordinary care is exercised in manufacture, in alignment of the samples, and in performing the experiments [12].

ASTM C297 describes specimen configurations used to determine the flatwise-tensile strength of sandwich constructions in the out-of-plane orientation. This test configuration and loading are similar to that of ASTM D897.

Other standard tensile tests, such as ASTM D2095, use bar or rod specimens that are easier to manufacture than those of ASTM D897. These test specimens, shown in Fig. 4, are loaded by pins through 4.76-mm-diameter holes. Although ASTM D2095 describes a fixture to assist in specimen alignment, eccentric forces still pose significant problems in alignment, resulting in relatively large scatter for such tests.

The exact stress distribution for all of these adhesive tensile tests is dependent on the relative moduli of the adhesive and adherends, the ratio of adhesive thickness to the dimensions of the joint, variations in adhesive thickness

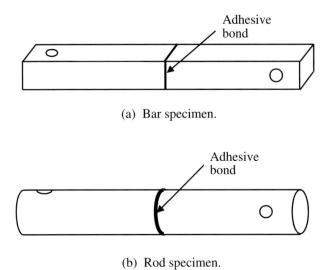

(a) Bar specimen.

(b) Rod specimen.

Fig. 4. Bar and rod specimens used for adhesive tensile testing (ASTM D2094).

within the joint, specimen alignment, and other factors. Caution must be exercised when comparing the 'tensile strength' data from one test with data obtained from another tensile test in which any of these factors differ. For most joints, the stress distribution can only be determined by numerical analysis (e.g. finite element analysis). Even with these methods, points of stress singularities usually exist for linear elastic analyses in regions where the elastic properties are discontinuous.

The results from all of these tensile tests are reported as the force at failure divided by the cross-sectional area of the adhesive. Such average stress information can be misleading. A note has already been made on the importance of alignment. Even when good alignment is obtained and the adhesive bond is of uniform thickness over the complete bond area, the maximum stresses in the bond line can differ markedly from the average stress as shown in Fig. 5 [12]. Furthermore, the stress distribution along the bond line is a strong function of details of the adhesive joint geometry. The family of curves shown in Fig. 5 was calculated using finite element analysis [13]. These calculations assume an elastic adhesive that is much less rigid than the steel adherends and with a Poisson ratio of 0.5. The different curves represent different adhesive thickness-to-diameter ratios, shown as parameters near the curves. The reader is referred to the work reported in [12–14], illustrating that significant shear stresses are developed in butt tensile joints. Later in this chapter it will be shown that analyses of this type can be used to predict variations in strength and the locations of the path for internal crack growth.

In conclusion to this section, it is emphasized that average tensile strength

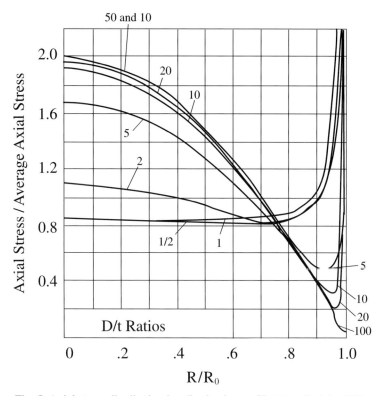

Fig. 5. Axial stress distribution in adhesive layer of butt tensile joint [12].

results, usually reported for tensile tests, must be used with extreme caution in predicting the strength of different, albeit superficially similar, joints.

2.3. Lap joint tests

The lap joint test is the most commonly used adhesive test, likely because test specimens are simple to construct and resemble the geometry of many practical joints. While this test is commonly referred to as the lap *shear* test, this is generally a misnomer, since failure is often more closely related to the induced tensile stresses than to the shear stresses. Further, it is conventional to report the 'apparent' adhesive strength as the load at failure divided by the area of overlap, even though the maximum stress will almost always differ markedly from this average value. Thus, while the lap joint test is commonly performed, the results from this test must be interpreted with caution. Issues associated with the single-lap joint test are discussed in ASTM D4896.

Although different lap specimen configurations are often used, the most common geometry is the single-lap configuration recommended in ASTM D1002 and

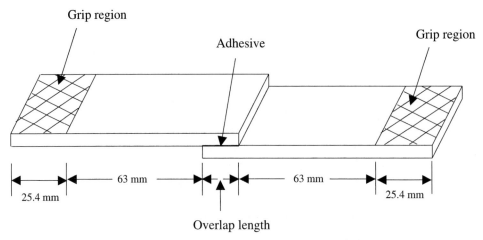

Fig. 6. Single lap test configuration (ASTM D1002).

shown in Fig. 6. The recommended specimen width is 25.4 mm. The flow of excess adhesive out of the edges of the overlap area during manufacture often poses a problem. As a result, single-lap specimens are often cut from two relatively large panels of the adherend that have been adhesively bonded together. This procedure, illustrated in Fig. 7, is described in ASTM D3165. In this case, the two adherends are bonded together over the entire length of the specimen. To produce an overlap length for adhesive testing, a notch is machined through each adherend. The notch depth is carefully controlled to avoid scoring the unnotched adherend. Under tensile loading, the applied load is transferred across the overlap length through the adhesive layer.

It has long been recognized that the lap specimen configuration will experience bending deformation under an applied tensile load such that the two lines of action of the applied forces will fall along the same line of action as shown in Fig. 8. This bending deformation induces tensile normal stresses near the ends of the bonded area. While the stress distribution in a lap joint is not amenable to accurate analytical solutions, there have been a number of efforts to determine the stresses. One of the earliest and best known of these is the now classical approach by Goland and Reissner [15] in which the joint is divided into parts and analyzed by mechanics of materials approaches. More recently, finite element techniques have been used. Work in this laboratory [12–14], by Guess et al. [16] and others (referenced in ASTM D4896), describe some of the difficulties in analyzing results from single-lap joint tests. Fig. 9, taken from ASTM D4896, shows the shear and normal stress distributions as a function of distance from the end of overlap as determined from finite element analysis. These results show that the shear stresses are not constant, increasing significantly in magnitude near the ends of the overlap. Additionally, loading of the lap joint induces normal stresses that vary along the

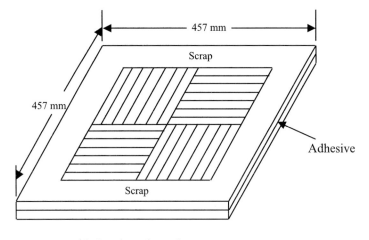

(a) Laminated panel.

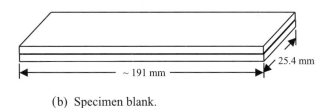

(b) Specimen blank.

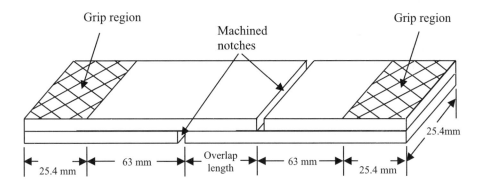

(c) Finished test specimen.

Fig. 7. Panel and specimen configuration for laminated panel testing (ASTM D3165).

length of the overlap region. The normal stresses are compressive in the central region but become tensile near the ends of the overlap. The magnitude of these tensile normal stresses becomes very large at the ends of the overlap. Careful observation of lap joint specimens during testing indicates that the first signs of

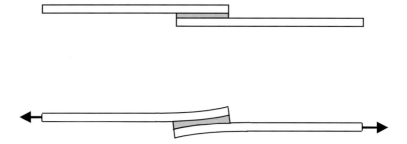

Fig. 8. Bending deformation of lap specimen under applied tensile load.

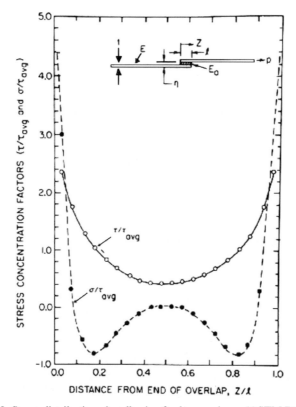

Fig. 9. Stress distributions in adhesive for lap specimen (ASTM D4896).

failure are typically cleavage at the ends of the overlap, consistent with the results from finite element analysis.

Efforts to alleviate these large tensile stresses, induced by the bending deformation, have led researchers to develop a double-lap specimen. The standardized

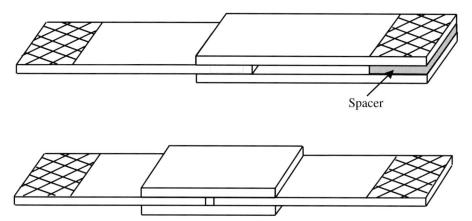

Fig. 10. Double-lap shear adhesive specimen configurations (ASTM D3528).

specimen configurations from ASTM D3528 are shown in Fig. 10. These efforts have not been completely successful, since experimental observations indicate that cleavage stresses play a dominant role in failure. Copper adherend/epoxy adhesive double-lap specimens investigated in this lab [17] showed cleavage crack opening at the ends of the overlap as the first visible signs of failure. Failed specimens shown in Fig. 11 exhibit extensive outward bending of the cover plates, providing graphic evidence that the adhesive and supporting plates were not only subjected to shear stresses. Finite element analysis of the double-lap specimen indicates that large tensile stresses exist at the ends of the overlap as in the single-lap

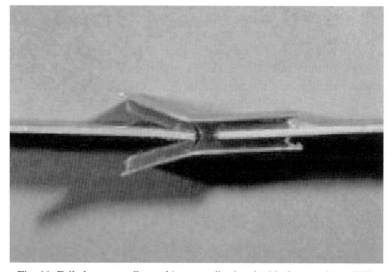

Fig. 11. Failed copper adherend/epoxy adhesive double-lap specimen [17].

configuration [13,17]. These results, as well as similar results by other researchers, indicate that failure of lap shear specimens is dependent more on the magnitude of the tensile normal stress near the ends of the overlap than on the maximum shear stress. It has also been demonstrated that the failure load of such joints is highly dependent on parameters such as adherend thickness, adhesive thickness, surface preparation techniques, method and time of cure, specimen width, and to a much smaller extent on the length of overlap.

Many handbooks and manufacturer's literature list the strength from adhesively bonded lap specimens as the 'apparent' shear strength of an adhesive. This usage has led some authors to conclude that the apparent shear strength is adequate for use as a design parameter. In fact, one popular and otherwise very good textbook on materials engineering includes the statement:

> "Thus in selecting an adhesive system, one must calculate the strength required. If, for example, you wish to design an adhesive-bonded lifting device for a 50 lb. (23 kg) machine component, you can use an adhesive with a 100 psi (0.689 MPa) shear strength and make the bond area 0.5 in² (3.2 cm²). You must use this type of joint-strength analysis in all adhesive-joint designs."

While this is certainly a logical method of design, it does not account for the non-uniform stress distribution within the lap joint nor the induced tensile stress, and in general may lead to incorrect predictions. For example, a logical extension of the above statement would be that when the weight of the machine component is increased to 100 lb. (46 kg), the bond area should be increased from 0.5 to 1.0 in² (3.2 to 6.4 cm²). This would be approximately true if the *width* of the overlap is doubled, but most certainly not true if the *length* of the overlap is doubled. In the latter case, the increase in joint strength would almost certainly be less than 100% and likely less than 30%. Furthermore, care must be exercised in referring to the load at failure divided by the area of overlap as the shear strength of an adhesive as if it were a material property even though these values are commonly reported in handbooks and manufacturer's literature. For a given adhesive, the value of shear strength value will depend on many factors as discussed previously. Thus, lap joint shear strengths should be used with caution in designing adhesive joints that differ in even seemingly minor details from the test specimens.

2.4. Peel tests

Nearly all of us have had at least some superficial, qualitative experience with an adhesive peel test when we removed a strip of adhesive tape from a flat surface by lifting and pulling on one end. Some interesting qualitative observations may be made by performing a simple test using a length of household cellophane tape. We suggest the reader to perform this experiment to experience the adhesive peel forces. Affix the length of tape to the smooth surface of a desk or tabletop,

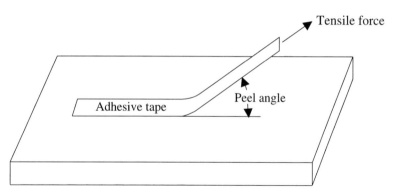

Fig. 12. Simple peel test to investigate peel forces.

applying sufficient pressure to attach it firmly. Next, lift one end of the tape to make an angle with the surface and apply a tension to the tape as shown in Fig. 12. Maintain a constant tape angle while peeling the tape from the surface, noting the force required. Next, reaffix the tape to the surface and peel it free at a different angle. Repeat this procedure for several peel angles ranging from a small angle such as 5° to an angle as large as 90°. Does the force that you apply to peel the tape from the surface change as the peel angle changes? Repeat the procedure using the same peel angle but with different peel rates. Is there a perceptible difference in the force required to sustain peeling for different peel rates?

Most adhesive tapes are composed of a flexible backing (paper, plastic, cloth, metal foil, etc.) to which a pressure-sensitive adhesive has been applied to one side (both sides for double-sided tapes). Pressure-sensitive adhesives typically consist of a rubbery material with a modifying tactifier that may be applied to the tape by a solvent system, hot melt, or by other means. One would expect such materials to be sensitive to the mode of stress (tensile versus shear) in the region where debonding occurs. Furthermore, since tacky rubbers of the type used in pressure-sensitive adhesive are viscoelastic, one would anticipate material properties to be time- and rate-dependent. Are these expectations consistent with the observations from your simple peel test?

Peel testing can be quantified with the aid of a universal testing machine. ASTM has formalized several peel tests, including: D1876, *Peel Resistance of Adhesives (T-Peel Test)*; D3167, *Floating Roller Peel Resistance of Adhesives*; D1781, *Climbing Drum Peel Test for Adhesives*; D903, *Peel or Stripping Strength of Adhesive Bonds*; and D2558, *Evaluating Peel Strength of Shoe Sole-Attaching Adhesives*. These test methods differ in the flexibility required of one or both of the adherends and the peel angle maintained during debonding. In all these tests, the resulting peel strength is generally defined as the force required to peel the adherends apart divided by the width of the peeled strip (N/m or lb./in.)

One of the more common peel tests is the T-Peel Test described in ASTM

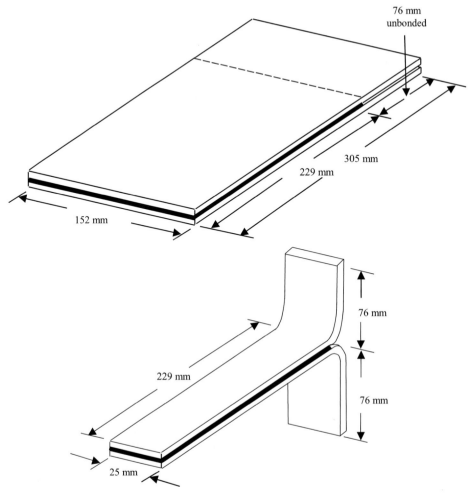

Fig. 13. T-peel test panel and test specimen (ASTM D1876).

D1876. The specimen and loading method for this test is illustrated in Fig. 13. In this test, two thin, flexible adherends measuring 152 mm by 305 mm are bonded over an area of 152 mm by 229 mm, as shown. Aluminum alloy plates, 0.81 mm in thickness, have been found to be well suited for many structural adhesives. The bonded panels are often cut into specimens of 25 mm by 305 mm for testing. At other times, the panel is tested as a single piece. The 76-mm-long unbonded regions of the two adherends are bent apart at right angles as shown in Fig. 13 for gripping with standard tensile testing grips. The peel force is recorded using a load cell on the testing machine as the specimen is loaded under a constant crosshead displacement rate of 254 mm (typically) per minute.

Another commonly performed peel test, the roller peel test, is described in

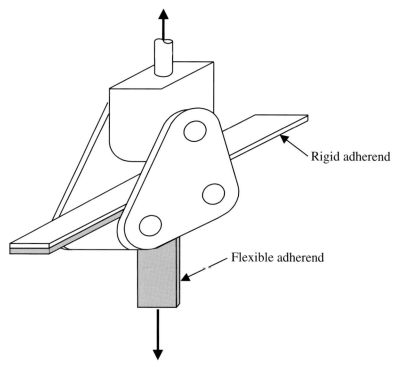

Rigid adherend

Flexible adherend

Fig. 14. Roller peel test configuration (ASTM D3167).

ASTM D3167. This test method, illustrated in Fig. 14, is used with specimens containing one adherend that is either relatively rigid or firmly attached to a rigid support. The other adherend is flexible, and is bent over a 25-mm-diameter roller as a tensile load is applied as shown. The stiffer adherend translates along the roller supports, maintaining the peel point in a fixed region between the grips. As a result, the fixture maintains a fixed peel angle as debonding occurs.

ASTM D1781 describes the climbing drum peel test that uses specimens with a relatively flexible adherend adhesively bonded to a rigid adherend. The test may also be used to test the adhesive bonds between the facesheets of sandwich structures. This test incorporates a hollow, flanged drum as illustrated in Fig. 15. Flexible straps are wrapped around the larger outer flanges of the drum and attached to the grips of a loading machine. The flexible part of the peel specimen is clamped to the central portion of the drum. Upon loading, the flexible straps unwind from the drum as the flexible adherend is wound around it and the drum moves up (hence the name 'climbing' drum peel test) thereby peeling the flexible adherend from the rigid adherend.

One of the simplest peel tests to conduct is the 180° peel test described in ASTM D903. In this test, one adherend must be flexible enough to allow a 180°

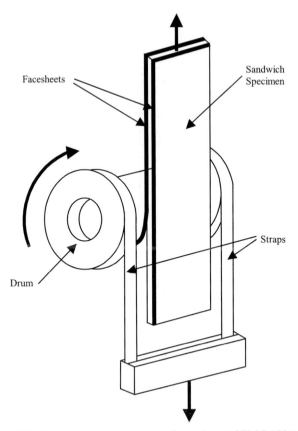

Fig. 15. Climbing drum peel test configuration (ASTM D1781).

bend to occur as shown in Fig. 16. The second adherend can be either flexible or rigid. The more rigid adherend is placed into one grip of the test machine and the more flexible member is bent back over the more rigid adherend and inserted into the other grip, producing a peel angle that is close to 180°.

The ASTM D-14 committee is currently working on a new 90° peel test in which a flexible adherend is bonded to a relatively rigid panel. The specimen is attached to a horizontally mounted, translating carriage that is attached to the lower crosshead of the testing machine as shown in Fig. 17. A tab on the end of the flexible material is attached to the upper grip of the testing machine. As the crossheads move apart during testing, the flexible member is peeled from the panel. The movable carriage maintains a peel angle of approximately 90°.

The stress distribution in a peel test is complex and dependent on the material properties of the adherends and adhesive as well as the specimen dimensions. In general, the stress distributions in peel tests are not well understood. A simple and easy to perform peel test using a length of household cellophane tape is again

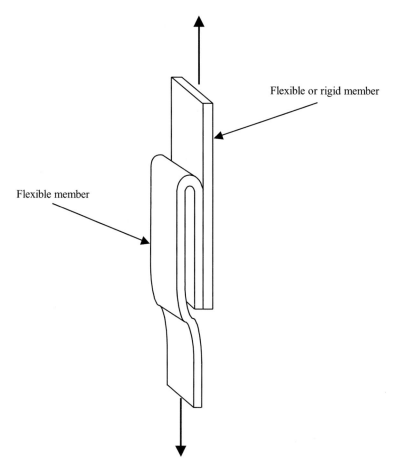

Flexible or rigid member

Flexible member

Fig. 16. 180° peel test configuration (ASTM D903).

used to illustrate this point. The reader is encouraged to obtain a long (~250 mm) strip of cellophane tape and make 25-mm 'tabs' by folding both ends over with the adhesive sides inward. Next, bring the tabs of the tape together to form a loop such that the adhesive sides of the tape are pressed together near the tabs with a loop below as shown in Fig. 18. Produce a short bonded section with a large unbonded loop below where the adhesive surfaces are facing each other but not bonded. Pull gently on the two tabs in opposite directions as shown in an attempt to peel apart the two segments of the tape. Intuitively, one might anticipate the state of stress at the bond line to be almost exclusively tensile (cleavage), perhaps with some induced shear stress. Would you anticipate any induced compressive stresses? Note what happens as the peel region approaches the loop of the tape where there is initially no bonding. It appears that there are compressive forces present so that the looped region in front of the failure is pressed together and

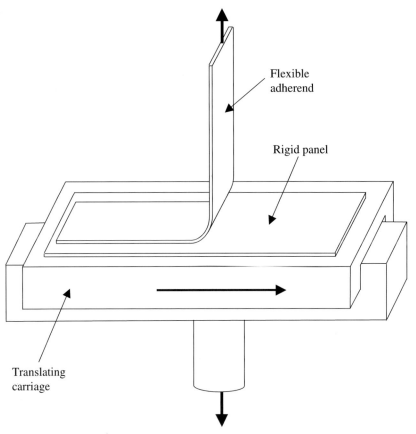

Fig. 17. 90° peel test currently under development.

bonded. With further peel, this newly bonded region progresses until the loop becomes very small and ultimate 'adhesive' failure occurs.

Clearly, the peel strength is not a 'fundamental' property for an adhesive. The value of force per unit width required to initiate or sustain peel is not only a function of the adhesive type, but also depends on the particular test method, rate of loading, thickness and stiffness of the adherend(s) and adhesive as well as other factors. Thus, peel tests generally do not yield results that may be used in quantitative design. This does not imply, however, that the peel test is not a useful test. Peel tests provide quantitative comparisons between different adhesive systems, insight into rate and temperature effects, etc. Additionally, peel tests can be used to provide fracture mechanics information as will be discussed in the next section. In the author's opinion, the latter aspect of peel tests has been perhaps most adroitly exploited by Gent and Hamed [18–20] who used peel tests in conjunction with fracture mechanics to obtain insights into time–temperature effects, the role of plasticity, and many other aspects of adhesive fracture.

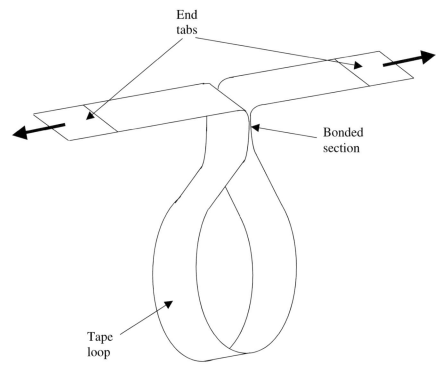

Fig. 18. Simple experiment to investigate peel stresses using cellophane tape.

2.5. Cleavage tests and tests based on fracture mechanics

As emphasized several times previously, the stress distribution in adhesive joints is generally complex and non-uniform, being dependent on the joint geometry, load alignment, and properties of the materials in the system. Furthermore, the maximum stresses in the bond often differ markedly from the average value even though the average stress is often the quantity reported as a test result. Cleavage tests are designed with an intentional, non-uniform distribution of stress. These tests differ from peel tests in that both adherends are relatively rigid, resulting in an approximately 0° peel angle. ASTM has standardized several cleavage tests for adhesives, including: D1062, *Cleavage Strength of Metal-To-Metal Adhesive Bonds*; D3433, *Fracture Strength in Cleavage of Adhesives in Bonded Joints*; D3807, *Strength Properties of Adhesives in Cleavage Peel by Tension Loading* (for plastics); and D5041, *Fracture Strength in Cleavage of Adhesives in Bonded Joints*.

The ASTM D1062 cleavage test uses the specimen configuration shown in Fig. 19. The specimen is fabricated from two identical metal pieces that are adhesively bonded to form the bond line to be tested. The load is applied off-axis

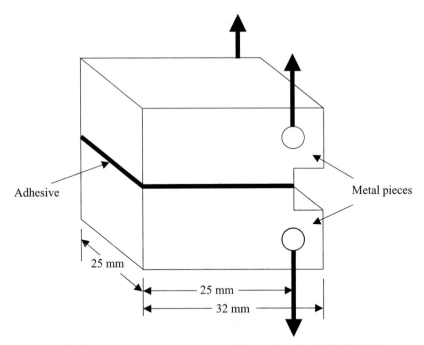

Fig. 19. Cleavage test for metal-to-metal adhesive bonds (ASTM D1062).

as shown, subjecting the adhesive specimen to a combined action of tension and bending. It is noted that this specimen geometry is very similar to that of the compact tensile specimen described in ASTM E399, used for determining the plane strain fracture toughness of metals. The specimen is pin-loaded to failure in tension through the loading holes. The ASTM D1062 standard requires that the breaking load be reported in force per specimen thickness. Additionally, the standard specifies that the percentages of the fracture surface experiencing cohesive, adhesive, and contact failures are to be reported.

ASTM D3807 uses a longer 'beam-like' cleavage peel specimen composed of two long slender rectangular strips bonded together over part of their length as shown in Fig. 20. The adherend strips are required to be 'semirigid', such that they can bend through an appreciable angle without failing. Near the ends of the unbonded lengths of these strips, a flexible wire is twisted about each of the strips. These wires are clamped using the grips of the tensile testing machine. Upon tensile loading, the wires separate the bonded strips, subjecting the bond line to combined tension and bending. Test results are reported as the force per specimen width required to propagate failure. It is noted that this test specimen is similar to those described in ASTM D3433. However, the analysis and reporting of test results for D3433 are quite different, as will be described subsequently.

ASTM D5041 uses specimens in the form of plates bonded together over one part of their length as in D3807, but uses a wedge forced between the unbonded

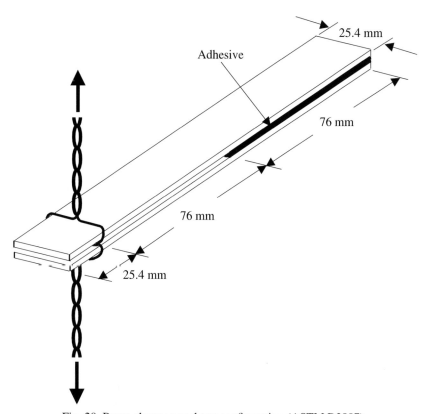

25.4 mm

Adhesive

76 mm

76 mm

25.4 mm

Fig. 20. Beam cleavage peel test configuration (ASTM D3807).

ends of the bonded plates to produce a cleavage load as illustrated in Fig. 21. ASTM D3762 uses a beam-like specimen configuration similar to D3807 and wedge loading similar to D5041 to assess environmental durability of adhesive bonds.

The primary result from these cleavage tests, the peak force per specimen width, is of little use in quantitative design of joints that differ in any significant detail from the test specimen and loading configuration. Similar to peel tests, however, these test results are of use for comparing adhesives and for qualitative evaluations. In contrast, the results from fracture mechanics testing do permit quantitative analysis of joint designs. The fracture mechanics approach views the adhesive joint as a system in which failure (typically the growth of a crack) requires that the stresses at the crack tip be sufficient to break bonds and its analysis involves an *energy balance*. In this approach, it is hypothesized that even if the stresses are very large (often theoretically infinite in the elastic case) a crack can grow *only* if sufficient energy is released from the stress field to account for the energy required to *create* the new crack surface as the fractured region enlarges. The critical value of this energy release rate (in joules per square meter of 'new' crack area) associated with fracture tests of adhesives is called by various

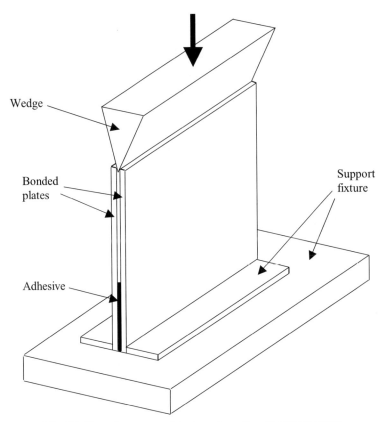

Fig. 21. Plate cleavage peel test configuration (ASTM D5041).

names, including the adhesive fracture energy, adhesive fracture toughness, and work of adhesion. Here G_c will be used to represent this critical energy release rate. The word *adhesion* is dropped from the comparable term when considering *cohesive* failure. This embodiment of fracture mechanics involves both a stress–strain analysis and an energy balance.

The analytical methods of fracture mechanics (both cohesive and adhesive) are described in a number of references [21–24] and will not be repeated here. However, a brief outline of one simple approach provides some insight into the concepts, principles, and methodologies involved for the reader who is not familiar with fracture mechanics. In the previous discussion of peel tests, it was noted that Gent and Hamed [18–20] had performed some extremely informative fracture mechanics tests using peel specimens. We consider a simplified fracture mechanic analysis of the 90° peel test shown in Fig. 17. Here we assume that the substrate is rigid and the 'peel adherend' is very flexible and perfectly elastic. The stress distribution in the vicinity of the 90° bend is complex and difficult to determine. If the material is perfectly elastic, however, this stress distribution is not required in the energy balance. As the flexible member is peeled from the rigid panel, the amount of strain energy associated with bending the curved portion of the beam remains constant. However, additional strain energy is associated with axial loading of a longer length of the flexible member. As the applied load P moves an amount δ during the peeling operation, it performs an amount of work equal to $P\delta$. In the energy balance, this input work is converted to fracture energy associated with both the formation of new surface area during the peel and new strain energy associated with axial loading as the vertical segment of the flexible adherend increases in length. This energy balance can be expressed as

$$P\delta = G_c \Delta A_a + \frac{\sigma^2}{2E} \Delta V$$

where P is the force required to sustain peel, ΔA_a is the new adhesive surface area produced by the peel, ΔV is the 'new volume' of flexible adherend subjected to axial loading, and E is the modulus of elasticity of the flexible adherend. This expression may be simplified by substituting for ΔA_a and ΔV using the relations

$$\Delta A_a = b\delta$$
$$\Delta V = A_s\delta$$

where b is the width and A_s is the cross-sectional area, respectively, of the flexible adherend. Substituting these relations and solving for G_c,

$$G_c = \frac{P}{b} - \frac{P^2}{2A_s E b}$$

Note that typical values of P/b observed in peel tests are on the order of 1 kN/m. Hence for relatively large value of $\Delta A_a E b$, the second term may be neglected and G_c may be approximated as the peel force per unit width, or

$$G_c = \frac{P}{b}$$

In principle, almost any test geometry can be used to determine the value of the adhesive fracture toughness. Some geometries, however, are easier to analyze or are more convenient from an experimental perspective. The split cantilever beam test, to be discussed in the next paragraph, is attractive from both standpoints. Once G_c is determined, the concepts of fracture mechanics can be used to determine the stress (or load) at failure for other loading geometries and crack sizes. Some geometries lend themselves much more readily to analysis and manufacture. To date, only ASTM D3433, *Fracture Strength in Cleavage of Adhesives in Bonded Joints*, is based on and analyzed by fracture mechanics. The specimen in this test, as noted above, is composed of two strips of dimensions 356 mm × 25 mm bonded together over all but 51 mm of their length as shown in Fig. 22. The specimen is loaded by pulling these two strips apart at their free 'cantilevered' ends in a universal testing machine.

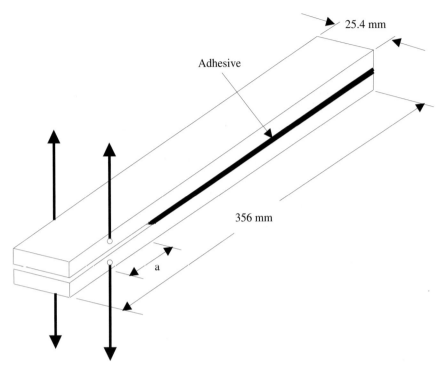

25.4 mm

Adhesive

356 mm

a

Fig. 22. Flat adherend specimen for determining fracture toughness (ASTM D3433).

If one assumes ideal cantilever boundary conditions, the analysis of this test configuration is very straightforward. The elastic energy stored in each of the cantilever beams is given by

$$U_{\mathrm{I}} = P\delta/2$$

where P is the applied load and δ is the free end deflection due to P. If we take the length of the cantilever to be a, its deflection is given by

$$\delta = Pa^3/3EI$$

where E is the modulus of elasticity of the strip material and I is the area moment of inertia of the strip. Using these relations, the total elastic strain energy stored in the two strips is

$$U = 2P^2 a^3/(6EI)$$

From this expression, the energy release rate, G_{c} may be calculated from

$$G_{\mathrm{c}} = \Delta U/\Delta A_a$$

Thus, the adhesive fracture toughness may be written as

$$G_{\mathrm{c}} = P^2 a^2/(bEI)$$

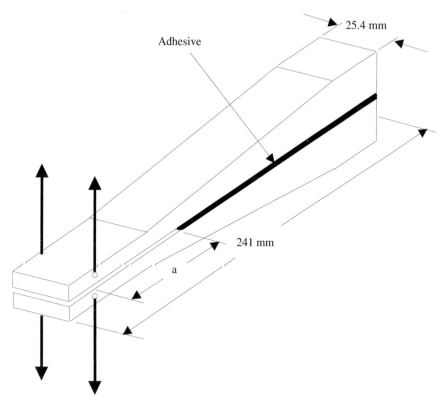

Fig. 23. Contoured adherend specimen for determining fracture toughness (ASTM D3433).

where A_a is the fracture area, b is the strip width and a is the distance from the line of action of P to the crack tip. Since $I = bh^3/12$, this expression can be written as

$$G_c = 12P^2a^2/(Eb^2h^3)$$

The expression for G_c given in ASTM D3433 may be written as

$$G_c = [12P^2a^2 + 4P^2h^2]/(Eb^2h^3)$$

The additional term in this expression is to account for strain energy due to shear stresses. Inspection of this equation indicates that this term is very small; the contribution to the energy release rate is 8% at $a = 2h$ and drops to only 2% for $a = 4h$. As will be discussed subsequently, numerical calculations show that other factors generally produce a much more significant contribution.

ASTM D3433 describes a second specimen geometry in which the adherends are thickness tapered as shown in Fig. 23. The taper is chosen to vary the beam compliance such that for a given load P at failure, the value of G_c is independent of the 'cantilever beam length', a. The two DCB configurations described in D3433 are convenient and widely used for evaluating adhesives as well as other

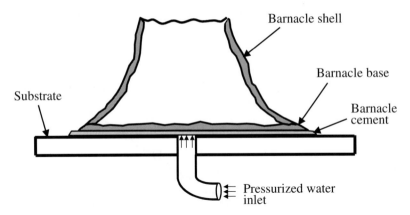

Fig. 24. Cross-sectional view of blister test for barnacle cement.

materials such as composite materials. As previously mentioned, however, fracture mechanics is not limited to this or any other specific testing geometry. In principle, any geometry for which an energy balance may be performed (or alternatively a stress intensity factor or J-integral calculated) may be used as an adhesive test.

Sometimes, specialized applications require the use of a non-standard testing geometry. As an example, the 'strength' of natural barnacle cement has been measured in the author's lab using a non-standard test. The test geometry to be tested was the natural joint between the base of the barnacle and plastic sheets that had been placed in the ocean. This configuration did not lend itself to tensile, lap, peel, or split cantilever testing. An array of holes were pre-drilled in the plates and filled with dental wax for conduction of the barnacle growth experiment. The wax was later removed at a moderately elevated temperature. The placement of a hole underneath the base of a barnacle formed a 'blister test' configuration as shown in Fig. 24. The hole was pressurized, causing failure of the barnacle cement and allowing for the determination of the adhesive fracture energy [23] using an analysis of the blister test by Williams [25].

The DCB test, the blister test, and several other geometries are somewhat amenable to the analytical analyses needed for fracture mechanics. As a consequence, most early fracture mechanics analyses focused on such geometries. Modern computational methods, particularly finite element methods (FEM), have lifted this restriction. A brief outline of how FEM might be used for this purpose may be helpful. Inherent in fracture mechanics is the concept that natural cracks or other crack-like discontinuities exist in materials, and that failure of an object generally initiates at such points [13,16,17,23–25]. Assuming that a crack (or a debonded region) is situated in an adhesive bond line, modern computation techniques can be used to facilitate the computation of stresses and strains throughout a body, even where analytical solutions may not be convenient or even possible.

The stresses and strains, which include the effects of a 'crack' in the bond line, are calculated throughout the adhesive and adjacent adherends. The strain energy, U_1, stored in the system for the particular crack size, A_1, is calculated. Next, the crack is allowed to grow to a slightly larger area, A_2, and the above process is repeated to determine the strain energy, U_2 corresponding to this increased crack area. The approximate energy release rate is given by

$$G = (U_1 - U_2)/(A_2 - A_1)$$

The critical energy release rate, $G_c = (\Delta U/\Delta A)_{crit}$, is that value of the energy release rate G that will cause the crack to grow in a particular material. Applied loads which result in energy release rates lower than this critical value, G_c, will not cause the given crack to propagate, while loads that produce energy release rates greater than this value will result in accelerated crack growth as noted previously. This critical energy release rate value G_c is also referred to as the adhesive fracture energy and the work of adhesion.

Researchers using numerical analysis in combination with fracture mechanics have developed efficient techniques to aid in the solution of problems of the type outlined above. While the calculation of the energy release rate described above is useful for visualizing the finite element approach, it is seldom used. Such approaches as the compliance method and the crack closure method are numerically easier [14,17,26,27]. The compliance method is analogous to the approach used previously for the DCB specimen. The deflection Δ due to a given load P is calculated at the point of load application for a given crack size. The work due to the loading ($P\Delta/2$ for an elastic system) is calculated. The work due to loading is calculated a second time for a slightly larger crack. The energy release rate is calculated as the difference in these two energy states divided by the corresponding difference in crack area. For the crack closure method, the specimen is meshed with finite elements, usually with smaller elements near regions of high stress gradients (e.g. near crack tips). Elements adjacent to the crack are given separate nodes to model the free surfaces of the crack. The energy release rate is determined from the nodal forces and nodal displacements associated with joining the first node from the crack tip with the matching node on the complementary surface, thus closing the crack by a length of one element length. The energy required to perform this crack closure operation divided by the area associated with the closure, is taken as the energy release rate. Our analyses have shown that with appropriate mesh refinement, the numerical results obtained by crack closure are in agreement with those obtained using the compliance method approach [12,14,17,26,27]. The crack closure method also facilitates a partition of the energy into components due to tensile (mode I) and shear (mode II) stresses at the crack tip. This partitioning is convenient in the analysis of adhesive joints for many adhesive materials where the critical energy release rate has been shown to be dependent on the mode of loading.

Most adhesive systems are not linearly elastic up to the failure point. Furthermore, the mechanisms associated with the creation of new surface area due to crack growth are not well understood. Surface 'creation' generally involves more than rupturing a plane of molecular bonds. Even taking surface roughness into consideration, the energy required to rupture molecular bonds of this increased area is a small fraction of the total 'fracture energy' for most practical adhesives. The total fracture energy also includes energy lost due to viscous, plastic, and other dissipation mechanisms at the tip of the crack and perhaps other mechanisms well removed from this immediate rupture point. Linear elastic stress analysis of fracture is, therefore, inexact.

Modern finite element analysis or other numerical methods have no problem in treating non-linear behavior. Our physical understanding of material behavior at such levels is lacking, however, and effective numerical analysis depends to a large extent on the experimental determination of these properties. Despite these limitations, many researchers have shown that elastic analyses of many adhesive systems can be very informative and useful. A number of adhesive systems are sufficiently linear, such that it is adequate to 'lump' the plastic deformation and other dissipative mechanisms at the crack tip into the adhesive fracture energy (critical energy release rate) term.

Analytical, numerical and experimental research based on adhesive fracture mechanics have been very informative and insightful. While we could highlight the work of researchers at the University of Akron, Virginia Tech, the University of Texas, Cal Tech, and a number of other university, industrial or government laboratories, we are most familiar with and will therefore draw on research from the Fracture and Adhesive Laboratory at the University of Utah for examples.

From finite element studies of butt joint tensile specimens and cone pull-out specimens, it was determined that the 'specific location' of the maximum energy release rate associated with a given size crack could be varied by making subtle changes in sample geometry [14,23]. The butt joints analyzed were composed of adherend materials that were very stiff compared to the adhesive. The adhesive was assumed uncompressible ($v = 0.5$). The analysis indicated that varying the adhesive thickness to diameter ratio (t/D) had a dramatic effect on the energy release rate. Additionally, varying the t/D ratio changed the location of the maximum energy release rate from the center of the butt joint to the outer edge.

Butt joint specimens were constructed using clear PMMA as the adherends and transparent polyurethane as the adhesive with a wide variety of thickness to diameter (t/D) ratios [12–14]. An experimental setup was constructed to facilitate observation of the bond line while the specimens were loaded in tension. Experimental observations were consistent with numerical predictions. Failure initiated at either the center or outside edge of the bonded surface, depending on the particular geometry tested. Further, the load required to sustain failure was

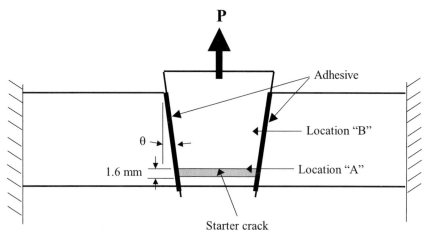

Fig. 25. Cross-sectional view of cone pull-out tests.

inversely proportional to the maximum energy release rate for specimens with varying t/D ratios.

The failure behavior of a series of cone pull-out tests was also explained with the aid of fracture mechanics and finite element analysis [23]. The specimens tested consisted of a truncated PMMA cone bonded into a matching hole in a PMMA plate using a clear polyurethane adhesive. A 1.6-mm-long starter crack extended completely around the periphery at the small end of the truncated cone as shown in Fig. 25. Two cone angles were tested: 0° (a cylinder) and 5°. When the 0° cone was loaded in tension, the failure always propagated from the starter crack (location 'A' in Fig. 25). When the cone angle was increased to 5°, failure never originated at this starter crack, but rather from a region that was roughly half the plate thickness (6.5 mm) away from the starter crack (location 'B' in Fig. 25). A finite element/fracture mechanics analysis of the two specimens indicated that an extremely short crack (~0.01 mm) located at the failure location for the 5° cone specimens had an energy release rate that was larger than that for the 6.5-mm-long starter crack located at the end. For the 0° cone specimen, the starter crack had the larger energy release rate. These experiments indicate that in both cases, crack growth initiated at the location of the largest energy release rate.

Double cantilever beam (DCB) specimens have also been analyzed using finite element/fracture mechanics techniques [28]. Results from these numerical analyses differ significantly from those obtained using the equations developed above and the equation found in ASTM D3433. The analytical equations found in D3433 assume "ideal cantilever boundary conditions" and that all of the work from the applied forces is stored as strain energy in the adherend beams, neglecting any strain energy in the adhesive. The finite element analysis raises questions as to the validity of both of these assumptions. The numerical results indicate that there is significant rotation of the beams at the assumed cantilever point. For shorter values of the split beam length a (but still within the D3433 recommended values), this end rotation can contribute 40% or more to the energy release rate. Neglecting the contribution of the adhesive to the strain energy might not appear to be significant since the adhesive thickness (and hence volume) is typically much less than that of the adherends. It must be noted, however, that the modulus for most adhesives is typically only a few percent of the modulus of the metal adherends and only 10 to 30% of the modulus of many composites. Therefore at comparable stresses, the strain energy density (strain energy per unit volume) of the adhesive can be much larger than for the adherends. Thus, the contribution of the adhesive to the strain may be significant in some cases.

Including the contribution of the adhesive to the strain energy in the finite element/fracture mechanics analysis facilitated a further observation [17,28]. Cracks were modeled at various locations through the thickness of the adhesive, and the energy release rate calculated for each of these locations. By varying the thicknesses of the adhesive and the adherends, it was possible to vary the paths of maximum energy release rate. That is, for most symmetric configurations cracks located near the centerline of the adhesive had larger energy release rates. For other configurations, the maximum energy release rate was near one of the adherends. It was also possible to devise specimen configurations for which the energy release rate was relatively independent of position through the adhesive thickness.

In an experimental investigation, DCB specimens similar to the geometries analyzed with variable split beam lengths a were constructed from aluminum using an epoxy adhesive [17,28]. The aluminum was selected so as to be in the elastic regime during the loads used in testing and an epoxy was chosen that exhibited relatively linear behavior up to its failure point. The DCB beam specimens were tested to failure in a universal testing machine. The crack length and the load required to propagate failure were used with the finite element/fracture mechanics analyses to determine the value of G_c for 13 different DCB specimen geometries. Additionally, values of G_c were obtained from the equation from ASTM D3433 using the same experimental results. The values of G_c determined using finite element analysis were very consistent, differing by less than 5% for those specimens where mode I loading dominated and by approximately 10% in the two cases where there was a significant mode II contribution to the loading at the crack. When using the equation from D3433, there was considerably more scatter, up to 50% in some cases. Experimental observations, shown in Fig. 26, confirmed that the adhesive cracks always followed the paths that the finite element/fracture mechanics analysis indicated were paths of maximum energy release rate. That is, if the FE/FM analyses indicated that this path was along the centerline of the adhesive, a very nearly planar crack was observed to propagate along this line as shown in Fig. 26a. For specimens where the analysis indicated that the maximum energy release rate was near the adhesive edge, the crack ran near this edge

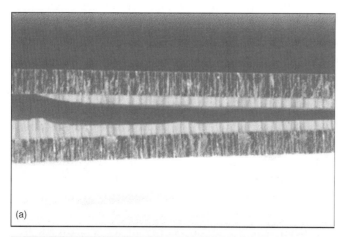

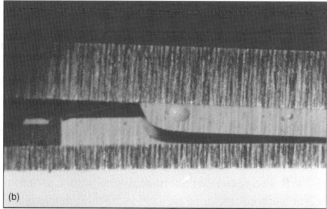

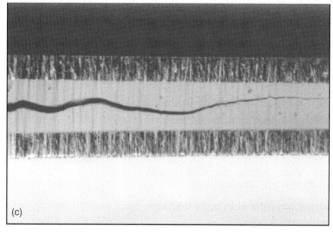

Fig. 26. Adhesive crack propagation in DCB specimens along paths of maximum G_c.

(leaving a thin but contiguous layer of adhesive on the aluminum adherend) as shown in Fig. 26b. Even when a starter crack was introduced somewhat removed from these locations of the maximum energy release rate, the crack propagated to these locations and after which followed the paths of maximum energy release rate. In the case where the energy release rate was relatively constant across the adhesive thickness, the planer nature of the crack was lost as the crack path was observed to 'wander' about the adhesive thickness (Fig. 26c), as its exact path was apparently determined by 'non-mechanical' aspects of the adhesive such as local changes in the chemistry or the physical nature of the adhesive.

It should be fairly obvious at this point that once the critical energy release rate G_c has been determined for a given adhesive/adherend system, fracture mechanics provides a very useful design tool. If the loading and the size of a typical crack-like discontinuity are known for a practical adhesively bonded joint, a stress analysis may be used to determine the strain energy in the system for the loading and an energy balance used to compute the energy release rate for the system to compare with the critical energy release rate, G_c. This procedure may be used to determine a factor of safety. For example, if the analysis indicated that the system of interest has an energy release rate, G, of M J/mm^2 and the given adhesive system is determined from experiments to have a G_c of N J/mm^2, the factor of safety can be defined as SF $= N/M$ for the joint. Modern numerical methods facilitate these computations for almost any joint geometry. It is important to note that the test geometry for determining G_c need not closely duplicate the practical geometry considered in the design. However, G_c may exhibit a dependence on the mode of stress at the crack tip, so this aspect should be very similar for the two geometries. It is also noted that alternate methods such as the stress intensity method, J-integral methods, etc. may be used in fracture mechanics design. These topics are described in the fracture mechanics texts referenced earlier.

Perhaps the most significant uncertainty in these finite element/fracture mechanics based approaches involves the value of crack length a used in the analyses. The crack length might be estimated, inferred from experiments, or determined in other ways. The authors have had some success in inferring the 'inherent crack size' by testing specimens with deliberately introduced cracks of varying sizes and extrapolating to the strength of joints for which no artificial crack was introduced. Using this inherent crack size in further analyses of other joint geometries fabricated with similar surface preparation, adhesive application, and cure process between the test and practical joints often produced excellent agreement between predictions and experimental strength results [12–14].

2.6. Thin-film versus bulk adhesive testing

It should be clear to the reader by now that certain adhesive test methods, such as the lap joint test, do not provide accurate determinations of adhesive strength. Furthermore, such tests do not provide a measure of the modulus of the adhesive, a required property for design purposes and for fracture mechanics analysis using finite element when the adhesive is to be modeled. The test methods presented above consider the adhesive in its 'in-situ' or 'thin-film' form. Another approach to determine the modulus and strength of an adhesive is through 'bulk' adhesive tests. Currently, there is some confusion in the literature as to whether mechanical properties obtained from bulk adhesive specimens may be used in the design and analysis of thin-film adhesive joints. Peretz [29] investigated the shear modulus and shear strength of adhesives for both thin adhesive layers and in bulk. Assuming a linear shear strain distribution throughout the adhesive thickness and uniform shear strain distribution throughout the bonded area, the in-situ shear modulus increased with increasing adhesive thickness up to the bulk material shear modulus. The shear strengths obtained from thin adhesive layers were similar to those obtained from bulk testing. Dolev and Ishai [30] conducted bulk and in-situ adhesive tests to compare mechanical properties under different states of stress. Good correlation between in-situ and bulk shear yield strength and elastic modulus was obtained. The authors concluded that elastic and strength properties of an in-situ adhesive may be determined by bulk adhesive testing. Lilleheden [31] performed a detailed experimental investigation of modulus variations in adhesives for differing adhesive thicknesses using a modified lap adherend specimen and moiré interferometry. These results showed no difference in the measured moduli of the adhesive between the thin-film and bulk forms. Discrepancies in moduli from previous studies [29,32] were attributed to factors such as variabilities in adhesive casting and curing conditions, lack of a well-defined state of stress, and inadequate methods of strain measurement. Swadener and Liechti [33,34] have conducted fracture mechanics analyses and experiments in which bond integrity is related to the global system properties well removed from the actual fracture locus.

Currently, no ASTM standard tests exist for bulk adhesive testing. However, since most adhesives are either plastic and/or elastomer-based, many of the standards included in ASTM Volume 8 (Sections 1 through 4) for plastics or Volume 9 (Sections 1 and 2) for rubbers might be adapted to test the properties of bulk adhesives. Tensile testing of bulk adhesive may be performed by casting or machining tensile specimens, either straight-sided or tapered specimens. Shear strength and shear modulus determinations of the bulk adhesive may be accomplished through several methods, including solid rod torsion testing or using the V-notched Iosipescu shear test method, commonly used for composite materials (ASTM D5379). The authors have had success in characterizing the tensile stress–strain response of structural adhesives using straight-sided tensile specimens, 150 mm long by 12.5 mm wide, machined from constant-thickness bulk adhesive panels. Additionally, the authors have characterized the shear stress–strain response of adhesives using V-notched specimens, 75 mm long by 19 mm wide, using the Iosipescu shear test method (ASTM D5379). Adhesive properties determined from these and other bulk adhesive tests are useful and informative in the design and analysis of adhesively bonded components.

2.7. Other standard tests

In the preceding sections, a number of tests have been described that are intended to provide information on the strength of adhesive joints. There are a number of other standard tests intended to provide other information on adhesives and adhesive joints that might relate to durability. A few of these standard tests will be reviewed briefly here.

The viscosity of an adhesive is related to its ability to flow and wet a surface and also to its ability to remain in place during the fabrication and cure of a joint. If the adhesive is too viscous, it may not make good contact and hence adequate mechanical connection with surface irregularities, asperities, etc. of a surface. If the adhesive is too 'runny' it may be difficult to manufacture a joint of consistent thickness. Changes in viscosity might also reflect separation of constituents or be evidence of premature partial curing of an adhesive. Accordingly, several standard tests have been devised to measure the viscosity of adhesives including: ASTM D1084, *Standard Test Methods for Viscosity of Adhesives*; D2556, *Standard Test Methods for Apparent Viscosity of Adhesives Having Shear Rate Dependent Flow Properties*; and D4499, *Standard Test Method for Heat Stability of Hot-Melt Adhesives*.

It is observed that for many well designed and manufactured practical adhesive joints, the location of failure is somewhat removed from the so-called interface. Nevertheless, it is absolutely essential that their be good interaction between the adhesive and the surface of the adherend(s). As a consequence, appropriate surface preparation before bonding is very important. Indeed, surface preparation is an extremely important part of adhesive science and technology. Surface *cleaning* is very important and in some cases all the surface preparation required. Cleaning may involve water, detergents, solvents, as well as sanding to remove contaminants, debris, loose oxides, etc. Surface preparation often includes much more than just a cleaning of the surface(s), however. Surface treatments may also be used, including roughening, which increases surface area and facilitates mechanical interlocking or hooking as described previously. Some surface treatments serve to prepare appropriate oxides (tightly bonded and coherent) or other surface layers for receiving the adhesive and facilitating mechanical interlocking if porous. The previously discussed Venable model of the oxide on aluminum presents a graphic picture of how a carefully prepared oxide surface might contribute to strong mechanical interlocking. Techniques standardized by ASTM for surface cleaning and preparation for adhesive bonding include D2651 and D3933 for metals, and D2093 for plastics.

Exposure to the environment and various environmental agents can have pronounced effects on adhesive joint strength and durability. Water, sunlight, solvents, etc. can affect the integrity of a joint. A number of standard tests have been designed for quantifying environmental effects including; D1713, *Test Method for Bonding Permanency of Water- or Solvent-Soluble Liquid Adhesives*; D1828, *Practice for Atmospheric Exposure of Adhesive-Bonded Joints and Structures*; and D1879, *Practice for Exposure of Adhesive Specimens to High-Energy Radiation*.

Time plays an important role in various aspects of adhesive performance. Most adhesives are polymers, which are inherently viscoelastic, such that modulus, strength, elongation at failure, and other properties depend on time and rate of loading. Furthermore, there are often chemical-stability-related attributes that are time-related. The useful shelf-life of many adhesives (or their components) is limited, and generally depends on the storage conditions. ASTM D1337, *Test Method for Storage Life of Adhesives By Consistency and Bond Strength*, addresses this problem. For many multi-component adhesive systems it is commonly known that the reaction starts at the time of component mixing and that application of the mixed adhesive and construction of the joint can continue only for a limited time. Use of the adhesive after this time is apt to affect joint strength and other properties. ASTM D1338, *Standard Test Method for Working Life of Liquid or Paste Adhesives by Consistency and Bond Strength*, is intended to address this feature. The effects of sustained loads on the property of a joint are addressed by several standards. Creep and time-to-failure are measured in several standard tests, such as: ASTM D1780, *Conducting Creep Tests of Metal-to-Metal Adhesives*; D2293, *Creep Properties of Adhesives in Shear by Compression Loading*; D2294, *Creep Properties of Adhesives in Shear by Tension Loading*; D3929, *Stress Cracking of Plastics by Adhesives Using the Bent-Beam Method*; and D3930 *Adhesives for Wood-Based Materials for Construction of Manufactured Homes*.

The combined effect of sustained loading and environment (particularly high humidity) is addressed by ASTM D3762, *Test Method for Adhesive-Bonded Surface Durability of Aluminum*. In this test, a double cantilever beam similar to that described for cleavage testing is loaded by forcing a wedge between the beam (rather than pulling them apart in a tensile testing machine) as shown in Fig. 27. The sample is then placed in an environmental chamber and the length of the adhesive crack observed and recorded as a function of time. ASTM D3166, *Standard Test Method for Fatigue Properties of Adhesives in Shear by Tension Loading*, uses a lap joint and a testing machine capable of applying a sinusoidal cyclic axial load to measure the fatigue resistance of a joint.

Tests are also available that attempt to infer long-term behavior from short-term testing. While such accelerated aging tests are never perfect, they may be the only alternative to observing a part in actual service for decades. The latter alternative is, of course, unsatisfactory if it involves waiting for this period of time to put a

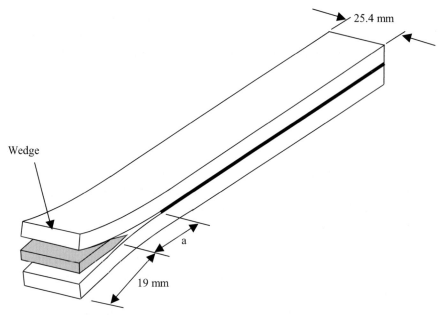

Fig. 27. Wedge test configuration for adhesive durability testing (ASTM D3433).

product on the market. As a case in point, consider the ASTM D3434, *Multiple-Cycle Accelerated Aging Test for Exterior Wet Use Wood Adhesives*, commonly referred to as the 'boiling test'. It is recognized that adhesives for laminating wood are generally not placed in boiling water during actual use. It is assumed, however, that resistance to boiling for a few hours or days may provide some evidence (or at least insight) into the durability of adhesively bonded wood panels under years of exposure to high humidity and elevated temperature. Another accelerated aging test is described in ASTM D3632, *Accelerated Aging of Adhesive Joints by the Oxygen-Pressure Method*. This test determines the deterioration in strength of adhesives exposed to high temperatures and high-pressure oxygen (typically 70°F and 300 psi, respectively). This test is used to identify adhesives that are likely candidates for more practical aging conditions.

In summary, there are many more tests that are used to explore additional properties of adhesives. This chapter has reviewed only the more commonly used tests for determining mechanical properties. Examples of other properties that may be tested for include impact strength, moisture absorption, corrosion, and attack by rodents and insects, just to name a few. The reader is referred to the ASTM Book of Standards Vol. 15.06 (revised and published annually) or the ISO counterpart for further details on the above and many other test standards.

Like most technical and scientific fields, the adhesives field has developed its own vocabulary, which at times may be somewhat unique from the use of similar words and phrases for even closely related fields. The usage of various words and terms used in adhesive science and technology is defined in ASTM D907, *Terminology of Adhesives*. Currently, ISO includes its adhesive terminology into the plastics terminology. However, an international committee is currently working on separating these two topics and, at the same time, correlating adhesives terminology with ASTM D907.

3. Conclusions

Adhesive joints are increasingly becoming the method of choice for joining component parts for a variety of subjective as well as objective reasons. Literally thousands of adhesives are available that differ in subtle as well as substantial ways. A problem faced by a designer is how to choose between the available adhesives. Mechanical strength is one of the most important attributes of an adhesive joint, and this in turn may be closely related to other features and properties. The nature of adhesion between materials is a complex subject to which the measurement of strength is intricately related. A variety of test methods have been devised with the purpose of determining the 'strength' of a joint. As useful as these test methods are, care must be exercised in interpreting the results from most tests. The secondary induced stresses can be as important in producing failure as the direct stresses. If the results from a particular adhesive strength test are to be used to design other adhesively bonded joints, it is important to exercise extreme care. The most direct, and likely safest, method would be to use a test geometry that duplicates the actual joint geometry in every essential detail. Test methods based on fracture mechanics have the potential of lifting this latter restriction. Here too, however, it is important to have an understanding of the details of the stress state near the adhesive crack. Modern computational techniques, primarily based on the finite element method, make nearly any joint geometry amenable to fracture mechanics analysis.

Standard test methods are available for the determination of adhesive and adhesive joint properties. These standard tests facilitate testing that can be duplicated and results that can be reported in consistent and understandable ways in different laboratories and settings.

References

1. Hartshorn, S.R. (Ed.), *Structural Adhesives*. Plenum Press, New York, NY, 1986.
2. Patrick, R.L. (Ed.), *Treatise on Adhesion and Adhesives*, Vols. 1–6. Marcel Dekker, New York, NY, 1966–1988.
3. Koelble, D.H., *Physical Chemistry of Adhesion*. Wiley, New York, NY, 1971.
4. Kinloch, A.J., *Adhesion and Adhesives*. Chapman and Hall, London, 1988.
5. Venables, J.D., McNamara, D.K., Chen, J.M., Sun, T.S. and Hopping, R.L., *10th Natl.*

SAMPE Tech. Conf. Proc., 362 (1978).

6.   Pluddemann, E.P., *Silane Coupling Agents*. Plenum Press, New York, NY, 1982.

7.   Fowkes, F.M., Hielscher, F.H. and Kelley, D.J., *J. Colloid Sci.*, **32**, 469 (1977).

8.   Mittal, K.L., *Pure Appl. Chem.*, **52**, 1295 (1980).

9.   Huntsburger, R.S., *Adhes. Age*, **21**, 32 (1978).

10. Padday, J.F. (Ed.), *Wetting, Spreading and Adhesion*. Academic Press, New York, NY, 1978.

11. Hammer, G.H. and Drzal, L.T., *Appl. Surf. Sci.*, **4**, 340 (1980).

12. Anderson, G.P., Chandapeta, S. and DeVries, K.L., Effect on removing eccentricity from button tensile tests. In: Johnson, S. (Ed.), *Adhesively Bonded Joints: Testing, Analysis, and Design*. ASTM, STP-981, 1988.

13. Anderson, G.P. and DeVries, K.L., Analysis of Standard Bond Strength Tests. In: Patrick, R. et al. (Ed.), *Adhesion and Adhesives*, Vol. 6. Marcel Dekker, New York, NY, 1988.

14. Anderson, G.P., DeVries, K.L. and Sharon, G., Evaluation of Adhesive Test Methods. In: Mittal, K.L. (Ed.), *Adhesive Joints, Formation, Characterization, and Testing*. Plenum Press, London, 1984.

15. Goland, M. and Reissner, E., *J. Appl. Mech.*, **11**, A-17 (1944).

16. Guess, T.R., Allred, R.E. and Gerstle, F.P., *J. Test. Eval.*, **5**, 84 (1977).

17. Borgmeier, P.R. and DeVries, K.L., Interpreting adhesive joint tests. *2nd Int. Symp. Adhesive Joints*, Pg 1, Newark, NJ, May 22–24, 2000.

18. Hamed, G.R., Energy conservation during peel testing. In: Patrick, R. et al. (Ed.), *Adhesion and Adhesives*, Vol. 6. Marcel Dekker, New York, NY, 1988.

19. Gent, A.N. and Hamed, G.R., *J. Appl. Polym. Sci.*, **21**, 2817 (1977).

20. Gent, N. and Hamed, G.R., *Polym. Eng. Sci.*, **17**, 462 (1977).

21. Griffith, A.A., *Poc. 1st Int. Congr. Appl. Mech.*, 55 (1924).

22. Broek, D., *The Practical Use of Fracture Mechanics*. Kluwer, Dordrecht, 1989.

23. Anderson, G.P., Bennett, S.J. and DeVries, K.L., *Analysis and Testing of Adhesive Bonds*. Academic Press, New York, NY, 1977.

24. Williams, J.G., *Fracture Mechanics of Polymers*. Ellis Horwood, Chichester, 1984.

25. Williams, M.L., *J. Adhes.*, **5**, 81 (1973).

26. Venables, J.D., McNamara, D.K., Chen, J.M., Sun, T.S. and Hopping, R.L., *Appl. Surf. Sci.*, **3**, 88 (1979).

27. Wang, S.S., Mandell, J.F. and McGarry, F.J., *Int. J. Fract.*, **14**, 39 (1978).

28. Devries, K.L. and Borgmeier, P.R., Fracture mechanics analyses of the behavior of adhesion test specimens. In: Van Ooij, W.J. and Anderson, H. (Ed.), *Adhesion Science and Technology — Mittal Festsschrift*, VSP Publ., Zeist, 1998, pp. 615–640.

29. Peretz, D., *J. Adhes.*, **9**, 115 (1978).

30. Dolev, G. and Ishai, O., *J. Adhes.*, **12**, 283 (1981).

31. Lilleheden, L., *Int. J. Adhes. Adhes.*, **14**, 31 (1994).

32. Brinson, H.F., *Composites*, **13**, 377 (1982).

33. Swadener, J.G. and Liechti, K.M., *J. Appl. Mech.*, **65**, 25 (1998).

34. Swadener, J.G., Liechti, K.M. and de Lozanne, A.L., *J. Mech. Phys. Solids*, **47**, 223 (1999).

Chapter 7

Measurement and analysis of the fracture properties of adhesive joints

M.D. THOULESS [a,b,*] and Q.D. YANG [a,1]

[a] *Department of Mechanical Engineering, University of Michigan, Ann Arbor, MI 48109, USA*
[b] *Department of Materials Science and Engineering, University of Michigan, Ann Arbor, MI 48109, USA*

1. Introduction

1.1. General concepts of fracture

Propagation of a crack involves processes that occur at different scales in a 'cohesive zone', a 'process zone', and a 'far-field zone' (Fig. 1). At the crack tip, the cohesive zone is associated with the material that actually separates during crack growth. As this material is deformed, it exerts forces across the putative crack surfaces until failure occurs, and the forces drop to zero as the crack advances. In other words, the deformation and failure of the material in the cohesive zone establish cohesive tractions that act across the crack plane. It should be appreciated that the relationship between the stresses and displacements of these cohesive tractions defines the fundamental fracture parameters of the system. In general, there are three possible modes of deformation at the crack tip — an opening mode, an in-plane shear mode, and an out-of-plane shear mode (Fig. 2). These are known as mode-I, mode-II and mode-III, respectively. In many geometries of practical interest, mode-III deformation is negligible, and only mode-I and mode-II deformations are considered in this paper. For each mode of deformation there are two parameters of primary importance: the peak stress supported across the crack plane during cohesive failure, and the area under the stress–displacement curve that characterizes the cohesive tractions. The former quantity is the 'cohesive strength' which is designated as $\hat{\sigma}$ in mode-I and $\hat{\tau}$ in

---

[*] Corresponding author. E-mail: thouless@engin.umich.edu
[1] Now with Rockwell Science Center, Thousand Oaks, CA 91360, USA.

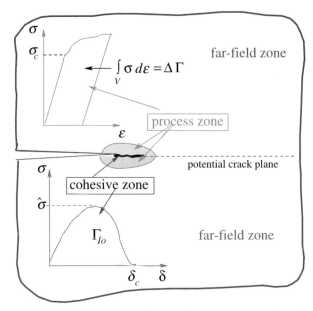

Fig. 1. Schematic illustration of the 'cohesive zone', a 'process zone', and a 'far-field zone' associated with fracture.

mode-II. The latter quantity is the energy associated with the creation of unit area of new crack surface. It is referred to as the 'intrinsic toughness', and is designated Γ_{Io} in mode-I and Γ_{IIo} in mode-II.[2] It should be appreciated that the cohesive tractions in all three modes may depend on loading rate [32–34], and on external factors such as the constraint acting on the cohesive zone [4,5] or the environment [28]. Therefore, in general, both the cohesive strength and intrinsic toughness should not be considered as unique material properties, except under carefully prescribed conditions.

  If the cohesive tractions are sufficiently large (typically greater than about three times the uniaxial stress required to trigger permanent deformation in the material under plane-strain conditions), they may induce non-linear, irreversible deformation in a process zone immediately surrounding the cohesive zone [48–50]. The energy dissipated by the irreversible deformation in the process zone gives rise to 'extrinsic' toughening. The total energy, Γ, required to create unit area of new crack surface (the 'toughness') consists of the intrinsic toughness associated with the cohesive zone plus the extrinsic toughness associated with the process zone, $\Delta\Gamma$. So that (omitting the subscripts identifying the deformation modes),

$$\Gamma = \Gamma_{\text{o}} + \Delta\Gamma. \tag{1.1}$$

---

[2] There are also two mode-III parameters, but these are not discussed in this paper.

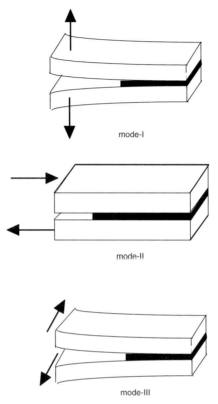

Fig. 2. Three possible modes of deformation at the crack tip — an opening mode (mode I), an in-plane shear mode (mode II), and an out-of-plane shear mode (mode III).

While non-linear deformations can occur in a process zone around a stationary crack, the energy losses are only realized when the crack begins to propagate and material in the process zone begins to unload. As a crack propagates, the extent of unloading increases and more of the energy dissipated in the process zone is realized as an increase in toughness. This is known as R-curve behavior. The magnitude of the extrinsic toughening increases from approximately zero for an initial crack with no process zone in its wake, to a fully toughened level when a wake consisting of a fully unloaded process zone extends far behind the crack tip (Fig. 3). If the process zone in the wake of a crack is removed, the toughness falls towards its intrinsic value [6]. In contrast to this type of behavior, it is possible for crack growth to occur without a process zone being induced if the cohesive tractions are small. There is then no R-curve behavior, and the energy associated with unit area of crack growth is independent of crack length, and remains at the value of the intrinsic toughness.

Beyond the boundary of any process zone that may exist is the far-field zone.

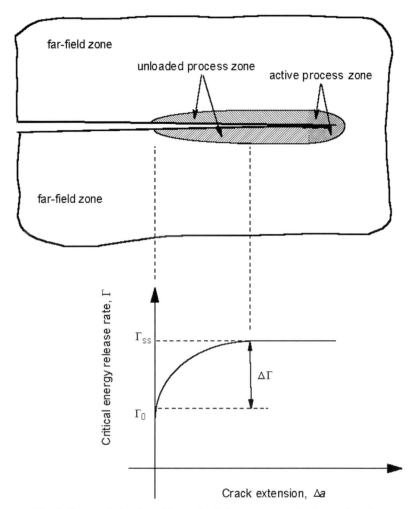

Fig. 3. R-curve behavior with a unloaded process zone in the crack wake.

In this region, the deformations are basically controlled by the macroscopic loads and geometry. (If there is no process zone, then the far-field zone extends up to the boundary of the cohesive zone.) The far-field zone provides the driving force for crack propagation. The difference between the work done by external loads as a crack propagates and any energy stored (or dissipated) in the far-field zone is associated with the applied crack-driving force (or applied energy-release rate). This quantity provides a crack-driving force to the process zone, or directly to the cohesive zone in the absence of a process zone).

It should be appreciated that the division of a material or system into the three regions of a cohesive, process and far-field zone is done for the convenience of analysis. There is a certain degree of flexibility in how these zones, that all interact with each other, are defined. To some extent, the definitions may be varied according to the scale at which fracture is analyzed, and according to how the fracture process is perceived. While an appropriate division may sometimes be physically obvious, there are other times when the arbitrariness of the choice is very apparent. If an understanding of the fracture process at the crack-tip scale is required, perhaps to predict the precise crack path or the mechanisms of crack-tip damage, then the cohesive zone has to be defined at a suitably fine scale. If a broader degree of understanding is desired, such as when predicting strength at an engineering scale without worrying about the finer details of fracture, then a coarser definition of the cohesive zone is acceptable in return for a reduced computational load.

A case where the divisions seem physically intuitive is the fracture of a fairly brittle crystalline material with a limited degree of plasticity at the crack tip. The cohesive zone is associated with the inter-atomic region along the crack plane in which material separation takes place. The cohesive tractions are provided by the inter-atomic potential as the atoms are separated from their initial equilibrium positions. These cohesive tractions induce plasticity within the process zone, which in turn is embedded in the linear-elastic region of the far-field zone. When the elastic region dominates the overall deformation of the system, this provides the classical picture of small-scale yielding in linear-elastic fracture mechanics. The inter-atomic traction–separation relationship gives Γ_o, while the energy dissipation realized in the process zone is $\Delta\Gamma$. However, the problem could also be considered by merging the cohesive and process zones into an enlarged cohesive zone. The cohesive traction–separation law then becomes the relationship between the stresses and displacements along a plane roughly corresponding to the elastic–plastic boundary. The cohesive tractions would be much smaller, but the displacements would be much larger, resulting in a large 'intrinsic' toughness that is the sum of the original intrinsic toughness and the original toughness increment. Now, the cohesive zone would be considered to be embedded directly in an elastic matrix (consistent with the smaller cohesive stresses); there would be no extrinsic toughening, and no R-curve behavior. [3]

An example for which the division into cohesive and process zones is not obvious is the problem of crack advance by void nucleation and coalescence in a ductile material [51]. Here, the cohesive zone might be taken to encompass all the cavities that contribute to crack growth. However, whether any cavitation off the fracture plane is associated with a process zone or with an extended cohesive zone is a matter of choice. Alternatively, the cohesive zone might be limited

---

[3] This alternative view is consistent with the notion that the surface-energy term in the Griffith approach [12] is replaced by a term incorporating the surface energy and the energy dissipated by plastic deformation [15].

to the actual plane along which separation occurs. Another ambiguous case is provided by a polymer in which crack growth occurs by chain scission associated with chain disentanglement and pullout. The cohesive zone might incorporate all the molecules involved with separation, so that its boundaries would be defined by the extent of material associated with these chains. Alternatively, the flow at the crack tip associated with this molecular motion might be viewed as crack-tip blunting and, consequently, be treated as belonging to a process zone. While these examples illustrate the flexibility associated with separating what are physically interconnected fracture processes, it should be emphasized that such divisions are necessary to simplify analyses of fracture unless full-scale atomic-level calculations are resorted to.

1.2. Cohesive-zone models of fracture

1.2.1. Introduction to cohesive zones

The discussion of the three zones associated with a crack provides the background for a very powerful approach to analyze fracture — the use of cohesive-zone models. It has been shown that this technique, which has its origins in the Dugdale [8] and Barenblatt [2] analyses of fracture, is particularly powerful when the cohesive tractions of the cohesive zone are incorporated into a finite-element model of the system [14,37,38,48–50]. Special elements that describe the cohesive traction–separation laws (and, in particular, incorporate the appropriate values of the cohesive strength, $\hat{\sigma}$ and $\hat{\tau}$, and intrinsic toughness, Γ_{Io} and Γ_{IIo}) replace the volume of material associated with the cohesive zone. These elements interact with the surrounding material during deformation, introducing a process zone if appropriate. Crack growth occurs when the elements in the cohesive zone reach the critical displacement for failure. One of the strengths of this approach is that the analysis does not require a sharp crack to be present, so it can be used for prediction of crack initiation as well as crack propagation [35].

As discussed earlier, the area under the traction–separation curve (Γ_{Io} and Γ_{IIo}) and the peak stress ($\hat{\sigma}$ and $\hat{\tau}$) are the important parameters that describe the cohesive tractions. The precise shape of the traction–separation law does not strongly influence the behavior of the system.[4] For example, one generally useful form of a mode-I traction–separation law is shown schematically in Fig. 4. It should be appreciated that while the area and peak stress are the two important parameters from a mechanics point-of-view, they may not necessarily represent the fundamental parameters from a materials perspective. In some ways, the peak

---

[4] Some caution needs to be taken to ensure that the compliance of the elements is not so extreme that it influences the overall deformation of the system.

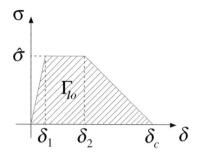

Fig. 4. Schematic mode-I traction–separation law that is generally useful in cohesive-zone models.

stress and critical displacement might be considered to be more fundamental parameters. However, from a mechanics point-of-view, a change in critical displacement affects fracture only through a change in the intrinsic toughness. This gives some insight into possible effects of rate and crack-tip constraints on fracture of adhesive joints [17,18]. For example, increased rates may raise the cohesive strength but decrease the critical displacement — whether the intrinsic toughness is decreased or increased by this change depends on how the total area under the traction–separation curve is affected. Similarly, an increased constraint on an adhesive layer may also raise the cohesive strength of the layer, but decrease the critical displacement. Again, short of experimental measurements, there is no way of predicting the net effect on toughness.

1.2.2. Mixed-mode fracture

In general, the cohesive traction–separation relationships for normal and shear deformation may well be completely unrelated, and a general approach for analyzing problems that involve different amounts of normal and shear deformation in the cohesive zone (mixed-mode problems) is required. The extent of mode-mixedness depends on the details of the mode-I and mode-II cohesive tractions, and on the geometry and loads. A general approach that accommodates unrelated traction–separation laws for opening and shear deformation involves recognizing that the total traction–separation work absorbed by the cohesive zone up to an arbitrary point in the deformation process, $\mathcal{G}$, (which is identical to $\mathcal{G}^\infty$, the applied energy-release rate, when there is no process zone), can be separated into the opening (mode-I) and shear (mode-II) components, $\mathcal{G}_I$ and $\mathcal{G}_{II}$, so that,

$$\mathcal{G} = \mathcal{G}_I + \mathcal{G}_{II}. \tag{1.2}$$

These two components can be calculated by integrating the appropriate traction–

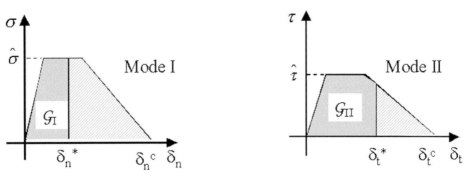

Fig. 5. Mixed-mode traction separation laws.

separation laws [57]

$$\mathcal{G}_I = \int_0^{\delta_n} \sigma(\delta_n)\,d\delta_n; \qquad \mathcal{G}_{II} = \int_0^{\delta_t} \tau(\delta_t)\,d\delta_t, \tag{1.3}$$

where δ_n and δ_t denote the normal and tangential displacements corresponding to the conditions of interest (Fig. 5).

A mixed-mode criterion for failure of the cohesive zone under different extents of mode-mixedness is also required. Under pure mode-I and pure mode-II conditions, the failure criterion appears to be simple: it is given by the conditions at which the tractions of the relevant cohesive law are reduced to zero ($\mathcal{G}_I = \Gamma_{Io}$ and $\mathcal{G}_{II} = \Gamma_{IIo}$, respectively). Mixed-mode criteria are much less obvious since the two modes of deformation may both contribute to failure together. An empirical relationship that has some physical appeal, and appears to capture some of the characteristics of experimental observations in this area is given by Wang and Suo [52]

$$\frac{\mathcal{G}_I^*}{\Gamma_{Io}} + \frac{\mathcal{G}_{II}^*}{\Gamma_{IIo}} = 1 \tag{1.4}$$

where Γ_{Io} and Γ_{IIo} are the total areas under the opening and shear traction–separation laws, and $\mathcal{G}_I^*$ and $\mathcal{G}_{II}^*$ are the values of $\mathcal{G}_I$ and $\mathcal{G}_{II}$ when the condition of Eq. 1.4 is met. In a mixed-mode scheme such as this, $\mathcal{G}_I$ and $\mathcal{G}_{II}$ evolve independently as the opening and shear displacements in the cohesive zone develop in response to the applied loads. Once the failure criterion of Eq. 1.4 is met, the normal and shear tractions across the crack plane are assumed to drop to zero instantaneously, and the crack advances. This scheme reduces to the correct results for pure mode-I and mode-II fracture ($\mathcal{G}_I = \Gamma_{Io}$ with $\mathcal{G}_{II} = 0$, and $\mathcal{G}_{II} = \Gamma_{IIo}$ with $\mathcal{G}_I = 0$, respectively). It also recovers the expected result that fracture is mode-I dominated in systems where in-plane shear rather than fracture occurs in response to mode-II loading, such as when dislocation emission

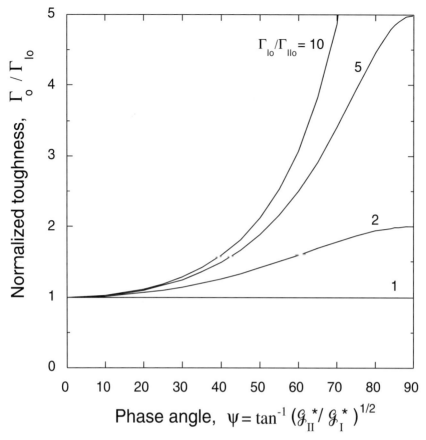

Fig. 6. A plot of the fracture toughness, given by $\Gamma_{\mathrm{o}} = \mathcal{G}_{\mathrm{I}}^* + \mathcal{G}_{\mathrm{II}}^*$, assuming the criterion of Eq. 1.4.

occurs easily along the interface [40], or when bonding across the interface is provided by the capillary forces of a liquid. In the latter case, the mode-I nature of fracture has been demonstrated experimentally [44]. The total energy-release rate, $\Gamma_{\mathrm{o}} = \mathcal{G}_{\mathrm{I}}^* + \mathcal{G}_{\mathrm{II}}^*$ required for fracture, assuming the criterion of Eq. 1.4, is plotted in Fig. 6. While this figure has been drawn for positive values of $\mathcal{G}_{\mathrm{I}}^*$, it should be noted that there appears to be evidence that crack growth can occur when the normal tractions are compressive [45,46].

1.3. Application of cohesive-zone models to the fracture of adhesive joints

So far, the discussion has outlined the general features of fracture that have to be understood before applying them to the specific problem of fracture of adhesive joints. It should be noted that since there are two physically distinct entities in an adhesive joint — the adhesive layer and the adherends — an obvious analytical approach is to associate the boundary between the different fracture zones discussed earlier with the boundaries between the two materials in the joint. In particular, it is often convenient to associate all the deformation in the adhesive layer with a cohesive zone. In other words, the role of the adhesive layer is assumed to be merely one of providing cohesive tractions across the interface between the adherends. As the adhesive layer deforms and fails, it exerts tractions on the adhesives and the forms of these normal and shear tractions lead directly to the traction–separation laws required for a cohesive-zone model. Therefore, in numerical analyses, the entire adhesive layer is replaced by cohesive-zone elements that mimic the deformation of the entire layer. In this approach, details such as whether crack propagation occurs by void nucleation, whether there is a plastic zone embedded in a larger elastic zone in the adhesive layer, or whether failure is interfacial or within the adhesive, are not addressed. All that matters is that the values of the effective cohesive strengths and intrinsic toughnesses of the adhesive layer have been correctly incorporated in the numerical model. However, it should be appreciated that differences in fracture mechanisms will physically affect fracture because of their influence on the effective cohesive tractions that act across the interface. However, as discussed earlier for geometrical and rate effects, the details can be ignored provided the appropriate cohesive traction–separation law is known.

There are other ways in which adhesive fracture can be modeled, and it is certainly a valid mode of enquiry to investigate what factors determine failure mechanisms and crack paths. While there is a substantial body of work based on continuum approaches dedicated to this topic, this type of study is not the emphasis of this chapter. Here, the problem of how to measure the properties of engineering joints, and how to use these data in a fashion that provides a predictive capability for their strength is addressed. Therefore, the adhesive layer is associated uniquely with the cohesive zone,[5] and attention is focused on techniques that measure the appropriate parameters of the traction–separation laws required for predictive purposes. However, it must be appreciated that a physical understanding of the actual fracture process can only be obtained by further investigations using cohesive-zone models at a finer scale. In particular, extrapolation of experimental results to conditions vastly different from those of the test configuration used to deduce the cohesive tractions of the adhesive layer may be unreliable if changes in the failure mechanism occur [4].

---

[5] This is actually a fairly common approach in the literature, even if it is not specifically recognized as such.

2. Fracture of joints with elastic adherends

2.1. Introduction to linear-elastic fracture mechanics

If the geometry and strength of an adhesive joint are such that the relationship between the applied loads and their displacements are essentially linear up to the point of crack growth, then the concepts of linear-elastic fracture mechanics (LEFM) can be used. For example, in mode-I geometries, this requires the non-dimensional parameter $E\Gamma_{\mathrm{Io}}/\sigma_y^2 h$ (where σ_y is the yield stress of the adherend, and h is a characteristic in-plane dimension of the joint) to be sufficiently low that fracture occurs before macroscopic plastic deformation of the adherends. It also requires the adhesive layer to have a negligible effect on the overall deformation of the system. This means that the non-dimensional parameter $E\Gamma_{\mathrm{Io}}/\hat{\sigma}^2 h$ must also be low; in addition, the adhesive layer thickness must be very small compared with h. Under these conditions, mode-I crack growth can be described using a single fracture parameter — the intrinsic toughness, Γ_{Io}, of the adhesive layer. The other fracture parameter, the cohesive strength of the layer, $\hat{\sigma}$, can be neglected. It should be repeated that the intrinsic toughness (as well as the cohesive strength) of an adhesive layer is not a unique parameter for one particular adhesive/adherend system. It may depend on the thickness of the adhesive layer, the locus of failure, the environment, and the crack velocity [9,11,30–34].

As discussed in Section 1.2.2, the normal and shear components of the intrinsic toughness, Γ_{Io} and Γ_{IIo} will often have different values. If LEFM conditions are satisfied, the applied energy-release rate, $\mathcal{G}^\infty$, can be calculated from the geometry and loading using linear elasticity. $\mathcal{G}^\infty$ is the value of energy-release rate that would act on a crack in a perfectly elastic body with a sharp crack (and cohesive tractions that are capable of supporting singular stresses). It is also possible to use linear-elastic calculations to deduce the nominal mode-I and mode-II components of $\mathcal{G}^\infty$, $\mathcal{G}_I^\infty$ and $\mathcal{G}_{II}^\infty$. The ratio of these components gives the nominal phase angle defined as

$$\psi^\infty = \tan^{-1}\left(\sqrt{\frac{\mathcal{G}_{II}^\infty}{\mathcal{G}_I^\infty}}\right) \tag{2.1}$$

The square root is required for consistency with the standard definition of the phase angle using applied stress–intensity factors [13].[6] A phase angle of $0°$ corresponds to a nominally pure mode-I configuration; a phase angle of $90°$ corresponds to a nominally pure mode-II configuration.

---

[6] If the adherends have different elastic properties, then a rigorous definition of the nominal phase angle should include a length scale to account for the oscillatory nature of the elastic stresses at the crack tip.

In the absence of a process zone, the applied energy-release rate, $\mathcal{G}^\infty$, equals the energy-release rate, $\mathcal{G}$, acting on the crack. The magnitude of the energy-release rate does not depend upon the nature of the cohesive zone tractions; however, the ratio of the two components of the energy-release rate at the crack tip, $\mathcal{G}_I$ and $\mathcal{G}_{II}$, do depend on the details of the cohesive tractions. In other words, $\mathcal{G}_I$ and $\mathcal{G}_{II}$ always equals the sum of $\mathcal{G}_I^\infty$ and $\mathcal{G}_{II}^\infty$ (since $\mathcal{G} = \mathcal{G}^\infty$), but $\mathcal{G}_I/\mathcal{G}_{II}$ does not equal $\mathcal{G}_I^\infty/\mathcal{G}_{II}^\infty$. Therefore, a rigorous analysis of mode-mixedness requires the cohesive-zone tractions to be considered. A continuum linear-elastic calculation only gives the nominal phase angle. However, despite the fact that the actual value of the mode-mixedness at the crack tip is generally different from the nominal value, the nominal phase angle has been suggested as a useful way of characterizing mixed-mode fracture [13]. While there is no physical basis for the assumption, if Eq. 1.4 is assumed to describe the relationship between the nominal components of the energy-release rate at fracture, it could be re-expressed in terms of the nominal phase angle

$$\frac{\Gamma_o}{\Gamma_{Io}} = \left[1 + \tan^2 \psi^\infty\right]\left[1 + \frac{\Gamma_{Io}}{\Gamma_{IIo}} \tan^2 \psi^\infty\right]^{-1} \tag{2.2a}$$

where $\Gamma_o = \mathcal{G}_I^* + \mathcal{G}_{II}^*$. A simple empirical form that captures the essential features of Eq. 2.2a, given by Evans and Hutchinson [10]

$$\Gamma_o/\Gamma_{Io} = \left\{1 + \tan^2\left[(1-\lambda)\psi^\infty\right]\right\}, \tag{2.2b}$$

where λ represents the relationship between Γ_{Io} and Γ_{IIo}, has been found to be quite useful for the purposes of analysis and prediction of the failure of interfaces in elastic systems [13,45,46].

2.2. Beam and plate geometries

There are many designs of test geometries to assess the strength of adhesive joints, However, the class of specimens based on plate and beam geometries are particularly important because it includes not only the obvious variants of the double-cantilever beam, but also the lap-shear, peel, blister and buckling geometries. A general analytical framework for the fracture mechanics of these types of specimens has been developed. This framework depends on the assumption that the crack is much longer than the thickness of the thinnest adherend, so simple beam theory applies. However, provided this assumption is satisfied, the analysis can incorporate arbitrary geometries and elastic mismatches between the adherends, and the effects of any general from of axial and bending loading, including non-uniform stresses and residual stresses [13]. At the heart of the analysis is the generic, steady-state form of the geometry that is shown in Fig. 7. This figure shows the effective loads acting at the crack tip: an axial load (per unit

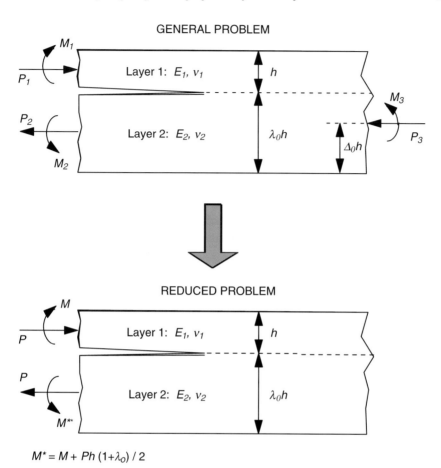

GENERAL PROBLEM

REDUCED PROBLEM

$M^* = M + Ph\,(1+\lambda_0)\,/\,2$

Fig. 7. The generic steady-state form of the beam and plate geometry.

width), P, and a bending moment (per unit width), M. Shear loading is neglected. The energy-release rate for this configuration is given by

$$\mathcal{G} = \frac{1}{2\hat{E}_2}\left\{\frac{P^2}{Uh} + \frac{M^2}{Vh^3} + 2\frac{PM}{\sqrt{UV}h^2}\sin\gamma\right\} \tag{2.3}$$

where

$$\frac{1}{U} = \frac{1}{\Sigma} + \frac{1}{\lambda_o} + \frac{6[1+\lambda_o]^2}{\lambda_o^3},$$

$$\frac{1}{V} = 12\left(\frac{1}{\Sigma} + \frac{1}{\lambda_o^3}\right),$$

$$\gamma = \sin^{-1}\left\{\frac{6\sqrt{UV}(1+\lambda_{o})}{\lambda_{o}^{3}}\right\},$$

$$\Sigma = \hat{E}_1/\hat{E}_2$$

$\hat{E}_i = E_i/(1 - v_i^2)$ in plane strain, and $\hat{E}_i = E_i$ in plane stress, E_1, E_2 are the Young's moduli of layers 1 and 2, v_1 and v_2 are Poisson's ratio for layers 1 and 2, and the thickness of layer 1 is h while the thickness of layer 2 is $\lambda_o h$. The nominal phase angle is given by

$$\psi^{\infty} = \tan^{-1}\left\{\frac{\eta\sin\omega - \cos(\omega + \gamma)}{\eta\cos\omega + \sin(\omega + \gamma)}\right\} \tag{2.4}$$

where,

$$\eta = \sqrt{\frac{V}{U}}\frac{Ph}{M},$$

and ω is a parameter that depends on the geometry and elastic properties of the adherends, which is tabulated in Suo and Hutchinson [42]. A value of $\omega = 52°$, can often be used with acceptable accuracy for practical purposes. While the use of this nominal phase-angle to characterize mixed-mode fracture has only been looked at in a limited sense, it does appear to provide a useful geometry-independent measure of the loading [39,45,46]. A reason why the concept may be useful is that all these geometries are essentially identical with respect to the crack tip — beams loaded by a combination of bending moment and axial load.[7]

The use of Eqs. 2.3 and 2.4 to analyze a particular configuration requires the applied loads and any residual stresses be reduced to the form shown in Fig. 7. Details of how this is done are given in Hutchinson and Suo [13]. It should also be noted that, since the quantities P and M are defined in terms of unit width of crack-front, Eqs. 2.3 and 2.4 can also be used directly when the crack width varies with crack length, as in the axisymmetric blister [3,24,25]. Furthermore, this analysis can be used when the depth of the specimens, h or λ_o, vary with crack length (such as in a tapered-double-cantilever beam geometry). It is the dimensions at the crack tip that are used in Eqs. 2.3 and 2.4. The values of P and M can often be found from simple plate or beam theory (assuming built-in boundary conditions at the crack tip). In Fig. 8, the actual geometry and loading of some common configurations are given along with the equivalent reduced loading.

---

[7] However, it should be noted that the use of P and M to describe the mixed-mode nature of the loading is not appropriate because Eq. 2.3 shows that these parameters are not orthogonal, unless λ_o is infinitely large. Therefore, their effects can not be added together in a linear fashion.

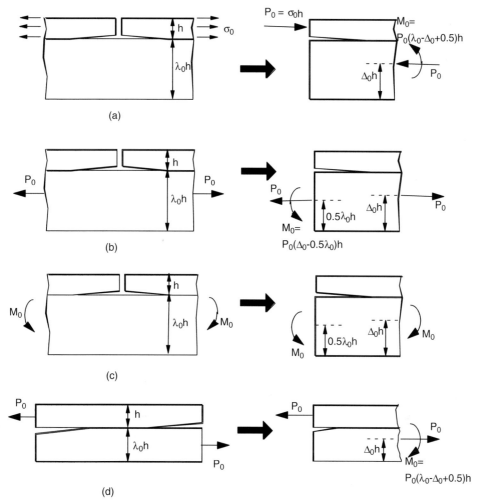

Fig. 8. Some common configurations based on beam and plate geometries.

However, there are geometries (such as buckles and axisymmetric blisters) for which numerical techniques have to be employed because of the non-linear deformations that occur in the linear regime [13,25,26]. Numerical techniques must also be used to evaluate energy-release rates for short cracks [16], and the effects of shear loading or non-isotropic adherends [43].

2.3. Double-cantilever-beam geometries

One of the most useful geometries of the type discussed above are those based on bonded cantilever beams. For example, the symmetrical double cantilever beam

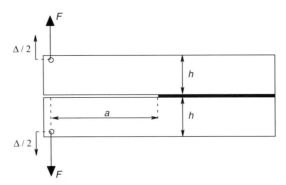

Fig. 9. Double-cantilever beam geometry.

(Fig. 9) is particularly useful for determining the mode-I toughness of a joint. When the crack is very long, the tip is loaded by a pure bending moment, and, under plane-stress conditions (appropriate when the height of the arms, h, is much greater than the width), Eq. 2.3 reduces to

$$\mathcal{G} = \frac{12a^2}{Eh^3} F^2, \tag{2.5}$$

where F is the applied force per unit width of the specimen. However, as shown in Fig. 10, the force F also loads the adherends in shear as well as in bending. The shear results in additional deformation at the crack tip beyond what would occur under an equivalent pure bending moment [29]. For many practical geometries, the ratio a/h is sufficiently low that the extra contribution to the energy-release rate from the shear force is important. Under these conditions, numerical, rather than analytical schemes need to be employed for rigorous calculations of the energy-release rate. Finite-element results for the energy-release rate are described

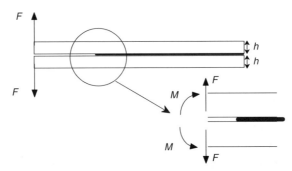

Fig. 10. An applied load provides a moment and a shear force at the tip of a crack.

by the empirical fit [43]

$$\mathcal{G} = \frac{12F^2}{Eh} \left(\frac{a}{h} + 0.677 \right)^2. \tag{2.6}$$

This is the appropriate equation to use when analyzing experimental double-cantilever beam (DCB) results in the LEFM regime (for $a/h > 1$). In this context, it should be noted that the expression

$$\mathcal{G} = 12a^2 F^2 \left(1 + h^2/3a^2 \right) / Eh^3, \tag{2.7}$$

which is the basis for the DCB specimen in the ASTM standards [1] is an approximate analytical result for the effects of shear [36,41]. However, the form of the correction factor for shear that this equation provides is incorrect (compare it with the correct form given in Eq. 2.6). Furthermore, Eq. 2.7 underestimates the energy-release rate for the double-cantilever beam geometry, and should not be used to analyze experimental results despite it being the basis of an ASTM standard.

The use of Eq. 2.6 to evaluate the energy-release rate requires the crack length to be measured during the course of the experiment. With transparent adherends, this can be done optically through the adherends so that the profile of the crack front can be observed. With other materials, the crack length must be monitored from the side of the sample either optically or using electrical resistance techniques. It is then desirable to use post-fracture observations to determine the crack-front profile and ensure that surface effects are not influencing the results unduly. It is also a matter of sound experimental procedure to monitor the displacements of the loading points, Δ (Fig. 9). This provides a means of calibrating the other parameters. Furthermore, measurement of time during the experiments allows the effects of crack velocity to be determined. In particular, it should be noted that values of toughness should be quoted in conjunction with the appropriate crack velocity. During a constant-displacement-rate test, there is a competition between the tendency for the applied energy-release rate to increase as the crack opening is increased, versus the tendency for the applied energy-release rate to decrease as the crack length increases. Whether the applied load increases or decreases depends on how fast the crack mouth is opened compared to the crack velocity. The rate of change of the energy-release rate, $\dot{\mathcal{G}}$, is approximately given by

$$\frac{\dot{\mathcal{G}}}{\mathcal{G}} = 2\frac{\dot{\Delta}}{\Delta} - 4\frac{\dot{a}}{a} \tag{2.8}$$

where $\dot{\Delta}$ is the rate at which the crack mouth is opened, and $\dot{a}$ is the crack velocity. If $\dot{\Delta}$ is kept constant, the energy-release rate will increase when $\dot{a} < a/2t$, where t is the time elapsed, but will decrease when $\dot{a} > a/2t$. Since the crack velocity itself may depend on the energy-release rate, a fairly rich variety of behavior can be expected that needs to be carefully interpreted [7].

If the crack length, a, load per unit width, F, and displacement, Δ, are continuously monitored, it is also possible to calculate the energy-release rate from the general expression [28]

$$\mathcal{G} = \frac{F^2}{2} \frac{dC}{da} \tag{2.9}$$

where $C = \Delta/F$ is the compliance of the specimen. This provides an excellent way of measuring the energy-release rate since it does not rely on analytical or numerical analyses of the geometry. Empirical fits to numerical analyses of the DCB specimen indicate that this compliance is approximately given by [54]

$$C = \frac{8}{Eh} \left(\frac{a}{h} + 0.677\right)^3, \tag{2.10}$$

to within about 10% if $a/h > 1$, where the second term in the bracket is the correction to the beam-bending equation caused by shear.

Eq. 2.10 forms the basis for a third type of test based on the double-cantilever beam geometry — the wedge test. In the wedge test, a wedge of known thickness is inserted into the crack mouth, and the resulting crack length is measured. If the wedge size, Δ, and crack length, a, are known, the applied energy-release rate can be deduced from Eqs. 2.6 and 2.10 as

$$\mathcal{G}^\infty = \frac{3E\Delta^2 h^3}{16(a + 0.677h)^4} \tag{2.11}$$

for $a/h > 1$. If the crack grows and is then arrested, the value of the energy-release rate when the crack is stationary gives the critical energy-release rate for crack arrest (which can be taken to be the value of toughness at zero crack velocity). If the crack length can be measured as a function of time, it is possible to deduce information about crack-growth kinetics [44]. Since wedge tests can be monitored over long periods of time and can be used in hostile environments, they form a good basis for investigating the effects of environment on statically loaded cracks, as in the 'Boeing wedge-test' [21]. For accurate results, the effects of shear should be included in the analysis, and Eq. 2.11 should be used to deduce the energy-release rate.

2.4. Effects of large deformation in adhesive layer

The discussion of the previous sections has been based on the assumptions of linear-elastic fracture mechanics. This requires that the adherends do not deform plastically, and that the deformation of the adhesive layer does not contribute significantly to the overall deformation of the system. Plastic deformation of the

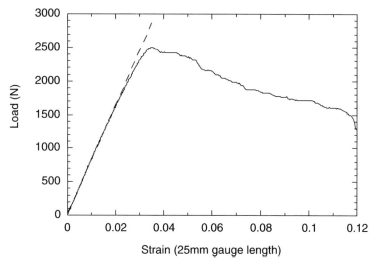

Fig. 11. An experimental load–strain curve for a DCB specimen (Al 6061-T6 adherends) fabricated with a 0.2-mm-thick layer of XD4600 (Ciba Specialty Products). For this specimen, $h = 34.0$ mm, $a = 73.0$ mm, $b = 9.53$ mm. This specimen was tested at a displacement rate of 1 mm/min [4].

adherends is addressed in Section 3. In this section, the effect of the deformation of the adhesive layer is considered. If the compliance of the adhesive layer is large relative to that of the adherends, or if the cohesive strength of the adhesive layer is relatively small, then deformation of the adhesive may contribute substantially to the overall deformation. This latter effect is illustrated in Fig. 11 which shows experimental data for a DCB specimen bonded with a commercial adhesive [4]. The constraint exerted by the adherends in this system was sufficiently great to cause cavitation and subsequent development of a large-scale bridging zone that supported a relatively low level of stress. This results in significant non-linearity of the load–displacement curve before fracture, and some care is required for the interpretation of the data.

Elastic-foundation models have been used to analyze the effects of both the compliance [23,55] and the plastic deformation of an adhesive layer [55]. Cohesive-zone models in which the adhesive layer is replaced by cohesive-zone elements provide powerful techniques to analyze the phenomena and to couple them to the fracture process. Using the trapezoidal traction–separation law shown in Fig. 4, normalized load–displacement curves for a DCB specimen are shown in Fig. 12 [4]. The maximum loads supported by the DCB joints are of the form [29]

$$\frac{F_{max}}{\Gamma_{Io}} = \frac{1}{\sqrt{12}} \sqrt{\frac{Eh}{\Gamma_{Io}}} f\left(\frac{a}{h}, \frac{E\Gamma_{Io}}{\hat{\sigma}^2 h}\right). \qquad (2.12a)$$

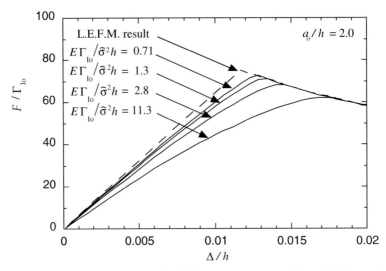

Fig. 12. Non-dimensional plot of the load–displacement curves for a DCB specimen with different values of the fracture parameters [4].

The LEFM result of Eq. 2.6,

$$\frac{F_{\max}}{\Gamma_{\mathrm{Io}}} = \frac{1}{\sqrt{12}}\sqrt{\frac{Eh}{\Gamma_{\mathrm{Io}}}}\, f\left(\frac{a}{h}\right),\tag{2.12b}$$

is appropriate only when the normalized parameter $E\Gamma_{\mathrm{Io}}/\hat{\sigma}^2 h$ is very small (i.e., a large cohesive strength, a low intrinsic toughness of the adhesive layer, or very thick adherends). In particular, it should be noted that if the peak load is used with Eq. 2.6 to calculate the energy-release rate at fracture, this value will result in an underestimate of the intrinsic toughness, Γ_{Io}, when $E\Gamma_{\mathrm{Io}}/\hat{\sigma}^2 h$ is greater than about one.[8] Accurate values for the fracture parameters of an adhesive system under these conditions have to be found by matching numerical predictions for both the shape of the load–displacement curve and the peak load to experimental measurements. Similar corrections need to be made to Eq. 2.10 when analyzing the results of a wedge test under conditions unless $E\Gamma_{\mathrm{Io}}/\hat{\sigma}^2 h$ is very small.

The discussion of the previous paragraph can also be illustrated following the approach of [55] in which the cohesive zone is viewed as providing an effective increase in the crack length. This can be seen from Fig. 11 because the results for the geometries with an initial crack and a cohesive zone all merge into the LEFM

---

[8] However, interpretation of such data in terms of extrinsic toughening and R-curves does provide an acceptable alternative approach. See Cavalli and Thouless [4] for a more detailed discussion of this point.

results for longer cracks. Empirical fits to the results of the numerical calculations shown in Fig. 11 [4,29] suggest that when the cohesive zone is fully developed (at the point of crack growth), the effective crack length, $a_{\text{effective}}$, is approximately

$$\frac{a_{\text{effective}}}{h} = \frac{a_o}{h} + f\left(\frac{E\Gamma_{\text{Io}}}{\hat{\sigma}^2 h}\right)$$

$$\approx \frac{a_o}{h} + \left(0.33\frac{E\Gamma_{\text{Io}}}{\hat{\sigma}^2 h} + 0.21\right)^{1/4}, \tag{2.13}$$

where a_o is the actual crack length (measured to the point where there is no bonding across the interface). This expression provides the basis for a technique to determine the fracture parameters from experimental data for a DCB geometry without resorting to numerical calculations. Substituting the effective crack length for a in Eq. 2.10 an approximate expression for the compliance can be found as

$$C = \frac{8}{E}\left\{\frac{a_o}{h} + \left(0.33\frac{E\Gamma_{\text{Io}}}{\hat{\sigma}^2 h} + 0.21\right)^{1/4}\right\}^3, \tag{2.14}$$

where the second term in the brackets is a measure of the additional compliance caused by deformation of the adhesive, and the third term is a measure of the additional compliance caused by shear. This equation can be used as the basis of an approximate analysis of either the wedge test or the DCB test when there is extensive deformation of the adhesive. For example, matching this compliance to the crack-opening displacement at the peak load, $F_{\max}$, allows an estimate for $E\Gamma_{\text{Io}}/\hat{\sigma}^2 h$ to be obtained. An estimate for the intrinsic toughness can then be found from an expression derived from Eq. 2.6

$$\Gamma_{\text{Io}} = \frac{12F_{\max}}{Eh}\left\{\frac{a_o}{h} + \left(0.33\frac{E\Gamma_{\text{Io}}}{\hat{\sigma}^2 h} + 0.21\right)^{1/4}\right\}^2. \tag{2.15}$$

3. Fracture of plastically deforming adhesive joints

3.1. Introduction

Linear-elastic fracture mechanics can only be used to characterize fracture of an adhesive joint when the two parameters $E\Gamma_{\text{Io}}/\hat{\sigma}^2 h$ and $E\Gamma_{\text{Io}}/\sigma_y^2 h$ (in mode-I) are sufficiently small that linear-elastic deformations dominate the behavior of the joint. The effect of the parameter $E\Gamma_{\text{Io}}/\hat{\sigma}^2 h$ was discussed in the previous section. In this section, the effects of large-scale plastic deformation in the adherends will be discussed.

In many practical applications, gross plastic deformation of the bonded materials occurs before fracture [27]. As a simple example, consider two adherends

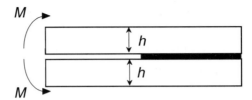

Fig. 13. A symmetrical, double-cantilever beam with each arm of thickness h, subjected to a pure bending moment M.

of thickness h bonded by a layer of adhesive and subjected to a remote bending moment, as shown in Fig. 13. Provided that $\hat{\sigma}/\sigma_y < 3$, so no process zone is introduced in the adherends, it can be easily shown [20] that if

$$\frac{E\Gamma_{\text{Io}}}{\sigma_y^2 h} > \frac{1}{3}, \tag{3.1}$$

macroscopic plastic deformation will occur before the crack starts to grow. If $\hat{\sigma}/\sigma_y > 3$, plastic deformation will occur in the adherends owing to the process zone, even if Eq. 3.1 is not met. There have been several energy-balance analyses based on classical beam- or plate-bending theories to investigate the effect of adherend plasticity on the fracture of adhesive joints [20,22,47]. At the heart of all these analyses lie two critical assumptions: (1) that bending is the dominant mode of deformation; and (2) that simple beam-bending theory can be used to describe the deformation of the adherends. As will be discussed below, these are invalid for most practical configurations. Therefore, the general notion of developing analytical approaches based on beam-bending theory for describing the fracture of plastically deforming adhesive joints is inappropriate. Rigorous, numerical calculations such as cohesive-zone models are required.

3.1.1. Pure bending

The problem with the second assumption — the use of simple beam-bending theory — can be demonstrated by considering the simplified geometry shown in Fig. 13 where a symmetrical double-cantilever beam consisting of two beams of thickness h and bonded by an adhesive layer (with an intrinsic toughness of Γ_{Io} and a cohesive strength of $\hat{\sigma}$) are separated by a pure bending moment, M. If the relationship between the stress, σ, and strain, ε, in the adherends is given by a power-law hardening relationship of the form

$$\sigma = A\varepsilon^n \tag{3.2}$$

where A and n are material constants, then an energy-balance calculation results in a prediction of a critical bending moment

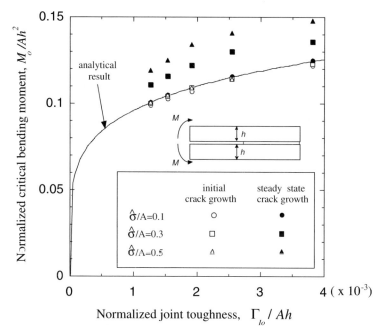

Fig. 14. The critical bending moment plotted as a function of joint toughness for three different values of $\hat{\sigma}/A$ for a symmetrical double-cantilever beam. The solid line represents the analytical expression (Eq. 3.3) [58].

$$\frac{M_0}{Ah^2} = \left(\frac{2}{\sqrt{3}}\right)\left(\frac{n+1}{4n}\right)^{\frac{n}{n+1}}\left[\frac{1}{2(n+2)}\right]^{\frac{1}{n+1}}\left(\frac{\Gamma_{Io}}{Ah}\right)^{\frac{n}{n+1}} \tag{3.3}$$

being required to separate the two adherends [58]. However, this calculation neglects the effects of $\hat{\sigma}$ which need to be explored by the use of a cohesive-zone model. Such an analysis shows that the cohesive stresses introduce a triaxial stress state that inhibits plastic deformation in the beams and invalidates the assumption of symmetrical bending used in the derivation of Eq. 3.3 [58]. Fig. 14 shows how the cohesive stress acts to raise the steady-state value of the critical bending moment above that predicted analytically. This triaxiality effect occurs even at relatively low values of $\hat{\sigma}$, well below that required to induce a process zone. (However, it should be noticed from Fig. 14, that the analytical result correctly predicts the initiation of crack growth, because the bending of the arms behind the crack tip are unaffected by the cohesive zone at this stage in the fracture process.)

3.1.2. Effect of shear

The second assumption invariably used in analytical derivations of the strength of plastically deforming joints is that the crack propagates under pure bending. This assumption is at the heart of several analyses that have been developed for the peel test [20,22]. However, as discussed in Section 2.3, if a specimen is loaded by an applied force rather than by a moment, there will be both a bending component and a shear component acting on the adherend at the crack tip (Fig. 10). When an applied force is transmitted through a film or laminate to the crack tip as a bending moment, bending deformations can lead to extensive plastic dissipation. In other words, the bending component of an applied force is not transmitted efficiently to the crack tip if yield occurs. In contrast, the shear component of an applied force is transmitted very efficiently to the crack tip, and does not result in extensive plastic deformation. The work done by the shearing component of the applied load efficiently contributes to overcoming the cohesive energy of an interface. The work done by the bending contribution is partially dissipated by plasticity. Numerical analyses [58,59] indicate that, for power-law plasticity, the shear component is dominant for small values of Ah/Γ_{Io} (where h is the adherend thickness), while the bending component is dominant at larger values of Ah/Γ_{Io}.

3.2. Peel test

The effect of shear is illustrated in an analysis of the symmetrical 90°-peel test [59]. Fig. 15 shows the configuration, and Fig. 16 shows how the steady-state peel force varies with the normalized thickness of the adherends, Ah/Γ_{Io}. While the bending assumption is valid for very large values of the parameter Ah/Γ_{Io}, as discussed in Section 3.1.1, large cohesive stresses introduce a constraint that elevates the bending moment (and hence, peel force) required for fracture above that predicted by pure bending analyses [59]. At lower values of Ah/Γ_{Io}, the effects of shear at the crack tip become important and decrease the peel force below that predicted by bending analyses. In the limit of $Ah/\Gamma_{\text{Io}} = 0$, only shear acts at the crack tip. Therefore, no energy is dissipated in bending, and the peel force tends to the thermodynamic value of $\Gamma_{\text{Io}}/2$. The competing effects of the cohesive stresses and the shear force cannot be treated analytically, and only numerical techniques should be used to analyze them.

To complete the discussion of the peel test, the effect of a finite yield stress for the adherends should be discussed. Fig. 16 was derived for a system with power-law hardening from zero yield strength. If the adherends have a yield strength of σ_y, there will be a critical thickness, similar to the form given in Eq. 3.1, above which macroscopic plasticity does not occur. While this regime has been explored in a numerical study by Wei and Hutchinson [53], no full analysis that covers both of the regimes explored by Yang et al. [59] and Wei and Hutchison [53] has yet been completed. However, a schematic sketch of the expected behavior is shown in Fig. 17. When the thickness of the adherends is small, shear forces will dominate so the energy dissipated in bending will become negligible and the peel force will tend to the thermodynamic limit of $\Gamma_{\text{Io}}/2$. The peel force is expected to

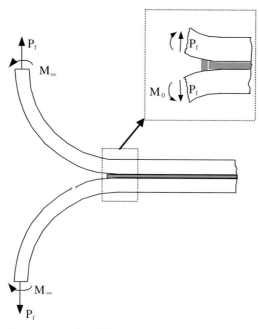

Fig. 15. Configuration for a symmetrical 90°-peel test, showing how, in general, both a bending moment and a shear force act at the crack tip.

rise as the effects of plastic bending become important for thicker adherends. For very thick adherends, bending will be dominated by elasticity. The peel force will then depend on the magnitude of the cohesive stress. No crack-tip plasticity will be induced if the cohesive stress is less than about $3\sigma_y$ [53], and the peel force will fall to an asymptotic limit of $\Gamma_{\text{Io}}/2$. The resultant peak in the peel force shown in Fig. 18 is similar to that suggested by Gent and Hamed [11]. If the cohesive stress is greater than about $3\sigma_y$, a process zone in the adherends will be induced by crack-tip plasticity, and the asymptotic peel force for very thick adherends will be substantially elevated [53].

3.3. General mixed-mode problems

To model geometries that are not symmetrical, the traction–separation laws for modes I and II need to be determined (a third law would be required for mode-III problems). In particular, values for Γ_{Io}, Γ_{IIo}, $\hat{\sigma}$ and $\hat{\tau}$ need to be determined from mode-I and mode-II tests. These are then incorporated into the traction–separation

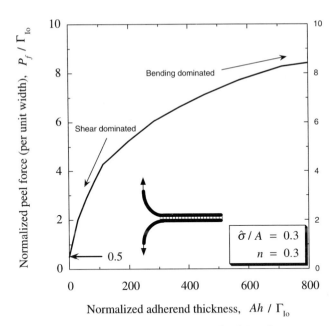

Fig. 16. Steady-state peel force plotted as a function of Ah/Γ_0 for a symmetrical 90°-peel specimen with power-law hardening. Superimposed on this plot is the analytical beam-bending solution [59].

laws of cohesive-zone elements that interact with the surrounding adherends in a finite-element model. An approach for determining the relevant parameters and incorporating them in numerical models has been recently established [17,57–60]. This will be reviewed here using, as an example, one particular material system consisting of aluminum (5754 alloy) and a 0.25-mm-thick commercial adhesive (XD4600 from Ciba Specialty Products).

3.3.1. Determination of mode-I parameters

The mode-I fracture parameters can be determined using a wedge-loaded double-cantilever beam specimen. The test configuration is sketched in the inset of Fig. 18. The contact force between the wedge and the beams provides a combination of a bending moment and shear loading that opens the crack. The use of different sized wedges and adherend thicknesses allows different combinations of bending moments and shear forces to act at the crack tip [56,58]. The bending moment causes a plastic radius of curvature, R_p, to be developed. This can be can be measured and compared to numerical predictions from a cohesive-zone model to deduce the mode-I fracture parameters [58]. This procedure was performed on the 2.0-mm-thick specimens and, when the crack propagated within the adhesive layer, the mode-I fracture parameters were found to be $\Gamma_{I0} = 1.4 \text{ kJ m}^{-2}$ and $\hat{\sigma} =$

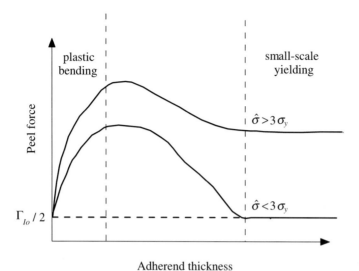

Fig. 17. Schematic plot of how the peel force varies with adherend thickness for a power-law hardening material with a finite yield strength σ_y.

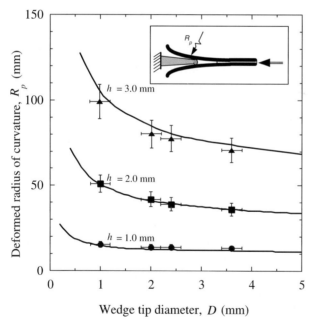

Fig. 18. Average values of the steady-state radii of curvature for symmetrical double-cantilever beams (1.0, 2.0, 3.0 mm thick) separated by wedges of different sizes being inserted into the interface. A comparison is made between the experimental observations and the predictions of the cohesive-zone model using values of $\Gamma_{Io} = 1.4$ kJ m^{-2} and $\hat{\sigma} = 100$ MPa. The dotted line represents the analytical predictions based on beam-bending theory [58].

100 MPa. These parameters were then used, without any further modifications, to predict the radii of curvature of DCB specimens with different adherend thickness. The excellent agreement between the numerical and experimental radii of curvature are shown in Fig. 18 as a function of the wedge-tip diameter, D. Furthermore, it has been shown that these mode-I fracture parameters can be used without modification to provide excellent agreement for the same combination of materials in other mode-I configurations [58,59].

It should be noted that the fracture parameters are sensitive to the constraint exerted on the adhesive layer by the deforming adherends. Both the cohesive strength and the critical displacement for fracture can be affected by the local stress state. A detailed analysis [17] has shown that these effects are not significant for the range of adherend thicknesses shown in Fig. 18. (Although they are responsible for the small discrepancies that can be seen between the predictions for the different thicknesses in this figure.) Furthermore, it has also been shown [17] that the traction–separation law was not very different when the same adhesive was used to bond steel rather than aluminum. However, a huge difference was observed when very thick aluminum adherends that deformed elastically during fracture were used [4]. Under these conditions, the large constraint altered the failure mechanism of the adhesive by causing void nucleation and growth. This led to large-scale bridging by the adhesive layer, a large increase in the critical displacement to failure of the adhesive layer, and a substantially enhanced level of Γ_{Io} — about 3.4 kJ m^{-2}. Finally, it should be observed that the mode-I fracture parameters depended on the locus of failure. When the crack was driven close to an interface with aluminum, Γ_{Io} dropped to 1.0 kJ m^{-2} and $\hat{\sigma}$ dropped to 60 MPa. These are the values of the mode-I parameters that have to be used when analyzing mixed-mode fracture with aluminum adherends, because under these conditions the crack invariably propagates at an interface.

3.3.2. Determination of mode-II parameters

The mode-II fracture parameters can be found by matching numerical and experimental results for the loads, displacements and the extent of crack propagation in three-point bending tests of adhesively bonded, end-notched flexure (ENF) specimens (see the inset of Fig. 19). These were determined to be $\Gamma_{\mathrm{IIo}} = 5.4$ kJ m^{-2} and $\hat{\tau} = 35$ MPa [60]. [9] Fig. 19 shows the resultant comparisons between the numerical and experimental load–displacement curves for ENF specimens fabricated from various thicknesses of aluminum adherends and a 0.25-mm-thick

---

[9] In this case, an independent check on the magnitude of $\hat{\tau}$ was found by determining the stress–strain curve of the adhesive layer when constrained between two aluminum plates in a napkin-ring test [60].

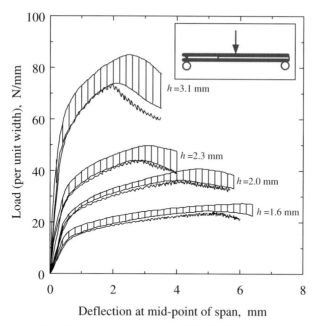

Fig. 19. A comparison between the experimental load–displacement curves for ENF specimens, and the numerical predictions using values of $\Gamma_{IIo} = 5.4$ kJ m^{-2} and $\hat{\tau} = 35$ MPa. The results for different values of adherend thickness, h, are plotted on this figure [60].

layer of XD4600 adhesive. The deformed shapes of these samples during and after fracture were also modeled very accurately using these parameters [60].

3.3.3. Mixed-mode fracture

Once the appropriate mode-I and mode-II fracture parameters for the system have been established, they can be combined using the mixed-mode failure criterion of Eq. 1.4 and the definitions of the energy-release rates given in Eqs. 1.3. By incorporating cohesive-zone elements developed using these equations into finite-element calculations of mixed-mode geometries, the fracture of a variety of different adhesive joints can be predicted quantitatively. Results for two specific geometries are illustrated here: an asymmetric T-peel joint and a single lap-shear joint. The predicted load–displacement curve of an asymmetrical T-peel joint having a thickness combination of 1.3/2.0 mm is shown in Fig. 20 by the dashed lines. It can be seen that the numerical prediction does an excellent job of reproducing the entire deformation history of the asymmetrical T-peel joint, including the onset of instability (where the numerical calculations ceased). The response of specimens with different thickness combinations were equally well predicted by the model [57]. Moreover, comparisons of the numerically predicted

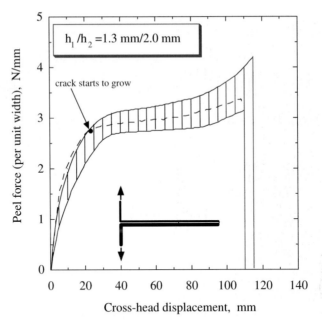

Fig. 20. A comparison between the experimental load–displacement curves for an asymmetrical T-peel specimen with one adherend which is 1.3 mm thick and the other 2.0 mm thick. The shaded areas in this figure indicate the range of experimental data from five specimens [57].

and experimentally observed deformations clearly demonstrate the excellent capability of this model to capture the detailed deformation of the sample (Fig. 21). All the features of the deformation, including the large strains and rotations, the extent of fracture, and the asymmetry of bending, are quantitatively captured. The predicted load–displacement data of a single lap-shear joint with an adherend thickness of 2.3 mm is plotted in Fig. 22, and compared to the experimental data. The deformation of the specimen is predicted accurately by the analysis (Fig. 23). Furthermore, the simulations do an excellent job of predicting the behavior of lap-shear joints of different geometries [18,57].

4. Summary

This paper has provided an in-depth discussion on the fracture of adhesive joints with a focus on how to deduce the 'intrinsic' fracture parameters of the adhesive layer from experimental results. The approach of using these parameters in cohesive-zone models to predict the fracture of adhesive joints has been outlined. By using this approach, many important issues that are frequently encountered in the fracture testing of adhesive joints, yet are usually ignored by traditional fracture analyses, have been fully addressed and the results are presented in the

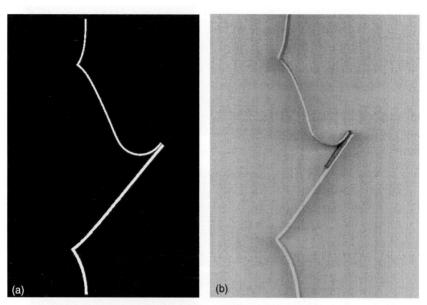

Fig. 21. A comparison of the numerically predicted (a) and experimentally observed (b) deformation of a 1.3/2.0 mm asymmetric T-peel specimen [57].

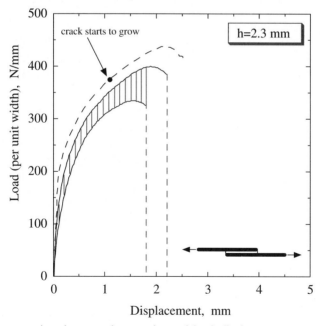

Fig. 22. A comparison between the experimental load–displacement curves for an a single lap-shear specimen with 2.3-mm-thick adherends. The shaded areas in this figure indicate the range of experimental data from five specimens [57].

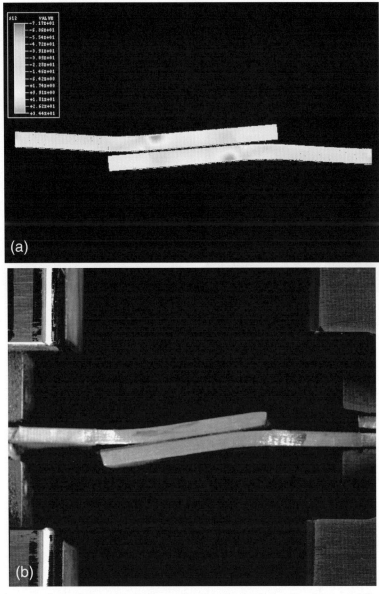

Fig. 23. A comparison of the numerically predicted and experimentally observed deformation of a 2.3-mm-thick single lap-shear joint [57].

context of this paper. These issues include shear loading of the adherends, the influence of adhesive compliance (large deformation in adhesive layers before failure), and the effects of gross plasticity in the adherends. A brief review of the key points is given below.

It is of crucial importance to recognize that there are generally two equally important normalized parameters that characterize the fracture of an adhesive joint. They are the joint toughness, $E\Gamma_{\mathrm{Io}}/\sigma_y^2 h$, and the joint strength, $E\Gamma_{\mathrm{Io}}/\hat{\sigma}^2 h$. Note that this is in contrast to the LEFM concepts that recognize only the role of joint toughness. The normalized joint toughness determines whether gross plasticity will occur in the bonded adherends, and the normalized joint strength characterizes the influence of adhesive compliance on the global deformation of the joint. *Equations derived from LEFM (Eqs. 2.3–2.5) are valid only when both the parameters are sufficiently small.*

A sufficiently small value of $E\Gamma_{\mathrm{Io}}/\sigma_y^2 h$ indicates that the adherends exhibit only elastic deformation during the fracture. However, this does not mean that LEFM results can be used without any modifications. Instead, one needs to examine further the effects of loading the adherends in shear, and the influence of large deformations in the adhesive layer. While numerical approaches are required for a rigorous analysis of these effects, approximate approaches based on the concept of replacing the actual crack length, a_o, with an effective crack length, $a_{\mathrm{effective}}$, are possible (Eqs. 2.12–2.15).

If $E\Gamma_{\mathrm{Io}}/\sigma_y^2 h$ exceeds a critical value that depends on the joint geometry (e.g., Eq. 3.1), gross plasticity will occur in the adherends before the crack starts to grow. Under this condition, local plasticity in the adherends is coupled to the fracture process in the adhesive. Local plasticity in the adherends is controlled by two stresses: the bending stress from the external loading, and the cohesive stress from the fracture process. *It is emphasized here that the traditional beam-bending type of analyses are inappropriate* because they neglect the effects of the cohesive stress, which is very significant even when it is relatively small (Fig. 14) because of the stress-sensitive nature of the post-yield behavior of the adherend materials. Moreover, the effects of shear loading of the adherends and adhesive compliance are all coupled to the fracture event and it is impossible to study them individually owing to the strongly non-linear nature of the problem. Therefore, rigorous full-scale numerical simulations are required.

The general-purpose mode-dependent cohesive zone modeling approach outlined in this paper (Eqs. 1.2–1.4) provides a powerful tool for analyzing this type of problem. The four fracture parameters required by the model can be obtained from two simple fracture tests: a symmetrical wedge test for Γ_{Io} and $\hat{\sigma}$, and an edge-notched flexure test for Γ_{IIo} and $\hat{\tau}$. While these parameters are sensitive to large changes in the constraint exerted on the adhesive layer by the adherends, they can be used to make accurate predictions about the strength of other adhesive joints made of the same material system.

Moreover, these mode-dependent cohesive-zone calculations reveal some very important mechanics associated with plastically deformed adhesive joints. *The most important is to appreciate that two competing modes, shear-dominated and bending-dominated, contribute to fracture.* Shear fracture is dominant when the normalized joint thickness, Ah/Γ, is relatively small and bending fracture dominates at large values. Competition between the two modes results in the wedge-tip size dependence of the measured radius of curvature (Fig. 18) in a wedge test, and in the thickness dependence of the measured peel force in a symmetrical 90°-peel test (Fig. 16). Note that Fig. 16 also shows that the transition zone from shear dominant to bending dominant fracture is fairly large and covers the thickness range of engineering application interests ($\sim$ mm). This emphasizes again the necessity of using full-scale CZM numerical calculations. These effects also occur in the fracture of mode-II and mixed-mode joints but they are less obvious because of the complicated stress state. However, the excellent agreements between the numerically predicted and experimentally observed deformations (Figs. 21 and 23) and load–displacement curves (Figs. 19, 20 and 22) demonstrate the point that all the important mechanisms are well represented in the CZM numerical calculations.

Acknowledgements

The authors would like to thank Dr. S.M. Ward, Dr. J. Hill, Mr. M.S. Kafkalidis and Mr. M.N. Cavalli for their assistance in this work. This work was partially supported by NSF Grant CMS 9624452 and the Ford Motor Company.

References

1. ASTM, Standard test method for fracture strength in cleavage of adhesives in bonded joints. Standard D 3433–3493, 1999.
2. Barenblatt, G.I., The mathematical theory of equilibrium cracks in brittle fracture. *Adv. Appl. Mech.*, **12**, 55–129 (1962).
3. Bennett, S.J., Devries, K.L. and Williams, M.L., Adhesive fracture mechanics. *Int. J. Fract.*, **10**, 33–43 (1974).
4. Cavalli, M.S. and Thouless, M.D., The effect of damage evolution on adhesive joint toughness. *J. Adhes.*, **76**, 75–92 (2001).
5. Chai, H., Micromechanics of shear deformation in cracked joints. *Int. J. Fract.*, **58**, 223–239 (1992).
6. Du, J.D., Thouless, M.D. and Yee, A.F., Development of a process zone in rubber-modified polymers. *Int. J. Fract.*, **92**, 271–286 (1998).
7. Du, J.D., Thouless, M.D. and Yee, A.F., Rate effects on the fracture of rubber-modified epoxies. *Acta Mater.*, **48**, 3581–3592 (2000).
8. Dugdale, D.S., Yielding of steel sheets containing slits. *J. Mech. Phys. Solids*, **8**, 100–108 (1960).
9. Evans, A.G., Rühle, M., Dalgleish, B.J. and Charalambides, P.G., The fracture energy of bi-material interfaces. *J. Mater. Sci. Eng. A*, **126**, 53–64 (1990).
10. Evans, A.G. and Hutchinson, J.W., Effects of non-planarity on the mixed mode fracture resistance of bimaterial interfaces. *Int. J. Solids Struct.*, **20**, 455–466 (1989).

11. Gent, A.N. and Hamed, G.R., Peel mechanics for an elastic–plastic adherend. *J. Appl. Polym. Sci.*, **21**, 2817–2831 (1977).

12. Griffith, A.A., The phenomena of rupture and flow in solids. *Phil. Trans. R. Soc. Lond. A*, **221**, 163–198 (1920).

13. Hutchinson, J.W. and Suo, Z., Mixed mode cracking in layered materials. *Adv. Appl. Mech.*, **29**, 63–191 (1992).

14. Hutchinson, J.W. and Evans, A.G., Mechanics of materials: top-down approaches to fracture. *Acta Mater.*, **48**, 125–135 (2000).

15. Irwin, G.R., Fracture dynamics, *Fracturing of Metals*, American Society for Metals, Cleveland, 1948, pp. 147–166 .

16. He, M.Y., Evans, A.G. and Hutchinson, J.W., Convergent debonding of films and fibers. *Acta Mater.*, **45**, 3481–3489 (1997).

17. Kafkalidis, M.S., Thouless, M.D., Yang, Q.D. and Ward, S.M., Deformation and fracture of an adhesive layer constrained by plastically deforming adherends. *Int. J. Adhes. Sci. Technol.*, **14**, 1593–1607 (2000).

18. Kafkalidis, M.S., Ph.D. Dissertation, University of Michigan, 2001.

19. Kanninen, M.F., An augmented double cantilever beam model for studying crack propagation and arrest. *Int. J. Fract.*, **9**, 83–91 (1973).

20. Kim, K.S. and Kim, J., Elasto-plastic analysis of the peel test for thin film adhesion. *J. Eng. Mat. Technol.*, **110**, 266–273 (1988).

21. Kinloch, A.J., Welch, L.S. and Bishop, H.E., The locus of environmental crack growth in bonded aluminum-alloy joints. *J. Adhes.*, **16**, 165–177 (1984).

22. Kinloch, A.J., Lau, C.C. and Williams, G.J., The peeling of flexible laminates. *Int. J. Fract.*, **66**, 45–70 (1994).

23. Krenk, S., Energy release rate of symmetric adhesive joints. *Eng. Fract. Mech.*, **43**, 549–559 (1992).

24. Jensen, H.M., On the blister test for interface toughness measurement. *Eng. Fract. Mech.*, **40**, 475–486 (1991).

25. Jensen, H.M., Analysis of mode mixity in blister test. *Int. J. Fract.*, **94**, 79–88 (1998).

26. Jensen, H.M. and Thouless, M.D., Effects of residual stresses in the blister test. *Int. J. Solids Struct.*, **30**, 779–795 (1993).

27. Lai, Y.-H. and Dillard, D.A., Using the fracture efficiency to compare adhesion tests. *Int. J. Solids Struct.*, **34**, 509–525 (1997).

28. Lawn, B., *Fracture of Brittle Solids*, 2nd edn. Cambridge Solid State Science Series, 1993.

29. Li, S., Wang, J. and Thouless, M.D., manuscript in preparation.

30. Liechti, K.M. and Chai, Y.S., Asymmetric shielding in interfacial fracture under in-plane shear. *J. Appl. Mech.*, **59**, 295–304 (1992).

31. Liechti, K.M. and Hanson, E.C., Nonlinear effects in mixed-mode interfacial delaminations. *Int. J. Fract. Mech.*, **5**, 114–128 (1990).

32. Lim, W.W., Hatano, Y. and Mizumachi, H., Fracture toughness of adhesive joints I: Relation between strain energy release rates in three different fracture modes and adhesive strength. *J. Appl. Polym. Sci.*, **52**, 967–973 (1994).

33. Lim, W.W. and Mizumachi, H., Fracture toughness of adhesive joints II: Temperature and rate dependencies of mode-I fracture toughness and adhesive tensile strength. *J. Appl. Polym. Sci.*, **57**, 55–61 (1995).

34. Lim, W.W. and Mizumachi, H., Fracture toughness of adhesive joints III: Temperature and rate dependencies of mode-II fracture toughness and adhesive shear strength. *J. Appl. Polym. Sci.*, **63**, 835–841 (1997).

35. Mohammed, I. and Liechti, K.M., Cohesive zone modeling of crack nucleation at bimaterial corners. *J. Mech. Phys. Solids*, **48**, 735–764 (2000).

36. Mostovoy, S., Crosley, P.B. and Ripling, E.J., Use of crack-line loaded specimens for measuring plane-strain fracture toughness. *J. Mater.*, **2**, 661 (1967).

37. Needleman, A., A continuum model for void nucleation by inclusion debonding. *J. Appl. Mech.*, **54**, 525–531 (1987).

38. Needleman, A., Numerical modeling of crack growth under dynamic loading conditions. *Comput. Mech.*, **19**, 463–469 (1997).

39. Papini, M., Fernlund, G. and Spelt, J.K., The effect of geometry on the fracture of adhesive joints. *Int. J. Adhes. Adhes.*, **14**, 5–13 (1994).

40. Rice, J.R. and Thomson, R.M., Ductile versus brittle behavior of crystals. *Philos. Mag.*, **29**, 73–97 (1974).

41. Ripling, E.J., Mostovoy, S. and Corten, H.T., Fracture mechanics: a tool for evaluating structural adhesives. *J. Adhes.*, **3**, 107–123 (1971).

42. Suo, Z. and Hutchinson, J.W., Interface crack between two elastic layers. *Int. J. Fract.*, **43**, 1–18 (1990).

43. Suo, Z., Bao, G., Fan, B. and Wang, T.C., Orthotropy rescaling and implications for fracture in composites. *Int. J. Solids Struct.*, **28**, 235–248 (1991).

44. Thouless, M.D., Mixed-mode fracture of a lubricated interface. *Acta Metall. Mater.*, **40**, 1281–1286 (1992).

45. Thouless, M.D., Hutchinson, J.W. and Liniger, E.G., Plane-strain, buckling-driven delamination of thin films: model experiments and mode-II fracture. *Acta Metall. Mater.*, **40**, 2639–2649 (1992).

46. Thouless, M.D., Jensen, H.M. and Liniger, E.G., Delamination from edge flaws. *Proc. R. Soc. Lond. A*, **447**, 271–279 (1994).

47. Thouless, M.D., Adams, J.L., Kafkalidis, M.S., Ward, S.M., Dickie, R.A. and Westerbeek, G.L., Determining the toughness of plastically deformation joints. *J. Mater. Sci.*, **33**, 189–197 (1998).

48. Tvergaard, V. and Hutchinson, J.W., The relation between crack growth resistance and fracture process parameters in elastic–plastic solids. *J. Mech. Phys. Solids*, **40**, 1377–1397 (1992).

49. Tvergaard, V. and Hutchinson, J.W., The influence of plasticity on the mixed mode interface toughness. *J. Mech. Phys. Solids*, **41**, 1119–1135 (1993).

50. Tvergaard, V. and Hutchinson, J.W., Toughness of an interface along a thin ductile joining elastic solids. *Philos. Mag. A*, **70**, 641–656 (1994).

51. Tvergaard, V. and Hutchinson, J.W., On the toughness of ductile adhesive joints. *J. Mech. Phys. Solids*, **44**, 789–800 (1996).

52. Wang, S.S. and Suo, Z., Experimental determination of interfacial toughness using Brazilnut-sandwich. *Acta Mater.*, **38**, 1279–1290 (1990).

53. Wei, Y. and Hutchinson, J.W., Interface strength, work of adhesion and plasticity in the peel test. *Int. J. Fract.*, **93**, 315–333 (1998).

54. Wiederhorn, S.M., Shorb, A.M. and Moses, R.L., Critical analysis of the theory of the double cantilever method of measuring fracture–surface energies. *J. Appl. Phys.*, **39**, 1569–1572 (1968).

55. Williams, G.J., Fracture in adhesive joints — the beam on elastic foundation model. In: *Proceedings of ASME International Mechanical Congress and Expositions*, 1995, pp. 1112–1117.

56. Yang, Q.D. and Thouless, M.D., Reply to comments on determining the toughness of plastically-deforming joints. *J. Mater. Sci. Lett.*, **18**, 2051–2053 (1999).

57. Yang, Q.D. and Thouless, M.D., Mixed-mode fracture analyses of plastically-deforming adhesive joints. *Int. J. Fract.*, **110**, 175–187 (2001).
58. Yang, Q.D., Thouless, M.D. and Ward, S.M., Numerical simulations of adhesively-bonded beams failing with extensive plastic deformation. *J. Mech. Phys. Solids*, **47**, 1337–1353 (1999).
59. Yang, Q.D., Thouless, M.D. and Ward, S.M., Analysis of the symmetrical 90°-peel test with extensive plastic deformation. *J. Adhes.*, **72**, 115–132 (2000).
60. Yang, Q.D., Thouless, M.D. and Ward, S.M., Elastic–plastic mode-II fracture of adhesive joints. *Int. J. Solids Struct.*, **38**, 3251–3262 (2001).

Chapter 8

The mechanics of peel tests

A.J. KINLOCH * and J.G. WILLIAMS

Department of Mechanical Engineering, Imperial College of Science, Technology and Medicine, Exhibition Rd., London SW7 2BX, UK

1. Introduction

The analysis of peeling of laminates and adhesive joints has a long history because peeling is important in many industrial processes and products. The evaluation of adhesive performance via the use of peel tests is a natural extension of this situation and the analysis of such tests is highly developed. A major attraction of peel tests is their apparent simplicity both practically and in their analysis. This is generally borne out in practice but, in both testing and data interpretation, attention to detail is important if reliable and useful results are to be obtained.

Some workers [1–10] have analysed the peel test by considering the stress distribution around the peel, or crack, front. This approach has met with little success, mainly due to the very complex stress distributions around the peel front, especially when the crack is located at the bimaterial interface, and the difficulty of defining a failure criterion. Others have used a fracture-mechanics method and adopted a stress-intensity factor approach which is based upon a stress-singularity argument [11,12]. This method has not proved to be very rewarding, mainly due to there being little physical basis for this approach in adhesive joints. For example, this approach predicts [13] oscillating stresses at the crack tip located at a bimaterial interface, which leads to the prediction that the crack face displacements also oscillate and, very near the crack tip, interfere. This is clearly a physically impossible solution. Mathematical ways around this problem have been proposed [14–17], but these typically lead to further complications. Also, in the case of adhesive joints where thin, constrained layers are often involved, the extent of the singular-field region is very localised and is often far smaller in extent than any plastic/viscoelastic process zone at the crack tip [18,19]. This is especially the case when the joint is subjected to bending loads [18]. Thus, the fundamental

---
* Corresponding author. E-mail: a.kinloch@ic.ac.uk

requirement of a singular-dominated zone is not met, and this invalidates any approach based on the assumption of a near-tip singularity field.

It is therefore not surprising that most workers [20–38] have adopted an approach to analysing the mechanics of the peel test which is *not* based upon considering either the stress distribution around the peel front or the determination of the stress-intensity factors. The approach they have adopted is one based upon applying a fracture-mechanics method using an energy-balance approach. A value of the adhesive fracture energy, G_c, is ascertained, which is the energy needed to propagate a crack through unit area of the joint, either cohesively through the adhesive layer or along the bimaterial interface [39]. The value of G_c should be characteristic of the joint and, ideally, independent of geometric parameters such as the applied peel angle, the thickness of the flexible substrate arm(s) being peeled and the thickness of the adhesive layer. However, it is recognised that, since the value of G_c includes plastic and viscoelastic energy dissipation which occurs locally at the crack tip, it will be a function of the rate and temperature at which the peel test is conducted. (Only if such energy losses are reduced to virtually zero and the locus of joint failure is exactly along the bimaterial interface, will the value of G_c be equivalent to the thermodynamic work of adhesion [22,23,39].)

Thus, the present chapter will concentrate upon the analysing the mechanics of the peel test by applying a continuum fracture-mechanics method using an energy-balance approach. The theoretical methods, via both analytical and numerical techniques, which have been reported to ascertain values of the adhesive fracture energy, G_c, will first be discussed. As will be seen, the current challenge is to model accurately any extensive plastic deformation which may occur in the flexible peeling arm, since if this is not accurately modelled then the value of G_c deduced may suffer a high degree of error. The latest developments in both the analytical and numerical approaches will be reviewed and the use of 'cohesive zone models' will be covered. The use of these approaches to interpret experimental data from peel tests will also be considered.

2. Analytical approaches

2.1. Introduction

The peel test in its most simple form is shown in Fig. 1 and is a thin peeling strip which is infinitely stiff in the axial direction but is completely flexible in bending. The strip is peeled away from a rigid substrate at an angle θ by a force P. When a critical value of P is reached the strip debonds from the surface and, if we assume that this occurs over a length da, then the force moves through a distance $da(1 - \cos\theta)$ as shown in Fig. 1. This performs work of $P\,da\,(1 - \cos\theta)$ and creates a new surface area of $b\,da$, where b is the width of the strip. Hence, the

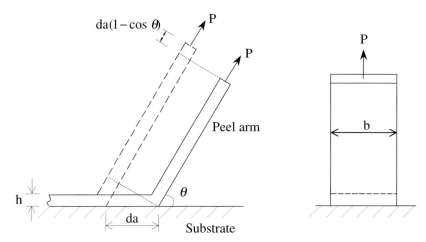

da(1−cos θ)

P

P

Peel arm

θ

h

da

Substrate

P

b

Fig. 1. Schematic of the basic peel test.

energy, $G_c^{\infty E}$, per unit area to create the new area is:

$$G_c^{\infty E} = \frac{P\,da\,(1 - \cos\theta)}{b\,da} = \frac{P}{b}(1 - \cos\theta) \tag{1}$$

This is the basis of all the analyses used in peeling and has an attractive simplicity. Note that the thickness of the strip, h, does not appear in the result, nor do any material properties. This is because we have assumed that the strip is perfectly flexible in bending and also is inextensible in tension (i.e. the peel arm behaves as a piece of 'string' which is infinitely rigid in axial tension; hence the superscript '∞E'), so that it simply transfers the external work to the surface in an non-prescribed way.

The test is an example of a steady-state process in that the debonding region remains the same as it grows. One would expect that, for a fixed value of $G_c^{\infty E}$, the load would remain constant and peeling would proceed at whatever speed was prescribed by the load point. If the rate of movement of the load point proceeds at a velocity V, then the rate of peeling is at $V(1 - \cos\theta)^{-1}$. Such a situation is an example of stable propagation and is very convenient for studying the adhesion of flexible substrates.

The assumptions outlined above are not always present and to understand them and cope with the consequences it is useful to set the test in the context of a more general fracture mechanics approach. An energy-based analysis is usually written in terms of the energy release rate, G, for a system where the adhesive fracture energy, G_c, is given by:

$$G_c = \left(\frac{dU_{\text{ext}} - dU_s - dU_d - dU_k}{dA} \right) \tag{2}$$

and dA is the increment of area created and is $b\,da$ for the peel test, and dU_{ext} is the increment of external work performed and is $P\,da\,(1-\cos\theta)$ in the peel test. The term dU_s is the change of stored strain energy, dU_d is the increment of the dissipated energy other than that in creating the surface and dU_k is the increment of kinetic energy change. All these last three terms are assumed to be zero in the simple analysis, so giving Eq. 1.

2.2. Tensile deformation of the arm

In the above, the terms dU_s and dU_d are both taken as zero since the strip is assumed to be infinitely flexible and inextensible, so the arm neither stores nor dissipates energy. Of course, this is never completely true, but is a reasonable assumption for many test specimens, and the errors incurred are small. However, such errors may arise in two ways.

The first is from the stretching of the peeling arm, or strip, by the force, P. The tensile stress, σ, acting on the peel arm is given by:

$$\sigma = \frac{P}{bh} \tag{3}$$

and clearly this is an important term to consider, since if h is small then the stress, σ, can exceed some limiting value and give failure of the arm before the peeling load is reached.

Secondly, there is also an effect on the change in the external work, dU_{ext}. Since the stress, σ, causes a strain, ε, which increases the distance through which the force, P, moves to $(1+\varepsilon-\cos\theta)\,da$. Thus, we have:

$$\frac{dU_{ext}}{b\,da} = \frac{P}{b}(1+\varepsilon-\cos\theta) \tag{4}$$

and there will also be an accompanying change in the sum of dU_s and dU_d,

$$\frac{d}{b\,da}(U_s+U_d) = h\int_0^\varepsilon \sigma\,d\varepsilon \tag{5}$$

Thus, we have for the value of G, but where elastic bending is still assumed to occur:

$$G_c^{eb} = \frac{P}{b}(1-\cos\theta)+h\left(\sigma\varepsilon-\int_0^\varepsilon \sigma\,d\varepsilon\right) \tag{6}$$

and Fig. 2 shows the stress versus strain curve and the two terms involving σ and ε. The term $\sigma\varepsilon$ is the work per unit volume done in the increment of growth and $\int_0^\varepsilon \sigma\,d\varepsilon$ is that taken up by the strip. The difference is shown shaded in Fig. 2 and is the addition to G_c^{eb}. (Here it is not necessary to distinguish between U_s and U_d, since no unloading occurs. However, it should be noted that a similar

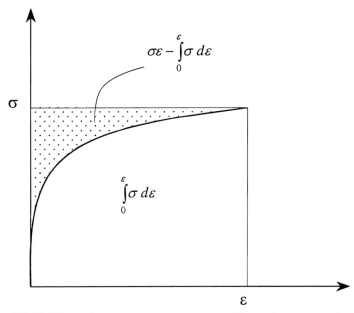

Fig. 2. Schematic stress versus strain curve of the peel-arm material.

scheme is used later in bending analysis where such a distinction is necessary.) This relationship may be rewritten as:

$$G_c^{eb} = \frac{P}{b}\left(1 - \cos\theta + \frac{1}{\sigma}\int_0^\sigma \varepsilon\, d\sigma\right) \tag{7}$$

For a stress versus strain curve of the form $\sigma \alpha \varepsilon^n$ we have:

$$\frac{1}{\sigma}\int_0^\sigma \varepsilon\, d\sigma = \left(\frac{n}{1+n}\right)\varepsilon \tag{8}$$

and for the linear case the right-hand side is equal to $\varepsilon/2$. Such corrections are included in the latest test protocols for the peel test [40].

2.3. Kinetic energy effects

It will be noted that kinetic energy term, U_k, has been ignored here and is determined entirely by the test speed V. Generally the speeds are slow so there is little change in kinetic energy but such effects may be easily included since the velocity components, å, of the peeled arm are:

$$\text{å}(1 - \cos\theta) \quad \text{and} \quad \text{å}\sin\theta \tag{9}$$

and:

$$\frac{dU_k}{b\,da} = \frac{1}{2}\rho h \mathring{a}^2 \left\{(1-\cos\theta)^2 + \sin^2\theta\right\} = \rho h\mathring{a}^2(1-\cos\theta) \tag{10}$$

where ρ is the arm density and $\mathring{a}$ is the peeling speed. Thus, using Eq. 2, and ignoring any strain, ε, in the peel arm and assuming only elastic bending, we have:

$$G_c^{\infty E} = \frac{P}{b}(1-\cos\theta) - \rho h\mathring{a}^2(1-\cos\theta) \tag{11}$$

i.e.:

$$G_c^{\infty E} = Eh(1-\cos\theta)\left[\varepsilon - \left(\frac{\mathring{a}}{C}\right)^2\right] \tag{12}$$

where $C^2 = E/\rho$; and C is the elastic wave speed and E is the tensile modulus of the peel arm. Thus, kinetic effects are likely to be important for $\mathring{a}$ values of the order of $\sqrt{\varepsilon}C$; i.e. for a polymer with $\varepsilon \sim 0.01$ and $C \sim 1500$ m/s and $\mathring{a} \sim 150$ m/s.

2.4. Other test configurations

Variations of the simple configuration considered above of an arm of material being peeled away from a rigidly supported substrate are often made for convenience or encountered in component design and the same form of analysis may be applied.

2.4.1. 'T-peel' tests

The most common related form of test piece is the 'T-peel' shown in Fig. 3 in which two thin strips are peeled apart. The forces are the same on each arm but for unequal thicknesses the bonded 'tail' finds an equilibrium angle. For any

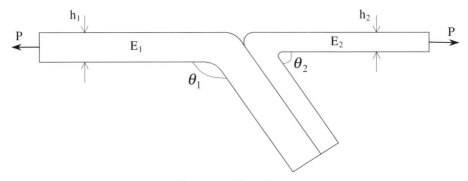

Fig. 3. The 'T-peel' test.

combination, of course, the applied force, P, moves $2\mathrm{d}a$ for a peel length of $\mathrm{d}a$ so, assuming $\varepsilon = 0$ and only elastic bending:

$$G_c^{\infty E} = 2\frac{P}{b} \tag{13}$$

When bending corrections are made, see Section 2.5, the values of the angles θ_1 and θ_2 must be known.

For linear materials, the strain correction may be included for each arm, i.e.:

$$G_c^{eb} = \left(2 + \frac{\varepsilon_1 + \varepsilon_2}{2}\right)\frac{P}{b} \tag{14}$$

where:

$$\varepsilon_1 = \frac{P}{bh_1 E_1} \quad \text{and} \quad \varepsilon_2 = \frac{P}{bh_2 E_2} \tag{15}$$

2.4.2. Debonded strip tests

The debonded strip configuration [40,41] is shown in Fig. 4 and is usually used to debond a central section. The angle θ changes here and is governed by the strain since:

$$\varepsilon = \left(\frac{1 - \cos\theta}{\cos\theta}\right) \tag{16}$$

By analogy with Fig. 1 we have:

$$P = \frac{F}{2\sin\theta} \tag{17}$$

and hence:

$$G_c^{eb} = \frac{F}{2b\sin\theta}\left(1 - \cos\theta + \frac{\varepsilon}{2}\right) \tag{18}$$

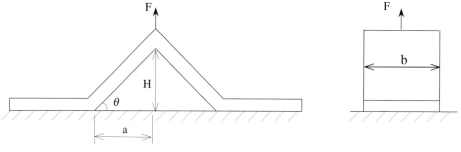

Fig. 4. The debonded strip test.

i.e.:

$$G_c^{eb} = \frac{F}{2b\sin\theta} \frac{(1-\cos\theta)(1+2\cos\theta)}{2\cos\theta} \tag{19}$$

which for small values of θ becomes [41,42]:

$$G_c^{eb} = \frac{3F\theta}{8b} \tag{20}$$

Such solutions are often written in the form of the load point displacement, H:

$$H = a\tan\theta \tag{21}$$

and we now have:

$$G_c^{eb} = \frac{FH}{2ba} \frac{\cos\theta}{\sin^2\theta} \frac{(1-\cos\theta)(1+2\cos\theta)}{2\cos\theta} = \frac{F}{2b} \frac{H}{a} \left[\frac{\sqrt{1+\left(\dfrac{H}{a}\right)^2}+\dfrac{1}{2}}{\sqrt{1+\left(\dfrac{H}{a}\right)^2}+1} \right] \tag{22}$$

2.4.3. Two-dimensional blister tests

A similar method is used if the load is applied using pressure as shown in Fig. 5. For this two-dimensional blister test the strip forms an arc of a circle of radius R giving:

$$H = R(1-\cos\theta) \quad \text{and} \quad a = R\sin\theta \tag{23}$$

The peeling force at each end is:

$$P = \frac{pba}{\sin\theta} = pRb \tag{24}$$

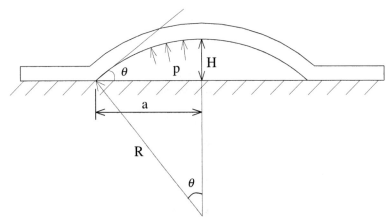

Fig. 5. The one-dimensional blister test.

and using the peeling analogue again:

$$G_c^{eb} = \frac{P}{b}\left(1 - \cos\theta + \frac{\varepsilon}{2}\right) = pR\left(1 - \cos\theta + \frac{\varepsilon}{2}\right) \tag{25}$$

The strain is given by:

$$\varepsilon = \frac{\theta - \sin\theta}{\sin\theta} \tag{26}$$

And hence:

$$G_c^{eb} = \frac{pH}{(1 - \cos\theta)}\left[(1 - \cos\theta) + \frac{1}{2}\left(\frac{\theta - \sin\theta}{\sin\theta}\right)\right] \tag{27}$$

For small values of θ, we have:

$$G_c^{eb} = pH\frac{7}{6} \tag{28}$$

2.4.4. Axisymmetric blister tests

The axisymmetric blister test is rather more complex [40,43], but the peeling analysis plus the assumption of a circular profile gives a reasonably accurate result. The peeling force per unit length along the circumference is given by:

$$\frac{P}{b} = \frac{\pi a^2 p}{2\pi a \sin\theta} = \frac{pa}{2\sin\theta} = \frac{pR}{2} \tag{29}$$

The strain is now biaxial and can be found from the cap area of the sphere:

$$2\varepsilon = \frac{2\pi R^2(1 - \cos\theta)}{\pi R^2 \sin 2\theta} - 1 \tag{30}$$

i.e.:

$$\varepsilon = \frac{(1 - \cos\theta)}{2(1 + \cos\theta)} \tag{31}$$

and:

$$G_c^{eb} = \frac{pR}{2}(1 - \cos\theta + \varepsilon) \tag{32}$$

in the biaxial case. For small values of θ this becomes:

$$G_c^{eb} = pH\frac{5}{8} \tag{33}$$

An exact solution [40] gives the constant varying with Poisson's ratio and is 0.652 at $\nu = 0.3$ and 0.645 at $\nu = 0.5$, i.e. differences of only about 4%. Such solutions are useful practically since only pressure and height need be measured. Further details of other cases may be found in [40].

2.5. Plasticity corrections

Peel tests have the considerable advantages of practical simplicity and ease of analysis but the latter assumes that there is no energy dissipation other than that creating the new surface, i.e. G_c. The peeling strip, however, undergoes intense bending near the contact point and this can lead to local energy dissipation proportional to the area created and thus falsely high values of G_c. For elastic, reversible, deformations the bending has no effect, since no energy is dissipated and Eqs. 6 or 7 may be employed. However, when plastic energy dissipation accompanies the peeling process then the term dU_d (see Eq. 2) needs to be extended to take such energy losses into account.

The onset of plasticity can be determined from an elastic analysis, since the moment at which the stress at the outer surface reaches the yield stress is given by [26,44]:

$$M_1 = \frac{1}{6}bh^2\sigma_y \tag{34}$$

where σ_y is the yield stress of the peel arm. The input G may be converted to a local elastic moment and hence:

$$G_c^{\infty E} = \frac{P}{b}(1 - \cos\theta) = \frac{6M_1^2}{Eb^2h^3} = \frac{h\sigma_y^2}{6E} = \frac{Eh}{6}\varepsilon_y^2 \tag{35}$$

where ε_y is the yield strain, i.e. σ_y/E. The local radius, R_1, of curvature at this point for when plastic yielding first occurs is given by:

$$\frac{1}{R_1} = \frac{12M_1}{Ebh^3} = \frac{2\varepsilon_y}{h} \tag{36}$$

The geometry near the contact point is shown in Fig. 6 and this derivation of M_1 assumes that at this point the local angle is zero and all of the input G is transmitted via bending. If the angle at the contact point is not zero but has a value of θ_0, then this is not so and only a proportion is transmitted in this way. The effect of θ_0 may be calculated by considering the moment at any point in the strip of co-ordinates v and x as shown in Fig. 6:

$$M = P\left[(x_0 - x)\sin\theta - (v_0 - v)\cos\theta\right] \tag{37}$$

where x_0 and v_0 are the co-ordinates of the load point. If ϕ is the local angle at the point with coordinates x and v, then large displacement elastic beam theory gives [30] for a beam of arc length s:

$$\sin\phi = \frac{dv}{ds}, \quad \cos\phi = \frac{dx}{ds} \quad \text{and} \quad \frac{d\phi}{ds} = \frac{1}{R} \tag{38}$$

and R is the local radius of curvature. Noting the elastic relationship:

$$\frac{1}{R} = \frac{12M}{Ebh^3} \tag{39}$$

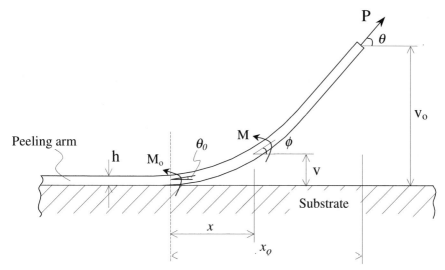

Fig. 6. Local moments in the peel test.

we have [28–30,44]:

$$\frac{1}{R^3}\frac{\mathrm{d}R}{\mathrm{d}\phi} = \frac{12P}{Ebh^3}\sin(\theta - \phi) \tag{40}$$

This equation may be integrated between the load point where $\phi = \theta$ and $R \to \infty$ and the contact point where $\phi = \theta_0$ and $R = R_0$, where R_0 is the local radius of curvature at the peel contact front itself, i.e.:

$$\frac{1}{2R_0^2} = \frac{12P}{Ebh^3}[1 - \cos(\theta - \theta_0)] = \frac{1}{2}\left(\frac{12M_0}{Ebh^3}\right)^2 \tag{41}$$

i.e.:

$$M_0^2 = \frac{Ebh^3}{b}P[1 - \cos(\theta - \theta_0)] \tag{42}$$

where M_0 is the local moment at the peel front.

Note that if $\theta_0 = 0$ then Eq. 35 is retrieved and for $\theta_0 = \theta$, $M_0 = 0$ we have no induced bending. It should also be observed that in steady state peeling M_0 is applied at the base and the subsequent deformation in the strip is all unloading to the load point where $M = 0$ and is mostly elastic.

When $M_0 > M_1$ then elastic–plastic bending occurs which may be analysed using plastic bending theory, and it is convenient to describe the analysis via the non-work hardening case [28–30,44] in which any bending moment $M > M_1$ may be written as:

$$M = M_\mathrm{p}\left[1 - \frac{1}{3}\left(\frac{c}{h}\right)^2\right] \tag{43}$$

where M_p is the fully plastic moment and is given by:

$$M_p = \frac{1}{4}bh^2\sigma_y = \frac{3}{2}M_1 \tag{44}$$

and $c/2$ is the distance of the elastic–plastic interface from the neutral axis and may be written in terms of the local radius of curvature, R:

$$\frac{1}{R} = \frac{2\varepsilon_y}{c} \tag{45}$$

and hence:

$$\frac{h}{c} = \frac{R_1}{R} = k \tag{46}$$

and:

$$m = \frac{M}{M_p} = 1 - \frac{1}{3k^2} \tag{47}$$

The deformation in the peeling process is rather complex and is illustrated in Fig. 7 via the relationship between m and k. The line OAC is the loading which occurs on debonding and for $m < 2/3$ and $k < 1$, the relationship is elastic, and:

$$m = \frac{2}{3}k \tag{48}$$

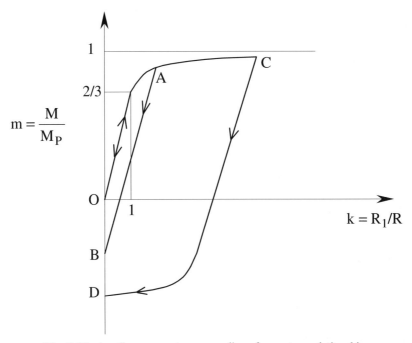

Fig. 7. The bending moment versus radius of curvature relationship.

Thus the unloading is on the same line and there is no energy dissipation from plastic bending of the arm. For $2/3 < m < 1$, the loading line is given by Eq. 47; but unloading is elastic for $1 < k < 2$ and the energy dissipation is given by the area OAB. For $k > 2$, some plastic unloading occurs and the dissipation is given by the area OACD. Note that some of this 'lost' energy is actually elastic because the strip does not unload in the steady state. These areas are the $dU_d/b\,da$ term with respect to bending energy losses in Eq. 2 and may be written as [28,29,44]:

Case (i) for $1 < k_0 < 2$:

$$\left(\frac{dU_d}{b\,da}\right)_{(i)} = \hat{G}\left(\frac{k_0^2}{3} + \frac{2}{3k_0} - 1\right) \tag{49}$$

Case (ii) for $k_0 > 2$:

$$\left(\frac{dU_d}{b\,da}\right)_{(ii)} = \hat{G}\left(2k_0 + \frac{10}{3k_0} - 5\right) \tag{50}$$

where $\hat{G}$ is the maximum elastic energy (per unit width per unit length) which can be stored in the peeling arm and for a non-work hardening material is given by:

$$\hat{G} = \frac{1}{2}Eh\varepsilon_y^2 \quad \text{and where} \quad k_0 = \frac{R_1}{R_0} \tag{51}$$

where R_0 is the radius of curvature at the peel front.

The input energies may be derived from Eq. 41 for a non-work hardening peeling arm as:

Case (i) for $1 < k_0 < 2$:

$$\frac{k_0^2}{3} = \frac{G_c^{\infty E}}{\hat{G}}\left[\frac{1 - \cos(\theta - \theta_0)}{1 - \cos\theta}\right] \tag{52}$$

Case (ii) for $k_0 > 2$:

$$2k_0 + \frac{8}{3k_0} - 4 = \frac{G_c^{\infty E}}{\hat{G}}\left[\frac{1 - \cos(\theta - \theta_0)}{1 - \cos\theta}\right] \tag{53}$$

(In this case, Eq. 50 is modified because of plastic unloading [30].) Similar relationships can be derived to include work hardening and the forms for the bi-linear model can be found in [30] and for a power law in [45]. In general, work hardening has not been found to be a significant effect in many practical cases. The true adhesive energy, G_c, for both solutions is given by:

$$G_c = G_c^{eb} - \left(\frac{dU_d}{b\,da}\right) + \hat{G} \tag{54}$$

The nature of the solution is apparent by considering the case of $k_0 \gg 2$ and $\theta = \pi/2$; which reduces to:

$$G_c \approx G_c^{eb}\sin\theta_0 + \hat{G} \quad \text{and} \quad k_0 = \frac{1}{2}\frac{G_c^{eb}}{\hat{G}} \tag{55}$$

As $\theta_0 \to 0$, $G_c \to \hat{G}$ and for $\theta_0 \to \pi/2$, then $G_c \to G_c^{eb}$. (The addition of $\hat{G}$ arises from the large k_0 assumption which is not true as $\theta \to \theta_0$.) Thus θ_0 is crucial to finding G_c and if it can be measured directly the correction for plastic bending of the peel arm can be made by finding k_0 from Eq. 53 and hence $dU_d/b\,da$ from Eq. 50. However, often θ_0 is small and difficult to measure and, since G_c^{eb} is generally much greater than $\hat{G}$, its value is vital and must be estimated.

2.6. Root rotations

If the adhesive energy dissipation is very local then θ_0 arises mostly from the elastic deformation of the adhered part of the strip as shown in Fig. 8. The deformation, v, may be deduced from a beam on an elastic foundation model in which the stiffness k_s arises from the half height of the beam, i.e.:

$$k_s = \frac{2bE}{h} \tag{56}$$

The deformation is given by [44,46]:

$$\frac{d^4v}{dx^4} = \frac{-12k_s v}{Ebh^3} = -4\lambda_1^4 v \tag{57}$$

where:

$$\lambda_1^4 = \frac{3k_s}{Ebh^3} = \frac{6}{h^4}, \qquad \text{i.e.} \quad \frac{1}{\lambda_1 h} = \frac{1}{6^{1/4}} \tag{58}$$

For a bending moment, M_0, applied at the peel front, the slope $(dv/dx)_{x=0}$, is given by:

$$\left(\frac{dv}{dx}\right)_{x=0} = -\frac{12M_0}{Ebh^3}\frac{1}{\lambda_1} = -\theta_0 \tag{59}$$

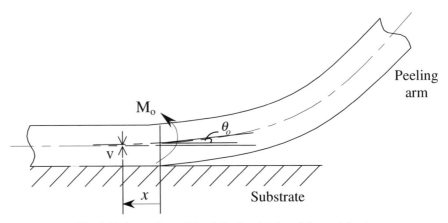

Fig. 8. Deformation and loads in the vicinity of the peel front.

and noting that:

$$\frac{1}{R_0} = \frac{12M_0}{Ebh^3} \tag{60}$$

we have:

$$\theta_0 = \frac{1}{\lambda_1 R_0} = \frac{1}{6^{1/4}} \frac{h}{R_0} = 1.28\varepsilon_y k_0 \tag{61}$$

The effective length over which the bending occurs, λ_1^{-1}, is short compared to h, so shear deformations are often important [47] and should be included. If μ is the shear modulus, then the inclusion of shear effects gives:

$$(\lambda_1 h)^{-2} = 0.24 + \frac{1}{10}\left[\frac{E}{\mu} - 2\nu\right] \tag{62}$$

It should be also noted that the stiffness, k_s, is determined by the transverse modulus, E_2, of the substrate and also by the presence of any adhesive layer. For an adhesive layer of thickness h_a, and modulus E_a, final expressions are [47]:

$$\frac{b}{k_s} = \frac{7}{40}\frac{h}{E_2} + \frac{h_a}{E_a} \tag{63}$$

and:

$$(\lambda_1 h)^{-2} = \left[\frac{0.175 + \dfrac{h_a E}{h E_a}}{\dfrac{3E_2}{E}}\right]^{1/2} + \frac{1}{10}\left[\frac{E}{\mu} - 2\nu\right] \tag{64}$$

Returning to Eq. 55, and noting that $\sin\theta_0 \approx \theta_0$ for small angles we have:

$$G_c \approx 0.64\varepsilon_y \frac{G_c^{eb2}}{\hat{G}} + \hat{G} \tag{65}$$

for the simple elastic root-rotation correction. Thus for a test in which G_c is constant and h is varied, we may note that:

$$\frac{G_c^{eb}}{G_c} = \frac{1.25}{\varepsilon_y^{1/2}}\left[\frac{\hat{G}}{G_c}\left(1 - \frac{\hat{G}}{G_c}\right)\right]^{1/2} \tag{66}$$

and:

$$\frac{\hat{G}}{G_c} = \frac{\sigma_y^2}{2E}\frac{h}{G_c} \tag{67}$$

Thus $G_c^{eb} \propto h^{1/2}$ for low values of h where $G_c^{eb} \to G_c$, which is not captured by the large k_0 solution used here. Similarly as:

$$\frac{\hat{G}}{G_c} \to 1, \quad \text{i.e. as:} \quad \hat{h} \to \frac{2EG_c}{\sigma_y^2} \quad \text{then:} \quad G_c^{eb} \to G_c \tag{68}$$

while the exact limit is given by Eq. 35 and is $3\hat{h}$. There is a maximum in G_c^{eb} at $\hat{G}/G_c = 1/2$; i.e. at $h = \hat{h}/2$, where:

$$\left(\frac{G_c^{eb}}{G_c}\right)_{max} = \frac{0.625}{\varepsilon_y^{1/2}} \tag{69}$$

2.7. Cohesive stresses

The model of deformation of the bonded section of the peel test results in a maximum stress, $\bar{\sigma}$, at the debond, or peel, front and for the simple elastic case this is given by:

$$\frac{k_s}{b} = \frac{\bar{\sigma}^2}{2G_c} = \frac{2E}{h} \tag{70}$$

i.e.:

$$\bar{\sigma} = 2\sqrt{\frac{EG_c}{h}} \tag{71}$$

which may be written as:

$$\frac{\bar{\sigma}}{\sigma_y} = \sqrt{\frac{2G_c}{\hat{G}}} \tag{72}$$

and since:

$$\frac{\hat{G}}{G_c} < 1, \quad \text{then:} \quad \frac{\bar{\sigma}}{\sigma_y} > \sqrt{2} \tag{73}$$

and the latter term increases for lower thickness values.

It is sensible to argue that the value of $\bar{\sigma}$ must be limited, and hence for low h values the stiffness may be adjusted to keep $\bar{\sigma}$ constant, and hence:

$$\lambda_1^4 = \frac{3k_s}{Ebh^3} = \frac{3}{2}\frac{\bar{\sigma}^2}{EG_ch^3} \tag{74}$$

i.e.:

$$\lambda_1^{-1} = h\left(\frac{2}{3}\frac{EG_c}{\bar{\sigma}^2h}\right)^{1/4} \tag{75}$$

and from Eq. 61 we now have:

$$\theta_0 = \frac{2\varepsilon_y k_0}{3^{1/4}} \left(\frac{G_c}{\hat{G}}\right)^{1/4} \left(\frac{\sigma_y}{\bar{\sigma}}\right)^{1/2} \tag{76}$$

and Eq. 66 becomes:

$$\frac{G_c^{eb}}{G_c} = \frac{1.15}{\varepsilon_y^{1/2}} \left(\frac{\bar{\sigma}}{\sigma_y}\right)^{1/4} \left[\left(\frac{\hat{G}}{G_c}\right)^{5/4} \left(1 - \frac{\hat{G}}{G_c}\right)\right]^{1/2} \tag{77}$$

In this case, $G_c^{eb} \propto h^{5/8}$ for small thicknesses and is a maximum at $5/9\hat{h}$ with:

$$\left(\frac{G_c^{eb}}{G_c}\right) = \frac{0.53}{\varepsilon_y^{1/2}} \left(\frac{\bar{\sigma}}{\sigma_y}\right)^{1/4} \tag{78}$$

The value of $\bar{\sigma}/\sigma_y$ is usually taken to be in the range of 1–4, so the value of the term G_c^{eb}/G_c is not greatly different from the elastic case [48].

Cohesive stress models are used in finite element simulations of the peel test, see [45,49–53] and are discussed below. The problem is quite demanding since it involves large deformations and plasticity. The results are generally similar in form to the analytical approach outlined here, and in particular the variation of $G_c^{\infty E}$ (or G_c^{eb}) with thickness. There is little experience with this approach to date and there is a difficulty in that the value of the term $\bar{\sigma}$ must be known. This may be important additional information but the analytical solution suggests that the solution is insensitive to its value, i.e. a dependence on $\bar{\sigma}$ of a quarter power over a limited range of values. The use of such an approach is potentially useful but is not yet well developed.

3. Experimental results

3.1. Introduction

Obviously, key questions which now arise are how good are these various analytical and finite element analysis methods at yielding a value of the adhesive fracture energy, G_c, (a) which is independent of the details of the peel test geometry, for example, independent of the peel angle and thickness of the peel arm; and (b) which agree with results from other test methods, for example, with values of G_c from standard linear-elastic fracture-mechanics (LEFM) tests.

3.2. Effect of peel angle

Kinloch et al. [30] studied the failure of peel tests of polyethylene/aluminium-foil laminates where the aluminium foil was bonded down to a rigid substrate and the

Table 1

Results for polyethylene/aluminium foil laminates for various peel angles

Peel angle (°)	$G_c^{\infty E}$ (J/m^2)	θ_0(theory) (°)	θ_0(expt.) (°)	G_c (J/m^2)
45	183	20.4	24–30	236
90	333	34.5	40–47	228
120	373	41.7	48–58	218
135	412	46.1	50–60	223
150	467	51.7	55–62	236

The thickness of the PE peel arm was 35 μm.

polyethylene film was peeled away at various angles. The results are shown in Table 1.

The values of $G_c^{\infty E}$ were determined using Eq. 1, and are similar in value to those of G_c^{eb} from Eq. 7, since the strain, ε, in the peel arm during the test was relatively low. Note that the values of $G_c^{\infty E}$ are highly dependent upon the value of the peel angle, θ, employed. The values of G_c were ascertained by allowing for the plastic energy dissipation that occurs in the peel test from using Eqs. 50 and 2, but modelling the peel arm as a material which work hardens according to a bi-linear elastic–plastic stress versus strain relationship [30]. Clearly, when the plastic deformation is taken into account, the values of G_c obtained are not significantly dependent upon the peel angle used and, as commented above, the differences in the values of $G_c^{\infty E}$ and G_c are largely due to the energy dissipated in plastic bending of the peel arm during the test. The values of the *local* peel angle, θ_0, determined from both the analytical theory and by direct experimental measurements are also quoted in Table 1. As may be seen, there is very good agreement between the values of θ_0 from the different methods and this acts as a direct check on the soundness of the analytical approach outlined above. Thus, the analytical approach does indeed yield a characteristic value of the adhesive fracture energy, G_c, which is independent of the peel angle.

Similar encouraging results have been reported by Moore and Williams [55] who led a 'round-robin' series of tests using, for example, a five-layer structure which formed a polypropylene/tie-layer/poly(ethylene vinyl alcohol)/tie-layer/polypropylene laminate approximately 120 μm thick, as used widely in packaging applications. They also reported that a characteristic value of the adhesive fracture energy, G_c, independent of the peel angle, θ, was obtained.

The results for the values of G_c discussed above were all derived using the analytical approach [30] described in previous sections. However, the peel test data shown in Table 1 have also been analysed [51,53] employing a finite element analysis embodying a cohesive zone model (CZM). In the CZM model, a maximum stress, $\bar{\sigma}$, is defined and the area under the stress versus displacement

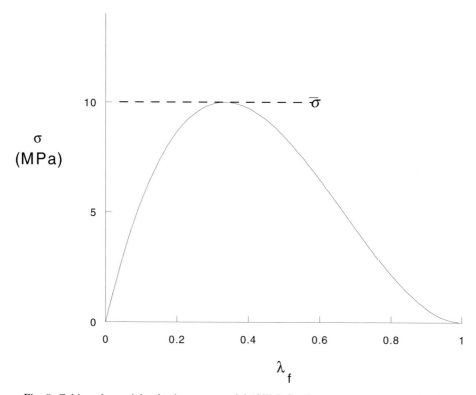

Fig. 9. Cubic polynomial cohesive zone model (CZM) for the stress versus separation law.

curve, which models the crack tip region, is taken to be equivalent to the adhesive fracture energy, G_c, see Section 2.7. In the present work, the stress versus separation law used in the CZM has been defined by a cubic polynomial, see Fig. 9 [54]. In this model, u_n and u_t are the normal and tangential components of the relative displacement of the crack faces, and the terms δ_n^c and δ_t^c are the critical values of these displacements. A single parameter, λ_f, based on these displacements may be defined by:

$$\lambda_f = \sqrt{\left(\frac{u_n}{\delta_n^c}\right)^2 + \left(\frac{u_t}{\delta_t^c}\right)^2} \tag{79}$$

such that the stress drops to zero at $\lambda_f = 1$. (In earlier work [19,56] it has been shown that by adopting the above approach the work of separation is equivalent to G_c, regardless of the combination of normal and tangential displacement taking place in the separation zone. Also, that there are no significant effects due to any mode-mix variations in the peel test, as discussed in more detail below.) Using this approach, it was shown that an elastic–plastic finite element analysis, employing a value of $\bar{\sigma} = \sigma_y$ (= 10 MPa) for the CZM, gave a value of G_c for the 90° peel

Table 2

Results for polyethylene/aluminium foil laminates with various thicknesses, h, of the polyethylene peel arm

Thickness, h (μm)	$G_c^{\infty E}$ (J/m^2)	θ_0(theory) (°)	θ_0(expt.) (°)	G_c (J/m^2)
30	195	59.5	54–66	69.8
45	205	50.0	41–49	62.3
60	240	46.0	38–46	69.3
75	260	43.4	38–45	71.5
105	260	36.1	24–32	67.3
135	225	29.1	22–28	59.5
165	240	27.1	22–28	65.4
215	220	21.9	17–21	68.2

The peel angle, θ, was 180°.

test of 230 ± 40 J/m^2. Thus, there is excellent agreement between the analytical (see Table 1) and the FEA CZM approaches to analysing the peel test data given in [30], and thereby deriving a characteristic value of G_c.

3.3. Effect of thickness of the peel arm

Firstly, the thickness, h, of the peel arm may influence the measured peel force, as shown [30] by the variation of the values of $G_c^{\infty E}$ as a function of h in Table 2. However, when the above analytical theory is employed, the resulting value of G_c is independent of the value of h, and furthermore there is good agreement between the values of the measured and theoretically calculated local peel angle, θ_0, at the crack front.

   Secondly, results for a rubber-toughened epoxy-adhesive bonding aluminium-alloy substrates, where a 'T-peel' test has been employed, have been reported [45]. The thicknesses of the aluminium-alloy arms was either 1 or 2 mm. When the reported peel loads, together with the work-hardening characteristics of the stress versus strain curve for the aluminium-alloy, are analysed [51] using the above analytical approach [30], which of course allows for the effect of peel arm thickness, then it is found that a very significant plasticity correction term ($dU_d/b\,da$) has to be applied. If such a correction is applied, then the values of G_c are in the range of 2900 ± 400 J/m^2 for the 1 mm and 2800 ± 300 J/m^2 for the 2 mm thick test specimens. Thus, again there is no effect of the thickness, h, of the arm of the peel specimen on the value of G_c.

3.4. Comparisons of test geometries

A comparison of the results of G_c evaluated from different test geometries is of particular interest, since if G_c has the same value from very different test geometries this gives confidence in its use for (a) material development and (b) component design and life-prediction studies.

Firstly, Moore and Williams [55] have reported results from 'T-peel' tests using the five-layer structure laminate discussed above. The values of G_c were obtained from 'T-peel' tests via the above analytical method, allowing for plastic bending of the peel arms. (Thus, measurement of the angles θ_1 and θ_2, see Fig. 3, was also undertaken since they are required for the plastic-bending analysis.) The results from the 'T-peel' and the fixed-arm peel test, where the peel angle was varied as discussed above, were not significantly different.

Secondly, the same rubber-toughened epoxy adhesive as was used for the 'T-peel' tests [45] discussed in Section 3.3 has also been studied [51,57] using a LEFM test specimen, i.e. the standard tapered-double cantilever-beam specimen [58]. At the same rate of test and for the same locus of joint failure, a value of G_c of 2750 ± 100 J/m^2 was determined using the LEFM test, compared with a value 2900 ± 400 J/m^2 from the 'T-peel' tests. Thus, here we have completely different test geometries giving the same value of G_c. So, again, a cross-check indicates the robustness of the above analytical approach for modelling the peel test.

Thirdly, earlier work [21] employing a crosslinked rubbery layer adhering a polymeric films has also demonstrated that the value of the adhesive fracture energy, G_c, is independent of the exact geometry of the test specimen. In this study, no tensile or plastic bending energy dissipation occurred in the peel test, and Eq. 1 was therefore used to deduce the value of G_c, since under these conditions $G_c \equiv G_c^{\infty E}$. The results from three very different adhesive test geometries (i.e. (a) a centrally cracked tensile sheet, (b) an edge-cracked rectangular sheet and (c) a peel test specimen) were found to give very similar values of G_c over a wide range of test rates and temperatures.

Thus, all the above studies reinforce the suggestion that the value of the adhesive fracture energy, G_c, characterises the toughness of the interface (or, more generally, the joint) and that the value of G_c is independent of the exact overall details of the geometry of the test specimen employed. Although, of course, the value of G_c may well be a function of the rate and temperature of test, as discussed below, and affected by the presence of any environment.

Finally, it is noteworthy that, from (a) the results shown in Table 1 (where the value of G_c is shown to be independent of peel angle), (b) the above discussions on the CZM approach in Section 3.2, and (c) the good agreement between values of G_c from different geometries referred to directly above, there appears to be no need to invoke the proposition that the ratio of mode I : mode II loading (i.e. tensile opening to in-plane shear loading) varies dramatically as the applied peel angle, θ, is changed. Such an effect might be expected to influence significantly the value of G_c. One major reason for this is undoubtedly the fact that the local angle, θ_0, at the peel front never actually experiences the extreme range of value used for the applied peel angle, θ. For example, from Table 1, it may be seen that, whilst the value of θ changes from 45° to 150°, the value of θ_0 merely changes from only about 30° to 60°. Hence, the actual range of any mode I : mode II ratio experienced by the crack tip regions during the peel test would be relatively very limited. It is of interest to note that the independence of the value of G_c upon the way the forces are applied to a joint has also been reported by De and Gent [59] from their work on joints which consist of a rubbery layer bonded to rigid substrates.

3.5. Experimental methods using rollers

An alternate strategy for calculating the increment of energy dissipated, dU_d, other than in creating new surface is to devise an experimental scheme for either removing it or measuring it directly. Some standard tests go some way in this direction and such schemes have been described by Gent and co-workers [26,27]. Fig. 10 shows a possible method of modifying the 90° peel test by forcing a roller, of radius r, onto the debonding point by applying a force D at an angle φ. The value of G_c arises from the difference between P and D, where D forces the strip to conform to the roller. In the work of Gent and Hamed [26], the value of r was made sufficiently large to avoid plastic bending and a relatively large force,

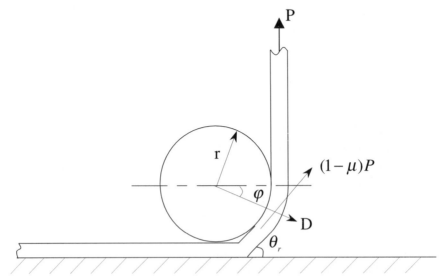

Fig. 10. The roller-assisted 90° peel test.

D, was applied to conform the peeling strip to the roller. A force, P, was then applied to an unbonded strip to determine the value of P necessary to simply bend the strip and overcome friction. When a bonded strip was then peeled, the value of G_c was determined by the difference in the applied peel forces. This scheme has recently received a good deal of attention [60–62] and has been extended to include plastic bending around the roller and a more complete calibration scheme. The analysis given here is a somewhat extended version of that given in [60].

For any values of P and D there will be an angle, θ_r, at the debond point as shown in Fig. 10. For an applied force P only a fraction, i.e. $(1 - \mu)P$, is applied at the peeling front; since friction arises in the roller, where μ is the coefficient of friction. The energy release rate, G, for this roller-assisted peel test is given by:

$$G = (1 - \mu)\frac{P}{b}(1 - \cos\theta_r) \tag{80}$$

and equilibrium gives:

$$\frac{D}{b}\cos\varphi = (1 - \mu)\frac{P}{b}\cos\theta_r \tag{81}$$

i.e.:

$$G = (1 - \mu)\frac{P}{b} - \frac{D}{b}\cos\varphi \tag{82}$$

and:

$$\cos\theta_r = \frac{D\cos\varphi}{(1 - \mu)P} \tag{83}$$

For the unbonded strip which is used in the calibration tests, then $G_c = 0$ and for elastic bending $\theta_r = 0°$. However, for the case of plastic bending then, again $G_c = 0$, but now G will increase as D is increased until it becomes constant at a value determined by r, say G_d. Thus, the calibration curve becomes:

$$\frac{P}{b} = \frac{1}{(1 - \mu)}\left(\frac{D}{b}\cos\varphi + G_d\right) \tag{84}$$

and is shown as line (i) in Fig. 11. For the bonded strip at $D = 0$, then $\theta_r = \theta = 90°$ and P/b is the 90° peel force. However, as D is increased, then the value of θ_r decreases and so does the value of P. Eventually, $\theta_r = \theta_0$ and there is no local bending and the only dissipation is G_d, but P/b is now increased compared to the case for the unbonded strip by $G_c/(1 - \mu)$. The line for a bonded strip is shown as line (ii) in Fig. 11 and the value of G_c may be found as shown. Fig. 12 shows data of this form taken from the work of Breslauer and Troczynski [60] which used nickel foils for the peel arm of 127 mm in thickness and $\varphi = 0$. The appropriate values are $\mu = 0.30$, $G_d = 268$ J/m^2 and $G_c = 230$ J/m^2.

To summarise, such schemes do show promise. However, they are more complex experimentally and introduce an extra source of energy dissipation via

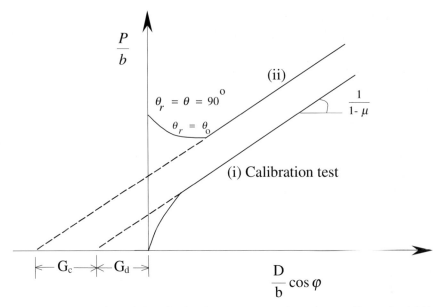

Fig. 11. The determination of the adhesive fracture energy, G_c, using a roller-assisted 90° peel test, and a calibration curve.

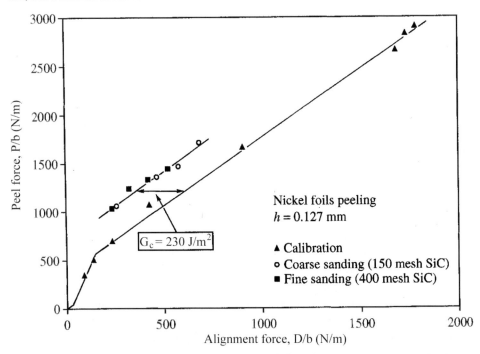

Fig. 12. Roller-assisted 90° peel test results for nickel foils bonded to aluminium alloy using an epoxy adhesive, after roughening of the foils with 150 mesh SiC grains (coarse sanding) or 400 mesh SiC grains (fine sanding) [60].

frictional losses, and possible errors may arise in the extrapolation that is required. Nevertheless, they have the attraction of direct experimental access to ascertaining the value of G_c and certainly should be explored, alongside analytical and numerical methods, for allowing for the plastic energy dissipation in the measured values of the peel force, P.

3.6. Effect of adhesive thickness

If an adhesive layer is present between the substrates, as for example in the case of the epoxy bonded aluminium-alloy peel tests referred to above, then the thickness of the adhesive layer, h_a, affects the value of G_c via two routes.

Firstly, the adhesive thickness, h_a, and adhesive modulus, E_a, both appear in Eq. 63 and when the adhesive layer is taken into account, this typically results in a relatively small decrease in the value of $dU_d/b\,da$; and once this effect is taken into account the value of G_c should be independent of the value of h_a.

Secondly, however, whilst the value of G_c may be independent of the overall geometry of the peel test, it is well established experimentally [25,39] that the value of G_c may be a function of the thickness, h_a, of the adhesive layer. This arises not because of any inaccuracies in the basic equations but due to the relative dimensions of the plastic–viscoelastic at the crack tip and the value of the thickness, h_a, of the adhesive layer. The typical form of the relationship between G_c and h_a for a rubbery adhesive bonding a polymeric substrate which is being peeled away is shown in Fig. 13. It may be seen that the value of G_c increases significantly as the thickness of the adhesive layer increases, and that this behaviour is especially pronounced when the value of h_a is low. However, the value of G_c eventually reaches a plateau value for relatively thick adhesive layers. These observations may be attributed to the energy dissipation that occurs within the adhesive layer in the plastic–viscoelastic zone at the peel front. It will be recalled that such localised energy dissipation ahead of the crack tip is included in the value of G_c. Essentially, as the thickness of the adhesive layer is increased in the peel test specimen, a larger volume of adhesive is subjected to deformation per unit area of detachment so that the energy expended increases. However, at large thicknesses the energy dissipated in the zone at the crack tip then becomes independent of the overall thickness of the adhesive layer, since the dissipation process zone no longer involves the entire layer of adhesive. This dependence of G_c upon the thickness of the adhesive layer has also been reported [39] for LEFM standard test specimens and arises from the same effects. (It should be noted that a similar effect could arise with respect to the thickness, h, of the peeling arm, assuming that the plastic–viscoelastic zone ahead of the peel front develops in the peel arm. If the value of h is decreased such that it becomes relatively thin compared to the diameter of the 'fully developed' plastic–viscoelastic zone ahead of the peel front, then this zone might well be prevented from fully developing at

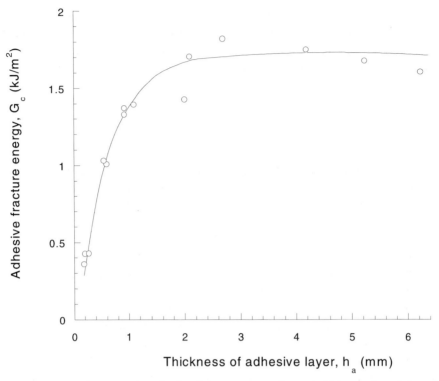

Fig. 13. Relation between the adhesive fracture energy, G_c, and thickness, h_a, of the adhesive layer for a poly(ethylene terephthalate) substrate peeling from a thermoplastic rubber adhesive layer at a peel angle of 180° [25].

low values of h, and hence a reduced value of G_c would be recorded. This effect has not apparently been reported in the literature, possibly because many peel tests employ an adhesive layer with a far lower value of the yield stress, σ_y, than the substrate arm. Hence, the plastic–viscoelastic zone at the peel front is confined to the adhesive layer; and this effect is thus seen to be a function of h_a, rather than h.)

3.7. Effect of test rate and temperature

It is noteworthy that the values of G_c from Tables 1 and 2 are measured for similar laminates and that the differences in the values of G_c arise from the fact that the tests were undertaken at different rates of test. The rate being significantly lower for the results shown in Table 2. The value of G_c does of course encompass (a) the intrinsic energy needed to rupture the inter-atomic and molecular bonds acting across the interface (or in the materials forming the interface, if cohesive failure in these materials occurs), *and* (b) the plastic and viscoelastic energy dissipated locally in a process zone ahead of the peel front. This latter energy loss will be rate and temperature dependent. Indeed, workers [21–23] have examined the rate and temperature dependence of the value of G_c as measured from peel test, and other test geometries and used the Williams, Landel and Ferry rate-temperature super-positioning relationship to form a 'master curve' for the dependence of the value of G_c upon the test rate and temperature.

4. Concluding remarks

The peel test is an attractive test method to assess the performance of a wide range of flexible laminates and adhesive joints. However, although it is a relatively simple test to undertake, it is often a complex test to analyse and thus obtain a characteristic measure of the toughness of the laminate, or adhesive joint.

The most successful approach that has been adopted is one based upon applying a fracture-mechanics method using an energy-balance approach. A value of the adhesive fracture energy, G_c, is thereby ascertained, which is the energy needed to propagate a crack through unit area of the joint, either cohesively through the adhesive layer or along the bimaterial interface. The value of G_c may be obtained via an analytical or a numerical analysis of the peel test. However, both of these theoretical approaches still need to be further refined in order to deal accurately with peel tests that involve large degrees of plastic deformation of the peel arm(s).

It has been shown that the value of G_c is essentially a characteristic of the laminate or joint. It is independent of geometric parameters such as the applied peel angle and the thickness of the flexible substrate arm(s) being peeled. Furthermore, the values of G_c obtained from peel tests agree well with values ascertained from very different joint geometries. However, it must be recognised that the value of G_c includes plastic and viscoelastic energy dissipation which occurs locally at the crack tip. Thus, for relatively tough adhesives, and when the thickness of the adhesive layer is relatively low, G_c will be a function of the thickness of the adhesive layer. Further, such energy dissipation will lead to the value of G_c being dependent upon the rate and temperature at which the peel test is conducted.

Finally, the fact that the value of G_c includes plastic and viscoelastic energy dissipation means that the peel test does not usually directly measure the intrinsic adhesion, even when the locus of failure is exactly at the interface between the materials forming the laminate, or joint. (Except in those rare cases when the test is conducted under conditions such that the plastic and viscoelastic energy dissipation tends to zero [23,39].) Also, if the measured peel force is used as the only parameter to be calculated, rather than the value of G_c, then it should always be borne in mind that the measured peel force will reflect changes in the geometric details of the peel test. Thus, inevitably, interpretation of the results from peel tests based simply upon knowing the values of only the measured peel forces will very often be open to some degree of doubt. This reinforces the need for (a) more physically realistic and accurate models and (b) improved test methods, such as the roller-assisted techniques, for the peel and associated tests.

Acknowledgements

We would like to thank Dr. H. Hadavinia for useful discussions and for preparation of the figures.

References

1. Kaelble, D.H., *Trans. Soc. Rheol.*, **3**, 161 (1959).
2. Kaelble, D.H., *Trans. Soc. Rheol.*, **4**, 45 (1960).
3. Kaelble, D.H., *Trans. Soc. Rheol.*, **9**, 135 (1965).
4. Kaelble, D.H. and Ho, C.L., *Trans. Soc. Rheol.*, **18**, 219 (1974).
5. Bikerman, J.J., *J. Appl. Phys.*, **28**, 1484 (1957).
6. Inoue, Y. and Kobatake, Y., *Appl. Sci. Res. A*, **8**, 321 (1959).
7. Gardon, J.L., *J. Appl. Polym. Sci.*, **7**, 625 (1963).
8. Gardon, J.L., *J. Appl. Polym. Sci.*, **7**, 643 (1963).
9. Spies, G.J., *Aircraft Eng.*, **30(March)**, 2 (1953).
10. Jouwersma, J., *J. Polym. Sci.*, **45**, 253 (1960).
11. Crocombe, A. and Adams, R.D., *J. Adhes.*, **12**, 127 (1981).
12. Thouless, M.D. and Jensen, H.M., *J. Adhes.*, **38**, 185 (1992).
13. Williams, M.L., *Bull. Seism. Soc. Am.*, **49**, 199 (1959).
14. Comninou, M., *J. Appl. Mech.*, **44**, 631 (1977).
15. Comninou, M., *J. Appl. Mech.*, **45**, 287 (1978).
16. Comninou, M. and Dundurs, J., *Res. Mech.*, **1**, 249 (1980).
17. Achenbach, J.D., Keer, L.M., Khetan, R.P. and Chen, S.H., *J. Elasticity*, **9**, 397 (1979).
18. Charalambides, M., Kinloch, A.J., Wang, Y. and Williams, J.G., *Int. J. Fract.*, **54**, 269 (1992).
19. Tvergaard, V. and Hutchinson, J.W., *J. Mech. Phys. Solids*, **41**, 1119 (1993).
20. Lindley, P.B., *J. Inst. Rubber Ind.*, **5**, 243 (1971).
21. Gent, A.N. and Kinloch, A.J., *J. Polym. Sci.*, **9(A-2)**, 659 (1971).
22. Andrews, E.H. and Kinloch, A.J., *Proc. R. Soc. Lond. A*, **332**, 385 (1973).
23. Andrews, E.H. and Kinloch, A.J., *Proc. R. Soc. Lond. A*, **332**, 401 (1973).
24. Gent, A.N. and Hamed, G.R., *J. Adhes.*, **7**, 91 (1975).
25. Gent, A.N. and Hamed, G.R., *Polym. Eng. Sci.*, **17**, 462 (1977).
26. Gent, A.N. and Hamed, G.R., *J. Appl. Polym. Sci.*, **21**, 2817 (1977).
27. Gent, A.N. and Kaang, S.Y., *J. Adhes.*, **24**, 173 (1987).
28. Kim, K.S. and Aravas, N., *Int. J. Solids Struct.*, **24**, 417 (1988).
29. Aravas, N., Kim, K.S. and Loukis, M.J., *Mater. Sci. Eng. A*, **107**, 159 (1989).
30. Kinloch, A.J., Lau, C.C. and Williams, J.G., *Int. J. Fract.*, **66**, 45 (1994) (See also: http://www.me.ic.ac.uk/materials/AACgroup/index.html, where software is available.).
31. Moidu, A.K., Sinclair, A.N. and Spelt, J.K., *J. Test. Eval.*, **23**, 241 (1995).
32. Gent, A.N., *Langmuir*, **12**, 4492 (1996).

33. de Gennes, P.G., *Langmuir*, **12**, 4497 (1996).
34. Thouless, M.D., Adams, J.L., Kafkalidis, M.S., Ward, S.M., Dickie, R.A. and Westerbeek, G.L., *J. Mater. Sci.*, **33**, 189 (1998).
35. Kinloch, A.J. and Williams, J.G., *J. Mater. Sci. Lett.*, **17**, 813 (1998).
36. Yang, Q.D. and Thouless, M.D., *J. Mater. Sci. Lett.*, **18**, 2051 (1999).
37. Kinloch, A.J. and Williams, J.G., *J. Mater. Sci. Lett.*, **18**, 2049 (1999).
38. Rahulkumar, P., Jagota, A., Bennison, S.J. and Saigal, S., *Int. J. Solids Struct.*, **37**, 1873 (2000).
39. Kinloch, A.J., *Adhesion and Adhesives: Science and Technology*. Chapman and Hall, London, 1987.
40. Williams, J.G., *Int. J. Fract.*, **87**, 265 (1997).
41. Lai, Y.-H. and Dillard, D.A., *J. Adhes.*, **56**, 59 (1996).
42. Gent, A.N. and Kaang, S.Y., *J. Appl. Polym. Sci.*, **32**, 4689 (1986).
43. Gent, A.N. and Lewandowski, L.H., *J. Appl. Polym. Sci.*, **33**, 1567 (1987).
44. Williams, J.G., *J. Adhes.*, **41**, 225 (1993).
45. Yang, Q.D., Thouless, M.D. and Ward, S.M., *J. Adhes.*, **72**, 115 (2000).
46. Kanninen, M.F., *Int. J. Fract.*, **9**, 83 (1973).
47. Williams, J.G. and Hadavinia, H., *Proceedings of the 14th European Fracture Conference*, Poland, September, 2002, to be published.
48. Hadavinia, H. and Williams, J.G., to be published.
49. Wei, Y. and Hutchinson, J.W., *Int. J. Fract.*, **93**, 315 (1998).
50. Rahulkumar, P., Jagota, A., Bennison, S.J. and Saigal, S., *Int. J. Solids Struct.*, **37**, 1873 (2000).
51. Hadavinia, H., Kinloch, A.J. and Williams, J.G., to be published.
52. Kinloch, A.J., Hadavinia, H., Blackman, B.R.K., Ring-Groth, M., Williams, J.G. and Busso, E.P., *Proceeding of the 23rd Adhesion Society Meeting*. Adhesion Society, 2000, p. 25.
53. Kinloch, A.J., Hadavinia, H., Blackman, B.R.K., Paraschi, M. and Williams, J.G., *Proceedings of the 24th Adhesion Society Meeting*. Adhesion Society, 2001, p. 44.
54. Chen, J., Crisfield, M., Kinloch, A.J., Busso, E.P., Matthews, F.L. and Qui, Y., *Mech. Compos. Mater.*, **6**, 301 (1999).
55. Moore, D.R. and Williams, J.G., In: Moore, D.R., Pavan, A. and Williams, J.G. (Eds.), *Fracture Mechanics Testing Methods for Polymers, Adhesives and Composites*. Elsevier Science, Amsterdam, 2001, p. 203.
56. Wei, Y. and Hutchinson, J.W., *Int. J. Fract.*, **95**, 1 (1999).
57. Kinloch, A.J., Blackman, B.R.K., Taylor, A.C. and Wang, Y., *Proceedings of Adhesion '99*. Inst. Materials, London, 1996, p. 467.
58. Blackman, B.R.K. and Kinloch, A.J., In: Moore, D.R., Pavan, A. and Williams, J.G. (Eds.), *Fracture Mechanics Testing Methods for Polymers, Adhesives and Composites*. Elsevier Science, Amsterdam, 2001, p. 225. (See also: *http://www.me.ic.ac.uk/materials/AACgroup/index.html*, where spreadsheets to calculate G_c are available.)
59. De, D.K. and Gent, A.N., *Rubber Chem. Technol*, **71**, 84 (1998).
60. Breslauer, E. and Troczynski, T., *J. Adhes. Sci. Technol.*, **12**, 367 (1998).
61. Sexsmith, M., Troczynski, T. and Breslauer, E., *J. Adhes. Sci. Technol.*, **11**, 141 (1997).
62. Breslauer, E. and Troczynski, T., *Mater. Sci. Eng. A*, **302**, 168 (2001).

Chapter 9

The mechanics of coatings

M. PAPINI [a,*] and J.K. SPELT [b]

[a] *Ryerson University, Department of Mechanical, Aerospace, and Industrial Engineering,*
350 Victoria Street, Toronto, ON, M5B 2K3, Canada
[b] *University of Toronto, Department of Mechanical and Industrial Engineering,*
5 King's College Road, Toronto, ON, M5S 3G8, Canada

1. Introduction

Coatings find use in a wide variety of industries, normally to serve one or more of the following purposes: (1) to protect the surface from corrosion; (2) to control friction and wear; and (3) to alter physical properties, such as reflectivity, colour, conductivity, etc. Though there are many different types of coatings, most can be classified as either organic or inorganic. Organic coatings are usually applied in liquid form with a spray gun or brush. A notable exception is powder coatings that are applied in solid form using electrostatic equipment and then sintered to form a continuous film. Inorganic coatings are usually metallic or ceramic based. Common examples include conversion coatings (e.g. phosphate), thin film coatings (e.g. Titanium nitride), which are normally applied via physical or chemical vapour deposition (PVD or CVD), and plasma sprayed coatings.

Coatings can fail in one of several modes: delamination, fracture, erosive wear, and general yield. Delamination of a coating refers to the loss of adhesion of the coating from the substrate, and, if the coating stresses are compressive, usually involves blistering. Residual stresses, thermal mismatch stresses, environmental attack, and impact or contact stresses are often cited as causes of delamination. Fracture refers to crack propagation within the coating itself, rather than crack propagation along the interface between the coating and substrate (i.e. delamination). Fracture in coatings can often be attributed to impact or contact phenomena.

---

[*] Corresponding author. Tel. +1 (416) 979-5000, ext. 7655; Fax: +1 (416) 979-5265, E-mail: mpapini@ryerson.ca

Organic coatings are typically quite soft and are often applied to hard surfaces. Inorganic coatings, on the other hand, are usually quite hard, and often applied to protect softer substrates from wear or the environment. On this basis, it is possible to make certain generalizations in the stress or fracture analysis of coatings. The aim of this chapter is to introduce the reader to the most important aspects of the mechanics of coatings. In keeping with the focus of this volume on adhesion, the emphasis will be on the interfacial cracking of coatings, and the determination of the stresses that drive the delamination process.

2. Measurement of mechanical properties

There are five properties that are necessary to perform mechanical analysis: hardness, strength, stiffness, coating fracture toughness and interfacial fracture toughness. In addition, we shall also discuss methods to determine coating thickness, as this can have an effect on some of the material properties, as discussed below. Hardness, both dynamic and static, is of interest in improving wear resistance. Stiffness and strength are required in stress analysis, while fracture toughness is of importance in the assessment of delamination.

Coating manufacturers are often unable to supply mechanical properties, such as Vickers hardness, Young's modulus, or yield strength. Manufacturers usually describe their products qualitatively using words such as 'hard', 'stiff', 'soft', or 'tough'. In fact, many of the current industry standard techniques for coating evaluation are qualitative, and of value only in comparisons of coating systems. Two examples of this are the often-quoted pencil hardness tests and adhesion scratch tests (see Section 2.3.2), which give only semi-quantitative measurements of hardness or adhesion.

Coating mechanical properties can be affected by the degree of cure, aging effects, and ambient conditions (e.g. temperature and humidity), making it difficult to establish a useful database. Moreover, organic coatings are viscoelastic, so that their properties depend on strain rate and temperature. A complete treatment of viscoelastic theory is beyond the scope of this chapter, and the interested reader is referred to Chapter 12 of this volume for a more complete discussion. The reader may also wish to consult refs. [1,2], which are review articles of transient and dynamic methods of coating characterization.

The measurement of coating properties is further complicated by possible differences between the bulk and in situ (i.e. when applied to a substrate) properties, which may also be a function of coating thickness. For example, the measurement of the in situ hardness of a soft coating on a hard substrate has been found to approach the bulk coating hardness for thick coatings, and approach the hardness of the substrate for thin coatings.

With these issues in mind, we shall review the methods currently available to measure coating mechanical properties. Most of the methods can be applied to both organic and inorganic coatings.

2.1. Measurement of coating stiffness

It is normally necessary to determine the Young's (tensile) modulus, E, of coatings in order to perform elastic stress and fracture analyses. Being, in general, a function of strain rate and temperature, E, is often measured using dynamic testing equipment, where measurements are made as a function of time-varying loads with loading frequency (strain rate) and temperature as controlled parameters. Fortunately, for many viscoelastic materials, there exists equivalence between mechanical behaviour with time and temperature [3]; increasing frequency corresponds to decreasing temperature, and vice versa. Thus, if a fixed frequency is used and the coating is characterized at a variety of temperatures, it is possible to use 'time–temperature superposition' to predict the behaviour at a fixed temperature over a range of frequencies.

The mechanical behaviour of coatings can also be measured using transient methods, such as creep and stress relaxation testing. Creep tests record strain under constant load while stress relaxation measurements record stress under constant strain. In both cases, the modulus as a function of time is calculated.

In dynamic coating tests, an oscillating force is applied to the coating, and the resulting strain is measured as a function of time. The modulus is then expressed in complex terms, according to standard viscoelastic theory. A variety of dynamic tests are available (see ref. [2]) for determining the complex modulus, the most common of which are the torsional pendulum and dynamic mechanical analysis. In torsional pendulum tests, a coating sample is twisted and allowed to oscillate freely. The amplitude decay due to energy dissipation is used to give the loss modulus, and the frequency of the oscillation is related to the elastic properties and to give the storage modulus. The main drawback of this method is that it gives the complex modulus at only one frequency. In dynamic mechanical analysis, a sample is subject to a forced displacement, and the resulting amplitude, along with the phase difference between stress and strain are measured. Instruments that allow control of amplitude, temperature and frequency are commercially available, allowing for measurements over a wide variety of temperatures and strain rates (i.e. frequencies).

If the elastic properties of the coating are required at very high rates of strain such as is the case in impact analysis, an ultrasonic time-of-flight method can be used to estimate E. The underlying assumption here is that at very high strain rates, the modulus versus strain rate curve reaches a plateau, and the response is elastic rather than viscoelastic. By measuring the time it takes for transverse and longitudinal ultrasonic waves to traverse a coating thickness, the transverse wave speed, c_t, and longitudinal wave speed, c_l can be determined. Knowing the density of the coating, ρ, both the Young's modulus, E, and Poisson's ratio, ν, can then be determined according to [4]:

$$E = 4\rho c_t^2 \frac{0.75 - \left(\dfrac{c_t}{c_l}\right)^2}{1 - \left(\dfrac{c_t}{c_l}\right)^2} \tag{1}$$

$$\nu = \frac{0.5 - \left(\dfrac{c_t}{c_l}\right)^2}{1 - \left(\dfrac{c_t}{c_l}\right)^2} \tag{2}$$

The method allows measurements to be made with the coating in situ, and has been used for the analysis of coatings subject to solid particle impact [5,6]. A drawback of the method is the need for a liquid couplant between the ultrasonic probe and the coating. Concerns about the effects of liquid absorption into the coating can be eliminated by first applying a very thin metal film using sputtering [7]. In ref. [7], high frequency ultrasonic surface waves (Raleigh) were used to measure the modulus of an epoxy layer in a region extending only a few microns from the free surface.

Inorganic coatings are usually not subject to viscoelastic behaviour unless the operating temperature is relatively close to the melting point. Difficulties do arise, however, because such coatings are in general very thin (often sub-micron), and cannot be easily separated from the substrate, or may fail under very low loads. A variety of techniques for stiffness determination are available, such as the bulge test, flexural test, tensile test, elastic wave methods, and indentation testing. In the bulge test, gas pressure is applied to a free-standing thin film mounted over a circular hole. The height of the bulge is measured as a function of the applied gas pressure and the stiffness is then extracted from this information [8]. In flexural tests, coatings are tested in situ by subjecting coated specimens to three-point (e.g. [9]) or four-point bending (e.g. [10]). The modulus of the coating is extracted from either load versus deflection or, if strain gauges are used, stress–strain curves in conjunction with composite beam theory.

If the coating is relatively thick, then a standard tensile test may be performed on a free film, with the Young's modulus determined from the load–displacement curve. For very thin free films, a thin metallic foil is coated on both sides. As the coated foil is loaded, strain gauges on either side record the superposition of the effects of the coating, foil, and strain gauges, which are then separated using composite beam theory to yield the coating Young's modulus [11]. An analysis of the errors in the flexure measurement of Young's modulus and Poisson's ratio associated with uncertainties in experimental parameters can be found in ref.

[12]. The authors conclude that a combination of the flexure, torsion, and thermal bending tests can be used to accurately determine the Young's modulus, residual stress, and thermal expansion coefficient, if the Poisson's ratio can be estimated. Combinations of experimental parameters that give accurate results are presented [12].

The stiffness of inorganic films may also be determined from the measurement of the propagation velocity of surface acoustic waves, C, which, for a homogeneous material is related to the Young's modulus and density as follows:

$$C = \frac{0.87 + 1.12\nu}{1 + \nu} \sqrt{\frac{E}{2\rho(1+\nu)}} \tag{3}$$

where ν is Poisson's ratio, E is Young's modulus, and ρ is the density of the material. Surface acoustic waves propagate very close to the surface, making them ideal for thin film characterization, even for films much thinner than the penetration depth of the wave. No special preparation of the sample is required, and the measurements are very quickly performed on in situ coatings. In principle, the wave can be induced by any number of standard piezoelectric transducers, but a novel and promising approach to this method is the use of pulsed lasers to induce local heating in an in situ coating, and thus an acoustic surface wave [13]. For a wave travelling in a layered material, the relationship is more complicated than Eq. 3, and a solution of the wave propagation equation must be found that satisfies the boundary conditions that stresses vanish at the surface and that displacements and stresses are continuous at the interface [4]. The result is that the wave propagation speed is related to the frequency of the waves, the film thickness, and the Young's moduli, Poisson's ratios, and densities of both the coating and substrate [14]. By comparing the measured surface wave velocity obtained over a wide variety of frequencies to theoretical wave velocity versus frequency curves (obtained by solving the above-mentioned wave equation), Young's modulus, density, and film thickness can all be determined simultaneously, provided that the coating is relatively thick. For very thin coatings, the information obtained from the measurement only allows one of the three parameters (i.e. Young's modulus, coating thickness, density) to be determined. Usually, the density and film thickness are taken as known, and Young's modulus is derived [14].

A commonly used method for mechanical characterization of inorganic coatings is quasi-static indentation. By monitoring the load and displacement of an indenter (usually a pyramidal diamond) as it penetrates the coating on a substrate, the Young's modulus, yield strength, and hardness can be determined. Commercially available nano-indenters typically provide indentation loads in the range of 0.2–30 mN. Of major concern in such measurements is the influence of the substrate. Substrate deflection is of concern in the case of hard coatings on softer substrates, while hard substrates constrain the indentation of soft coatings. The

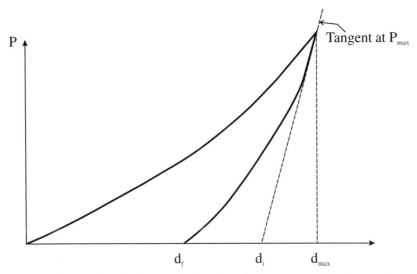

Fig. 1. A typical indentation load versus indentation depth curve resulting from measurements made with a nano-indenter.

traditional guideline to minimize such effects is that the indentation depth should be at most 1/10th of the coating thickness [15]. Recently, however, investigators have found that the critical thickness ratio is a function of both coating/substrate yield strength ratio and indenter geometry [16–19] and that the 1/10th rule overestimates the required critical thickness ratio in many applications. Of additional concern is the effect of the elasticity of the indenter. We shall now outline a method that takes into account all of these effects and was used for elastic modulus characterization of the interphase in an epoxy aluminium system [20].

A typical indentation load–depth plot is shown in Fig. 1. For measurement of Young's modulus, the initial slope of the *unloading* portion of a plot of indentation load versus penetration should be used. This is because the loading portion includes both elastic and plastic effects, whereas the unloading portion is always elastic (e.g. [21]). The unloading curve can be fitted using the power law relationship suggested in [22]:

$$P = \delta(d - d_{\mathrm{f}})^{m} \tag{4}$$

where P is the indentation load, d is the indentation depth, and d_{f} is the depth at which the indenter load returns to zero. The parameters δ, m, and d_{f} are determined by fitting the portion of the unloading curve between 40% and 90% of the maximum load. This has been found to reduce the effect of creep during unloading [23]. The slope of the tangent at maximum load on the unloading curve gives the uncorrected reduced modulus, E_{ru} which does not account for elasticity of the substrate and indenter, as:

$$\frac{dP}{dd} = \beta^* E_{ru} \sqrt{A} \tag{5}$$

where A is the cross-sectional area of the indenter as a function of the distance to the indenter tip, and β^* is a parameter determined from an elastic analysis of the punch problem based on the pioneering work of Chen and Engel [24] and has values of 1.129, 1.142, and 1.167, for circular, square, and triangular punches, respectively [21]. To correct for the elasticity of the indenter and substrate, the following relation can be used:

$$\frac{1}{E_{ru}} = \frac{1}{E_r}\left(1 - e^{-\frac{\alpha^*}{a}h}\right) + \frac{1}{E_{rs}}\left(e^{-\frac{\alpha^*}{a}h}\right) + \frac{1}{E_{ri}} \tag{6}$$

where the reduced modulus $E_r = E/(1 - v^2)$ and E_{rs} and E_{ri} are the reduced moduli of the substrate and indenter, respectively, h is the thickness of the coating, and a is the square root of the indenter cross-sectional area at maximum load. The parameter α^* is a function of a/h determined by analysis of the elastic punch problem and can be found in ref. [21]. Eq. 6 can then be used to extract the reduced modulus of the coating, E_r from indentation data.

2.2. Measurement of coating hardness and yield strength

The wear properties of coatings are closely related to hardness. Coatings are often referred to as 'stiff' or 'hard', and many times these terms are used interchangeably; however, they mean completely different things. Stiffness and compliance are related to the coating's resistance to deformation under *elastic* conditions, while hardness is a measure of a coating's resistance to indentation under *plastic flow* conditions. Hardness is thus closely related to the yield strength of the bulk coating material. In other words, the material properties which control indentation depth per unit applied force are hardness under plastic flow, and Young's modulus under elastic conditions.

A commonly used quick characterization of relative hardness is provided by the pencil scratch test [25] in which a series of standardized, sharp pencils are drawn across the coating at a specific angle to find the 'softest' one that causes a scratch. Thus, coating hardness is reported as that of the scratching pencil (e.g. 4B–4H). The test is obviously subject to a certain amount of subjectivity, since it depends on the force applied to the pencil, the pencil consistency (i.e. different manufacturers of pencils make pencils of slightly different hardness), pencil sharpness and inclination. For this reason, pencil hardness should only be used as a comparative measure of coating hardness, when the same person performs the test on a variety of coatings with the same pencils.

The micro- and nano-indentation techniques discussed in the previous section can also be used to characterize the hardness of both inorganic and organic coatings. In hardness testing, the four commonly used tests for micro-hardness, Rockwell, Vickers, Brinell, and Knoop, are all based on applying a known load to an indenter and measuring the size of the indentation:

$$H = \frac{F}{A_i} \tag{7}$$

where H is the hardness, F is the applied load, and A_i is the indented area [26]. Because the indentation is measured after the indenter has been removed, elastic recovery is not considered, and hardness is only associated with plastic flow. This is in contrast to the nano-indentation techniques described in the previous section, which use the continuous load–penetration curve to determine both elastic and plastic coating properties. The principal difference between the four hardness tests is the shape of the indenter; a sphere for the Brinell test, a cone for the Rockwell test, and a pyramid for the Vickers and Knoop. The latter two tests have the advantage that the measured hardness is independent of the indenter load.

As was the case with the measurement of coating modulus (Section 2.1), viscoelastic behaviour and the influence of the substrate can affect hardness measurements. With organic coatings, it is common to distinguish between conventional quasi-static hardness and dynamic hardness, p_d, defined as the resistance to plastic indentation at very high strain rates, where it is assumed that both viscoelastic and elastic effects are negligible.

Dynamic hardness (also called plastic flow pressure) is often used in the rigid-plastic analysis of solid particle impacts on both coatings and bare metals, where it is assumed that, because of the high velocities involved, the material response is a fully plastic one (e.g. [27]). The dynamic hardness, assumed constant during an impact, can be used to estimate the trajectory of a particle as it plows the target by the use of Eq. 7, with F being the retarding force, and A_i being the instantaneous contact area between the impacting particle and the target. Good success in predicting the crater size in single spherical particle impacts on bare targets has been obtained by various authors [27–31], and the technique has shown promise in predicting erosion behaviour for coated substrates, if elastic rebound effects are considered [5]. The method has been extended to predict coating erosion due to arbitrarily shaped angular particles [32,33].

Indentation hardness data can also give the yield stress via analysis of the stresses in the coating due to the indenter. For an elastic–perfectly plastic coating response, plastic flow occurs at the indentation pressure corresponding to the hardness as defined in Eq. 7, so that an elastic analysis of the stresses below the indentation at these conditions, in conjunction with an appropriate yield criterion results in the relationship between hardness and yield stress. The resulting relationship is a function of the indenter shape. For example, for the dynamic indentation of a urethane coating on an epoxy primed aluminium substrate by a spherical indenter, the authors applied an approximate elastic theory [34] in conjunction with the Tresca yield theory and obtained good results with the simple condition $\sigma_y = p_d$; i.e. a perfectly plastic response [5].

The yield stress of a coating can also be measured using standard tensile testing of free films to obtain the stress–strain curve. The major difficulties are the fabrication of free films having uniform thickness and the gripping of the thin specimens. As in all mechanical testing of organic (viscoelastic) coatings, the temperature, humidity, and strain rate should all be carefully controlled and noted.

2.3. Measurement of coating fracture toughness and interfacial adhesion strength

Coating fracture usually occurs either in the coating itself or along the interface between the coating and substrate. For cracks propagating within the coating, the crack path followed will be the one that maximizes the mode I stress intensity, whereas interfacial cracks may be mixed mode (see Section 4.2). We must therefore distinguish between the mode I fracture toughness of the coating, K_{IC}, and the mixed-mode toughness of an interface, $K(\psi)$. These concepts are covered in detail in Section 4, and the reader may want to skip forward and read that section at this point.

2.3.1. Measurement of coating toughness

Nano- or micro-indenters can been used to characterize the fracture toughness of both the coating and the coating–substrate interface. The fracture toughness, K_{IC}, of coatings has been related to the propagation of radial cracks in the coating due to Vickers indentation:

$$K_{IC} = \frac{BP}{c^{3/2}} \tag{8}$$

where B is a geometrical factor determined by experimental calibration, P is the peak indentation load, and c is a characteristic dimension of the radial crack [35–37]. According to [36], $B = 0.08$ gives a good estimate for polycrystalline inorganic coatings when $c/a > 2.5$, where a is the radius of the plastic zone below the indentation. The authors claim accuracy within about 10% for commonly encountered inorganic coatings.

Another commonly used method to determine fracture toughness of coatings is through the use of compliance fracture specimens, whereby the change in compliance of the specimen is measured or calculated as the crack propagates, and related to the mode I critical strain energy release rate, G_{IC}. Elementary fracture mechanics gives the required relationship [38]:

$$G = \frac{P^2}{2w} \frac{dC}{da} \tag{9}$$

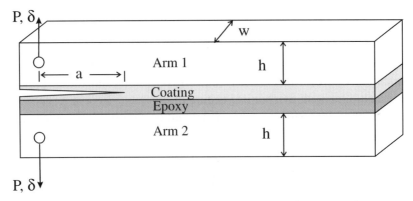

Fig. 2. The modified double cantilever beam (DCB) specimen for coating fracture toughness measurement.

where P is the load on the specimen, w and C are the width and compliance of the specimen, and a is the crack length. The most commonly used specimen is the modified double cantilever beam (DCB) specimen shown in Fig. 2 (e.g. [39]). For evaluation of coating strength, one of the arms of the specimen (the arms are usually made of a stiffer material such as steel or aluminium) is coated and attached to the other arm using an epoxy. To ensure that crack propagation occurs in the coating, rather than any of the interfaces or in the epoxy, both the interfaces, and the epoxy should be of higher toughness than the coating. This is not a stringent condition, however, since crack propagation can be directed toward the coating by employing a mixed-mode loading on the DCB. This is accomplished simply by applying a greater load to the coated arm than to the arm that is bonded with the epoxy adhesive [39,40].

As an example, consider the *homogeneous* DCB (i.e. no coating or epoxy applied). For this specimen, elementary beam theory gives the compliance as [38]:

$$C(a) = \frac{\delta}{P} = \frac{8a^3}{Eh^3w} \tag{10}$$

so that the energy release rate is given by

$$G = \frac{12P^2a^2}{Eh^3w^2} \tag{11}$$

The compliance change as the crack propagates can either be calculated or obtained experimentally. If an analytical expression exists for $C(a)$, then the critical strain energy release rate, G_{IC}, can be obtained by noting the load at which the crack propagates, P_c, together with the crack length, a, usually measured via an optical microscope. For example, in the case of a homogeneous beam, substitution of P_c and a into Eq. 11 gives G_{IC} for the case of a homogeneous beam. G_{IC} can then be related to K_{IC} via Eq. 18.

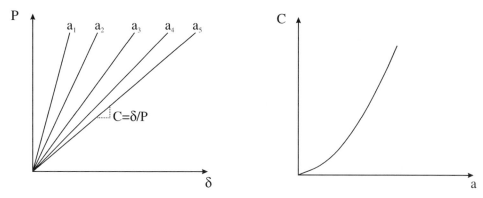

Fig. 3. Experimental calculation of compliance for DCB specimens.

For the DCB with a coating and epoxy layer shown in Fig. 2, the presence of these two layers affects the compliance, and a beam-on-an-elastic-foundation model can be used to calculate $C(a)$ [41]. It should be noted that if the relative toughness of the two interfaces, the coating, and the epoxy is such that the crack propagates along the coating/arm interface, then the measurement of G_C is no longer the mode I toughness of the coating, but a mixed-mode interfacial toughness (i.e. $G_C(\psi)$, see Section 4).

If the compliance is to be determined experimentally, the displacement of the arms, δ, is noted as a function of the applied load, P, and plotted for different crack lengths (Fig. 3). The slope of the lines in the left panel of Fig. 3 gives the compliance, $C = \delta/P$. If the compliance is then plotted as a function of crack length, and curve fitted (Fig. 3, right panel), G can be determined using Eq. 9. This experimental procedure was used successfully in the measurement of G_{IC} for plasma-sprayed coatings [42].

2.3.2. Measurement of interfacial fracture toughness

A common mode of coating failure is delamination at the coating–substrate interface. The strength of this interface is measured using 'adhesion' tests that range from the simple adhesion scratch test to the more sophisticated tests based on fracture and/or contact mechanics. Many of the qualitative tests, such as the Adhesion Tape Test, are covered in the ASTM standards (e.g. [43–45]). These should be used only for comparative purposes, where a quick evaluation of adhesion is required.

Unfortunately, different techniques of adhesion measurement often give contradictory results. For example, the adhesion strength of TiN films deposited on annealed steel as a function of sputtering time was assessed using the following tests [46]: (a) adhesion scratch; (b) four-point bend; (c) cavitation; (d) impact;

(e) laser-acoustic; (f) scanning acoustic microscope; and (g) Rockwell hardness. The seven methods failed to provide consistent data, and some of the methods even indicated a reduction in film adhesion for samples expected to have the best film quality (the highest presputtering time is expected to yield the highest adhesion strength). Of these techniques, the micro-indentation (Rockwell, Vickers, etc.) and scratch tests are the most common, and are discussed below, together with methods based on interfacial fracture mechanics specimens.

Because coating stiffness affects the stresses that cause delamination, strain rate and temperature can have a significant effect on results with organic coatings. With this restriction, the techniques for adhesion measurement work equally well for organic and inorganic coatings.

The adhesion scratch test involves drawing a diamond tip across the coating surface while gradually increasing the normal load on the tip until the coating is removed. The minimum load required to initiate removal is termed the 'critical load' and is often quoted as a measure of coating adhesion. The critical load by itself is useful only for comparative assessments of adhesion between coatings of the same thickness; however, a variety of authors have suggested methods of obtaining more quantitative adhesion data from the test. Early attempts focussed on the calculation of the interface shearing force [47], or the critical value of the ploughing stress required to remove the coating [48]. More recent attempts use energy-based or fracture mechanics criteria to estimate the interfacial fracture toughness [49–51], with varying degrees of success. More accurate measurements of interfacial fracture strength can be obtained using tests based on bimaterial fracture mechanics (e.g. indentation tests, or interfacial fracture specimens, as discussed in Section 4). It should be noted that the modified DCB specimen described in Section 2.3.1 can also be used to measure interfacial fracture toughness by choosing the loadings on the arms so that crack propagation is directed toward the interface.

Indentation tests can also be used for measurement of the interfacial fracture strength. The coating is indented by either a ball or Vickers indenter until an interface crack is initiated at a critical load; further loading causes the interface crack to propagate in a stable fashion. It is usually assumed that indentation causes a plastic zone to form below the indenter, and that this constant-volume process causes a compressive radial stress in the coating adjacent to the indenter that drives the delamination. The delamination may or may not be assisted by buckling of the coating due to the compressive stresses. Interfacial fracture mechanics techniques are then used to relate the depth of indentation at the onset of delamination to the fracture toughness and mode mix of the interface [6,52–56]. This will be discussed further in Section 4.2.

Interface toughness measurement can be accomplished using the blister test, in which a hole is drilled from the uncoated side of the substrate to the film-substrate interface. Air pressure is then applied through the hole, causing the coating to delaminate from the substrate. Alternatively, the coating can be made to delaminate by applying a normal point load to the coating through the hole, so that the coating is pushed off the substrate. The interfacial fracture toughness can be related to the internal pressure or point load by coupling plate theory to a mixed-mode fracture analysis (see Section 4.2.3). Jensen has analyzed both the point and pressure loading cases, and gives solutions for energy release rates and mode mix [57]. Interfacial cracking is covered in more detail in Section 4.

It should be noted that these methods work well for inferring the interfacial crack *propagation* behaviour, but they cannot measure the conditions under which a crack will *initiate*. Estimates of the interfacial shear stress to initiate debonding can be obtained from indentation tests using an approximate stress analysis in which the coating is assumed very thin, so that stresses are approximately constant through the thickness [34]. In this model [58], the plastic zone below the indentation is modelled as a cylinder, which is replaced conceptually by a hole under an internal pressure (see Fig. 4), p, which is related to the indentation hardness of the material:

$$p = \frac{2H}{3} \tag{12}$$

The approximate elastic stress analysis of the remaining coating then gives the maximum interfacial shear stress, τ, at crack initiation at the film-hole boundary:

$$\tau = \frac{-\left(\frac{2}{3}\right)H}{\dfrac{K_1'\left(\dfrac{\xi c}{h}\right)}{K_1\left(\dfrac{\xi c}{h}\right)} + \dfrac{vh}{\xi^2 c}} \tag{13}$$

where H is the hardness of the coating, $K_1'(x) = \mathrm{d}K_1(x)/\mathrm{d}x$, $K_1(x)$ is the first-order modified Bessel function of the second kind, v is Poisson's ratio of the coating, c is the contact radius during indentation, h is the coating thickness, and

$$\xi = \left[\frac{6(1-v)}{4+v}\right]^{1/2} \tag{14}$$

The normal stress across the interface is predicted to be zero. It should be noted that this approximate analysis is based on normal stresses and strains which are averaged through the thickness of the coating, and thus the resulting interfacial shear stress is not suitable for the calculation of the energy release rate and mode ratio. Nevertheless, this nominal value of the critical interfacial shear stress was found to be useful in predicting initiation for an alkyd coating on a steel substrate indented by a rigid sphere [6,56]. A promising method for predicting crack initiation based on the use of the stress concentration in the vicinity of free edges of bonded materials [59,60] is discussed in Section 4.4.

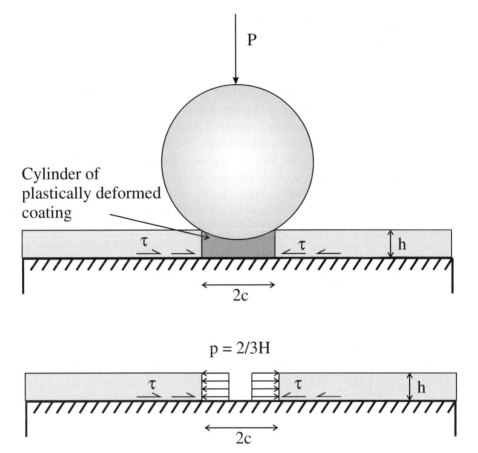

Fig. 4. Scheme used to estimate interfacial shear stresses leading to crack initiation in indented coatings.

2.4. Measurement of coating thickness

As mentioned above, coating thickness is found to affect many of the properties of coatings, and therefore, must be measured accurately. There is a variety of methods to measure both dry and wet film thickness, both non-destructive and destructive. An in-depth discussion of all the possible methods is beyond the scope of this chapter, so we shall only mention the main techniques and direct the reader to the appropriate references. A comprehensive review article covers many methods for organic coating thickness measurement [61], and many techniques are covered in the ASTM standards.

For wet film thickness measurement of paints and varnishes (i.e. before drying of the freshly applied coating), it is common to use mechanical wet film thickness gauges, which involve pressing or rolling a calibrated scale onto a wet film.

Examples include the wheel, Pfund, notch, and comb gauges. The appropriate ASTM standards can be found in refs. [62,63].

Direct measurements of dry organic film thickness can be made using micrometers or dial gauges to obtain the difference in thickness between the non-covered and covered substrate. Films should, of course, be hard and stiff enough to resist indentation of the gauges. The ASTM standards for these tests are in ref. [64]. Dry film thickness can also be measured by observing, under a microscope, precision angular cuts made using drill bits, groove cutting instruments, or grinding instruments. The substrate should be sufficiently rigid to prevent deformation of the coating during the cutting process and this method is recommended for coatings of thickness up to 1.3 mm [65].

For inorganic coatings, an ASTM guide [66] covers coating measurement using commercially available gauges using the following principles: (a) magnetic attraction (nonmagnetic coatings on magnetic substrates); (b) eddy-current (metallic coatings on nonconductive substrates), (c) X-ray fluorescence (for metallic coatings up to 15 μm thickness); (d) beta backscatter (for coatings having an atomic number sufficiently different than that of the substrate); (e) Coulometric (amount of electricity consumed in dissolving the coating in an electrolyte is measured and related to thickness; suitable for many different coating/substrate combinations); (f) double-beam interference microscope (for thin or transparent coatings); (g) microscopic (thickness measured in magnified image of coating cross-section); and (h) strip and weigh (difference in substrate-coating weight before and after stripping).

In addition, ultrasonic methods may be used to measure film thickness for both organic and non-organic coatings. Time-of-flight methods (Section 2.1) may be used if the density and either the Young's modulus or Poisson's ratio of the coating are known. Raleigh surface waves can also be used to measure film thickness (e.g. [67]). The velocity of these waves, being a function of the coating thickness and material properties of coating and substrate, is measured as a function of frequency and compared to the theoretical dispersion curve. The use of this method is of particular interest because the effect of surface irregularities and roughness is minimized.

3. Analysis and measurement of stresses in coatings and thin films

Stresses in coatings usually arise from one or more of the following sources: (a) residual stresses due to thermal mismatch as the coating is cured or thermally processed; (b) residual stresses due to solvent evaporation or change in volume; and (c) contact as the coating is indented. These stresses may lead to yielding and permanent deformation of the coating, or fracture along the interface or in the coating itself (Section 4). In all cases, the stresses arise because coating displacement is inhibited by the adhesion of the coating to the substrate. In some cases, the residual stresses are extremely high, as in the processing of ceramic coatings, where the stresses can reach values as high as 1 GPa [55].

A large literature exists on the numerical analysis of coated systems. In most cases, either boundary value or finite element analysis is used to determine residual and contact stresses. The reader is referred to ref. [68], which contains a bibliography of 259 papers dealing with the use of numerical methods in the analysis of coated systems published between 1996 and 1998 alone.

3.1. Residual stresses due to thermal mismatch

Consider a thin coating on a substrate having coefficients of thermal expansion α_T^c and α_T^s, respectively. Since the coating is thin, the isotropic normal stress in the coating plane can be considered approximately constant through the thickness and, in the case of a coating–substrate system, cooled from temperature T_2 to T_1, is given by:

$$\sigma_t = \int_{T_2}^{T_1} \frac{E}{1-\nu} \left(\alpha_T^c - \alpha_T^s \right) \mathrm{d}T \tag{15}$$

In the case of $\alpha_T^s > \alpha_T^c$ (or $T_1 > T_2$), the coating stress is compressive, and the coating might be subject to buckling and blistering. If $\alpha_T^c > \alpha_T^s$, the tendency is for the coating to contract more than the substrate thereby generating a tensile stress which may crack the coating. The use of Eq. 15 in conjunction with interfacial fracture analysis is explained in detail in Section 4.2.2.

In practice, the ability of a coating to carry stresses is significantly reduced when the coating is heated above its glass transition temperature, T_g, so that T_g may replace T_2 in the above expression. The residual normal stress given by Eq. 15 often leads to shear stresses and 'peeling' or normal stresses perpendicular to the coating–substrate interface. Typically, for the case where the coating is very thin compared to the substrate, the ratio of shearing (mode II) to peeling (mode I) stress intensity factors for a crack at the interface is in the range 1.0–1.7.

There is a variety of methods available to coating, substrate and interface stresses. For metallic films, X-ray diffraction, in which the diffraction angle of X-rays is correlated to the internal strain in the crystalline structure of the coating, can be used to characterize the residual stress state. A common method is the so-called '$\sin^2 \Psi$' method, normally used when anisotropy is expected in the residual stress. The elastic lattice strain is measured via X-ray diffraction with respect to Euler angles, φ and ψ as $\epsilon_{\varphi,\psi}$, related to the Cartesian strain tensor, ϵ_{ij}, and finally to the residual stress tensor (see, for example, [69,70]).

A method that is commonly used with vacuum deposited thin films is the cantilever technique. A thin slab is cantilevered and, as a coating is deposited on it, strains are expected to develop in it due to thermal mismatch, recrystallization processes, and phase transformations. Using elementary beam theory, the average stress in the coating can be related to the radius of curvature before, R_b, and after, R_a the film is deposited according to the well-known Stoney [71] relationship:

$$\sigma = \frac{1}{6} \frac{E_s h_s^2}{(1 - \nu_s)h_f} \left(\frac{1}{R_a} - \frac{1}{R_b} \right) \tag{16}$$

where E, ν, and h are the Young's modulus, Poisson's ratio, and thickness with the subscripts s and f referring to substrate and film. Either the radius of curvature or the cantilever end deflection can be measured in real time to monitor the development of residual stresses. Often, optical methods such as laser reflection fibre optic strain gauges, or laser interferometry are used to measure cantilever end deflection, which can then be related to beam curvature [72]. A recently published article reviews residual stress measurement and analysis as it applies to vacuum deposited thin films, and contains descriptions of these techniques [72]. Eq. 16 can be used to monitor the average stress (sum of intrinsic stresses, thermal mismatch stresses and external stresses) in thin films during the deposition process, and is accurate for cases in which $h_f \ll h_s$. A more general formula that incorporates a correction for thicker coatings is [72]:

$$\sigma = \left[\frac{1}{6} \frac{E_s h_s^2}{(1 - \nu_s)h_f} \left(\frac{1}{R_a} - \frac{1}{R_b} \right) \right] \frac{\left[1 + \frac{E_c(1 - \nu_s)}{E_s(1 - \nu_c)} \left(\frac{h_s}{h_c} \right)^3 \right]}{1 + \frac{h_s}{h_c}} \tag{17}$$

A more generally applicable method is based on the use of nano- or micro-indenters. The indenter load–displacement curve is recorded over both loading and unloading. By assuming that the contact pressure does not change in the presence of elastic residual stresses, equi-biaxial residual stresses can be determined from indentation tests for an elastic coating and substrate [73].

Residual stresses can also be determined analytically by treating both the coating and substrate as beams (e.g. [74]). The results can be used to estimate the likelihood of cracks occurring in the coating itself or of cracks initiating at the interface (as opposed to the propagation of existing flaws at the interface which can be analyzed using the methods of Section 4).

3.2. Contact stresses

The wide use of micro- or nano-indenters to determine coating material properties and residual stresses often relies on elastic contact analysis. In addition, coating contact stress is associated with erosive failure of coatings due to particle impact. Erosive behaviour is often modeled by assuming that impact velocities are high, so that coating response is fully plastic, and elastic stress analysis can be used to estimate the conditions under which this is the case. As will be shown in Section 4.2.3, impact or indentation of a coating can lead to buckling and delamination in coatings and elastic contact analysis can be used to estimate the interfacial stresses that lead to the initiation of interfacial cracks.

There is a variety of techniques available to analyze contact stresses in coating systems. The problem is analytically complex, and simplifying assumptions must often be made to obtain solutions. For example, in many cases it is not possible to satisfy all of the equilibrium, boundary, and surface displacement conditions exactly. In addition, sharp indenters sometimes lead to stress singularities at edges.

Solutions for simplified cases in which it is assumed that plane sections of the coating remain plane after indentation-induced compression, can be found in the book by Johnson [75]. Johnson's approach was subsequently extended by Jaffar [76]. An early work [24] assumed frictionless contact and used elastic continuum mechanics to estimate the stress distribution in layered materials. In this analysis, force and displacement continuity at the interfaces was satisfied exactly, while the surface displacement condition (profile of the indenter), was only approximated. These authors analyzed indentation stresses created by circular flat-ended and parabolic indenters numerically on both single and multilayer systems, with good results.

A similar, but further simplified approach is that of Matthewson [34]. The assumptions are that the indenter is rigid, and that the coating is thin enough so that the stresses in the coating may be adequately described by the average of each stress through the thickness. This leads to an over determined problem in which not all of the continuity conditions can be met, but can be satisfied approximately. The approach, however, matches experimental observations quite well, and can be used to obtain estimates of the stress distribution due to contact of axisymmetric profiles.

Another approach is to extend the classical contact theory for indenters on elastic half-spaces developed by Hertz [77] and Huber [78] to the case of layered materials. An example of such an approach is ref. [79], in which the authors modify the Hertz/Huber analysis by considering the coating material properties as a function of indentation depth. Mathematically, the authors treat the transition from coating to substrate as a discontinuity in Young's modulus and Poisson's ratio represented by a Heaviside step function, and re-derive the appropriate Hertzian equations. The results match FEA calculations well.

4. Analysis of adhesion and interfacial fracture toughness of coatings

Cracking, whether in the coating itself, or along the coating/substrate interface (delamination) is one of the most common modes of failure of coatings and results from stresses that can be created by a variety of effects, such as thermal mismatch, coating swelling or shrinkage, and indentation or particle impact. A coating on a substrate is a bimaterial system, and as such is often characterized by sharp gradients in properties in the region of the interface. The problem is further complicated by the sensitivity of coating–substrate interface adhesion to pretreatment (or contamination), the properties of both the coating and substrate, and the thickness of the coating, resulting in a large amount of scatter in experimental data. The aim of this section is to introduce some of the more important techniques available for the analysis of interfacial fracture of coatings. Discussion will be detailed enough to provide a basic understanding of interfacial cracking, but will omit some of the mathematical details of the analyses.

Much of what is summarized here is presented in more detail in the excellent article on mixed-mode cracking in layered materials by Hutchinson and Suo [55]. In that reference, the authors consider not only interfacial fracture, but also cracking within coatings and cracking in substrates below coatings. Only interfacial cracking (delamination) will be considered in detail here. Before proceeding however, we will first discuss some basic concepts of fracture mechanics.

4.1. Mixed-mode cracking

It is usual to characterize the fracture behaviour of materials by the use of either stress intensity factors, usually denoted by K, which describe the intensity of stress near a crack tip, or energy release rates, usually denoted by G or J, which give the amount of elastic energy the system provides to extend a crack a unit area. There is a great body of literature on the use of these parameters to analyze coating systems. For example, fracture mechanics techniques have been used to study debonding of thin films due to residual stresses (e.g. [53,55,80–84]), due to indentation [6,52–56,85], and to calibrate various specimens used to evaluate the adhesion strength of coatings [51,55,57,86,87].

For homogenous, isotropic materials, K and G are related through the well-known Irwin relationship, which for planar problems is:

$$G = G_I + G_{II} = \frac{K^2}{\bar{E}} = \frac{K_I^2 + K_{II}^2}{\bar{E}} \tag{18}$$

where G, G_I, G_{II} are the total, mode I, and mode II components, respectively, of the strain energy release rate. K, K_I, and K_{II} are the corresponding total, mode I, and mode II stress intensity factors, and $\bar{E}$ depends on whether the conditions at the crack tip are plane stress or plane strain, according to:

$$\bar{E} = E \qquad \text{plane stress}$$
$$\bar{E} = \frac{E}{1 - v^2} \qquad \text{plane strain} \tag{19}$$

Mode I crack growth occurs when tensile stresses act perpendicular to the crack plane, tending to split the material. A mode II loading results from shear stresses acting on the opposing crack faces in the direction perpendicular to the line describing the crack front. Thus, working in terms of stress intensity or strain energy release rates is essentially equivalent, keeping Eq. 18 in mind. The strain energy release rate depends on the geometry, materials and loads on the system, and it is assumed that crack propagation in homogeneous systems occurs when the applied strain energy release rate, G, equals the critical strain energy release rate G_{IC} (a material property) in mode I (crack opening mode). Because the crack is free to propagate in any direction in a homogenous material, and because critical strain energy release rates are usually lowest in mode I, cracks tend to follow the path that maximizes G_I (and makes $G_{II} = 0$); therefore G_{IC} is sufficient to describe the fracture strength of the material. In contrast to this homogenous case, cracks along bimaterial interfaces are often 'trapped' by the interface and forced to travel along it. The cracks cannot, in these cases, propagate along the path that maximizes mode I, and are thus referred to as mixed-mode cracks.

If we restrict ourselves to 2D problems, a crack along a bimaterial interface will generally be in a state of combined mode I, in which stresses ahead of the crack tip tend to open the crack, and mode II, in which stresses ahead of the crack tip tend to shear the crack faces (there are some limitations to this interpretation of mode mix for cases where the materials have large differences in mechanical properties, and this will be discussed in the following section). Following the work of Hutchinson and Suo, the mode mix is often expressed as a phase angle, ψ, defined as [55]:

$$\psi = \tan^{-1}\left(\frac{K_{II}}{K_I}\right) \tag{20}$$

so that $\psi = 0°$ and $\psi = \pm 90°$ refer to pure mode I, and mode II, respectively.

The general procedure to predict fracture in mixed-mode problems is similar to traditional linear elastic fracture mechanics (LEFM) procedures; that is, the applied stress intensity, K_I, is compared to the critical stress intensity at crack propagation K_{IC} (a material property). The crack propagates when $K_I = K_{IC}$. The difference with mixed-mode problems is that the critical stress intensity or strain energy release rate is a function of the mode mix, so that in order to characterize the strength of the interface, a 'fracture envelope' that describes the dependence of the total critical stress intensity, K_C, on the mode mix must be constructed experimentally:

$$K_C = K_C(\psi) \tag{21}$$

An example of such a fracture envelope, in this case developed for the case of mixed-mode cracking of an adhesive joint system consisting of 7075-T6 aluminium adherends bonded with a 0.4 mm thick structural epoxy appears in

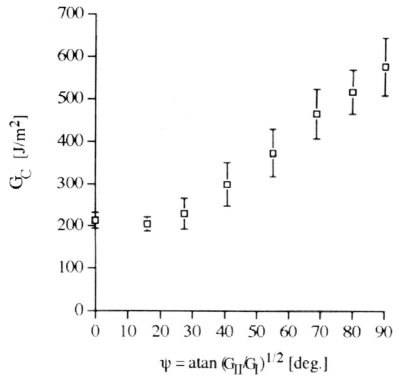

Fig. 5. Fracture envelope, G_c versus phase angle, ψ, for FPL-etched 7075-T6 aluminum adherends bonded with 0.4 mm Cybond 4523GB epoxy adhesive cured at 150°C for 45 min [88].

Fig. 5 [88]. In this case, the fracture parameter is expressed as J_c, the energy release rate, rather than K_c. Note that a relationship similar to that of Eq. 18, relates the energy release rate, J, to the stress intensity, K. In ref. [88], J was used rather than G to emphasize that some of the specimens studied behaved in a non-linear elastic manner (and thus the J integral was used to calculate the energy release rate). For linear elastic behaviour, G is identical to J. The mode I fracture toughness (i.e. $\psi = 0°$) is significantly lower than the mode II fracture toughness ($\psi = \pm 90°$). It is important to note that a fracture envelope is defined for a specific material system and method of manufacture. For example, for the system of Fig. 5, fracture occurred within the bond line, and thus the relevant parameters that affect fracture toughness, and hence must be specified, are the adherend and adhesive material, the bond line thickness, the adherend pretreatment, and the adhesive cure cycle. For coatings that fail due to delamination along the interface, specification of the pretreatment procedure on the substrate is of particular importance, although fracture toughness may also depend on parameters such as the coating application procedures, coating thickness and coating age. The procedure used to generate

Fig. 5 for an adhesive system is essentially the same as that used for an epoxy coating [39]. In this case, a secondary bond was used to join the coating to another adherend thereby creating a DCB fracture specimen. The mixed-mode fracture toughness of the coating was then determined using an approach related to Eq. 9. This method of analysis is slightly less accurate than that described in the next section, but may be adequate for many applications.

Once the particular coating system is characterized, any geometry utilizing this system can be described by the fracture envelope. Of course, the applied stress intensity or strain energy release rate must be found in terms of the mode I and II components, for comparison with Eq. 21; i.e. the crack propagates when

$$K(\psi) = K_c(\psi) \tag{22}$$

where ψ here refers to the particular mode mix at the crack tip.

In order to perform mixed-mode analyses, it is thus necessary to analytically determine the applied stress intensity, K (or equivalently the applied strain energy release rate, G), and applied phase angle, ψ, as a function of the particular loading and geometry, so that the left side of Eq. 22 is fully determined. The right side of Eq. 22 is determined experimentally using interfacial fracture specimens, as will be discussed below. The discussion in the next section will focus on the determination of K and ψ for coating systems in which fracture occurs along the interface.

4.2. Interfacial cracking: method of Suo and Hutchinson

4.2.1. Fracture mechanics concepts

Suo and Hutchinson developed a general method for the analysis of mixed-mode interfacial cracking [89]. Because of its wide applicability, it is worth reviewing it in some detail here. The notation for the fracture mechanics analysis that follows is the one introduced by Rice [90], and used by Hutchinson and coworkers [55,89], and is chosen because the expressions reduce to their LEFM equivalents if both materials are the same.

An interface crack in a bimaterial system (Fig. 6) involves elastic mismatch, which is described in terms of the dimensionless Dundurs parameters [91]. For a coating on a substrate, these are:

$$\alpha = \frac{\mu_C(\kappa_S + 1) - \mu_S(\kappa_C + 1)}{\mu_C(\kappa_S + 1) + \mu_S(\kappa_C + 1)} \tag{23}$$

and

$$\beta = \frac{\mu_C(\kappa_S - 1) - \mu_S(\kappa_C - 1)}{\mu_C(\kappa_S + 1) + \mu_S(\kappa_C + 1)} \tag{24}$$

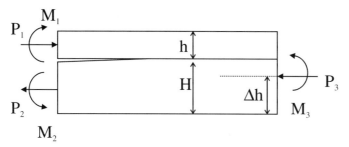

Fig. 6. The definition of geometry and loading for a bimaterial system with an interface crack used in the Suo and Hutchinson analysis [89].

where the subscripts C and S refer to the coating and substrate. Assuming plane strain conditions,

$$\kappa_C = 3 - 4\nu_C \quad \text{and} \quad \kappa_S = 3 - 4\nu_S \tag{25}$$

where ν is the Poisson's ratio and μ is the shear modulus. The stresses on the bimaterial interface directly in front of the crack at a distance r, are in the form [90]:

$$\sigma_{22} + i\sigma_{12} = \frac{(K_1 + iK_2)}{\sqrt{2\pi r}} r^{i\varepsilon} \tag{26}$$

where the parameter ε is given by

$$\varepsilon = \frac{1}{2\pi} \ln\left(\frac{1-\beta}{1+\beta}\right) \tag{27}$$

and the complex stress intensity factor $K = K_1 + iK_2$ has real and imaginary parts which are similar to the conventional mode I and mode II stress intensity factors in LEFM. In fact, were it not for the $r^{i\varepsilon}$ term, Eq. 26 would give the mode I (σ_{22}) and mode II (σ_{12}) components of stress ahead of the crack, and K_1 and K_2 would reduce to the K_I and K_{II} of traditional LEFM. However, for $\beta \neq 0$, the $r^{i\varepsilon}$ term in introduces an oscillating singularity and complicates the usual notions of stress intensity and mode mix. Unfortunately, commonly used coatings and substrates often have a combination of material properties resulting in nonzero β. Hutchinson and Suo suggested, in this case, a definition of mode mix in terms of a characteristic length, l, as follows [55]:

$$\Psi = \tan^{-1}\left[\frac{\text{Im}(Kl^{i\varepsilon})}{\text{Re}(Kl^{i\varepsilon})}\right] = \tan^{-1}\left(\frac{K_2}{K_1}\right) = \tan^{-1}\left(\frac{\sigma_{12}(r=l)}{\sigma_{22}(r=l)}\right) \tag{28}$$

where K is the complex stress intensity factor. A mode I crack then becomes one that has zero shear stress, σ_{12} at a distance l ahead of the crack tip, and a mode II crack one that has zero normal stress, σ_{22} at that point. In the case of $\varepsilon = 0$, or where elastic mismatch is small, Eq. 28 reduces to the traditional LEFM value given in Eq. 20. The choice of the reference length, l, is arbitrary, and, in the thin film case, the coating thickness, h, is normally used. The mode mix depends on coating thickness, but Eq. 28 shows that data at differing coating thickness can be compared using the transformation law [55]:

$$\psi_2 = \psi_1 + \varepsilon \ln\left(\frac{h_2}{h_1}\right) \tag{29}$$

where h_1 and h_2 are the two coating thicknesses being compared.

Eq. 28 shows that ψ can be in the range $\pm 90°$ when $K_1 \geq 0$ since K_2 can be positive or negative depending on the sign of the shear stress, σ_{12}. The case of $K_1 < 0$, implying crack face compression and friction, is strictly not allowed by this analysis, and will be discussed below.

The relationship between G and K for interfacial cracking is:

$$G = \frac{(1 - \beta^2)}{2}\left(\frac{1}{\bar{E}_1} + \frac{1}{\bar{E}_2}\right)(K_1^2 + K_2^2) \tag{30}$$

where the subscripts in the Young's moduli refer to the materials on either side of the bimaterial interface, and the over-bars refer to the definition given in Eq. 19. This again reduces to the traditional LEFM value given in Eq. 18 for zero elastic mismatch.

An artifact of the oscillating singularity in Eq. 26 is that the crack faces are predicted to contact at some distance behind the crack tip. In most cases, the size of the region in which the crack faces are in contact is very small compared to the process zone (plastic zone ahead of crack tip), and thus the usual argument of LEFM can be invoked (i.e. the behaviour of the material in the process zone is assumed to be characterized by the state at some distance from the crack tip where the stress state, as given by Eq. 26, is well-defined [92]). Hutchinson and Suo give an estimate of the range of ψ over which the crack faces are expected to be open as a function of two parameters, l (characterizing the size of the process zone), and L (which characterizes an in-plane length of the cracked geometry being considered). For $\varepsilon > 0$, the condition for which the crack is open is [89]:

$$-\frac{\pi}{2} + 2\varepsilon < \psi < \frac{\pi}{2} + 2\varepsilon - \varepsilon \ln\left(\frac{1}{10}\frac{L}{l}\right) \tag{31}$$

For a coated system, typical values of l and L are, respectively, the coating thickness, and the crack length. In many cases, this range is sufficiently large to avoid the complications of stress oscillation. For example, for an alkyd paint on a steel substrate, undergoing impact-induced buckling and delamination, the range was $-84° < \psi < 90°$ [6], which is almost the full range of possible phase angles.

Let us now proceed to the actual determination of the stress intensity factors and phase angles necessary for interfacial fracture analysis of coatings. As mentioned earlier, Hutchinson, Suo and their coworkers solved the problem of an interfacial crack between two elastic layers in a very general form, so that many different types of interfacial cracking problems could be treated [89]. Their solution is in terms of the beam theory reactions (i.e. forces and bending moments) at the edges of the layers on either side of the crack, as shown in Fig. 6, and is valid for any combination of coating and substrate thickness. We shall now review this general case, and then the specific case applicable to thin coating–substrate systems.

4.2.2. Solution for general loading

This section will present the work done by Suo and Hutchinson on a generally applicable analysis of interfacial fracture. The details of the analysis will not be considered here, only the final outcome. The problem was originally solved in 1990 [89], but we shall follow the convention of the authors in their more compact reformulation of the problem in 1992 [55].

Consider the two elastic layers with an interfacial crack parallel to the layers shown in Fig. 6. All loads at the edges of the layers are per unit width and applied at the centroids of their respective beam elements. The strain energy release rate, stress intensity factor, and mode mix for this crack can be calculated in terms of the reactions at the edges of the layers. The assumption here is that stress fields in the layers can be approximated by the beam-theory approximations, and that shear deformations are negligible. The portion to the right of the crack tip in Fig. 6 can be considered a composite beam, and the following geometric parameters are useful for deriving fracture parameters:

$$\Sigma = \frac{\bar{E}_1}{\bar{E}_2}, \qquad \eta = \frac{h}{H}, \qquad \Delta = \frac{1 + 2\Sigma\eta + \Sigma\eta^2}{2\eta(1 + \Sigma\eta)} \tag{32}$$

The strain energy release rate, G, can be calculated by taking the difference in strain energy per unit length per unit width stored in the layers far behind and ahead of the crack tip (this can also be derived by a J-integral formulation, as was done, for example, in [88]), giving:

$$G = \frac{1}{2\bar{E}_1}\left(\frac{P_1^2}{h} + 12\frac{M_1^2}{h^3}\right) + \frac{1}{2\bar{E}_2}\left(\frac{P_2^2}{H} + 12\frac{M_2^2}{H^3} - \frac{P_3^2}{Ah} - \frac{M_3^2}{Ih^3}\right) \tag{33}$$

where A and I are the dimensionless composite beam area and moment of inertia given by:

$$A = \frac{1}{\eta} + \Sigma,$$

$$I = \Sigma\left[\left(\Delta - \frac{1}{\eta}\right)^2 - \left(\Delta - \frac{1}{\eta}\right) + \frac{1}{3}\right] + \frac{\Delta}{\eta}\left(\Delta - \frac{1}{\eta}\right) + \frac{1}{3\eta^3} \tag{34}$$

The corresponding complex stress intensity factor, which is more useful because it gives mode mix information, is:

$$K = h^{-i\varepsilon} \sqrt{\left(\frac{1-\alpha}{1-\beta^2}\right)} \left(\frac{P}{\sqrt{2hU}} - ie^{i\gamma}\frac{M}{\sqrt{2h^3V}}\right) e^{i\omega} \tag{35}$$

where the loads P and M are given by

$$P = P_1 - C_1 P_3 - \frac{C_2 M_3}{h}, \qquad M = M_1 - C_3 M_3 \tag{36}$$

with

$$C_1 = \frac{\Sigma}{A}, \qquad C_2 = \frac{\Sigma}{I}\left(\frac{1}{\eta} + \frac{1}{2} - \Delta\right), \qquad C_3 = \frac{\Sigma}{12I} \tag{37}$$

and

$$\frac{1}{U} = 1 + \Sigma\eta\left(4 + 6\eta + 3\eta^2\right), \qquad \frac{1}{V} = 12\left(1 + \Sigma\eta^3\right),$$

$$\frac{\sin\gamma}{\sqrt{UV}} = 6\Sigma\eta^2(1+\eta) \tag{38}$$

By separating the real and imaginary parts of Eq. 35, the phase angle describing the mode mix, as described by Eq. 28 with $l = h$, is:

$$\psi = \tan^{-1}\left(\frac{\sqrt{\dfrac{V}{U}}\dfrac{Ph}{M}\sin\omega - \cos(\omega+\gamma)}{\sqrt{\dfrac{V}{U}}\dfrac{Ph}{M}\cos\omega + \sin(\omega+\gamma)}\right) \tag{39}$$

With the exception of the unknown parameter, ω, the entire solution up to this point was determined by arguments associated with linearity, geometry, and dimensionality. The determination of ω, however, requires rigorous solution of the elasticity problem. This was done by Suo and Hutchinson using integral equation methods, and a tabulation of ω, which depends on α, β, and η, appears in ref. [89]. The variation of ω with these parameters for two specific cases is shown in Fig. 7, from ref. [55].

The preceding analysis is valid for any combination of edge loads, as long as the interface crack propagates parallel to the two elastic layers. All that is required is that the stresses in the layers be expressed in terms of the edge loads in Fig. 6. It is thus possible to study a multitude of interfacial cracking problems by combining analyses based on basic strength of materials approaches (to determine the edge loads) with the above equations. In this manner, the applied stress intensity factor, energy release rate, and mode mix can be determined as a function of loads applied far from the crack tip. The applied parameters are then matched to corresponding experimentally determined values (e.g. [88]) in order to predict fracture loads.

It is important to note at this point, that combinations of edge loads that result in a negative K_1 (i.e. the denominator of Eq. 39) always result in crack closure,

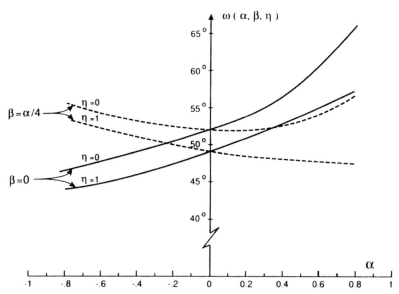

Fig. 7. The function $\omega(\alpha,\beta,\eta)$ for two commonly occurring cases of β [55].

and therefore render the analysis invalid because friction on the crack faces has not been taken into account. However, it is not unreasonable to expect that a system in which the crack tends to close would be more resistant to crack propagation than one in which the crack tends to open, so that a system giving negative K_1 can be considered desirable from a design point of view. If the substrate is much thicker than the coating, the results of the simplified analysis in the following section can be used.

Ref. [55] reviews the use of this method to model interfacial cracks in pre-stressed coatings propagating from edges and holes. In these cases, the only loads are on the top layer (i.e. the prestress load), but no restriction is put on the thickness of the substrate. To find the proper direction of the edge load corresponding to a coating under residual stress, superposition can be used, as described in refs. [55,89], and commonly referred to as the 'cut and paste' procedure. Consider a coating–substrate system initially unstressed at a temperature T_1 that is cooled to a temperature T_2. The difference in thermal expansion coefficients will cause a biaxial misfit stress in the coating given by Eq. 15. If there is an interfacial crack, the question that arises is what is the equivalent edge load on the coating in Fig. 6. If $\alpha_T^c > \alpha_T^s$, then the coating tends to contract more than the substrate, as shown in Fig. 8a. If a tensile load given by Eq. 15 is applied to the coating, the misfit strain is recovered, and if the two layers are bonded as shown in Fig. 8b, then the interface is stress free, and $G = 0$. In order to obtain the original problem (i.e. misfit stress across interface, but no edge load), it is required to add on a compressive edge stress equal to the misfit stress, as shown in Fig. 8c. Because

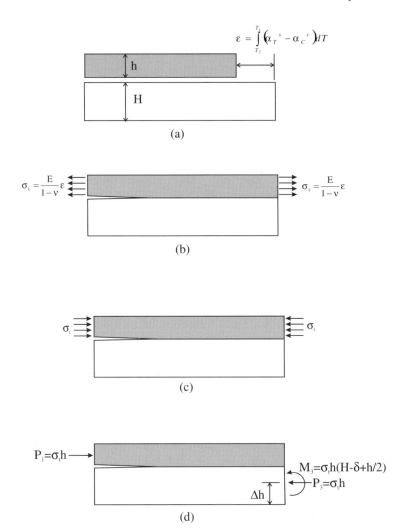

Fig. 8. Superposition to obtain edge loading for residual stresses [55,89] for case where coating–substrate system is cooled from T_1 to T_2, with $\alpha_c > \alpha_s$. (a) Misfit strain when coating and substrate not bonded. (b) Tensile stress required to cancel misfit strain; $G = 0$ in this case. (c) Compressive stress required to obtain original loading of no edge load, but with misfit strain. (d) Equivalent loading for use in method of Suo and Hutchinson [89].

$G = 0$ for Fig. 8b, the equivalent edge load P_1 for use in Eqs. 33–39 is the one corresponding to the loading in Fig. 8c (i.e. that shown in Fig. 8d). The whole procedure can be repeated for the case of $\alpha_T^s > \alpha_T^c$, in which case the resulting misfit stress is compressive, and the resulting equivalent edge load is tensile.

The results of the analysis can also be used to interpret data from interface fracture toughness specimens. As a simple illustration, consider the case of the

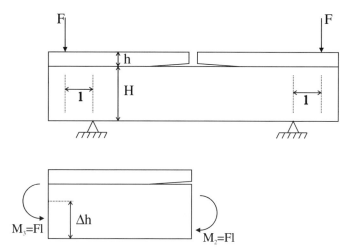

Fig. 9. The four-point bend specimen for interfacial toughness measurement, and equivalent edge loading.

commonly used four-point bend flexure specimen shown in Fig. 9. A notch is introduced in the coating at the centre of the specimen, and upon loading, an interfacial crack will propagate as shown. In this case, the only load is applied to the bottom layer (i.e. the substrate) and, using the conventions of Fig. 6, has magnitude $M_2 = Fl$. The corresponding load on the composite beam side is, of course, $M_3 = Fl$, as shown in Fig. 9. Specializing Eq. 33 to this case results in:

$$G = \frac{(Fl)^2}{\bar{E}_2 h^3}\left(6\eta^3 - \frac{1}{2I}\right) \tag{40}$$

where $\bar{E}_2$ refers to the lower material (i.e. the substrate). Eq. 35 and 39 can be similarly specialized and used to calculate the interfacial stress intensity factor and phase angle. An advantage of such a specimen is that once a steady-state crack length has been reached, the strain energy release rate does not depend on crack length, making it well suited for measuring the toughness of an interface. Both G and ψ are functions of the thickness ratio, η, and this fact can be used to obtain the interfacial strength of a coating over a wide range of phase angles simply by changing the coating and/or substrate thickness. A solution also exists for a modified specimen [93] where there also is a tensile load applied at the neutral axis of the composite beam [89]. It is found that by varying the magnitude of the tensile load and the bending moment (Fl), the full range of phase angles can be obtained. The four-point bend specimen can also been used to determine the interfacial fracture strength in multi-layer systems [94]. Comparisons of calculated G using composite beam theory and finite element analysis reveal good agreement.

Other examples of specimens that may be used to quantify interfacial fracture toughness as a function of mode mix include the blister test, the Brazil

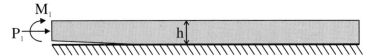

Fig. 10. Simplified edge loading for a case where substrate is much thicker than coating.

nut sandwich [95], the indentation test and the modified DCB specimen (see Section 2.3.1).

4.2.3. Thin coatings and films with loading on coating only

In many common cases (e.g. thin films), the coating is much thinner than the substrate, and the solution of the previous section is simplified by setting $\eta = 0$. If we further restrict our discussion to cases where the only load is on the thin layer (Fig. 10), we can further simplify the expressions for strain energy release rate, stress intensity, and phase angle. This is suitable for discussion of the thin film problems of buckling delamination, blistering, and delamination due to thermal mismatch or other residual stresses. In all of these situations, the mismatch in stresses can be expressed as an edge load on the thin coating alone, with no loading of the substrate on the edge. In this case, for $h \ll H$, $P_2 = M_2 = \eta = 0$, Eqs. 33, 35, and 39 become, respectively,

$$G = \frac{6\left(1 - v_1^2\right)}{E_1 h^3} \left(M_1^2 + \frac{h^2 P_1^2}{12}\right) \tag{41}$$

$$K = h^{-i\varepsilon} \sqrt{6\left(\frac{1 - \alpha}{1 - \beta^2}\right) \frac{1}{h^{3/2}} \left(\frac{h P_1}{\sqrt{12}} - i M_1\right) e^{i\omega}} \tag{42}$$

$$\tan \psi = \frac{\mathrm{Im}(K h^{i\varepsilon})}{\mathrm{Re}(K h^{i\varepsilon})} = \frac{\sqrt{12} M_1 \cos\omega - h P_1 \sin\omega}{-\sqrt{12} M_1 \sin\omega - h P_1 \cos\omega} \tag{43}$$

where E_1 and v_1 refer to the coating. Note that the assumed directions of M_1 and P_1 are as in Fig. 10.

This simpler case has been used with good success in the analysis of indentation-induced delamination, as shown in Fig. 11 [53]. In that study, the indentation was assumed to produce a fully plastic deformation just below the point of contact, resulting in a uniform biaxial compressive elastic stress in the coating adjacent to this zone of plastically deformed coating. By treating a de-laminating coating as a clamped disk under biaxial compressive stress, it was possible to predict the critical load under which the coating would begin to buckle and delaminate. In the case of a coating indented such that buckling occurred, an asymptotic postbuckling analysis was used to determine the loads at the edge of

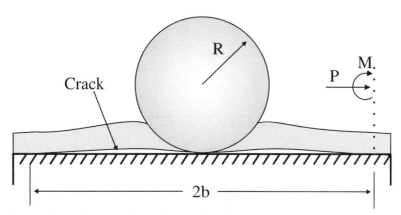

Fig. 11. Indentation induced buckling delamination of a coating. Equivalent edge loading on thin layer also shown.

the coating (see Fig. 11) that drove the delamination. Coupling the postbuckling loads at the edge of the delamination with a strain energy analysis similar to that reviewed above, resulted in expressions for the strain energy release rate and mode mix for indentation-induced interfacial cracks. Note that even in the absence of buckling, the compressive load may drive delamination with a phase angle equal to ω (Eq. 43 with $M_1 = 0$), which depends on the Dundurs parameters α and β, and can be found tabulated in ref. [89].

The above approach was modified to model the delamination of alkyd paint on a steel substrate by the impact of a spherical particle. The buckling analysis accounted for the presence of the indenting particle by setting coating deflection to zero over the area at which the sphere contacts the substrate at the point of maximum penetration [6,56]. This yielded the bending moment and compressive force needed for evaluation of the strain energy release rate, Eq. 41, and mode ratio, Eq. 43. It should be noted that, although the material properties were such that ε was nonzero, so that the difficulties associated with the oscillating singularity explained in Section 4.2.1 existed, a comparison of measured and predicted delamination radii revealed excellent agreement in all cases. It is also worth noting that the preceding analyses model interfacial crack propagation, not initiation, so that an existing flaw is assumed in the coating. There has been little discussion in the literature regarding conditions leading to initiation of interface cracks under indentation. For spherical particles impacting alkyd paint on steel, it was found that buckling initiation corresponded to the penetration of the coating to the substrate, generating a critical interfacial shear stress [6,56].

A number of authors have studied coating blistering due to biaxial compressive residual stresses arising from, for example, thermal mismatch between coating and substrate [53,55,84,96,97]. Blistering can also be driven by water absorption in coatings and resulting osmotic effects. In ref. [98], it was found that blisters grew in adhesive joints subject to environmental aging due to osmotic pressure

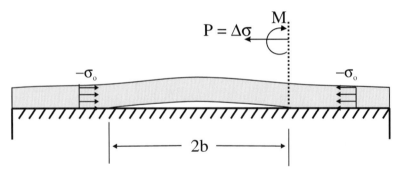

Fig. 12. Blistering of coating due to compressive residual stress inducing buckling. Note that edge normal load, the change in stress due to buckling, $\Delta\sigma$, is positive.

originating in water clusters at microscopic voids. Blistering arises from the coating buckling (see Fig. 12), and an analysis similar to that used to study indentation induced buckling can be used to find the critical compressive stress. A delamination of length $2b$ is assumed present in the interface, and, in the absence of buckling, there is no difference in strain in the coating on either side of the delamination so that crack propagation cannot occur. If buckling is found to occur, then a postbuckled analysis similar to that previously used in indentation-induced buckling analyses can be used to find the edge loads (normal and bending moment) on the top layer. Eqs. 41–43 may then be used to calculate the energy release rate, stress intensity factor and phase angle of the interfacial crack. It is important to note once again that, if buckling does not occur, the crack cannot propagate; therefore, the normal edge load should be the *change* in stress from the unbuckled to buckled state, $\Delta\sigma$, and not the total stress in the coating. Thus, the edge condition on the crack is tensile, as shown in Fig. 12, since buckling relaxes the compressive stress state in the coating. For a straight sided blister, coupling of the buckling solution (in this case the coating is treated as a wide beam under compressive stress and the buckling solution is exact in closed form) to the interfacial crack solution, Eqs. 41–43, give the strain energy release rate and phase angle of loading as [97]:

$$G = \frac{(1 - \nu_c^2)h\sigma_o^2}{2E_c}\left(1 - \frac{\sigma_c}{\sigma_o}\right)\left(1 + \frac{3\sigma_c}{\sigma_o}\right) \tag{44}$$

$$\psi = \frac{\text{Im}(Kh^{i\varepsilon})}{\text{Re}(Kh^{i\varepsilon})} = \tan^{-1}\left[\frac{2 + \tan\omega\sqrt{\dfrac{\sigma_o}{\sigma_c} - 1}}{-2\tan\omega + \sqrt{\dfrac{\sigma_o}{\sigma_c} - 1}}\right] \tag{45}$$

where the material properties are of the coating, σ_o is the compressive residual stress in the coating, and σ_c is the critical compressive stress required for buckling, which is equal to:

$$\sigma_c = \frac{\pi^2 E_c \left(\dfrac{h}{b}\right)^2}{12(1 - v_c^2)} \tag{46}$$

where b is half of the delamination length (Fig. 12). This gives the expected result that a thin coating is more likely to buckle than a thick one. This result also has important implications for the propagation of a blister, as was discussed at length in [97]. As b increases in Eq. 46, σ_c decreases, and therefore Eq. 44 predicts that the crack driving force, G, increases as the delamination grows. This would lead to unstable growth of the blister (i.e. the blister would grow without limit), which contradicts experimental observation. However, examination of Eq. 20 shows that, as b increases, the mode II contribution increases. For example, in the absence of elastic mismatch, $\omega = 52.1°$, and at $\sigma_0 = \sigma_c$, $\psi = -37.9°$; whereas when $\sigma_0 = 7.6\sigma_c$, $\psi = -90°$ (i.e. pure mode II). Because the fracture toughness in mode II is generally significantly higher than that in mode I, a propagating blister encounters increasingly higher fracture toughness and growth is stable.

Eq. 45 indicates that K_1 will be negative for

$$\frac{\sigma_0}{\sigma_c} > 1 + 4\tan^2 \omega \tag{47}$$

so that some account of friction between crack faces must be made in these cases [97].

Similar solutions exist for the more commonly encountered case of circular blisters in coatings [53,55,96]. In this case, the buckling analysis is performed on a clamped circular disk of radius $2b$ (Fig. 12) and is more complex, so that either an asymptotic solution is used, similar to that used for indentation induced buckling presented earlier in this section, or a numerical analysis is necessary. It is found that both the phase angle and energy release rate do not vary as sharply with σ_0/σ_c as with the straight sided blister, and that under certain conditions, the interfacial crack is found to kink upwards into the coating, producing a circular spall [53,55]. As before, analyses of this type describe only the growth of a delamination, and not its initiation. Because stresses do not develop at the interface in compressed films in the absence of buckling, a relatively large interface flaw is necessary for crack propagation to occur. Contaminants at the interface can produce such flaws and the minimum flaw size necessary for the coating to buckle can be derived from Eqs. 44–46, knowing the critical energy release rate $G_c(\psi)$.

This simplified analysis in which the substrate is considered much thicker than the coating has been used to analyze the pressurized blister test for interfacial toughness [57]. By using non-linear von Karman plate theory, it was possible to derive the generic coating end loads in Eqs. 41–43 (i.e. P_1 and M_1) in terms of the applied blister pressure or concentrated loads, and thus calculate the energy

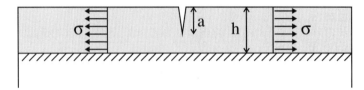

Fig. 13. Surface cracking due to tensile coating residual stress.

release rate and mode mix for the blister test. The results demonstrate the validity of the blister test for measuring interface toughness.

4.3. Other cracking patterns

In the above analyses, we have only considered interfacial fracture. Several other cracking patterns have been observed in coated systems subject to residual stresses. The more common examples include coating surface cracks (Fig. 13), coating channelling cracks (Fig. 14), substrate cracks (Fig. 15), and spalling

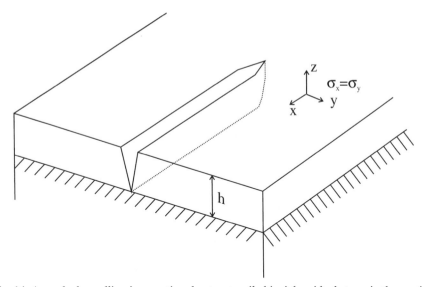

Fig. 14. A crack channelling in a coating due to a tensile biaxial residual stress in the coating.

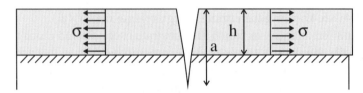

Fig. 15. A crack with its tip in the substrate due to a tensile residual stress in the coating.

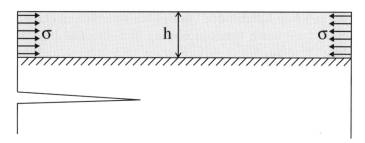

Fig. 16. Substrate spalling due to compressive residual stresses in the coating.

(Fig. 16). Restricting ourselves to the case where the coating is much thinner than the substrate, some general observations can be made regarding the cracking patterns that are most likely to appear. The energy release rate and mode mix for all cracking mechanisms are found to depend on the elastic properties of both the coating and the substrate, the fracture toughness of the coating, substrate, and interface, and the coating thickness. In this section, the aim is to demonstrate the relative likelihood of a crack in a given configuration propagating as a function of the above parameters.

It has been shown [55,99] that, regardless of whether cracking occurs in the substrate, coating, or interface, the energy release rate driving crack propagation for a thin film on a semi-infinite substrate with a residual stress, σ, can be expressed in the following form:

$$G = \frac{Z\sigma^2 h}{\bar{E}_c} \tag{48}$$

where Z is a dimensionless number which depends on the geometry of the cracking configuration and the elastic properties of both the coating and the substrate, h is the coating thickness, and $\bar{E}_c$ is defined (for the coating) in Eq. 19. It follows that, for a given cracking pattern, cracking is more likely if the coating is thicker and more compliant. Intuitively, this makes sense owing to the fact that the strain energy per unit coating area stored in the coating increases with increasing coating thickness and compliance. At crack propagation, $G = G_C$, the appropriate fracture toughness depends on whether the crack is in the coating, substrate, or along the interface, so that Eq. 48 may be used in design situations to predict the critical coating thickness below which crack propagation is not expected to occur [55]:

$$h_c = \frac{G_C \bar{E}_c}{Z\sigma^2} \tag{49}$$

For cracking in either the substrate or coating, usually $G_C = G_{IC}$, because the cracks are free to follow the path of maximum mode I. However, as has been shown in previous sections, if the crack is interfacial, then G_C in Eq. 49 must be taken at the proper phase angle for the particular loading and geometry.

Hutchinson and Suo tabulated Z values for different cracking patterns assuming no elastic mismatch between coating and substrate [55]. It was found that the highest Z occurs for a surface crack in a coating and the lowest for substrate spalling. This does not necessarily mean that one is more likely to see surface cracks than substrate spalling, as this also depends on the relative fracture toughness of the coating and substrate, and the existence of a flaw in either the coating or the substrate (the Z values quoted are for steady-state propagation in the presence of a flaw, and not crack initiation).

Table 1 summarizes the dependence of Z on the elastic properties of the coating and substrate for some commonly encountered cracking patterns that have not been covered in the previous sections, together with the relevant articles to be consulted for further information. It should be noted that: (a) The trends are extrapolated for the case of a thin coating on a substrate (i.e. $\eta = 0$); (b) where Z depends on β, a representative value of $\beta = \alpha/4$ has been chosen, as is the case when the Poisson's ratio of the coating and substrate are equal (a common approximation); (c) for interfacial cracking, the crack is inherently mixed mode, so the dependence of the phase angle is also shown; and (d) where possible, trends such as E_s/E_c approaches infinity and zero have been shown.

Table 1 shows only steady-state values, with an assumed cracking geometry already in existence. As mentioned earlier, the appearance of a particular cracking geometry will largely depend on the relative fracture toughness of the coating, substrate, and interface, and the mode mix, since G_{IIC} is usually greater than G_{IC}. In cases where the substrate is relatively brittle, cracking may occur there especially since it is governed only by G_{IC} in homogeneous materials. In fact, cracks that begin in the coating or along the interface have a tendency to grow into the substrate and propagate parallel to the interface (spalling) if the substrate is sufficiently brittle [55,100,107]. Similar arguments can be made that one should expect surface cracks or channel networks in coatings that are very brittle, and that interfacial cracks should be expected when the interface is of low toughness. One final note is necessary before proceeding to a discussion of Table 1.

The coating stiffness is in the denominator in the crack driving force in Eq. 48, which means that, as mentioned previously, cracking of any type is more likely to be seen with compliant coatings than with stiff ones. However, the Z values generally depend on the ratio of the substrate to coating stiffness, and some important observations can be made from the entries in Table 1 if one treats the coating stiffness as fixed or known. For coating surface cracks, when the coating is very stiff compared to the substrate, Table 1 shows that the crack driving force increases as the crack approaches the interface, so that a crack may well continue into a brittle substrate. When the substrate is very stiff compared to the coating, the opposite is expected, and the crack will likely arrest at the interface. If the substrate and interface are relatively tough, and the coating is much stiffer than the substrate, the surface crack may channel across the surface. The entries in Table 1

Table 1
Summary of Z values for thin coatings

Cracking pattern	Z dependence on E_s/E_f			Relevant fracture toughness	References
Surface cracking in coating (Fig. 13)	$E_s/E_c = 0.11 \begin{cases} Z \to \infty & \text{for} & a/h \to 1 \\ Z = 3 & \text{for} & a/h = 0.5 \\ Z \to 0 & \text{for} & a/h \to 0 \end{cases}$			Coating mode I	[80,103–105]
	$E_s/E_c \to \infty \begin{cases} Z \to 0 & \text{for} & a/h \to 1 \\ Z = 1.2 & \text{for} & a/h = 0.5 \\ Z \to 0 & \text{for} & a/h \to 0 \end{cases}$				
Cracks channeling in coating (Fig. 14)	$E_s/E_c \to 0 \quad Z \to \infty$			Coating mode I	[80,103–105]
	$E_s/E_c \to \infty \quad Z \to 1.1$				
Crack in film with tip in substrate (Fig. 15)	$E_s/E_c = 1 \begin{cases} Z = 4.2 & \text{for} & a/h = 1.2 \\ Z = 2.5 & \text{for} & a/h = 3.5 \end{cases}$			Substrate mode I	[106]
	$E_s/E_c = 19 \begin{cases} Z = 0.25 & \text{for} & a/h = 1.2 \\ Z = 0.1 & \text{for} & a/h = 3.5 \end{cases}$				
Substrate spalling (Fig. 16)	$E_s/E_c = 9 \quad Z = 0.44$			Substrate mode I	[55,107]
	$E_s/E_c = 0.25 \quad Z = 0.25$				
Interfacial from edge or from hole (Fig. 17)	$Z = 0.5$ irrespective of E_s/E_c			Interfacial mixed mode	[55,89]
	$E_s/E_c = 9 \quad \psi = 55.5°$				
	$E_s/E_c = 0.11 \quad \psi = 0.57°$				

for channelling seem to support this hypothesis, as the crack driving force is high for channelling in the case of a coating stiffer than the substrate.

In the case of a crack in a coating with the tip in the substrate (which can occur if the substrate is brittle), Table 1 shows that if the coating is stiffer than the substrate then the crack is more likely to continue propagating in the substrate perpendicular to the interface. It can also be seen that, regardless of the relative stiffness of the coating and substrate, the crack driving force decreases with increasing crack length so that the crack propagation is expected to occur in a stable manner, until a point where it may be more energetically favourable for the crack to deflect to a path parallel to the interface (i.e. spalling). Table 1 shows that the relative substrate–coating stiffness does not have a effect on the likelihood of this happening. It should be noted that spalling under tensile edge loads (i.e. compressive residual stresses) is unlikely to occur, given that the tensile load induces crack closure.

For interfacial cracks (Table 1), the crack driving force does not depend on the ratio of substrate/coating stiffness for thin films (but does increase with increasing coating compliance, Eq. 48). However, the amount of mode II generally increases as the ratio of substrate to coating stiffness decreases. Given that the mixed-mode fracture toughness of interfaces generally increases with increasing mode II component, this would seem to indicate that a crack is more likely to propagate when the substrate is very stiff compared to the coating. If the residual stress in the coating is compressive, then a tensile end condition results on the layer, which produces crack closure, and some account of crack face friction must be made, as noted in Section 4.2.1.

The above results are for the case of single cracks in a particular configuration, and neglect the effects of multiple crack interactions. A discussion of recent work on crack spacing in the case of multiple channelling cracks can be found in [101,102]. It should be noted that more complicated cracking patterns such as spiral cracks have been reported, and a recent paper [105] outlines a theory based on the interaction of the different types of cracks described here, that appears to be able to predict more complicated cracking patterns.

4.4. Summary of possible cracking patterns and coating design

The results of the previous sections are brought together in the flowchart shown in Fig. 18. This chart shows the likelihood of certain cracking patterns being observed depending on whether the stresses in the coating are tensile or compressive, the interfacial toughness is high or low, the relative toughness of the coating and substrate, and the relative stiffness of the coating and substrate. It can be used as a guide in determining which cracking patterns are of concern for a particular coating/substrate system. For example, an epoxy based coating on an aluminium substrate is an example of a brittle coating on a much tougher

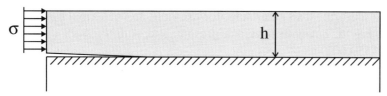

Fig. 17. A crack emanating from an edge or hole and propagating along a coating/substrate interface.

substrate, with $E_s > E_c$. If there is a tensile residual stress in the coating and the interfacial toughness is high, Fig. 18 indicates that the most likely cracking pattern is surface cracking, with the crack arresting at the interface. On the other hand, if the interfacial toughness is low, and the residual stress is tensile, surface cracking is still likely to occur, but with the crack proceeding along the interface. There is also the additional concern that interfacial cracking might occur from edge or hole defects in the coating.

In the case that the stress in the coating is compressive, Fig. 18 shows that regardless of relative coating/substrate/interface toughness, blistering (Fig. 12) is the only cracking pattern likely to be seen (the bending moment due to buckling tends to open the crack). For cracks emanating from an edge along the interface or in the substrate, compressive coating residual stresses result in a tensile edge condition (see Fig. 8), which, in turn results in crack closure. For example, in the case of interfacial cracking (Fig. 17), for a positive edge stress, K_1 is negative (i.e. the denominator of Eq. 43), meaning that the crack has a tendency to close. The equation describing mode mix for substrate spalling exhibits similar behaviour under a tensile edge stress (i.e. results in $K_1 < 0$), so that substrate spalling in the case of high interfacial toughness is also unlikely (see [55,107]).

It was noted in Section 2.3.2 that most of the current interfacial fracture mechanics methodologies describe *steady-state* crack propagation, but not the *initiation* of interfacial cracks. A recent approach to the prediction of initiation is based on the calculation of the singular stress field at the free edge of a bimaterial system loaded on the top layer [59,60]. Because the crack is assumed not to exist initially in this analysis, a very different singular field is predicted, and the results can be used to predict initiation of cracks in residually stressed coatings. Because the predictions of this theory sometimes contradict the predictions of the Suo and Hutchinson approach, we shall briefly review it as a final note.

It has been established in the literature that, for the case of two bonded semi-infinite layers, the interface stresses in the vicinity of the edge exhibit singular behaviour (e.g. [108–110]). A number of authors have developed this theory to investigate the singularity for different edge angles under different thermal and mechanical loading (e.g. [111,112]). Recently, this method has been used to study the stress intensity near a free edge in layered materials loaded on the top layer only by a uniform stress σ, arising, for example, from thermal mismatch residual stresses (Fig. 6, but with no crack present, and with $P_2 = P_3 = M_1 = M_2 = 0$; $P_1 = \sigma h$) [59,60].

The free-edge solution gives interface stresses in the following form [59]:

$$\sigma_{\theta\theta}^{fe} = K_{fe} r^{\lambda-1} + \sigma_0 \tag{50}$$

$$\sigma_{r\theta}^{fe} = K_{fe} r^{\lambda-1} f_{r\theta}(\theta = 0) \tag{51}$$

where $\sigma_{\theta\theta}^{fe}$ and $\sigma_{r\theta}^{fe}$ are the free-edge normal and shear interfacial stresses, and K_{fe} is the free-edge stress intensity factor, defined as [59]:

$$K_{fe} = \lim_{r \to 0} \frac{\sigma_{\theta\theta}^{fe}(\theta = 0)}{r^{\lambda-1}} \tag{52}$$

λ and $f_{r\theta}$ can be found analytically, depend only on material mismatch (i.e. Dundurs parameters α and β), and can be found in refs. [59] and [113,114], respectively, while σ_0 is [59]:

$$\sigma_0 = \frac{-\sigma}{\bar{E}_1} \left[\frac{8\mu_1\mu_2}{\mu_1(\kappa_2 - 3) - \mu_2(\kappa_1 - 3)} \right] \tag{53}$$

and is given in ref. [59]. The subscripts indicate the upper (1) and lower (2) layers. Comparison of Eqs. 50 and 51 with Eq. 26 shows that the form of the singularity for the free-edge and steady-state cracking solutions is quite different. Notably, in the commonly encountered case of $\beta = 0$, the exponent of the singularity, λ, depends on the material properties in the case of the free-edge solution, while it is always a square root dependence in the case of the steady-state solution. Also, the interfacial stresses depend only on one stress intensity factor in the case of the free-edge solution, while they depend on two stress intensity factors in the case of the steady-state solution. The free-edge stress intensity factor, K_{fe}, must be determined via numerical methods, and for the geometry presently considered, was calculated via finite element analysis in ref. [59]. The authors compared their results to the steady-state interfacial cracking solution of Suo and Hutchinson, and found that the singularity behaved, in some cases, quite differently [60]. Notably, for $\beta = 0$, when α is positive, K_{fe} is always negative, meaning that compressive singular normal interface stresses appear for $\alpha > 0$ (Eq. 50). Based on the assumption that negative stress intensity induces crack closure in both the steady-state and free-edge solutions, the authors [60] suggest design criteria for avoiding both crack initiation (free-edge solution) and propagation (steady-state solution). As noted previously, negative values of K_1 are strictly not admissible in the steady-state solution, owing to the fact that crack face contact and friction have not been considered. But it is reasonable to expect, for design purposes, that situations resulting in crack closure are less likely to promote crack propagation than those resulting in opening of the crack. Fig. 19 is a plot of $\eta = \eta_c$, versus α for $\beta = 0$, where η_c is defined as the critical thickness ratio that gives $K_1 = 0$, forming the boundary between positive and negative K_1 [60]. The boundary between positive and negative K_{fe} is at $\alpha = 0$, as described above. In interpreting the results of Fig. 19, it should be noted that the following relationships were used [60] to normalize both the steady-state and free-edge stress intensity factors:

$$K_1^* = \frac{K_1}{\sigma \sqrt{h + H}} \tag{54}$$

$$K^* = \frac{K_{fe}(h + H)^{\lambda - 1}}{\sigma} \tag{55}$$

where K_1^* refers to the steady-state mode I stress intensity, K^* the free-edge stress intensity, and σ is the applied stress on the top layer. If a coating–substrate system can be designed such that α and η are in quadrant I, the system will be resistant to both initiation and propagation of cracks, i.e. the best scenario possible.

Unfortunately, coatings are often very thin, resulting in small η, so that the worst-case scenario of quadrant III is often a reality. In this case, it is desirable to have the stiffest possible coating on the most compliant possible coating, making the value of α large, so that quadrant IV governs, where at least crack initiation is inhibited.

It should be noted that the free-edge singularity approach to predicting crack initiation is relatively new, and has yet to be conclusively proven by experiment. A major criticism of this approach is that crack often initiate from defects and voids, so that the free-edge singularity might not be dominant. In fact, the present authors have found that crack initiation and propagation in structural adhesive joints was unaffected by the presence of either an adhesive spew fillet or a starter crack [115]. In those experiments, however, the crack propagated within the adhesive layer, with an associated 'damage zone' consisting of voids ahead of the macrocrack. It is unclear whether such behaviour would be observed in interfacial cracking. Recent work by various workers (e.g. [116,117]) based on cohesive zone modelling of crack initiation might answer such questions. In this approach, the interface is modelled numerically as a distinct region with its own constitutive behaviour described by a 'traction–separation' relation that links the coating and substrate. This permits the simulation of loading events leading to crack initiation and then propagation at the interface.

5. Summary

This chapter contains a review of the most important concepts and tools regarding the analysis and design of coating systems. Methods were presented to test and analyze the failure of coatings in all of the common modes of failure: delamination, fracture, erosive wear, and general yield in both organic and inorganic coatings. Despite there often being major differences in the behaviour of organic and inorganic coatings, a significant amount of overlap in testing methodologies and analysis techniques does exist.

Section 2 focussed on the measurement of the principal mechanical properties of coatings: stiffness, hardness, yield strength, fracture toughness, and coating thickness. It was demonstrated that measurement of these properties for coatings is often complicated by a variety of parameters, including differences between bulk and in situ properties, coating thickness effects, aging effects, degree of cure, and viscoelastic effects.

For coating stiffness tests, it was necessary to distinguish between those that worked well for organic coatings and those that worked well for inorganic coatings. For viscoelastic coatings (i.e. organic), either transient (e.g. creep, stress relaxation) tests or dynamic (e.g. torsional pendulum or dynamic mechanical) tests can be used to characterize coating stiffness. For coatings that are relatively insensitive to viscoelastic effects, or for viscoelastic coatings whose properties are needed at very high rates of strain, the ultrasonic time-of-flight method can be used. For inorganic coatings, standard tensile tests, bending tests, or measurement of surface acoustic wave speeds can be used to measure coating stiffness. An alternate method, based on indentation of the coating, can also be used if certain restrictions on indentation depth are met in order to avoid errors introduced by interference from the substrate below the coating.

Hardness tests measure the resistance of a coating to indentation under conditions of plastic flow. The most commonly used test for organic coatings is the pencil scratch test, a very subjective test that should only be used to assess the relative hardness of coatings. Micro- and nano-indentation tests can also be used for determination of hardness and yield strength of inorganic coatings, provided that the aforementioned limitations on coating penetration depth are observed. For assessments of the erosion resistance of coatings, a dynamic hardness or plastic flow pressure is often required, and this can be measured via single particle impact tests.

Coating fracture toughness is most often measured via compliance methods using modified double cantilever beam (DCB) specimens in which the coating is adhesively bonded to a second substrate. The method can also be used to evaluate interfacial fracture toughness by a suitable choice of loading on the two arms of the specimen, so that crack propagation can be directed to occur along the coating/substrate interface.

Interfacial fracture toughness can be estimated using a number of qualitative methods such as the adhesion tape test, or scratch test. In addition to DCB tests, more quantitative measurements can be made using indentation delamination tests, blister tests, or various interfacial fracture specimens (Section 4). It is important to note that, in the case of organic (viscoelastic) coatings, the tests should be performed at the proper ambient conditions and strain rate.

Many coating parameters (e.g. hardness, interfacial fracture strength, etc.) have been found to depend on coating thickness. For this reason, a wide variety of tests exist to measure this important coating parameter. The choice of method depends on the type of coating (i.e. organic or inorganic), whether a dry or wet thickness is desired, the desired accuracy, and whether a non-destructive evaluation is necessary.

Section 3 dealt with the analysis and measurement of stresses in coatings and thin films. Residual stresses were found to be very important (especially those due to thermal mismatch), and basic methods for the estimation of these stresses were presented. For metallic films, residual stresses are most commonly measured using X-ray diffraction techniques. For vacuum deposited thin films, the most commonly used approach is the cantilever method, which can also be used to measure residual stresses in organic coatings. Indenter load–displacement curves from nano- or micro-indentation tests sometimes can be used to measure residual stresses in both organic and inorganic coating, if the assumption is made that contact pressure does not change in the presence of elastic residual stresses.

Knowledge of coating contact stress is important because of the wide use of micro- or nano-indentation, and also for estimating erosive wear characteristics of coatings. Though the problem of contact between an indenter and a layered material is inherently very complex, simplifying assumptions can be made leading to closed form solutions for many common geometries.

In Section 4, the analysis of coating adhesion and interfacial fracture toughness was reviewed. Because of the bimaterial nature of coating systems, the resulting interfacial cracks are often mixed-mode, and can exhibit very complex and unusual behaviour (e.g. oscillating crack tip singularity). Fortunately, a generally applicable analysis of interfacial fracture phenomenon has been developed, making it possible to estimate the fracture strength of interfaces in coating systems using fracture data generated from standard interfacial fracture specimens. This analysis can be used to model interfacial crack propagation in a wide variety of systems, including arbitrarily loaded coated beams, coating blistering phenomenon due to compressive residual stresses, and impact-induced buckling delamination of coatings.

Analyses similar to that described above for interfacial fracture exist for a variety of cracking patterns. The results of these analyses can be used to predict what type of cracking pattern is most likely for a given coating system, and thus can aid in the design of coating systems. Table 1 and Fig. 18 in particular can be used as a guide to coating system design.

As a final note, it is important to realize that the generally applicable analysis of interfacial fracture described in Sections 4.1, 4.2 and 4.3 only works well in predicting crack propagation behaviour. In Section 4.4, a recent analysis based on a solution of the free-edge singularity problem (i.e. no pre-existing crack is assumed) is presented, and can be used to predict crack initiation behaviour. The design requirements based on the free-edge solution are sometimes radically different from those based on interfacial fracture mechanics (Fig. 19), so that a

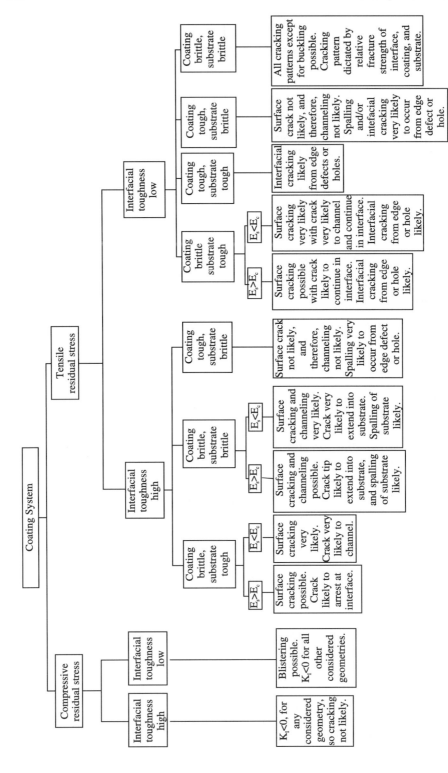

Fig. 18. Summary of likely cracking patterns as a function of interface, coating, and substrate properties.

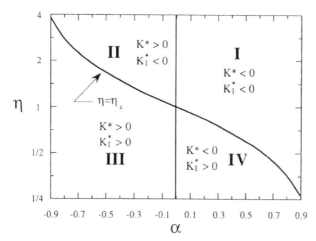

Fig. 19. Dependence of critical coating thickness ratio, η_c on α, for $\beta = 0$ [60].

design that avoids crack propagation does not always result in a design that avoids crack initiation.

References

1. Zozel, A., *Prog. Org. Coat.*, **8(1)**, 47 (1980).
2. Ikeda, K., *Kolloid-Z.*, **160**, 44 (1958).
3. Williams, M.L., Landel, R.F. and Ferry, J.D., *J. Am. Chem. Soc.*, **77**, 3701 (1955).
4. Achenbach, J.D., *Wave Propagation in Elastic Solids*, Elsevier North Holland, Amsterdam, 1984.
5. Papini, M. and Spelt, J.K., *Wear*, **222**, 38–48 (1998).
6. Papini, M. and Spelt, J.K., *Int. J. Mech. Sci.*, **40(10)**, 1061–1068 (1998).
7. Ardebili, V., Sinclair, A.N. and Spelt, J.K., *IEEE Trans. Ultrason. Ferroelectr. Freq. Contr.*, **44(1)**, 102–107 (1997).
8. Wang, W., Tsakalakos, T. and Hilliard, J., *J. Appl. Phys.*, **48**, 879 (1977).
9. Jämting, A.K., Bell, J.M., Swain, M.V. and Schwarzer, N., *Thin Solid Films*, 304–309 (1997).
10. Zaouali, M. and Lebrun, J.L., *Proceedings of the 9th International Conference on Experimental Mechanics*, Vol. 3, 1990, pp. 1281–1290.
11. Hollman, P., Larson, M., Hedenqvist, P. and Hogmark, S., *Surf. Coat. Technol.*, **90**, 234–238 (1997).
12. Schwarzer, N., Richter, F. and Hecht, G., *Surf. Coat. Technol.*, **60**, 396–400 (1993).
13. Schneider, D., Schwartz, T. and Schultrich, B., *Thin Solid Films*, **219**, 92 (1992).
14. Schneider, D., Schwartz, T., Scheibe, H. and Panzer, M., *Thin Solid Films*, **295**, 107–116 (1997).
15. Peggs, G.N. and Leigh, I.C., *Report MOM 62, UK National Physical Laboratory*, 1983.
16. Lebouvier, D., Gilormini, P. and Felder, E., *J. Phys. D: Appl. Phys.*, **18**, 199–210 (1985).
17. Lebouvier, D., Gilormini, P. and Felder, E., *Thin Solid Films*, **172**, 227–239 (1989).
18. Sun, Y., Bell, T. and Zheng, S., *Thin Solid Films*, **258**, 198–204 (1995).

19. Cai, X. and Bangert, H., *Thin Solid Films*, **264**, 59–71 (1995).
20. Safavi-Ardebili, V., Sinclair, A.N. and Spelt, J.K., *J. Adhes.*, **62**, 93–111 (1997).
21. King, R.B., *Int. J. Sol. Struct.*, **23(12)**, 1657–1664 (1987).
22. Oliver, W.C. and Pharr, G.M., *J. Mater. Res.*, **7**, 1564 (1992).
23. Baker, S.P., *Proc. Mater. Res. Soc. Symp.*, **308**, 209 (1993).
24. Chen, W.T. and Engel, P.A., *Int. J. Solids Struct.*, **8**, 1257–1281 (1972).
25. ASTM D3363-92, American Society for Testing of Materials, Philadephia, PA, 1992.
26. Williams, J.A., *Engineering Tribology*, Oxford University Press, 1994.
27. Hutchings, I.M., Winter, R.E. and Field, J.E., *Proc. R. Soc. Lond.*, **A348**, 379 (1976).
28. Hutchings, I.M., *Erosion: Prevention and Useful Applications, ASTM STP664*, W. In: Adler, F. (Ed.), American Society for Testing and Materials, 1979, pp. 59–76.
29. Rickersby, D.G. and MacMillan, N.H., *Int. J. Mech. Sci.*, **22**, 491–494 (1980).
30. Sundararajan, G. and Shewmon, P.G., *Int. J. Impact Eng.*, **6(1)**, 3–22 (1987).
31. Tirupataiah, Y., Venkataraman, B. and Sundararajan, G., *Mater. Sci. Eng.*, **A24**, 133–140 (1990).
32. Papini, M. and Spelt, J.K., *Int. J. Mech. Sci.*, **42**, 991–1006 (2000).
33. Papini, M. and Spelt, J.K., *Int. J. Mech. Sci.*, **42**, 1007–1025 (2000).
34. Matthewson, M.J., *J. Mech. Phys. Solids*, **29(2)**, 89–113 (1981).
35. Anstis, G.R., Chantikul, P., Lawn, B.R. and Marshall, D.B., *J. Am. Ceram. Soc.*, **64**, 533 (1981).
36. Evans, E.G. and Charles, E.A., *J. Am. Ceram. Soc.*, **59**, 371 (1976).
37. Niihara, K., Morena, M.R. and Hasselman, D.P.H., *J. Mat. Sci. Lett.*, **1**, 13 (1982).
38. Broek, D., *Elementary Engineering Fracture Mechanics*, 4th edn. Martinus Nijhoff, Dordrecht, 1986, pp. 136–140.
39. Wylde, J.W. and Spelt, J.K., *Int. J. Adhes. Adhes.*, **18(4)**, 237–246 (1998).
40. Fernlund, G. and Spelt, J., *Comp. Sci. Technol.*, **50**, 441–449 (1994).
41. Troczynski, T. and Camire, J., *Eng. Fract. Mech.*, **51(2)**, 327–332 (1995).
42. Guo, D.Z. and Wang, L.J., *Surf. Coat. Technol.*, **56**, 19–25 (1992).
43. ASTM D3359-97, American Society for Testing of Materials, Philadephia, PA, 1997.
44. ASTM B571-97, American Society for Testing of Materials, Philadephia, PA, 1997.
45. ASTM D4541-95e1, American Society for Testing of Materials, Philadephia, PA, 1995.
46. Ollendorf, H. and Schneider, D., *Surf. Coat. Technol.*, **113**, 86–102 (1999).
47. Benjamin, P. and Weaver, C., *Proc. R. Soc. Lond. A*, **254**, 163–176 (1960).
48. Burnett, P.J. and Rickerby, D.S., *Thin Solid Films*, **101**, 243 (1983).
49. Laugier, M.T., *Thin Solid Films*, **117**, 243–249 (1984).
50. Venkataraman, S., Kohlstedt, D.L. and Gerberich, W.W., *J. Mat. Sci.*, **7**, 1126–1132 (1992).
51. Thouless, M.D., *Eng. Fract. Mech.*, **61**, 75–81 (1998).
52. Marshall, D.B. and Evans, A.G., *J. Appl. Phys.*, **56(10)**, 2632–2638 (1984).
53. Evans, A.G. and Hutchinson, J.W., *Int. J. Solids Struct.*, **20(5)**, 455–466 (1984).
54. Rosenfeld, L.G., Ritter, J.E., Lardner, T.J. and Lin, M.R., *J. Appl. Phys.*, **67(7)**, 3291–3296 (1990).
55. Hutchinson, J.W. and Suo, Z., *Adv. Appl. Mech.*, **29**, 63–191 (1992).
56. Papini, M. and Spelt, J.K., *Int. J. Mech. Sci.*, **40(10)**, 1043–1059 (1998).
57. Jensen, H.M., *Eng. Fract. Mech.*, **40(3)**, 475–486 (1991).
58. Matthewson, M.J., *Appl. Phys. Lett.*, **49(21)**, 1426–1428 (1986).
59. Klingbeil, N.W. and Beuth, J.L., *Eng. Fract. Mech.*, **66**, 93–110 (2000).
60. Klingbeil, N.W. and Beuth, J.L., *Eng. Fract. Mech.*, **66**, 111–128 (2000).
61. Kämpf, G., *Prog. Org. Coat.*, **1**, 335–350 (1973).

62. ASTM D1212-91(1996)e1, American Society for Testing of Materials, Philadephia, PA, 1996.
63. ASTM D4414-95, American Society for Testing of Materials, Philadephia, PA, 1995.
64. ASTM D1005-95, American Society for Testing of Materials, Philadephia, PA, 1995.
65. ASTM D4138-94(1999), American Society for Testing of Materials, Philadephia, PA, 1999.
66. ASTM B659-90(1997), American Society for Testing of Materials, Philadephia, PA, 1997.
67. Lakestani, F., Coste, J. and Denis, R., *NDT E. Int.*, **28(3)**, 171–178 (1995).
68. Makerle, J., *Finite Elem. Anal. Des.*, **34**, 113–124 (2000).
69. Bacmann, J.J., In: Hecht, G., Richter, F. and Hahn, J. (Eds.), *Thin Films*. Informationsgesellschaft, Oberursel, 1994.
70. Perry, A.J., Sue, J.A. and Martin, P.J., *Surf. Coat. Technol.*, **81**, 17–28 (1996).
71. Maissel, L. and Glang, R. (Eds.), *Handbook of Thin Film Technology*, McGraw-Hill, New York, 1970.
72. Tamulevicius, S., *Vacuum*, **52(2)**, 127–139 (1998).
73. Suresh, S. and Giannakopoulus, A.E., *Acta Mater.*, **46(16)**, 5755–5767 (1998).
74. Chiu, C. and Liou, Y., *Thin Solid Films*, **268**, 91–97 (1995).
75. Johnson, *Contact Mechanics*, Cambridge University Press, Cambridge, 1985.
76. Jaffar, M.J., *Int. J. Mech. Sci.*, **31(3)**, 229–235 (1989).
77. Hertz, H., *J. Reine Angew. Math.*, **92**, 156–173 (1881).
78. Huber, M.T., *Ann. Phys.*, **14**, 153–163 (1904).
79. Schwarzer, N. and Richter, F., *Surf. Coat. Technol.*, **74/75**, 97–103 (1995).
80. Hu, M.S. and Evans, A.G., *Acta Metall.*, **37(3)**, 917–925 (1989).
81. Jorgenson, O., Horsewell, A., Sorenson, B.F. and Leisner, P., *Acta Metall. Mater.*, **43(11)**, 3991–4000 (1995).
82. Chiu, C. and Liou, Y., *Thin Solid Films*, **268**, 91–97 (1995).
83. Choi, S.R., Hutchinson, J.W. and Evans, A.G., *Mech. Mat.*, **31**, 431–447 (1999).
84. Chai, H., *Int. J. Fract.*, **46**, 237–256 (1990).
85. Oliveira, S.A.G. and Bower, A.F., *Wear*, **198**, 15–32 (1996).
86. Cao, H.C. and Evans, A.G., *Mech. Mat.*, **7**, 295–304 (1989).
87. Babu, M.V. and Kumar, R.K., *Eng. Fract. Mech.*, **55(2)**, 235–248 (1996).
88. Fernlund, G., Papini, M., McCammond, D. and Spelt, J.K., *Comp. Sci. Technol.*, **51**, 587–600 (1994).
89. Suo, Z. and Hutchinson, J.W., *Int. J. Fract.*, **43**, 1–18 (1990).
90. Rice, J.R., *J. Appl. Mech.*, **55**, 98–103 (1988).
91. Dundurs, J., *J. Appl. Mech.*, **36**, 650–652 (1969).
92. Oliveira, S.A.G. and Bower, A.F., *Wear*, **198**, 15–32 (1996).
93. Charalambides, P.G., Lund, J. and McMeeking, R.M., *J. Appl. Mech.*, **56**, 77–82 (1989).
94. Klingbeil, N.W. and Beuth, J.L., *Eng. Fract. Mech.*, **56(1)**, 113–126 (1997).
95. Wang, J.S. and Suo, Z., *Acta Met.*, **38**, 1279 (1990).
96. Yin, W., *Int. J. Solids Struct.*, **21(5)**, 503–514 (1985).
97. Thouless, M.D., Hutchinson, J.W. and Liniger, E.G., *Acta Metall. Mater.*, **40(10)**, 2639–2649 (1992).
98. Tu, Y. and Spelt, J.K., *J. Adhes.*, **72**, 359–372 (2000).
99. Evans, E.G., Drory, M.D. and Hu, M.S., *J. Mater. Res.*, **3**, 1043–1049 (1988).
100. He, M. and Hutchison, J.W., *J. Appl. Mech.*, **56**, 270–278 (1989).
101. Thouless, M.D., *J. Am. Ceram. Soc.*, **73**, 2144–2146 (1990).
102. Strawbridge, A. and Evans, H.E., *Eng. Failure Anal.*, **2(2)**, 85–103 (1995).
103. Beuth, J.L., *Int. J. Sol. Struct.*, **29**, 1657–1675 (1992).

104. Beuth, J.L. and Klingbeil, N.W., *J. Mech. Phys. Solids*, **44**, 1411–1428 (1996).
105. Xia, Z.C. and Hutchinson, J.W., *J. Mech. Phys. Solids*, **48**, 1107–1131 (2000).
106. Ye, T., Suo, Z. and Evans, A.G., *Int. J. Solids Struct.*, **29**, 2639–2648 (1992).
107. Suo, Z. and Hutchinson, J.W., *Int. J. Solids Struct.*, **25(11)**, 1337–1353 (1989).
108. Bogy, D.B., *Int. J. Solids Struct.*, **6**, 1287–1313 (1970).
109. Bogy, D.B., *J. Appl. Mech.*, **35**, 460–466 (1968).
110. Bogy, D.B., *J. Appl. Mech.*, **38**, 377–386 (1971).
111. Suga, T., Mizuno, K. and Miyazawa, K., *MRS Int. Meet. Adv. Mater.*, **8**, 137–142 (1989).
112. Yang, Y.Y. and Munz, D., *Int. J. Solids Struct.*, **34(10)**, 1199–1216 (1997).
113. Okajima, M., Ph.D. Thesis, Carnegie Mellon University, 1985.
114. Yang, Y.Y., Ph.D. Thesis, University of Karlsruhe, 1992.
115. Papini, M., Fernlund, G. and Spelt, J.K., *Comp. Sci. Technol.*, **52**, 568–570 (1994).
116. Mohammed, I. and Liechti, K.M., *J. Mech. Phys. Solids*, **48**, 735–764 (2000).
117. Tvergaard, V. and Hutchinson, J.W., *J. Mech. Phys. Solids*, **41**, 1119–1135 (1993).

Chapter 10

Stresses and fracture of elastomeric bonds

N. SHEPHARD [*]

Dow Corning Corporation, Midland, MI, USA

1. Introduction

This chapter will provide an overview of the important differences in the stress states within bonded joints involving elastomeric adhesives and sealants. Elastomeric adhesives require special consideration because their mechanical behavior is significantly different than most engineering materials. Elastomers are generally very flexible but nearly incompressible. In addition, the physical properties of elastomers are rate and temperature dependent. Quantifying the mechanical properties of elastomers is generally more difficult than typical engineering materials. However, unique engineering solutions to complicated design problems can be obtained when elastomeric adhesives are used with a thorough understanding of their unusual mechanical response. To achieve that understanding, we will first review the mechanical properties of elastomeric adhesives. Then we will explore how the mechanical properties of elastomeric adhesives influence the stress in the adhesive joint. Specific methods for measuring the strength of elastomeric adhesives under simple loading conditions will be reviewed. Followed by methods for predicting the stresses in more complicated real world engineering applications by computer modeling techniques and advanced analytical techniques. Finally, methods for designing with elastomeric adhesives will be discussed. But, before we begin, let us consider a few notable applications of elastomeric adhesives in our world today.

A large complex structure such as a building must be designed to be flexible. Naturally occurring gradients in the building's temperature would result in stresses caused by thermal expansion and contraction leading to cracks and eventual destruction of the building. Therefore, predefined cracks are built into the design of the building. Such cracks are more commonly called expansion joints. Each joint is designed to accommodate the change in size of the building's walls and floors during temperature changes. An elastomeric adhesive commonly called a

[*] Corresponding author. E-mail: n.shephard@dowcorning.com

sealant protects the joint. The sealant prevents water, air and dirt from infiltrating the building. The sealant also prevents larger objects from being pinched in the joint, thus preventing damage to the adjoining substrates. In addition to movement caused by temperature changes, wind, relative humidity and seismic events can lead to movement of the joints. Similar joints protected by sealants can be found in concrete roads, bridges, ships and aircraft.

In contrast, the pacemaker leads that are imbedded in the wall of a cardiac patient's heart are sealed to the polyurethane body of the pacemaker using a similar sealant. Without the sealant, the sensitive electronic components contained within the pacemaker could be exposed to the patient's body fluids leading to failure of the pacemaker and great risk to the patient.

Extremely small components also benefit from the use of elastomeric adhesives. Common epoxy adhesives are rigid and concentrate the stress leading to device failure. However, elastomeric adhesives can be used to dissipate such stress and yield a more durable electronic device. This technology enables the development of inexpensive but durable electronic components, such as, cellular telephones, portable computers and automotive applications. For example, the computer command control module, which regulates the operation of modern automobiles, can survive the harsh conditions under the hood because it is encapsulated in elastomeric adhesives known as potting compounds.

Vibration is a major destructive force in many types of machinery. Vibration can lead to wear and fatigue failure of rigid materials like alloys and plastic components. The helicopter benefits from the unique flexibility and vibration dampening characteristics of elastomeric adhesives. Elastomers are used to transfer torque to the rotor blades while providing flexibility and vibration control. Rubber tires are another example of elastomeric adhesive use. Consumers have enjoyed a steady increase in the useful life expectancy of rubber tires due to improvements in adhesion between the elastomer, tire cords, and the filler. The rubber tire industry has driven a great deal of the fundamental research on elastomeric materials in engineering design.

The above examples illustrate the key design features of elastomeric adhesives: elasticity, low tensile and shear modulus, great compressive strength, and unique vibration and dampening control. Now let us take a closer look.

2. Mechanical properties of elastomeric adhesives contrasted to rigid adhesives and substrates

2.1. Small strain conditions

By definition, elastomeric (often termed rubbery or just rubber) adhesives are viscoelastic solids. In order to describe their behavior it is useful to start with the

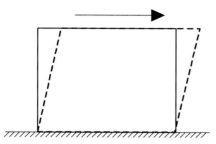

Fig. 1. Simple shear condition.

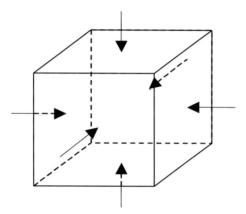

Fig. 2. Triaxial compression.

simpler components of viscoelastic materials, namely, perfectly viscous behavior and perfectly elastic behavior, as outlined in Ferry's classic text [1]. A viscoelastic material comprises attributes of both the liquid and the elastic solid. A perfectly viscous material obeys the relationship defined by Newton whereby the stress is directly proportional to the rate of strain but independent of the strain itself. A perfectly elastic solid obeys Hooke's law that states that the stress is directly proportional to the strain and not affected by the rate of strain. For isotropic materials undergoing small strains the above relationships are generally true.

The above relationships can be measured using simple experiments in shear or compression. In the case of the shear experiment (Fig. 1), the shape of the material changes without changes in the volume of the material and Eq. 1 is followed.

$$\tau = \eta \dot{\gamma} \tag{1}$$

Conversely a hydrostatic compression experiment results in changes in volume and with no change in shape of the material as seen in Fig. 2 and Eq. 2.

$$p = -B\Delta V \tag{2}$$

When a material is subjected to simple extension, changes in both shape and in

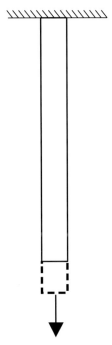

Fig. 3. Simple tension.

volume will usually occur as noted in Fig. 3.

$$\sigma = E\varepsilon \qquad (3)$$

The Young's modulus, bulk compressibility and shear modulus can be related through the addition of a fourth constant, Poisson's ratio v.

$$E = 3B(1 - 2v) = 2(1 + v)G \qquad (4)$$

where Poisson's ratio quantifies the ratio of the transverse strain to the longitudinal strain.

$$v = \frac{-\varepsilon_y}{\varepsilon_x} \qquad (5)$$

Rearrangement of Eq. 4 leads to a convenient method for evaluating the value of v by measuring E and B.

$$1 - 2v = \frac{E}{3B} \qquad (6)$$

A value of Poisson's ratio equal to 0.5 indicates that the material deforms without volume change. From Table 1, we note that rubber-like materials all have a value of 0.49 or greater for Poisson's ratio, but not 0.5. It should be noted that a value of 0.5 for Poisson's ratio would result in a bulk modulus of infinity. That

Table 1

Elastic constants for various classes of materials

	Young's modulus, E (GPa)	Bulk modulus, K (GPa)	Poisson's ratio, ν	Coefficient of thermal expansion $(10^{-6}/°C)$
(a) Ceramic				
Alumina	400	238	0.22	
Silicon carbide	396	236	0.22	
Type E glass	72.4	43	0.22	9
(b) Metal				
Aluminum	70.3	76	0.345	23.6
Copper	130	138	0.343	16.9
Mild steel	212	171	0.293	11.7
(c) Plastic				
Epoxy	3.5	3	0.33	70
Polystyrene	3.2	3	0.33	70
PMMA	3.3	4.1	0.37	70
(d) Rubber				
Polyisoprene	0.001	2	0.4999	90
Poly(isobutene-co-isoprene)	0.001	1.97	0.4995	80
(e) Liquids				
Mercury		25		3.3
Water		2.1		
Dodecane		1.3		

is clearly never the case in the real world. However, it is often useful to use the value 0.5 for Poisson's ratio when discussing elastomers. This unique feature of elastomeric materials can be explained by considering the molecular arrangement of atoms in an elastomer.

All elastomeric materials have several key features in common. Firstly, the molecules must be polymeric. In addition, the polymer must be composed of atoms arranged such that long-range motion between the atoms can readily occur. Polymers composed of linear chains are a necessary first condition. However, the polymer chains must also be composed of bonds that have low rotational energy so that the linear chains can easily coil and uncoil. However, the polymer must not be highly crystalline or once again the chain mobility will be too low. It is the flexing of the polymer coil, which provides the spring-like response of the rubber. Finally, the polymer chains need to be chemically or physically connected to each other to provide long-range interaction. When subjected to a tensile or shear stress, an elastomeric material deforms by slippage of the polymer chains past each other to relieve the mechanical stress. The actual distance between individual atoms remains essentially unchanged. Therefore, the volume of the material does not significantly change during the deformation. Conversely, rigid materials such as metal alloys undergo deformation by the direct stretching of atomic bonds. This results in much higher modulus as well as significant volume change. However, an elastomeric material can only respond by chain slippage if the chains can undergo long-range motion at a similar or faster speed that the imposed external strain. The ratio of the average mobility of the polymer to the speed of the deformation is a useful relationship known as the Deborah number, De. The mobility of the polymer chains can be reduced by reducing the temperature of the material or by constraining the polymer chains by crystallization or excessive cross-linking. Alternatively, the rate of strain can be increased beyond the speed of the polymer. Explosions, bullets, and high-speed collisions are examples of high-speed strain rates, which can result in brittle fracture of normally rubbery materials.

2.2. Large strains

Up to this point, we have only been discussing small strain experiments. However, we know that elastomeric materials can undergo very large deformation. The complete stress–strain curves for several types of materials can be seen in Fig. 4. Fig. 4a depicts the stress–strain curve for an elastic–brittle material like tempered steel, note that the strain to break is only 0.004. Deformation occurs by the reversible stretching of atomic bonds until breaking occurs. The response for a plastic material is initially similar to an elastic material, but with an additional region beyond the reversible linear response. This new region is the area of plastic yielding. Deformation now occurs by the large-scale movement of atoms within the matrix. This deformation is not reversible. Lead, copper and brass are examples of metals that undergo plastic yielding (Fig. 4b). Such metals are commonly called ductile. Many polymeric materials undergo plastic yielding such as nylon (Fig. 4c). We note that the mechanical response for an elastomeric material is quite different. The stress–strain curve is not linear and no significant yield point is detected (Fig. 4d). The strain to break is greater than 6 and quite reversible for an unfilled rubber. As mentioned before, the deformation occurs by the slippage of polymer chains past each other, the deformation of atomic bonds does not occur until the breaking strain is reached.

The relative amount of polymer chain slippage during a deformation is described by the Deborah number. As mentioned earlier, this ratio can be affected by changes in temperature or rate of deformation. Because of this, the mechanical response of a rubber is dependent on time and temperature. A typical response can be seen in Fig. 5. At low temperature or high strain rate, the rubbery material behaves like a brittle solid.

Conversely, at high temperature or slow strain rate, the rubber behaves more

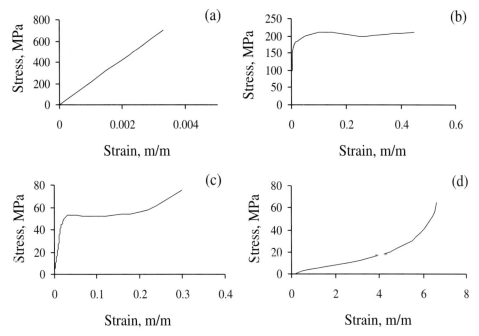

Fig. 4. Stress versus strain curves for (a) annealed steel, (b) copper, (c) nylon and (d) natural rubber.

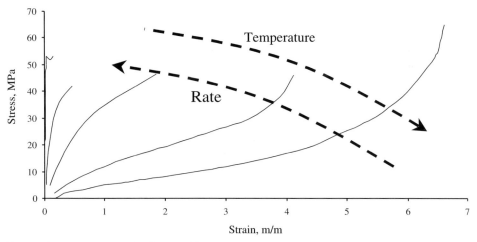

Fig. 5. Stress versus strain for a rubbery material. Arrows indicate measurements taken at different rates or test temperatures.

like a low modulus solid. The modulus of the rubbery material is a good indicator of the relative chain mobility. As seen in Fig. 6, the modulus increases as the temperature of the material decreases or the speed of the deformation increases.

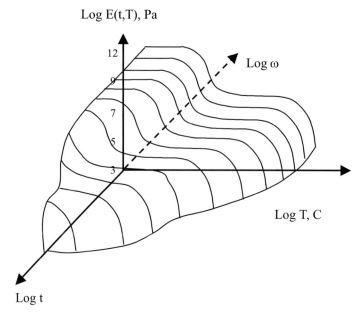

Fig. 6. Modulus as a function of temperature and test speed plotted on log scale.

2.3. Viscoelasticity

2.3.1. Phenomenological treatment of viscoelasticity

The viscoelastic response of the rubbery material can be modeled using simple mechanical models representing combinations of Eq. 1 and the time derivative of Eq. 3. A simple Dashpot model represents the viscous component and the spring model represents the elastic component. Note that E and η represent the appropriate modulus and viscosity based on the state of stress being modeled.

2.3.1.1. Transient loading patterns.   Creep relaxation, stress relaxation, constant rate of strain and constant rate of stress are all examples of transient loading patterns. The Maxwell model (Fig. 7) is used to represent a viscoelastic liquid and is especially useful for stress relaxation experiments. This simple model can be generalized to give good approximations for real material systems by combining numerous elements in parallel. This is known as a Maxwell–Wiechert model.

The Voigt model is good for modeling viscoelastic solids in creep experiments. The more generalized version is the Voigt–Kelvin model, which is a series expansion of the Voigt model (Fig. 8).

Sometimes it is advantageous to combine Voigt and Maxwell models. The series combination known as the Burgers model contains four experimental constants, which describe the creep behavior of a material. Numerous other com-

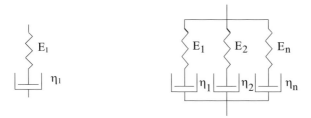

Maxwell Model   Maxwell-Wiechert Model

Fig. 7. Mechanical models useful for modeling stress relaxation.

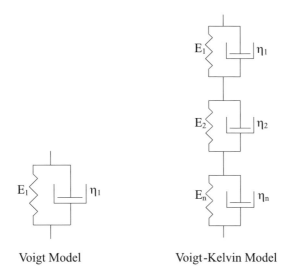

Voigt Model   Voigt-Kelvin Model

Fig. 8. The Voigt model can be used to model creep.

binations have been used to describe viscoelastic materials. The above examples take advantage of the Boltzmann superposition principle, which states that all the stresses add independently and the resultant strains are additive.

The proportionality constant between modulus and viscosity is known as the relaxation time, τ_r. For creep experiments, the proportionality constant is known as the retardation time. For the generalized models, there is a spectrum of relaxation or retardation times to account for the various viscoelastic processes occurring at different time scales in the material.

$$E = \tau_r \eta \tag{7}$$

$$E_i = \tau_{r_i} \eta_i \tag{8}$$

A summary of the important equations for the Maxwell element and the Voigt element can be found in Table 2. A detailed derivation of the equations can be found in the classic texts [1,2].

Table 2

Summary of viscoelastic models

	Maxwell	Voigt
Equation of motion	$\dfrac{d\varepsilon}{dt} = \dfrac{1}{E}\dfrac{d\sigma}{dt} + \dfrac{\sigma}{\eta}$	$\sigma(t) = \varepsilon(t)E_0 + \eta\dfrac{d\varepsilon(t)}{dt}$
Stress relaxation	$E(t) = \dfrac{\sigma(t)}{\varepsilon_0} = E\,e^{-t/\tau_r}$	$D(t) = E$ but physically impossible
Generalized	$E(t) = \displaystyle\int_0^{\infty} E(\tau_r)e^{-t/\tau_r}\,d\tau$	
	$E(t) = \displaystyle\int_{-\infty}^{\infty} \bar{H}(\ln\tau_r)e^{-/\tau_r}\,d\ln\tau_r$	
Creep	$D(t) = \dfrac{\varepsilon(t)}{\sigma_0} = D + \dfrac{t}{\eta}$	$D(t) = D(1 - e^{-t/\tau_r})$
Generalized	$D(t) = \displaystyle\int D(\tau_r)(1 - e^{-t/\tau_r})d\tau_r$	
	$D(t) \cong \displaystyle\int_{-\infty}^{\infty} \bar{L}(\ln\tau_r)(1 - e^{-t/\tau_r})d\ln\tau_r$	
Stress at constant strain rate	$\sigma(t) = \dot{\varepsilon}\displaystyle\int_{-\infty}^{\infty} \tau_r\bar{H}(\ln\tau_r)(1 - e^{-1/\tau_r})d\ln\tau_r$	
Steady state flow viscosity	$\eta = \displaystyle\int \tau_r E(\tau_r)d\tau_r = \int \psi_r\bar{H}(\ln\tau_r)d\ln\tau_r$	
Where:	$\left[\bar{H}(\ln\tau_r)\right]_{\tau_r=t} \cong \dfrac{dE(t)}{d(\ln t)}$	$\left[\bar{L}(\ln\tau_r)\right]_{\tau_r=t} \cong -\dfrac{d\varepsilon(t)}{d(\ln t)}$

2.3.1.2. Dynamic experiments. Often a viscoelastic material is subjected to stresses or strains, which change over time. When experiments are set up to mimic such conditions, they are referred to as dynamic experiments or dynamic mechanical analysis. For transient experiments, the time scale is qualitatively proportional to the inverse of the test frequency used for dynamic experiments. We will consider the above equations with respect to a simple sinusoidal stress as is often imposed by common commercial test machines.

When a sinusoidal stress is applied to a linear viscoelastic material, a corresponding sinusoidal strain will result, but it will be out of phase with the stress as seen in Fig. 9. The components of the stress vector can be decomposed into a component in phase and a component 90° out of phase as depicted in Fig. 10. The complex modulus is obtained by dividing the complex stress by the complex

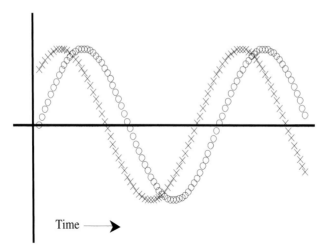

Fig. 9. Strain response, ○, of a viscoelastic material to a dynamic stress, ×.

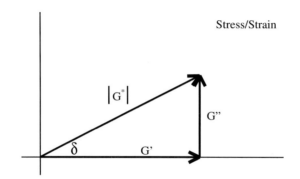

Fig. 10. Vectorial projection of storage and loss components.

strain. For a shear stress, the equations for complex modulus are given below:

$$G^* = G' + iG'' \tag{9}$$

$$|G^*| = \sqrt{G'^2 + G''^2} \tag{10}$$

$$\tan \delta = G''/G' \tag{11}$$

$$G' = |G^*| \cos \delta \tag{12}$$

$$G'' = |G^*| \sin \delta \tag{13}$$

where G^* is the complex shear modulus, G' is the storage modulus also known as the elastic modulus, G'' is the loss modulus which represents the viscous response of the material. δ is the phase angle between the storage and loss components. Dynamic measurements are useful when a wide range of strain rates

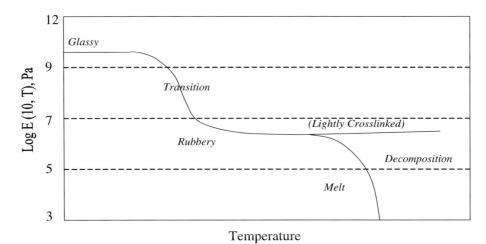

Fig. 11. Regions of viscoelastic response for an uncrosslinked and a crosslinked rubber. Redrawn from ref. [3].

are of experimental interest. Various methods can be used to yield test data at frequencies from 10^{-5} to 10^{12} cycles/s.

2.3.2. Time–temperature correspondence

Fig. 11 depicts the various regions of viscoelastic response. By plotting the modulus as a function of temperature or frequency as seen in Fig. 12. one can easily see this phenomenon [3]. The transition from glassy to rubbery response is known as the glass transition temperature T_g. The rubbery condition exists for all the reasons covered in Section 2.2. Flexibility of the polymer chain was one of those conditions. The polymer flexibility directly relates to temperature. At the point when polymer motion is limited to less that a few repeat units, the material becomes glassy. The glass transition can be described by a specific temperature for a given time or it can be described as a specific time at a given temperature in accordance with the Deborah number previously defined.

The relationship between time and temperature can be useful for extrapolating to conditions that may not be experimentally convenient. This can be demonstrated by the following creep experiment. Modulus is measured as a function of time at a constant temperature, and repeated for several other temperatures. The resulting data is graphed on the left side of Fig. 12, then all the data is assembled on the right side of the graft by horizontally sliding the individual curves to create a smooth continuous curve spanning many decades of time for one specific temperature. Such a curve is known as a master curve. The relative amount of horizontal shifting is known as the shift factor, a_T. Williams, Landel and Ferry first made this method public in the mid 1950s [4]. For many polymers, a simple equation,

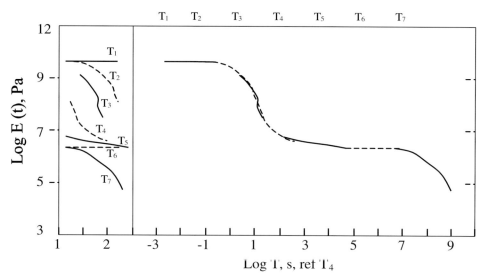

Fig. 12. Creep compliance data, left, taken at several temperatures and horizontally shifted to generate a master curve, right.

known as the WLF equation, adequately describes the shift factor function:

$$\log a_T = \frac{-C_1(T - T_g)}{C_2 + T - T_g} \tag{14}$$

where C_1 and C_2 are constants, T is the temperature during the measurement and T_g is the glass transition temperature. It has been shown that the WLF equation follows the same form as the Doolittle equation used to describe the viscosity of a liquid. This method of extrapolation is often referred to as time–temperature superposition. It is more generically known as the method of shifting variables when it is used for variables other than temperature. This is a very powerful method for predicting the performance of a polymer over a very wide range of response times. However, it should be noted that this method is only valid under specific conditions. A plot of the shift factor versus temperature should result in a smooth monotonic function. Any kinks or discontinuities are an indication of a change in the mechanism as a function of temperature and invalidates the experiment. Crystallinity and thermal decomposition can also represent changes that would invalidate the applicability of the method of shifting variables.

3. Stress distributions in butt joints

An elastomeric joint is subjected to mechanical stresses that are the result of several possible sources, which often occur in combinations. The most commonly

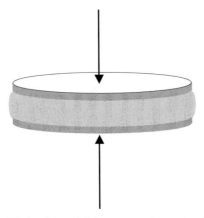

Fig. 13. A rubber disk in compression or tension.

recognized source of stress is due to changes in the joint dimensions caused by applied loads. Stresses can also occur, without significant substrate movement, due to changes in the volume occupied by the elastomer. Such changes can be the result of temperature changes, swelling by water or other solvents, or due to evaporation of volatile ingredients. Stresses are not uniform. The stresses are usually a complex combination of tension, compression and shear. For relatively simple cases, closed form analytical solutions exist and will be discussed below. But, for more complex conditions some form of computer-aided numerical analysis is usually needed.

We will first consider the simple case of a rubber disk confined between two rigid plates and subjected to tension or compression as seen in Fig. 13. There are several solutions based on different boundary conditions and simplifying assumptions. We will see that the extent of constrainment of the elastomer is a key property governing the stress distribution and apparent stiffness of the joint. The amount of constrainment is given by the height to diameter ratio, the Poisson's ratio of the elastomer, and the friction at the substrate–elastomer surfaces as seen in Fig. 14.

Closed form solutions have been derived which neglect outer edge effects, but can account for full adhesion or high friction, zero adhesion and intermediate levels of adhesion at the substrate. The aforementioned derivations are made possible by the simplifying assumption that Poisson's ratio for the elastomer is 0.5. Such solutions are useful for determining apparent stiffness and stress distributions in the interior of the joint, as we shall see in the following paragraphs.

Eq. 3 describes the mechanical response of a perfectly lubricated circular disk. On the other hand, Gent et al. have estimated the normal and shear stresses for small compressions or extensions of perfectly bonded blocks of radius, a, and thickness, h [5–7]. It was shown that the shear stress obeys the same differential equation as the torsion of prismatic shafts or the pressurized membrane problems.

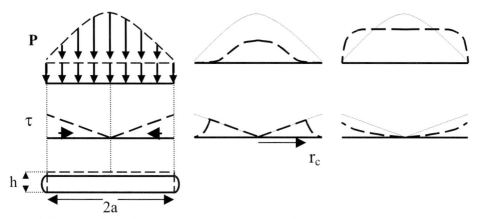

Fig. 14. Stress conditions in the button joint. Left: good adhesion, $\nu = 0.5$. Center: intermediate adhesion, $\nu = 0.5$. Right: good adhesion, $\nu < 0.5$.

The uniform compressive or tensile stress resulting in the initial strain is given by Eq. 3 and the shear stress acting at a radius, r, is given by:

$$\tau = 3G\left(\frac{r}{h}\right)\varepsilon \tag{15}$$

The corresponding normal stress is given as a pressure, P related to the shear stress by:

$$\tau = \left(\frac{h}{2}\right)\left(\frac{\partial P}{\partial r}\right) \tag{16}$$

Integration yields the pressure associated with the given shear stresses:

$$P = 3G\varepsilon\left(\frac{a^2}{h^2}\right)\left[1 - \left(\frac{r^2}{h^2}\right)\right] \tag{17}$$

Integrating the normal stresses resulting from the normal and shear displacements gives the total normal force:

$$F = 3\pi^2 G\varepsilon\left(1 + \frac{a^2}{2h^2}\right) \tag{18}$$

From the above derivation, it is important to note that the effect of constraint increases the stiffness of the component by the factor $(1 + a^2/2h^2)$. As seen in Fig. 15, for large thin blocks of rubber, the effective modulus approaches that of the bulk modulus, B, for compression when a/h is greater than about 10. Thornton et al. analyzed the interfacial stresses for the intermediate case when some amount of friction or adhesion occurs between the rubber disk and the rigid blocks [8]. A new variable, μ, defines the static coefficient of friction for which a critical radius exists at which the shear stresses cannot overcome the coefficient

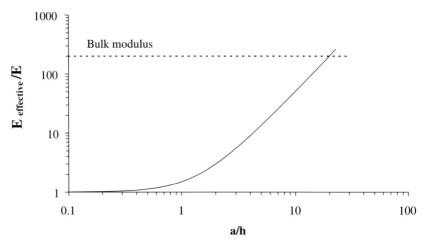

Fig. 15. The effect of constraint, a/h, on the apparent stiffness of a thin elastomeric disk. Redrawn from ref. [7].

of friction and Eq. 20 holds. Beyond that radius, the shear stress is proportional to the normal stress times the coefficient of friction. The following equations were derived for the three boundary conditions:

For the perfectly lubricated condition, $\mu \to 0$:

$$F = \pi a^2 3G\varepsilon \tag{19}$$

For the perfectly adhering condition, $\mu \geq a/h$:

$$F = 3\pi^2 G\varepsilon \left(1 + \frac{a^2}{2h^2}\right) \tag{20}$$

For the intermediate condition, $0 \prec \mu \prec a/h$:

$$F = 6\pi G\varepsilon \left\{ \frac{r_c^4}{4h^2} + \frac{r_c^3}{2\mu h} + \frac{r_c^2}{2\mu^2} + \frac{r_c h}{4\mu^3} - \frac{ha}{2\mu} - \frac{h^2}{4\mu^2} \right\} \tag{21}$$

In Fig. 14, center, we see that the shear stresses are equivalent to the solution by Gent (Fig. 14, left), up to the point that slippage occurs, known as the critical radius. In addition, the normal stresses are lower and vanish beyond the critical radius, as compared to the perfectly adhering solution. The above derivation assumed that Poisson's ratio equals 0.5. Of course, real elastomers are not incompressible. Lai et al. have studied the effect of compressibility on the stress distributions in thin elastomeric blocks and annular bushings [9]. For the special cases of an infinite strip of finite width the following closed form solution for the effective modulus was obtained:

$$E_{\text{eff}} = K \left[1 - \frac{\tanh \beta}{\beta} \right] \tag{22}$$

$$\beta = \frac{3}{\eta}\sqrt{\frac{2(1-2v)}{1+v}} \tag{23}$$

where K is the bulk modulus and $\eta = 1/S$ = total free area/area of one loaded surface (inverse of the classical shape factor, S). The solution agrees well with numerical solutions so long as the substrates are completely rigid. For highly constrained joints, $h/a < 0.1$, the substrates may deflect and reduce the predicted stresses in the elastomer. Fig. 14 (right) describes the shape of the stress field when Poisson's ratio is less than 0.5. The pressure distribution is flattened as the constraint increases. We see that the incompressible assumption leads to significant errors in the estimation of the pressure distribution within the block. For an elastomer with $v = 0.4995$, the incompressible assumption leads to an effective stiffness error of 10%, 47%, and 187% for shape factors of 5, 10 and 20, respectively.

Numerical solutions to the above joint conditions have been provided which include the effect of compressibility and are more accurate near the edge of the joint [10,11]. Gent and Hwang used a linear elastic model to demonstrate a 40% reduction in apparent modulus when Poisson's ratio was changed from 0.4999 to 0.49 [12]. Chang and Peng have studied the butt joint using nonlinear finite element analysis [13]. The material was described by the nonlinear elastic strain energy function of Ogden–Tschoegl, modified to include a bulk modulus term, which accounts for specific values of Poisson's ratio. Thus, the effect of strain softening or strain hardening could be studied along with the effect of compressibility. For small strains, their calculations agreed well with Lindsey, Shapery and Williams so long as a/h was larger than 10 [14]. But for values of $a/h < 6$, the approximate solutions of Gent and Lindley were better matched [5]. Fig. 16 depicts the expected changes in the normal stress distribution as a function of strain hardening or softening.

Numerical techniques can provide insight into the stresses in the elastomeric butt joint. However, the correct choice for Poisson's ratio is essential. To further complicate the situation, measurement of Poisson's ratio is not trivial. The measurement of Poisson's ratio was first performed using classical dilatometry techniques [15]. Poisson's ratio is also time dependent. The WLF equation has been used to describe the temperature and rate dependence of Poisson's ratio [16,17]. Furthermore, Poisson's ratio can be a function of filler–polymer interactions and strain induced crystallization. For a highly filled composite, greater than 50 vol% glass beads, the value of Poisson's ratio dropped from nearly 0.5 to 0.25 over a strain of 0–40%, respectively [18]. The measurement of bulk modulus and shear or tensile modulus affords an easier way to estimate Poisson's ratio by utilizing Eq. 6 [19,20]. More recently, photoelastic techniques, ultrasonics and contact strain gages have been used to measure Poisson's ratio [21–23]. Finally, as noted by Kakavas, defects in the sample itself, such as bubbles, can significantly

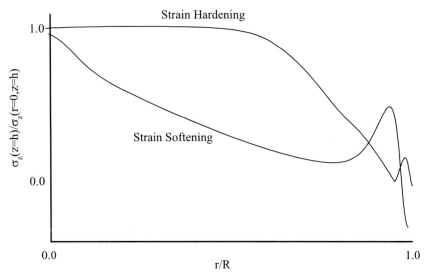

Fig. 16. Normalized stress versus radial position for a strain hardening and a strain softening model. Redrawn from ref. [13].

reduce the apparent Poisson's ratio [24]. A comparison of the effective modulus calculated by the finite element method and predicted by Gent's equation led Kakavas to conclude that the Poisson's ratio used in his calculations was too high. Further investigation leads to a correction term for effective Poisson's ratio based on the true Poisson's ratio and the volume of voids in the sample.

4. Measuring fracture energy

The interfacial strength of an elastomeric material is best defined using an energy balance approach as discussed in Chapter 2. The strain energy release rate, $\mathcal{G}_c$, is simply the energy needed to generate a new surface by propagating a crack though the material as originated by Griffith while studying brittle materials [25]. Pocius has recently reviewed the derivation leading to the following for the special case of a completely brittle material as seen in Fig. 17 [26]:

$$\frac{F^2}{2w}\frac{\partial C}{\partial a} \geq \mathcal{G}_c \tag{24}$$

where C is the compliance of the material $\mathcal{G}_c$ is the critical strain energy release rate. Rapid crack growth will occur if $\mathcal{G}_c$ is exceeded. The above analysis is generalized to various geometric configurations by discussing the resistance to crack growth in terms of the total energy available to drive a crack through the material. For the specific case of elastomeric materials, energy losses occur near

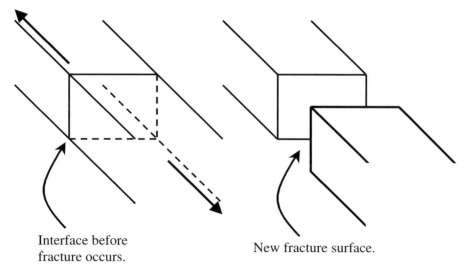

Interface before fracture occurs.

New fracture surface.

Fig. 17. The calculation of the fracture energy based on the work needed to create two new surfaces in the material.

the vicinity of the crack tip which are included in the total energy measured by fracture tests. The magnitude of the losses are generally related to the material properties of the elastomer being strained to failure. Frequently, the energy losses make up the majority of the energy being measured. Thus, an energy balance approach can generate fracture measurements that are independent of the test geometry. One need only measure or calculate the elastic energy needed to drive the crack in terms of the total strain energy per unit crack surface formed:

$$\mathcal{G} = -\left(\frac{\partial W}{\partial A}\right)_l \tag{25}$$

4.1. Displacement versus load-controlled measurement conditions

As noted in Section 2.3, the deformation of elastomeric adhesives is a function of the speed and temperature. Correspondingly, the measured strain energy release rate is also a function of the test speed and the material temperature. Another important consideration is the loading condition. There are two ways that a joint can be stressed, by separating the components at a specific speed and measuring the resulting force or by applying a specific force and measuring the speed of the crack growth. Often both occur at the same time, but we will discuss them separately for now.

Many commercial test machines measure the strength of the test specimen by subjecting them to a specific displacement condition and then measuring the stresses that occur. This is known as a displacement controlled test condition.

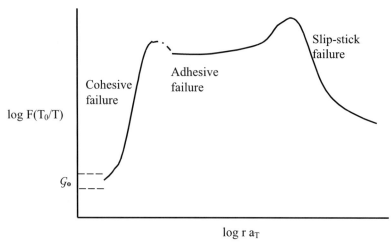

Fig. 18. Strain energy as a function of test speed. Redrawn from ref. [27].

Typical crack speed versus applied strain energy for displacement controlled testing can be found in Fig. 18 [27].

Other test methods can be conducted which apply a load on the sample and measure the speed of crack growth. Such methods are commonly called load controlled. A typical curve of crack growth versus applied strain energy can be seen in Fig. 19.

Note the existence of a rate-independent threshold fracture energy, $\mathcal{G}_0$, which occurs under very slow strain rate conditions. The fracture energy is generally separated into two components, the classical work of adhesion and the bulk dissipative term. The classical work of adhesion is the thermodynamically reversible and rate-independent energy required to separate two surfaces. It is described in detail in Volume II "Surfaces, Chemistry and Applications". It is important to note that the thermodynamic work of adhesion is usually much smaller than the measured fracture energy. That is because the energy needed to propagate a crack in the material, the fracture energy or strain energy release rate, also includes the energy spent in viscous response, inertia, and in the generation of heat; all of which are rate-dependent properties. In other words, the thermodynamic work of adhesion is associated with the processes occurring along the interface that is being separated. While, the fracture energy measures the thermodynamic work of adhesion along with all the other rate-dependent energy consuming mechanisms, which are occurring in the bulk of the materials being tested. The threshold fracture approaches the value of the theoretical work of adhesion and under ideal conditions should be equivalent. Several researchers have proposed the following equation, which attempts to separate the measured strain energy release rate into

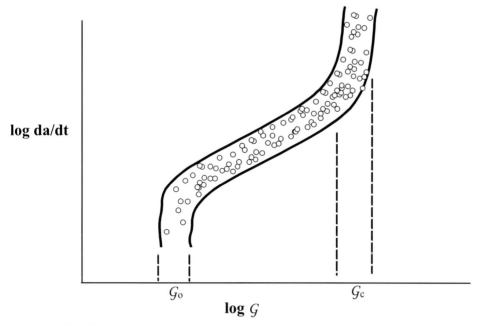

Fig. 19. Generalized crack speed as a function of applied fracture energy.

contributions from the interface and contributions from the bulk [28–30].

$$\mathcal{G} = \mathcal{G}_0 + f(\mathcal{G}_0)\zeta \qquad (26)$$

where ζ is the rate-dependent contributions occurring in the bulk of the adhesive. When conducting experiments on various adhesives, all with different viscoelastic response, it is necessary to measure the strain energy release rate at many different speeds or temperatures in order to distinguish between mechanisms associated with the bulk elastomer versus mechanisms associated with changes to the interface. Furthermore, the interpretation of the results should be done with the expected use conditions in mind. For example, if the adhesive needs to survive small stresses for many years, then the slow-rate data are most important. But, if the adhesive needs to be used on explosive shielding or tornado resistant buildings, then the high-speed data is very important.

4.2. Common fracture energy test methods

Starting with rather simple geometric shapes and loading conditions facilitate the measurement of the fracture energy. The following methods represent some of the more common techniques for measuring the crack speed versus strain energy release rate.

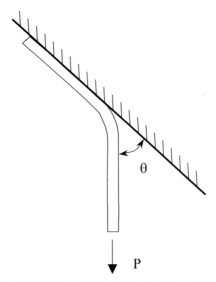

Fig. 20. Geometry of the peel test. Redrawn from ref. [31].

4.2.1. Peeling

Kinlock and Williams describes the peel test in detail in Chapter 8. The test geometry can be seen in Fig. 20. This discussion of the peel test will mainly address thick elastomeric specimens. The derivation of $\mathcal{G}$ for the peel test was recently reviewed by Maugis [31]. $\mathcal{G}$ can be expressed as follows:

$$\mathcal{G} = \frac{P}{w}(1 - \cos\theta) + \frac{P^2}{2w^2 Eh} \tag{27}$$

where w is the peel width, θ the peel angel, P the applied load, E the modulus of the peeling member and h the thickness of the peeling member. The second term of this equation accounts for the strain energy used to stretch the peeling member. It is common to embed a wire cloth, fiberglass or similar material in the elastomer before cure, then E becomes very large and the second term can be neglected.

The peel test is most often conducted under constant displacement rate and with a peeling angle of 180°. Under this condition, the displacement rate is predetermined and the peeling force is measured. The crack speed is 1/2 the displacement speed.

Conversely, a load-controlled experiment can be conducted by hanging a known mass off the peeling member. However, the peeling speed is more difficult to measure. An alternative method for load-controlled peeling can be accomplished by using a creep frame and lever arm to apply a load to the peeling specimen. The deflection of the lever arm can be easily measured with electronic displacement transducers and the crack speed deduced by simple geometric relationships [32].

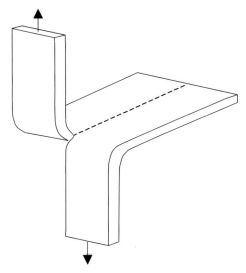

Fig. 21. Test geometry for the trouser tear method.

The effect of thickness can be corrected by plotting the strain rate instead of the peeling rate. This is analogous to measuring strain rate instead of crosshead speed for a uniaxial tension experiment. The strain rate can be estimated by the following equation [33]:

$$\dot{\varepsilon} = \frac{\dot{a}}{b} \qquad (28)$$

where $\dot{\varepsilon}$ is the strain rate and $\dot{a}$ is the peeling rate. The gage length is assumed to be the same as the strained thickness, b. At constant peeling rate, the strain rate increases as the strained thickness, b, decreases.

The value of $\mathcal{G}$ is a function of the peeling angle, width and force as described by Eq. 27. In addition, it is rate dependent as described by Eqs. 26 and 28. It is important to note these dependencies when attempting to compare adhesion test data from peel tests conducted under different conditions.

4.2.2. Trouser tear test

The trouser tear test gets its name from the shape of the test specimen, which resembles a pair of men's trousers as seen in Fig. 21. The test is conducted by pulling each leg in opposite directions, thus propagating a crack down the center of the specimen. The test is most commonly conducted in displacement control whereby the two legs are pulled apart at a specified speed while a load cell measures the force. The test can also be conducted in load control by applying a constant load to the legs of the specimen usually with a hanging weight or lever actuated creep frame. The fracture energy can be obtained from Eq. 29 [34,35].

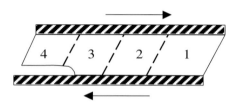

Fig. 22. The simple shear test geometry.

$$\mathcal{G} = 2F/w \tag{29}$$

The experiment can be further simplified by embedding a strip of fiberglass or wire cloth on both sides of the crack parallel to the trouser legs, effectively eliminating stretching in the legs during testing. The crack length is then determined by measuring the length of the legs or by optical means.

4.2.3. Simple shear, pure shear

4.2.3.1. Simple shear. Many tests configurations can be analyzed by comparing the stored strain energy in the elastomer as a crack passes through the region. In Fig. 22, we note that the stored strain energy per unit volume, U, is released in region 4 because the rubber has debonded. Region 3 is in the process of debonding. Regions 2 and 1 are still strained and have potential energy for driving the crack tip. The strain energy per unit volume can be determined from stress–strain measurements. Thus, for the simple shear case, the strain energy release rate is:

$$\mathcal{G} = Uh \tag{30}$$

This test is often conducted at a constant shear strain condition. Under such conditions, the strain energy release rate is constant as a function of crack length. This is very helpful in the analysis. However, it is also important to point out that changes in the modulus of the elastomer will change the stored strain energy. If this occurs over the time interval of the test, then U will need to be reevaluated. Such changes are very common during environmental aging.

4.2.3.2. Pure shear. The pure shear test is a special case of the tensile test in which the length is very long compared the width of the elastomer thus preventing the elastomer from contracting along that length of the specimen. This results in plane strain conditions that result in the principal stress occurring at a 45° angle out of the paper as drawn in Fig. 23. The analysis of the fracture energy is identical to the simple shear case [36]. At constant strain and modulus, the strain energy release rate is independent of crack length. This is good for the same reasons as

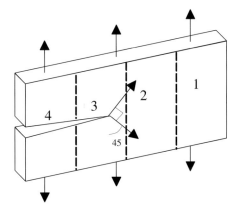

Fig. 23. The pure shear test specimen.

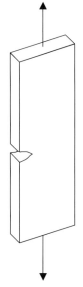

Fig. 24. The tensile strip fracture test specimen.

stated for the simple shear case. Again, modulus changes during the test must be monitored independently in order to correct the strain energy density.

4.2.4. Tensile strip specimen

The tensile strip specimen, described in Fig. 24 was studied by Rivlin and Thomas [36]. The strain energy release rate is described by the following equations:

$$\mathcal{G} = kU_b a \tag{31}$$

$$k = 2\pi \sqrt{1 + \varepsilon_b} \approx \frac{3}{\sqrt{\lambda}} \tag{32}$$

where U_b is the strain energy per unit volume, and k is a function of the extension. This equation could also be used in the pure shear specimen when the crack length is short and the strain energy is a function of crack length. Like the previous three methods, measurement of the crack length is the most difficult part of the experiment. Unlike the previous methods, the strain energy release rate is a function of crack length. Therefore, the strain energy release rate changes during the experiment.

4.3. Methods for measuring the threshold fracture energy

The threshold fracture energy is very interesting for two important reasons. Firstly, it is the energy most closely associated with changes in the interface properties. Secondly, it is relevant for long-term durability measurements. Referring back to Eq. 26, we recall that the threshold fracture energy often only represents a small fraction of the total energy measured. The function ζ is the major contribution to the energy measured. Therefore, it is necessary to try to reduce the size of ζ until it is nearly zero, thus, revealing the hidden value of $\mathcal{G}_0$. This is accomplished by conducting the tests at very slow speeds. One quickly notes that practical test speeds are still too fast and ζ still too large to see the true size of $\mathcal{G}_0$. The method of time–temperature superposition is a useful way to obtain data at extremely slow speeds. Further reduction in the value of ζ is accomplished by using plasticizers to increase the mobility of the polymer chain. However, this method introduces a new variable, which may or may not be applicable to one's real world application. If the bulk material does not change as part of the independent variables, then it may be possible to assume that the function ζ does not change from sample to sample. A simple change in surface preparation without changes in the elastomer composition is one example in which ζ may be assumed constant. In this case, a log–log plot of crack speed versus strain energy release rate will provide a family of curves, one for each surface condition. The relative change in $\mathcal{G}_0$ can be estimated by the vertical distance between each curve as seen in Fig. 25.

The method of time–temperature superposition has been used extensively for the study of the threshold strain energy release rate. A classic example can be found in Figs. 26 and 27 redrawn from the work of Chun and Gent [37]. In this work, the trouser tear test was used to generate the master curve spanning 15 decades of time. The threshold strain energy release rate was estimated at a crack speed of 10^{-18} m/s. Such extremely slow rates are commonly obtained using master curves. The shift factor was obtained using the WLF equation with $C_1 = 17.6$ and $C_2 = 52$. Gent and Lai have shown that many hydrocarbon

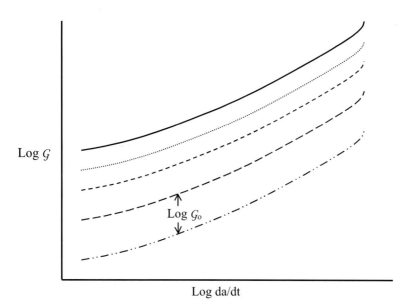

Fig. 25. Estimation of $\mathcal{G}_0$ when ζ is reasonably constant for a given set of measurements. Each line represent a different surface treatment. The distance between them is $\mathcal{G}_0$.

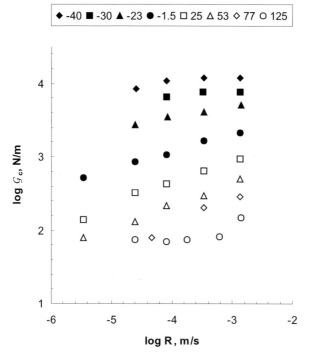

Fig. 26. Strain energy release rate data measured at various temperatures and test speeds. Redrawn from ref. [37].

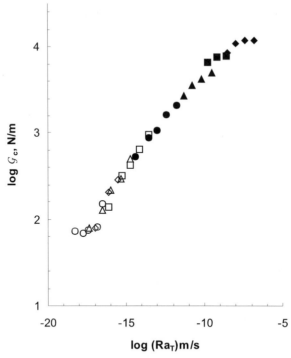

Fig. 27. The master curve referenced to $-55°C = T_g$. Redrawn from ref. [37].

polymers have very similar master curves when referenced to the T_g [38]. Gent and Tobias have generated master curves using the trouser tear test with the polymer under a swollen condition to further reduce the value of ζ [39].

Mazich et al. utilized the pure shear fracture test to generate master curves of threshold fracture energy for polydimethylsiloxane polymers as a function of crosslink density [40]. Their data compared well with recent data from Gent and Tobias [39].

In all cases the threshold fracture energy is still about 10–20 times larger than the theoretical work of adhesion predicted from thermodynamic estimates. This discrepancy has been explained by the theory of Lake and Thomas [41]. The work to break a polymer chain includes the work needed to pull the chain taunt in addition to the work needed to break a the chemical bond between two individual atoms in the chain. Thus, the fracture energy scales with the square root of the molecular weight between crosslinks as defined below:

$$\mathcal{G}_{c,o} = (3/8)^{1/2} \rho A U q^{1/2} l M_o^{1/2} M_c^{1/2} \tag{33}$$

where ρ is the polymer density, A is Avogadro's number, U is the dissociation energy of the weakest main chain bond, q is the number of main chain bonds per free jointed segment, l is the projected length of a main chain bond, M_o is the average molecular weight per main chain atom and M_c is the average molecular weight between crosslinks.

4.4. Techniques for obtaining interfacial fracture energy

While it has often been considered good to have cohesive failure, it is of little value when interface strength is the desired measurement. Once cohesive failure occurs, all we can say about the interface is that the stressed in the joint in the region of the interface were less than the strength of the interface at the time of testing. If one is studying different surface treatments or adhesion promoter additives, then cohesive failure prevents differentiation of the variables controlling the strength of the interface in particular. This problem is further complicated by many government and industry tests that are not ideally useful for measuring interfacial strength. For example, the 180° peel test conducted at 2 inches per minute constant peel rate tends to favor cohesive failure as seen in the work by Shephard and Wightman [42]. The failure mode during peeling could be changed by adjusting the peel test conditions. Slower speed, higher modulus and thinner test specimens favored adhesive failure as seen in Fig. 28.

Interfacial fracture energy is often difficult to obtain when working with elastomeric materials. This is mainly due to the relatively low tear strength of the elastomer. If the stresses in the joint exceed the tear strength of the elastomer, then the fracture will be cohesive in the elastomer. Therefore, the stresses in the joint during testing must be designed to never exceed the tear strength of the elastomer. Such a condition could lead to extremely slow failure rates. In such cases, the interfacial failure can be facilitated by applying and environmental stress, such as water, heat or ultraviolet light to degrade the strength of the interface.

4.5. Methods for fatigue fracture energy

The calculation of fracture energy for fatigue experiments is identical to previously discussed approaches. The main difference is that the strain energy release rate is now a varying function of time. The resulting data generally takes on the shape of the curved in Fig. 19. Of course the y-axis is da/dn and the x-axis is some measure of strain energy per cycle. The variable, n, denotes each cycle. Typically, the time function is a simple sinusoidal function with a specific frequency and amplitude. However, a more complicated driving function can be used to simulate a specific use application such as a road sealant undergoing displacements due to road traffic [43].

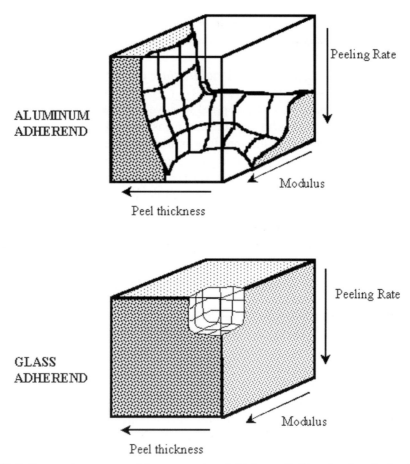

Fig. 28. Failure mode versus modulus, rate and peel thickness. Solid region is cohesive failure, clear regions are adhesive failure. Redrawn from ref. [42].

4.5.1. Uniaxial tension

This method is based on the tensile strip specimen. Therefore the fatigue cycles can never go into compression because the specimen would bend rather than support a significant compressive load. Lake and Lindley used this method to generate data for natural rubber [44]. During that work, they noted the three distinct events in the crack growth curves. $\mathcal{G}_o$ the minimum energy necessary to initiate crack growth, the power law region and $\mathcal{G}_c$ when crack growth is rapid completely analogous to the previous discussion. The following equation has been used to describe the power law region [45]:

$$\frac{\mathrm{d}a}{\mathrm{d}n} = \frac{1}{K}\mathcal{G}^{\beta} \tag{34}$$

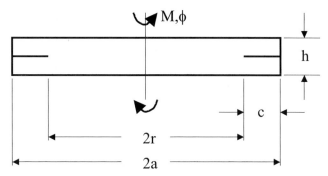

Fig. 29. Geometry of the torsion test specimen. Redrawn from ref. [47].

in which K and β are constants associated with a specific elastomer. Integration leads to N, the number of cycles to failure based on an initial crack length, a_o:

$$N = \frac{K}{(\beta - 1)(2k(\lambda)U_b)^\beta} \cdot \frac{1}{a_o^{\beta-1}} \tag{35}$$

where k, λ and U are defined by Eqs. 31 and 32. Recently, Lacasse et al. measured the cut growth in a silicone sealant and found good agreement between experimental data and Eq. 35 [46].

4.5.2. Torsion

A description of the torsion test can be seen in Fig. 29. The derivation of the strain energy release rate has been recently reviewed [47].

$$\mathcal{G} = \left(r + \delta^*\right)^2 G\phi^2/2h \tag{36}$$

where δ is a correction factor, which adjusted the crack length to account for frictional forces during torsion. This factor could be removed if the sample was also placed in tension during the test. We note that with this test, the more the crack grows, the lower the strain energy release rate for a given strain amplitude. This is very useful because the crack growth is predicted to arrest. The arrest fracture energy is often assumed to be the same as the threshold fracture energy. Therefore, one needs to simply run the test at constant amplitude and wait for the crack to stop growing in order to find the threshold fracture energy. In a follow up paper, Gent reports that data obtained in torsion compare well with previously published data generated using a tensile specimen [48].

4.5.3. Compression

Very little published fracture mechanics data has been found utilizing a compression test. However, Stevenson measured compression data that agreed well

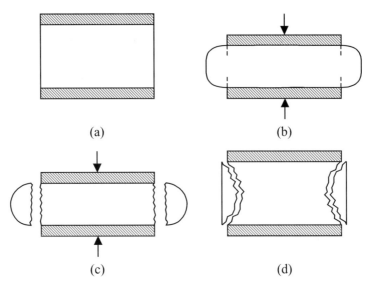

Fig. 30. Crack growth in a compression test specimen. Redrawn from ref. [49].

with tensile data, once again, demonstrating the power of the energy approach with fracture mechanics [49]. The compression test geometry and resulting crack growth can be seen in Fig. 30. Crack growth occurs such that a parabolic cross-section is removed from the outer radius of the specimen. The strain energy release rate is given by:

$$G = 0.5Uh \tag{37}$$

$$U = 0.5E_c\varepsilon_c^2 \tag{38}$$

$$E_c = E(1 + \kappa s^2) \tag{39}$$

$$s = \text{one loaded area/force-free area} \tag{40}$$

where E_c is the effective compression modulus, s is a shape factor, and κ is a numerical factor. A comparison of the experimental data with the theoretical data generated from Eq. 37 can be seen in Fig. 31. The corresponding crack growth data for tensile testing was also plotted with reasonable agreement.

The measurement of crack growth under compression as described above is a very experimentally tedious problem. It is no longer necessary to collect such data. Instead, the same fracture data in tension can be readily obtained followed by conversion of the data to a particular compression problem.

The value of the energy approach to fracture measurement is that it can be readily used on many very analytically difficult problems. Researchers have demonstrated the ability to directly correlate data measured in one test geometry with data obtained with completely different test geometry. Simple test pieces can

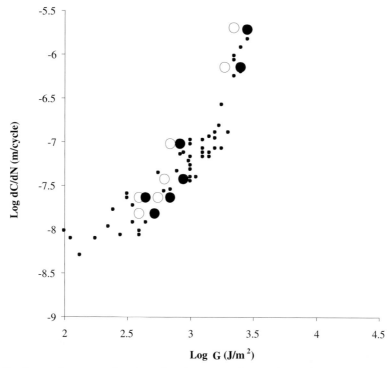

Fig. 31. Crack growth rate as a function of strain energy: ●, experimental results for compression, ○ theoretical results for compression, • experimental results for simple tension. Redrawn from ref. [49].

be used to generate the characteristic fracture energy versus crack speed curves which can then be applied to very complicated real world parts by using numerical methods to map the strain energy in the part and using that data to determine if crack growth is likely.

5. Designing with elastomeric adhesives

5.1. Design approach

When designing an elastomeric joint, the design approach is based on measurement of material properties using relatively simple test methods and joint shapes. This is followed with analytical or more likely numerical solutions to determine the state of stress in the more complicated real world part. Then the strain energy density information is compared to the fracture energy data to determine the likelihood of crack growth within the part. Finally, the optimized part should be prepared and tested to validate the joint design.

Using the fracture mechanics approach automatically accounts for the occasional non-ideal part. The assumption that a flaw will exist enables the ability to determine if the flaw will propagate or simply remain at its current size. A joint that can tolerate occasional unintentional defects will be far easier to manufacture and result in fewer field failures.

5.2. Conditions to avoid

Careful analysis of the butt joint has led to several generalizations regarding the stability of small defects in the elastomer. The theory of large elastic deformation developed by Rivlin and outlined by Gent predicts that small voids in the rubber will expand without bound when a critical value of hydrostatic tension of about 2.5 G is exceeded [50,51]. Using the equations from Gent's analysis of the butt joint, one can predict when catastrophic failure will occur due to rupture of the elastomer. For thin blocks, this equates to a mean tensile stress of about 1.25 G. In compression, failure occurs at the edges where the maximum shear deformation should not exceed about 100%, which is an average compressive stress of about 3 G or strain of $h/3a$.

In addition, compressibility is an important factor when the constraint on the elastomer is high. Under such conditions, it is important to obtain an accurate measurement of Poisson's ratio. Otherwise, the stresses used to calculate the strain energy density could be off by 50% or more.

Conditions where three-sided adhesion occurs leads to premature failure by subjecting the elastomer to triaxial stresses. A few examples of good and bad joint design can be seen in Fig. 32. Frequently, it is necessary to prevent adhesion on one or more surfaces in order to eliminate triaxial stress.

Synergistic degradation effects can also lead to early joint failure. Fatigue causes faster crack growth than do steady-state stresses. Mechanical stress in combination with environmental stresses like standing water, heat, ultraviolet light, acid rain, ozone or solvents can reduce service life many fold.

5.3. The isotropic assumption is often false

Throughout this chapter, all the equations have assumed homogeneous isotropic materials. In many cases, the material properties of the elastomer are not uniform. Some common problems leading to anisotropic material properties are poor mixing of components leading to trapped air, undispersed filler particles, or poor crosslinking. Some elastomers cure by diffusion of moisture into the elastomer. Such processes lead to gradients in modulus according to the moisture diffusion path in the joint. Temperature gradients during cure also can lead to modulus gradients. Shrinkage during cure can introduce non-uniform pre-stresses. Filler settling can also lead to heterogeneity. Solvent diffusion can lead to heterogeneous

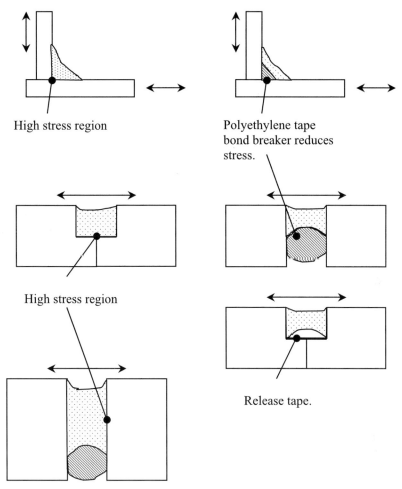

High stress region

Polyethylene tape
bond breaker reduces
stress.

High stress region

Release tape.

Fig. 32. Left side demonstrates poor joint designs that contain high stresses. Right side demonstrates better designs.

swelling. Finally, an elastomer that was isotropic to begin with may undergo degradation on the surface or in the center of the bulk at a different rate. This can lead to large changes in material properties.

Acknowledgements

The recent reviews by Gent and the classic text by Ferry were invaluable during the preparation of this manuscript. The support of the Dow Corning Corporation for time and materials is also appreciated.

References

1. Ferry, J.D., *Viscoelastic Properties of Polymers*. John Wiley and Sons, New York, 1961.
2. Aklonis, J.J. and MacKnight, W.J., *Introduction to Polymer Viscoelasticity*, 2nd edn. John Wiley and Sons, New York, 1983.
3. Aklonis, J.J. and MacKnight, W.J., *Introduction to Polymer Viscoelasticity*, 2nd edn. John Wiley and Sons, New York, 1983, p. 37.
4. Williams, M.L., Landel, R.F. and Ferry, J.D., *J. Am. Chem. Soc.*, **77**, 3701 (1955).
5. Gent, A.N. and Lindley, P.B., *Proc. Inst. Mech. Eng. (Lond.)*, **173**, 111 (1959).
6. Gent, A.N. and Meinecke, E.A., *Polym. Eng. Sci.*, **10**, 48 (1970).
7. Gent, A.N., Henry, R.L. and Roxbury, M.L., *Trans. ASME Appl. Mech.*, 855 (1974).
8. Thornton, J.S., Montgomery, R.E., Thompson, C.M. and Dillard, D.A., *Polym. Eng. Sci.*, **28(10)**, 655 (1988).
9. Lai, Y.H., Dillard, D.A. and Thornton, J.S., The effect of compressibility on the stress distribution in thin elastomeric blocks and annular bushings. *J. Appl. Mech.*, **59**, 902–908 (1992).
10. Messner, A.M., Stress distributions in poker chip tensile specimens, *Bulletin of Working Group on Mechanical Behavior*. CPIA Publ., **27(109)** (1963).
11. Anderson, G.P., DeVries, K.L. and Sharon, G., Evaluation of Tensile Tests for Adhesive Bonds. In: Johnson, W.S. (Ed.), *Delamination and Debonding of Materials, ASTM STP 876*. American Society for Testing and Materials, Philadelphia, PA, 1985, pp. 115–134.
12. Gent, A.N. and Hwang, Y.-C., *Rubber Chem. Technol.*, **61**, 630 (1988).
13. Chang, W.V. and Peng, S.H., *J. Adhes. Sci. Technol.*, **6(8)**, 919 (1992).
14. Lindsey, G.H., Schapery, R.A., Williams, M.L. and Zak, A.R., Galcit report on sm 63–6, Technical report, Caltech, 1963.
15. Smith, T.L., *Trans. Soc. Rheol.*, **3**, 113 (1959).
16. Kruse, R.B., *CPIA Publ.*, **2**, 337 (1962).
17. Waterman, H.A., *Rheol. Acta*, **16**, 31 (1977).
18. Smith, T.L., *Trans. Soc. Rheol.*, **3**, 113 (1959).
19. Smith, T.C., *Polym. Eng. Sci.*, **16**, 394 (1976).
20. Schwarzl, F.R. In: Erigen, A.C., Liebowitz, H., Koh, S.L. and Crowley, T.M. (Eds.), *Mechanics and Chemistry of Solid Propellants*. Pergamon, Oxford, 1967.
21. Richard, T.G., *J. Compos. Mater.*, **9**, 108 (1975).
22. Knollman, G.C., Martinson, R.H. and Bellin, T.L., *J. Appl. Phys.*, **50**, 111 (1979).
23. Fedors, R.F. and Hong, S.D., *J. Polym. Sci., Polym. Phys. Ed.*, **20**, 777 (1982).
24. Kakavas, P.A., *J. Appl. Polym. Sci.*, **59**, 251 (1996).
25. Griffith, A.A., *Phil. Trans. R. Soc.*, **A221**, 163 (1920).
26. Pocius, A.V., *Adhesion and Adhesives Technology*, Carl Hanser Verlag, New York, 1997, Chapt. 4.
27. Kaelble, D.H., *J. Adhes.*, **1**, 102 (1969).
28. Gent, A.N. and Kinloch, A.J., *J. Polym. Sci.*, **9(A-2)**, 659 (1971).
29. Gent, A.N. and Schultz, J., *J. Adhes.*, **3**, 281 (1972).
30. Andrews, E.H. and Kinloch, A.J., *Proc. R. Soc. Lond. A.*, **332**, 401 (1973).
31. Maugis, D., *Contact, Adhesion and Rupture of Elastic Solids*. Springer-Verlag, Berlin, 2000, Chapt. 5.
32. Shephard, N.E. and Wightman, J.P., A simple device for measuring adhesive failure to sealant joints. In: Klosowski, J.M. (Ed.), *Science and Technology of Building Seals, Sealants, Glazing, and Waterproofing, Seventh Volume, ASTM STP 1334*. American Society for Testing and Materials, Philadelphia, PA, 1998.

33. Cochrane, H. and Lin, C.S., *Rubber Chem. Technol.*, **66**, 48 (1993).
34. Mueller, H.K. and Knauss, W.G., *Trans. Soc. Rheol.*, **15**, 217 (1971).
35. Ahagon, A. and Gent, A.N., *J. Polym. Sci. Polym. Phys. Ed.*, **13**, 1903 (1975).
36. Rivlin, R.S. and Thomas, A.G., *J. Polym. Sci.*, **10**, 291 (1953).
37. Chun, H. and Gent, A.N., *Rubber Chem. Technol.*, **69**, 577 (1996).
38. Gent, A.N. and Lai, S.M., *J. Polym. Sci., Polym. Phys. Ed.*, **19**, 1619 (1994).
39. Gent, A.N. and Tobias, R.H., *J. Polym. Sci. Polym. Phys. Ed.*, **20**, 2051 (1982).
40. Mazich, K.A., Samus, M.A., Smith, C.A. and Rossi, G., *Macromolecules*, **24(10)**, 2766 (1991).
41. Lake, G.J. and Thomas, A.G., *Proc. R. Soc. Lond. A*, **300**, 108 (1967).
42. Shephard, N.E. and Wightman, J.P., An Analysis of the 180° Peel Test for Measuring Sealant Adhesion. In: Lacasse, Michael, A. (Ed.), *Science and Technology of Building Seals and Sealants: Fifth Volume, ASTM STP 1271*. American Society for Testing and Materials, Philadelphia, PA, 1995.
43. Al-Qadi, I.L., Imad, L. and Abo-Qudais, S.A., Test method for evaluating pavement sealants under simultaneous cyclic shear and normal deflections. In: Klosowski, J. (Ed.), *Science and Technology of Building Seals, Sealants, Glazing, and Waterproofing: 3rd Vol., ASTM STP 1254*. American Society for Testing and Materials, Philadelphia, PA, 1994.
44. Lake, G.J. and Lindley, P.B., *J. Polym. Sci.*, **9**, 1233 (1965).
45. Gent, A.N., Lindley, P.B. and Thomas, A.G., *J. Polym. Sci.*, **8**, 707 (1964).
46. Lacasse, M.A., Margeson, J.C. and Dick, B.A. Static and dynamic cut growth fatigue characteristics of silicone based elastomeric sealants. *RILEM Proceedings, 28, Durability of Building Sealants: Proceedings of the International RILEM Symposium on Durability of Building Sealants*, Garston, Watford, UK, 10/11/94, pp. 1–16, 1996 (ISBN:0419210709) (NRCC-38996).
47. Aboutorabi, H., Ebbott, T., Gent, A.N. and Yeoh, O.H., *Rubber Chem. Technol.*, **71**, 76 (1997).
48. Gent, A.N. and De, D.K., *Rubber Chem. Technol.*, **71**, 84 (1997).
49. Stevenson, A., *Rubber Chem. Technol.*, **59**, 208 (1986).
50. Rivlin, R.S., In: Eirich, F.R. (Ed.), *Rheology, Theory and Applications*, Vol. 1. Academic Press, New York, 1956, Chapter 10.
51. Gent, A.N. (Ed.), *Engineering with Rubber*. Hanser Publishers, New York, 1992. Chapt. 3.

Chapter 11

Crack path selection in adhesively bonded joints

BUO CHEN [a,*] and DAVID A. DILLARD [b]

[a] *Cooper Tire and Rubber Company, Findlay, OH, USA*
[b] *Professor of Engineering Science and Mechanics, Virginia Tech, Blacksburg, VA, USA*

1. Introduction

1.1. Background

A wide variety of both strength tests, measuring some average stress at break, and fracture tests, measuring the energy required to propagate a debond, have been introduced to test adhesive joints to failure. In addition to quantifying the stresses or energy release rates required for mechanical separation, a meaningful evaluation of the mechanical performance often requires accurate identification and interpretation of the locus of failure and the crack propagation behavior. Perhaps based on the old adage that chains always break at the weakest link, conventional wisdom suggests that bonded joints should also break at the weakest location. The weakest link criterion is applicable to discrete systems such as chains, but is not directly applicable to continuous systems, including bonded joints, where complex stress distributions interact with spatially varying material properties. Fracture of an adhesive bond involves complex interactions between loading conditions, geometry, and material property variations causing the locus of failure to vary even within the same material system [1–3].

The final locus of failure and fracture trajectory of an adhesive bond is the result of the interactions among the material properties such as the tensile strength of all the components, quality of adhesion at the interface, fracture toughness of the bonds, and the stress state at the crack tip of existing flaws or debonds [4]. To predict the locus of failure in an adhesively bonded joint loaded in an arbitrary manner such as illustrated in Fig. 1, the mechanisms controlling the

---
* Corresponding author. E-mail: bhchen@coopertire.com

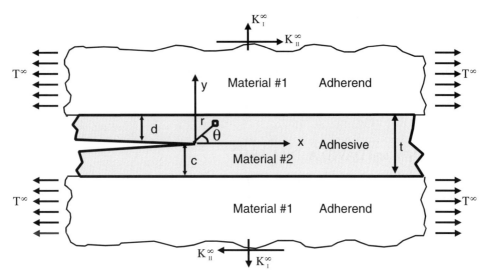

Fig. 1. A crack within the adhesive layer in an adhesive bond. The adherend is designated as material 1 and adhesive is designated as material 2. The coordinate system is set at the crack tip. If $d = 0$, then the crack becomes an interface crack located at the interface between the adherend and the adhesive.

direction of debonding must be understood. Fracture mechanics, sometimes used in conjunction with finite element methods, has proven to be a powerful tool in modeling and predicting failure modes [5,6]. The foundations for fracture mechanics provided in Chapter 2, including the fundamental relationships for stresses, displacements, and fracture parameters for both cohesive cracks and interfacial debonds, will serve as the basis for this discussion.

This chapter examines the fracture mechanics concepts that influence crack path selection behavior in adhesively bonded joints, and reviews a recent study aimed at understanding and predicting crack path selection. The discussion briefly reviews relevant fracture mechanics and interface mechanics theory, illustrates results from experimental studies, and highlights both numerical and analytical modeling efforts. Specifically, the effects of mode mixity and the T-stress, both of which can be affected by specimen geometry, material properties, loading mode, and residual stress state, will be provided. The concepts and results summarized herein should be useful in interpreting the role mechanics plays in determining the locus of failure in adhesive joints, and has implications for test specimen selection, failure mode analysis, and predictive modeling.

1.2. Direction of crack propagation and mixed mode fracture

Over the years, several fracture mechanics criteria have been developed to determine the direction of crack propagation for cracks in brittle, homogeneous, isotropic solids. Among these criteria, three primary ones have been widely discussed in the literature; namely,

(1) *Maximum opening stress criterion*: This criterion, proposed by Ergodan and Sih [7], dictates that the direction of cracking is perpendicular to the direction of maximum opening stress at the crack tip.

(2) *Maximum energy release rate criterion*: Palaniswamy and Knauss [8] suggested this criterion, whereby the direction of crack propagation can be obtained by maximizing the energy release rate as a function of the angle of crack kinking.

(3) *Mode I fracture criterion*: Goldstein and Salganik [9] and Cotterell and Rice [10] proposed that a crack will propagate along a path such that pure mode I fracture is maintained at the crack tip, i.e. $K_{II} = 0$ at the growing crack tip.

Although these three criteria specify different aspects, they all yield similar results and no experimentally distinguishable differences have been observed [2,3,11,12]. Consequently, the choice of criterion in practical applications depends on convenience. The maximum opening stress and the maximum strain energy release rate criteria are often used in analytical studies, whereas the mode I fracture criterion is usually more convenient to use in numerical analysis since standard stress intensity factor extrapolation schemes are usually available in most commercial finite element analysis (FEA) codes. All three criteria are consistent with the mechanics notion that cracks tend to grow perpendicular to the largest tensile stress.

These criteria, although developed primarily for cracks in homogeneous materials, can be extended to bi-materials systems, such as adhesively bonded joints, coatings, and laminated materials, provided the debond is propagating within one of the (homogeneous) layers. Care should be used when applying these criteria to determine the direction of cracking for cracks located near the bi-material interface, however, due to differences in fracture toughness in the vicinity of the interface [13] as well as the more complex stress states that exist. According to these criteria, a crack in an adhesive bond can be steered to different locations if the local stress state at the crack tip is mixed mode. Consequently, various failure locations can result and failure does not necessarily occur at the weakest site within the material. Anyone who has tried to tear a perforated page from a notepad is familiar with this paradox. To illustrate the concept, Fig. 2 suggests that by applying shear, one may be able to steer the debond towards one interface or the other, depending on the direction of the shear. Cracks will have a tendency to run perpendicular to the principal diagonal [1], at least until they reach a fracture-resistant substrate in the case of adhesive bonds.

---

[1] The principal diagonal, defined by the line connecting tip to tip of a shear stress state, is the diagonal that is tensile rather than compressive.

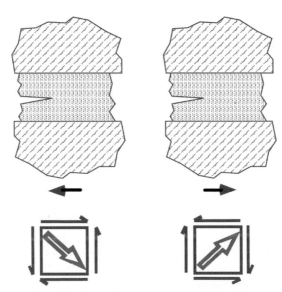

Fig. 2. The presence of shear can steer growing cracks to one interface or the other depending on the direction of the shear. Cracks tend to propagate perpendicular to the principal diagonal.

The fracture mode mixity at the crack tip in homogeneous materials is defined as the ratio between the mode II and mode I stress intensity factors.

$$\psi = \tan^{-1}\left(\frac{K_{\mathrm{II}}}{K_{\mathrm{I}}}\right) \tag{1}$$

where ψ is usually referred as the phase angle, and is analogous to (Eq. 33 in Chapter 2) that was defined for an interface crack. The global fracture mode mixity can also be defined based on the strain energy release rate ratio, as

$$\eta = \frac{G_{\mathrm{II}}}{G} \tag{2}$$

where G is the total strain energy release rate available and G_{II} is the strain energy release rate available for mode II crack propagation. When a debond is propagating cohesively within an adhesive layer, these relationships will prove useful in predicting the direction of cracking. For simple geometries, analytical solutions for the stress intensity factors, energy release rates, and mode mixities are possible, but for more complex configurations, numerical techniques are often required [14].

If mode II loading is present, the crack can be steered away from a weaker region and actually propagate in a tougher region of the joint, altering the locus of failure, and violating the weakest link perspective. The specimen geometry and loading mode can favor failure at one interface or the other, depending on the stress state that is established throughout the adhesive layer and at the debond

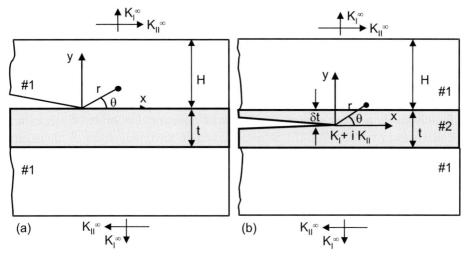

Fig. 3. Cracks in adhesive bonds: (a) interface crack; (b) sub-interface crack. K_{I}^{∞}, and K_{II}^{∞} are taken as external loading, which can also be described as far field stress intensity factors.

tip. One should distinguish between the globally applied mode mix, which results from considering the loads applied to an identical monolithic geometry and given by K_{I}^{∞} and K_{II}^{∞}, and the local mode mixity at the crack tip: K_{I} and K_{II}, for cohesive failures, or K_1 and K_2, for interfacial cracks, as discussed in Chapter 2. Crack path selection is ultimately controlled by the local stress field at the crack tip, as shown in Fig. 3, rather than by the far field values, although both are related [15].

1.3. Directional stability of a crack and the T-stress

Another important factor in predicting the locus of failure in adhesively bonded joints is the directional stability of the crack propagation. In investigating slightly curved or kinked cracks in linear elastic, homogeneous materials under mode I loading, Cotterell and Rice [10] concluded that the T-stress given in Eq. 1 of Chapter 2 plays an important role in the directional stability of the crack propagation. The crack is directionally stable if the T-stress is negative, but is directionally unstable if the T-stress is positive. A similar trend has also been found in three-dimensional crack propagation studies [16]. Although the T-stress is a linear elastic fracture mechanics concept and is calculated from the linear elastic material properties of the same solid containing the crack, the T-stress plays a similar role in inelastic materials such as the elastic–plastic materials [17,18]. The T-stress is a non-singular stress that is parallel to the crack, and passes around the crack.

Through considering higher order terms in Williams [19] asymptotic stress expansion, given in (Eq. 1 of Chapter 2) for a crack in homogeneous materials, Chao et al. further investigated the effect of the T-stress on the crack propagation manner and indicated that the transition point between directionally stable and unstable cracks is slightly more complex. For studies discussed in this chapter, however, the criteria developed by Cotterell and Rice is still found to be satisfactory and therefore is used.

In adhesively bonded systems, the issue of directional stability of cracks was first discussed by Chai [20–22], who observed an intriguing form of alternating crack trajectory in the mode I delamination failure of graphite reinforced epoxy composite laminates and aluminum/epoxy bonds. The crack periodically alternated between the two interfaces with a characteristic length of 3–4 times the thickness of the adhesive layer. More specifically, as the crack advanced, the crack propagated along one interface and then gradually deviated away with an increasing slope until the other interface was approached. An abrupt kink then occurred as the crack approached the opposite interface. The crack then propagated near the interface for a distance of about 2–3 times the thickness of the adhesive layer before deviating from the interface again. As a result, a characteristic length of 3–4 times the thickness of the adhesive was observed in the crack trajectory, reflecting very directionally unstable crack propagation.

Fleck et al. [15] and Akisanya and Fleck [2,3] investigated this directional stability issue in adhesive bonds and concluded that, similar to the situation in homogeneous materials, the directional stability of cracks in adhesive bonds also depends on the magnitude of the T-stress when the bonds are under predominantly mode I loading. The crack propagation in an adhesive bond is directionally stable if the T-stress is negative and is directionally unstable if the T-stress is positive. This argument revealed the threshold of the transition of the directional stability of cracks in adhesively bonded joints and provided an important foundation for later studies.

In their analyses for the sandwich geometry shown in Fig. 4, the adherends were assumed to be semi-infinite, the adhesive was assumed to be linear elastic, and a semi-infinite straight crack was present within the adhesive layer. The

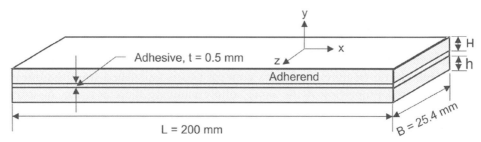

Fig. 4. Geometry of adhesively bonded double cantilever beam (DCB) specimens. For symmetric specimens, $h = H$, and for asymmetric specimens, $h \neq H$.

T-stress can be calculated as [15]

$$T = \frac{1-\alpha}{1+\alpha}T^{\infty} + \sigma_0 + C_{\mathrm{I}}(c/t,\alpha,\beta)\frac{K_{\mathrm{I}}^{\infty}}{\sqrt{t}} + C_{\mathrm{II}}(c/t,\alpha,\beta)\frac{K_{\mathrm{II}}^{\infty}}{\sqrt{t}} \qquad (3)$$

where σ_0 is the residual stress in the adhesive and t is the thickness of the adhesive layer. $C_{\mathrm{I}}(c/t,\alpha,\beta)$ and $C_{\mathrm{II}}(c/t,\alpha,\beta)$ in Eq. 3 are tabulated, non-dimensional functions [15], c is defined as shown in Fig. 1, and α and β are the Dundurs' parameters defined in Chapter 2. K_{I}^{∞}, K_{II}^{∞}, and T^{∞} in Eq. 3 are solutions for the case of monolithic material obtained by neglecting the adhesive layer and here they are used as far-field loading. The manner in which K_{I}^{∞} and K_{II}^{∞} are related to a specific applied loading can be found [23], as can T^{∞} for several commonly used testing geometries [24]. For specimens of a monolithic material, the *T*-stresses, which is the T^{∞} in Eq. 3, are positive for several commonly used testing geometries such as compact tension and double cantilever beam (DCB) specimens [24]. When an adhesive layer with lower modulus than the adherends is introduced, then the *T*-stress in the specimen may become negative due to the material mismatch, stabilizing the crack as will be demonstrated later.

Since the above analyses were based on the assumption that the adherends were semi-infinite, the effect of adherend bending on the *T*-stress was excluded. Since in most of the adhesively bonded joints in reality, the thickness of the adherends is finite, the *T*-stress is found to be higher than that predicted by Eq. 3, and will increase when the thickness of the adherends decreases due to the adherend bending. In fact, the *T*-stress obtained using Eq. 3 represents the lower bound of the *T*-stress in real adhesive bonds with sandwich geometry according to Chen and Dillard [25]. They also reported that this adherend bending effect on the *T*-stress induces a mild influence of the thickness of the adherends on the directional stability of the cracks in the DCB specimens as will be discussed later in this Chapter.

To calculate the *T*-stress for real and complex adhesive bond geometries, many numerical methods have been proposed [24,26,27] over the years and the FEA method is the most direct one. According to Eq. 1 of Chapter 2, the *T*-stress can be calculated by substituting the stress σ_{xx}, σ_{yy}, and the stress intensity factors obtained from finite element analysis of the equation [28,29]. The calculations can be further simplified and the analysis applied to most commonly used adhesive bond geometries [25].

When the fracture is pure mode I, as with symmetric DCB specimens, $K_{\mathrm{II}} = 0$. Then the *T*-stress along the crack plane ($\theta = 0$ and $\pm\pi$) can be easily obtained using

$$T = \sigma_{xx} - \sigma_{yy} \qquad (4)$$

If the fracture is mixed mode, as with asymmetric DCB specimens in which one adherend is thicker than the other, K_{II} does not equal to zero. Along the crack plane ahead of the crack tip ($\theta = 0$), the T-stress can be still calculated using Eq. 4. However, behind the crack tip ($\theta = \pm\pi$), the T-stress is given by

$$T = \sigma_{xx} + K_{\mathrm{II}}\sqrt{\frac{2}{\pi r}} \tag{5}$$

where σ_{xx} is obtained from the finite element analysis, and K_{II} can be calculated from fitting the σ_{xy} data, which is also obtained from the finite element analysis, as

$$K_{\mathrm{II}} = \left(\sigma_{xy}\sqrt{2\pi r}\right)\Big|_{\theta=0} \tag{6}$$

If an advanced finite element package, such as ABAQUS® [30] version 6.1-1 or later, is used, then direct extrapolation of the T-stress can also be made.

If the residual stress in the adhesive is not zero, the T-stress is then given by

$$T = T_0 + \sigma_0 \tag{7}$$

where T_0 is the T-stress under zero residual stress state calculated using the FEA method and σ_0 is the residual stress in the adhesive layer. According to Eq. 7, the T-stress is linearly related with the residual stress in the adhesive layer, indicating the magnitude of the T-stress can be altered through varying the residual stress state. The relationship presented in Eq. 7 was a key factor for successfully demonstrating the dependence of the directional stability of cracks on the T-stress level in adhesively bonded joints, as will be discussed in the next section.

Together, the mode mixity and T-stress play critical roles in determining crack path selection and locus of failure in adhesively bonded joints. The tendencies induced by these two mechanics principles, combined with the spatial variation in mechanical properties within the adherends, adhesive, and interphase region, determine the ultimate failure mode. For bonded systems involving large spatial variations in mechanical properties, debonds may favor a weak interface rather than obey the tendencies imposed by the mechanics principles outlined in these two sections. For systems having reasonably adequate mechanical properties throughout the bond, however, these mechanics principles may control the failure event.

2. Experimental studies

To demonstrate the T-stress effect and to understand other factors affecting crack path selection, the authors and their coworkers [13,25,31,32] carried out a series of experimental studies with adhesively bonded joints to determine the effects of T-stress, specimen geometry, external loading conditions, surface pretreatment, and material properties on the observed failure mode. These results demonstrated that the crack path of a crack can be determined based on the material properties, specimen dimensions, residual stress state of the bond, and the external loading conditions. Therefore, the crack path selection behavior could be controlled and the locus of failure accurately predicted for a series of model epoxies (with different rubber toughener content) used to bond aluminum adherends. Some of the results are repeated herein to demonstrate the role that mechanics plays in determining the failure mode in adhesively bonded joints. The results presented are believed to be representative of what might be seen with other structural adhesive systems, and with some care, these results may have relevance for non-structural and elastomeric bonds as well.

2.1. Material system and specimen preparation

2.1.1. Material and specimen

The model adhesive used was Dow Chemical epoxy resin D.E.R. 331 mixed with an M-5 silica filler, a dicyandiamide ('DICY') curing agent, a tertiary amine accelerator (PDMU), and various amounts of rubber toughener (Kelpoxy G272-100), details of which have been listed elsewhere [33]. The final products, according to the rubber concentration level, were designated as adhesives A (0% rubber), B (4.1%), C (8.1%), and E (15%). Adherends were cut from 25-mm-wide 6061-T6 aluminum alloy bar stock.

Beam-type fracture specimens were fabricated, with a width of 25 mm and length of 200 mm. The thickness of the adhesive layer was controlled using metal shims; adhesive thickness was controlled to be 0.5 mm in most tests discussed in this chapter unless otherwise stated. However, the thickness of the adherends was varied, and specimens with both symmetric and asymmetric adherend thicknesses were prepared. Before bonding, the surfaces of the adherends were treated with one of three different surface preparation methods: acetone wipe, which was used simply to provide surface uniformity among specimens; base-acid etch, and P2 etch. The base-acid etch procedure is a deep cleaning procedure, and a new aluminum oxide surface was generated after the preparation [34,35]. The treatment was carried out by immersing aluminum in 5% (weight ratio) aqueous sodium hydroxide solution at 50°C for 5 min; rinsing the specimen in de-ionized (DI) water; neutralizing residual surface sodium hydroxide in dilute nitric acid; rinsing the adherend in DI water again; air drying the specimen; and placing the adherend in a desiccator until bonding was carried out. The P2 surface treatment was employed to develop a robust oxide surface and avoids the use of toxic chromium(VI). In the procedure an Fe(III) solution was used to oxidize the aluminum surface. The P2 etch method can greatly improve the surface morphology and chemistry of the aluminum substrates and therefore can

Table 1

Material characterization results for the model epoxy adhesive formulations used in the study

Adhesive designation	Rubber concentration (%)	CTE (10^{-6}/°C)	T_g (°C)	Modulus (GPa)	Calculated residual stress (MPa)
A	0	58	125	3.10	14.8
B	4.1	59.5	119	3.06	14.38
C	8.1	62	112	2.97	13.8
E	15.0	65	106	2.85	13.4

significantly improve adhesion [34,36]. The specimens were cured at 170°C for 90 min, cooled to room temperature, and then stored in a desiccator prior to testing.

The material properties of the cured adhesive were characterized using differential scanning calorimetry (DSC) (for the glass transition temperature, T_g), thermal mechanical analysis (TMA) (for coefficient of thermal expansion (CTE), α_2), and room temperature dogbone tensile tests (for Young's modulus, E_2). The results shown in Table 1 indicate that as the rubber concentration increases, the CTE of the material increases but the modulus and the glass transition temperature decrease slightly. Poisson's ratios for all the adhesives were estimated as $\nu_2 = 0.33$ at room temperature. The material properties for the aluminum 6061-T6 substrate are Young's modulus $E_1 = 70$ GPa, Poisson's ratio $\nu_1 = 0.33$, and CTE $\alpha_1 = 26 \times 10^{-6}$/°C.

Due to the mismatch of the coefficients of thermal expansion, an equal biaxial residual stress σ_0 was induced throughout the adhesive layer after curing of the specimens. If the adherends are assumed to be relatively thick and rigid, as compared with the adhesive, the residual stress is accurately approximated by

$$\sigma_0 = \frac{E_2}{1 - \nu_2} (\alpha_2 - \alpha_1)(T_{\text{sft}} - T) \tag{8}$$

T_{sft} is the stress free temperature of the adhesive, which was measured using a curvature measurement technique [37] for each adhesive, and was very close to the glass transition temperature of the respective adhesive (the results are also listed in Table 1). T is the test temperature (room temperature in this study). Since the coefficients of thermal expansion of the adhesives increased with rubber concentration and meanwhile the modulus and the glass transition temperature decreased slightly, as shown in Table 1, the calculated residual stresses induced in the specimens during curing were very similar for all the adhesives. On the other hand, as will be discussed later, the rubber toughener enhanced the fracture toughness of the adhesive bonds significantly.

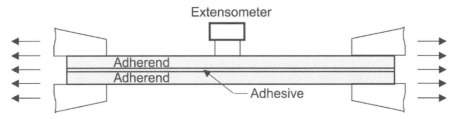

Fig. 5. Schematic of DCB specimens loaded in tension in a universal test machine until the adherends were plastically deformed in order to alter the residual stress state in the adhesive layer.

2.1.2. Altering the T-stress

As discussed earlier with Eq. 7, the T-stress can be varied through altering the residual stress in the adhesive bonds. Although prior studies have achieved variations in residual stress through altering the cure procedure or adherends chosen [38], in this work a mechanical method was introduced to alter the residual stress (and consequently, the T-stress levels) in adhesively bonded joints. In these studies, specimens were axially loaded in an Instron machine until the substrates were plastically deformed as shown in Fig. 5. An MTS extensometer was attached to the specimens to monitor the strain. Because the adherends and adhesive are loaded in parallel, the strain measured through the extensometer was the strain for the adherends as well as the adhesive. Upon unloading, the plastic deformation remaining in the substrate, ε_p, was recorded from the strain–stress curve as shown in Fig. 6. Tensile tests of dogbone specimens made of the neat adhesive were also conducted, and a typical stress–strain curve for adhesive C is also shown in Fig. 6

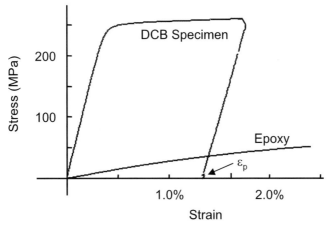

Fig. 6. The stress–strain curve for the DCB specimens under uniaxial tension and neat adhesive C dogbone specimens.

along with a typical curve for the sandwich specimens. According to the tensile test results for the neat adhesive, the adhesive did not yield at the strain levels investigated. Due to the plastic deformation in the adherends, the residual stress in the adhesive layer was increased and so was the T-stress level according to Eq. 7. Setting the coordinate system as shown in Fig. 4, the total residual stress in the adhesive now becomes [39]

$$\sigma_x = \frac{E_2}{1 - v_2}(\alpha_2 - \alpha_1)(T_{\text{sft}} - T) + \frac{E_2}{1 - v_2^2}(1 - v_1 v_2)\varepsilon_{\text{p}}$$

$$\sigma_z = \frac{E_2}{1 - v_2}(\alpha_2 - \alpha_1)(T_{\text{sft}} - T) + \frac{E_2}{1 - v_2^2}(v_2 - v_1)\varepsilon_{\text{p}} \tag{9}$$

Ink was used to determine if any microcracks were induced in the adhesive layer during the stretching procedure, which could influence the crack path. No evidence of microcracks was observed before the strain exceeded 2.1%, which is beyond the strain levels reported herein.

As compared with other methods, the stretching method provides a convenient and more direct way to alter the residual stress and consequently, the T-stress level, in the adhesive bonds. With delicate control of the testing frame achieved through operating the GPIB interface using LabVIEW [40] software, the expected plastic deformation level could be achieved within 3% error; therefore, the desired T-stress level can also be achieved rather precisely. Using this method, the T-stress can be continuously varied over a wide range (of 47 MPa for the material system used in the studies). Since the stretching method is purely a mechanical method, no material properties have been altered, permitting direct comparisons of the test results. The availability of tensile test frames provides a convenient way to perform the stretching and alter the residual stress in adhesive bonds.

2.2. Testing procedure

The fracture testing and post-failure analysis methods used were quasi-static DCB and end notch flex (ENF) tests, and low-speed impact tests on DCB specimens. Quasi-static fracture tests were conducted using a procedure similar to that described in ASTM D3433, using crosshead speeds appropriate to induce failure in about 1 min. Data were analyzed using procedures described by Blackman et al. [41], and Parvatareddy et al. [42]. Low speed impact tests were conducted on DCB specimens using a falling wedge test setup described by Xu et al. [43], using analysis techniques recommended by Blackman et al. [44–46]. Post-failure analyses were conducted to determine the locus of failure, using X-ray photo-electron spectroscopy (XPS), scanning electron microscopy (SEM), atomic force microscopy (AFM), and Auger depth profiling. Details for the test methods and post-failure analyses procedures can be found in the literature [13,25,31,32].

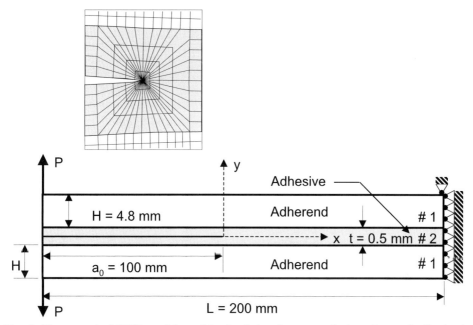

Fig. 7. The numerical DCB model used in the finite element analysis; a layer of adhesive is sandwiched between two adherends.

2.3. Experimental results and discussion

2.3.1. T-stress and directional stability of cracks

Finite element analysis was first conducted to determine quantitatively the T-stress level in the DCB specimens using the numerical method discussed earlier in this chapter. The DCB model analyzed is shown in Fig. 7. An adhesive layer (material 2) is sandwiched between two adherends (material 1). The thicknesses of both the adhesive and the adherends were varied in the analysis to examine the effects of specimen geometry. A straight crack is located in the middle of the adhesive layer, and the displacements of one end of the model are totally constrained. Eight-node, plane-strain elements with reduced integration were used to mesh the geometry. Quarter point singular elements were constructed around the crack tip to properly capture the singularity, as shown in the inset in Fig. 7. Both the adherend and adhesive were modeled as linear elastic materials with material constants $E_1 = 70$ GPa, $E_2 = 2.97$ GPa, and $\nu_1 = \nu_2 = 0.33$. In order to determine the magnitude of loading in the finite element analysis, the adhesive bonds were assumed to have a single fracture toughness value of 310 J/m^2, which is referred as the iso-fracture toughness model, since the measured fracture toughness of the adhesive bonds appeared to be independent of the crack propagation behavior for this material

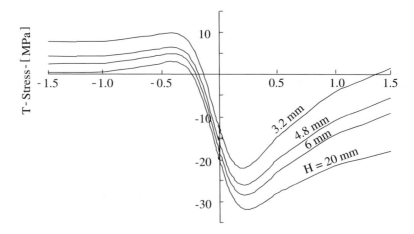

Distance from Crack Tip - x [mm]

Fig. 8. The T-stresses for DCB specimens with different adherend thicknesses and zero residual stress.

system [25]. The single fracture toughness value was obtained based on the quasi-static testing results of DCB specimens using the testing procedure mentioned above.

Fig. 8 shows the T-stress distribution along the x-axis, obtained from the FEA analysis for symmetric DCB specimens with zero residual stress. The adhesive thickness used was 0.5 mm and various adherends thicknesses ($H = 20$, 6, 4.8, and 3.2 mm) were analyzed. The T-stress converges to the bending stress of the composite beam as x decreases behind the crack tip and converges to zero as x increases in front of the crack tip, which is beyond the range in Fig. 8. This figure also shows that the T-stress is non-singular at the crack tip and increases as the adherend thickness decreases. Although the T-stress in the specimen is negative when the residual stress is zero, increasing the residual stress can induce a positive T-stress state, according to Eq. 7, thereby destabilizing the propagating debond. T-peel specimens, with their thin adherends, are notorious for producing alternating debonds for adhesives with only modest toughness.

To confirm the theoretical predictions that the T-stress affects the directional stability of cracks, DCB specimens with adhesive C and adherend thicknesses of 4.8 mm were prepared. The surface preparation for the adherends was simply an acetone wipe, and various levels of residual stress were achieved among the specimens using the stretching method. Specimens were then tested quasi-statically according to the procedure discussed earlier. The resulting fracture surfaces of the specimens were carefully examined and three representative specimens were selected as shown in Fig. 9, from which, the effect of the T-stress on the directional stability of cracks can be inferred. The initial residual stress

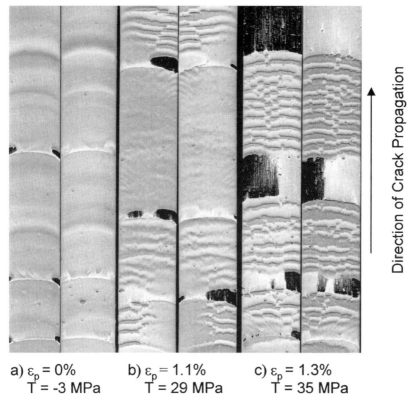

Direction of Crack Propagation

a) $\varepsilon_p = 0\%$ b) $\varepsilon_p = 1.1\%$ c) $\varepsilon_p = 1.3\%$
   T = -3 MPa   T = 29 MPa   T = 35 MPa

Fig. 9. The observed fracture surfaces in symmetric ($H = 4.8$ m) DCB specimens with different levels of plastic deformation. From left to right, the failures are cohesive with directionally stable crack, cohesive with oscillatory crack trajectory, and interfacial (or very close to the interface) with alternating crack trajectory.

based on thermal mismatch in all three specimens shown in Fig. 9 was 13 MPa as calculated using Eq. 8. However, each specimen was stretched to a different level of plastic deformation in the adherends before testing. As a result, the final residual stress and the T-stress varied among these three specimens according to Eq. 7. Specimen a was an as-produced specimen; no plastic deformation was introduced in the adherends, and the consequent T-stress was -3 MPa obtained using the FEA results and Eq. 7. The failure surfaces of specimen a appeared cohesive (except for a few spots along the edges where the debond had arrested) and the crack was directionally stable. On the other hand, specimen b was stretched to about 1.1% plastic deformation. Consequently, the T-stress increased to 29 MPa and an oscillatory crack trajectory was observed on the failure surfaces, indicating a tendency toward directionally unstable crack propagation. When the plastic deformation reached the level $\varepsilon_p = 1.3\%$ as with specimen c, the corresponding T-stress was 35 MPa. The crack in this specimen alternated between the two

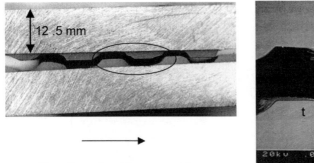
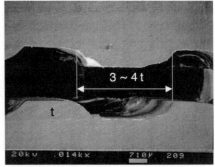

Direction of crack propagation

Fig. 10. The cross-section of the failed specimen with $T = 35$ MPa and alternating crack trajectory. The picture on the right is the scanning electron microscopy (SEM) micrograph of the circled portion of the cross-section.

interfaces and failure occurred at or very close to the interfaces. This alternating crack trajectory, illustrated in the micrographs in Fig. 10, represents directionally unstable crack propagation, and is very similar to Chai's original observations [20,22]. As the crack advanced, the crack propagated near one interface, then gradually deviated away with an increasing slope until the other interface was approached. An abrupt kink then occurred, allowing the crack to propagate near the interface for a distance about 2–3 times the thickness of the adhesive layer before deviated from the interface gradually again.

The testing results demonstrate that the magnitude of the T-stress controls the directional stability of crack propagation in adhesively bonded joints and the magnitude of the oscillation of the crack trajectory appeared to increase with the T-stress in the tests. On the other hand, for this material system, the fracture toughness of the bonds was not significantly affected by the amount of stretching; average fracture energies for all three specimens ranged between 310 and 320 J/m^2.

As obtained from the finite element analysis mentioned above, Fig. 11 shows the relation between the T-stress and the thickness of the adhesive layer for the DCB specimens with zero residual stress and various thicknesses of adherends. In the figure, the T-stresses in the finite element analysis were taken at the crack tip. Along with Fig. 8, Fig. 11 indicates that the T-stress obtained by Fleck et al. [15] is the lower bound, since the adherends were assumed to be semi-infinite in their analysis. As the adherend thickness decreases, the T-stress increases and the difference is not negligible if the adherend thickness is less than 6 mm for the configuration studied. Therefore, for this particular material system, when the adherend thickness is less than about 10 times the thickness of the adhesive, the effect of adherend bending on the T-stress level is no longer negligible and the crack propagation is predicted to be more directionally unstable as the adherend

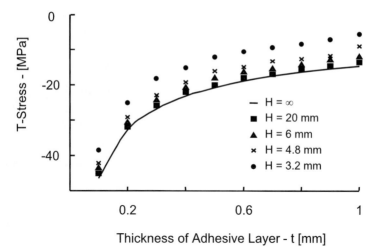

Fig. 11. The specimen geometry dependence of the *T*-stress in DCB specimens for specimens with no residual stress. Solid line represents Fleck et al. [15] solution for semi-infinite adherends.

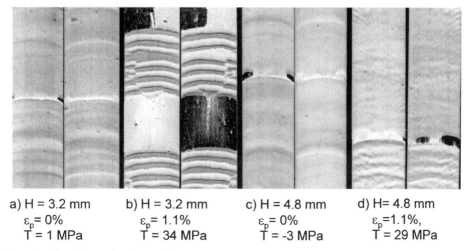

Fig. 12. The effect of adherend thickness on the *T*-stress level and the directional stability of cracks. The crack tends to be more directionally unstable when the thickness of adherend decreases.

thickness decreases. Fig. 12 illustrates the difference in directional stability for representative specimens selected from two groups of DCB specimens with adhesive C and adherend thicknesses of 4.8 and 3.2 mm. Directional instability is greatest in the specimens with the smaller adherend thickness. We note that the interfacial failure region, clearly seen in specimen b, occurred during slow debond propagation, whereas the alternating debond occurred while the debond was propagating more rapidly.

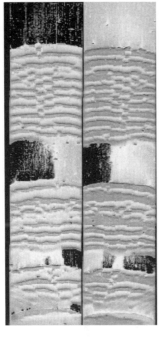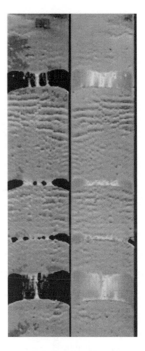

a) $\varepsilon_p = 1.3\%$ b) $\varepsilon_p = 1.3\%$
   t = 0.5 mm   t = 0.25 mm

Fig. 13. The effect of adhesive thickness on the directional stability of cracks in DCB specimens.

Fig. 11 also shows that the adhesive thickness has a significant effect on the T-stress in DCB specimens. As the adhesive thickness decreases, the T-stress decreases, which indicates that directionally unstable cracks are less likely to occur in specimens with thinner adhesive layers. Fig. 13 illustrates representative specimens selected from two groups of symmetric DCB specimens with adhesive C and adhesive thickness of 0.5 and 0.25 mm. The specimens were subjected to mechanical stretching until 1.3% of plastic deformation occurred in the adherends before testing to increase the residual stress and consequently the T-stress. The adhesive thickness was 0.5 mm for specimen a and was 0.25 mm for specimen b. Both specimens have a final residual stress of 38.6 MPa. However, because the adhesive layer in specimen b is thinner, the T-stress in specimen b (26 MPa) is lower that the T-stress in specimen a (35 MPa). The alternating debonding in specimen a represents more directional instability than the predominantly oscillating debond seen in specimen b, as predicted in Fig. 12.

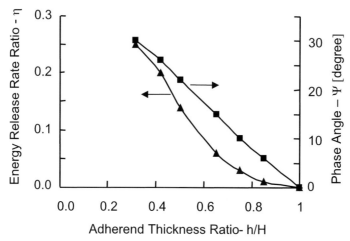

Fig. 14. The global fracture mode mixity in asymmetric DCB specimens.

2.3.2. Fracture mode mixity and the locus of failure

As outlined in Section 1.2, mode mixity can significantly alter the locus of failure, even driving debonds away from weaker regions of the joint. To investigate the effect of fracture mode mixity on the locus of failure in adhesive bonds, quasi-static DCB and ENF tests were conducted. Specimens were made of adhesive C with acetone wipe surface preparation and they were all as-produced, therefore the T-stresses were all negative. For the ENF tests, specimens were symmetric and for the DCB tests, both symmetric and asymmetric specimens were used with three different adherend thickness ratios, i.e. $h/H = 0.5$, 0.75, and 1. Finite element analysis was used to quantify the mode mixity and the results are shown in Fig. 14; adherend thickness ratios of $h/H = 0.5$, 0.75, and 1 correspond to fracture mode mixity of $\psi = 22°$, 10°, and 0° or $\eta = 14\%$, 3%, and 0%, respectively.

After failure, the failure surfaces were first visually examined. The failure surfaces of the specimens tested in ENF and of the DCB specimens with $h/H = 0.5$ all appeared to be interfacial. On the other hand, a visible layer of epoxy film was found on the failure surfaces of the DCB specimens with $h/H = 0.75$ and 1, which indicates that the failures were cohesive. To identify the locus of failure quantitatively, one typical specimen from each test was selected, based on the visual examination, for XPS and Auger depth profile analyses. The analyses were carried out on two representative areas on the failure surfaces of each specimen selected and the average values are reported. Table 2 shows the XPS analysis results for all the tests. Five elements were detected on the failure surfaces and their concentrations varied for each test. Of significance, the carbon concentration decreases as the fracture mode mixity increases whereas the aluminum and oxygen concentrations increase. Since carbon is mainly from the epoxy adhesive

Table 2

XPS elemental analysis results for the typical specimens selected from each test

Analysis results	Test method			
	Symmetric DCB ($h/H = 1$)	Asymmetric DCB ($h/H = 0.75$)	Asymmetric DCB ($h/H = 0.5$)	ENF
Phase angle (Ψ)	0°	10°	22°	90°
SERR ratio (η)	0%	3%	14%	100%
C%	76.4	76.3	53.0	44.5
Al%	0.2	0.3	9.3	14.8
O%	18.7	18.8	31.8	36.7
N%	2.6	2.5	2.6	2.0
Si%	2.1	2.1	3.6	2.0

Table 3

Auger depth profile results for the typical specimens selected from each test

Analysis results	Test method			
	Symmetric DCB ($h/H = 1$)	Asymmetric DCB ($h/H = 0.75$)	Asymmetric DCB ($h/H = 0.5$)	ENF
Phase angle (Ψ)	0°	10°	22°	90°
SERR ratio (η)	0%	3%	14%	100%
Adhesive thickness	250 μm	50 μm	4 nm	3.5 nm

and aluminum is from the aluminum adherend, Table 2 indicates that failure tends to be more interfacial as the mode mixity increases. This indication is further verified by the Auger depth profile data, which precisely quantifies the locus of failure as shown in Table 3. In the mode I tests, the adhesive layer left on the failure surfaces was approximately 250 μm thick, which indicated that failure occurred in the middle of the adhesive layer since the total thickness of the adhesive in the specimens was 0.5 mm. As the mode mixity increases, Table 3 shows that the locus of failure shifts toward the interface as indicated by the Auger surface analysis data. In the asymmetric DCB test with $h/H = 0.75$, which contains 3% of mode II fracture component, a 50-μm-thick polymer film was detected on the failure surfaces. As the mode mixity increased to 14% ($h/H = 0.5$), the residual polymer film thickness decreased to 4 nm. In the ENF tests, the polymer film thickness was 3.5 nm, which indicated that the failure is very close to the interface. Thus, even when failures visually appear to be interfacial, surface analysis techniques are able to support the predicted trends that increased amounts of mode II loading drive the debond closer to the interface.

The results of XPS analyses and Auger depth profile clearly identify the

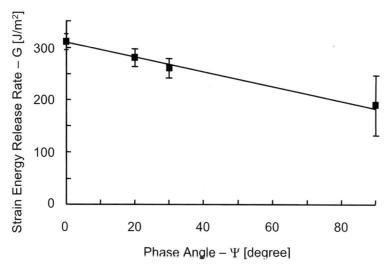

Fig. 15. Fracture toughnesses of the adhesive bonds measured in different tests. Error bars represent ±1 standard deviation.

locus of failure and demonstrate the effect of fracture mode mixity. The results also showed that the locus of failure in asymmetric DCB tests with fracture mode mixity of 14% is very close to the locus of failure in the mode II ENF tests. These results suggest that, at least for this system, the asymmetric DCB test with adherend thickness ratio of 0.5 or higher can be used rather than the ENF test to produce interfacial failures. Since the asymmetric DCB test is conducted under predominantly opening mode, the onset of fracture and the crack propagation sequence are much easier to observe. As shown in Fig. 15, however, the toughnesses measured at these different mode mixity ratios are different. Other details of asymmetric DCB analysis and testing can be found in Sundararaman and Davidson [47] and Xiao et al. [48] where the fracture behavior of asymmetric DCB specimens prepared using adherends of dissimilar materials were discussed.

The critical strain energy release rates measured in each test are shown in Fig. 15. The fracture toughness measured decreases as the mode II fracture component increased in the tests for this particular material system. This mode mixity dependence of the fracture toughness of adhesively bonded joints apparently is in contrast with the observations of other researchers for other material systems [49–54]. This contradiction can be explained through analyzing the locus of failure. As discussed in Swadener and Liechti [52] and Swadener et al. [53], the locus of failure in their studies was independent of the fracture mode mixity, and the size of the plastic deformation zone at the crack tip increased with the fracture mode mixity. This increased plastic zone was shown to be responsible for a shear-induced toughening mechanism, which consequently, caused the fracture toughness to increase with the mode II components in their studies. In this study, however, as

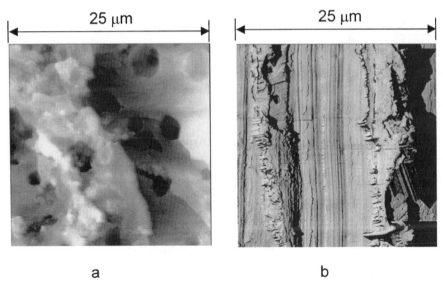

Fig. 16. Atomic force microscopy images (tapping mode) for the failure surfaces. Image a was taken from the typical specimen tested under pure mode I, and image b was taken from the typical specimen tested under pure mode II.

the XPS and Auger depth profile analyses indicated, the failure was cohesive in the mode I tests and became more interfacial as the mode II component in the fracture increased. AFM was also conducted on the adhesive side of the failure surfaces of the typical specimens tested under pure mode I and pure mode II, respectively, and digital images were obtained using a Nanoscope IIIa controller as shown in Fig. 16. In image a, which was taken from the typical specimen tested under pure mode I, evenly distributed rubber toughener particles are visible, and the surface morphology indicates a ductile failure. On the other hand, in image b, which was taken from the typical specimen tested under pure mode II, the failure surface reflects the shape of the aluminum substrate surfaces and no rubber toughener particles are evident. These AFM images suggest that rubber toughener particles were bypassed when the failure was interfacial, whereas the bond was effectively toughened when the failure was cohesive due to the presence of the rubber particles. This strong fracture mode mixity dependency of the locus of failure explains why, for this particular material system, the fracture toughness of the specimen decreased as the mode II fracture component increased. In addition, the crack trajectories were very tortuous locally in the mode I tests, which can be observed under a microscope, whereas were much more straight in the mixed mode tests because the failure occurred at or near the interface and the adherends provided rigid constraints. Consequently, more energy was consumed when cracks propagate within the adhesive than along the interface. Therefore, the apparent critical strain energy release rate measured in mode I tests is higher than that measured in mixed-mode tests, as has also been reported by Parvatareddy and Dillard [55].

2.3.3. Rate of debonding and the locus of failure

The rate of debonding can have a significant effect on the locus of failure. For the model adhesive system studied, rapid failures have resulted in cohesive failures, whereas slow propagation often leads to interfacial failures. Differences in viscoelastic properties of the bulk adhesive and the interphase region may be responsible in part for these trends. The complex dynamics of a rapidly propagating debond is expected to play a role in crack path selection. To investigate the effect of the rate of crack propagation on the locus of failure in adhesively bonded joints under mode I fracture loading, both quasi-static and low-speed impact tests on DCB specimens prepared using adhesive C and acetone wipe surface cleaning were conducted. The specimens tested were symmetric and with various levels of T-stress introduced using the mechanical stretching method. The rate of crack propagation was obtained for both tests using the high-speed camera system. For this particular material system, the rate of crack propagation in the low-speed impact DCB tests was relatively constant throughout the whole specimen and the magnitude was estimated to be 1 m/s. However, in quasi-static tests, the debond rate varied dramatically. During the test, as a loading cycle started, the crack first propagated slowly at a rate with a magnitude of 10^{-5} m/s until the maximum load value was reached, at which time the crack jumped ahead for a certain distance at a rate with magnitude of 1 m/s, prior to crack arrest. After failure, the failure surfaces were first visually examined to determine the crack propagation behavior in each specimen, and specimens with typical failure surfaces were selected to illustrate the fracture processes.

The failure surfaces from specimens a and b in Fig. 17 were two typical as-produced specimens and the T-stress in both specimens was negative. Specimen a was tested under quasi-static loading and specimen b was tested under low-speed impact. The magnitude of the rate of crack propagation for each region in the specimens is marked schematically in the figure. Since the T-stress was negative and the cracks were directionally stable, the failures in both specimens appeared to be within the adhesive layer (except for a few limited locations along the edge in specimen a regardless of the rate of crack propagation. This observation suggests that the effect of debond rate on the locus of failure is not significant under the negative T-stress state. As the T-stress increased, the cracks tended to be directionally unstable and the effect of debond rate on the locus of failure became more pronounced.

Fig. 18 shows two typical specimens selected from the DCB tests and the T-stress in both specimens was 35 MPa. Specimen a was tested under quasi-static loading and specimen b was tested under low-speed impact. The magnitude of

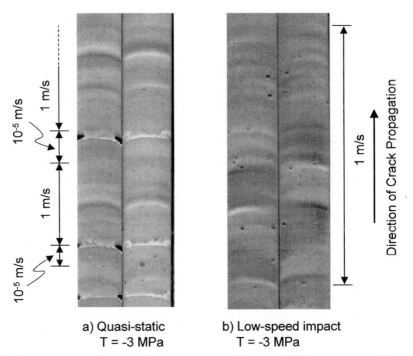

a) Quasi-static
T = -3 MPa

b) Low-speed impact
T = -3 MPa

Fig. 17. The rate dependence of the locus of the failure in DCB specimens with negative T-stresses.

the rate of crack propagation for each region in the specimens is also marked in the figure according to the information recorded using the high-speed camera. In specimen a, the locus of failure visually appeared to be interfacial when the crack propagated slowly, and appeared to be cohesive in the region of fast crack propagation. In specimen b, since the rate of crack propagation is relatively constant and is comparable to the fast region in the quasi-static test, the failure was cohesive and the crack propagation behavior was relatively uniform throughout the specimen. The association of the failure mode with the rate of crack propagation in Fig. 18 indicated that in the DCB specimens with high T-stress levels, the locus of failure varied with the rate of crack propagation. These results are very consistent with other observations in dynamic fracture experiments by Ravi-Chandar and Knauss [56–59].

To further identify the locus of failure and to correlate with the effect of debond rate, both XPS and SEM were conducted on both the fast region and the slow region of specimen a shown in Fig. 18. The surface analyses were conducted on two representative areas (analysis area 1×3 mm on the slow region and 1×1 mm on the fast region due to the alternating feature of the crack) of each region and the average values were reported.

The XPS results indicate that in the region with fast crack propagation, the

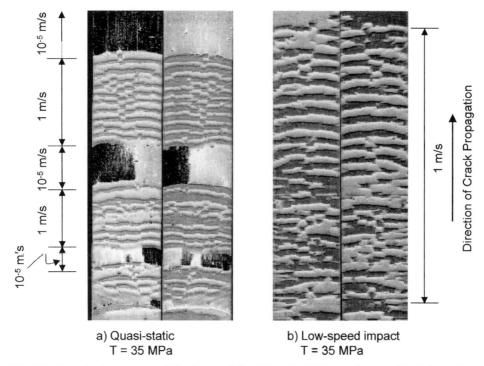

a) Quasi-static
T = 35 MPa

b) Low-speed impact
T = 35 MPa

Fig. 18. The rate dependence of the locus of the failure in DCB specimens with high positive *T*-stresses. The *T*-stress in both specimens was 35 MPa.

carbon concentration on the failure surface (76.4%) was almost equal to the carbon concentration in the bulk epoxy adhesive, which is 76.6%, while the aluminum concentration was 0.4%. On the other hand, in the region with slow crack propagation, significantly more aluminum was detected (2.1%) and the oxygen concentration was higher than that in the fast region. The trend in the variation of the elemental concentrations on the failure surfaces verified that failure was more interfacial when the rate of crack propagation was low (high aluminum and oxygen concentrations). SEM was conducted on the adhesive side of the failure surface and the failure surface morphology was noted for each region. In the fast region as shown in Fig. 19b, the failure surface was relatively rough with visible evidence for evenly distributed rubber toughener particles and polymer drawing. In the slow region, on the other hand, the failure surface is relatively smooth reflecting the shape of the aluminum substrate surfaces and no rubber toughener is evident. These SEM photomicrographs support the results obtained in the XPS analysis and further demonstrate the effect of the rate of crack propagation on the locus of failure in specimens with high *T*-stress levels.

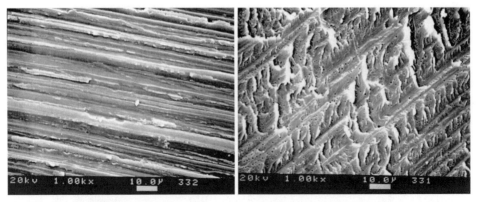

a) Region with slow crack propagation   b) Region with fast crack propagation

Fig. 19. The SEM micrographs for regions with different rate of crack propagation.

2.3.4. Fracture mode mixity and the directional stability of cracks

As experimentally demonstrated in previous sections, under globally mode I loading, the directional stability of cracks in adhesive bonds depends on the T-stress level. On the other hand, the XPS and Auger depth profile analyses results reveal that the locus of failure in adhesive bonds is controlled by the fracture mode mixity. However, concerns have been raised about the effect of mode mixity on the directional stability of cracks. To investigate the mode mixity effect on the directional stability of cracks, low-speed impact tests on DCB specimens and adherend thickness ratios of $h/H = 0.5$, 0.75, and 1 were conducted. All the specimens were prepared with adhesive C and acetone wipe cleaning and were subjected to mechanical stretching until the adherends were plastically deformed before the tests to achieve positive T-stress states in the specimens. After debonding, failure surfaces were carefully examined to determine the crack propagation behavior.

Fig. 20 shows the failure surfaces of three typical specimens selected from each specimen group of different adherend thickness ratio. The plastic deformation in the adherends introduced before the tests in order to alter the T-stress was 1.3% for all the three specimens. As a result, the T-stresses for all the three specimens are positive according the FEA results in Section 2.3.1 and Eq. 7, and their values are shown in Fig. 20. Due to the positive T-stress level (35 MPa), the crack trajectory in specimen a, in which the fracture is mode I since the specimen is symmetric ($h/H = 1$), is alternating, highly directionally unstable. In specimen b, the T-stress has increased slightly to 38 MPa due to the low-level fracture mode mixity with $G_{II}/G = 3\%$ introduced by the asymmetric adherends ($h/H = 0.75$). However, the crack trajectory is predominantly directionally stable except in limited locations where alternating cracks were observed. This effect

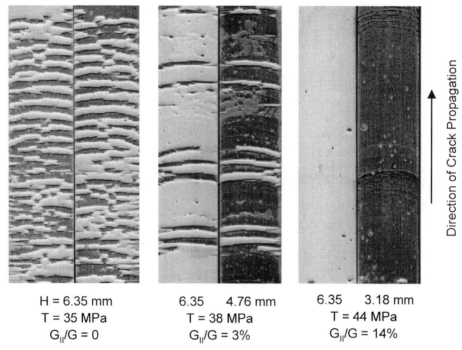

Direction of Crack Propagation

H = 6.35 mm	6.35 4.76 mm	6.35 3.18 mm
T = 35 MPa	T = 38 MPa	T = 44 MPa
$G_{II}/G = 0$	$G_{II}/G = 3\%$	$G_{II}/G = 14\%$

Fig. 20. The effect of fracture mode mixity on the directional stability of cracks in DCB specimens with high positive T-stress levels tested under low-speed impact.

of mixed mode fracture on the directionally stability of cracks becomes more pronounced in specimen c, in which the T-stress has increased even more to 44 MPa due to the high fracture mode mixity ($G_{II}/G = 14\%$) introduced from the adherend asymmetry ($h/H = 0.5$). However, the crack trajectory in specimen c is very directionally stable with a locus of failure occurring at the interface between the adhesive and the thin adherend by visual examination.

The results suggest that although the fracture mode mixity will cause the T-stress in the specimen to increase, the crack propagation will be stabilized very rapidly as the mode mixity increases regardless of the T-stress state. On the other hand the results also indicate that directionally unstable cracks can only be observed in predominantly mode I fracture tests with mode mixity $G_{II}/G < 3\%$ for this particular material system. Beyond this point, the mode mixity forces the debond to propagate along the preferred interface, preventing directional instability.

2.3.5. Surface preparation and mixed mode fracture tests

Effective surface pretreatments are known to improve both the initial and long-term performance of many substrate/adhesive systems. Surface treatments may alter the surface roughness, oxide stability, surface energetics, and other factors that can all lead to improved adhesion. To investigate the effect of surface preparation on the locus of failure in adhesive bonds under mixed mode fracture tests, both quasi-static DCB and ENF tests were conducted on specimens with adherend surfaces prepared with acetone wipe, base-acid etch, and P2 etch. The adhesive used was adhesive C and all the specimens tested were as-produced and hence with negative T-stress levels. For the ENF tests, specimens were symmetric and for the DCB tests, specimens were both symmetric and asymmetric with three different adherends thickness ratios, i.e. $h/H = 0.5, 0.75$, and 1, which gives rise a fracture mode mixity of $\psi = 22°, 10°$, and $0°$, respectively, as discussed earlier. The failure surfaces were first examined visually, and one typical specimen was selected from each test for subsequent XPS and Auger depth profiling to identify the locus of failure. The XPS analysis was carried out on two representative areas on both the aluminum and the adhesive sides of each failed specimen. On the other hand, the Auger depth profiling was only conducted on two small areas on the aluminum side of the failure surfaces. For both tests, only the average values are reported.

Table 4 shows the XPS results for each test. Five elements, carbon, aluminum, nitrogen, silicon, and oxygen were detected on the failure surfaces and their concentrations varied with the testing conditions. Since the major sources for each element are already clear, variations in the concentrations of these elements imply changes in the locus of failure in the specimens. Carbon is the major element of the epoxy adhesive; nitrogen is from the DICY curing agents and is usually present on the surface at a very low level; silicon is from the filler; and aluminum is exclusively from the aluminum adherend. Although both the adhesive and the aluminum surface contain oxygen, the oxygen concentration in the aluminum oxide layer is much higher. According to Table 4, the carbon concentration in the failure surfaces of the symmetric DCB specimens is very high while the aluminum concentration is apparently below the detecting limit of the XPS (about 0.2%). As the fracture mode mixity increases, the carbon concentration on the failure surfaces decreases while the aluminum and oxygen concentrations increase. On the failure surfaces of the ENF specimens, the aluminum and oxygen concentrations are relatively high and the carbon concentration is relatively low. In addition, this carbon is unlikely from the epoxy according to its chemical nature shown in the XPS spectrum, but is more likely from the air contamination or aluminum extrusion. Since the high aluminum and oxygen concentrations suggest the failure location is within the aluminum oxide layer, these results indicate that failure tended to be more interfacial as the mode mixity increased as discussed in Section 2.3.2.

On the other hand, as the surface preparation method varies, the element concentrations, especially for carbon and aluminum, also vary significantly in the tests with mode mixity G_{II}/G higher than 14%, and the trend of the variation

Table 4

XPS elemental analysis results for typical specimens selected from each test

Analysis results	Test method			
	Symmetric DCB $(h/H = 1)$	Asymmetric DCB $(h/H = 0.75)$	Asymmetric DCB $(h/H = 0.5)$	ENF
Phase angle (ψ)	0°	10°	22°	90°
SERR ratio	0%	3%	14%	100%
C%				
Acetone	76.4	76.3	53.0	44.5
B/A	76.5	76.5	73.1	45.5
P2	76.4	76.4	74.0	52.8
Al%				
Acetone	0.2	0.3	9.3	14.8
B/A	0.3	0.5	1.9	13.7
P2	0.1	0.2	0.4	9.1
O%				
Acetone	18.7	18.8	31.8	36.7
B/A	18.6	17.9	19.0	34.4
P2	18.8	18.5	18.1	31.9
N%				
Acetone	2.6	2.5	2.6	2.0
B/A	2.4	2.7	2.0	2.4
P2	2.7	2.6	3.0	2.6
Si%				
Acetone	2.1	2.3	3.6	2.0
B/A	2.2	2.4	4.0	4.0
P2	2.0	2.0	4.5	36

The adherends of the specimens were prepared using an acetone wipe, a base/acid etch, or a P2 etch.

suggests an effect of interface properties on the locus of failure. For instance, when a more advanced surface preparation method was used, the carbon and silicon concentrations increased and the aluminum and oxygen concentrations decreased. These results suggest that advanced surface preparation methods enhance adhesion and displace failure from the interface.

To further quantify the locus of failure, the epoxy film thicknesses on the failure surfaces of each specimen were measured. On the failure surfaces of the specimens tested under mode I loading or under mixed mode loading with a phase angle of 10°, a visible layer of adhesive was observed. For these specimens, a Nikon Measurescope 2305 was used to measure the epoxy film thickness. On the other hand, for the specimens tested under mode II loading or under mixed mode loading with a phase angle of 22°, the failure surfaces visually appeared to be

Table 5

Auger depth profile results for typical specimens selected from each test. The adherends of the specimens were prepared using an acetone wipe, a base/acid etch, or a P2 etch

Analysis results	Test method			
	Symmetric DCB ($h/H = 1$)	Asymmetric DCB ($h/H = 0.75$)	Asymmetric DCB ($h/H = 0.5$)	ENF
Phase angle Ψ	0°	10°	22°	90°
SERR ratio η	0%	3%	14%	100%
Depth profile				
Acetone	250 μm	50 μm	4 nm	3.5 nm
B/A	250 μm	55 μm	12 nm	6.0 nm
P2	250 μm	57 μm	100 nm	26.5 nm

interfacial. The Auger depth profiling method was then used for those specimens, in which the failure appeared to occur at or near the interface.

As shown in Table 5, in the mode I test, the thicknesses of the residual adhesive layer on the failure surfaces were about 250 μm for all the specimens with different surface preparations, which indicated that the failures all occurred in the middle of the adhesive layer in the test regardless of the surface preparation method since the total thickness of the adhesive of the specimens was 0.5 mm. When the phase angle increased as in the asymmetric DCB test with $h/H = 0.75$, which contains 3% of mode II fracture component, a layer of epoxy film with a thickness of around 50 μm was detected on the failure surfaces of all the specimens. Although the failure was still cohesive, the decrease in the film thickness on the metal side of the failure surfaces indicated that the locus of failure shifted toward the interface due to the increase in the mode mixity. On the other hand, because the failure was still cohesive, no significant effect of interface properties on the locus of failure was observed. When the mode mixity increased to 14% as in the asymmetric DCB test with $h/H = 0.5$, where the mode mixity strongly forced the crack toward the interface, the effect of interface properties on the locus of failure became pronounced. In the specimen with adherends prepared with acetone wipe, a 4-nm-thick epoxy film was detected on the failure surfaces; in the specimen with adherends treated with base/acid etch, the film thickness was 12 nm; and in the P2 etched specimen, a visible layer of film, which was estimated to be about 100 nm, was observed on the failure surfaces. This increasing trend in the measured film thickness from the failure surfaces suggested that the advanced surface preparation methods enhance adhesion and displace failure from the interface, which also confirmed the indications obtained from the XPS analyses. In the ENF test, a similar trend in the variation of film thickness was observed.

The XPS and the Auger depth profile analyses clearly identified the locus of failure and verify the analytical prediction made through applying the criteria of direction of cracking to adhesive bonds. Through testing specimens prepared with different surface preparation techniques, these results also demonstrated the effect of interface properties on the locus of failure and verify that crack path selection in adhesive bonds is a result of interactions between external loads and material properties [4]. The results also indicated that since the locus of failure is very sensitive to the interface properties in the asymmetric DCB tests with fracture mode mixity of 14% or higher, the asymmetric DCB test can be employed rather than the ENF test to evaluate the interface fracture properties in adhesively bonded joints. As discussed earlier, since the asymmetric DCB test is conducted under predominantly opening mode, the onset of fracture and the crack propagation sequence are much easier to observe; this substitution can greatly simplify the testing procedure.

2.3.6. Surface preparation and the rate dependence of the locus of failure

In Section 2.3.3, the rate dependence of the locus of failure in adhesive bonds was studied. The results showed that when the T-stress was negative, the failures were all cohesive and the cracks were directionally stable regardless of the debond rate; as the T-stress increased, the cracks became directionally unstable and a very pronounced effect of debond rate on the locus of failure was observed. The failure was more interfacial when the debond rate was low. To investigate the influence of the interface properties on the rate dependence of the locus of failure in specimens with high T-stresses, two groups of symmetric DCB specimens with adherend surfaces prepared using either an acetone wipe or P2 etch were prepared and tested under quasi-static loading condition. The adhesive used was adhesive C and the Kodak EktaPro high-speed camera system was used in the same manner as before to monitor the fracture sequence and to obtain the rate of crack propagation. After failure, a representative specimen from each group of specimens was selected based on visual examination for the XPS analyses to identify the locus of failure.

In Fig. 21, the failure surfaces of the two representative specimens are shown. The adherend surfaces of specimen a were prepared using an acetone wipe, and a P2 etch was used in preparing the substrates for specimen b. The T-stresses in both specimens were 35 MPa and the magnitude of the crack propagation rate for different regions of the specimens, which were obtained using the high-speed camera, were marked along both specimens to quantify the effect of debond rate. Through visual examination of the failure surfaces of specimens a and b, the influence of surface preparation on the rate dependence of the locus of failure can be observed. In specimen a, in the region of slow crack propagation, the failure surface was clear, indicating interfacial failure. On the other hand, in the region of fast crack propagation, the failure appeared to be cohesive since a visible layer

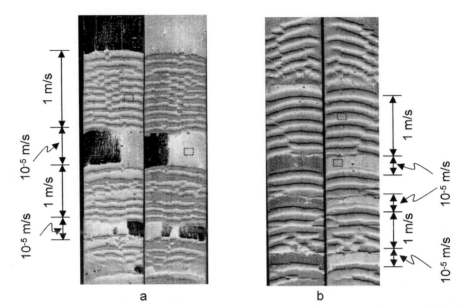

Fig. 21. The failure surfaces of the DCB specimens prepared using acetone wipe (a) and P2 etch (b), respectively. The dotted boxes indicate the areas where XPS analyses were conducted.

of adhesive film was observed on the failure surface. However, in specimen b, the difference in the locus of failure between the slow and fast crack propagation regions was not as pronounced. As a matter of fact, a visible layer of adhesive film was observed in both regions. Another noticeable difference in the failure surfaces between specimens a and b is that the regions of slow crack propagation in specimen b were considerably smaller than in specimen a, indicating that the rate dependence of the locus of failure was significantly reduced due to the variation in interface properties.

The XPS analyses were conducted on the areas in both the slow and fast crack propagation regions of each specimen as schematically shown in Fig. 21. The XPS data further identified the locus of failure and supported the results of visual examinations. As shown in Table 6, for specimen a, the major element concentrations on the failure surfaces, especially carbon and aluminum, varied significantly between the slow and the fast crack propagation regions and indicated that the failure was more interfacial in the region of slow crack propagation. On the other hand, for specimen b, although the variations of the major element concentrations between the slow and the fast regions also indicated a similar trend in the rate dependence of the locus of failure, the magnitude of the variation suggested that this debond rate effect in specimen b was not as pronounced as in specimen a.

Overall, the comparison of the rate dependence of the locus of failure between

Table 6

XPS elemental analysis results of the symmetric DCB specimens with either an acetone wipe or a P2 etch surface preparation

Specimen	Surface treatment	Region analyzed	XPS atomic percentage				
			C%	Al%	O%	N%	Si%
A	Acetone	Fast	76.1	0.4	16.8	2.5	4.3
		Slow	72.3	2.1	20.9	2.6	2.1
B	P2	Fast	76.5	0.2	16.9	4.3	2.1
		Slow	75.9	0.9	15.7	4.1	3.4

the two representative specimens with different surface preparations revealed that the interface properties significantly affect the crack propagation behavior. Advanced surface preparation techniques enhance the adhesion between the adhesive and the substrates and consequently, the rate dependence of the locus of failure is reduced. Of particular significance, however, is that the debond did not propagate exclusively at the weaker interface. The stress state destabilized the debond, leading to a crack that alternated between two interfaces with different properties.

2.3.7. Asymmetric surface preparation and the directionally unstable cracks

As discussed earlier in this chapter, the direction of crack propagation is stabilized very rapidly as the mode mixity increases; when the mode mixity G_{II}/G is more than 3%, cracks in the specimens were all directionally stable regardless of the T-stress levels examined. In this section, the effect of asymmetric surface preparation on the directionally unstable cracks is of interest. Geometrically symmetric DCB specimens with one adherend prepared using an acetone wipe and the other using a P2 etch were prepared using adhesive C and they were tested under mode I low-speed impact conditions. The low-speed impact was chosen due to the reason that the rate of crack propagation was relatively constant in this test, and therefore, the effect of debond rate on the locus of failure was minimized. Before the tests, each specimen was mechanically stretched to achieve a high residual stress state and consequently to achieve a high positive T-stress state such that alternating crack propagation was observed in the specimen. After failure, post-failure analyses via XPS and Auger depth profile were conducted on the failure surfaces of a typical specimen to identify the locus of failure and the crack propagation trajectory.

The failure surfaces of the typical specimen selected are shown in Fig. 22. The T-stress in the specimen was 36 MPa and the crack trajectory alternated between the two interfaces, which can be observed in the side-view photograph. To further identify the locus of failure, both the XPS and Auger depth profile analyses were conducted on representative areas of both sides of the specimen as schematically

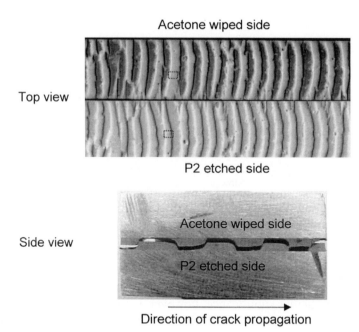

Acetone wiped side

Top view

P2 etched side

Side view

Acetone wiped side

P2 etched side

Direction of crack propagation

Fig. 22. The failure surfaces and the crack trajectory of the DCB specimen with asymmetric surface preparation. The dotted boxes indicate the areas where XPS and Auger analyses were conducted.

Table 7

The post failure analysis results on the failure surfaces of the DCB specimen with asymmetric surface preparation

Surface treatment	XPS atomic percentage					Film thickness
	%C	%Al	%O	%Si	%N	(μm)
Acetone	75.8	0.9	17.2	3.8	2.3	0.6
P2	76.7	0.1	16.5	4.5	2.2	1.4

shown in Fig. 22 and the results are listed in Table 7. The XPS results show that the carbon and silicon concentrations on the acetone wiped adherend surface are lower than on the P2 etched adherend surface, whereas the aluminum and oxygen concentrations are much higher. This trend of the variation of the major element concentrations on the failure surfaces indicates that the locus of failure on the acetone wiped adherend side was more interfacial than on the P2 etched adherend side. The exact locations of the failure on both adherends surfaces were revealed by the Auger depth profile data as shown in the Table 7. On the surfaces of the adherend prepared using the acetone wipe, a layer of adhesive film of 0.6 μm thick was detected. However, on the surfaces of the adherend prepared using the

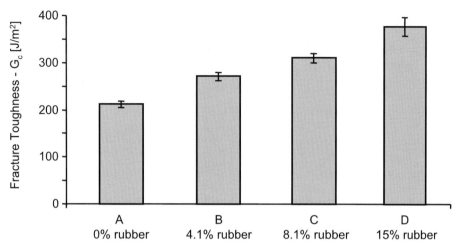

Fig. 23. The fracture toughness of the DCB specimens using adhesives with different levels of rubber concentrations. Error bars represent ±1 standard deviation.

P2 etch, the adhesive film detected was much thicker (1.4 μm), which indicated that the failure was more cohesive.

2.3.8. Toughness of the adhesive and the directional stability of cracks

As pointed out by Pocius [60], the directional stability of cracks is significantly affected by the fracture toughness of adhesive bonds. An energy balance model used to analyze crack propagation predicted that directionally unstable cracks are more unlikely to occur as the fracture toughness of the adhesive bonds increase [32]. This energy balance model will be discussed in this chapter later.

To verify the prediction, DCB specimens using adhesives A, B, C and E were prepared. Due to the various levels of rubber concentration in the adhesives, the fracture toughness of the DCB specimens varied with the adhesive. The critical fracture toughnesses measured in the quasi-static tests for the DCB specimens using different adhesives are shown in Fig. 23, which indicates that the fracture toughness of the bonds increased significantly with rubber concentration level. After the specimens were prepared, they were subjected to mechanical stretching to achieve various levels of the T-stress in the specimens. The specimens were then tested under low-speed impact loading in order to minimize the effect of debond rate in the tests. After failure, the failure surfaces in each specimen were carefully examined to determine crack trajectory and the crack propagation manner.

Typical failure surfaces observed in as-produced specimens are shown in Fig. 24. Specimen a was bonded using adhesive A, which contains no rubber toughener and is the most brittle adhesive in the series. The failure surfaces

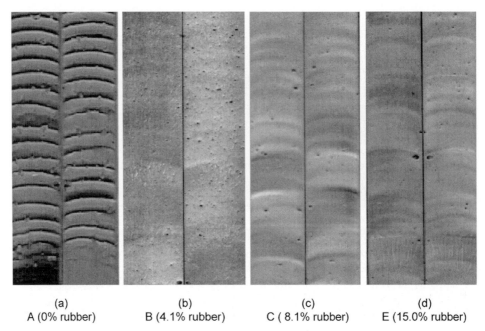

(a)	(b)	(c)	(d)
A (0% rubber)	B (4.1% rubber)	C (8.1% rubber)	E (15.0% rubber)

Fig. 24. The failure surfaces of the as-produced DCB specimens prepared using different adhesives.

of this specimen revealed an alternating crack trajectory, which indicated that the crack propagation was directionally unstable. As the rubber concentration increased in the adhesive as in specimens b, c and d, the failures all appeared to be cohesive with directionally stable crack trajectory. A similar trend had been observed in stretched specimens. All three specimens in Fig. 25 contained 1.1% plastic deformation in the adherends and from specimens (a–c), the rubber concentration in the adhesive increased from 4.1% to 15.0%. Examinations of the failure surfaces of theses specimens indicate that the crack was directionally unstable in specimen a and became more and more stable in specimens b and c. Figs. 24 and 25 reveal that the directional stability of the crack is significantly affected by the rubber concentrations in the adhesives.

3. Analytical studies

Fundamental understanding of the mechanics of the crack path selection behavior and predictive capabilities for both directionally stable and unstable crack propagation in adhesive bonds can be gained through analytical models of the bonded system. Through these studies, important insights were gained and the trajectories for directionally unstable crack propagation observed in the tests were

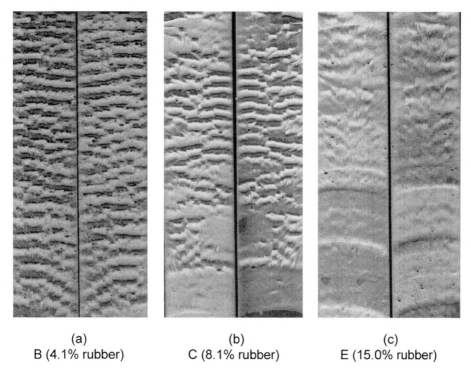

(a)
B (4.1% rubber)

(b)
C (8.1% rubber)

(c)
E (15.0% rubber)

Fig. 25. The failure surfaces of the DCB specimens prepared using different adhesives. All the specimens contained 1.1% plastic deformation in the adherends

simulated. In the following sections, the discussions mainly focus on an energy balance model analyzing the probability of directional unstable crack propagation, analysis of the alternating crack using the interface mechanics, and simulations of the directionally unstable cracks for various systems.

3.1. Energy balance and directional stability of crack propagation

Because of adhesive shrinkage during cure or the thermal mismatch between adherends and adhesive, a residual stress is often induced during the fabrication of adhesive bonds. If the residual stress is released during bond failure, the stored strain energy will be reduced, and is potentially available to assist in driving crack propagation. For a relatively straight crack propagating anywhere within an adhesive layer between two relatively stiff adherends, the stored energy associated with the residual stress is not available to drive cracking because the fracture plane is parallel to the biaxial stress state. On the other hand, anytime a crack moves with a substantial angle with respect to the bond plane, stored elastic energy is relieved as the residual stresses are locally reduced. Compared to a directionally stable crack, where the crack trajectory is fairly straight, Fig. 10 shows that due

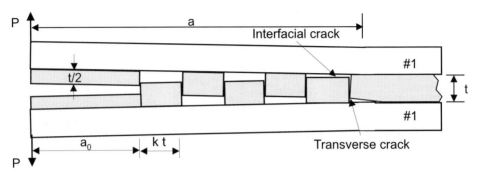

Fig. 26. A double cantilever beam (DCB) specimen with simplified crack trajectory for directionally unstable crack propagation.

to the alternating feature, more residual stress is released in the directionally
unstable crack propagation, and consequently, a greater amount of energy is
available to propagate the crack. On the other hand, because the crack trajectory
is more tortuous, the crack propagates along a longer crack path when the crack
is directionally unstable, which indicates that more energy is consumed during the
fracture process. This close relationship between the manner of crack propagation
and the energy flow in the system suggests that the directional stability of cracks
in adhesive bonds can be predicted through a rather simple analysis of the energy
balance within a bonded system.

Fig. 26 shows a DCB model with an alternating crack in the bondline that is
idealized as a 'square wave' similar to that used in Akisanya and Fleck [2,3].
Although the simplification of the crack trajectory is rather crude since many
details are ignored, the error induced is negligible as far as energy flow is
concerned because the model only contains about a 5% error in the total length
of the crack trajectory and the total area of the additional free surfaces associated
with the release of residual stress. Aluminum adherends bonded with a 0.5-mm-
thick bondline of adhesive C were analyzed, using aforementioned properties.
To further simplify the analysis, several other assumptions were made: according
to the results measured in the quasi-static tests, the measured fracture toughness
of the aluminum/epoxy bonds is approximately independent of the directional
stability of cracks for this materials system studied based on the plane area of
the specimens; therefore, the specimens are assumed to have a single critical
fracture toughness value $G_c = 310 \, \text{J/m}^2$ as before. Because the adherends are stiff
compared to the adhesive; the adherend bending due to the residual stress in the
adhesive is negligible in the debonded region. For the same reason, the effect of
the adhesive layer on the flexural rigidity of the specimens is negligible.

For the crack trajectory shown in Fig. 26, the energy available during the
fracture process consists of two parts, U_1 and U_2. U_1 is the strain energy available
when the crack propagates along the interfaces and U_2 is the strain energy

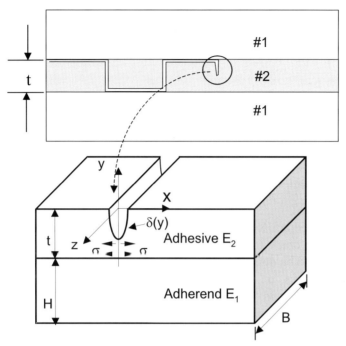

Fig. 27. The geometry and the stress state of a transverse crack in the DCB specimen with simplified crack trajectory for directionally unstable crack propagation.

available for the transverse crack propagation. U_1 is the same as the energy available during the directionally stable crack propagation and can be obtained using the conventional compliance analysis as

$$U_1 = \frac{a^3 P^2}{3EI} \tag{10}$$

where a is the crack length, EI is the effective flexural rigidity of the DCB specimen, and P is the load. For comparison purposes, an iso-fracture toughness is assumed, resulting in P being given by

$$P = \frac{\sqrt{G_c B EI}}{a} \tag{11}$$

To calculate the strain energy U_2, a single transverse crack is isolated as shown in Fig. 27 and the strain energy u available for this particular crack is given by

$$u = \frac{\sigma}{2} \int_0^t \delta(y) \, dy \tag{12}$$

where $\delta(y)$ is the crack opening displacement profile and σ is the stress distributed along the crack surfaces prior to cracking. If the crack is assumed to be a Griffith crack as Suo [61] and Nairn [62] suggested in studying tunneling cracks in coatings and composites, $\delta(z)$ under plane-strain conditions is given by

$$\delta(y) = 4\sigma \frac{1-v_2^2}{E_2} \sqrt{y^2 - z^2} \tag{13}$$

The choice of $\delta(z)$ was also verified by the finite element analysis conducted to investigate the geometry of a vertical opening crack in an adhesively bonded DCB specimen. Substituting Eq. 13 into Eq. 12, the strain energy available for transverse crack propagation U_2 is then obtained as

$$U_2 = \frac{\pi}{2} \left(\frac{a - a_0 + kt}{2kt} \right) \left(\frac{1-v_2^2}{E_2} \right) B\sigma^2 t^2 \tag{14}$$

where a_0 is the precrack length as shown in Fig. 26, kt is the characteristic length of the alternating crack trajectory, and the opening stress applied at the crack surfaces σ is still to be determined.

The normal stress, σ, depends on the final residual stress in the adhesive, specimen geometry, and the external loads. Setting the coordinate system as shown in Fig. 27, and recognizing that $\sigma_y = 0$ following debonding, one may approximate σ as

$$\sigma = \frac{Pa(t/2+d)}{I} + \sigma_0 + \sigma_p \tag{15}$$

where $I \approx BH^3/12$, and d is given by

$$d = \frac{nH^2 - t^2}{2(t+nH)} \tag{16}$$

The first term in Eq. 15 is the normal stress induced at the crack surfaces due to the bending of the adherends prior to cracking; the second term σ_0 is the thermal residual stress induced from curing; and the third term σ_p is the residual stress induced from the mechanical stretching used to alter the T-stress state in the specimen as discussed earlier. The second and third term in Eq. 15 combined is given by Eq. 9. Substituting Eq. 15 into Eq. 14, the strain energy available for the transverse crack propagation is then obtained. Dividing the total strain energy $U = U_1 + U_2$ by the total crack area, the average strain energy release rate for the crack propagation is obtained.

Shown in Fig. 28 is the result of the analysis. The ordinate of the figure is the normalized average strain energy release rate with respect to the critical strain energy release rate G_c (= 310 J/m^3) of the DCB specimens made of adhesive C, and the abscissa represents the thickness of the adherends. The dashed line represents the critical strain energy release rate for the alternating crack propagation and therefore, also represents the threshold of the transition of the directional stability of the crack propagation. If the available strain energy release

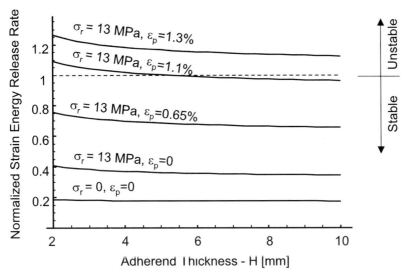

Fig. 28. Available strain energy release rate for directionally unstable crack propagation in DCB specimen with various conditions.

rate is higher than the critical strain energy release rate for the alternating crack trajectory, cracks in the specimen are more likely to be directionally unstable since there is enough energy available for cracks to propagate along the alternating path. On the other hand, if the available strain energy release rate for the alternating crack propagation is lower than the critical strain energy release rate, cracks are more likely directionally stable. Fig. 28 shows that when the plastic deformation in the adherends ε_p is less than 1.1%, the curves are all below the dashed line indicating a likelihood for directionally stable crack propagation. As the plastic deformation ε_p in the adherends increases (consequently, the T-stress increases), more strain energy becomes available and the curves shift upwards, suggesting an increase in the probability of directionally unstable cracks. This result is quantitatively consistent with the predictions made by Fleck et al. [15] using the T-stress argument and the experimental results discussed in Section 2.3.1.

All curves in Fig. 28 are monotonically decreasing with increasing adherend thickness, indicating an effect of adherend bending on the directional stability of cracks. For instance, the curve with $\varepsilon_p = 1.3\%$ and $\sigma_0 = 13$ MPa is higher than the dashed line when H is less than 4 mm and is lower than the dashed line when H is greater than 4 mm. Since the dashed line represents the threshold of the directional stability of crack propagation, this curve indicates that for specimens with $\varepsilon_p = 1.3\%$ and $\sigma_0 = 13$ MPa, the directional stability of cracks varies with the thickness of the adherends. This prediction is also consistent with the experimental results discussed in Section 2.3.1 where DCB specimens with adherend thicknesses H of mm and 3.2 mm, respectively, were tested to investigate the effect of adherend bending. The experimental results indicated that cracks in specimens with $H = 3.2$ mm were more directionally unstable than the cracks in specimens with $H = 4.8$ mm when the residual stresses and the plastic deformation levels were the same.

In addition, Fig. 28 also indicates an effect of the toughness of the adhesive bonds on the directional stability of cracks. When the toughness of the bond increases, all the curves will shift down vertically, indicating that the transition between the directionally stable and unstable crack is less likely to occur. This result is again consistent with the experimental observations discussed in Section 2.3.8, which showed that as the rubber concentration in the adhesive increases, the fracture toughness of the bonds increases and consequently, the transition from directionally stable cracks to directionally unstable cracks is more unlikely to occur.

As seen, even relatively crude analytical energy balance models are helpful in understanding crack stability in adhesive layers. This energy-based perspective provides an intuitive understanding of the effect of in-plane stresses within an adhesive layer on crack path selection. Indeed, residual stresses have been shown to induce some very interesting failure modes in bonded joints, including spiral crack patterns [63].

3.2. Interface mechanics and the prediction of crack trajectories

In this section, discussion focuses on the interface fracture mechanics and the details of crack trajectory predictions that are possible with numerical implementation of these concepts. According to the interface fracture mechanics theory discussed in Chapter 2, a crack at the interface between the adherend and adhesive can be represented by a sub-interface crack lying a small distance (δt) below the interface and the complex stress intensity factors K_1 and K_2 for the interface crack are related to the conventional stress intensity factors K_I and K_{II} for the sub-interface crack as

$$K_1 + i K_{II} = q e^{i\phi} (K_1 + i K_2) \delta t^{i\varepsilon} \tag{17}$$

where $q = \sqrt{(1 - \beta^2)/(1 + \alpha)}$ is a real quantity and $\phi(\alpha, \beta)$ is a dimensionless function of the elastic moduli listed in Hutchinson and Suo [12] for different material combinations. In addition, a similar relationship also exists between the local complex stress intensity factors, K_1 and K_2, for the interface crack and the far field stress intensity factors, K_I^∞ and K_{II}^∞:

$$K_1 + i K_2 = p \left(K_I^\infty + i K_{II}^\infty \right) t^{-i\varepsilon} e^{i\omega(\alpha, \beta)} \tag{18}$$

where $p = \sqrt{(1 - \alpha)/(1 - \beta^2)}$, $\omega(\alpha, \beta)$ is a dimensionless functions of material constants and is also listed in Hutchinson and Suo [12], and K_I^∞ and K_{II}^∞ is directly related to the external loads applied on the specimen and can be found in Tada et al. [23]. Specifically, for symmetric DCB specimens, which are under predominantly mode I loading, the relationship is given by

$$K_I^\infty + i K_{II}^\infty = K_I^\infty = f \frac{Pa}{B} H^{-3/2} \left(2\sqrt{3} + 2.315 \frac{H}{a} \right) \tag{19}$$

where P is the external load, B is the width of the specimen, a is the crack length, and f equals 1 for plane stress and $1/\sqrt{1 - v_1^2}$ for plane strain. Substituting Eq. 19 into Eq. 18, the local complex stress intensity factors for an interface crack in a DCB specimen can be expressed as a function of the external load as

$$K_1 + i K_2 = p f \frac{Pa}{B} H^{-3/2} \left(2\sqrt{3} + 2.315 \frac{H}{a} \right) t^{-i\varepsilon} e^{i\omega(\alpha,\beta)} \tag{20}$$

If the crack is at the sub-interface, the local stress intensity factor K_I and K_{II} can also be expressed as a function of the external load by substituting Eq. 20 into Eq. 17 as

$$\begin{aligned} K_I &= q \, |K| \cos \left[\omega + \phi + \varepsilon \ln (\delta t / t) \right] \\ K_{II} &= q \, |K| \sin \left[\omega + \phi + \varepsilon \ln (\delta t / t) \right] \end{aligned} \tag{21}$$

where, $|K|$ is given by

$$|K| = p f \sqrt{\frac{G E_1}{12}} \left(2\sqrt{3} + 2.315 \frac{H}{a} \right) \tag{22}$$

and $G = 12 P^2 a^2 / E_1 B^2 H^3$ is the applied strain energy release rate in the specimen.

Fig. 29 shows that due to the material mismatch, when the distance between the sub-interface crack and the interface δt approaches zero, the corresponding K_{II} component at the crack tip is very high and is acting in such a direction that the crack tends to deviate away from the interface toward the centerline of the bond. As the distance δt increases, the corresponding K_{II} value drops drastically, which suggests that the sub-interface crack will deviate from the interface in a rather gradual fashion. This prediction is consistent with crack trajectory shown in the SEM micrograph of the DCB specimens with directionally unstable cracks in Fig. 10. Since differences in the material mismatch will result in variations in the stress distribution, Fig. 29 also indicates that the crack propagation behavior will also be different for different materials systems.

The analysis of the interface mechanics provides useful insights into crack propagation behavior in adhesively bonded joints. A finite element model for the DCB specimen was constructed using Franc2D/L [64], a convenient code for this task because of its capability for automatic remeshing in the vicinity of a growing crack. An adhesive layer (material 2) with thickness of $t = 0.5$ mm is sandwiched between two adherends (material 1) with thickness of $H = 6$ mm, and

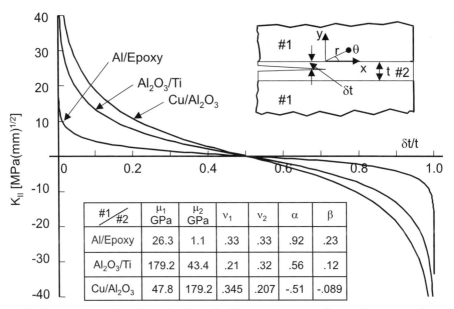

Fig. 29. Parametric study of the local mode II stress intensity factor K_{II} versus the non-dimensional location of the sub-interfacial crack for different material combinations. The applied strain energy release rate used in the calculation was $310\,\text{J/m}^2$.

a straight crack is located at the interface between the adhesive and the adherend. At the crack tip, there is a small crack kink into the adhesive with a length of 0.01 mm and an angle of $30°$ with respect to the interface. The purpose of the crack perturbation is to start the crack deviation quickly and reduce simulation time. According to the interface mechanics discussed earlier, an interface crack will deviate away from the interface automatically and no crack perturbation if necessary. The displacements of one end of the model were constrained. Moments of opposite direction were applied on the other end of the adherends to simulate the external loads. A horizontal tensile stress T^{∞} was applied to achieve the desired T-stress level. Three types of elements were used in the analysis. Eight-node, plane-strain elements were used with reduced integration in the area away from the crack tip; right around the crack tip, the elements used were quarter-point singular elements; and in the area in between, triangle elements were used for the convenience of remeshing during the crack propagation. Both the adherends and adhesive were modeled as linear elastic materials with material constants $E_1 = 70$ GPa, $E_2 = 2.97$ GPa, and $\nu_1 = \nu_2 = 0.33$. The residual stress in the adhesive layer was estimated as 13 MPa and the adhesive bond was assumed to have an iso-fracture toughness value of $310\,\text{J/m}^2$ as discussed earlier. The interface crack with a small perturbation was assumed to be present originally in the specimen and T^{∞} was adjusted to such a value that $K_{II} = 0$ at the crack tip. This analysis intended to predict the crack trajectory as the crack advances through the following procedure:

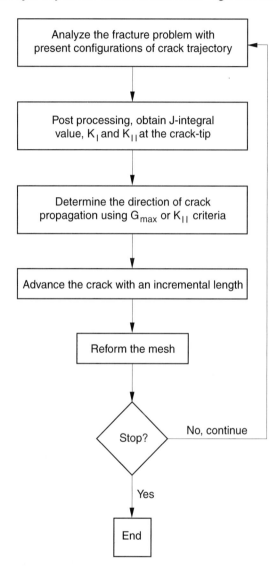

In the finite element analysis, the stress singularity at the crack tip was simulated using the quarter-point singular element, which provides a singularity with an order of $r^{-1/2}$. However, according to Cook and Ergodan [65], the order of the stress singularity of a crack perpendicular to a bi-material interface approaches to $r^{\lambda-1}$ as the crack propagates toward the interface, where λ is determined by the material mismatch at the interface. Consequently, certain errors existed in stresses at the crack tip in the finite element analysis when the crack approached

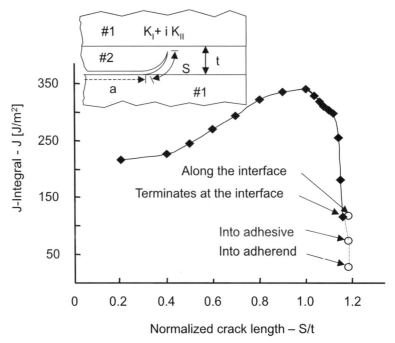

Fig. 30. Strain energy release rate (J-integral value) available at the crack tip versus the non-dimensional crack length obtained using the finite element analysis.

the interface. However, since only the direction of the crack propagation was of interest in this study, which was determined by the phase angle instead of the stresses, errors in the magnitude of stress singularity should have minimal affect on the resulting crack trajectory.

The strain energy release rate (J-integral value) versus the normalized kinked crack length S/t is plotted in Fig. 30. The result shows that the energy available for the crack increases as the crack advances thickness-wise in the adhesive layer and decreases drastically as the crack approaches the opposite interface due to the rigid boundary of the adherend. If the crack continues to grow, the crack could possibly propagate into the adherend, kink into the interface, or reflect back to the adhesive layer. The final direction depends on the amount of energy release rate available relative to the fracture toughness of the adhesive bond in that particular direction according to the criteria of cracking direction reviewed earlier. Fig. 30 shows that the most likely scenario is for the crack to kink into the interface since more energy is available in that direction than for the other possibilities. Supportive information can also be found in He et al. [66]. A similar tendency is found in Fig. 31, where the phase angle $\psi = \tan^{-1}(K_{II}/K_I)$ at the crack tip versus the normalized kinked crack length S/t is plotted. As the crack propagates, the phase angle is negative, which indicates that the crack would propagate along a

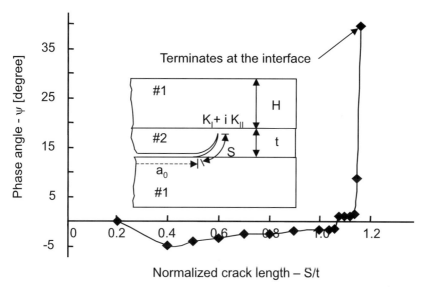

Fig. 31. Phase angle at the crack tip versus the non-dimensional crack length obtained by the finite element analysis.

trajectory with an increasing slope. When the crack approaches the top interface, ψ increases rapidly and becomes positive at the interface, which suggests an abrupt kink for the crack. Furthermore, the fracture toughness of adherends in adhesive bonds is often much higher than that of the adhesive. Consequently, the crack will likely kink into the interface due the restriction of the adherends. Supportive information can also be found in He and Hutchinson [67] where the deflection behavior of a straight crack located close to the interface of dissimilar materials is analyzed.

After kinking occurs, the crack grows along the interface. Meanwhile, the phase angle at the crack tip increases. Fig. 32 shows the phase angle $\psi = \tan^{-1}(K_{II}/K_I)$ versus the normalized interfacial crack length L/t, where K_{II} and K_I are obtained using the sub-interface crack concept and by assuming $\delta t \ll 1$. The results are for several values of the Dundur's parameter, with $\alpha = 0.92$ corresponding to the aluminum/epoxy case discussed herein. For this case, Fig. 32 suggests that after propagating along the interface for 2–3 times the thickness of the adhesive layer, the crack will start to leave the interface due to the increase in phase angle, which is consistent with the experimental observations of the crack trajectory shown in Fig. 10. These results are also consistent with the results discussed in Akisanya and Fleck [2,3] theoretically. Fig. 32 also indicates that the characteristic spacing of the alternating cracks depends on the moduli, as has been observed experimentally.

Fig. 33 shows the final results of the FEA simulation of the crack trajectory for

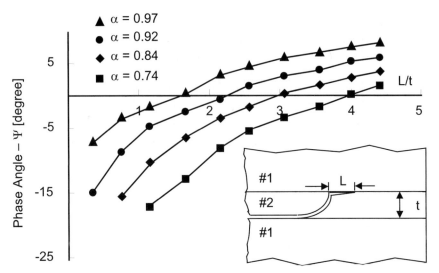

Fig. 32. The phase angle at the crack tip versus the normalized kinked crack length s/t for different materials combinations obtained from the parametric study.

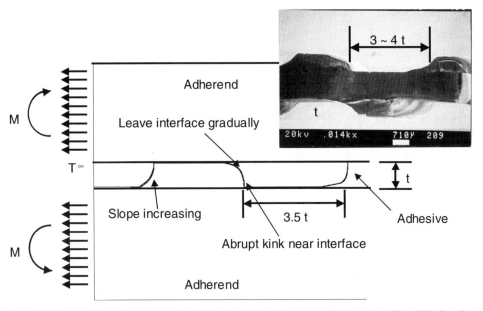

Fig. 33. The crack trajectory predicted by the finite element analysis using Franc2dl for the adhesive C system studied. The result reflects the overall characteristics of the actual crack trajectory such as the characteristic length of the crack as shown in the SEM micrograph.

the C system studied (the FEA mesh is hidden for a clear observation of the crack path). Compared to the SEM micrograph of the actual crack trajectory shown in the inset, the numerical simulation accurately reproduces the characteristic length of the alternating crack of 3–4 times the thickness of the adhesive layer and the overall characteristic shape of the actual crack trajectory.

Clearly, interface fracture mechanics coupled with finite element analysis is a powerful tool for modeling failure in adhesive bonds. Perhaps the biggest limitation is the difficulty in knowing the spatial variation in material properties within the adhesive layer and at the interfaces. Further understanding in this area could lead to an even greater understanding and ability to accurately predict the failure mode and locus of failure in adhesive joints of various types.

4. Implications for selecting and interpreting adhesion tests

Although the focus of this chapter has been drawn from a study of a series of model epoxies used to bond aluminum adherends, the implications of the role that mechanics plays in the failure of adhesive bonds are much broader. If one interface of a bond is particularly weak, no amount of mode mixity may be sufficient to steer the debond away from the weak zone. On the other hand, for many practical bonds having reasonably good properties throughout the bondline, mode mixity and T-stress can significantly affect the failure mode. Because of this significant effect, test method selection and interpretation benefit significantly from an understanding of the mechanic principles that will affect the crack trajectory.

From a mode mixity standpoint, greater amounts of mode II loading will tend to steer the debond to the mechanically favored interface. Mode II loading can be achieved through shearing the specimen, but is also encountered when dissimilar adherends are used. Mode mixity favors debonds propagating near the interface of the more highly strained adherend, as may be induced by increased loading, decreased adherend thickness, or reduced modulus. These principles have been employed, both intentionally and unintentionally, for specimens reported in the literature. Kramer [68], for example, has long advocated the use of asymmetric DCB specimens for their ability to produce interfacial failures with the more compliant adherend. The inverted blister specimen has been used successfully for brittle polymer [69] and ice [70] adhesion. By using a thin blister adherend bonded to a stiff section of the brittle adhesive, failure is driven along the interface rather than rupturing the brittle material. Bonding studs to core-sawn plugs of thick coatings can lead to cohesive failures within the coating material, whereas pulling thin adherends away from the relatively thick coatings result in interfacial failures. Failures in single lap joints often reveal the effect of mode mixity on the locus of failure, as illustrated in Fig. 34. These failures are consistent with the tendency of cracks to propagate perpendicular to the largest tensile stress (in homogeneous materials) if debonds grow inward from the two edges, as discussed by Adams et al. [71]. Coatings, loaded primarily by a tensile residual stress state,

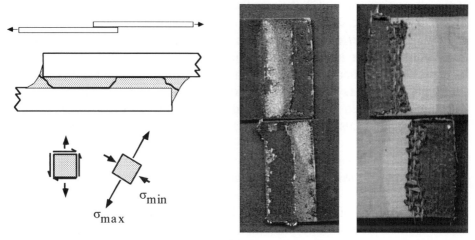

Fig. 34. Mode mixity helps explain the locus of failure in single lap joints, where debonds tend to propagate as shown.

will typically fail at the interface because the mode mixity drives the debond to the interface.

From the T-stress standpoint, increased residual stress, thinner adherends, thicker adhesive layers, and less ductile adhesives contribute to directional instability, leading to cracks that may alternate from one adherend to another. Material properties also play a role; cracks within adhesives that are significantly softer than the adherends are trapped [15], and more likely to appear as cohesive failures. This is consistent with the observation of Gent [72] that failures in bonds of rubber to a rigid adherend are almost invariably cohesive, leaving behind a visible layer of rubber. As the moduli of the adherends and adhesive become more similar, this crack trapping effect is less pronounced, so interfacial failures are more likely. The effect of increased T-stress are also seen in T-peel specimens, where bending of the thin adherends tends to fragment the adhesive at a characteristic spacing.

While all of us can cite counterexamples to each of these tendencies, they nonetheless remain useful in selecting and interpreting adhesion tests. If one is interested in probing interfacial failures, selecting test specimen configurations that favor this failure mode may prove useful. For example, accelerated characterization techniques are often desired to investigate the long-term performance of adhesive bonds. Such methods can be quite useful provided that the failure modes produced under the service and accelerated conditions are identical. Since long-term failures often occur at the interface where the complexities of surfaces, material gradients, and stress anomalies occur, selection of accelerated characterization test specimens that favor interfacial failures may prove useful. This choice must be balanced against the need to select test methods that will provide useful insights into the response under the intended applications.

5. Summary

The theoretical background, experimental results, and analytical or numerical studies provided in this chapter support the fact that the final locus of failure in adhesive bonds is strongly affected by the stress state that exists within the bondline and at the tip of a growing debond. Although chains (and other discrete systems) break at the weakest link, adhesive bonds and other continuous systems do not always fail in such a simplistic fashion. Instead, the failure path is determined by the interaction of a complex stress field with spatially varying mechanical properties of the material system. Variations in stoichiometry, crystallinity, molecular weight distributions, weak boundary layers, and other parameters may render substantial variations in strength and toughness throughout a bondline. On the other hand, mode II loading and the T-stress are two mechanics parameters that can, in many cases, determine, or at least influence, the failure mode. Visual observations and surface analysis techniques, such as XPS and Auger depth profiling, have consistently demonstrated this effect. An understanding of these principles and the complex interactions between stress state and material properties can prove useful in selecting and interpreting test methods for adhesive bonds.

References

1. Cao, H.C. and Evans, A.G., An experimental study of the fracture resistance of bimaterial interfaces. *Mech. Mater.*, **7**, 295–304 (1989).
2. Akisanya, A.R. and Fleck, N.A., Analysis of a wavy crack in sandwich specimens. *Int. J. Fract.*, **55**, 29–45 (1992).
3. Akisanya, A.R. and Fleck, N.A., Brittle fracture of adhesive joints. *Int. J. Fract.*, **58**, 93–114 (1992).
4. Dillard, D.A., Chen, B., Parvatareddy, H., Lefebvre, D. and Dillard, J.G., Where does it fail, and what does that mean? *J. Adhes.*, in preparation.
5. Freund, L.B. and Kim, K.S., *Mater. Res. Soc. Symp. Proc.*, **226**, 291 (1991).
6. Meguid, S.A. and Papanikos, P., Fatigue crack growth behavior of interacting holes in airframe alloys. In: S.A. Meguid (Ed.), *Mechanics in Design*, 1996, pp. 731–739.
7. Ergodan, V.F. and Sih, G.C., On crack extension in plates under plane loading and transverse shear. *Trans. ASME J. Bas. Eng.*, **85**, 519–527 (1963).
8. Palaniswamy, K. and Knauss, W.G., On the problem of crack extension in brittle solids under general loading. In: Nemat-Nasser, S. (Ed.), *Mechanics Today*, Vol. 4. Pergamon Press, New York, 1978, pp. 87–148.
9. Goldstein, R.V. and Salganik, R.L., Brittle fracture of solids with arbitrary cracks. *Int. J. Fract.*, **10(4)**, (1974).
10. Cotterell, B. and Rice, J.R., Slightly curved or kinked cracks. *Int. J. Fract.*, **16**, 155–169 (1980).
11. Geubelle, P.H. and Knauss, W.G., Crack propagation at and near bimaterial interfaces: linear analysis. *J. Appl. Mech.*, **61**, 560–566 (1994).
12. Hutchinson, J.W. and Suo, Z., Mixed mode cracking in layered materials. *Adv. Appl. Mech.*,

29, 63–191 (1992).

13. Chen, B. and Dillard, D.A., Numerical analysis of the directionally unstable crack propagation in adhesively bonded joints. *Int. J. Solids Struct.*, **38**, 6907–6924 (2001).

14. Lin, C. and Liechti, K.M., Similarity concepts in the fatigue fracture of adhesively bonded joints. *J. Adhes.*, **24**, 101–121 (1986).

15. Fleck, N.A., Hutchinson, J.W. and Suo, Z., Crack path selection in a brittle adhesive layer. *Int. J. Solids Struct.*, **27(13)**, 1683–1703 (1991).

16. Xu, G., Bower, F. and Ortiz, M., An analysis of non-planar crack growth under mixed mode loading. *Int. J. Solids Struct.*, **31**, 2167–2193 (1994).

17. Lasson, S.G. and Carlsson, A.J., Influence of non-singular stress terms and specimen geometry on small-scale yielding at crack tips in elastic–plastic materials. *J. Mech. Phys. Solids*, **21**, 263–277 (1973).

18. Chao, Y.J., Liu, S. and Broviak, B.J., Brittle fracture: constraint effect and crack curving under model I conditions. *Proceedings of Pressure Vessel and Piping Conference*, American Society of Mechanical Engineers, July 2000, Seattle, WA, 2000.

19. Williams, M.L., On the stress distribution at the base of a stationary crack. *J. Appl. Mech.*, **24**, 109–114 (1957).

20. Chai, H., The characterization of mode I delamination failure in non-woven, multidirectional laminates. *Composites*, **14(4)**, 277–290 (1984).

21. Chai, H., On the correlation between the mode I fracture of adhesive joints and laminated composites. *Eng. Fract. Mech.*, **24(3)**, 413–431 (1986).

22. Chai, H., A note on crack trajectory in an elastic strip bounded by rigid substrates. *Int. J. Fract.*, **32**, 211–213 (1987).

23. Tada, H., Paris, P.C. and Irwin, H.R., *The Stress Analysis of Cracks Handbook*. Del Research, Hellertown, Pennsylvania, PA, 1985.

24. Lasson, S.G. and Carlsson, A.J., *J. Mech. Phys. Solids*, **21**, 263 (1973).

25. Chen, B., Dillard, D.A., Dillard, J.G. and Clark, R.L. Jr., Crack path selection in adhesively bonded joints: the roles of external loads and specimen geometry. *Int. J. Fract.*, **114**, 167–190 (2002).

26. Ayatollahi, M.R., Smith, D.J. and Pavier, M.J., *Key Eng. Mater.*, **83**, 145–149 (1998).

27. Leevers, P.S. and Radon, J.C., *Int. J. Fract.*, **19**, 311 (1982).

28. Sladek, J. and Sladek, V., Regularized integral representation of thermal stresses. *Eng. Anal. Bound. Elem.*, **8**, 224 (1991).

29. Sherry, A.H., France, C.C. and Goldthoppe, M.R., Compendium of T-stress solutions for two and three dimensional cracked geometries. *Fatigue Fract. Eng. Mater. Struct.*, **18**, 141 (1995).

30. *ABAQUS Theory Manual*, Version 6.1, Hibbitt, Karlsson, and Sorensen, Pawtucket, RI, 2000.

31. Chen, B. and Dillard, D.A., The effect of the T-stress on the crack path selection in adhesively bonded joints. *Int. J. Adhes. Adhes.*, **21**, 357–368 (2001).

32. Chen, B., Dillard, J.G., Dillard, D.A. and Clark, R.L. Jr., Crack path selection in adhesively bonded joints: the roles of material properties. *J. Adhes.*, **75**, 405–434 (2001).

33. Vrana, M.A., Dillard, J.G., Ward, T.C., Rakestraw, M.D. and Dillard, D.A., The influence of curing agent content on the mechanical and adhesive properties of dicyandiamide cured epoxy systems. *J. Adhes.*, **55**, 31–42 (1995).

34. Wegman, R.F., *Surface Preparation Techniques for Adhesive Bonding*. Noyes Publications, Park Ridge, NJ, 1989.

35. Minford, J.D., *Handbook of Aluminum Bonding Technology and Data*. Marcel Dekker, New York, 1993.

36. Rogers, N.L., *Proceedings of the 13th National SAMPE Technical Conference.* Society for the Advancement of Material and Process Engineering, 1981, p. 640.
37. Dillard, D.A., Park, T.G., Zhang, H. and Chen, B., Measurement of residual stresses and thermal expansion in adhesive bonds, in preparation.
38. Daghyani, H.R., Ye, L. and Mai, Y.W., Effect of thermal residual stress on the crack path in adhesively bonded joints. *J. Mater. Sci.*, **31**, 2523–2529 (1996).
39. Dillard, D.A., Chen, B., Chang, T. and Lai, Y.H., Analysis of the notched coating adhesion test. *J. Adhes.*, **69**, 99 (1999).
40. *LabVIEW – Graphical programming for instrumentation*, National Instruments, Version 5.0, 1998.
41. Blackman, B.R.K., Dear, J.P., Kinloch, A.J. and Osiyemi, S., The calculation of adhesive fracture energies from double-cantilever beam test specimens. *J. Mater. Sci. Lett.*, **10**, 253–256 (1991).
42. Parvatareddy, H., Dillard, J.G., McGrath, J.E. and Dillard, D.A., Environmental aging of the Ti-6Al-4V/FM-5 polyimide adhesive bonded system: implications of physical and chemical aging on durability. *J. Adhes. Sci. Technol.*, **12(6)**, 615–637 (1998).
43. Xu, S., Ohanehi, D.C., Yu, J.-H., Dillard, D.A., Lefebvre, D. and Schultz, R.A., Characterization of Impact Resistance of Electrically Conductive Adhesives. *Proceedings of the 24th Annual Meeting of the Adhesion Society*, Williamsburg, VA, February 25–28, 2001, pp. 321–323.
44. Blackman, B.R.K., Dear, J.P., Kinloch, A.J., Macgillivray, H., Wang, Y., Williams, J.G. and Yayla, P., The failure of fibre composites and adhesively bonded fibre composites under high rates of test, part I mode I loading – experimental studies. *J. Mater. Sci.*, **30**, 5885–5900 (1995).
45. Blackman, B.R.K., Kinloch, A.J., Wang, Y. and Williams, J.G., The failure of fibre composites and adhesively bonded fibre composites under high rates of test, part II mode I loading – dynamic effects. *J. Mater. Sci.*, **31**, 4451–4466 (1996).
46. Blackman, B.R.K., Dear, J.P., Kinloch, A.J., Macgillivray, H., Wang, Y., Williams, J.G. and Yayla, P., The failure of fibre composites and adhesively bonded fibre composites under high rates of test, part III mixed-mode I/II and mode II loadings. *J. Mater. Sci.*, **31**, 4467–4477 (1996).
47. Sundararaman, V. and Davidson, B.D., An unsymmetric double cantilever beam test for interfacial fracture toughness determination. *Int. J. Solids Struct.*, **34(7)**, 799–817 (1997).
48. Xiao, F., Hui, C.Y. and Kramer, E.J., Analysis of a mixed mode fracture specimen: the asymmetric double cantilever beam. *J. Mater. Sci.*, **28**, 5620–5629 (1993).
49. Liechti, K.M. and Liang, Y.M., Toughening Mechanisms in Mixed-Mode Interfacial Fracture. *Int. J. Solids Struct.*, **32(67)**, 957–978 (1995).
50. Liechti, K.M. and Freda, T., On the use of laminated beams for the determination of pure mixed-mode fracture properties of structural adhesives. *J. Adhes.*, **28**, 145–169 (1989).
51. Liechti, K.M. and Hanson, E.C., Nonlinear effects in mixed-mode interfacial delaminations. *Int. J. Fract.*, **36**, 199–217 (1988).
52. Swadener, J.G. and Liechti, K.M., Asymmetric shielding mechanisms in the mixed-mode fracture of a glass/epoxy interface. *J. Appl. Mech.*, 65(1), 25–29 (1998).
53. Swadener, J.G., Liechti, K.M. and de Lozanne, A.L., Intrinsic toughness and adhesion mechanisms of glass/epoxy interface. *J. Mech. Phys. Solids*, **47(2)**, 223–258 (1999).
54. Thouless, M.D., Fracture of a model interface under mixed-mode loading. *Acta. Metall. Mater.*, **38**, 1135–1140 (1990).
55. Parvatareddy, H. and Dillard, D.A., Effect of mode mixity on the fracture toughness of Ti-6Al-4V/FM-5 adhesive joints. *Int. J. Fract.*, **96**, 215–228 (1999).

56. Ravi-Chandar, K. and Knauss, W.G., An investigation into dynamic fracture. I – Crack initiation and arrest. *Int. J. Fract.*, **25**, 247–262 (1984).

57. Ravi-Chandar, K. and Knauss, W.G., An investigation into dynamic fracture. II – Microstructural aspects. *Int. J. Fract.*, **26**, 65–80 (1984).

58. Ravi-Chandar, K. and Knauss, W.G., An investigation into dynamic fracture. III – On steady-state crack propagation and branching. *Int. J. Fract.*, **26**, 141–154 (1984).

59. Ravi-Chandar, K. and Knauss, W.G., An investigation into dynamic fracture. IV – On the interaction of stress wave with propagating cracks. *Int. J. Fract.*, **26**, 189–200 (1984).

60. Pocius, A., Verbal communication, 1998.

61. Suo, Z., Failure of brittle adhesive joints. *Appl. Mech. Rev.*, **43(5)**, S276–S279 (1990).

62. Nairn, J.A., The strain energy release rate of composite microcracking: a variational approach. *J. Compos. Mater.*, **23**, 1106–1129 (1989).

63. Dillard, D.A., Spiral tunneling cracks induced by environmental stress cracking in LARC™-TPI adhesives. *J. Adhes.*, **44**, 51–67 (1994).

64. *Franc2D/L: A crack propagation simulator for plane layered structures*, Version 1.4, Swenson, D. and James, M., Kansas State University, 1999.

65. Cook, T.S. and Ergodan, F., Stresses in bonded materials with a crack perpendicular to the interface. *Int. J. Eng. Sci.*, **10**, 677–697 (1972).

66. He, M.Y., Bartlett, A., Evans, A.G. and Hutchinson, J.W., Kinking of a crack out of an interface: role of in-plane stress. *J. Am. Ceramic Soc.*, **74(4)**, 767–771 (1991).

67. He, M.Y. and Hutchinson, J.W., Crack deflection at an interface between dissimilar elastic materials. *Int. J. Solids Struct.*, **25(9)**, 1053–1067 (1989).

68. Kramer, E.J., Jiao, J., Gurumurthuy, C.K., Sha, Y., Hui, C.Y. and Borgesen, P., Measurement of Interfacial Fracture Toughness Under Combined Mechanical and Thermal Stresses. *J. Electr. Packaging*, **120(4)**, 349–353 (1998).

69. Kinloch, A.J. and Fernando, M., Use of the inverted-blister test to study the adhesion of photopolymers. *Int. J. Adhes. Adhes.*, **10(2)**, 69–76 (1990).

70. Penn, L.S. and Jiang, K.R., Use of the blister test to study the adhesion of brittle materials. Part II. Application. *J. Adhes.*, **32**, 217–226 (1990).

71. Adams, R.D., Comyn, J. and Wake, W.C., *Structural Adhesive Joints in Engineering*, 2nd edn. Chapman and Hall, 1997.

72. Gent, personal communication, 1998.

Chapter 12

Rheology for adhesion science and technology

A. BERKER [*]

3M Company, St. Paul, MN, USA

1. Introduction

Rheology deals with the deformation and flow of matter. In its practice, understanding the mechanical response of non-Newtonian materials to stresses and strains constitutes a large portion of its focus. In adhesion science and technology, one encounters many materials whose mechanical response can be non-Newtonian. These responses are encountered at the manufacturing, processing and dispensing stages of the adhesive. Furthermore, one frequently finds that end-use adhesive properties depend on those molecular parameters which also dictate the materials' rheological behavior–molecular architecture, average molecular weight, molecular weight distribution, glass transition temperature and plateau modulus being prime examples. Thus, it is not surprising to find that the adhesive and rheological properties of materials can be related.

In this introductory monograph, it will not be possible to touch upon many of the facets of the broad topic of rheology. Likewise, we will not be able to get into much detail of the specific areas that are covered. Fortunately, there are numerous references that the reader can consult and we will try to point these out as we go along. The purpose of this chapter is to give the reader some familiarity with the essentials of rheology, to provide useful references for further study, and to point out connections to adhesion whenever possible.

The basic building blocks of rheology are dynamics and kinematics and the relation between them, for given classes of materials, is the mechanical constitutive equation. Constitutive equations contain material functions which can be obtained from controlled testing — rheometry. Under suitable conditions, a

---

[*] Corresponding author. E-mail: aberker@mmm.com

certain class of materials may behave like a viscous liquid, an elastic solid, or they may exhibit characteristics of both. Such viscoelastic materials, if subjected to small enough mechanical stimuli, are seen to have moduli and compliances that are independent of the magnitude of the applied stimulus. Under this so-called 'linear viscoelastic regime' a universal constitutive relation holds, while the materials' individual response shows a great sensitivity to its molecular peculiarities. Linear viscoelasticity thus serves two important purposes: (1) it offers an array of analytical tests that probe the molecular structure and build up a mechanical 'fingerprint'; and (2) it provides a limiting form which the myriad constitutive relations must reduce to under sufficiently small deformations.

Processing conditions, on the other hand, generally involve high deformations where the material response is non-linear. Thus we need to be able to formulate constitutive relations that are valid beyond the linear viscoelastic regime. Here, certain principles come to our aid to assure that the relation being proposed is free of gross defects such as violating the laws of thermodynamics. Both continuum mechanics and molecular modeling have been used to generate constitutive relations. Since even a rudimentary discussion of this topic requires tensor algebra and calculus, we will not go into the details of how constitutive equations are generated. Instead we will focus on a few relations that are able to account for viscoelastic behavior, viscoplasticity and thixotropy. Among the simpler of these is the generalized Newtonian fluid (GNF) that is a widely used constitutive relation for viscous fluids that show shear thinning. It is primarily used in engineering calculations for pressure drops and flow rates. The most widely used member of this class of fluids is the so-called power-law fluid (PLF).

To understand the processing and dispensing behavior of adhesives we often need to construct flow models of the process. While all models are approximations, they still have to satisfy continuity of mass and the balance of momentum and thermal energy. The constitutive equation, plus the appropriate initial and/or boundary conditions of the problem at hand, provide closure to these balance laws. While any realistic solution to a particular processing problem will generally involve numerical computations, several generic problems of interest may be amenable to analytical development. For instance, when a pressure-sensitive adhesive is pressed down unto a surface, we have an example of squeeze flow. Similarly, if a paste is spread onto a surface via a knife, we have an example of wedge flow. These two flows will be discussed for the PLF.

2. Kinematics and dynamics

In continuum mechanics, a *body* is considered to be a continuous distribution of *material points*. A *deformation* is a smooth, one-to-one mapping which is subject to certain restrictions and which carries each material point into a point in Euclidean 3-space. The deformation is a vector, so that if we apply the gradient operator on it we get a tensor [1] called the deformation gradient, F. A given deformation is volume preserving, or isochoric, if the determinant of the deformation gradient, $\det(F)$, is equal to one.

Now let us introduce the element of time. At time t, in Euclidean space, the body occupies a region that is called its configuration. Let τ and t refer to some past time and the present, respectively. We would like to track the motion of a material particle, P, as the body moves from its configuration at τ to its present configuration at t. Thus, if we let the vector $X(\tau)$ be the place occupied by the material point P at τ, we can use its position at time t as a 'marker' to identify P, viz.

$$X_t = X(t) \tag{1}$$

The trajectory of all such material points is called the motion of the body

$$X(\tau) = f(X_t, \tau) \tag{2}$$

The particle velocity is the rate of change of its trajectory

$$v = \frac{\mathrm{d}}{\mathrm{d}\tau} X(\tau) = \dot{X}(\tau) = \frac{\partial}{\partial \tau} f(X_t, \tau) \tag{3}$$

Now consider two neighboring material points X_t and $X_t + \mathrm{d}X_t$ which were located at neighboring places $X(\tau)$ and $X(\tau) + \mathrm{d}X(\tau)$ at time τ. The relative deformation gradient tensor F is defined by

$$\mathrm{d}X(\tau) = F \, \mathrm{d}X_t \tag{4}$$

Thus, the relative deformation gradient tensor describes the change of the local configuration that occurs between times τ and t in the neighborhood of P. Since F depends on both τ and t, it is a relative tensor and we write $F_t(\tau)$ to denote this. From Eqs. 2 and 4, we note that the relative deformation gradient tensor is the gradient of the motion, i.e.

$$F_t(\tau) = \nabla f \tag{5}$$

where ∇ is the gradient operator. Using the relative deformation gradient tensor, we next define the Cauchy tensor

$$C_t(\tau) = F_t^{\mathrm{T}}(\tau) F_t(\tau) \tag{6}$$

and the Green tensor

$$B_t(\tau) = F_t(\tau) F_t^{\mathrm{T}}(\tau) \tag{7}$$

---

[1] Tensors and vectors will be given bold-face symbols throughout this chapter.

where F^T denotes the transpose of F. The inverse of the Cauchy and Green tensors are called the Finger and Piola tensors, respectively. The Cauchy and Piola tensors relate the separation of two neighboring material points between the two states (times) τ and t, while the Finger and Green tensors relate the way small material areas change upon deformation between the two states τ and t. Note that if the motion consists solely of a rigid rotation then $C = B = I$. Since strain is a measure of deformation which vanishes in the absence of stretch, we can therefore also define relative strain tensors as $C_t(\tau) - I$ (Cauchy strain tensor), etc.

Turning next to deformation rates, if we let Y_t $(= X_t + dX_t)$ and X_t be two neighboring material particles, using Eq. 4 we can form

$$Y(\tau) - X(\tau) = F_t(\tau)(Y_t - X_t) \tag{8}$$

Taking $d/d\tau$ of Eq. 8, using Eq. 3, and setting $\tau = t$, we get

$$\nabla v(t) = \dot{F}_t(t) \tag{9}$$

To investigate the way ∇v changes along the particle's flow path, we introduce – via Eqs. 3 and 4 – a tensor that is independent of the reference state, viz.

$$L_t(\tau) \equiv \nabla v(f(X_t, \tau)) = \dot{F}_t(\tau) F_t^{-1}(\tau) \equiv L(\tau) \tag{10}$$

Next, using the relative Cauchy tensor, we take $d/d\tau$ and then set $\tau = t$, leading to

$$\dot{C}_t(t) = \nabla v + \nabla v^T \equiv 2D \tag{11}$$

where $D(t)$ is called the rate of deformation tensor. It is symmetric, i.e. $D = D^T$. The velocity gradient tensor can be uniquely decomposed into symmetric plus skew parts via

$$\nabla v = \frac{1}{2}\left(\nabla v + \nabla v^T\right) + \frac{1}{2}\left(\nabla v - \nabla v^T\right) \tag{12}$$

so that we have

$$D(t) = \frac{1}{2}\left(\nabla v + \nabla v^T\right), \qquad W(t) = \frac{1}{2}\left(\nabla v - \nabla v^T\right) \tag{13}$$

where $W(t)$ called the vorticity tensor and is skew, i.e $W = -W^T$.

Eq. 11 identifies $\dot{C}_t(t)$ with twice the rate of deformation tensor. To get accelerations of higher order we can use Eq. 6 to show that (omitting the argument τ and the reference time t)

$$\frac{\partial^n}{\partial \tau^n} C_t(\tau) = \frac{\partial^{n-1}}{\partial \tau^{n-1}}\left[(LF)^T F + F^T(LF)\right] = \frac{\partial^{n-2}}{\partial \tau^{n-2}}\left[F^T A_2 F\right] = \cdots$$
$$= \frac{\partial}{\partial \tau}\left[F^T A_{n-1} F\right] = F^T A_n F \tag{14}$$

where

$$\boldsymbol{A}_{n+1} = \overset{\triangle}{\boldsymbol{A}_n} \equiv \dot{\boldsymbol{A}}_n + \left(\boldsymbol{A}_n \boldsymbol{L} + \boldsymbol{L}^{\mathrm{T}} \boldsymbol{A}_n\right);$$
$$n = 0, 1, 2, \ldots, \quad (\boldsymbol{A}_0 = \boldsymbol{I}, \ \boldsymbol{A}_1 = 2\boldsymbol{D}, \ldots) \tag{15}$$

defines the Rivlin–Ericksen tensors, $\boldsymbol{A}_n$, and the time derivative operation known as the 'lower convected' derivative, $\overset{\triangle}{\boldsymbol{A}_n}$. Similarly, if we start with the Finger tensor, $\boldsymbol{C}_t^{-1}(\tau)$, we can generate

$$\frac{\partial^n}{\partial \tau^n} \boldsymbol{C}_t^{-1}(\tau) = -\boldsymbol{F}^{-1} \boldsymbol{B}_n \boldsymbol{F}^{-T} \tag{16}$$

where

$$\boldsymbol{B}_{n+1} = \overset{\triangledown}{\boldsymbol{B}_n} \equiv \dot{\boldsymbol{B}}_n - \left(\boldsymbol{L}\boldsymbol{B}_n + \boldsymbol{B}_n \boldsymbol{L}^{\mathrm{T}}\right);$$
$$n = 0, 1, 2, \ldots, \quad (\boldsymbol{B}_0 = -\boldsymbol{I}, \ \boldsymbol{B}_1 = 2\boldsymbol{D}, \ldots) \tag{17}$$

are called the White–Metzner tensors, $\boldsymbol{B}_n$, and the time derivative operation known as the 'upper convected' derivative, $\overset{\triangledown}{\boldsymbol{B}_n}$.

Time derivatives play a central role in rheology. As seen above, the upper and lower convected derivatives fall out naturally from the deformation tensors. The familiar 'partial' derivative, $\partial/\partial t$, corresponds to an observer with a fixed position. The 'total' derivative, $\mathrm{d}/\mathrm{d}t$, allows the observer to move freely in space, while if the observer follows a material point we have the 'material', or 'substantial' derivative, denoted variously by the symbols $\mathrm{d}_{(m)}/\mathrm{d}t$, $\mathrm{D}/\mathrm{D}t$ or $(\dot{\ })$. We could expect that these different expressions could find their way into constitutive relations (see Section 5) as time rates of change of quantities that are functions of spatial position and time. However, only certain rate operations can be used by themselves in constitutive relations. This will depend on how two different observers who are in rigid motion with respect to each other measure the same quantity. The expectation is that a valid constitutive relation should be invariant to such changes in observer. This principle is called 'material frame indifference' or 'material objectivity', and constitutes one of the main tests that a proposed constitutive relation has to pass before being considered admissible.

Kinematics deals with time and distances. To measure time one needs to have a particular instant as a reference. Similarly, the motion of a body remains ambiguous unless it is relative to something else, e.g. to two objects, or points, with a fixed distance separating them. In effect, such a collection of points can be thought of as an observer who is watching the motion, or as a 'frame of reference'. Since such a frame is constructed out of points with fixed separations, it can only undergo rigid body motions, i.e. translations and rotations. Thus, let $X(t)$, $c(t)$ and o be points referred to an 'old' frame and $X^*(t^*)$ be the transform of $X(t)$ into a 'new' frame such that

$$\boldsymbol{X}^* = \boldsymbol{c} + \boldsymbol{Q}(\boldsymbol{X} - \boldsymbol{o}), \quad t^* = t + t_0 \tag{18}$$

where t_0 is a constant and $\boldsymbol{Q} = \boldsymbol{Q}(t)$ is a rotation. The transform given by Eq. 18 is called a change of frame. The difference of two points is a geometric vector and would transform as

$$X^* - Y^* = u^* = \boldsymbol{Q}(t)[X - Y] = \boldsymbol{Q}(t)u \tag{19}$$

Vectors which transform according to Eq. 19 under a change of frame are called 'indifferent' or 'objective' vectors. A tensor that acts on an indifferent vector to produce another indifferent vector is called an indifferent tensor. Indifferent vectors and tensors can appear in constitutive equations by themselves without violating the principle of material objectivity. Summarizing, objective scalars, vectors and tensors undergo a change of frame according to the rules given respectively by

$$\alpha^* = \alpha, \quad \boldsymbol{a}^* = \boldsymbol{Q}(t)\boldsymbol{a}, \quad \boldsymbol{A}^* = \boldsymbol{Q}(t)\boldsymbol{A}\boldsymbol{Q}^{\mathrm{T}}(t) \tag{20}$$

Next, let us briefly focus on dynamics. In specifying the dynamic state of a body we will be concerned with *contact* and *body forces*. Contact forces arise by the interaction of material particles in different parts of the body and by the interaction of the body with its surroundings across its boundary. Body forces are those that the surroundings exert on the interior of the body. The primary example of a body force is gravity.

We follow Cauchy and assume ([16], p. 97) the existence of *a surface force density* $s(\boldsymbol{n},\boldsymbol{x},t)$ defined as a force per unit area for each unit normal $\boldsymbol{n}$ at every $(\boldsymbol{x},t)$ along the trajectory of the body's motion. The vectors s and $\boldsymbol{n}$ can have, in general, different directions and the 'surface' can be any oriented surface cut through the body. The two sides of the surface are designated as 'negative' and 'positive' in such a way that $\boldsymbol{n}$ points from the negative into the positive side. $s(\boldsymbol{n},\boldsymbol{x},t)$ is exerted across the surface by the material on the positive side upon that on the negative side. Thus, if we visualize a part of the body enclosed by a surface with unit *outward* normal $\boldsymbol{n}$, then s is the positive (tensile) force per unit area exerted by the surroundings unto the part. For points $\boldsymbol{x}$ that are on the boundary of the body, s is called the surface traction. For body forces we define the field $\boldsymbol{b}(\boldsymbol{x},t)$ as the force per unit volume exerted by the environment on the interior points $\boldsymbol{x}$ of the body.

Cauchy's theorem states ([16], p. 101) that given a body in motion under the action of surface (i.e. contact) and body force fields s and $\boldsymbol{b}$, respectively, then the momentum balance laws are satisfied if and only if there exists a *symmetric* spatial tensor field $\boldsymbol{\sigma}$, called the Cauchy stress, such that

$$s(\boldsymbol{n}) = \boldsymbol{\sigma}\boldsymbol{n} \tag{21}$$

for each unit vector $\boldsymbol{n}$, where $\boldsymbol{\sigma}$ satisfies the balance of linear momentum ('equation of motion' — see Section 3). The symmetry of $\boldsymbol{\sigma}$ is equivalent to the balance of angular momentum ([16], p. 105). With the tension-positive sign convention, a hydrostatic pressure is negative. It is useful to separate out an isotropic part from the total stress tensor that is associated with a 'pressure', p, which for incompressible fluids is undetermined to within an arbitrary scalar

$$\sigma = -p\boldsymbol{I} + \boldsymbol{S} \tag{22}$$

For each given isochoric flow problem, p must then be determined by the balance laws and the particular boundary conditions of that problem, while the extra stress tensor $\boldsymbol{S}$ must be supplied by a constitutive equation for the material. The Cauchy and extra stress tensors are both objective.

3. Balance laws [2]

Let $\boldsymbol{r}$, t, $\boldsymbol{v}$ denote the position vector, time and the velocity vector respectively, and take the density $\rho(\boldsymbol{r},t)$ of the body $\mathcal{B}(t)$ to be a non-negative field. Conservation of mass requires that the mass be constant over time. Using the General Transport Theorem ([28], p. 19) to differentiate the integral, we thus must have

$$\frac{\mathrm{d}}{\mathrm{d}t} \int_{\mathcal{B}(t)} \rho \, \mathrm{d}V = \int_{\mathcal{B}(t)} [\dot{\rho} + \rho \operatorname{tr}(\boldsymbol{L})] \, \mathrm{d}V = 0 \tag{23}$$

or, locally,

$$\dot{\rho} + \rho \operatorname{tr}(\boldsymbol{L}) = \dot{\rho} + \rho (\nabla \cdot \boldsymbol{v}) = 0 \tag{24}$$

Eq. 24 is usually referred to as the 'continuity equation'. If $s(\boldsymbol{n},\boldsymbol{r},t)$ is the surface traction and $\boldsymbol{b}(\boldsymbol{r},t)$ the body force per unit volume, then we need to satisfy the balance of linear momentum and the balance of angular momentum. The former can be stated as

$$\frac{\mathrm{d}}{\mathrm{d}t} \int_{\mathcal{B}(t)} \rho \boldsymbol{v} \, \mathrm{d}V = \int_{\partial \mathcal{B}(t)} s \, \mathrm{d}A + \int_{\mathcal{B}(t)} \boldsymbol{b} \, \mathrm{d}V \tag{25}$$

where the traction is related to the unit normal $\boldsymbol{n}$ to the body's surface $\partial \mathcal{B}(t)$ via Eq. 21. Application of the divergence theorem on the integral balance in Eq. 25 yields the local form for the balance of linear momentum

$$\rho \dot{\boldsymbol{v}} = \nabla \cdot \sigma + \boldsymbol{b} \tag{26}$$

Eq. 26 is commonly referred to as the 'equation of motion'. The balance of angular momentum, in local form, leads to the requirement that the total stress tensor be

---

[2] I am grateful to W.E. VanArsdale for providing me with a compact summary of the balance laws from which this section is taken.

symmetric, i.e.

$$\boldsymbol{\sigma} = \boldsymbol{\sigma}^{\mathrm{T}} \tag{27}$$

Next, the balance of energy requires that

$$\dot{E} = H + W \tag{28}$$

where $E(t)$ is the energy, $W(t)$ the mechanical power and $H(t)$ the thermal power of the body. The local form of Eq. 28, usually called the 'energy equation', is given by

$$\rho \dot{u} = \boldsymbol{\sigma} \cdot \boldsymbol{L} + \phi - \nabla \cdot \boldsymbol{q} \tag{29}$$

where u is the specific internal energy, $\phi(\boldsymbol{r}, t)$ is the body *heat source density* and $\boldsymbol{q}(\boldsymbol{r}, t)$ is the outward heat flux vector. We generally assume further that

$$u = cT \quad \text{and} \quad \boldsymbol{q} = -k\nabla T \tag{30}$$

where c is the specific heat, k the thermal conductivity and $T(\boldsymbol{r}, t)$ (≥ 0) the temperature field. Next, letting s denote the specific entropy, the Clasius–Duhem inequality is given, in local form, by the relation

$$\rho \dot{s} \geq (\phi / T) - \left[\nabla \cdot (\boldsymbol{q} / T) \right] \tag{31}$$

Eq. 31 is the recognition that there is a limit to the rate at which heat may be converted into energy in the absence of mechanical work. If we introduce the specific Helmholtz free energy, a, given by

$$a = u - sT \tag{32}$$

into Eq. 31 and use the energy balance, we get the alternate expression

$$-\rho \dot{a} - \rho s \dot{T} + \boldsymbol{\sigma} \cdot \boldsymbol{L} - (\boldsymbol{q} \cdot \nabla T) / T \geq 0 \tag{33}$$

This inequality is satisfied if

$$-(\boldsymbol{q} \cdot \nabla T) \geq 0 \quad \text{(Fourier's inequality)} \tag{34}$$

and

$$\mathcal{D} \equiv -\rho \left(\dot{a} + s \dot{T} \right) + \boldsymbol{\sigma} \cdot \boldsymbol{L} \geq 0 \quad \text{(Planck's inequality)} \tag{35}$$

The first of these inequalities says that heat flow is in the opposite direction of the temperature gradient, while the second says that the internal dissipation rate $\mathcal{D}$ should be non-negative. In an isothermal process we are left only with the requirement that

$$-\rho \dot{a} + \boldsymbol{\sigma} \cdot \boldsymbol{L} \geq 0 \quad \text{('dissipation inequality')} \tag{36}$$

The equations of continuity, motion and energy are used routinely in solving flow problems. The Planck or dissipation inequalities are used to ensure that a proposed constitutive relation does not violate thermodynamics.

4. Linear viscoelasticity

Between the extremes of viscous fluids and elastic solids are materials that seem to exhibit both traits. These are called viscoelastic materials or memory fluids, and their dual nature becomes most evident when we subject them to time-dependent (unsteady) tests. The three major types of unsteady tests are the so-called relaxation, creep and dynamic tests. In the previous sections, we gave definitions and descriptions for stress, strain and deformation rates. These quantities are now used in defining the various unsteady tests. Thus, in a relaxation test the sample is subjected to a sudden, constant, strain. The stress shoots up in response and then gradually decays ('relaxes'). In the creep test, a sudden stress is applied and held constant. Now the strain picks up quickly and then, while continuing to increase, slows down on its rate of increase. We say the material 'creeps' under the constant stress. In dynamic tests, one confining wall is made to move periodically with respect to another. One monitors both the strain and the stress as a function of time.

In the above unsteady tests, if one keeps the level of imposed stress and strain low enough, the measured material functions show an independence from these applied stimuli levels, exhibiting only a dependence on time (or frequency). This type of response indicates linear viscoelastic behavior. The primary modes of deformation employed in these tests are either *shear* or *extension*. If there is no volume change accompanying the deformation, a single *modulus* or *compliance*, whether real or complex, but a function of time (or frequency) and temperature only, suffices to characterize the material behavior. We will define moduli and compliances further below. Let us now start examining these and other key topics in linear viscoelasticity.

4.1. Relaxation and creep tests

Without any differentiation between shear and elongation, let σ and γ denote stress and strain, respectively. We are thus ignoring the tensorial nature of these quantities for the time being. One way to present the relaxation and creep functions is via the spring/dashpot mechanical analog models of viscoelasticity. Of these, the so-called Maxwell model consists of a spring with a constant elastic modulus G, hooked up in series with a dashpot having a constant viscosity μ. When we pull at the ends of the assembly, a common stress, σ, is experienced by the two elements in series. In response to this, the spring undergoes a strain γ_s while the dashpot undergoes a strain γ_d. The time derivative of the total strain is

$$\dot{\gamma} = \dot{\gamma}_s + \dot{\gamma}_d = \dot{\sigma}/G + \sigma/\mu \tag{37}$$

In a relaxation test, we impose an instantaneous strain that is then held constant, i.e.

$$\gamma(t) = \gamma(0) = \text{constant} \quad \Rightarrow \quad \dot{\gamma} = 0 \tag{38}$$

so that Eqs. 37 and 38 can be solved to give the stress in a single Maxwell element in relaxation as

$$\sigma(t) = G\gamma(0)e^{-t/\lambda} = \sigma(0)e^{-t/\lambda}, \qquad \lambda \equiv \mu/G = \text{constant} \tag{39}$$

where λ is called the characteristic relaxation time of the assembly. From Eq. 39, we see that λ is the time required for the stress to decay to $1/e$ of its initial value. For the creep test where $\sigma(t) = \sigma(0) = \text{constant}$, it is easy to show that Eq. 37 now leads to the following strain in a single Maxwell element in creep

$$\gamma(t) = \frac{\sigma(0)}{G}\left(1 + \frac{t}{\lambda}\right) = \gamma(0)\left(1 + \frac{t}{\lambda}\right) \tag{40}$$

A quite different response than the single Maxwell model may be obtained from the so-called single element 'Voigt model' which now consists of a spring and dashpot in parallel. When this element is pulled, its components share a common strain, γ, which leads to different stresses σ_s and σ_d in the spring and dashpot, respectively. Now the total stress is

$$\sigma = \sigma_s + \sigma_d = G\gamma + \mu\dot{\gamma} \tag{41}$$

It is straightforward to show that Eq. 41 leads to the stress in single Voigt element in relaxation as

$$\sigma(t) = G\gamma(0) \tag{42}$$

and to the strain in creep as

$$\gamma(t) = \frac{\sigma(0)}{G}\left(1 - e^{-t/\lambda}\right) \tag{43}$$

In the Voigt creep test, the initial value of the strain, $\gamma(0)$, is zero since the dashpot will prevent the spring from moving instantaneously. At long times, Eq. 43 predicts that the strain will saturate at a value

$$\gamma(\infty) = \sigma(0)/G \tag{44}$$

i.e. there is a finite ultimate deformation. The time constant λ, again given by Eq. 39_3, now represents the time required for the strain to grow within $1/e$ of its ultimate value and is referred to as the characteristic retardation time. Fig. 1 depicts the two single element models in relaxation and creep.

The single element Maxwell in relaxation and the single element Voigt in creep display some of the features of real viscoelastic materials subjected to these tests. However, the single element model predictions are usually too crude — actual responses tend to be more gradual. It is possible to improve the model predictions by combining the best features of both models and by adding more and/or new types of elements (e.g. mass for inertia and a friction block for yield stress).

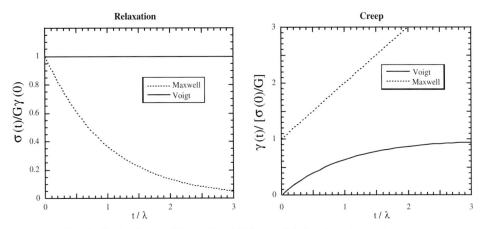

Fig. 1. Single element Maxwell and Voigt models in relaxation and creep.

4.2. Moduli and compliances

In a relaxation test, we would expect a response in the form

$$\sigma(t)/\gamma_0 = \varphi(t) + G_\infty \tag{45}$$

where $\varphi(t)$ is a non-increasing function called the 'relaxation function', G_∞ is the 'equilibrium modulus' ($= 0$ for liquids) and γ_0 is the applied (constant) initial strain. The factor $\sigma(t)/\gamma_0$ is the 'relaxation modulus'. The equilibrium modulus accounts for solid-like response, hence we can let $\varphi(\infty) = 0$. At time $t = 0$, when the step strain γ_0 is imposed, the material responds with an 'instantaneous modulus' $[\varphi(0) + G_\infty]$ which is characteristic of an elastic material. In a creep experiment, the response would be of the form

$$\gamma(t)/\sigma_0 = J_0 + \psi(t) + (t/\eta) \tag{46}$$

where $\psi(t)$ is a non-decreasing function called the 'creep function', J_0 is the 'instantaneous compliance', σ_0 is the applied (constant) initial stress and η is the viscosity. The factor $\gamma(t)/\sigma_0$ is the 'creep compliance'. Since the initial response is incorporated into the instantaneous compliance we can let $\psi(0) = 0$. The equilibrium compliance, J_∞, which corresponds to the maximum recoverable (i.e. elastic) deformation, $\gamma(\infty)$, is given by

$$J_\infty = J_0 + \psi(\infty) \tag{47}$$

In a dynamic experiment whose oscillations are sinusoidal we can again have a relaxation or creep test. If we express the imposed strain (relaxation test) as

$$\gamma(t) = \gamma_0 e^{i\omega t} \tag{48}$$

where ω is the radial frequency and $i = \sqrt{-1}$, then the stress response will have a part that is in phase with the strain — the elastic part — and one which is out of phase with the strain (or, alternatively, in phase with $\dot{\gamma}(t)$ — the viscous part)

$$\sigma(t) \equiv G_\infty \gamma(t) + \hat{\sigma}_0 e^{i(\omega t + \delta)} \tag{49}$$

where δ (referred to as the 'loss angle') is the phase difference between the stress and strain. The quantity σ/γ is known as the 'complex modulus', G^*

$$G^* = \sigma(t)/\gamma(t) = G_\infty + (\hat{\sigma}_0/\gamma_0) e^{i\delta} \equiv (G_\infty + G') + iG'' \tag{50}$$

where

$$G' = (\hat{\sigma}_0/\gamma_0) \cos\delta = \text{' storage modulus'},$$
$$G'' = (\hat{\sigma}_0/\gamma_0) \sin\delta = \text{'loss modulus'} \tag{51}$$

so that

$$\tan\delta = G''/G' = \text{'loss tangent'} \tag{52}$$

The reader should note that some authors introduce the storage modulus differently — by burying the G_∞ term into the definition of G' — in which case, Eq. 50 assumes the form

$$G^* \equiv G' + iG'' \tag{50a}$$

Next, if we impose a sinusoidal stress (dynamic creep test)

$$\sigma(t) = \sigma_0 e^{i\omega t} \tag{53}$$

then the response will be

$$\gamma(t) \equiv J_0 \sigma(t) + \hat{\gamma}_0 e^{i(\omega t - \delta)} + \int_0^t [\sigma(t')/\eta] \, dt' \tag{54}$$

where we have again taken the out-of-phase component in such a way that the strain lags the stress. The last term in Eq. 54 is the viscous flow component that originates from a viscous fluid response only. Thus we can define the 'complex compliance' as

$$J^* = \gamma(t)/\sigma(t) = J_0 + (\hat{\gamma}_0/\sigma_0) e^{-i\delta} - i(1/\omega\eta)$$
$$= (J_0 + J') - i[(1/\omega\eta) + J''] = 1/G^* \tag{55}$$

where

$$J' = (\hat{\gamma}_0/\sigma_0) \cos\delta = \text{'storage compliance'},$$
$$J'' = (\hat{\gamma}_0/\sigma_0) \sin\delta = \text{'loss compliance'} \tag{56}$$

and

$$\tan \delta = J''/J' \tag{57}$$

We can also define the 'complex viscosity' by

$$\eta^* = \sigma(t)/\dot{\gamma}(t) = \left(\sigma_0 e^{i\omega t}\right) \bigg/ \frac{\mathrm{d}}{\mathrm{d}t}\left[\gamma_0 e^{i(\omega t - \delta)}\right] = G^*/(i\omega)$$
$$= \left[\sigma_0/(\gamma_0\omega)\right](\sin \delta - i\cos \delta) \equiv \eta' - i\eta'' \tag{58}$$

where

$$\eta' = \left[\sigma_0/(\gamma_0\omega)\right]\sin \delta = G''/\omega, \quad \eta'' = \left[\sigma_0/(\gamma_0\omega)\right]\cos \delta = G'/\omega \tag{59}$$

In Eq. 59, η' is also called the dynamic viscosity. In Eq. 55, J_0 is the instantaneous compliance (sometimes called the glassy compliance and denoted by J_g) and is associated with solid-like elastic behavior while $1/\omega\eta$ is associated with liquid-like viscous behavior, which leaves J' and J'' as the terms that are thus associated with viscoelastic behavior. Again, the reader should note that some authors introduce the dynamic creep components differently, by burying these terms into the definitions of J' and J''. If one goes along with the latter definitions, Eq. 55 becomes

$$J^* \equiv J' - iJ'' \tag{55a}$$

and we get the component interrelations as

$$\left.\begin{array}{ll} G' = J'/\left(J'^2 + J''^2\right), & G'' = J''/\left(J'^2 + J''^2\right) \\ J' = G'/\left(G'^2 + G''^2\right), & J'' = G''/\left(G'^2 + G''^2\right) \end{array}\right\} \tag{60}$$

In the definitions given by Eqs. 50a and 55a, the liquid or solid cases are to be implicitly understood from the context. Thus, if we are dealing with a solid (e.g. a cross-linked rubber), the $1/\omega\eta$ term would drop out of the compliance relations and G' will have a part due to G_∞.

4.3. Boltzmann superposition principle

We can interrelate the relaxation and creep functions and the dynamic moduli and compliances via Boltzmann's superposition principle which states that all effects of past history can be considered independently in their contributions to the present state of the (linear) viscoelastic material. Thus, if one subjects the material to, say, incremental strains $(\gamma_0 - 0)$, $(\gamma_1 - \gamma_0)$, . . . , $(\gamma_n - \gamma_{n-1})$ at times $t_0, t_1, \ldots,$ t_n, then the total stress $\sigma(t)$ at time t will be

$$\sigma(t) = \sigma(t - t_0) + \sigma(t - t_1) + \cdots + \sigma(t - t_n) \tag{61}$$

Generalizing, one has

$$\sigma(t) = \int_{\gamma(-\infty)}^{\gamma(t)} [\varphi(t-t') + G_\infty] d\gamma = \int_{-\infty}^{t} [\varphi(t-t') + G_\infty] \dot{\gamma}(t') dt' \tag{62}$$

and similarly

$$\gamma(t) = \int_{\sigma(-\infty)}^{\sigma(t)} \left[\psi(t-t') + \frac{t-t'}{\eta} + J_o\right] d\sigma$$

$$= \int_{-\infty}^{t} \left[\psi(t-t') + \frac{t-t'}{\eta} + J_o\right] \dot{\sigma}(t') dt' \tag{63}$$

Eqs. 62 and 63 are known as the Boltzmann–Volterra integral equations and are usually given in the form

$$\sigma(t) = \int_{-\infty}^{t} G(t-t')\dot{\gamma}(t') dt', \qquad \gamma(t) = \int_{-\infty}^{t} J(t-t')\dot{\sigma}(t') dt' \tag{64}$$

where $G(t)$ and $J(t)$ are simply called the steady relaxation modulus and steady creep compliance, respectively.

4.4. Time–temperature superposition

For a homogeneous viscoelastic material that is confined to a temperature range within which there is no phase change, the effects of time and temperature can be interchangeable. In other words, if we apply a stress or strain to the material, we can choose either to wait for it to relax or creep at the fixed test temperature or we can get the same response faster by raising the test temperature. This principle is called time–temperature superposition and is widely used in linear viscoelasticity. Taking into account any small variation of the plateau modulus, we can express the principle mathematically (say, in steady shear) as

$$G(t,T) = \left[G_N^0(T)/G_N^0(T_{\text{ref}})\right] G(t/a_T, T_{\text{ref}}) \equiv b_T G(t/a_T, T_{\text{ref}}) \tag{65}$$

where T_{ref} is a reference temperature and a_T is called the 'shift factor'. Now if we take

$$G_N^0 = \rho R T/M_e \tag{66}$$

where M_e is the entanglement molecular weight, ρ the density and R the universal gas constant, we have

$$b_T = G_N^0(T)/G_N^0(T_{\text{ref}}) = \rho T/\rho_{\text{ref}} T_{\text{ref}} \tag{67}$$

The function b_T is called the 'vertical shift factor' and is usually close to one. If the temperature range is very wide or if the material has a peculiar architecture (e.g. highly branched polymers) vertical shifting may become important. According to Eq. 65, isothermal data taken at different temperatures and plotted as $G(t)$ vs. t, can be shifted horizontally along the time axis to produce a single 'master curve' at T_{ref}. As we shall see later, the same thing can also be done when working with dynamic data to produce a master curve of the components of the complex modulus (say). In the latter case, one would start out with isothermal frequency (ω) sweep data in the form of $G'(\omega)$, $G''(\omega)$ vs. ω taken at different temperatures and the curves would be shifted horizontally along the ω-axis to produce the master curve at T_{ref}.

If $T_g < T < T_g + 100$ K then one finds that most amorphous polymers obey the Williams–Landel–Ferry (WLF) equation ([13], p. 303)

$$\log a_t = \frac{-c_1 (T - T_{ref})}{c_2 + (T - T_{ref})} \tag{68}$$

where c_1 and c_2 are constants. If we choose the reference temperature to be the glass transition temperature, T_g, then the constants assume almost universal values of

$$c_1 \simeq 17.4, \quad c_2 \simeq 51.6 \text{ K} \quad \text{at} \quad T_{ref} = T_g \tag{69}$$

At temperatures greater than approximately $T_g + 100$ K, one frequently finds that the shift factor data can be fitted better via the Arrhenius form

$$\ln a_T = \frac{\Delta E_{ref}}{R} \left(\frac{1}{T} - \frac{1}{T_{ref}} \right) \tag{70}$$

where the activation energy difference, ΔE_{ref}, is generally only a weak function of temperature. The shift factor is usually obtained experimentally, via horizontal shifting, without having to rely on the universal constants of WLF. It is defined as

$$a_T = \lambda_i (T) / \lambda_i (T_{ref}) \tag{71}$$

where λ_i refers to the i-th relaxation time of the material. In different regimes (e.g. dilute solution, melts) the ratio in Eq. 71 may be given by different expressions. For instance, for dilute polymer solutions

$$a_T = \lambda_i (T) / \lambda_i (T_{ref}) = (T_{ref} / T) \left\{ [(\eta_0 - \eta_s)/c]_T / [(\eta_0 - \eta_s)/c]_{T_{ref}} \right\} \tag{72}$$

where c is polymer concentration, η_0 is the zero-shear viscosity of the polymer and η_s is the solvent viscosity. In a concentrated polymer–solvent system, on the other hand, molecular theories suggest that

$$a_T = \lambda_i (T) / \lambda_i (T_{ref}) = (T_{ref} / T) \left\{ [a^2 \varsigma_0]_T / [a^2 \varsigma_0]_{T_{ref}} \right\} \tag{73}$$

where a^2 denotes the root-mean-squared end-to-end distance between monomer units and ς_0 stands for the monomeric friction coefficient. In fact, the key concept

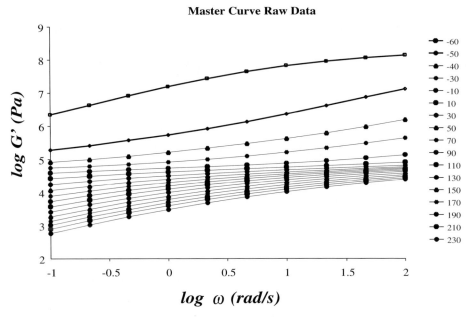

Fig. 2. Storage modulus, isothermal frequency sweep data.

behind the shift factor is that the monomeric friction factor scales all motions of
the macromolecule. If we recall that η_o scales with the longest relaxation time
and is also proportional to the monomeric friction coefficient, then the ratio of
relaxation times at two different temperatures will essentially scale as the ratio of
the respective monomeric friction coefficients at those two temperatures, i.e. the
dynamics at one temperature will scale as the shift factor times the dynamics at
the other temperature.

The linear viscoelastic master curve of a material serves as an important
'fingerprint' for its mechanical behavior and the fine features of these master
curves correlate with the particular materials' molecular details. For these reasons,
master curves are widely generated in practice. Below, we illustrate an example
[38] of master curve generation where the original data were taken under dynamic
testing. Fig. 2 shows the data, Fig. 3 shows the master curve obtained by means of
the shift factor calculated from the data in Fig. 2. Finally, Fig. 4 shows a plot of
the shift factor that is seen to display WLF-type behavior.

Since time–temperature shifting, as illustrated in Figs. 2–4, enjoys widespread
use in linear viscoelasticity, it is important to have a consistent set of criteria
for its validity. It is advisable that one have nearly exact matching of shapes of
adjacent curves with over more than half-range overlap, that the shift factor have
a reasonable form (e.g. WLF, Arrhenius) and possess the same value for all of
the viscoelastic functions. A sharp test of how well one has time–temperature

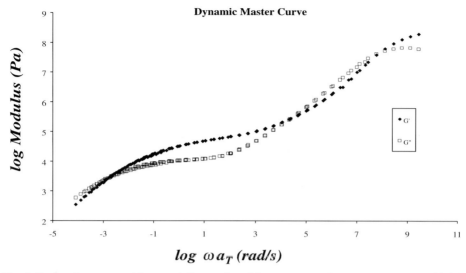

Fig. 3. Reduced storage and loss moduli vs. reduced frequency at reference temperature of 30°C.

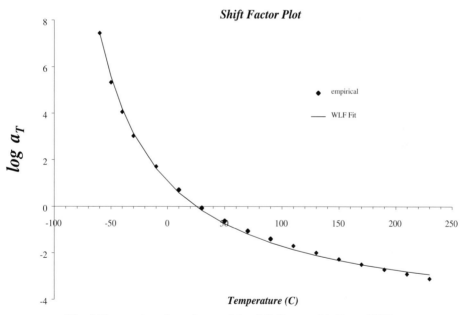

Fig. 4. Temperature dependence of the shift factor, with $T_{\mathrm{ref}} = 30°C$.

shifting is perhaps to re-plot the (say) usual $\log(G', G'')$ vs. $\log(\omega a_T)$ data in the alternate form of $\tan\delta$ $(= G''/G')$ vs. $\log|G^*|$. While the former plots may look convincing, the latter eliminates the horizontal shift and should yield a temperature-independent curve. Furthermore, one can now directly see the amount of vertical shifting, while this information lays hidden in the $\log(G',G'')$ vs. $\log(\omega a_T)$ plot.

4.5. Relaxation and retardation spectra

The Maxwell and Voigt models contain one, constant, characteristic time given by $\lambda = \mu/G$. If we construct a model with N (>1) elements, each with their own characteristic times $\lambda_i = \mu_i/G_i$, $i = 1, 2, \ldots, N$, then we might expect better fits to data. Experience shows that viscoelastic materials may have a large number of characteristic times that could also be approximated as a continuous spectrum. In the latter case, letting the single element Maxwell relaxation behavior to be our guide, we can define

$$\varphi(t) \equiv \int_0^\infty F(\lambda)e^{-t/\lambda}\,d\lambda \equiv \int_{-\infty}^\infty H(\ln\lambda)e^{-t/\lambda}\,d\ln\lambda, \quad (H(\ln\lambda) = \lambda F(\lambda)) \quad (74)$$

where $F(\lambda)$ and $H(\ln\lambda)$ are called the relaxation spectrum and the logarithmic relaxation spectrum, respectively. Similarly, taking the single element Voigt creep behavior as our guide, we have

$$\psi(t) \equiv \int_0^\infty \Phi(\lambda)\left[1 - e^{-t/\lambda}\right]d\lambda \equiv \int_{-\infty}^\infty L(\ln\lambda)\left[1 - e^{-t/\lambda}\right]d\ln\lambda,$$

$$(L(\ln\lambda) = \lambda\Phi(\lambda)) \quad (75)$$

where $\Phi(\lambda)$ and $L(\ln\lambda)$ are called the retardation spectrum and the logarithmic retardation spectrum, respectively. Using the relaxation and retardation spectra we can establish exact relations between the various moduli and compliances.

4.6. Exact relationships

When characterizing a material in the linear viscoelastic range, one may not always be able to perform all the individual tests needed for the direct evaluation of the desired material functions. For instance, suppose we are interested in the long-time creep behavior, i.e. we would like to obtain the steady creep compliance, $J(t)$. This is ideally measured in a torque-controlled device by applying a constant stress and measuring the resulting deformation. Let us assume, on the other hand, that all we have available is a displacement-controlled device from which we have obtained the steady relaxation modulus, $G(t)$. Using the exact relationship between these two material functions, we would be in a position to calculate the desired creep compliance from the given relaxation modulus without necessarily having to do a creep test.

Another frequently encountered situation is where one may have dynamic data in the frequency domain and wishes to calculate material functions in the time domain, or vice versa. For instance, suppose a material is sensitive to long thermal exposures that might be necessary to measure a steady property. We could instead perform a series of short frequency sweeps during which we may be reasonably assured of the material's thermal stability, and then use that dynamic data to calculate the desired steady property.

As alluded to above, in linear viscoelasticity, given one material function, it is theoretically possible to calculate all the other material functions. Since the theory is well established and available in numerous references (see, e.g. [13], Chapt. 3) we will not give a listing of these relations, but encourage the reader to make use of them as needed.

5. Constitutive equations

Everyday experience with common liquids teaches us that they have no preferred equilibrium configuration so that, at rest, they take the shape of the container they happen to be in. If we stir such a fluid, the drag will depend on how rapidly it is stirred and not on how 'much' we deform it (e.g. how many revolutions one stirs). Conversely, the resistance one feels in stretching a rubber band depends on how far it is stretched and not on how fast this is done. Furthermore, there is a preferred shape — the unstretched length — that the material likes to return to.

The liquid is an example of a viscous fluid while the rubber band is an example of an elastic solid. There are materials with mechanical responses that span the entire range between these extremes. Mechanical constitutive equations are relations between the *dynamics* (stresses and their time rates of change) and *kinematics* (deformations and their time rates of change) of materials. As such they provide closure to the balance equations (see Section 3), thereby allowing the solution to a specific mechanical problem involving a specific material.

It is unrealistic to expect one 'universal' constitutive equation to work satisfactorily in all classes of mechanical problems for all materials. In fact, even a single material may require employing different constitutive equations for solving different problems. Hence, one tries to come up with constitutive equations that address a sufficiently broad class of problems and/or are applicable for sufficiently broad classes of idealized material behavior.

A further perspective into such classifications may be gained by considering the element of time. In the case of the familiar 'silly putty' which bounces like a rubber ball, but flows when left alone, the crucial factor that decides between its solid or liquid-like response is the time scale of the experiment. This leads naturally to the definition of a dimensionless parameter called the Deborah number, viz.

$$N_{\mathrm{De}} \equiv \lambda/t_{\mathrm{p}} \tag{76}$$

where λ is a characteristic time of the material and t_{p} is a characteristic time scale for the process during which a significant change in kinematics occurs. Hence, when we bounce the silly putty we might expect that $t_{\mathrm{p}} \ll \lambda$, and when we let it flow, $t_{\mathrm{p}} \gg \lambda$. This identifies $N_{\mathrm{De}} \gg 1$ with elastic and $N_{\mathrm{De}} \ll 1$ with viscous behavior, respectively. The characteristic time of a fluid can range from the order of tens of microseconds (e.g. multigrade mineral oils under engine operation temperatures) to seconds (e.g. polymer melts). It decreases with increasing temperature and increases with increasing pressure and molecular weight. Linear viscoelasticity yields a quantity which has the dimensions of time and which also has been experimentally correlated with the primary normal stress coefficient Ψ_1 (see further below) and the shear viscosity η at vanishing shear rates. This quantity is given by Eq. 77, below, and one could certainly use it as a characteristic time of the material, viz.

$$\lambda = \int_0^\infty sG(s)\,\mathrm{d}s \Bigg/ \int_0^\infty G(s)\,\mathrm{d}s = \left(\eta''/\omega\right)/\eta'\big|_{\omega\to 0} = \Psi_1/(2\eta)\big|_{\dot\gamma\to 0} \tag{77}$$

Another possible candidate for λ may be the reciprocal of the shear rate at which the shear viscosity attains 80% of the zero-shear value in a shear viscosity vs. shear rate plot (the so-called 'flow curve'). This has been shown [24] to be the point of maximum curvature of the flow curve for the Cross model (see further below).

Most polymers will have a spectrum of relaxation times, of which the longest ones will play a major role in determining how much the material can relax in a given time interval after the imposition of a loading. One can look at this relaxation as the materials' way of 'erasing' or 'forgetting' the imposed disturbance. In that sense, the larger the λ, the longer is the 'memory' of the material. Thus, in a viscous liquid with a short relaxation time there would be hardly any memory at all and the stress might be expected to depend entirely on the current rate of deformation. In contrast, the rubber band mentioned earlier, which always tries to get back to its original length may be thought of as having a very long memory. We could then classify the 'in-between' materials as possessing differing degrees of memory and refer to them, collectively, as memory fluids. For such materials we would expect the response to depend on the history of the stimuli.

Of the many continuum models available, the simplest non-Newtonian ones are the viscous fluids. These are liquids for which the stress depends exclusively on the *current* value of the rate of deformation tensor. The so-called 'generalized Newtonian fluid' (GNF) is a sub-class of viscous fluids. The GNF is used primarily in steady shear flows that are important in processing equipment. Thus, the GNF models try to accurately capture the details of the way the shear viscosity depends on the shear rate, i.e. the materials' *flow curve*, to get reliable estimates of the volumetric flow rate under given pressure drops or vice versa. Several GNF models will now be covered since they are fairly important in engineering calculations ([4], Chapt. 4). However, we will limit our discussion of how other constitutive equations based on continuum mechanics can be constructed (using some of the ideas mentioned above) to merely a verbal sketch of the mathematical process. We will also not discuss constitutive equations based on molecular modeling, since that too would be beyond the scope of this chapter. Finally, we will briefly look at complex materials such as those with a yield strength and time-dependency.

5.1. Viscous models

A fairly simple model of a liquid can be constructed by assuming that the stress depends only on the current value of the rate of deformation

$$\boldsymbol{\sigma} = \boldsymbol{\sigma}(\boldsymbol{A}_1); \quad \boldsymbol{A}_1 = 2\boldsymbol{D} \tag{78}$$

where $\boldsymbol{A}_1$ is the first Rivlin–Ericksen kinematic tensor (see Section 2). Material objectivity (see Section 3) requires that $\boldsymbol{\sigma}$ be an isotropic tensor function, so that it can be represented by ([16], p. 233)

$$\boldsymbol{\sigma}(\boldsymbol{A}_1) = \mu_0 \boldsymbol{I} + \mu_1 \boldsymbol{A}_1 + \mu_2 \boldsymbol{A}_1^2 \tag{79}$$

where the coefficients can depend on the principal invariants of $\boldsymbol{A}_1$ which are given by

$$i_1(\boldsymbol{A}_1) = \mathrm{tr}(\boldsymbol{A}_1) = 0; \quad i_2(\boldsymbol{A}_1) = -\frac{1}{2}\mathrm{tr}(\boldsymbol{A}_1^2); \quad i_3(\boldsymbol{A}_1) = \frac{1}{3}\mathrm{tr}(\boldsymbol{A}_1^3) \tag{80}$$

Further, the dissipation inequality (see Section 3) requires that

$$\mu_1 \mathrm{tr}(\boldsymbol{A}_1^2) + \mu_2 \mathrm{tr}(\boldsymbol{A}_1^3) \geq 0 \tag{81}$$

The isotropic term in Eq. 79 can be lumped with the pressure, giving the Reiner–Rivlin fluid (RR)

$$\boldsymbol{\sigma} = -p\boldsymbol{I} + \mu_1 \boldsymbol{A}_1 + \mu_2 \boldsymbol{A}_1^2 \tag{82}$$

with μ_1 and μ_2 subject to Eq. 81 and depending only on the invariants $i_2(\boldsymbol{A}_1)$ and $i_3(\boldsymbol{A}_1)$. In steady simple shear flow with the magnitude of the shear rate given by $\dot{\gamma}$, the RR fluid yields the following so-called 'viscometric functions' (note: 1 is the flow direction, 2 is the direction along the velocity gradient and 3 is the neutral direction)

$$\left.\begin{array}{l} \eta = \sigma_{12}/\dot{\gamma} = \mu_1(i_2, i_3) = \mu_1(-\dot{\gamma}^2, 0) \\ \Psi_1 = (\sigma_{11} - \sigma_{22})/\dot{\gamma}^2 = 0 \\ \Psi_2 = (\sigma_{22} - \sigma_{33})/\dot{\gamma}^2 = \mu_2(i_2, i_3) = \mu_2(-\dot{\gamma}^2, 0) \end{array}\right\} \tag{83}$$

From Eq. 83 we observe that the viscometric functions are insensitive to the direction of shear and that the primary normal stress coefficient is zero. Hence this is not a realistic model for most shear sensitive fluids. Eq. 82, with $\mu_1 = \mu =$ constant and $\mu_2 = 0$ is the Newtonian fluid. If we keep μ_1 shear rate dependent and set $\mu_2 = 0$, we then have the GNF. Several special cases of the GNF are discussed below.

5.1.1. Elbirli–Yasuda–Carreau (EYC) model

This is a five-parameter model given by the expression

$$(\eta - \eta_\infty)/(\eta_0 - \eta_\infty) = \left[1 + (\lambda\dot{\gamma})^a\right]^b \tag{84}$$

where all the parameters are positive constants. η_0, η_∞ are the zero shear rate and 'infinite' shear rate viscosities, λ is a time constant, b is a constant related to the slope of the flow curve, i.e. the power-law exponent n (see also, below) and a is a constant which determines the breadth of the transition from the Newtonian plateau to the power-law region. In order to have the same high shear asymptote as the power-law model one sets $b = (n-1)/a$. An example of a flow curve for polymer melts and the fit using Eq. 84 (with $\eta_\infty = 0$) is presented in Fig. 5.

As can be seen from Fig. 5, the EYC model allows excellent fits to typical melt flow curves. For this reason, the model has found widespread use in calculations such as in the modeling of flows in molds, runners and dies. We can show that the flow rate-pressure drop relation for the EYC fluid (with $\eta_\infty = 0$) under steady laminar flow in a pipe of radius R is given by

$$Q = \frac{\pi R^3}{3\lambda}\left[\Psi_R^{1/a} - \frac{1}{a}\left(\frac{\eta_0}{S_R\lambda}\right)^3 \int_0^{\Psi_R} x^{\frac{4}{a}-1}(1+x)^{\frac{3(n-1)}{a}}\,dx\right]; \quad \left(\Psi_R = (\lambda\dot{\gamma}_R)^a\right) \tag{85}$$

where $\dot{\gamma}_R$ denotes the shear rate at the wall.

5.1.2. Power law (PL) model

The PL model is a non-Newtonian relation that is widely used in engineering calculations. It is given by

$$\eta = m\dot{\gamma}^{n-1} \tag{86}$$

where the parameters m and n are positive constants, called the consistency index and power law index, respectively. When $n < 1$ the fluid is called pseudoplastic and when $n > 1$, it is called dilatant. The PL has no inherent time constant. We note that on a $\log\eta$ vs. $\log\dot{\gamma}$ plot, Eq. 86 gives a straight line with slope $(n-1)$. Looking at Fig. 5 we see that at the higher shear rates, the flow curve does indeed present a straight line. It is only this portion of the flow curve that Eq. 86 is

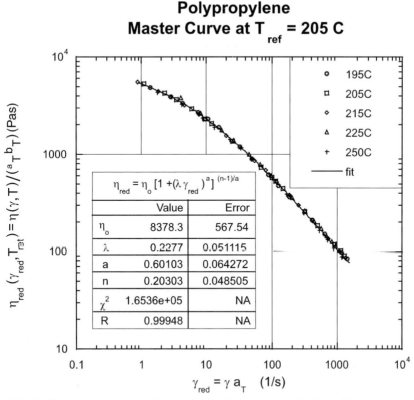

Fig. 5. Flow curve data for polypropylene and the fit using Eq. 84 (with $\eta_\infty = 0$).

intended for. In principle, this restriction can lead to problems if one wishes to use the PL relation in flow modeling of process equipment, since there will invariably be regions away from walls where the shear rate drops down to low values where Eq. 86 no longer applies. If we fit a materials' flow curve by Eq. 84 (with $\eta_\infty = 0$) and let τ_w denote the shear stress at the wall of a particular flow geometry, the criterion for the validity of using the PL relation in modeling the flow is

$$\tau_w \gg \frac{\eta_0}{\lambda} \tag{87}$$

5.1.3. Truncated power law (TPL) model

This is a three-parameter model given by

$$\eta = \begin{cases} \eta_0; & \dot{\gamma} \leq \dot{\gamma}_0 \\ \eta_0 \, (\dot{\gamma}/\dot{\gamma}_0)^{n-1}; & \dot{\gamma} \geq \dot{\gamma}_0 \end{cases} \tag{88}$$

A relatively 'quick-fix' for the low shear rate limitation of the PL is to use Eq. 88. On a $\log \eta$ vs. $\log \dot{\gamma}$ plot, Eq. 88 gives two straight lines with slopes equal to zero and $(n-1)$ which cross at $\dot{\gamma} = \dot{\gamma}_0$. The parameter $\dot{\gamma}_0$ thus serves as a value of the shear rate above which 'significant' shear thinning sets in and its inverse serves as a time constant. If we fit a materials' flow curve by Eq. 88, the criterion for the validity of using the PL relation in modeling the flow would now be

$$\tau_w \gg \eta_0 \dot{\gamma}_0 \tag{89}$$

5.1.4. Ellis model

This fluid is given by

$$\eta_0/\eta = 1 + (\tau/\tau_{1/2})^{\alpha-1}; \quad \tau = \sqrt{(\boldsymbol{\tau} \cdot \boldsymbol{\tau})/2} \tag{90}$$

The Ellis fluid parameters allow the construction of a time constant given by $(\eta_0/\tau_{1/2})$. One of the advantages of the Ellis model is that the low to high shear rate range can be spanned continuously, without the need to 'patch together' two separate equations (cf. TPL model). On the other hand, the shear rate dependence of viscosity now becomes implicit (via the magnitude of the shear stress, τ).

5.1.5. Cross model

If we set $b = -1$ in Eq. 84 (EYC), we obtain the four parameter Cross model. Except for those applications involving dilute solutions, one rarely has a need for the high shear rate asymptote, η_∞, of the shear viscosity. Indeed, most polymer melts will undergo melt fracture before this second plateau is ever reached. For polymer melts one therefore has the simpler three parameter Cross model which can still span the Newtonian plateau to the power law region in one equation

$$\eta/\eta_0 = \left[1 + (\lambda\dot{\gamma})^a\right]^{-1}; \quad a = 1 - n \tag{91}$$

Various isothermal and non-isothermal flow problems have been solved with the above model fluids. Bird et al. [4] (p. 229) lists some of these solutions for rectangular slits and circular pipes.

5.2. Elastic and viscoelastic models

We will now give a brief sketch of some of the other continuum models beyond the simple viscous ones considered above. The purpose here is to show how these models are interconnected and allow the reader to gain a broader scope of the subject without going into the mathematical details.

First, staying within the confines of a viscous fluid, the next level of generality beyond that of the RR fluid (see Eqs. 78 and 82) would be to allow the stress to depend on higher rates of deformation tensors while maintaining the restriction that these tensors be evaluated at the current time. Including the first n Rivlin–

Ericksen tensors in this way leads to the so-called Rivlin–Ericksen (RE) fluid. Part of the complexity of the RE constitutive equation (say, staying with the case of A_1 and A_2 dependency, only) is due to keeping the full representation of the stress tensor as an isotropic tensor-valued function in A_1 and A_2. It is possible to simplify this relation by considering an alternative approach. Now one essentially looks at the problem as a perturbation expansion for slow flows. Thus, at rest, the stress tensor is given by the isotropic hydrostatic pressure only. The first order 'correction' includes an additional term proportional to A_1, which gives us the Newtonian fluid. At second order, we would include the 'square' terms only, viz.

$$\boldsymbol{\sigma} = -p\boldsymbol{I} + \mu_0 \boldsymbol{A}_1 + \mu_1 \boldsymbol{A}_1^2 + \mu_2 \boldsymbol{A}_2 \tag{92}$$

Eq. 92 is the so-called 'second order fluid'. Proceeding in this fashion yields the class of constitutive relations known as order fluids. They are meant to account for memory effects in an incremental way as one goes to higher orders. Being a subset of the RE fluid, they suffer similar shortfalls. A special member of this class of fluids is the Criminale–Ericksen–Filbey Fluid ([4], p. 503) that is specifically designed for viscometric flows, with three material functions taken as the three viscometric functions themselves, i.e. η, Ψ_1 and Ψ_2 that are functions of the shear rate.

At the other extreme end from a viscous liquid we have the isotropic, elastic solid. Now, instead of the deformation rate, we would expect the stress to depend on deformation. As a deformation measure we can take $\boldsymbol{C}$ or $\boldsymbol{B}$ or their inverses, or any combination of them. Working with $\boldsymbol{B}$, say, we let

$$\boldsymbol{\sigma} = \boldsymbol{\sigma}(\boldsymbol{B}); \quad \boldsymbol{B} = \boldsymbol{F}\boldsymbol{F}^{\mathrm{T}} \tag{93}$$

Applying the principle of material objectivity plus the dissipation inequality, we are led to the constitutive equation for an incompressible *elastic solid*

$$\boldsymbol{\sigma} = -p\boldsymbol{I} + 2\rho \frac{\partial \hat{a}}{\partial i_1} \boldsymbol{B} - 2\rho \frac{\partial \hat{a}}{\partial i_2} \boldsymbol{B}^{-1}; \quad a(\boldsymbol{B}) \equiv \hat{a}(i_1(\boldsymbol{B}), i_2(\boldsymbol{B}), i_3(\boldsymbol{B})) \tag{94}$$

where a is the specific Helmholtz free energy (see Section 3), ρ the density and i_1, i_2 are the first and second principal invariants of $\boldsymbol{B}$. In the reference configuration $\boldsymbol{B} = \boldsymbol{I}$, and the stress reduces to a pressure so that one can choose the unstressed state of the material as the reference configuration.

In the isotropic solid, the unstressed state is chosen as the *fixed* reference configuration. One way to allow for fluid-like behavior is then to let the unstressed configuration *evolve* in time. VanArsdale [31] has developed constitutive equations for suspensions using this concept of elastic fluids. Working with the deformation tensor $\boldsymbol{B}^{-1} = \boldsymbol{F}^{-1\mathrm{T}}\boldsymbol{F}^{-1}$, VanArsdale uses a new deformation tensor b that is similar to $\boldsymbol{B}^{-1}$ but which, instead, characterizes the deformation from the evolving unstressed state. The evolution is expressed by the relation for the lower convected derivative of b as

$$\overset{\Delta}{b} = 2\Sigma(b); \quad \Sigma = \Sigma^{\mathrm{T}} \tag{95}$$

where $\Sigma(b)$ is a symmetric tensor called the slippage tensor. For the elastic solid $\Sigma = 0$. Now the constitutive equation for an elastic fluid can be given as

$$\sigma = -pI + S(b); \quad a = a(b); \quad \overset{\Delta}{b} = 2\Sigma(b) \tag{96}$$

where $S(b)$, $a(b)$ and $\Sigma(b)$ must be specified for each material. Material objectivity requires that b be an objective tensor, so that $S(b)$, $a(b)$ and $\Sigma(b)$ are isotropic functions. The dissipation inequality also places restrictions on these functions. We let

$$a(b) \equiv \hat{a}(i_1(b), i_2(b), i_3(b)) \tag{97}$$

and introduce another symmetric isotropic tensor $d(b)$, which is the rate of deformation tensor for the evolving unstressed configuration such that

$$\Sigma(b) = bd(b) \tag{98}$$

Several well-known constitutive equations can now be generated from the above equations. Thus, if one picks

$$S(b) = G_0\left(b^{-1} - I\right); \quad d = \frac{1}{2G_0\lambda}Sb \tag{99}$$

where G_0 and λ are positive constants, then we get

$$\overset{\triangledown}{S} + \frac{1}{\lambda}S = G_0 A_1 \tag{100}$$

where A_1 is the first Rivlin–Ericksen tensor. Eq. 100 is the constitutive equation for the upper convected Maxwell fluid, (UCM). The material constants G_0 and λ are the modulus and characteristic time, respectively. Recalling the simple spring and dashpot in a series model (the Maxwell element) from linear viscoelasticity, we see that Eq. 100 is a properly 'upgraded' tensorial version of the scalar Maxwell model where the indifferent upper convected time derivative has replaced the partial derivative with respect to time — an operation which does not preserve objectivity. It is possible to integrate Eq. 100 to obtain the integral version of the same constitutive relation as

$$S(t) = -\int_{-\infty}^{t} M\left(t - t'\right) \gamma_0\left(t, t'\right) \mathrm{d}t' \tag{101}$$

with

$$\gamma_0\left(t, t'\right) = -F(t)F^{-1}\left(t'\right)F^{-1\mathrm{T}}\left(t'\right)F^{\mathrm{T}}(t) = -F(t)C^{-1}\left(t'\right)F^{\mathrm{T}}(t) \tag{102}$$

and

$$M\left(t-t'\right) = \frac{\partial}{\partial t'}G\left(t-t'\right), \quad G\left(t-t'\right) = G_0 e^{-(t-t')/\lambda} \tag{103}$$

$M(t-t')$ is called the memory function. Eq. 101 is the constitutive relation for the Lodge rubber-like liquid (LRL).

If we assume that the initial stress $S(-\infty)$ is a pressure, then, to within isotropic terms, we can generalize Eq. 102 to include incompressible fluids. These obey the so-called the Rivlin–Sawyers model given by

$$S(t) = \int_{-\infty}^{t} \left\{ M_1\left[t-t',i_1\left(\gamma_0\right),i_2\left(\gamma_0\right)\right]\gamma_0\left(t,t'\right) \right.$$
$$\left. + M_2\left[t-t',i_1\left(\gamma_0\right),i_2\left(\gamma_0\right)\right]\gamma_0^{-1}\left(t,t'\right)\right\} dt' \tag{104}$$

Wagner [32–34] assumed that

$$M_1\left[t-t',i_1\left(\gamma_0\right),i_2\left(\gamma_0\right)\right] = -M\left(t-t'\right)h\left[i_1\left(\gamma_0\right),i_2\left(\gamma_0\right)\right]; \quad M_2 = 0 \tag{105}$$

where h is called the damping function. The separability of the memory function indicated by Eq. 105_1 has been observed experimentally for a variety of polymer melts. The memory function $M(t-t')$ can be identified with the linear viscoelastic memory function and the damping function accounts for finite strain non-linear behavior. When $h = 1$, the Wagner model reduces to the LRL. Various semi-empirical forms have been proposed for the damping function [22], as, for instance

$$h = \sum_{k=1}^{K} f_k \exp\left(-n_k\sqrt{\alpha i_1+(1-\alpha)i_2-3}\right); \quad \sum_{k=1}^{K} f_k = 1 \tag{106}$$

where K is a positive integer and the parameters f_k, n_k and α have to be extracted from large strain data.

The LRL can also be obtained via network theories. Since $h = 1$ can thus be given a network interpretation, it is natural to consider the Wagner fluid in this light, as well. In this view, as the strain increases, temporary junctions in the network disentangle. The probability of an entanglement to survive a relative deformation between the instant of creation, t', and the instant of observation, t, can then be equated with the damping function. If one further stipulates that junctions lost in an increasing deformation are not re-formed when the deformation is decreased — say, during a recovery experiment — then the strain-dependent disentanglement becomes an irreversible process. Wagner's irreversible model expresses this irreversibility constraint on the damping function as

$$h\left[i_1,i_2\right] = \min_{t'\leq t''\leq t}\left\{h\left[i_1\left(t'',t\right),i_2\left(t'',t\right)\right]\right\} \tag{107}$$

where i_1 and i_2 are the invariants of $\boldsymbol{C}_t^{-1}(t')$. In Eq. 104, by analogy with elastic solids, if we set

$$M_1 = -2\rho\frac{\partial}{\partial i_1}U\left[i_1(\gamma_0),i_2(\gamma_0)\right]; \quad M_2 = 2\rho\frac{\partial}{\partial i_2}U\left[i_1(\gamma_0),i_2(\gamma_0)\right] \tag{108}$$

where $U\left[i_1(\gamma_0),i_2(\gamma_0)\right]$ is a specific energy rate, then we obtain the Kaye–Bernstein, Kearsley, Zapas (K-BKZ) constitutive equation.

In the limit of small elastic deformations one can define the strain tensor

$$2\boldsymbol{e} = -\ln(\boldsymbol{b}) = \sum_{n=1}^{\infty}\frac{(\boldsymbol{I}-\boldsymbol{b})^n}{n}; \quad |\boldsymbol{I}-\boldsymbol{b}| < 1 \tag{109}$$

Again making use of the slippage tensor, this time we are led to an evolution equation involving the corotational derivative

$$\overset{\circ}{\boldsymbol{e}} = \dot{\boldsymbol{e}} + \boldsymbol{e}\boldsymbol{W} - \boldsymbol{W}\boldsymbol{e} = \boldsymbol{D} - \boldsymbol{\Sigma} - \boldsymbol{e}\boldsymbol{\Sigma} - \boldsymbol{\Sigma}\boldsymbol{e} + o(\boldsymbol{e}); \quad \boldsymbol{W} = skw(\boldsymbol{L}) \tag{110}$$

and, for incompressible materials, to the constitutive equation

$$\boldsymbol{S} = 2G_1\boldsymbol{e}; \quad \boldsymbol{\Sigma} = \frac{\boldsymbol{e}}{\lambda_1}; \quad \mathrm{tr}(\boldsymbol{e}) = 0 \tag{111}$$

where G_1 and λ_1 are constants. If we eliminate $\boldsymbol{e}$, Eq. 111 leads to the corotational Maxwell model

$$\overset{\circ}{\boldsymbol{S}} + \frac{1}{\lambda_1}\boldsymbol{S} = 2G_1\boldsymbol{D} \tag{112}$$

For a viscous fluid we let the stress depend on $\boldsymbol{D}$ and for the elastic fluid we let it depend, via the slippage tensor, on $\boldsymbol{b}$. It would therefore be natural to explore the consequences of assuming

$$\boldsymbol{\sigma} = -p\boldsymbol{I} + \boldsymbol{S}(\boldsymbol{b},\boldsymbol{D}); \quad a = a(\boldsymbol{b},\boldsymbol{D}); \quad \overset{\triangle}{\boldsymbol{b}} = 2\boldsymbol{\Sigma}(\boldsymbol{b},\boldsymbol{D}) \tag{113}$$

This should yield a viscous, elastic fluid. Material objectivity leads to the requirement that $\boldsymbol{S}(\boldsymbol{b},\boldsymbol{D})$, $a(\boldsymbol{b},\boldsymbol{D})$ and $\boldsymbol{\Sigma}(\boldsymbol{b},\boldsymbol{D})$ be isotropic functions in both arguments. If deformation rates and elastic deformations are small, for isochoric deformations one can take

$$\left.\begin{array}{l} \boldsymbol{S} = 2G\boldsymbol{e} + 2\mu\boldsymbol{D} \\ \boldsymbol{\Sigma} = \dfrac{\boldsymbol{e}}{\lambda} - \beta\boldsymbol{D} \end{array}\right\}; \quad \mathrm{tr}(\boldsymbol{D}) = 0; \quad \mathrm{tr}(\boldsymbol{e}) = 0 \tag{114}$$

where G, μ, λ and β are constants. The deformation obeys the evolution equation

$$\overset{\circ}{\boldsymbol{e}} = (1+\beta)\boldsymbol{D} - \frac{\boldsymbol{e}}{\lambda} + o(\boldsymbol{e}) \tag{115}$$

This leads to the constitutive relation

$$\overset{\circ}{\boldsymbol{S}} + \frac{1}{\lambda}\boldsymbol{S} = 2\mu\overset{\circ}{\boldsymbol{D}} + \frac{2[G\lambda(1+\beta)+\mu]}{\lambda}\boldsymbol{D} \tag{116}$$

which is called the corotational Jeffreys model. Note that for $\mu = 0$, Eq. 116 reduces to the corotational Maxwell model.

So far we have started with the viscous fluid and elastic solid limiting cases and tried to generate constitutive equations that bring in the fluids' memory in an incremental way. Another approach would be to tackle the full memory effects by starting out with sufficient generality. The incompressible simple fluid (ISF) of Noll [11] does this by allowing the stress to depend on not just the current value, but on the entire history of the local deformation. It is 'simple' in the sense that only the deformation history of the immediate neighborhood of a material particle serves to determine the state of stress of that particle. Otherwise, the constitutive equation itself can hardly be called simple; on the contrary, it is far too general to be used in most flow problems, so that one has to resort to simplifying assumptions — via flow classifications — to make it tractable. The ISF is given by the relation

$$\left.\begin{array}{l} \boldsymbol{\sigma} = -p\boldsymbol{I} + \overset{\infty}{\underset{s=0}{\mathcal{F}}} \left[\boldsymbol{F}_t(t-s) \right] \\ \det[\boldsymbol{F}_t(t-s)] = 1 \end{array}\right\} \tag{117}$$

where

$$\boldsymbol{F}^t(s) \equiv \boldsymbol{F}_t(t-s); \quad s \geq 0 \tag{118}$$

is called the 'history' of the relative deformation gradient tensor, s is the time lag extending out into the past from the current time t and $\mathcal{F}_{s=0}^{\infty}[\boldsymbol{F}_t(t-s)]$ is a tensor-valued functional whose argument is the family of all possible tensorial history functions. The histories are subject to the constraint given by Eq. 117$_2$ which ensures isochoric deformations.

Since the ISF is a very general constitutive relation, it should be applicable to a large class of materials under different flows. Any rigorous solutions to flow problems using it are therefore likely to have relevance to a wide class of materials. For instance, if we make the flow problem trivially simple by letting it to be a fluid at rest, then Eq. 117 predicts that the stress is hydrostatic. We know that this is true for most materials. It is not true, however, for other materials such as those with a yield stress that are rigid at rest or for granular media; such 'complex' materials do not fall under simple fluids and we will discuss them separately, later. Fortunately, there are non-trivial flow problems where rigorous solutions using Eq. 117 are possible, and we shall look at some of these next.

In steady shear flow (also called simple shear flow) it can be shown that the ISF predicts the components of the stress tensor to be of the form

$$[\boldsymbol{\sigma}] = \begin{bmatrix} \sigma_{11} & \sigma_{12} & 0 \\ \sigma_{12} & \sigma_{22} & 0 \\ 0 & 0 & \sigma_{33} \end{bmatrix} \tag{119}$$

We can define the so-called viscometric functions by

$$
\left.
\begin{aligned}
\sigma_{12} &= \tau(\dot{\gamma}) = \eta(\dot{\gamma})\dot{\gamma} \\
\sigma_{11} - \sigma_{22} &= N_1(\dot{\gamma}) = \Psi_1(\dot{\gamma})\dot{\gamma}^2 \\
\sigma_{22} - \sigma_{33} &= N_2(\dot{\gamma}) = \Psi_2(\dot{\gamma})\dot{\gamma}^2
\end{aligned}
\right\}
\tag{120}
$$

where τ, N_1, N_2 are called the shear stress and the primary and secondary normal stress differences, while η, Ψ_1, Ψ_2 are called the shear viscosity, primary and secondary normal stress coefficients, respectively. Furthermore, we can also show that

$$
\left.
\begin{aligned}
\tau(-\dot{\gamma}) &= -\tau(\dot{\gamma}); & \eta(-\dot{\gamma}) &= \eta(\dot{\gamma}) \\
N_1(-\dot{\gamma}) &= N_1(\dot{\gamma}); & \Psi_1(-\dot{\gamma}) &= \Psi_1(\dot{\gamma}) \\
N_2(-\dot{\gamma}) &= N_2(\dot{\gamma}); & \Psi_2(-\dot{\gamma}) &= \Psi_2(\dot{\gamma})
\end{aligned}
\right\}
\tag{121}
$$

so that, if the direction of the flow is reversed, the shear stress should change sign but the normal stress differences should be insensitive to the direction of flow. While it is important in its own right, simple shear flow belongs to a larger class of flows, called viscometric flows, which are widely used in rheological characterization of polymers and other materials. It can be shown that the stress tensor in viscometric flows is again of the form given by Eq. 119 and that the rheological state of the material under these flows is completely determined by specifying η, Ψ_1, Ψ_2.

Viscometric flows, in turn, are a sub-class of a larger class of flows for which it is possible to generate rigorous solutions using the ISF. These are called motions with constant stretch history (MCSH). In a MCSH the stretch history is independent of the instant of observation, t, but instead depends solely on the time lag, $s = t - \tau$, $\infty < \tau \le t$. Noll has shown that a flow is a MCSH if and only if the relative deformation gradient, relative to some fixed time $t = 0$, has the form

$$
\boldsymbol{F}_0(\tau) = \boldsymbol{Q}(\tau)e^{\tau M}; \quad \boldsymbol{Q}^{\mathrm{T}}\boldsymbol{Q} = \boldsymbol{I}; \quad \boldsymbol{Q}(0) = \boldsymbol{I}
\tag{122}
$$

where $\boldsymbol{M}$ is a constant tensor. Defining

$$
\boldsymbol{L}_1(t) \equiv \boldsymbol{Q}(t)\boldsymbol{M}\boldsymbol{Q}(t)^{\mathrm{T}}
\tag{123}
$$

one can classify MCSH into three broad categories

$$
\left.
\begin{aligned}
\text{I)} &\quad \boldsymbol{L}_1^2 = \boldsymbol{0}, \quad \boldsymbol{L}_1 \ne \boldsymbol{0} \\
\text{II)} &\quad \boldsymbol{L}_1^3 = \boldsymbol{0}, \quad \boldsymbol{L}_1^2 \ne \boldsymbol{0} \\
\text{III)} &\quad \boldsymbol{L}_1^n \ne \boldsymbol{0}, \quad n = 1, 2, \dots
\end{aligned}
\right\}
\tag{124}
$$

Flows in category I are called viscometric flows; they include simple shear, Poiseuille and plane Poiseuille, Couette, helical, torsional and cone-and-plate flows. Category II flows are called fourth order flows; they include superposition of category I flows in curvilinear orthogonal coordinate systems when the components of the metric tensor do not change along the path line of each particle. Category III contains extensional flows.

Extensional flows are important in polymer characterization (e.g. sensitivity of extensional viscosity to long chain branching), polymer processing (e.g. fiber spinning, film blowing, coating, converging/diverging sections in dies, etc.) and in constitutive modeling for being a severe test on the predictive powers of constitutive equations. Slattery [29] has provided an analysis of *unsteady* extensional flows for the ISF that also allows one to classify *steady* extensional flows. Following that analysis, we define an unsteady extensional flow by the velocity field

$$v_i = (a_i x_i + V_i) A(t); \qquad \text{(no sum on } i) \tag{125}$$

where x_i are the Cartesian components of the position vector at time t, v_i the corresponding components of the velocity vector, $A(t)$ is an arbitrary function of time and a_i, V_i are constant vectors. For an incompressible fluid undergoing the above flow, the continuity and momentum equations lead to the pressure field

$$-p = \rho\phi + \rho \sum_{i=1}^{3} \left[\left(\frac{1}{2} a_i x_i^2 + V_i x_i \right) \left(\frac{dA}{dt} + a_i A^2 \right) \right] + f(t) \tag{126}$$

where $f(t)$ is an arbitrary function of time and the body force potential ϕ is given by

$$g = -\nabla\phi \tag{127}$$

when the body forces, ρg, are conservative (e.g. gravitational). The particle paths are obtained by solving

$$\left. \begin{aligned} -\frac{d\xi_i}{ds} &= (a_i \xi_i + V_i) A(t-s) \\ \xi_i|_{s=0} &= x_i \end{aligned} \right\}; \qquad \text{(no sum on } i) \tag{128}$$

which, upon integration, yields

$$\xi_i = -\frac{V_i}{a_i} + \left(x_i + \frac{V_i}{a_i} \right) \exp\left[a_i \int_0^s A(t-u)\,du \right]; \qquad \text{(no sum on } i) \tag{129}$$

The particle paths allow one to calculate the relative Cauchy tensor which leads to the stress tensor

$$\left. \begin{aligned} \sigma_{ii} &= \rho \sum_{i=1}^{3} \left[\left(\frac{1}{2} a_i x_i^2 + V_i x_i \right) \left(\frac{dA}{dt} + a_i A^2 \right) \right] + \alpha_0 + \alpha_1 a_i \int_0^s A(t-u)\,du \\ &\quad + \alpha_2 \left[a_i \int_0^s A(t-u)\,du \right]^2 + \rho\phi + f(t) \\ \sigma_{ij} &= 0; \quad i \neq j \end{aligned} \right\} \tag{130}$$

To classify steady extensional flows, we note that Eq. 130 can be expressed as

$$\left.\begin{array}{l} \sigma_{ii} + p = a_i H + a_i^2 L; \quad (\text{no sum on } i) \\ \sigma_{ij} = 0; \quad i \neq j \end{array}\right\} \tag{131}$$

where the isotropic terms have been merged and H and L are material functions of the invariants given by

$$I_1 = \sum_{i=1}^{3} a_i = 0, \quad I_2 = \sum_{i=1}^{3} a_i^2, \quad I_3 = \sum_{i=1}^{3} a_i^3 \tag{132}$$

For Newtonian fluids with constant viscosity μ

$$H = 2\mu, \quad L = 0 \tag{133}$$

Starting from the above expression, one can classify steady extensional flows into three categories, as shown below.

5.2.1. Uniaxial extension ('simple elongation')

An example would be stretching a cylindrical rod along its longitudinal axis (1-axis) at a constant stretch rate of $\dot{\varepsilon}$, while letting the diameter shrink uniformly. The a_i are given by

$$a_i = \{a_1, a_2, a_3\} = \{\dot{\varepsilon}, -\dot{\varepsilon}/2, -\dot{\varepsilon}/2\} \tag{134}$$

Inserting Eq. 134 into Eq. 131, we get

$$(\sigma_{11} - \sigma_{22})/\dot{\varepsilon} = 3H/2 + 3\dot{\varepsilon}L/4 \equiv \eta_e(\dot{\varepsilon}), \quad \sigma_{22} - \sigma_{33} = 0 \tag{135}$$

Note that substituting the Newtonian values of H and L from Eq. 133 into Eq. 135 yields

$$\eta_e(\dot{\varepsilon})|_{\text{Newtonian}} = 3\mu \tag{136}$$

Furthermore, we observe that under vanishing stretch rates, the uniaxial extensional viscosity, $\eta_e(\dot{\varepsilon})$, tends to its Newtonian value given by

$$\eta_e(\dot{\varepsilon})|_{\dot{\varepsilon} \to 0} = 3\mu \tag{137}$$

5.2.2. Biaxial extension

An example would be stretching a sheet along the 1- and 2-directions at the same rate $\dot{\varepsilon}_b$, while letting its thickness shrink in the 3-direction. The a_i are given by

$$a_i = \{a_1, a_2, a_3\} = \{\dot{\varepsilon}_b, \dot{\varepsilon}_b, -2\dot{\varepsilon}_b\} \tag{138}$$

Substituting this into Eq. 131, we get

$$(\sigma_{11} - \sigma_{33})/\dot{\varepsilon}_b = 3H - 3\dot{\varepsilon}_b L \equiv \eta_b(\dot{\varepsilon}_b), \qquad \sigma_{11} - \sigma_{22} = 0 \tag{139}$$

Now the limiting cases become

$$\eta_b(\dot{\varepsilon}_b)|_{\dot{\varepsilon}_b \to 0} = \eta_b(\dot{\varepsilon}_b)|_{\text{Newtonian}} = 6\mu \tag{140}$$

5.2.3. Planar extension

An example would be stretching a sheet that is clamped at one end along the 1-direction, with constraints along the sides in the 3-direction, while letting it shrink in the 2-direction at the same rate. The a_i are given by

$$a_i = \{a_1, a_2, a_3\} = \{\dot{\varepsilon}_p, -\dot{\varepsilon}_p, 0\} \tag{141}$$

Substituting this into Eq. 131, we get

$$(\sigma_{11} - \sigma_{22})/\dot{\varepsilon}_p = 2H \equiv \eta_p(\dot{\varepsilon}_p), \qquad \sigma_{33} = -p \tag{142}$$

The limiting cases become

$$\eta_p(\dot{\varepsilon}_p)|_{\dot{\varepsilon}_p \to 0} = \eta_p(\dot{\varepsilon}_p)|_{\text{Newtonian}} = 4\mu \tag{143}$$

Beyond the few categories discussed above, it is difficult to work with the ISF in its full form. However, by either assuming slow flows or small deformations, it is possible to obtain approximate expressions in the form of expansions. When the ISF is expanded for slow flows one obtains the order fluids we discussed earlier. When we expand for small deformations, the ISF yields integral relations which account for the fading memory. At the 0th order we again have the hydrostatic pressure, but now at the 1st order we get the constitutive equation for linear viscoelasticity, viz.

$$\boldsymbol{\sigma} = -p\boldsymbol{I} + \int_0^{\infty} f(s)\boldsymbol{G}^t(s)\,\mathrm{d}s; \qquad \boldsymbol{G}^t(s) = \boldsymbol{C}_t(t-s) - \boldsymbol{I} \tag{144}$$

At the 2nd order, one has the constitutive equation for second order viscoelasticity

$$\boldsymbol{\sigma} = -p\boldsymbol{I} + \int_0^{\infty} f(s)\boldsymbol{G}^t(s)\,\mathrm{d}s$$

$$+ \int_0^{\infty}\int_0^{\infty} \{\alpha(s_1, s_2)\,\boldsymbol{G}^t(s_1)\,\boldsymbol{G}^t(s_2) + \beta(s_1, s_2)\,\mathrm{tr}[\boldsymbol{G}^t(s_1)]\,\boldsymbol{G}^t(s_2)\}\,\mathrm{d}s_1\,\mathrm{d}s_2 \tag{145}$$

Eqs. 144 and 145 are the initial terms in the class of constitutive equations known as memory integral expansions. If we now expand $G^t(s)$ in a power series about the instant of observation, t, using

$$G^t(s) = C_t(t-s) = \sum_{n=0}^{\infty} \frac{(-1)^n}{n!} A_n(t) s^n \tag{146}$$

then we recover the retarded motion expansions. If we neglect coupling effects between the deformations experienced by a fluid element at past times $t', t'', \ldots,$ insofar as to their effects on the stress at t is concerned, then one obtains the Rivlin–Sawyers Fluid which we had encountered earlier as a generalization of the LRL. It is equally possible to carry out the above developments in terms of the history of the Finger tensor.

Other useful constitutive relations can be generated by the inclusion of non-linear terms in stress. Thus, if we were to start with the UCM (Eq. 100), and add a term that is quadratic in stress we get

$$S + \lambda_1 \overset{\triangledown}{S} + \frac{\alpha}{G_0} S^2 = \eta A_1 \tag{147}$$

where α is a constant with a value between zero and one. Eq. 147 is a special case of the Giesekus model which considerably improves the predictive powers of the UCM model. Another nonlinear model which has proven useful is the Phan–Thien Tanner (PTT) constitutive relation which is based on a network model. In its multiple relaxation times (multi-mode) version, the i-th stress tensor of the PTT obeys the relation

$$\lambda_i \left\{ \dot{S}_i - \mathcal{L} S_i - S_i \mathcal{L}^T \right\} + Y_i S_i = 2\lambda_i G_i D; \quad \mathcal{L} = L - \xi D;$$

$$\text{(no sum on } i, i = 1, 2, \ldots, N) \tag{148}$$

where λ_i, are the relaxation times, G_i are the moduli, ξ is a constant that is usually between zero and one and the extra stress (for an N-mode model) is given by

$$S = \sum_{i=1}^{N} S_i \tag{149}$$

Two forms have been used for the function Y_i appearing in Eq. 148

$$Y_i = \begin{cases} 1 + (\varepsilon/G_i)\,\mathrm{tr}(S_i) \\ \exp\left[(\varepsilon/G_i)\,\mathrm{tr}(S_i)\right] \end{cases}; \qquad \text{(no sum on } i) \tag{150}$$

where ε is another constant. When ε is equal to zero, the PTT model becomes identical to the so-called Johnson–Segalman constitutive relation. The extensional flow response of the PTT is mainly governed by ε. The model predicts a limiting uniaxial extensional viscosity with stretch rate, in line with observations.

5.3. Complex materials

There are many materials of practical importance, such as suspensions, pastes, gels, powders, etc., whose mechanical behavior show peculiarities beyond those of a typical viscoelastic polymer. Typically, these 'complex' materials are two-phase, solid-fluid mixtures of relatively rigid particles in a suspending fluid and may display time- as well as rate-dependency, yield, compressibility and wall-slip. In analyzing such materials, one can model each phase separately and then combine the information via mixture theories. This approach, while being relatively rigorous, can also be complicated. In composites, with specialized end-use properties that depend critically on the properties and distribution of the individual phases, one may be forced to confront the full complexity of the problem [9] in order to come up with realistic 'effective' property predictions. For most flow problems, a simpler approach is to describe the material as rigid prior to yield, and as viscous during flow [5]. A *yield criterion* is typically introduced to determine the onset of flow. The stress during flow is specified in terms of the deformation rate. One or more scalar variables can be introduced to describe changes in the microstructure that affect flow behavior. However, such scalar measures are incapable of predicting elastic response and normal stress effects for which one may need to employ some tensor measure of particle interactions. Hand [17] introduced such a tensor to describe microscopic structure for suspensions. We will restrict the present discussion to a review of some existing models for two-phase materials within the context of a simple constitutive framework [1] that contains several current models for viscoplastic and thixotropic materials.

Two-phase mixtures of relatively rigid particles in a Newtonian fluid have a limited ability for storing elastic energy. Consequently, these materials may be modeled as strictly *dissipative* during flow [15]. In its simplest form, the stress for such a material depends on the current value of the rate of deformation tensor D. We assume that the stress in a deforming two-phase material is given by

$$\sigma = -pI + ME,$$
$$\left(E \equiv \mathrm{dev}(D)/|\mathrm{dev}(D)|; \quad \mathrm{dev}(D) = D - \mathrm{tr}(D)I/3 \right) \tag{151}$$

where M is a modulus, E is a normalized measure of deformation rate and where 'dev' stands for the deviatoric part of a tensor. The dissipation inequality is satisfied if

$$M \geq 0 \quad \text{and} \quad p\,\mathrm{tr}(D) \leq 0 \tag{152}$$

for all possible D, where the second constraint vanishes for incompressible materials. The modulus M typically depends on $|D|$ for models of viscoplastic materials. An additional dependence on the history of this invariant is characteristic of models for thixotropy. E does not vanish in the absence of flow since its principal values are constrained [1]. In addition, the stress ME does not vanish provided the modulus remains finite as $|\text{dev}(\boldsymbol{D})| \to 0$. This limiting behavior excludes any possibility of the stress vanishing upon the removal of applied loads. A yield criterion can be specified to denote the inception of flow. For example, a material yields according to the von Mises criterion [26] if

$$\lim_{|\text{dev}(\boldsymbol{D})| \to 0} |\text{dev}(\sigma)| = Y_0 \tag{153}$$

in terms of the yield strength Y_0. This condition implies a limiting value

$$\lim_{|\text{dev}(\boldsymbol{D})| \to 0} |M| = Y_0 \tag{154}$$

Several models for viscoplastic materials are contained in the expression

$$M = M_0 \left[\Gamma + (2\lambda |\boldsymbol{D}|)^m \right]^n \tag{155}$$

where the equilibrium modulus M_0 is a constant that must be positive to satisfy the dissipation inequality. The exponents m and n are typically positive to insure realistic limiting values for M at low and high rates of deformation. The gel strength Γ is often normalized to have a value between zero and one. This parameter and the characteristic time λ are usually constant, but can vary with the magnitude $|\boldsymbol{D}|$. The material parameters in Eq. 155 are restricted to values that insure the modulus remains finite as $|\boldsymbol{D}|$ goes to zero. The value of M in this limit determines the yield strength of the material according to the constraint given by Eq. 154. This limiting value depends on the gel strength for materials with a constant characteristic time λ_0. In this context, materials without gel strength do not exhibit yield. This behavior is typical of viscous fluids, such as the Newtonian and Power Law models, given, respectively, by

$$M = 2M_0\lambda_0 |\boldsymbol{D}| \tag{156}$$

$$M = M_0 (2\lambda_0 |\boldsymbol{D}|)^n \tag{157}$$

where the product $M_0\lambda_0$ is a viscosity and n is the power law index. Even a material with gel strength will not yield if Γ vanishes as $|\boldsymbol{D}|$ goes to zero. We expect Γ to decrease as the material's internal structure is disrupted by flow at higher rates. Many of the early models for viscoplastic materials are characterized by a unit gel strength. These include: Bingham [3] ($m = n = 1$), Herschel and Bulkley [19] ($n = 1$), Casson [6] ($m = 0.5$, $n = 2$), and Robertson and Stiff [27] ($m = 1$). In each case, the equilibrium modulus M_0 is equivalent to the yield strength. The various models referred to above are plotted in Fig. 6.

Since Eq. 151 is meant for engineering calculations, it is useful to know the resulting expressions for volumetric flow rates in common geometries. For steady rectilinear flow in the z-direction in a slit of width w and gap h, the volumetric

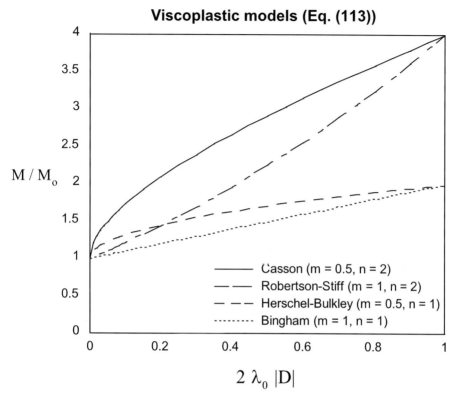

Fig. 6. Several viscoplastic models as given by Eq. 155. Γ is unity, M_0 and λ_0 are constant.

flow rate is

$$Q_{\text{slit}} = -\frac{wh^2}{2}\frac{\mathrm{d}p}{\mathrm{d}z}\left[\alpha + \frac{h}{2H^3 Y_0}\int_1^H \frac{\Gamma^{1/m}}{\lambda}\varsigma\left(\varsigma^{1/n}-1\right)^{1/m}\mathrm{d}\varsigma\right];$$

$$H = \frac{h}{h_p}; \quad |\boldsymbol{v}_{\text{slip}}| = \alpha\tau_{\text{wall}} \tag{158}$$

where α denotes the slip coefficient and $|\boldsymbol{v}_{\text{slip}}|$ is the slip speed tangent to the boundaries while τ_{wall} denotes the magnitude of the shear stress at the wall. Under the special conditions of constant gel strength and characteristic time, it is possible to obtain closed form solutions, with or without wall slip, for both the steady slit and pipe flows of materials obeying Eq. 155 [2]. These slit flow results are summarized in graphical form in Fig. 7 for specific values of the parameters.

For steady flow in a circular cylindrical pipe of radius a, the general expression

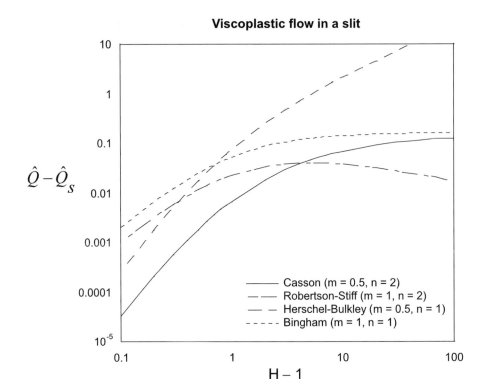

Fig. 7. Volumetric flow rate for viscoplastic flow in a slit. $\hat{Q} = Q\left[\frac{2\lambda_0 Y_0}{wh^3(-dp/dz)}\right]$,

$\hat{Q}_s = \alpha\frac{\lambda_0 Y_0}{h}$ ('s' = slip), $H = \frac{h}{h_p}$.

for the volumetric flow rate is [2]

$$Q_{\text{pipe}} = -\frac{\pi a^3}{2}\frac{dp}{dz}\left[\alpha + \frac{a}{R^4 Y_0}\int_1^R \frac{\Gamma^{1/m}}{\lambda}\varsigma^2\left(\varsigma^{1/n} - 1\right)^{1/m}d\varsigma\right];$$

$$R = \frac{a}{a_p} \tag{159}$$

As in the slit case, special cases can be worked out for the circular cylindrical pipe which are summarized in graphical form in Fig. 8 for specific values of the parameters.

Many of the materials described as viscoplastic also exhibit time-dependent effects associated with a change in structure. This behavior is characterized by a *reversible* decrease in shear viscosity with time under isothermal conditions. Materials that fit this description are called thixotropic and one can describe them using the same constitutive equation suggested for incompressible viscoplastic materials. However now the modulus, Eq. 155, evolves with time through the gel strength Γ and the characteristic time λ. Evolution equations are supplied in the

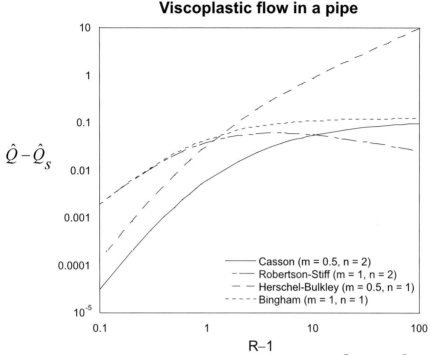

Fig. 8. Volumetric flow rate for viscoplastic flow in a pipe. $\hat{Q} = Q \left[\frac{2\lambda_0 Y_0}{\pi a^4 (-\mathrm{d}p/\mathrm{d}z)} \right]$,

$\hat{Q}_s = \alpha \frac{\lambda_0 Y_0}{a}$ ('s' = slip), $R = \frac{a}{a_p}$.

form

$$\left. \begin{array}{l} \dot{\lambda} = f(\lambda, \Gamma, \boldsymbol{D}) \\ \dot{\Gamma} = g(\lambda, \Gamma, \boldsymbol{D}) \end{array} \right\} \tag{160}$$

In this context, the parameters Γ and λ are scalar measures of internal structure that depend on the history of $\boldsymbol{D}$. Models based on these measures are strictly dissipative if Γ and λ do not contribute to the material's free energy. Cheng and Evans [7] study similar constitutive relations in which the viscosity evolves with time. Their model can be expressed as

$$M = 2M_0 \lambda |\boldsymbol{D}|, \quad \dot{\lambda} = \frac{\lambda_\infty (1+a) - \lambda}{\lambda_1} + 2b|\boldsymbol{D}|(\lambda_\infty - \lambda);$$

$(a, b, \lambda_1, \lambda_\infty \text{ positive constants})$ \hfill (161)

Steady values of λ decrease from $\lambda_\infty (1+a)$ at rest ($|\boldsymbol{D}| = 0$) to λ_∞, as $|\boldsymbol{D}|$ becomes unbounded. Hence, the viscosity $M_0 \lambda$ decreases as structure breaks down at larger deformation rates. This material does not exhibit yield because the gel strength is zero. Cheng [8] has generalized this approach to allow for additional scalar measures of structure. Fredrickson [14] suggest a similar model for thixotropic suspensions based on an evolution equation for the inverse of viscosity. This model can be expressed in the form

$$M = 2M_0\lambda|\boldsymbol{D}|, \quad \dot{\lambda} = \frac{\lambda}{\lambda_0\lambda_1}(\lambda_0 - \lambda) + \frac{a\lambda^2|\boldsymbol{D}|^2}{\lambda_\infty}(\lambda_\infty - \lambda);$$

$$(a, \lambda_0, \lambda_1, \lambda_\infty \text{ positive constants}) \tag{162}$$

Steady values of λ vary from λ_0 at rest to λ_∞ as $|\boldsymbol{D}|$ becomes unbounded. This change corresponds to thixotropic behavior for $\lambda_0 > \lambda_\infty$, while $\lambda_0 < \lambda_\infty$ implies the formation of structure associated with antithixotropy. This material exhibits a yield stress for $\lambda_0 = 0$ since λ goes like $|\boldsymbol{D}|^{-1}$ as $|\boldsymbol{D}|$ goes to 0. The characteristic time becomes equivalent to a gel strength for this choice of parameter values. Slibar and Paslay [30] developed a model to describe drilling mud. The gel strength in this model is given as an integral of the magnitude $|\boldsymbol{D}|$ over time. This integral can be differentiated to obtain the following form of the model

$$M = M_0(\Gamma + 2\lambda_0|\boldsymbol{D}|), \quad \dot{\Gamma} = \frac{\Gamma(1 - \Gamma)}{\lambda_1} - a|\boldsymbol{D}|\Gamma^2;$$

$$(a, \lambda_1 \text{ positive constants}) \tag{163}$$

Steady values of Γ decrease from $\Gamma = 1$ at rest to $\Gamma = 0$ as $|\boldsymbol{D}|$ becomes unbounded. Consequently, this model exhibits yield and the time-dependent behavior characteristic of thixotropic materials. Harris [18] proposes a continuum theory for time-dependent behavior that has some similarities with the Slibar and Paslay model. This theory is based on an integral expression for the difference between viscosity and its rest value following a long time period. This approach can be formulated in terms of the equations

$$M = 2M_0\lambda|\boldsymbol{D}|, \quad \lambda - \lambda_0 = \int_{-\infty}^{t} h(|\boldsymbol{D}|)M(t - t')\,dt' \tag{164}$$

where the characteristic time λ has a rest value λ_0 and $M(t - t')$ is a memory function. The dimensionless function h is determined by $|\boldsymbol{D}|$ evaluated at t'. Eq. 164 is equivalent to models studied by Cheng and Evans [7] for any memory function $M(t - t')$ that corresponds to an evolution equation for λ. For example, the equation

$$\dot{\lambda} = h(|\boldsymbol{D}|) + (\lambda_0 - \lambda)/\lambda_1 \tag{165}$$

is obtained for the exponential memory function $M(t - t') = \exp[(t' - t)/\lambda_1]$, where λ_1 is a constant relaxation time. In these examples, the evolution equation typically depends on the current value of the variable and the magnitude $|\boldsymbol{D}|$. Material objectivity implies that the rate of deformation can only appear in such equations through invariants like $|D|$. The only other possible invariant for an incompressible material vanishes in shearing flows. Consequently, any evolution equation of the form of Eq. 160 would be insensitive to changes in the sign of D during such a flow. For example, the magnitude of the shear stress would be unaffected by reversing the shear rate. This experiment tests the validity of models based on scalar measures of structure for a given material. Models based on tensorial measures of microstructure are discussed in [31].

6. Rheometry

Rheological measurements are performed so as to obtain a test fluid's material functions. Under viscometric flows we have seen that the shear viscosity and the primary and secondary normal stress differences suffice to rheologically characterize the fluid. If the flow field is extensional and the material is able to attain a state of dynamic equilibrium, then one measures the extensional viscosity; otherwise, we measure the extensional viscosity growth or decay functions. In this section, we will examine steady and dynamic shear plus uniaxial extensional tests, since these make up the majority of routine rheological characterization.

6.1. Shear rheometry

To measure shear viscosity one has to place the test fluid in between confining walls that are then subjected to a shearing motion relative to one another. As long as the fluid adheres to these walls, one can control the flow by manipulating the walls and achieve a shear field within the fluid. For steady shear we wish to measure the three viscometric functions. The cone and plate geometry is the preferred arrangement for steady shear under moderate to low shear rates. At higher shear rates this flow becomes unstable and is susceptible to edge failures. At these and higher shear rates, more common to those encountered in industrial processes, one usually relies on capillary rheometry to measure the shear viscosity. Under dynamic shear one measures the components of the complex shear modulus. These dynamic tests are usually carried out in a parallel plate arrangement.

6.1.1. Cone and plate steady shear flow

The test fluid is placed between a fixed lower plate and a rotating upper cone. The cone angle, θ_0, is usually very small ($\sim$2–4°). The radius is R and the constant angular velocity of the cone is W.

The flow is assumed to be steady and laminar. The velocity field, with respect to a spherical coordinate system $\{r, \theta, \phi\}$ with origin at the cone apex, is postulated to be

$$v = \{v_r, v_\theta, v_\varphi\} = \{0, 0, v_\varphi(\theta)\} \tag{166}$$

Neglecting inertial and body forces, the equation of motion reduces to

$$r\text{-component:} \quad 0 = \frac{1}{r^2}\frac{\partial}{\partial r}\left(r^2\sigma_{rr}\right) - \frac{1}{r}\left(\sigma_{\theta\theta} + \sigma_{\varphi\varphi}\right) \left.\begin{array}{c} \\ \\ \\ \\ \\ \end{array}\right\}$$

$$\theta\text{-component:} \quad 0 = \frac{\partial\sigma_{\theta\theta}}{\partial\theta} + \left(\sigma_{\theta\theta} - \sigma_{\varphi\varphi}\right)\cot\theta \qquad (167)$$

$$\varphi\text{-component:} \quad 0 = \frac{\partial\sigma_{\theta\varphi}}{\partial\theta} + 2\sigma_{\theta\varphi}\cot\theta$$

Integrating Eq. 167₃ we have

$$\sigma_{\theta\varphi} = c/\sin^2\theta; \quad c = \text{constant} \tag{168}$$

The torque exerted on the plate is

$$\mathcal{T} = \int_0^{2\pi}\int_0^R \sigma_{\theta\varphi}\big|_{\theta=\frac{\pi}{2}} r^2 \mathrm{d}r\mathrm{d}\varphi = \frac{2\pi R^3 c}{3} \tag{169}$$

so that

$$\sigma_{\theta\varphi} = \frac{3\mathcal{T}}{2\pi R^3\sin^2\theta} \simeq \frac{3\mathcal{T}}{2\pi R^3} \tag{170}$$

since the cone angle is small (i.e. $\theta \sim 90°$). Eq. 170 says that the shear stress is essentially constant throughout the flow field. It follows that the shear rate will also be constant, and we can show that it is given by

$$\dot{\gamma} = -\dot{\gamma}_{\theta\phi} = W/\theta_\mathrm{o} \tag{171}$$

From Eqs. 170 and 171, the shear viscosity is

$$\eta(\dot{\gamma}) = \frac{3\theta_\mathrm{o}\mathcal{T}}{2\pi R^3 W} \tag{172}$$

The fact that both the shear stress and shear rate are independent of position in the gap is what makes the cone and plate arrangement so desirable for steady shear characterization. To get the normal stresses we note that, since $\cot\theta \approx 0$, Eq. 167₂ indicates $\sigma_{\theta\theta} = \sigma_{\theta\theta}(r)$. Furthermore, since the viscometric functions

$$N_1 = \sigma_{\varphi\varphi} - \sigma_{\theta\theta}, \quad N_2 = \sigma_{\theta\theta} - \sigma_{rr} \tag{173}$$

depend solely on the shear rate — which is a constant — they too must be independent of position, so that

$$\frac{\partial N_2}{\partial r} = \frac{\partial}{\partial r}\left(\sigma_{\theta\theta} - \sigma_{rr}\right) = 0 \tag{174}$$

Thus, via Eqs. 167₁ and 174

$$r\frac{\partial \sigma_{rr}}{\partial r} = \frac{\partial \sigma_{rr}}{\partial \ln r} = \frac{\partial \sigma_{\theta\theta}}{\partial \ln r} = \left(\sigma_{\varphi\varphi} + \sigma_{\theta\theta} - 2\sigma_{rr}\right) = N_1 + 2N_2 \tag{175}$$

Since $\sigma_{\theta\theta}$ corresponds to what a flush-mounted pressure transducer would read on the surface of the cone or plate, a plot of the measured pressure vs. $\ln r$ should yield a straight line with slope equal to the rhs of Eq. 175. Similarly, the total thrust on the plate can be calculated from

$$F = -\int_0^{2\pi}\int_0^R \sigma_{\theta\theta} r \, dr \, d\varphi - p_a \pi R^2 \tag{176}$$

where p_a is the ambient pressure. If we neglect surface tension, then

$$\sigma_{rr}|_R = -p_a \tag{177}$$

so that Eq. 176 reduces to

$$F = \frac{\pi R^2}{2}\left(\sigma_{\theta\theta} - \sigma_{\varphi\varphi}\right) = \frac{\pi R^2 N_1}{2} \tag{178}$$

Hence pressure and thrust measurements, via Eqs. 175 and 178, allow one to calculate N_1 and N_2.

6.1.2. Capillary rheometry

While the cone and plate geometry is the preferred arrangement to obtain the steady viscometric functions, it is limited to low shear rates — usually, to those less than 10 s^{-1}. At higher shear rates encountered in processing (~ 10–10^6 s^{-1}), it is customary to resort to capillary rheometry to measure the shear viscosity. Unfortunately, the normal stress differences cannot be obtained from this test. To get N_1 at high shear rates one can, however, employ a slit device based on the so-called 'hole pressure' effect [21].

In a typical capillary rheometer, one has a temperature-controlled barrel into which the test material (usually in powder or pellet form) is packed. Directly downstream of the barrel is a cylindrical die with known length and radius. A piston is programmed to force the molten material through the die at a constant rate. A pressure transducer located near the die entry records the pressure drop. The capillary rheometer is widely employed and has been analyzed in detail (see Macosko [25], who gives a thorough discussion). The expression for the shear viscosity is

$$\eta(\dot{\gamma}_R) = S_R/\dot{\gamma}_R = \frac{S_R}{(Q/\pi R^3)}\left[3 + \frac{d\ln\left(Q/\pi R^3\right)}{d\ln S_R}\right]^{-1}, \quad \left(S_R = \frac{R\Delta p}{2L}\right) \tag{179}$$

where R is the die radius, Q is the volumetric flow rate, $\dot{\gamma}_R$ is the shear rate at the die wall, S_R is the shear stress at the die wall and Δp is the measured pressure drop across the die length L. For a PL fluid with index n, Eq. 179 reduces to

$$\eta(\dot{\gamma}_R) = S_R/\dot{\gamma}_R = \frac{S_R}{\left(Q/\pi R^3\right)} \left[3 + \frac{1}{n}\right]^{-1} \tag{180}$$

Eqs. 179 and 180 indicate that the shear rate at the die wall has a correction due to the non-Newtonian character of the viscosity. This correction on the shear rate is called the Weissenberg–Rabinowitch correction. The shear stress itself may also need to be corrected so as to account for pressure losses due to elastic effects at the die entrance and exit. This correction is known as the Bagley correction. The Bagley correction can be obtained by repeating the pressure drop measurements with different aspect ratio (L/R) dies. The pressure drop vs. L/R plot which results is called a Bagley plot; the intercept on the negative L/R axis represents the additional length of 'fictitious' capillary which corresponds to the extra losses.

6.1.3. Parallel plate oscillatory shear

This is the most commonly employed dynamic test for linear viscoelastic characterization. The amplitude of the oscillations is kept small to ensure linearity and angular frequencies and gaps are adjusted according to the material at hand to avoid inertial effects. One can have both of the plates or a single plate moving, depending on the rheometer. A widely used commercial rheometer keeps the lower plate fixed while the top one is driven. The torque to keep the plate fixed is then measured at the lower plate. Another variety has the lower plate driven while the top plate is allowed to move. Here the top plate is connected to an air-bearing rotor that is then linked to a transducer to measure the angular displacement. This, in turn, is hooked up with a torsion bar to measure the torque on the top plate. Here we will only examine the fixed bottom plate case. Stiffness issues in parallel plate oscillatory shear tests are discussed in Walters [35].

A circular cylindrical coordinate system (r, θ, z) is employed with origin at the center of the bottom plate and positive z-axis pointing to the top plate. The plates have a (common) radius R and separation (gap) H. The top plate is driven at an angular velocity of $W(t)$ such that its angular displacement is given by

$$\theta(t) = \theta_0 \Re \left\{ e^{i\omega t} \right\} \tag{181}$$

where θ_0 is real and $\Re\{z\}$ stands for the operation of taking the real part of the complex number z. The BCs are

$$\left. \begin{array}{lll} \text{At } z = 0 & \Rightarrow & v_r = v_\theta = v_z = 0 \\ \text{At } z = H & \Rightarrow & v_r = v_z = 0, \quad v_\theta = rW(t) = r\dot{\theta}(t) \end{array} \right\} \tag{182}$$

The BCs lead us to postulate a velocity field in the form

$$v = \{v_r, v_\theta, v_z\} = \{0, v_\theta(r, z, t), 0\} \quad \text{with} \quad v_\theta \equiv r\Re\left\{f(z)e^{i\omega t}\right\} \tag{183}$$

where $f(z)$ is complex, and in view of the BCs,

$$f(0) = 0 \quad \text{and} \quad f(H) = \theta_0 i\omega \tag{184}$$

From the assumed velocity field given by Eq. 183, the only non-zero component of the rate of deformation tensor is given by

$$\dot{\gamma}_{\theta z} = \frac{1}{r}\frac{\partial v_z}{\partial \theta} + \frac{\partial v_\theta}{\partial z} = \frac{\partial v_\theta}{\partial z} = r\Re\left\{f'(z)e^{i\omega t}\right\} \tag{185}$$

Hence the only non-zero component of the stress tensor for the general linear viscoelastic model is

$$S_{\theta z} = \int_{-\infty}^{t} G(t - t')\dot{\gamma}_{\theta z}(t')\,dt' = r\Re\left\{\eta^*(\omega)f'(z)e^{i\omega t}\right\} \tag{186}$$

where the complex viscosity is given by

$$\eta^*(\omega) = \eta'(\omega) - i\eta''(\omega) = \int_{0}^{\infty} G(s)e^{-i\omega s}\,ds \tag{187}$$

Neglecting inertia and gravity, the equations of motion are

$$\left.\begin{array}{ll}
r\text{-component:} & 0 = -\dfrac{\partial p}{\partial r} \\[2mm]
\theta\text{-component:} & 0 = -\dfrac{1}{r}\dfrac{\partial p}{\partial \theta} + \dfrac{\partial S_{\theta z}}{\partial z} \\[2mm]
z\text{-component:} & 0 = -\dfrac{\partial p}{\partial z}
\end{array}\right\} \tag{188}$$

From Eqs. 188$_1$ and 188$_3$ we have that $p = p(\theta)$ only. Combining $p = p(\theta)$ with (188)$_2$ and (186), we are led to $p = p_0 = $ constant, which again via (188)$_2$ implies that

$$\partial S_{\theta z}/\partial z = 0 \quad \Rightarrow \quad d^2 f/dz^2 = 0 \tag{189}$$

Eq. 189, subject to the BCs (Eq. 184) yield

$$f(z) = i\omega\theta_0 z/H \tag{190}$$

which, substituted into Eq. 186, gives

$$S_{\theta z} = (r\theta_0/H)\Re\left\{\eta^* i\omega e^{i\omega t}\right\} \tag{191}$$

The torque, $\mathcal{T}$, required to keep the bottom plate stationary is evaluated from

$$\mathcal{T} = 2\pi \int_0^R r \, \Re\{S_{\theta z}\}|_{z=0} r \, dr = \left(\frac{\pi R^4 \omega \theta_0}{2H}\right) \Re\left\{(i\eta' + \eta'') \, e^{i\omega t}\right\} \tag{192}$$

If we represent the torque as

$$\mathcal{T} = \mathcal{T}_0 \Re\left\{e^{i(\omega t + \delta)}\right\} \tag{193}$$

where $\mathcal{T}_0$ is the (real) amplitude and δ is the relative phase shift (= loss angle), then Eqs. 192 and 193 yield the components of the complex viscosity as

$$\eta' = \left(\frac{2H\mathcal{T}_0 \sin\delta}{\pi R^4 \omega \theta_0}\right), \quad \eta'' = \left(\frac{2H\mathcal{T}_0 \cos\delta}{\pi R^4 \omega \theta_0}\right) \tag{194}$$

6.2. Extensional rheometry

If we want to find out how a fluid behaves under extension, we have to somehow 'grip' and stretch it. Experimentally, this is much more difficult than the shear arrangement, especially if the fluid has a low viscosity. Earlier (see Section 5) we saw that it is possible to classify steady extensional flows under the categories of uniaxial, biaxial and planar flows. We will now examine uniaxial testing, since this mode is more commonly employed as a routine characterization tool. Here we encounter two approaches: the first seeks to impart a uniform extensional field and back out a true material function, while the second employs a mixed flow field that is 'rich' in its extensional component (e.g. converging flows) and use it to back out a measured property of the fluid which is somehow related to its extensional viscosity.

6.2.1. Uniaxial extension

With melts there are basically two experimental methods for achieving uniaxial flow. Both techniques start out with the sample below its softening point and with well-defined dimensions. In one, the sample is placed between two clamps which are then moved away from each other at a prescribed velocity, and in the other the sample is pulled between two counter-rotating gears. With the former technique, the clamp separation must increase exponentially with time so that it is difficult to generate Hencky strains (see Section 5) above ~3.5. With the latter technique, the sample length is fixed, allowing one to go to higher Hencky strains (~6–7). Supporting the sample while it gets heated up beyond its softening point to the test temperature can be achieved either by neutrally buoyant, inert, heated oils or by means of heated, inert gas cushions. For a sample of cross-sectional area $A(t)$ at time t that is undergoing uniaxial flow with a constant stretch rate $\dot{\varepsilon}_0$, the cross-sectional area is expected to diminish from its initial value A_0 according to

$$A(t) = A_0 \exp(-\dot{\varepsilon}_0 t) \tag{195}$$

Monitoring the tensile force $F(t)$ on the sample then gives the normal stress difference as a function of time ($= F(t)/A(t)$) from which the extensional viscosity can be evaluated.

6.2.2. Converging flow

The uniaxial extension tests mentioned above require specialized equipment and have limitations on how high a value of stretch and stretch rate one can achieve. In view of this, there have been efforts to back out extensional properties of melts from easier tests which, however, may be less well-controlled. The simplest of these situations arise while conducting capillary rheometry where there is a converging flow as the fluid advances from the large diameter of the barrel down to the much smaller diameter of the die. Cogswell [10] proposed an approximate method to extract the uniaxial viscosity from capillary testing. His approach is to take the total pressure drop associated with the entrance from the barrel to the die as being made up of contributions from shear and extension. These contributions are calculated for incremental cones via force balances, and the resulting shear and extensional pressure drops are then associated with the shear stress and the normal stress difference, respectively. Since the entry flow cone angle is unknown, the total entrance pressure drop is calculated by summing the incremental pressure drops and assuming that this sum is minimized for the true value. Viscous flow is assumed throughout, i.e. elastic behavior is neglected. The relationship for the uniaxial extensional viscosity, $\eta_e(\dot{\varepsilon})$, arising from Cogswell's analysis is

$$\eta_e(\dot{\varepsilon}) = \frac{9}{32} \frac{(n+1)^2}{\eta} \left(\frac{P_0}{\dot{\gamma}_a} \right)^2, \qquad \left(\dot{\gamma}_a = \frac{4Q}{\pi R^3} \right) \tag{196}$$

where the stretch rate is given by

$$\dot{\varepsilon} = \frac{4}{3} \frac{\eta \dot{\gamma}_a^2}{(n+1) P_0} \tag{197}$$

In Eqs. 196 and 197, P_0 is the exit pressure which can be well approximated by the pressure drop through an orifice of the same radius as the die and n is the PL exponent which is obtained from capillary data (with die radius and length given by R and L, respectively). Cogswell's method, as well as other alternative methods for obtaining the uniaxial viscosity, have been compared against direct measurements in an extensional rheometer by Laun and Schuch [23].

7. Additional examples

In this section we illustrate two rheology problems that have direct relevance to adhesion. When pressure is applied on a pressure-sensitive adhesive, the material will flow between the backing and the substrate. This is an example of so-called 'squeeze flow'. Likewise, if we have two plates that are placed at an angle to each other, with adhesive in between them at the joint, and proceed to press the plates together, then we have an example of 'wedge flow'. This case could also approximate the flow one gets when a clump of material, such as caulk, is spread on a surface with a spatula. We now examine these two flows.

7.1. Squeeze flow

Consider a fluid placed between two parallel circular disks of radius R. The gap $2h$ between the disks is assumed to be narrow, i.e. to be much less than R at all times. A force F is applied to each disk so as to push them together. We wish to know what the force needs to be in order to maintain a prescribed motion of the disks when the fluid is Newtonian or PL. The case where the material is a Bingham fluid with a yield stress is complicated (see, however, Wilson [36]).

A circular cylindrical coordinate system (r, θ, z) is taken at the mid-plane as shown in Fig. 9.

7.1.1. Newtonian fluid case

We will neglect gravity and inertial forces and assume axial symmetry. Let us postulate that the velocity field is of the form

$$\boldsymbol{v} = \{v_r, v_\theta, v_z\} = \{v_r(t, r, z), 0, v_z(t, z)\} \tag{198}$$

where t denotes time. The equation of continuity is

$$\frac{1}{r}\frac{\partial}{\partial r}(r v_r) + \frac{\partial v_z}{\partial z} = 0 \tag{199}$$

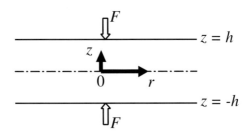

Fig. 9. Squeeze flow geometry and coordinate system.

Substituting Eq. 198 into 199 we see that the radial component of the velocity must be of the form

$$v_r = r f(t,z) \tag{200}$$

where the unknown function f satisfies

$$2f + (\partial v_z / \partial z) = 0 \tag{201}$$

The z-component of the equation of motion, without the inertial and body forces, reads

$$\rho \frac{\partial v_z}{\partial t} = -\frac{\partial p}{\partial z} + \frac{1}{r}\frac{\partial}{\partial r}(r S_{rz}) + \frac{1}{r}\frac{\partial}{\partial \theta} S_{\theta z} + \frac{\partial}{\partial z} S_{zz} \tag{202}$$

The third term on the rhs of Eq. 202 is zero due to axi-symmetry. We further restrict ourselves to a 'quasi' steady-state solution, i.e. we assume that at any given time t, the flow can be approximated as being steady. This would mean, for instance, that the impulsive loading involved in the start-up of the squeeze would not be covered by the solution. The quasi steady-state assumption allows us to discard the term on the lhs of Eq. 202. The simplified z-momentum equation is thus

$$0 = -\frac{\partial p}{\partial z} + \frac{1}{r}\frac{\partial}{\partial r}(r S_{rz}) + \frac{\partial}{\partial z} S_{zz} \tag{203}$$

The r-momentum equation, similarly simplified, reads

$$0 = -\frac{\partial p}{\partial r} + \frac{1}{r}\frac{\partial}{\partial r}(r S_{rr}) + \frac{\partial}{\partial z} S_{zr} - \frac{S_{\theta\theta}}{r} \tag{204}$$

For a Newtonian fluid with viscosity μ the extra stress components in the cylindrical coordinate system are given by

$$S_{rz} = S_{zr} = \mu \left(\frac{\partial v_r}{\partial z} + \frac{\partial v_z}{\partial r} \right) = \mu \frac{\partial v_r}{\partial z} \tag{205}$$

$$S_{rr} = 2\mu \frac{\partial v_r}{\partial r}, \quad S_{\theta\theta} = 2\mu \left(\frac{1}{r}\frac{\partial v_\theta}{\partial \theta} + \frac{v_r}{r} \right) = 2\mu \frac{v_r}{r}, \quad S_{zz} = 2\mu \frac{\partial v_z}{\partial z} \tag{206}$$

From the continuity equation, Eq. 199, we can get an order of magnitude estimate of the magnitudes of the two velocity components relative to each other, viz.

$$\underbrace{\frac{1}{r}\frac{\partial}{\partial r}(r v_r)}_{O(v_r/R)} + \underbrace{\frac{\partial v_z}{\partial z}}_{O(v_z/h)} = 0 \quad \Rightarrow \quad v_z = O\left(\frac{h}{R} v_r \right) \tag{207}$$

Eq. 207 confirms our intuitive sense that since the gap is narrow, the main flow component should be that in the r-direction, i.e. $v_r \gg v_z$. Now let us look at the order of magnitudes of the various terms in the momentum equations, viz.

$$0 = -\frac{\partial p}{\partial z} + \underbrace{\frac{1}{r}\frac{\partial}{\partial r}(r\,S_{rz})}_{O(\mu v_r/Rh)} + \underbrace{\frac{\partial}{\partial z}S_{zz}}_{O(\mu v_z/h^2)} \tag{208}$$

$$0 = -\frac{\partial p}{\partial r} + \underbrace{\frac{1}{r}\frac{\partial}{\partial r}(r\,S_{rr})}_{O(\mu v_r/R^2)} + \underbrace{\frac{\partial}{\partial z}S_{zr}}_{O(\mu v_r/h^2)} - \underbrace{\frac{S_{\theta\theta}}{r}}_{O(\mu v_r/R^2)} \tag{209}$$

Taking the r-momentum equation, Eq. 209, first, we note that retaining the dominant stress gradient only leads to

$$0 = -\frac{\partial p}{\partial r} + \frac{\partial}{\partial z}S_{zr} = -\frac{\partial p}{\partial r} + \mu\frac{\partial^2 v_r}{\partial z^2} \tag{210}$$

Eq. 210, in turn, gives us an estimate for the order of magnitude of the unknown pressure, viz.

$$p = O\left(\mu v_r R/h^2\right) \quad \Rightarrow \quad \partial p/\partial z = O\left(\mu v_r R/h^3\right) \tag{211}$$

Comparing Eq. 211$_2$ with the other terms in Eq. 208, we see that the axial pressure gradient dominates the other extra stress gradients in the z-momentum equation, leaving

$$0 = -\partial p/\partial z \tag{212}$$

The simplified equation set that needs to be solved is therefore Eqs. 200, 201, 210$_2$ with and 212. The boundary conditions that need to be satisfied are

$$\left.\begin{array}{ll} \partial f/\partial z = 0, & v_z = 0 \text{ at } z = 0 \quad \text{(symmetry)} \\ f = 0, & v_z = dh/dt = \dot h \text{ at } z = h \quad \text{(no slip)} \\ p = p_a & \text{at } r = R \end{array}\right\} \tag{213}$$

where p_a denotes ambient pressure. To solve the set, we first differentiate Eq. 210$_2$ with respect to z and use Eq. 212 to get

$$d^3 f/dz^3 = 0 \tag{214}$$

which, combined with the two boundary conditions on f, yield

$$f = c\left(h^2 - z^2\right); \quad c = \text{constant} \tag{215}$$

Next, integrating Eq. 201 with f given by Eq. 215 we find, using the BC for v_z at $z = 0$,

$$v_z = -2c\left(h^2 z - \frac{z^3}{3}\right) \tag{216}$$

Using the remaining BC for v_z at $z = h$ we find the constant c to be

$$c = \frac{3}{4} \frac{(-\dot{h})}{h^3} \tag{217}$$

Finally, going back to Eq. 210_2 and using Eq. 200, we integrate with respect to r and use the BC for p at R to get the pressure field as

$$p = p_a + \frac{3(-\dot{h})\mu R^2}{4h^3} \left[1 - \left(\frac{r}{R}\right)^2 \right] \tag{218}$$

Since $S_{zz} = 0$ at $z = h$ (see Eqs. 206_3 and 216), the force on the disk at $z = h$ is

$$F = \int_0^{2\pi} \int_0^R (p - p_a)_{z=h} r \, dr \, d\theta = \frac{3\pi R^4 \mu(-\dot{h})}{8h^3} \tag{219}$$

Eq. 219 is known as the Stefan equation and it tells us what the force $F(t)$ must be in order to sustain the prescribed motion $h(t)$. If we want to know what the disk motion will be under a constant applied force F, then we can integrate Eq. 219 with respect to time, while holding F constant, to get

$$\frac{1}{h^2} - \frac{1}{h_0^2} = \frac{16Ft}{3\pi \mu R^4}; \quad h_0 = h(0) \tag{220}$$

Note that in obtaining Eq. 220, we have implicitly assumed that, from the start, the fluid fills the entire area between $r = 0$ to $r = R$. It can be shown that if the fluid only partially fills the disk area then one gets

$$\frac{1}{h^4} - \frac{1}{h_0^4} = \frac{128\pi Ft}{3\mu V^2}; \quad V = \text{volume of fluid at } t = 0 \quad (< 2\pi R^2 h_0) \tag{221}$$

7.1.2. PL fluid case

We can proceed the same way as with the Newtonian fluid case up to the point where we need to express the stress components in terms of the velocity gradients. Next we need the magnitude of the rate of deformation tensor

$$\dot{\gamma} = \left[(2\dot{\gamma}_{rz}^2 + \dot{\gamma}_{rr}^2 + \dot{\gamma}_{\theta\theta}^2 + \dot{\gamma}_{zz}^2)/2 \right]^{1/2} = (12f^2 + r^2 f'^2)^{1/2}; \quad (f' = df/dz) \tag{222}$$

where we have used Eqs. 200 and 201. We could now proceed with the analysis, along the lines of the Newtonian case, by keeping the same dominant stress components in the equations of motion. Note that there are complications near $r = 0$ in approximating the shear rate by the second term only on the rhs of Eq. 222_2. A rigorous singular perturbation solution of the plane flow version of this problem has been given by Johnson [20]. An alternative approach is discussed in Bird et al. [4] (p. 189) where it is assumed that the instantaneous volumetric flow rate $Q(r)$ across the cylindrical surface between the disks at r is equal to that for flow through a slit of thickness $2h$ and width $2\pi r$. The latter can be easily obtained for the PL fluid and for a slit of thickness $2B$, width W and length L, is given by

$$Q_{\text{slit}} = \frac{2WB^2}{2+\frac{1}{n}}\left[\frac{(p_0-p_L)B}{mL}\right]^{1/n} \tag{223}$$

To switch back to the squeeze geometry, we consider the flow between the disks in the region r to $r+dr$ so that Eq. 223 can be re-written in the form

$$Q_{\text{squeeze}} = Q(r) = \frac{2(2\pi r)h^2}{2+\frac{1}{n}}\left[\left(-\frac{dp}{dr}\right)\frac{h}{m}\right]^{1/n} \tag{224}$$

Next, we can make an overall mass balance to give the relation

$$Q(r) = 2\pi r^2(-\dot{h}) \tag{225}$$

Equating Eqs. 224 and 225 one can solve for the pressure field and using that, get the force on the plate

$$p = p_a + \frac{m(-\dot{h})^n}{h^{2n+1}}\left(\frac{2n+1}{2n}\right)^n\frac{R^{n+1}}{n+1}\left[1-\left(\frac{r}{R}\right)^{n+1}\right] \tag{226}$$

$$F = \frac{(-\dot{h})^n}{h^{2n+1}}\left(\frac{2n+1}{2n}\right)^n\frac{\pi mR^{n+3}}{n+3} \tag{227}$$

Eq. 227 is called the Scott equation and is the PL counterpart of the Stefan equation to which it reduces when $n = 1$. If the fluid only partially fills the disk area then one gets

$$\frac{1}{h^\alpha}-\frac{1}{h_0^\alpha} = \left(\frac{3n+5}{2n+1}\right)\left[\frac{2(n+3)(2\pi)^{(n+1)/2}F}{mV^{(n+3)/2}}\right]^{1/n}t;$$

$$\alpha = \frac{3n+5}{2n}; \quad V = \text{fluid volume }(< 2\pi R^2 h_0) \tag{228}$$

By comparing Eqs. 221 and 228, it is possible to find out, for instance, what the effect of shear thinning would be when a given volume of adhesive drop is squeezed in between the two sides of a joint. In particular, one could judge whether the drop can be expected to spread to a wider or smaller area than a corresponding Newtonian drop of the same volume under the same force.

7.2. Wedge flow

Consider a fluid placed between two infinite planes that form a wedge of internal angle 2α. The plates are closing together with a constant angular velocity Ω. We

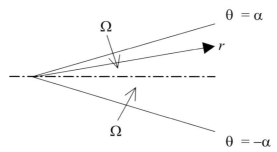

Fig. 10. Wedge flow geometry and coordinate system.

wish to know what the flow field will be when the fluid is Newtonian or PL. The case where the material has a yield stress is, again, more difficult to analyze (see [37]).

A circular cylindrical coordinate system (r, θ, z) is taken with origin at the mid-plane as shown in Fig. 10.

7.2.1. Newtonian fluid case

We will neglect gravity and inertial forces and assume angular symmetry about the mid-plane. Let us postulate that the velocity field is of the form

$$\boldsymbol{v} = \{v_r, v_\theta, v_z\} = \{\Omega r f'(\theta), -2\Omega r f(\theta), 0\}; \quad \left(f'(\theta) = \frac{df}{d\theta}\right) \tag{229}$$

where f denotes an unknown function of θ. The equation of continuity

$$\frac{1}{r}\frac{\partial}{\partial r}(r v_r) + \frac{1}{r}\frac{\partial v_\theta}{\partial \theta} = 0 \tag{230}$$

is automatically satisfied by the choice of the velocity components in Eq. 229. The r- and θ-components of the equation of motion, without the inertial and body forces, read respectively as

$$0 = -\frac{\partial p}{\partial r} + \frac{1}{r}\frac{\partial}{\partial r}(r S_{rr}) + \frac{1}{r}\frac{\partial}{\partial \theta}S_{r\theta} - \frac{S_{\theta\theta}}{r} \tag{231}$$

$$0 = -\frac{1}{r}\frac{\partial p}{\partial \theta} + \frac{1}{r^2}\frac{\partial}{\partial r}(r^2 S_{r\theta}) + \frac{1}{r}\frac{\partial}{\partial \theta}S_{\theta\theta} \tag{232}$$

For a Newtonian fluid with viscosity μ, the non-zero extra stress components in the cylindrical coordinate system are given by

$$S_{r\theta} = \mu\left[r\frac{\partial}{\partial r}\left(\frac{v_\theta}{r}\right) + \frac{1}{r}\frac{\partial v_r}{\partial \theta}\right] = \mu\Omega f'' \tag{233}$$

$$S_{rr} = 2\mu\frac{\partial v_r}{\partial r} = 2\mu\Omega f', \quad S_{\theta\theta} = 2\mu\left(\frac{1}{r}\frac{\partial v_\theta}{\partial \theta} + \frac{v_r}{r}\right) = -2\mu\Omega f' \tag{234}$$

Substituting Eqs. 233 through 234 into the equations of motion we get

$$0 = -\frac{\partial p}{\partial r} + \frac{\mu\Omega}{r}\left(f''' + 4f'\right) \quad \text{and} \quad 0 = -\frac{1}{r}\frac{\partial p}{\partial \theta} \tag{235}$$

Hence, the governing equation for the unknown function f is

$$f^{(iv)} + 4f'' = 0 \tag{236}$$

The boundary conditions that need to be satisfied are

$$\left. \begin{array}{l} f = 0, \quad f'' = 0 \ \text{at} \ \theta = 0 \quad \text{(symmetry)} \\ f = 1/2, \quad f' = 0 \ \text{at} \ \theta = \alpha \quad \text{(no slip)} \end{array} \right\} \tag{237}$$

The solution is

$$f(\theta) = \frac{1}{2}\left[\frac{2\theta\cos(2\alpha) - \sin(2\theta)}{2\alpha\cos(2\alpha) - \sin(2\alpha)}\right] \tag{238}$$

7.2.2. PL fluid case

We can proceed the same way as with the Newtonian fluid case up to the point where we need to express the stress components in terms of the velocity gradients. We need the magnitude of the rate of deformation tensor

$$\dot{\gamma} = \left[(2\dot{\gamma}_{r\theta}^2 + \dot{\gamma}_{rr}^2 + \dot{\gamma}_{\theta\theta}^2 + \dot{\gamma}_{zz}^2)/2\right]^{1/2} = \Omega\left(4f'^2 + f''^2\right)^{1/2} \equiv \Omega g(\theta); \quad (g > 0) \tag{239}$$

The non-zero stress components are given by

$$S_{r\theta} = \eta(\dot{\gamma})\dot{\gamma}_{r\theta} = (m\dot{\gamma}^{n-1})\dot{\gamma}_{r\theta} = m\Omega^n g^{n-1} f'' \tag{240}$$

$$S_{rr} = \eta(\dot{\gamma})\dot{\gamma}_{rr} = 2m\Omega^n g^{n-1} f', \quad S_{\theta\theta} = \eta(\dot{\gamma})\dot{\gamma}_{\theta\theta} = -2\Omega^n m g^{n-1} f' \tag{241}$$

Substituting Eqs. 240 through 241 into the equations of motion we get

$$0 = -\frac{\partial p}{\partial r} + \frac{4m\Omega^n}{r}g^{n-1}f' + \frac{m\Omega^n}{r}\left(g^{n-1}f''\right)' \tag{242}$$

$$0 = -\frac{\partial p}{\partial \theta} + 2m\Omega^n g^{n-1}f'' - 2m\Omega^n\left(g^{n-1}f'\right)' \tag{243}$$

Cross-differentiating Eqs. 242 and 243 to eliminate the pressure, we arrive at the governing equation for the unknown function f

$$\left(g^{n-1}f''\right)'' + 4\left(g^{n-1}f'\right)' = 0 \tag{244}$$

which is subject to the same BCs as in the Newtonian case, Eq. 237. Eq. 244 is a nonlinear ordinary differential equation that can be integrated once but, beyond that, has to be solved numerically. Since the problem is of the two-point boundary value type, a common numerical approach would be to use so-called 'shooting' techniques. Closed form solutions are possible [12] if the function f is represented by a quadratic in θ.

References

1. Berker, A. and VanArsdale, W.E., Phenomenological models of viscoplastic, thixotropic and granular materials. *Rheol. Acta*, **31**, 119–138 (1992).
2. Berker, A. and VanArsdale, W.E., Steady flow of viscoplastic materials in a rectangular slit. In: Siginer, D.A. et al. (Eds.), *Developments in Non-Newtonian Flows*. AMD-Vol. 175, ASME, New York, 1993, pp. 21–25
3. Bingham, E.C., *Fluidity and Plasticity*. McGraw-Hill, New York, 1922.
4. Bird, R.B., Armstrong, R.C. and Hassager, O., *Dynamics of Polymeric Liquids, Vol. 1: Fluid Mechanics*, 2nd edn. John Wiley and Sons, New York, 1987.
5. Bird, R.B., Dai, G.C. and Yarusso, B.J., The rheology and flow of viscoplastic materials. *Rev. Chem. Eng.*, **1**, 1 70 (1982).
6. Casson, N., A flow equation for pigment–oil suspensions of the printing ink type. In: Mill, C.C. (Ed.), *Rheology of Disperse Systems*. London, 1957, pp. 84–104.
7. Cheng, D.C.-H. and Evans, F., Phenomenological characterization of the rheological behavior of inelastic reversible thixotropic and antithixrotropic fluids. *Br. J. Appl. Phys.*, **16**, 1599–1617 (1965).
8. Cheng, D.C.-H., On the behavior of thixotropic fluids with a distribution of structure. *J. Phys. D: Appl. Phys.*, **7**, L155–L158 (1974).
9. Christensen, L.M., *Mechanics of Composite Materials*. John-Wiley, New York, 1979.
10. Cogswell, F.N., Converging flow of polymer melts in extrusion dies. *Polym. Eng. Sci.*, **12(1)**, 64–73 (1972).
11. Coleman, B.D., Markovitz, H. and Noll, W., *Viscometric Flows of Non-Newtonian fluids*. Springer Verlag, Berlin, 1966, p. 17.
12. Dong Chen, X., Slip and no-slip squeezing flow of liquid food in a wedge. *Rheol. Acta*, **32(5)**, 477–482 (1993).
13. Ferry, J.D., *Viscoelastic Properties of Polymers*, 2nd edn. John Wiley and Sons, New York, 1970.
14. Fredrickson, A.G., A model for the thixotropy of suspensions. *AIChE J.*, **16(3)**, 436–441 (1970).
15. Goddard, J.D., Dissipative materials as models of thixotropy and plasticity. *J. Non-Newtonian Fluid Mech.*, **14**, 141–160 (1984).
16. Gurtin, M.E., *An Introduction to Continuum Mechanics*. Academic Press, San Diego, 1981.
17. Hand, G.S., A theory for anisotropic fluids. *J. Fluid Mech.*, **13**, 33–46 (1962).
18. Harris, J., A continuum theory of time-dependent inelastic flow. *Rheol. Acta*, **6(1)**, 6–12 (1967).
19. Herschel, W.H. and Bulkley, R., Konsistenzmessungen von Gummi-Benzöllosungen. *Kolloid Z.*, **39**, 291–300 (1926).
20. Johnson, R.E., Power-law creep of a material being compressed between parallel plates: a singular perturbation problem. *J. Eng. Math.*, **18**, 105–117 (1984).
21. Kaye, A., Lodge, A.S. and Vale, D.G., Determination of normal stress differences in steady shear flow. *Rheol. Acta*, **7(4)**, 368–379 (1968).
22. Larson, R.G., *Constitutive Equations for Polymer Melts and Solutions*. Butterworths, Boston, 1988.

23. Laun, H.M. and Schuch, H., Transient elongational viscosities and drawability of polymer melts. *J. Rheol.*, **33(1)**, 119–175 (1989).
24. Lavallée, C. and Berker, A., More on the predictions of molecular weight distributions of linear polymers from their rheology. *J. Rheol.*, **41(4)**, 851–871 (1997).
25. Macosko, C.W., *Rheology: Principles, Measurements and Applications*. VCH Publishers, New York, 1994.
26. Mendelson, A., *Plasticity: Theory and Applications*. MacMillan, New York, 1970.
27. Robertson, R.E. and Stiff, H.A., An improved mathematical model for relating shear stress to shear rate in drilling fluids and cement slurries. *Soc. Pet. Eng. J.*, (February), 31–36 (1976).
28. Slattery, J.C., Momentum, energy and mass transfer in continua. Robert E. Krieger Publishing Co., Huntington, New York, 1978.
29. Slattery, J.C., Unsteady relative extension of incompressible simple fluids. *Phys. Fluids*, **7**, 1913–1914 (1964).
30. Slibar, A. and Paslay, P.R., On the analytical description of the flow of thixotropic materials. In: Reiner, M. and Abid, D. (Eds.), *International Symposium on Second-order Effects in Elasticity, Plasticity and Fluid Dynamics*. Haifa, Israel, 1962, pp. 314–330.
31. VanArsdale, W.E., The application of a theory for elastic fluids in suspension rheology. *J. Rheol.*, **26(5)**, 477–491 (1982).
32. Wagner, M.H., Analysis of time-dependent non-linear stress growth data for shear and elongational flow of a low density branched polyethylene melt. *Rheol. Acta*, **15**, 136–142 (1976).
33. Wagner, M.H., Prediction of primary normal stress difference from shear viscosity data using a single integral constitutive equation. *Rheol. Acta*, **16**, 43–50 (1977).
34. Wagner, M.H., Zur Netzwerktheorie von Polymer-Schmelzen. *Rheol. Acta*, **18**, 33–50 (1979).
35. Walters, K., *Rheometry*. Halsted Press, London, 1975.
36. Wilson, S.D.R., Squeezing flow of a Bingham material. *J. Non-Newtonian Fluid Mech.*, **47**, 211–219 (1993).
37. Wilson, S.D.R., Squeezing flow of a yield-stress fluid in a wedge of slowly-varying angle. *J. Non-Newtonian Fluid Mech.*, **50**, 45–63 (1993).
38. Yarusso, D.J., Private communication, 2000.

Chapter 13

Effect of rheology on PSA performance

DAVID J. YARUSSO [*]

3M Company, St. Paul, MN, USA

1. Introduction to pressure-sensitive adhesives

Pressure-sensitive adhesives (PSAs) have become familiar materials in our world. Most people have used tapes such as masking tape, electrical tape, or transparent tape, for example Scotch® Magic™ Transparent Tape. All of these have pressure-sensitive adhesives of various kinds coated on a paper or polymeric film backing. The adhesive strength of such materials can vary widely from the easily removable Post-It® note to the permanent bonds formed by the double-sided foam tapes such as Scotch® VHB™ tape and the tapes used to mount body side moldings on vehicles. The optimization of the performance of such materials in terms of the ability to bond and hold on a variety of surfaces as well as the ability to be removed from surfaces cleanly and without damage is an ongoing effort in the design of PSA products.

A pressure-sensitive adhesive must have the following characteristics:
- it is permanently tacky at room temperature (in dry, solvent free form)
- adheres to a variety of surfaces upon mere contact without the need of more than finger or hand pressure
- requires no activation by water, solvent, or heat in order to form a bond
- has sufficient cohesive strength and elastic nature that it can be handled with the fingers and removed from smooth surfaces without leaving a residue.

2. Interfacial and rheological requirements of adhesives

As one begins to read the literature on adhesives and especially pressure-sensitive adhesives, it is easy to become confused about the relative importance of the adhesive/substrate interface and the mechanical or rheological properties of the

---
[*] Corresponding author. E-mail: djyarusso@mmm.com

adhesive for adhesion. In fact, both features are critical to adhesion of PSAs and the two act in a multiplicative way.

As we examine the properties of adhesives, we will be looking at various rheological measurements, especially linear viscoelastic properties, and relating them to the performance of the PSA. The reader who is unfamiliar with rheological properties and measurements is encouraged to read Chapter 12 by Ali Berker before proceeding.

With regard to the interface, we need to be concerned about the role of interfacial energies in controlling wetting behavior as well as the strength of intermolecular forces acting across the bonded interface.

Rheological properties and interfacial properties interact in both the bonding and the debonding processes for adhesives including pressure-sensitive adhesives. In this section, we will examine some of the important requirements on both the interfacial properties and the adhesive rheology and how they are coupled.

2.1. General requirements of adhesives

Before considering the unique requirements of PSAs, it will be instructive to first consider the requirements of adhesives generally and then look at how PSAs meet those requirements.

2.1.1. The wetting and bonding process

All adhesives must obtain intimate wetting of the substrates to which they are applied. They must flow to allow molecular contact with as much of the substrate surface as possible, overcoming the roughness of the surface.

In the wetting and bonding process, the interfacial properties provide the driving force for the adhesive to spread on the substrate surface. The rheological properties control the resistance to the flow required for that spreading process. The ideal situation for bond formation is to have interfacial properties which provide a strong driving force for the adhesive to spread on the substrate and for the adhesive to provide minimal resistance to the flow required for it to spread.

The tendency of a liquid to spread on a solid surface is often characterized by measuring the contact angle. Picture a drop of a liquid on a surface as shown in Fig. 1. If the drop is small so that gravity forces are negligible, it has been shown [1] that, at equilibrium,

$$\cos\theta = \frac{\gamma_{SV} - \gamma_{SL}}{\gamma_{LV}}$$

The subscripts S, L, and V represent the solid, liquid, and vapor phases, respectively. Therefore, γ_{LV} represents the interfacial energy between the liquid and vapor phases.

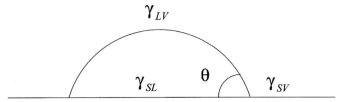

Fig. 1. A drop of liquid on a smooth, rigid solid. The interfacial energies associated with the solid–liquid, solid–vapor, and liquid–vapor interfaces are designated by the γ values. The equilibrium contact angle is given by θ.

If the adhesive is the liquid in this situation, we see that low surface energy of the adhesive, (γ_{LV}) high surface energy of the substrate (γ_{SV}), and low interfacial energy between the adhesive and the substrate (γ_{SL}) all favor low contact angle and spreading of the adhesive. In fact, ideally, an adhesive should have a contact angle of zero with the substrate. Such will be the case when

$$\gamma_{SV} - \gamma_{SL} \geq \gamma_{LV}$$

In other words, the overall system energy can be reduced by replacing the solid–vapor interface with the two interfaces between the solid and liquid and between the liquid and the vapor.

For most adhesives which are applied as liquids, the resistance of the adhesive to the deformation required of it to spread on the substrate is governed by its viscosity. The lower the viscosity, the better for enhancing the wetting and spreading process required for bond formation.

2.1.2. Resistance to debonding

Once the adhesive has spread on the substrate, there must be some kind of attractive interaction across the interface to allow the adhesive to bond to the surface. Generally, if the interfacial energies are favorable to spreading, there must be at least a weak attractive interaction. However, even adhesives with similar spreading behavior can have interfacial forces of very different strength.

The most common type of interactions are the dispersion forces. These are the forces which result from the random fluctuations of electron clouds in materials. These fluctuations create an instantaneous electrical dipole which then induces a dipole of opposite orientation on the other side of the interface. Although very weak, these interactions are present in all materials as long as there is intimate molecular contact at the interface. The interfacial forces in most PSAs are dominated by dispersion interactions.

For some adhesive–substrate combinations, stronger, specific interactions are possible. Most of these can be thought of as generalized Lewis acid-base interactions. This subject is treated in detail by Fowkes [2].

Of course, no matter how well the adhesive wets and adheres to the substrate, it would be of little use if the adhesive itself could not bear a load. Therefore, after bonding, the adhesive must behave like a solid to perform its function. Sometimes, an adhesive bond will have a specific stiffness requirement. Certainly, the adhesive can no longer behave like a liquid or it will simply separate (fail cohesively) under load.

Beyond bearing the load, the adhesive must also resist fracture and debonding. In this regard, fracture toughness is the important property. Stiffness is the resistance to deformation and is measured by the modulus, the ratio of stress to strain in the initial linear region of deformation. Fracture toughness has to do with the resistance of the material to propagating a crack leading to fracture. The two do not go hand-in-hand. In fact, often one must trade off one for the other. Materials which provide a mechanism for energy absorption in deformation have a much higher fracture toughness than those which do not. Consider metals and glasses, for example. Glass has a very high stiffness. The bonds resist deformation strongly. However, when the stress applied exceeds the value required to overcome the forces holding individual molecules in place, there is no means of slippage of molecules to a new location. A fracture crack propagates easily. In ductile metals, the molecules can slip to new stable positions in response to stress. This shear yielding process absorbs huge amounts of energy, preventing the propagation of the crack.

Similar considerations control the resistance to debonding of adhesives. In some cases, when the interface is very strong, the process is identical to that in bulk materials. The joint fails by cohesive failure within the adhesive layer. However, even when the crack propagates along an interface between an adhesive and a substrate, the ability of the system to resist the propagation of such a crack depends on the ability of the adhesive to absorb energy by a yielding deformation near the crack tip.

2.2. Comparison of PSAs to other adhesives

2.2.1. Physical or chemical state change for other adhesives

For many adhesives, the contradiction between the desire for liquid-like behavior for bonding and solid-like behavior for debonding resistance is resolved by the fact that the adhesive undergoes a change in physical state between the two events. For example, a hot melt adhesive, such as that used in a hot melt glue gun, becomes a liquid by heating to a temperature which exceeds the crystalline melting point of the component polymer. In the molten state, it has sufficiently low viscosity to flow and meet the bonding criteria. Upon cooling, it recrystallizes and becomes a solid which can resist deformation. An epoxy adhesive consists of two reactive liquid components. These are mixed and applied before the chemical curing reaction has progressed very far. Therefore, the adhesive is still a liquid for the bond-making process. As the reaction progresses, a polymer network is built up and the glass transition temperature increases until it exceeds room temperature, at which time the adhesive becomes a very hard solid. Solvent-based adhesives are liquid by virtue of the fact that the component polymer is dissolved in a solvent. The adhesive is applied in this state and then the solvent evaporates, leaving the solid polymer in the bond.

2.2.2. Only time scale sensitivity for PSAs

Pressure-sensitive adhesives are unique among adhesives in that they do not undergo a change in physical state between the bonding process and the performance life of the adhesive in which it must resist debonding. The adhesive material on a tape when the tape is applied is identical in properties to that which is there when the tape is doing its job or when it is being peeled off. Such adhesives manage to satisfy the requirements of liquid-like behavior for bond formation and solid-like behavior to resist debonding entirely as a result of their viscoelastic properties. Like all viscoelastic materials, their rheological responses have characteristics of both elastic solids and viscous liquids. They respond more like liquids when subjected to slow deformations and more like solids when we attempt to deform them rapidly. The apparent modulus increases continuously as the deformation rate increases. It so happens that the natural time scales for bonding for a pressure-sensitive adhesive product are longer than those for debonding in typical tack or peel tests. A properly designed PSA will exhibit a strong gradient in stiffness between these two time scales, allowing it both to bond quickly and to resist debonding, thus exhibiting tack and peel adhesion.

Pressure-sensitive adhesives were so named because they bond under very light pressure. However, since we name other adhesives based on what accomplishes their change in state (heat-activated, solvent-activated adhesives, etc.), perhaps it would be more descriptive to refer to these materials as 'time-scale sensitive' adhesives.

2.2.3. Surface properties and wetting in PSAs

PSAs must meet the same criteria as other adhesives with regard to surface energy and contact angle so that they have a driving force to spread on the substrate. However, because PSAs are not low viscosity liquids, it is difficult to measure contact angle or surface tension in the usual ways. Often we infer these properties from the behavior of chemically similar low molecular weight analogs to the PSA.

In general, PSAs behave more like very soft solids than like liquids during the wetting process. They are soft enough to deform sufficiently to achieve intimate contact with the surfaces to which they are applied but they retain some elastic memory, at least for a time after bonding. Therefore, it is not the viscosity which dictates the resistance to the wetting flow, but rather the modulus, with lower modulus promoting wetting.

2.2.4. Debonding in PSAs

An effective pressure-sensitive adhesive must provide resistance to debonding from the substrate. However, in many PSA applications, (masking tape, for example) it is desired that the adhesive be removable with some moderate force. Unlike structural adhesives, stronger is not always better. It is also generally desirable that the PSA separate from the substrate and not fail cohesively within the bulk of the adhesive. In tape applications, it is undesirable for the PSA to separate from the backing, leaving adhesive residue on the substrate.

PSAs are very soft compared to most other types of adhesives and therefore, they deform to a much greater degree when they debond. When subjected to peeling stresses, internal voids appear in the PSA because they are highly resistant to volume expansion, similar to rubbery polymers in this regard. Often, these voids coalesce, creating filaments of adhesive which then proceed to elongate. Other times, the adhesive takes on more of a curtain shape near the peel front. In either case, the deformation in the adhesive is primarily extensional and the elongation achieved before debonding is often as high as 10 times the initial adhesive thickness. For most PSAs, the failure is by detachment of these adhesive strands from the substrate rather than by cohesive failure within the strand. Several researchers have studied the visual appearance of the adhesive in the peel front under different conditions of peel speed and for different types of PSAs [3,4].

The interfacial interactions bond the adhesive molecules to those of the substrate, but stresses to debond this interface can only be applied by deforming the adhesive itself. When a tape is peeled, a force is applied through the backing which is transferred through the adhesive to the interface in the zone of the peel front. The strength of that interface determines how much the adhesive must deform before providing the condition at which the interface will separate. The mechanical properties of the adhesive determine how much work it takes to deform the adhesive to that state. The energy required to do this work manifests itself in the peel force. This is the essence of the coupling between the interface and the rheology in the debonding process.

3. Rheology and common performance tests

3.1. Time scale-dependent properties and PSA performance

The three most common tests of PSA performance (tack, peel, and shear) each have certain characteristic time scales and requirements of the rheological properties at those time scales. In this section, we will examine those requirements and the correlation between rheology and performance.

3.2. Tack

In the context of PSAs, the term 'tack' refers to the ability of an adhesive to form a bond rapidly and to provide significant resistance to debonding. One of the common means of testing tack is with a probe tack test. A probe, usually cylindrical with a flat or slightly convex face, is brought into contact with the adhesive surface, held for a certain dwell time, then pulled away. Although conventional tack tests measure only the maximum force required to pull the probe away, modern techniques allow measurement of the full force vs. displacement curve during debonding [5]. One can analyze the shape of this curve as well as identify the peak force and the area under the tack curve which is the work done to remove the probe. For typical tack tests, the dwell time of the probe is on the order of 1 s.

Dahlquist [6] noted that there seems to be a minimum value of the compliance of a PSA in order for it to exhibit tack or equivalently, a maximum value of the modulus. He analyzed data on rheological properties as a function of time (or frequency) and temperature as well as tack as a function of dwell time, separation rate and temperature. He observed that only those materials with sufficiently high compliance in the 1-s time scale at the temperature of use were tacky. This so-called 'Dahlquist criterion' is expressed as follows for various common rheological measurements

$J(1 \text{ s}) \geq 3 \times 10^{-6} \text{ Pa}^{-1}$ Creep compliance

$G(1 \text{ s}) \leq 3 \times 10^{5} \text{ Pa}$ Stress relaxation modulus

$|E^*|(1 \text{ s}^{-1}) \leq 1 \times 10^{6} \text{ Pa}$ Magnitude of complex (dynamic) tensile modulus

$|G^*|(1 \text{ s}^{-1}) \leq 3 \times 10^{5} \text{Pa}$ Magnitude of complex (dynamic) shear modulus

The choice of 1 s as the time scale is somewhat arbitrary and is based on the amount of time one typically would be willing to allow for a PSA to bond to judge it tacky. However, in some real applications, it may be necessary for a PSA to form a bond faster, such as in high speed splicing on a paper machine. In such cases, it appears that the PSA will be able to form a bond if the compliance at the actual required time scale meets this same numerical factor.

Much experimental work suggests that Dahlquist's criterion is necessary but not sufficient for tack. Meeting this compliance criterion ensures that the material can deform sufficiently to wet the surface and form the bond but says nothing about its ability to resist the debonding. Highly swollen gels are examples of materials which might meet this criterion but would not be judged to be tacky. Although they are sufficiently soft to wet the surface, these materials lack the ability to dissipate sufficient energy in the debonding process to provide a significant resistance to debonding.

The resistance to probe removal in a standard tack test is governed by the adhesive response to deformations in a much shorter time than the 1-s scale governing bonding. If we imagine the adhesive stretching in the direction perpendicular to the plane of the adhesive film as the probe is removed, its elongation rate is equal to the probe removal speed divided by the adhesive thickness. One can look at the magnitude of the dynamic tensile modulus, $|E^*|$, at an angular frequency equal to this elongation rate for an idea of how the material will respond at this removal speed. Because the thickness of PSA coatings is so small (on the order of 30 μm), these elongation rates are typically on the order of 100 s^{-1}. The deformation time scale is the reciprocal, about 0.01 s.

It has been observed that the adhesive cavitates under such deformation, often breaking up into filaments which are then elongated until they break away from the probe. The force vs. displacement curve during such a test looks like the stress vs. strain curve of the adhesive at the appropriate rate, but terminates at the point where the adhesive debonds rather than at the ultimate tensile strength of the adhesive. Such large strain properties cannot be directly predicted from the linear viscoelastic properties which are more commonly measured but they are correlated. The initial stiffness in the tensile test conducted at a particular elongation rate is strongly correlated with the storage modulus at an angular frequency numerically equal to the elongation rate. The overall area under the stress–strain curve is more complicated because it depends not only on the tensile properties of the adhesive, but also on the condition at which debonding occurs. In general, we can say that the energy under the tack curve will depend on the ability of the adhesive to absorb energy in tensile deformation at the elongation rate characteristic of debonding.

3.3. Peel

For most peel testing, one tries to ensure that the wetting and bonding process is complete. Sometimes the tape will be bonded to the surface and then aged to ensure equilibration of the bond. At the very least, the dwell time and conditions are specified and are considerably longer than the short dwell time used for a tack test. The ability of the adhesive to undergo viscous flow in the time scale allowed will be important in predicting the tendency for the adhesion to increase with dwell time, especially on rough surfaces. Of course, there are also interfacial chemistry mechanisms for adhesion build with time which are outside the scope of this chapter.

The time scale which is characteristic of the adhesive deformation in peeling of a PSA tape is governed by the peeling rate and the adhesive thickness. Fig. 2 shows the peel front for a PSA tape according to Kaelble [7]. The length of the

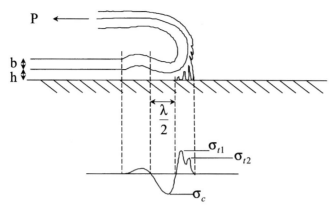

Fig. 2. Stylized drawing of typical PSA peel front showing local tensile and compressive stresses exerted by the adhesive on the substrate according to Kaelble [7].

stress wave, λ, is approximately the length in the direction of peel over which the adhesive is significantly deformed. Although the actual length depends in detail on the backing and adhesive thickness and properties, it is always of the order of the adhesive thickness. Therefore the time from beginning to end of deformation for any given adhesive element is of the order of the adhesive thickness divided by the peel front propagation rate. For a typical adhesive thickness of 30 μm and at a the moderate peeling rate of 30 cm/min, the characteristic time scale of adhesive deformation in peel is

$$t = \frac{30 \times 10^{-6} \text{ m}}{30 \times 10^{-2} \text{ m/min}} \cdot \frac{60 \text{ s}}{\text{min}} = 6 \times 10^{-3} \text{ s}$$

Thus, the peeling time scale is much shorter than the typical bonding time scale. A value of 0.01 s is often used as a representative value of the correct order although, of course, the actual value depends inversely on the peeling rate. This is the same order of magnitude as the debonding time scale for typical tack tests.

Just as in the probe tack experiment, the primary mode of adhesive deformation in peeling is in extension out of the plane of the tape. The amount of work done to deform the adhesive (which is proportional to the peel force) will depend on the stress–strain behavior of the adhesive at the rate dictated by the peeling and on the condition for debonding of the adhesive from the substrate.

The qualitative rule which is often quoted is that the adhesive should be stiff in the time scale of peel. However, it is probably more important for the adhesive to have high energy loss characteristics (i.e. high E'' or G'', viscous component of dynamic modulus) at the relevant frequency. However, if one couples the requirement for high modulus in the 0.01 s time scale with that of low modulus in the 1 s time scale to achieve wetting, high energy dissipation in the range between these time scales necessarily follows. In other words, whenever the

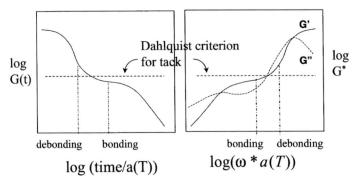

Fig. 3. Linear viscoelastic master curves of PSA in relation to Dahlquist criterion for tack.

storage modulus is changing rapidly with frequency, the loss modulus is relatively high. Although not obvious, this fact results from the interrelations among the viscoelastic functions as described by Ferry [8]. Elastic materials have little rate dependence and low loss characteristics. Viscous materials have strongly rate-dependent properties and high loss when the viscosity is high.

The linear viscoelastic properties of a typical PSA as a function of frequency and time scale are shown in Fig. 3 in relation to the Dahlquist tack criterion and the characteristic time scales for bonding and debonding for typical tack and peel tests. Note that the PSA cannot be characterized as liquid-like in the bonding time scale in the sense of having more viscous than elastic character. The G' value (elastic component of dynamic shear modulus) is typically greater than G'' (viscous component) because this time scale normally falls in the rubbery plateau region of the response. Indeed, the PSA might do a better job of wetting the surface and be tackier if the molecular weight were low enough so that the viscous character did dominate in this region. However, such a material would have little or no chance of being removable cleanly from the surface. Experience has shown that true liquid flow in this time scale is not necessary. It is sufficient for the modulus of this soft semi-solid to be low enough to allow sufficient deformation for the PSA to fully wet the substrate. Notice also that the PSA is not only stiffer in the debonding time scale than in the bonding time scale but the energy dissipation (as indicated by G'') is much higher.

As one might expect because of the role of viscoelasticity in the performance of PSAs, it has been found that if one measures peel force as a function of peeling speed at various temperatures, the curves can be reduced to a single master curve using the same shift factors which govern the superposition of the rheological properties (e.g. Kaelble [9], Derail et al. [10,11]). For uncrosslinked PSAs, such a master curve will have a shape like that of the solid curve in Fig. 4. At low rates (high temperatures), the PSA will split cohesively rather than separating cleanly from the substrate. As the peeling rate increases, the force increases in this

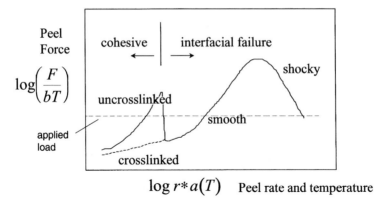

Fig. 4. General shape of peel force master curve for peel rate and temperature dependence.

zone. At a critical speed, the failure mode makes a transition to interfacial failure, often accompanied by a reduction in the peel force (although not always). As the peeling rate continues to increase, the peel force also increases in this region and the peel is smooth and steady. At some point, the peel will become unstable when tested using a constant pulling rate. The force will oscillate as the peel front sticks and slips. The average peel force will drop with increasing rate in this region. This unstable region is often referred to as 'shocky' peel. Gent and Petrich [12] showed that similar results are observed for rubbers which have been thermally bonded to a substrate and then crosslinked. The difference is primarily in the location of the transitions. For adhesives with a greater degree of crosslinking, the cohesive failure regime can be suppressed and the curve may look more like that of the dotted curve in Fig. 4.

The rate at which the transition from cohesive to interfacial failure occurs is strongly correlated with the frequency at which the transition from rubbery to flow behavior occurs in the linear viscoelastic properties. The second transition from smooth to unsteady peel is related to the frequency at which the transition from rubbery to glassy behavior occurs in the viscoelastic properties. More quantitative connections between the rheological master curves and the peel master curves will be explored in Section 5.

Although a multitude of adhesive performance tests exist, many of them are simply peel tests under different conditions of rate and temperature. All of these can be understood in the context of the peel master curve, including peel tests conducted under constant applied load where the peel rate is the measured quantity. Locating the peeling load on the vertical axis, one can read across to find the intersection of that load with the curve. As an example, consider the dotted line in Fig. 4. There are actually four intersections and therefore four possible peeling rates which can occur under this load. Normally, the peel will occur at one of the two intersections where the slope is positive as the peel is stable under those conditions. If the load is applied gently, it is likely that the peel will occur on the first intersection and the failure will be cohesive. A sudden jolt can cause a transition to the second state with much faster peel in the interfacial failure mode. From the two curves, it becomes clear why excessive crosslinking can reduce the resistance to peel under applied load conditions, although it will eliminate the transfer of adhesive to the substrate.

The resistance to peel under steady loads is important in many PSA applications. Failures in such cases can be by peeling at very low rates. For example, masking tape is used to hang masking paper from walls or sides of vehicles. The tape may be expected to hold this load for a day or more. If the tape peels at this load at a rate sufficient to cause significant loss of the drape over this time, its performance is unacceptable. Similarly, high performance foam tapes have been used to hold body side moldings on cars. If the molding initially has a flat shape and the car surface is curved, after bonding, the molding will be exerting a constant stress in the direction of peeling the adhesive at the ends of the piece. In this case, the PSA must resist that peeling stress for the lifetime of the vehicle. Even lifting a few centimeters per year is unacceptable.

Such performance characteristics cannot be predicted from the value of the peel force at the typical rates of standard peel tests. In fact, since the low rate regions of the peeling master curve are governed by very different molecular processes than those governing the high rate regions, there is almost no correlation between the two.

3.4. Shear

Performance in static load shear tests (see Fig. 5) can be related to the viscosity of the adhesive when dealing with uncrosslinked or lightly crosslinked (below the gel point) PSAs. If we assume that this material will creep governed by the zero shear rate limiting viscosity of the adhesive, η_0, and that the area determining the shear stress, τ, is the constantly decreasing overlap area: $w \cdot l(t)$, then

$$\tau = \frac{F}{wl(t)} = \frac{\eta_0}{h}\frac{dl}{dt}$$

Integration and solution of this equation from $l = l_0$ to $l = 0$ leads to the following relation of Dahlquist [13] between shear hang time, viscosity and geometry

$$t_c = \frac{l_0^2 w \eta_0}{2hF}$$

where t_c represents failure time, l_0 is the length of overlap of tape on panel in direction parallel to applied force, w is the tape width, h is the adhesive thickness, F is the applied force and η_0 is the limiting viscosity at zero shear rate.

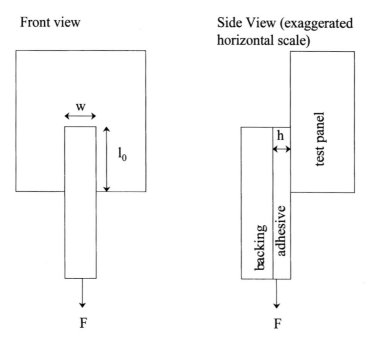

Fig. 5. Static shear test geometry. Load, F, is applied by hanging a weight from a sample adhered to a panel with a rectangular overlap area.

The value of η_0 can be obtained from a creep compliance test by waiting until the creeping flow reaches steady state. It can also be obtained from the dynamic mechanical data from the following relation

$$\eta_0 = \lim_{\omega \to 0} \frac{G''}{\omega}$$

For crosslinked adhesives (including the physically crosslinked block polymer PSAs), the static load shear test fails not by steady creeping flow of the PSA but by a 'pop-off' mechanism which is probably a very low angle peeling process. The failure times in these cases are very difficult to predict and not simply related to the rheology. Zosel [14] points out that the time scale and deformation rate of the PSA in a static load shear test depends on the failure time. Examining the rheological properties of the adhesive at the time scale dictated by the shear failure time, he found that the magnitude of the dynamic modulus, $|G^*|$, at the angular frequency defined by

$$\omega = \dot{\gamma}_s = \frac{l_0}{h \cdot t_c}$$

was approximately equal for a variety of adhesives. However, all the tests were done with the same load and geometry. Zosel states that a similar relation is obtained at other values of the shear stress, but with a different value of the dynamic modulus. Based on his data relating the maximum shear stress in a dynamic shear failure test to the dynamic modulus at the appropriate frequency, it would seem that a general relationship would be possible involving a ratio of the modulus at the failure time scale and the applied shear stress, but such a relationship has not been developed.

4. Viscoelastic windows of performance

4.1. Definition of viscoelastic window concept

The idea of identifying classes of PSAs by a 'viscoelastic window' within which they fall was proposed by Chang [15] and further discussed in a later publication [16]. We have talked of the need for liquid-like behavior over long deformation times and solid-like behavior in response to short time deformation. Chang chooses to look at the storage and loss components of the dynamic modulus at two frequencies: 0.01 and 100 s^{-1}, to characterize the viscoelastic window of an adhesive. These four values are used to construct the four corners of the window on a plot like that shown in Fig. 6.

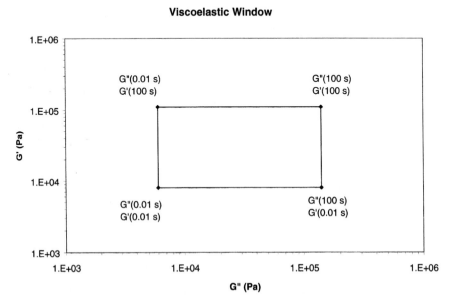

Fig. 6. Schematic diagram of a viscoelastic window of a PSA.

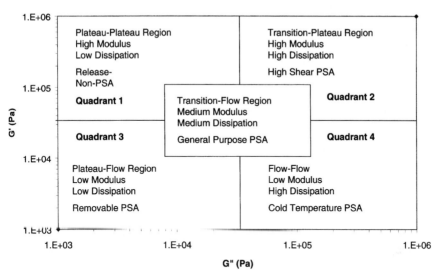

Classification of Materials

Fig. 7. Classification of materials by quadrants in which their viscoelastic windows appear.

4.2. Classification of adhesive types

Chang then proceeds to divide up the space into four quadrants and a central section as shown in Fig. 7. One can generalize the properties of different materials based on where in this diagram their viscoelastic windows primarily fall.

The first line of the description within each quadrant indicates the regions of the rheological master curve within which the debonding and bonding frequencies fall, respectively.

One of the advantages of this approach is that both the low and high frequency required by this characterization are accessible with available instruments. One does not need to use time–temperature superposition and construct a master curve to obtain these values. Since data at only two frequencies are needed and at a single temperature, data acquisition is very fast. Obviously, much detail from the master curve is not present here, but for a rapid evaluation of how materials compare to one another and assessment of their potential utility as PSAs, this is a very powerful tool.

5. From rheology to peel force

5.1. Introduction

Although we have discussed the rheological requirements of adhesives and some qualitative trends in peel, tack, and shear performance as a function of the rheology, we have not yet attempted to make a quantitative connection between the rheological properties and peel force. This is an active area of research and in this section we will examine some of the approaches which have been taken to address this issue.

5.2. Correlation between peel master curves and rheological master curves

Since the interface strength does not depend on the rate of peel, the shape of the peel master curve must arise from the rate dependence of the properties of the adhesives and we would like to be able to predict it from rheological measurements.

It is tempting to try to relate directly the peel master curve to one or another rheological function. However, such attempts have been unsuccessful. Although the $G''(\omega)$ function is similar in shape to the peel force vs. rate function from the onset of interfacial failure onward, the magnitude of the variations in G'' are much larger than the peel force variations. Of course, no such simple analysis could predict the location of the transition from cohesive to adhesive failure.

Many have recognized that the interface strength and the rheological properties are coupled and in a roughly multiplicative way. The data of Andrews and Kinloch [17] suggest that the fracture energy, G_c, can be expressed as a product of a viscoelastic function and the interface strength as given by an intrinsic adhesive failure energy, G_c^0

$$G_c = G_c^0 f(\dot{c} \cdot a(T))$$

where $\dot{c}$ represents the peel front propagation rate and f is a function. No simple relationship between measurable rheological properties and this viscoelastic function has been proposed.

Their data consist of master curves for fracture energy of a styrene–butadiene rubber from various surfaces along with the cohesive fracture energy of the rubber itself as a function of crack propagation rate. The results showed that the curves were the same shape, differing only by a vertical shift on the log axis, i.e. by a multiplicative factor. This factor is presumed to be associated with the different values of the intrinsic adhesive failure energy on these surfaces. The curves from different surfaces appear to be simply vertically shifted on a log axis, i.e. the values differ by a multiplicative constant. These data are not for PSAs, but for rubber samples in cohesive failure and peeling from surfaces against which the rubber was cured.

However, other data exist on PSAs which show that the peel master curves for the same adhesive on different surfaces are not simply shifted vertically from one another. Consider the data of Kaelble in Fig. 8 for an acrylic PSA peeling from glass and polytetrafluoroethylene (PTFE). Although the curves are similar and indeed the values on PTFE are consistently lower than on glass, the transition to

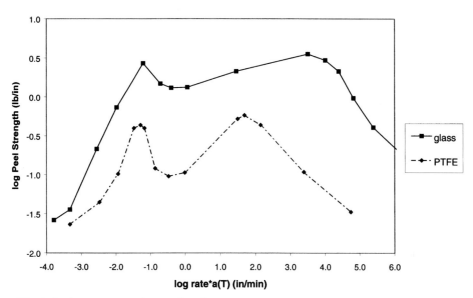

Fig. 8. Peel master curve for acrylic adhesive on various surfaces according to Kaelble [7].

shocky behavior occurs at a substantially lower rate on PTFE than on glass. Such a difference cannot be predicted by the simple multiplicative relationship.

5.3. Energy balance considerations

We will now examine attempts to more quantitatively model the peeling process and predict the peeling force from the mechanical properties and the deformation of the adhesive. In order to understand these approaches, we must first understand the relationship between energy absorption in the adhesive and peel force.

Consider the energy balance involved in steady state peeling of a pressure-sensitive tape at constant rate. For simplicity we will consider 90° peel so that the distance the peel front moves is equal to the distance of motion of the tape tab being peeled. For this simplified analysis, we will also assume an inextensible backing and a rigid substrate so that no energy is stored or dissipated in the deformation of the backing or the substrate. We apply a force F and peel a distance d. Therefore the work done is given by $F \times d$. Where did this energy go? Since neither the kinetic nor the potential energy of the system has changed, all that work must have been dissipated as heat. Where did that occur? The answer is in the adhesive as it is stretched and then relaxed again as it passes through the peeling nip. The volume of adhesive which has been deformed is given by $d \times w \times h$ where w is the tape width and h is the adhesive thickness. If the energy dissipation per unit volume of adhesive under this deformation cycle is given by U, then we have

$$U w h d = F d$$

or

$$U h = \frac{F}{w}$$

This simplified analysis assumed that the entire thickness of the adhesive takes part in the deformation to an equal degree. Although this may not always be true, a linear dependence of peel strength (i.e. peel force per unit width) on adhesive thickness is commonly observed within the range of thickness typical of adhesive tapes.

Often, one defines a fracture energy for peeling which is the energy absorbed per unit area of interface peeled. If we designate the fracture energy as G_c, then

$$G_c = \frac{F d}{d w} = \frac{F}{w}$$

Although it may not seem obvious at first, force per unit width is dimensionally equivalent to energy per unit area.

In reality, the backings are generally not inextensible and there are energy effects associated with bending of the backing as well. Kinloch [18] has published an improved analysis of peel which takes into account the effects of peel angle and the stretching of the film backing, allowing one to extract a fracture energy which is characteristic of the adhesive bond even for highly deformable backings and unifies results from various peel angles. In other words, it isolates the contribution to the fracture energy which comes from the adhesive deformation alone. This value is still not the thermodynamic work of adhesion, however, but reflects the large amount of energy absorbed in adhesive deformation. It is these corrected values of the fracture energy which should, ideally, be used for comparison to modeling predictions based on adhesive deformation alone.

5.4. Tensile elongation approximation

There seems to be a strong consensus that the dominant mode of deformation of the adhesive during peeling is extensional flow. Such a view is certainly supported by the peel visualization experiments which have been done [3,4]. Even though sometimes the adhesive breaks up into individual strands and other times deforms as sheets perpendicular to the peel front, it appears that approximating the complex deformation field as uniaxial extension is a reasonable approximation.

Therefore, let us imagine the adhesive as a set of independent packets of material, each of which is subjected to uniaxial tensile deformation in the direction perpendicular to the substrate surface as shown in Fig. 9. For 90° peel, the

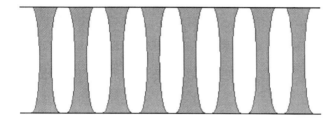

View looking into peel nip

Fig. 9. Picture of simplified model of adhesive deforming as individual tensile elements.

elongation rate of each strand is the peel rate divided by the adhesive thickness which is the same as the initial length of each strand. Under this approximation, the energy absorbed in each adhesive strand is given by the area under the force vs. distance curve of the adhesive up to the point at which it detaches from the surface. If we know the stress–strain curve of the adhesive material at the appropriate elongation rate and temperature, then the amount of energy absorbed per unit volume of adhesive is the area under the tensile stress vs. strain curve up to the detachment point. If detachment does not occur and the adhesive fails cohesively, then the peel force should be governed simply by the area of the stress–strain curve up to the point of fracture of the adhesive strand.

Although difficult, it is possible to measure stress vs. strain curves of PSAs. Examples of such work include that of Christenson et al. [3] and Piau et al. [23]. One can do this at various elongation rates and temperatures and create a material response function. Of course, it is much easier to obtain rheological data at small strains than to obtain tensile stress–strain data. One can assume a shape of the stress vs. strain function (i.e. a constitutive relationship) and then use the small strain data to assign values to the parameters in such a function. In order for a predictive model of peel to be useful, one should be able to use readily obtained rheological parameters like those obtained from linear viscoelastic master curve measurements and predict peel force master curves.

5.5. Constitutive equations for adhesive deformation

In order to avoid the necessity of measuring the stress–strain curves of the adhesive as a function of rate and temperature, we would like to have a constitutive equation for the adhesive, the parameters of which can be determined from relatively simple rheological measurements. Authors have taken multiple approaches to this problem. In some of the work, the adhesive stress–strain curves in uniaxial extension were measured which allows a direct comparison between the constitutive model and the data. In other cases, the model was used as a tool to allow prediction of the peel behavior from measurements of the linear viscoelastic properties but the stress–strain curves were not measured.

Yarusso [33] used a generalized Maxwell model, like that shown in Fig. 11, but with a larger number, n, of parallel Maxwell elements. For this model, the fundamental equation governing stress and strain in each element in tension is

$$\frac{d\varepsilon}{dt} = \frac{1}{E_i}\frac{d\sigma_i}{dt} + \frac{\sigma_i}{\theta_i E_i}$$

With a constant extension rate such that

$$\varepsilon(t) = Rt$$

the resulting equation for the tensile stress as a function of time of deformation is

$$\sigma(t) = R\sum_{i=1}^{n} E_i\theta_i\left(1 - e^{-t/\theta_i}\right)$$

In this work, the adhesive tensile properties were not measured directly. The equation was used to allow prediction of the peel results, with the material parameters being derived from the linear viscoelastic property measurements.

Christenson and McKinley [19] evaluated a generalized linear Maxwell model as well as the upper convected Maxwell model and the Giesekus model. These authors worked with the tensorial forms of these functions which are capable of correctly treating large strain deformations.

When the coordinate axes are chosen to align with the principal directions of the stress, the stress tensor in uniaxial elongation is

$$\boldsymbol{\tau} = \begin{bmatrix} \tau_{xx} & 0 & 0 \\ 0 & \tau_{yy} & 0 \\ 0 & 0 & \tau_{zz} \end{bmatrix}$$

and the strain rate tensor for uniaxial extension is

$$\dot{\boldsymbol{\gamma}} = \begin{bmatrix} -1 & 0 & 0 \\ 0 & -1 & 0 \\ 0 & 0 & 2 \end{bmatrix}\dot{\varepsilon}(t)$$

where $\dot{\varepsilon}$ represents the elongational strain rate. All three of these constitutive equations have a contribution from an element with only a viscous contribution. Its contribution to the stress tensor is

$$\boldsymbol{\tau}_\infty = -\eta_\infty\dot{\boldsymbol{\gamma}}$$

The contributions from the other elements of the relaxation time distribution for the three constitutive equations considered have the following forms.

$$\boldsymbol{\tau}_i + \theta_i \frac{\mathrm{d}}{\mathrm{d}t} \boldsymbol{\tau}_i = -\eta_i \dot{\boldsymbol{\gamma}} \quad \text{Generalized linear Maxwell model}$$

$$\boldsymbol{\tau}_i + \theta_i \boldsymbol{\tau}_{(1),i} = -\eta_i \dot{\boldsymbol{\gamma}} \quad \text{Upper convected Maxwell model}$$

in which $\boldsymbol{\tau}_{(1)}$ represents the upper convected time derivative of the stress tensor defined as follows:

$$\boldsymbol{\tau}_{(1)} = \frac{\partial}{\partial t} \begin{bmatrix} \tau_{xx} & 0 & 0 \\ 0 & \tau_{yy} & 0 \\ 0 & 0 & \tau_{zz} \end{bmatrix} - \begin{bmatrix} -\tau_{xx} & 0 & 0 \\ 0 & -\tau_{yy} & 0 \\ 0 & 0 & 2\tau_{zz} \end{bmatrix} \dot{\varepsilon}(t)$$

$$\boldsymbol{\tau}_i + \theta_i \boldsymbol{\tau}_{(1),i} - \alpha_i \frac{\theta_i}{\eta_i} (\boldsymbol{\tau}_i \cdot \boldsymbol{\tau}_i) = -\eta_i \dot{\boldsymbol{\gamma}} \quad \text{Giesekus model}$$

In the Giesekus model, the parameters α_i are greater than zero and are called the mobility factors.

For the generalized linear Maxwell model (GLM) and the upper convected model (UCM), the only material parameters needed are contained in the relaxation time spectrum of the material which can be obtained from simple linear viscoelastic measurements. For the Giesekus model, one needs in addition the mobility factors which Christensen and McKinley obtained by fitting the stress–strain curves of the adhesive. The advantage of the Giesekus model was that it provided them with a better description of the stress–strain curves. This, of course, is to be expected since those curves were used to deduce the parameters of the model.

The experimental stress–strain curves of the adhesive measured by these authors were bounded by the UCM and GLM predictions, with the UCM predictions exceeding the measured stress values and the GLM predictions falling below the measured values.

Derail et al. [10,11] used a non-linear integral constitutive equation of the KBKZ [20] type, which, for uniaxial extension, has the form

$$\sigma(t) = \int_{-\infty}^{t} m(t - t') h(\lambda) \left(\lambda^2 - \frac{1}{\lambda} \right) \mathrm{d}t'$$

in which $m(t)$ is the memory function and is related to the relaxation modulus by

$$m(t) = -\frac{\mathrm{d}G(t)}{\mathrm{d}t}$$

The strain is given by λ and the function $h(\lambda)$ is a damping function which corrects for the 'too large' strains which are predicted by the Lodge [21] elastic liquid constitutive equation which is identical to the equation above with $h(\lambda) = 1$.

The damping function chosen by these authors comes from Wagner [22] and is claimed to fit the non-linear shear and elongation data for a large number of linear polymers with this form

$$h(\lambda) = \left[\lambda^2 \exp(-b) + \lambda^a (1 - \exp(-b))\right]^{-1}$$

with $b = 4$ and $a = 0.3$. These authors did not measure the tensile properties of the adhesive directly but used this model to allow prediction of the peel master curve from the linear viscoelastic data.

In the work of Piau and colleagues [23,24], the Lodge elastic liquid constitutive model was used directly. These authors found that the Lodge model provided reasonable agreement with their uniaxial extension results, showing slightly more strain hardening than was observed in the data.

Therefore, with the exception of the Giesekus model, the parameters for all of these constitutive equations can be deduced from the relaxation time spectrum of the material which can be obtained from the small strain linear viscoelasticity measurements alone. There are various numerical methods in the literature which allow the determination of this spectrum from measured viscoelastic master curves, such as dynamic modulus, relaxation modulus, and creep compliance.

5.6. Modeling of peel force vs. rate and failure criteria

Several authors have shown that the peel force vs. rate can be calculated from the energy absorbed in elongation of the PSA using a suitable constitutive equation for the PSA along with measurements of the amount of strain in the adhesive strands at the point of detachment or strand breaking in the peel nip.

For example, Christenson et al. [3,19] performed a detailed study of polyisobutylene-based pressure-sensitive adhesives. Although these authors did not postulate a specific detachment criterion, they did extensive work characterizing the linear viscoelastic properties, the tensile stress–strain properties, and the peel force. In addition, they conducted detailed visualization of the deformation of the adhesive during peel and therefore, could assess the ability to predict the peel force from the mechanical properties of the adhesive and the visually observed detachment strain. In this work, the adhesive consisted of a blend of high and low molecular weight polyisobutylene. They showed that when they used the Giesekus model as the constitutive equation for the adhesive, they could accurately describe the stress–strain curves of the adhesive and the peel force was well predicted by the integral of the stress–strain curve up to the measured detachment strain. Their results are summarized in Table 1.

The three rows for each peel rate represent replicate tests. It is clear that the agreement between the calculated and the measured peel forces is quite good.

One of the things that Christenson points out in his work is that the contribution to the peel energy at a given peel rate comes from a broad spectrum of the

Table 1

Measurement and prediction results of Christenson and McKinley [19] on polyisobutylene pressure-sensitive adhesives

Peel rate (mm/min)	Time at detachment (s)	Strain at detachment	Measured peel force (N)	Calculated peel force (N)
50	3.08	1.57	1.3	1.15
50	3.00	1.37	1.0	1.00
50	3.18	1.47	1.2	1.12
100	1.81	1.75	1.9	1.67
100	1.89	1.68	1.7	1.67
100	1.83	1.55	1.4	1.51
200	0.76	1.57	1.9	1.52
200	0.99	1.73	2.0	2.05
200	1.03	2.00	2.4	2.43

relaxation times of the adhesive. The commonly held approximation that the behavior in peel at a given elongation rate is dominated by the viscoelastic properties at the equivalent frequency is shown to be crude at best and misleading at worst.

The work of Piau et al. [23] follows a similar pattern. Modeling is done using the measured stress–strain behavior of the adhesive, the measured elongation at detachment from visualization of the peel front and the measured peel force. They show that the peel force data can be calculated from the energy absorption in elongation of the adhesive strands but they do not identify a detachment criterion.

In the region of the peel behavior where the adhesive is splitting (i.e. the adhesive strands are breaking before detaching from the substrate), one would expect that the failure condition would be the same as when conducting tensile stress–strain testing on the PSA material. Such agreement was found in the results of Gent and Petrich [12], and Piau et al. [23,24], for example. Others who have not necessarily measured the fracture strain of the adhesive have nevertheless had good success in predicting the peel force vs. rate in the cohesive failure regime by assuming a constant value of the fracture strain for the adhesive and using that value along with the chosen constitutive equation. The work of Yarusso [33] and Derail et al. [10,11] are examples. Derail et al. chose a particular value of the fracture strain which they found to be roughly constant in some measurements of the stress–strain curves of their adhesives. Thus, it appears that the strain at which the adhesive strand will break is not strongly dependent on elongation rate or temperature.

When one reaches the peel rate/temperature conditions for interfacial debonding between the adhesive and the substrate, the appropriate failure criterion becomes a matter which has not been fully solved. Qualitative analysis of the curves

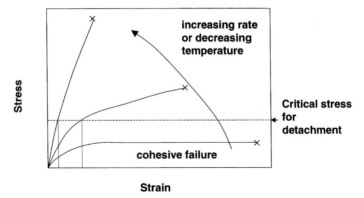

Fig. 10. Set of stress–strain curves of PSAs as a function of rate with critical stress and critical strain lines shown.

is sufficient to rule out certain possibilities. For example, an intuitively appealing criterion would be one of constant tensile stress at the point of debonding. This postulate can predict the transition from cohesive failure at low rate to interfacial failure at high rate as shown by Gent and Petrich [12]. However, one finds that if this postulate is used, the predicted peel force will always decrease with increasing rate in the interfacial failure zone. That this is so can be seen from inspection of Fig. 10. As the elongation rate increases, a viscoelastic material will behave as if it is stiffer, giving the family of stress–strain curves shown as a function to rate, with the modulus increasing and the break elongation decreasing as elongation rate is increased. At sufficiently low rate, the stress never reaches the critical value and the adhesive elongates up to its internal failure point and breaks. At higher rates, the critical stress is reached and debonding occurs where the stress–strain curve crosses the critical detachment stress level. However, because the curves rise more steeply at higher elongation rates, the area under the curve up to the debond point will necessarily decrease as the rate is increased. Therefore, the peel force will decrease with increasing peel rate for all rates beyond the transition from cohesive to interfacial failure, contrary to observations.

A second reasonable postulate might be that the detachment will occur at a critical strain or degree of deformation. Such a postulate would suggest that the peel force should increase with increasing rate but would not predict the transition from cohesive to interfacial failure as the rate increases.

One suggestion for an interfacial detachment failure criterion can be found in the work of Derail et al. [10,11]. The adhesives used in this work were blends of polybutadiene and tackifying resin. These authors also assume that the dominant deformation mode in the adhesive is elongational deformation. They employ a non-linear integral constitutive model for the adhesive stress–strain behavior of the KBKZ type [25]. This model describes fairly well the viscoelastic behavior of

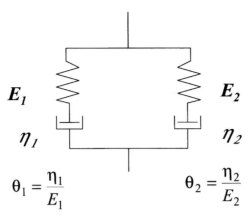

$$\theta_1 = \frac{\eta_1}{E_1} \qquad\qquad \theta_2 = \frac{\eta_2}{E_2}$$

Fig. 11. Two parallel Maxwell elements to model adhesive rheology in approach of Hata.

polymers in strong flows. The model contains a memory function of time which can be obtained directly from the relaxation modulus, $G(t)$ or from the complex shear modulus, $G^*(\omega)$ by an inverse Fourier transform. The model also contains a damping function. These authors chose a form for this function given by Wagner [26]. They use a detailed approach to account for the contributions to the peel force from the adhesive deformation as well as the bending of the backing, in this case an aluminum foil.

In the cohesive failure zone, they find a good fit to the data by assuming that failure occurs at a particular level of the Hencky strain. In the interfacial failure regime, they use a failure criterion which states that the product of the stress and the crack opening displacement at the crack tip is a constant, derived from work of de Gennes [27] and Hui et al. [28]. Their approach does predict the qualitative features of the measured peel master curves including the transitions from cohesive to adhesive failure and the transition to stick-slip peel at high rates. Furthermore, they obtain reasonably good quantitative agreement with the data with a single adjustable parameter which is the numerical value of the fracture criterion.

Another failure criterion postulate was proposed originally by Hata [29]. He suggests a debonding criterion based on a critical value of the stored elastic energy density in the adhesive. To facilitate calculation and understanding, he suggests using a simple mechanical model analogy to treat the rheological properties of the adhesive. In its simplest form, we model the adhesive as a set of two parallel Maxwell elements, as shown in Fig. 11. In order to approximate the behavior of the adhesive, we set one of the spring constants to a level approximating that of the glassy state and the other to a level characteristic of the rubbery plateau modulus. The viscosities of the elements are chosen so that the time constants of the two elements approximate the time scale of the transitions from glassy to rubbery behavior and from rubbery to flow behavior in the rheological master curves.

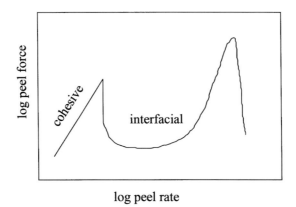

Fig. 12. Peel force vs. rate for two-element model.

Although this rheological model is a very crude representation of the behavior of any real PSA, it is sufficient to produce all the key qualitative features of PSA peel master curves when coupled with the stored elastic energy density debonding criterion. When the peel rate (and therefore the elongation rate of adhesive strands) is very low (i.e. the deformation time is long compared to the time scale for the transition from rubbery to flow behavior), nearly all of the deformation occurs in the viscous elements and the spring elements are stretched very little. As a result, the system never stores much elastic energy and the debonding criterion can never be met. Under these conditions, the model predicts cohesive failure. If one further postulates a maximum elongation to break, then one can calculate an energy of deformation up to the cohesive failure point and ultimately a peel force. In this zone, the peel force will increase in proportion to the peel rate. At a critical speed, the amount of energy stored in the springs will be sufficient to satisfy the stored elastic energy density criterion before the maximum elongation to break is reached and a transition to interfacial failure will occur. The model can predict the drop in peel force at this point. Further increases in speed result in greater energy dissipation in the stretching of the adhesive before reaching the debonding criterion over a wide range of rate and ultimately, a maximum is reached and then the energy (and peel force) begin to decrease with increasing rate. This maximum occurs at approximately the point where the deformation time becomes short compared to the shortest time constant in the rheological model, i.e. the time scale for the transition from rubbery to glassy behavior. A predicted peel vs. rate curve for such a simple two element model is shown in Fig. 12.

This approach was used in subsequent papers by Mizumachi and colleagues [30–32] to model behavior of PSAs in a rolling wheel tack test with controlled pulling rate. Hata recognized the oversimplification of the rheological properties which results from using a model with only two relaxation times and suggesting

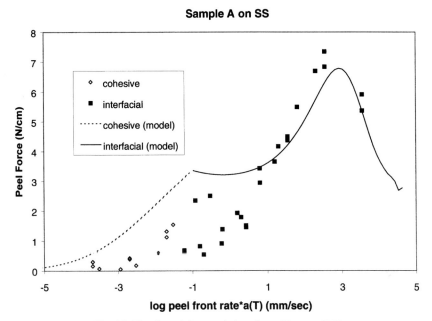

Fig. 13. Fit of model to peel data from Yarusso [33].

improving the approach by using a generalized Maxwell model with many parallel Maxwell elements. In a paper by Yarusso [33], the parameters of a generalized Maxwell model were obtained from the measured linear viscoelastic data for a set of PSAs and this approach was successfully used to model the peel master curves using only the failure criterion as a parameter of fit. An example of one of the model fits to the data is shown in Fig. 13. The model was unsuccessful, however, in predicting the shift in the time scale of the transition to shocky peel with a change from a high energy (high adhesion) substrate to a low energy release coating. As in Kaelble's data on PTFE, the transition to shocky peel occurred at a much lower rate than for the high-energy substrate. The model predicted a slight decrease in this rate as the critical stored elastic energy density was decreased to agree with the magnitude of the peel force, but the amount of shift was much larger in the experimental data than was predicted by the model.

It should be noted that for an adhesive which is behaving primarily as an elastic solid, the detachment failure criterion used by Derail et al. is equivalent to a stored elastic energy density criterion but under conditions where the deformation is primarily viscous, the two criteria are quite different. None of these authors has been able to successfully link the values of these failure criteria to fundamental interfacial properties or the thermodynamic work of adhesion. Clearly, much remains to be done to complete our understanding of the relationships among surface properties, adhesive rheology, and peel force.

6. Polymer structure and formulation effects on PSA rheology

Now that we have discussed the relationships between rheological properties and PSA performance, we will consider the control of the rheological properties through polymer structure and formulation. An excellent detailed review of many of these issues can be found in an article by Creton [34].

6.1. Regions of the master curve

All amorphous, single phase polymer systems have similarly shaped rheological functions. The general shape is shown in Fig. 14. There are four major regions of the response: glassy behavior, the glass-to-rubber transition region, the rubbery plateau, and the viscous flow region. Various polymers differ primarily in the location of the transitions and in the magnitude of the plateau modulus.

6.2. The glassy zone

All polymers have about the same modulus in the glassy region: of the order of 10^9 Pa. Not much about this region is important for the behavior of PSAs. Systems differ primarily in the location of the transition from glassy to rubbery behavior and that feature is extremely important in controlling the performance of PSAs.

6.3. Control of the glass transition temperature and its location in time scale

The glass transition is usually characterized by the temperature at which it is observed in a particular test. A common method involving mechanical properties measures the dynamic mechanical properties at a fixed frequency while scanning

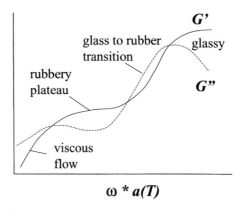

Fig. 14. Major regions of viscoelastic behavior.

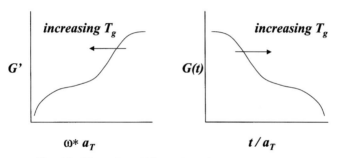

Fig. 15. Effect of T_g shift on viscoelastic master curve.

temperature. The location of the peak in the tan δ function is taken as the glass transition temperature, T_g. However, of more direct relevance to PSA performance is the location of the transition in the time or frequency domain of the rheological master curves. Increasing T_g will increase the transition position in time and decrease it in frequency as shown in Fig. 15.

6.3.1. Polymer structure

The chemical structure of the base polymer of a PSA determines its T_g. For an acrylic PSA which has no other additives, this is the only control of the transition position of the adhesive. Highly flexible chain backbones, unhindered in their rotation by the side groups tend to have low T_g values. The acrylic polymers used as PSAs contain at least one long chain alkyl acrylate monomer. These monomers act as a sort of internal plasticizer, diluting the chain backbone and contributing to the free volume of the system.

6.3.2. Copolymerization

The typical approach in synthesizing an acrylic polymer useful as a PSA is to use a variety of comonomers to control the glass transition temperature and the cohesive strength of the material. One can estimate the glass transition temperature of a copolymer based on the T_g values of the homopolymers produced by the individual monomers using the well known Fox equation. This empirical relationship states that

$$\frac{1}{T_g} = \sum_i \frac{w_i}{T_{gi}} \tag{1}$$

where T_g represents glass transition temperature of the copolymer in absolute temperature units (e.g. Kelvin), w_i is the weight fraction of component i, T_{gi} is the glass transition temperature of the homopolymer of component i in absolute temperature units, and the summation is taken over all the components in the copolymer. Of course, since the T_g is dependent on the testing rate, the same test conditions must be used for all the values in this equation.

6.4. Factors affecting the plateau modulus

6.4.1. Natural entanglement density of a polymer

According to the work of Ferry [8], every polymer has a natural density of entanglements with controls the magnitude of the modulus in the rubbery plateau region. This entanglement spacing can be expressed by an average molecular weight per entanglement. This number, coupled with the density, determines the plateau modulus. Polymers with long side chains tend to have a large molecular weight between entanglements and a low plateau modulus.

6.4.2. The role of diluents

When the polymer is diluted with low molecular weight molecules such as tackifiers and plasticizers, the spacing between entanglements increases and the plateau modulus goes down. The modulus decrease goes approximately as the second power of the polymer concentration, according to results of Graessley [35].

$$G_r = G_r^0 v_p^2$$

where G_r represents the rubbery plateau modulus of the mixture, G_r^0 is the plateau modulus of undiluted polymer, and v_p is the volume fraction of polymer in the system.

In block copolymer-based PSAs, there is an additional factor affecting the plateau modulus which is the filler effect introduced by the rigid microdomains of the glassy endblock. Kraus and Rollman [36] suggest a form which combines the Graessley effect with a stiffening effect following the Guth and Gold [37] equation as proposed by Holden [38] for block polymers.

$$G_r = G_r^0 v_p^2 \left(1 + 2.5 v_h + 14.1 v_h^2\right)$$

where v_p represents the volume fraction of polymer in the rubbery (midblock) phase alone and v_h represents the volume fraction of hard domain phase in the total system.

Dupont and De Keyzer [39], also used this equation to estimate the plateau modulus of PSAs formulated from the classical rubbery triblock copolymers such as styrene–isoprene–styrene and styrene–butadiene–styrene polymers commonly used in PSAs and their hydrogenated counterparts.

6.5. Formulation with tackifiers and plasticizers

Many PSAs are formulated by blending elastomers with tackifiers and plasticizers. The elastomers commonly used (such as natural rubber, polybutadiene, styrene-diene block copolymers, etc.) have plateau moduli which are higher than the Dahlquist criterion and therefore require dilution to be soft enough to wet the substrates. Furthermore, their glass transition temperatures are relatively low, resulting in low energy dissipation characteristics at normal debonding rates at room temperature. Blending of the elastomer with particular low molecular weight additives can address both of these issues and allow tailoring of the adhesive properties.

Tackifiers are low molecular weight materials with relatively high glass transition temperatures. Plasticizers are also low molecular weight materials but of low glass transition temperature. These definitions are not absolute, but rather are relative to the glass transition temperature of the system into which they are being blended.

The blending of a polymer with molecularly miscible tackifiers and plasticizers allows manipulation of the T_g of the system. The T_g can be roughly predicted by the Fox equation (Eq. 1). In this case the components in the summation are simply the different components in the mixture and the values needed are the glass transition temperatures for each of the materials: the polymer, the tackifier, and the plasticizer. Tackifiers generally have T_g values in the range from -30 to $+100°C$. Plasticizers have lower T_g values. Again, the transition position in time scale will be related to the T_g as described above for copolymers.

At the same time, the addition of these low molecular weight components dilutes the entanglements of the polymer and reduces the plateau modulus. The net result for the addition of a tackifier to an elastomer is shown in Fig. 16. The effect of plasticizer addition is shown in Fig. 17. Note that the combination of the shift in

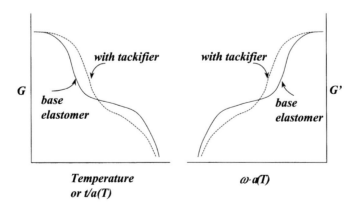

Fig. 16. Effect of tackifiers on rheology.

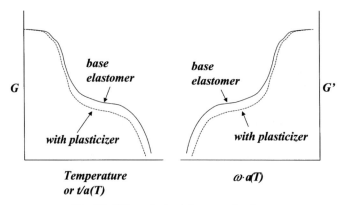

Fig. 17. Effect of plasticizers on rheology.

the glass-to-rubber transition and the depression of the plateau modulus can allow a tackified rubber to meet the criteria described earlier for a PSA, whereas the base elastomer is typically too stiff in the 1-s time scale and has little energy dissipation in the 0.01-s time scale. The use of plasticizer alone can soften the material enough to meet the Dahlquist criterion and allow wetting of the substrate, but it does not create the condition of high dissipation in the debonding time scale. Therefore, plasticized elastomers can be easily removed from surfaces with relatively little resistance. The use of a combination of tackifier and plasticizer allows one to manipulate the position of the glass-to-rubber transition and the magnitude of the plateau modulus independently over constrained ranges. Formulation of PSAs is largely a question of optimizing these two rheological characteristics to achieve the desired balance of performance properties.

6.6. Factors affecting the long time (low frequency) zone

6.6.1. Molecular weight and its distribution

If the molecular weight of a polymer is very low, i.e. less than its natural entanglement spacing, then the rubbery plateau region will not exist. For most polymers useful as PSAs, there will be a finite rubbery plateau and its length will increase as the average molecular weight increases. The transition to flow behavior will be more gradual for broad molecular weight distributions and sharper for a narrow distribution. The trend with molecular weight is shown in Fig. 18 along with the effect of entanglement density on the plateau modulus level.

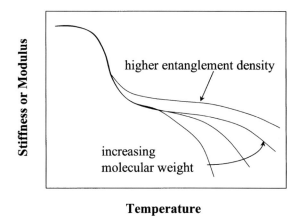

Temperature

Fig. 18. Effect of molecular weight and entanglement density.

6.6.2. Branching and crosslinking

Crosslinking in PSAs is normally conducted to enhance cohesive strength and prevent cohesive splitting of the PSA and the resulting adhesive residue left on the substrate. It is therefore desirable to crosslink to the point of creating a finite gel content in the material but not to the point of increasing the plateau modulus of the system. Therefore, crosslinking normally affects only the flow zone of the rheological master curve and not the plateau or glassy transition zones.

Sometimes, crosslinking processes stop short of the gel point, resulting in a high concentration of long chain branched molecules which are still soluble, but provide a greatly increased time scale for viscous flow.

If the material has a finite gel fraction, then there is no true flow regime. Instead, the relaxation modulus or the elastic part of the dynamic modulus will reach a limiting finite value at infinite time or infinitesimal frequency.

6.6.3. Behavior near the gel point

The behavior of G' and G'' with respect to frequency can be extremely instructive with regard to the state of gelation of the system as discussed by Winter [40]. The gel point is a critical point at which the first infinitesimal amount of gel of infinite (i.e. limited only by the macroscopic sample size) molecular weight. Below the gel point, G' is lower and falls faster with decreasing frequency than G'' in the flow region. When the crosslink density is higher than the critical level needed for gelation, G' remains higher than G'' and approaches a finite value as the frequency goes to zero. Right at the gel point, the two curves lie on top of one another over an extended range of frequency, as shown in Fig. 19.

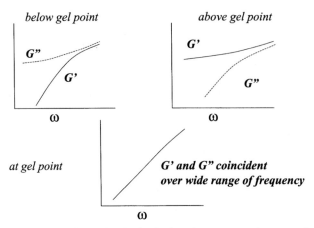

Fig. 19. Dynamic mechanical behavior in the low frequency regime near the gel point.

7. Summary

We have seen that there are many clearly identifiable rheological criteria controlling performance of PSAs in different tests and applications. Furthermore, the control of the rheological properties of PSA systems follows generally understood principles of polymer science. Detailed stress and energy absorption analysis methods have allowed us to understand the influences of adhesive contributions as well as those of the backing deformation to the apparent peeling strength. Recent efforts have begun to elucidate the quantitative relationship between rheological master curves and peel master curves. The single largest remaining unknown is the relationship between the strength of the interface as reflected in the thermodynamic work of adhesion and the detachment failure criterion associated with debonding of the PSA. If this link could be established, we would be in a position to predict most aspects of PSA performance based on the fundamental properties associated with interfacial chemistry and adhesive rheology.

References

1. Cherry, B.W., *Polymer Surfaces*. Cambridge University Press, Cambridge, 1981, pp. 24–25.
2. Fowkes, F.M., *J. Adhes. Sci. Tech.*, **1(1)**, 7–27 (1987).
3. Christenson, S., Everland, H., Hassager, O. and Almdal, K., *Int. J. Adhes. Adhes.*, **18**, 131–137 (1998).
4. Urahama, Y., *J. Adhes.*, **31**, 47–58 (1989).
5. Chuang, H.K., Chiu, C. and Paniagua, R., *Adhesives Age*. Sept. 1997, p. 18–23.
6. Dahlquist, C.A., *Proceedings of the Nottingham Conference on Adhesion*, Part III. MacLaren and Sons, London, 1966, p. 143.

7. Kaelble, D.H., *Trans. Soc. Rheol.*, **15(2)**, 275–296 (1971).
8. Ferry, J.D., *Viscoelastic Properties of Polymers*, 3rd edn., Wiley and Sons, New York, 1980, Chapters 3 and 4.
9. Kaelble, D.H., *J. Adhes.*, **1**, 102–123 (1969).
10. Derail, C., Allal, A., Marin, G. and Tordjeman, Ph., *J. Adhes.*, **61**, 123–157 (1997).
11. Derail, C., Allal, A., Marin, G. and Tordjeman, Ph., *J. Adhes.*, **68**, 203–228 (1998).
12. Gent, A.N. and Petrich, R.P., *Proc. R. Soc. Lond. A*, **310**, 433 (1969).
13. Dahlquist, C.A. In: Patrick, R.L. (Ed.), *Adhesion and Adhesives*, Vol. 2. Marcel Dekker, New York, 1969, p. 219.
14. Zosel, A., *J. Adhes.*, **44**, 1–16 (1994).
15. Chang, E.P., *J. Adhes.*, **34**, 189–200 (1991).
16. Chang, E.P., *J. Adhes.*, **60**, 233–248 (1997).
17. Andrews, E.H. and Kinloch, A.J., *Proc. R. Soc. Lond. A*, **324**, 301 (1971).
18. Kinloch, A.J., Lau, C.C. and Williams, J.G., *Int. J. Fract.*, **66**, 45–70 (1994).
19. Christensen, S. and McKinley, G.H., *Int. J. Adhes. Adhes.*, **18**, 333–343 (1998).
20. Bernstein, B., Kearsley, E.A. and Zapas, L.J., *Trans. Soc. Rheol.*, **7**, 391 (1963).
21. Lodge, A.S., *Elastic Liquids*. Academic Press, New York, 1964.
22. Wagner, M.H., *Rheol. Acta*, **18**, 681 (1979).
23. Piau, J.-M., Verdier, C. and Benyahia, L., *Rheol. Acta*, **36**, 449–461 (1997).
24. Verdier, C., Piau, J.-M. and Benyahia, L., *J. Adhes.*, **68**, 93–116 (1998).
25. Bernstein, B., Kearsley, E.A. and Zapas, L.J., *Trans. Soc. Rheol.*, **7**, 391 (1963).
26. Wagner, M.H., *Rheol. Acta*, **18**, 681 (1979).
27. de Gennes, P.G., *C. R. Acad. Sci.*, **307(II)**, 1949 (1988).
28. Hui, C.Y., Xu, D.B. and Kramer, E.J., *J. Appl. Phys.*, **72(8)**, 3294 (1992).
29. Hata, T., *J. Adhes.*, **4**, 161 (1972).
30. Mizumachi, H., *J. Appl. Polym. Sci.*, **30**, 2675 (1985).
31. Mizumachi, H. and Hatano, Y., *J. Appl. Polym. Sci.*, **37**, 3097 (1989).
32. Tsukatani, T., Hatano, Y. and Mizumachi, H., *J. Adhes.*, **31**, 59 (1989).
33. Yarusso, D.J., *J. Adhes.*, **70**, 299–320 (1999).
34. Creton, C., Materials science of pressure sensitive adhesives. In: Meijer, H.E.H. (Ed.), *Materials Science and Technology, Vol. 18, Processing of Polymers*, Wiley-VCH, Weinheim, Germany, 1997.
35. Graessley, W.W., *Adv. Polym. Sci.*, **16**, 1 (1974).
36. Kraus, G. and Rollmann, W., *J. Appl. Polym. Sci.*, **21**, 3311–3318 (1977).
37. Guth, E. and Gold, O., *Phys. Rev.*, **53**, 322 (1938).
38. Holden, G. In: Ceresa, R.J. (Ed.), *Block and Graft Copolymerization*, Wiley, New York, 1973.
39. Dupont, M. and De Keyzer, N., PSTC Tech XVII, Technical Sem. Proc. May, 1994.
40. Winter, H.H., *Polym. Eng. Sci.*, **27**, 1698–1702 (1987).

Chapter 14

Tack

COSTANTINO CRETON [a,*] and PASCALE FABRE [b]

[a] *Laboratoire PCSM, Ecole Supérieure de Physique et Chimie Industrielle, 10 Rue Vauquelin, 75231 Paris Cedex 05, France*
[b] *Centre de Recherche Paul Pascal, Av. Dr Schweitzer, 33600 PESSAC, France*

1. Introduction

In order to be effective, an adhesive must possess both liquid properties, to wet the surface when the bond is formed, and solid properties, to sustain a certain level of stress during the process of debonding. Structural adhesives accomplish this by a chemical reaction, typically a polymerization, which transforms a liquid mixture of oligomers into a crosslinked polymer. For pressure-sensitive adhesives (PSA), however, this transition must occur without any change in temperature or chemical reaction. This property is called tack and gives PSA the ability to form a bond of measurable strength by simple contact with a surface. It gives PSA their easy and safe handling, since the adhesive can be applied to the surface without the use of any solvent, dispersant or heat source.

Since a PSA must have some amount of tack to be considered as such, it is essential to understand what are the minimum requirements in terms of molecular structure of the components and in terms of formulation for a material to be tacky. However, because tackiness is a complex and not yet completely understood mechanism, it is relatively difficult to establish simple relevant criteria for a good adhesive material.

Traditionally, tack properties of a PSA have been correlated to their linear rheological behavior, such as elastic and loss modulus [1–3]. While this type of phenomenological analysis provides many clues for the practical design of PSA, it is intrinsically limited by the fact that a tack experiment involves large strains and transient behaviors of the PSA, which cannot be easily predicted by either viscosity (shear, elongational) or any other small strain steady-state dynamical property. The simple observation of the debonding of a PSA tape from a solid

* Corresponding author. E-mail: costantino.creton@espci.fr

reflects the complexity of the phenomena at work: final rupture often occurs through the formation of a fibrillar structure [4,5] and measured tack energies are much larger than the thermodynamic work of adhesion, W_a, characterizing the reversible formation of chemical bonds at the interface.

Moreover, to consider the adhesive alone is not sufficient to predict its behavior in a situation where surface effects can be important: the occurrence of bubbles or fibrils is not only a matter of the viscoelastic properties of the adhesive but depends also on the characteristics of the surface of the adherend: roughness, surface tension, and on the thickness of the adhesive film. In fact, surface and bulk effects are coupled and it would thus be more accurate to consider adhesive/substrate pairs than adhesives and substrates separately.

Despite these difficulties, recent experimental and theoretical work focusing on the microscopic mechanisms taking place during the debonding of the PSA from the substrate, have greatly enhanced our understanding of tackiness.

2. Experimental methods

Since by definition a tacky material must be sticky to the touch, all standardized testing methods of tackiness seek to reproduce in one way or another the test of a thumb being brought in contact and subsequently removed from the adhesive surface.

The main experimental methods to quantify tack can be divided into two categories: methods which provide essentially a single number are designed to be very close to the application and mainly aimed at quick comparisons between materials. Among those, described schematically in Fig. 1, are loop tack, rolling ball or rolling cylinder tack. In those methods, typically the contact time, debonding velocity, applied pressure, are reasonably reproducible but cannot be independently controlled. At the other extreme, probe methods are more difficult to implement (although standard instruments are commercially available) but provide much more information and allow the control of the main experimental parameters independently. Because these tests are more informative, they will be the focus of the rest of this paper.

All probe methods are based on the physical principle described in Fig. 2. A probe, with a flat or spherical tip, is brought into contact with the adhesive film, kept in contact for a given time and under a given average compressive pressure, and then removed at a constant velocity, V_{deb}. The result of the test is a force vs. time curve of the adhesive film in tension, which can be easily converted to stress vs. strain by a proper normalization [8].

Several variations of the test exist. Historically, the first referenced probe test is the Polyken probe tack test developed by Hammond [6]. In this case, the probe has a flat tip and is upside down. The compressive force is controlled by lifting a

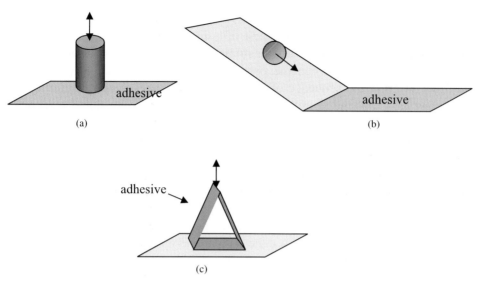

Fig. 1. Schematic of the different methods for the evaluation of tack properties. (a) Probe tack. (b) Rolling ball tack. (c) Loop tack.

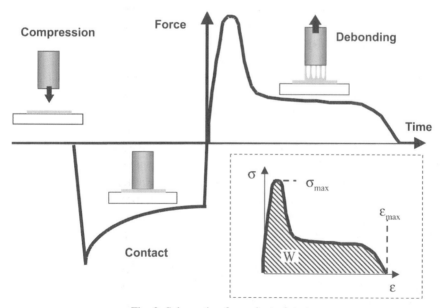

Fig. 2. Schematic of a probe tack test.

weight over which the adhesive is deposited. In this early version, only the maximum force of the debonding curve was recorded and taken as a measure of tack.

An improvement over this methodology was developed by Zosel [7], who used a displacement controlled instrument and pointed out the importance of considering the complete debonding curve rather than simply the maximum in tensile force. He also used for the first time in situ optical observations during the debonding of the probe, demonstrating that good adhesives are able to form bridging fibrils between the probe and the substrate [4]. However, it is only recently that the sequence of microscopic processes occurring during the debonding of a flat-ended probe from a soft adhesive were elucidated, again thanks to in situ optical observations and measurements [8,9].

In parallel to this development, other groups have been using spherical tip probes to test tackiness. Using a spherical tip has two important consequences. It solves the practical problem of good alignment between the probe and the film which gave rise to poor reproducibility of the results [10] and it allows, at least in the early stages of the debonding process, to study quantitatively the motion of a circular crack and to use the energy release rate concepts at the edge of the contact zone [11] to measure a value of $\mathcal{G}_c$, the critical energy release rate. However, this latter approach, often called the JKR method, is limited to relatively elastic PSA as discussed in detail in Chapter 15 of this book. For soft and highly viscoelastic systems, using a flat probe has the advantage of applying a uniform displacement field to the layer facilitating the analysis of the fibrillation process.

This chapter will therefore focus on results obtained with the flat probe with a particular emphasis on the recent theoretical and experimental developments which have shed some light on the microscopic mechanisms of debonding.

3. Analysis of the bonding and debonding mechanisms

The tackiness of a specific adhesive is dependent on its ability to bond under light pressure and short contact time, while forming a fibrillar structure upon debonding from the substrate. It is then natural to test tackiness with a quick bonding and debonding test such as those described in the introduction.

Unfortunately, such a test applies a rather complicated loading and unloading cycle to a highly deformable material. Therefore, a microscopic analysis of the sequence of events occurring during a tack test is necessary in order to attempt a detailed interpretation of a tack curve. Rather than presenting an exhaustive review of the most recent theories, we have chosen to present here a rather phenomenological picture which, while it leaves certain aspects unexplained, remains consistent with experimental results.

Starting from a flat probe tack test such as that described in Fig. 2, the sequence of events can be broken down into four main events [8]:
(1) Bonding to the surface of the adherend (compression).
(2) First stage of debonding: initiation of the failure process through the formation of cavities or cracks at the interface or in the bulk (tension, small deformation).

(3) Second stage of debonding: formation of a foamed structure of cavities elongated in the direction normal to the plane of the adhesive film (tension, large deformation).

(4) Final stage of debonding: separation of the two surfaces either by failure of the fibrils (cohesive failure, i.e. some adhesive remains on both surfaces) or by detachment of the foot of the fibrils from the surface (adhesive failure, i.e. there is no adhesive left on the probe surface). Note that we refer here to a visually observable presence or absence of adhesive on the surface. Surface analysis techniques like XPS almost always find molecular traces of adhesive on the adherent's surface.

The debonding part of this sequence of events is shown in Fig. 3 in parallel with a stress–strain measurement. While the exact sequence of events depends also on the geometry of the test and thickness of the adhesive layer as discussed in Section 3.2.4, the general features of a stress–strain curve obtained in a probe test of a PSA, remain the same and will be characterized typically with three parameters defined in Fig. 2: a maximum stress, σ_{max}, a maximum extension, ε_{max}, and a work of separation, W, defined as the integral under the stress–strain curve. This work of separation should not be confused with the thermodynamic work of adhesion, W_a, an interfacial equilibrium parameter calculated from the values of surface and interfacial energies or with $\mathcal{G}_c$, characterizing an energy dissipated during the propagation of a crack and which will be discussed in more detail in Section 3.2.2.

3.1. Bonding process

The requirements for a good bonding to the surface of the adherent have been discussed by Dahlquist some years ago and these requirements are still widely used in the industry today [12]. They specify that a PSA needs to have a tensile elastic modulus E' at 1 Hz lower than 0.1 MPa to bond properly to the surface. More recently, this criterion has been rationalized in terms of the contact between a rough surface against an elastic plane [13,14]. The key result of this study is that one expects good molecular contact under zero pressure when the surface forces exactly balance the elastic energy cost involved in deforming the adhesive film to conform to the rough surface. In terms of elastic modulus of the adhesive, this can be written as:

$$E < W_a \left(\frac{R^{1/2}}{\zeta^{3/2}} \right) \tag{1}$$

where R represents the average radius of curvature of the asperities of the model surface and ζ represents the average amplitude of the roughness. For $R \sim 50\ \mu m$, $\zeta \sim 2\ \mu m$ and $W_a \sim 50\ mJ/m^2$, one finds a threshold elastic modulus of the order of 0.1 MPa.

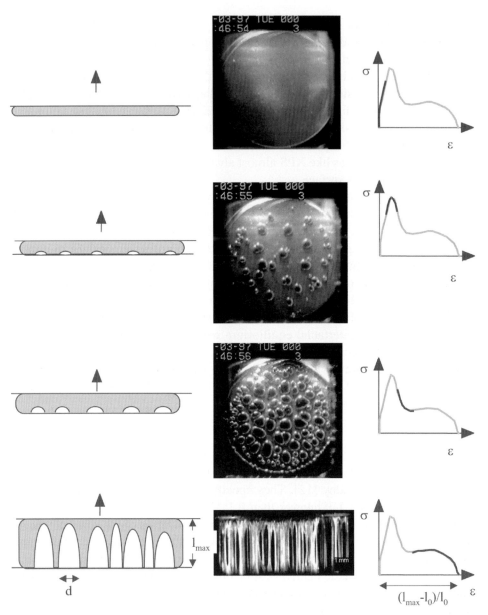

Fig. 3. Schematic of the deformation mechanisms taking place during a probe tack test and corresponding images of the stages of debonding. Images from [70].

3.2. Debonding mechanisms

A bonding model can predict whether the bonding stage will occur properly but is not sufficient to know the level of energy dissipation which will occur upon debonding (for example water would easily pass the test but is not a useful PSA). Therefore, we will now consider the details of the debonding process.

Normally after the contact is established during the bonding stage, a tensile force is applied to the adhesive film until failure and debonding occurs. A useful PSA will require a much larger amount of mechanical work to break the contact than to form it and this work is in particular done against the deformation of bridging fibrils which can extend several times the initial thickness of the film as shown in Fig. 3 [4,8,15,16].

3.2.1. Initiation of failure: cavitation

Since the adhesive material is rubbery, it deforms at nearly constant volume and the formation of fibrils can only occur through the prior formation of voids between them. Depending on the experimental system, these voids can first appear at the interface between the adhesive and the adherend or in the bulk of the adhesive layer but they invariably expand in the bulk of the adhesive layer as illustrated in Fig. 3 [8,17].

The occurrence of this cavitation process can be readily understood by noting that the elastic tensile modulus, E, of a typical PSA is about 4–5 orders of magnitude lower than its bulk compressive modulus. A mechanical analysis of the growth of an existing cavity in an elastic rubber shows that in such a medium, a preexisting cavity is predicted to grow in an unstable manner if the applied hydrostatic tensile pressure exceeds the tensile modulus, E, of the adhesive [18,19]. This expansion condition can be roughly written as:

$$\sigma > E \tag{2}$$

If the nucleation of cavities is indeed responsible for the decrease in the tensile force in Fig. 2, according to Eq. 2, one expects the measured σ_{max} to be directly related to the elastic modulus E of the adhesive and therefore to obey time–temperature superposition. This is indeed confirmed by experiments showing that for several PSA on steel surfaces, σ_{max} at a given reduced debonding rate is directly proportional to E' (the value of the elastic modulus measured with steady-state oscillatory shear measurements) at an equivalent reduced average deformation rate [8,17]. This result, shown in Fig. 4, implies therefore that, provided that Dahlquist's criterion is satisfied, the larger the elastic modulus of the PSA, the higher its tack force on a high energy surface. This statement may, however, no longer be true for low energy surfaces as discussed in Section 6.1.

Moreover, it should be kept in mind that such a simplistic model for the nucleation of cavities would predict a simultaneous expansion of all existing cavities at the same identical applied hydrostatic stress. This is contrary to experimental results which show that cavities appear sequentially at a range of applied stresses. Furthermore, a difficult outstanding question is that of the nature

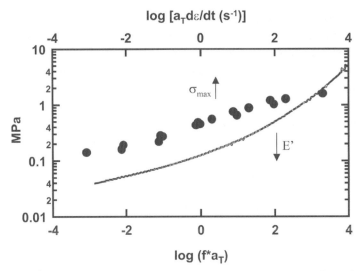

Fig. 4. σ_{max} and E' as a function of the reduced shear frequency $a_T f = a_T \omega/2\pi$, where a_T is the WLF shift factor, or of the reduced strain rate $a_T \, d\varepsilon/dt$. $\bullet$, σ_{max}; ——, E'. Data from [8].

of the defects able to expand into cavities. Gay and Leibler argued that, for rough surfaces or rough adhesives, cavities should expand from defects consisting of air bubbles trapped at the interface during the bonding process [15]. This may be true in certain cases, but cavities are just as easily nucleated on smooth surfaces and in the bulk. Better criteria for the expansion of a cavity, considering the existence of defects, are currently developed to obtain a quantitative prediction of the maximum tack stress [20].

3.2.2. Foam formation

Until now, the viscoelastic properties of the adhesive and the surface properties of the substrate (except for its roughness) have not played a significant role in controlling the mechanisms. They will be essential, however, in the subsequent process, i.e. the evolution of individually expanded cavities into an elongated microfoam structure.

Although the formation of the foam is a rather complicated process, it can be approximately characterized by two important parameters: the cell size, d, in the plane of the adhesive film and the maximum extension, l_{max}, of the cell walls in the direction normal to the plane of the adhesive film as shown in Fig. 3.

The final size of the cells d of the foam will clearly play a role in the macroscopic stress sustained by the walls. This final size of the cells will be controlled by three parameters which are characteristic of the behavior of the adhesive and of the interface [20–22]. Two of them are bulk parameters, the unrelaxed elastic modulus of the adhesive, E_o (typically at high frequency) and the average Deborah number at which the experiment is being conducted. This Deborah number is defined as the product of the initial macroscopic strain rate of the test, V_{deb}/h_o, by a relevant relaxation time of the adhesive, τ:

$$De = \tau V_{deb}/h_o \tag{3}$$

where h_o is the initial thickness of the film and V_{deb} is the probe velocity. At high values of De, one assumes that the adhesive behaves essentially elastically with a modulus E_o, while at low values of De, significant relaxation of the stresses can occur within the time frame of the experiment.

The third parameter is interfacial: it is the critical energy release rate $\mathscr{G}_c$ characteristic of an adherent/adhesive pair. It can be approximately defined as:

$$\mathscr{G}_c = \mathscr{G}_o(1 + a_T\phi(\dot{a})) \tag{4}$$

where $\mathscr{G}_o$ is the energy release rate extrapolated at a vanishing crack velocity. $\dot{a}$, a_T is the time–temperature shift factor and ϕ is a bulk dissipative function which depends on debonding rate [23,24]. $\mathscr{G}_c$ characterizes the amount of energy dissipated by a propagating crack at the interface between the PSA and the surface. This parameter is widely used to characterize the adhesion between a crosslinked elastomer and a hard surface [24,25] and, with some precautions, can also be used for viscoelastic PSAs [11].

Coming back to the foam formation, the simplest case is that of high values of $\mathscr{G}_c$ and high Deborah numbers. In this case, the characteristic lateral dimension of the cells, d, will only be controlled by the elastic modulus, E, and the thickness of the layer, h_o:

$$d \approx h_o\left(\frac{K}{E}\right)^{1/2} \tag{5}$$

where K is the bulk modulus of the adhesive [20,21]. The distance d is representative of the length over which the stress is relaxed by the expansion of the cavity. Forgetting about numerical constants, this means that one expects the lateral dimension of these cells to be directly proportional to the thickness of the film and inversely proportional to the square root of the shear modulus (the bulk modulus K does not vary much from one soft adhesive to another and is of the order of 1 GPa).

If the polymer can relax during the test ($De < 1$), the cavities can expand laterally causing what is analogous to a dewetting of the sample as shown in Fig. 5 [22].

3.2.3. Fibrillation and failure

Assuming that the cavities have expanded laterally as much as they can without coalescing, the last stage of the debonding process will start with the vertical

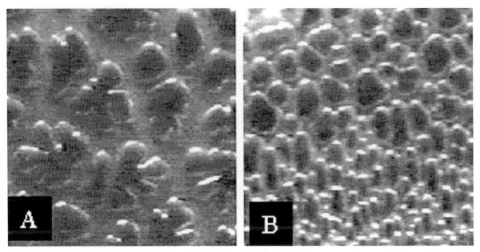

Fig. 5. Comparison of the observed cavities in a poly(2-ethylhexyl acrylate) adhesive debonded: (A) at 10 μm/s; and (B) at 100 μm/s. Images from [8].

elongation of the walls between cells. This mechanism implies that, at the molecular scale, there is a progressive orientation of the polymer chains in the direction of traction [26,27]. There is an interesting analogy between this process and the formation of craze fibrils in glassy and semi-crystalline polymers. However, while craze fibrils grow in length only by drawing fresh material from a reservoir of unoriented polymer [28], the situation is less clear for PSA fibrils where some of the extension is a result of fibril creep and some is due to the drawing of unoriented polymer from the foot. The respective weight of these two mechanisms will depend on the rheological properties of the adhesive in elongation: a weakly strain-hardening adhesive will favor fibril creep, while fibrils formed by a strongly strain-hardening adhesive will grow by drawing of unoriented polymer from the foot.

Once most of the polymer chains are well-oriented, the stress on the fibrils can increase again, causing either an instability and a fracture of the fibrils themselves (macroscopically, a cohesive fracture) or the detachment of the foot of the fibril from the surface of the adherent (macroscopically an adhesive fracture).

The occurrence of one or the other of these processes will depend on a delicate balance between the tensile properties of the adhesive and the interfacial parameter, $\mathcal{G}_c$. Despite the fact that the level of stress and the maximum extension that these fibrils will achieve often controls the amount of work necessary to debond the adhesive (the external work done during this process can sometimes represent up to 80% of the practical debonding energy), no quantitative analytical treatment of this extension and fracture process exists for such highly non-linear materials. Numerical methods have, however, been successful in predicting at least the extensional behavior if not the point of fracture [29,30].

3.2.4. Effect of geometry

While the above section oversimplifies the debonding process by separating it into individual stages which are not really independent of each other, it is, however, a first step towards a better understanding of the critical parameters controlling tackiness. The three stages of debonding described earlier, assumed that cavitation was the first mechanism in the failure process. While this is true for useful PSAs, it is worthwhile to briefly consider the limits where this may no longer be true. When a tensile stress is applied to a confined layer of arbitrary elastic modulus, one can envision two limiting cases: a very hard adhesive will not cavitate, but form a crack (and the probe tack will become the butt joint geometry) and a simple liquid will form a single filament (and this experiment is called a squeeze-flow test).

In order to understand in which conditions cavitation is likely to occur, it is necessary to consider the effect of the coupling between the experimental geometry and the mechanical and interfacial properties of the adhesive on the failure mechanisms.

The results of fracture tests of adhesive bonds are almost never independent of the experimental geometry because the presence of the interface with its discontinuity in elastic properties ensures that the stress field at the interface depends on both the external loading and the elastic properties mismatch as discussed in chapters on hard adhesives. However, soft adhesives have the added complication to dissipate energy, not only in a restricted plastic zone near the interface, but over a large volume, often the entire volume of the sample. This means that there is a very strong coupling between the boundary conditions of the test (thickness of the layer, size of the probe and stiffness of the probe) and the observed deformation mechanisms.

In a recent study, Crosby et al. [31] have discussed the different possible initial failure mechanisms of a thin adhesive elastic layer in a probe test and have extracted two geometrical parameters which couple with the two material parameters, E and $\mathcal{G}_c$: the degree of confinement of the adhesive layer (represented by the ratio of a lateral dimension over a thickness of the layer) and a characteristic ratio between the size of a preexisting internal flaw, a_c, and a lateral dimension of the system, a. They distinguished among three main types of initial failure: bulk cavitation, internal crack and external crack as shown in Fig. 6.

When the confinement is high, elastic instabilities such as cavitation in the layer are strongly favored, the more so of course if the elastic modulus of the adhesive is low and the interfacial energy release rate, $\mathcal{G}_c$, is high. On the other hand, weak adhesion, a higher elastic modulus and a lesser degree of confinement all favor crack propagation as shown in Fig. 7.

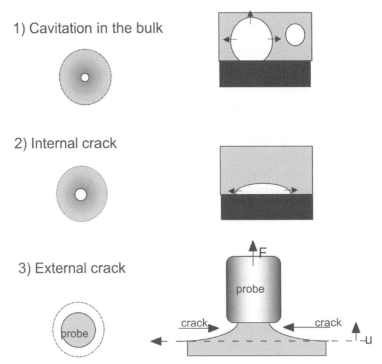

Fig. 6. Schematic of the initial failure mechanisms of the elastic layer. Top views (left) and side views (right). Arrows indicate the direction of expansion of the cracks or cavities. In the case of bulk cavitation, the nucleation can be at the interface or in the bulk.

4. Molecular structure and adhesive properties

A polymer (or copolymer) chain is characterized by the nature and distribution of its constitutive monomers along the chain, and molecular parameters, such as its number average molecular weight, M_n, and polydispersity, M_w/M_n, its average molecular weight between entanglements, M_e and its degree of branching. These factors have an important effect on the bulk properties of the material, such as the glass transition temperature, T_g, or the large scale organization. On the other hand, the molecular parameters also determine the rheological behavior, expressed by the different moduli and characteristic relaxation times for small (monomer) or large (chain or a part of a chain) scale motions. In Fig. 8, the typical tensile modulus, E, of a high molecular weight polymer with a narrow molecular weight distribution is represented. At a given temperature, for relaxation times shorter than a characteristic time, t_e, or frequencies of motion larger than $1/t_e$, the chain is unable to relax at a large scale: the high frequency value of the modulus is related to the nature of the monomers and their local mobility. Between t_e and t_d, the chain is relaxed on a typical scale corresponding to the distance between

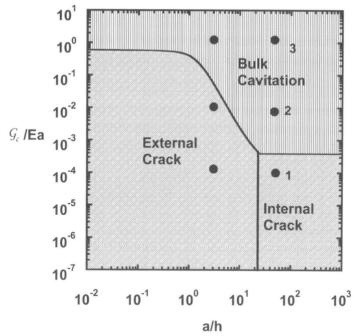

Fig. 7. Deformation map with $\mathcal{G}_c/Ea$ as a function of a/h. Points 1: low value of $\mathcal{G}_c/E$, transition from external crack to internal crack with increasing confinement. Points 2: intermediate value of $\mathcal{G}_c/E$, transition from external crack to cavitation with increasing confinement. Points 3: high values of $\mathcal{G}_c/E$, always failure by cavitation. Case 1 is typical of a crosslinked rubber on steel or of a block copolymer-based PSA on release paper, case 2 is typical of an acrylic PSA on release paper and case 3 is typical of a PSA on high energy surfaces.

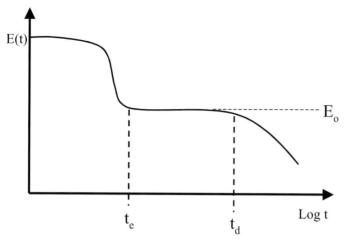

Fig. 8. Relaxation modulus, $E(t)$, as a function of time at a fixed temperature for a high molecular weight polymer with a narrow molecular weight distribution.

entanglements: one observes a plateau region for the modulus E_o, where its value is inversely proportional to the average molecular weight between entanglements M_e. Finally, at times larger than t_d, the polymer flows like a liquid: in this range, the value of the modulus is related to the polymer molecular weight, and its degree of branching [32,33].

It is helpful, in order to understand the adhesive behavior of a PSA, to compare, for the same temperature, the typical experimental times or frequencies involved in a quick tack experiment (contact time, separation rate) to the characteristic relaxation times defined above. For instance, when the bond formation step is realized in a time shorter than t_e, the very high modulus of the material will prevent it from making a good contact, thus leading to a poor adhesive behavior. In a similar way, the separation rate has to be within a range where fibrils can form in order to obtain large dissipation and thus a good adhesive behavior. In practical situations, the separation rate and temperature are usually imposed by the specific use that is being made of the adhesive, so that one needs to play with the molecular weight of the polymer, its degree of branching or crosslinking and the monomer friction coefficient in order to modify its relaxation times and thus its viscoelastic losses during the fibrillation stage. Thus, even if not sufficient by itself, the knowledge of the linear viscoelastic properties, E' and E'', versus frequency curve gives a strong indication of the suitability of a given material for adhesive purposes.

As a first approximation, a suitable molecular structure for a tacky material could be described as a nearly uncrosslinked network: a low plateau modulus will give a high compliance of the layer and therefore a good contact with the surface, and crosslinks and entanglements will give the necessary cohesive strength to form stable fibrils at debonding frequencies. In practice, this is often realized by combining two main ingredients: a partially crosslinked, low T_g high molecular weight component and a high T_g, low molecular weight component called a tackifier, but as we will see in the following, many types of structures can lead to the right balance of properties.

4.1. Influence of molecular parameters: nature of the monomers

4.1.1. Glass transition temperature (T_g)

Since the characteristic relaxation times of the base-polymer in the PSA are always decreasing with temperature, one would expect the properties of a PSA to be also monotonically dependent on temperature: experimentally, however, one observes the existence of a fairly sharp optimum of the tack properties of a material with temperature. This is illustrated in Fig. 9, for various polyacrylates [34] and will be discussed in more detail in Section 5.1.2. More generally, it is admitted that,

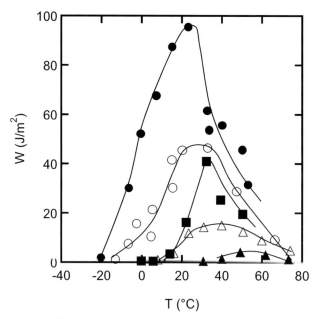

Fig. 9. Debonding energy, W, as a function of temperature for probe tack tests of different acrylic polymers. ●, Poly(2-ethylhexyl acrylate); ○, poly(n-butyl acrylate); ■, poly(isobutyl acrylate); △, poly(ethyl acrylate); ▲, poly(methyl acrylate). Contact time 0.02 s. Data from [34].

for the PSA properties of an adhesive to be optimal, its T_g should lie somewhere between 70 and 50°C below the use temperature for an acrylate or natural rubber based adhesive and between 30 and 50°C below the usage temperature for a styrenic block copolymer (SBC)-based adhesive. This temperature corresponds to a balance between an elastic modulus, E, lower than 10^5 Pa (Dahlquist's criterion) and a high level of dissipation upon debonding. It is, in fact, one of the roles of a tackifying resin, or of a comonomer in an acrylic PSA, to raise the T_g of the system when the T_g of the polymer alone is too low.

4.1.2. Monomer polarity

Since it was noticed that the incorporation of polar monomers, such as acrylic acid, could enhance the cohesive and adhesive strength of a material, several studies have been devoted to the role played by the polarity of the monomer in the adhesion process.

The influence of incorporating polar monomers is complex: it both modifies the surface tension and the bulk properties of the adhesive. However, these two effects are not generally dominant at the same time. In tack experiments, the most visible change brought about by acrylic acid is a sharp increase in the long relaxation times of the polymer [35]. As shown in Fig. 10, this causes a shift in the

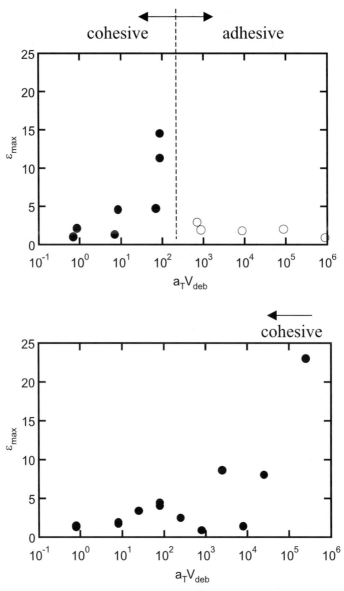

Fig. 10. Maximum extension of the fibrils, ε_{max}, as a function of reduced debonding rate, $a_T V_{deb}$, for PnBA with 2.5% acrylic acid (upper panel); and for PnBA without acrylic acid (lower panel). $T = 23°C$, $t_c = 1$ s. •, cohesive failure; ○, adhesive failure. Molecular weights and molecular weight distributions are identical for both polymers. Data from [17].

characteristic debonding rate at which the transition from cohesive failure of the fibril to adhesive failure is observed [17].

However, in peel tests, which typically involve long contact times, the dominant effect is reversed: acrylic acid moieties can slowly migrate to the interface with the substrate and cause a significant increase of the interactions which can switch the fracture mode back to cohesive [36]. More generally, the presence of monomers of different polarities can lead to specific time-dependent effects related to the kinetics of diffusion or reorientation of the different species at the surface during the experimental time. Finally, the presence of acrylic acid can suppress the time–temperature equivalence generally observed for adhesive tests of soft polymers [8,37].

4.2. Influence of molecular parameters: characteristics of the chain

4.2.1. Molecular weight of the chain (M_n)

Commercial PSAs have typically a very broad molecular weight distribution or alternatively a very broad distribution of terminal relaxation times. This polydispersity is essential to obtain good PSA properties.

   In order to understand why this is the case, it is useful to examine the results obtained with polymers with a narrow or very narrow molecular weight distribution [17,34,38]. As shown in Fig. 11 for a polyisobutylene, if the comparison is made for the same experimental conditions, reasonable tackiness is only obtained in a relatively narrow range of molecular weights. This can be understood in the following way: the increase of M_n increases the viscosity, η, since $\eta \propto M_n^{3.4}$, or, in terms of characteristic relaxation times, the terminal relaxation time, τ_d. In a tack experiment, this increase in viscosity leads to a larger cohesive strength of the

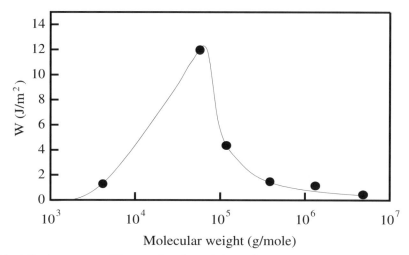

Fig. 11. Adhesion energy, W, as a function of the weight average molecular weight for a polyisobutylene with a low polydispersity. Data from [34].

material, which, in turn, causes a better stability of the PSA fibrils, once they are fully formed, and thus to a larger value of ε_{max}. On the other hand, the formation of the fibrils from the initially formed cavities requires a certain amount of flow so if the terminal relaxation time and the viscosity increase too much the fibrillar structure is never formed. As a general guide, experiments conducted at room temperature and relatively high debonding rates show a maximum tack energy at a molecular weight which is approximately 5–10 times the average molecular weight between entanglements, M_e.

When considering the molecular weight effects, one should, nevertheless, remember the influence of experimental parameters: the optimum molecular weight will depend strongly on the rate and temperature of the test with high temperatures and low debonding rates shifting the optimum towards higher molecular weights [17].

Conversely, the temperature window of optimum tack for a given molecular weight will be rather narrow (approximately two decades in frequency or 10°C based on Fig. 10). For the more common case of polymers which are not monodisperse but have a broad molecular weight distribution, the resulting broad distribution of terminal relaxation times circumvents this problem allowing both fibril formation and fibril stability for a range of experimental conditions. As a result, an optimum in tack energy with M_n is still observed but becomes also broader and is shifted to higher molecular weights relative to the monodisperse case.

4.2.2. Molecular weight between entanglements (M_e)

Entanglements are crucial in the behavior of PSA. Clearly unentangled polymers do not work well as PSA since they do not have enough cohesive strength to form stable fibrils. On the other hand, the terminal relaxation time of an entangled polymer, τ_d, is dependent on $(M_n/M_e)^3$ from the reptation model and plays a role in controlling the elongation of the fibrils as described in the preceding section. Additionally, the average molecular weight between entanglements also plays a major role in the early stages of the debonding process since it controls the elastic modulus in the plateau region. Indeed, since tack tests are usually conducted in a frequency range where E is in its plateau region (where $E' = E_o \propto 1/M_e$) a change in M_e has a direct consequence on the value of E' at the debonding frequency.

For polymers with molecular weights much larger than M_e, a transition in debonding mechanism is observed for M_e larger than approximately 10^4 g/mol [39]. At low values of M_e, adhesive failure by crack propagation is observed while for high M_e polymers, failure by cavitation and fibrillation is observed. This transition can be understood by a purely mechanical argument. The critical stress for cavitation in the bulk to occur is proportional to the elastic modulus E' of the polymer at the testing frequency. Therefore, when E' decreases, the critical stress for cavitation decreases, and it eventually replaces $\mathcal{G}_c$ as the failure criterion [31,40]. Therefore, for low M_e, the initial debonding is by crack propagation and cannot then evolve towards a fibrillar structure while for high M_e, the debonding occurs through cavitation and fibrillation and a larger amount of energy is dissipated in the process. This transition from fibrillation to homogeneous deformation, which can also be concomitant with a transition from cohesive to adhesive failure, is normally accompanied by a drop in tack energy. Zosel pointed out that this critical value of M_e corresponds in fact to the Dahlquist criterion for the elastic modulus, E, below which a material is not considered tacky: $E = 10^5$ Pa. M_e thus plays a very important role on the debonding stage in a PSA, since it influences the ability for fibrillation, the type of rupture, and consequently the debonding energy.

4.2.3. Chemical crosslinks and molecular weight between crosslinks (M_c)

Experiments on a series of polydimethylsiloxane (PDMS) adhesives, have shown that the tack energy is maximum, for a degree of crosslinking which is slightly above the gel point [41]. Below the gel point, the crosslinks' main effect is to increase the terminal relaxation time of the polymer. The presence of long branches will, in particular, play a large role on the elongational flow properties of the polymer and increase fibril stability. This, in turn, will lead to a higher value of ε_{max} and a higher tack energy.

Above the gel point, additional crosslinking will initially have an effect on the strain hardening which occurs in extension. Significant strain hardening in extension will occur in the polymer for an increasingly lower level of strain, causing early detachment of the fibrils and a lower measured ε_{max}. This effect is illustrated in Fig. 12 on a series of acrylic polymers crosslinked beyond the gel point [39]. One should note that the peak, and the height of the plateau, remain unchanged implying that the viscous dissipation in the fibrils and the plateau modulus, E_o, are much less affected by a modification of the crosslink structure. By extrapolating these measurements, one can argue that fibrillation should be suppressed altogether when the crosslink density becomes equal to the entanglement density. Following the map of Fig. 6, a change in mechanism from cavitation to crack propagation has occurred, mainly caused by a decrease in the interfacial dissipative term, $\mathcal{G}_c$, this time.

Therefore, in a similar way as entanglements do, crosslinks decrease the ability of the material to flow, and eventually lead to a transition to failure by crack propagation and coalescence, and thus to a lower tack energy. However, crosslinks are permanent and completely prevent large deformations; this is not the case for entanglements, provided that the deformation occurs at sufficiently low rates.

Crosslinking is widely used in the PSA industry to tune the properties of removable adhesives by reducing ε_{max}. An example of that type of effect, is the

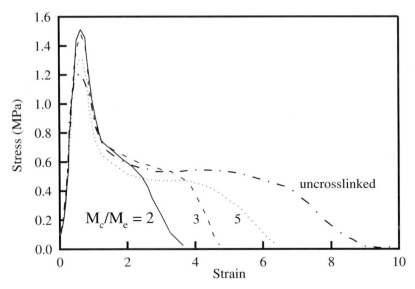

Fig. 12. Stress–strain curves of PnBA on steel, UV crosslinked to different degrees as indicated by the ratio M_c/M_e. Data from [39].

development of PSA for trauma-free removal patches, where in situ irradiation leads to crosslinking, and thus to highly reduced peeling forces for the adhesive [42].

4.3. Influence of the organization of the chains at a nanometric scale

The main polymers used as PSA can be divided in a few large classes according to their chemical structures: natural rubber based, styrenic block copolymer based (mainly triblocks and diblocks of styrene and butadiene or isoprene), acrylics and for a smaller part, silicones. From a microstructural point of view, however, the classification should be different. As already mentioned, a pressure-sensitive adhesive requires both a good shear resistance and some capability to flow in order to function properly. The resistance to shear is normally achieved by introducing crosslinks: chemical ones as in natural rubber or acrylics, but also physical ones. In the case of strongly immiscible comonomers, for example, and depending on the chain statistics, the architecture and monomer composition of the polymer chain can lead to microphase separation so that the microscopic domains act as physical crosslinks for the system: this is the case for styrenic block copolymers. In certain cases, in addition, there can be a long range ordered structure, such as a lamellar stacking. All these structures lead to different behaviors as far as tack is concerned. We list below several types of systems, as examples of different types of phase-separated structures, although their importance from an industrial point of view is not necessarily comparable.

4.3.1. Block copolymers

Block copolymers with incompatible blocks which are able to microphase separate are good candidates for PSA properties. Indeed, blends of ABA triblocks and AB diblocks, where the rubbery midblock of the ABA is the majority phase and the glassy endblocks self organize in hard spherical domains and form physical crosslinks, are widely used as base polymers for PSA. The actual adhesives are always compounded with a low molecular weight tackifier resin able to swell the rubbery phase and dilute the entanglement network. Linear styrene–rubber–styrene copolymers, with rubber being isoprene, butadiene, ethylene/propylene or ethylene/butylene, are the most widely used block copolymers in this category.

This class of PSA has unique properties which are related to their molecular superstructure. Indeed, compared to chemical crosslinks, physical crosslinks present several advantages, the first one being that they are reversible with temperature, thus leading to a large viscosity decrease above the glass transition temperature of the endblocks: this makes these systems very suitable for hot-melt processing, an increasingly popular processing method due to environmental regulations. A second advantage is that the crosslink density is naturally fixed by the composition of the system, provided that it is at equilibrium, and thus easier to control than chemical crosslinking. Finally, the physical crosslinks provided by the hard domains are quite fixed under low stresses, giving very good creep properties, but can be broken and reformed at high stresses, which is essential for the formation of the fibrillar structure.

As shown schematically in Fig. 13, one can distinguish two types of configuration for the triblock molecules: if the two endblocks of a molecule are incorporated in separate hard domains, the triblock will be described as a bridge. If, alternatively, both endblocks are incorporated in the same hard domain, the triblock will be described as a hairpin molecule.

Although this process is not fully understood, it is believed that the number of bridge molecules between hard domains controls the large strain behavior. An increase in the amount of tackifying resin in a pure triblock copolymer causes a change in the ratio of bridge to hairpin molecules, and consequently a change in the value of ε_{max} as shown in Fig. 14.

Similarly to the case of simpler homogeneous systems discussed in Section 4.2 the rheological properties of these PSA will be essential in controlling their adhesive behavior. However, the modifications of molecular structure to obtain these rheological properties will be different. For example, the apparent plateau modulus, E', will no longer be controlled only by M_e, but also by the molecular weight of the elastomeric midblock and by the amount of resin which is incorporated. The terminal relaxation time controlling fibril extension can be somewhat tuned by replacing some of the triblock in the adhesive by double the amount of diblock with one half the molecular weight, effectively modifying the ratio of hairpin

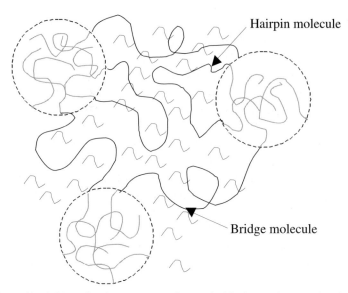

Fig. 13. Schematic of the molecular structure of styrenic block copolymers showing the hairpin and bridge configurations.

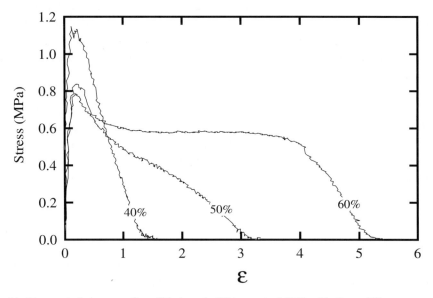

Fig. 14. Stress–strain curves of an SIS + resin PSA on steel [71] with three different amounts of resin in wt%. The base polymer was Vector 4311 (Exxon) and the resin was Regalite R101 (Hercules). $V_{deb} = 10 \, \mu m/s$, $T = 40°C$.

to bridges. It is important to note that this ratio can effectively be modified by the processing conditions (hot-melt, solution cast). In a recent work, Flanigan et al. [43] showed that the adhesive properties of acrylic triblock copolymers cast from two different solvents could be very different, even though their structure as determined by X-ray scattering was the same, i.e. the hard domains size and spacing were identical.

Other studies on the properties of block copolymers with more complicated architectures exist, although not all of them consider tack. Some studies comparing properties of radial versus simple block architecture allow to compare the effect of chemical versus physical crosslinks in the case of structured systems. The data shows that there is a decrease of melt viscosity together with better adhesive properties for star copolymers [44]. This is related to a point that was discussed earlier: the different architectures of the molecules lead to different physical crosslinking densities. In the same spirit, block copolymers with four different arms, two polyisoprene, and two polystyrene-*b*-poly(ethylene/butylene) arms were studied, and led to a better combination of shear strength and melt viscosity for adhesive applications [45], compared to the conventional linear SIS and SEBS triblocks.

4.3.2. Change in structure through phase transition

Previously discussed systems are always phase-separated at application temperatures and undergo a change in properties at high temperatures from elastomeric tacky to liquid for processing reasons: this change occurs over a wide temperature range. It is, however, possible to tune the molecular structure to obtain a transition from solid non-tacky to elastomeric tacky for a given temperature independent of debonding rate, by using a thermodynamic phase transition.

Recent experiments [46] have compared the tack properties of a given system below and above the phase transition between an organized smectic phase and an isotropic phase. The system used was a side-chain fluorinated copolymer, with two types of pendant groups both able to crystallize. Below the transition, in the smectic phase, the system possesses no tack due to its high elastic modulus that prevents the formation of a good bond with the surface. Above the transition, a soft, phase-separated region with possibly a continuous network of crystalline domains, allows the system both to achieve a good bond for short contact times and extended cavitation followed by large deformations. This type of system presents the interesting feature to have a very abrupt non tacky–tacky transition at a temperature which can, in principle, be varied at will by changing their monomer composition.

4.3.3. Heterogeneous particles

There are very few studies published on PSA films obtained from heterogeneous latexes: indeed, most PSA used in the form of latexes, such as styrene–butadiene or acrylic systems, are random copolymers. Although emulsion polymerization does not allow in general as much control over the molecular structure and therefore over adhesive properties, the use of solvent-free adhesives to replace their solvent-based counterparts is increasingly important, due to the recent environmental regulations. One example where the structure of the final adhesive film can be finely tuned at a scale of a few tenths of a nanometer is given by a recent study [47] which compares the tack properties of two acrylate copolymer samples synthesized either with a batch or continuous feed process, leading to particles of different heterogeneities. The tack properties of the two systems are markedly different, thus raising the question of the role of structural heterogeneities at the particle scale. Such heterogeneities may, in principle, be probed by AFM methods, effectively performing a 'nano-tack' experiment on isolated particles or on monolayers of latex particles [48]: this powerful tool could help to understand the correlation between the macroscopic properties of an adhesive and its microscopic response.

4.4. Role of other components

A formulated adhesive contains in general, in addition to the base polymer, some small molecule additives: a tackifier, and some fillers, plasticizers and antioxidants. Additionally if the adhesive has been obtained by emulsion polymerization it contains some surfactants. Tackifiers and plasticizers are added to tune the viscoelastic properties of the adhesive, while surfactants are generally unwanted residues of the polymerization process. Again, much technology, most of it proprietary, is involved in these formulation parameters and we only address here some generic points based on what is available in the open literature.

4.4.1. Tackifiers

The role of the tackifier is to adjust the viscoelastic properties of the system. It typically consists in short chain polymers of molecular mass between 300 and 3000 g/mol, with a softening temperature between 60 and 115°C, depending on the adhesive. The tackifying resin has a dual role: increasing the T_g of the system, which increases the viscoelastic losses at high frequencies, and decreasing the modulus at the low frequencies that are important at the bond formation stage [49,50]. The decrease in elastic modulus in the plateau region can be interpreted as a dilution of the entanglement structure. For a given T_g of the tackifier, there is an optimum weight fraction in a PSA: at low tackifier contents, the plateau modulus is too high to satisfy Dahlquist's criterion and at high tackifier contents, the T_g of the PSA becomes too high and again poor bonding occurs.

A simple way to determine the optimum amount of resin is to determine the amount which gives the minimum value of the elastic modulus of the system as shown in Fig. 15 [34].

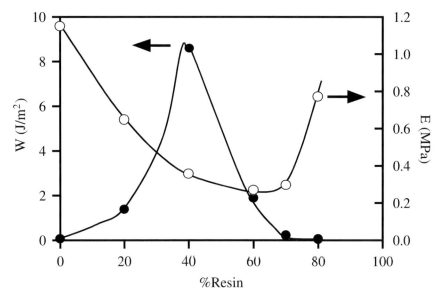

Fig. 15. Elastic modulus, E, in a relaxation experiment (○) and adhesion energy, W, in a probe tack test on steel (●) as a function of resin content for a natural rubber/glycerol ester of hydrogenated resin blend. Data from [34].

Of course, in order to be able to modify the viscoelastic properties of a system, and to change the entanglement density, a tackifier needs to be miscible with the adhesive, which implies that it has to be chemically adapted to the adhesive [50–54]. This is realized via different families of tackifiers, depending on the nature of the adhesive compound.

Giving an exhaustive list of tackifiers together with their compatibilities with adhesives would be beyond the scope of this review, since new families of tackifiers with better compatibility appear on a regular basis [55] but the interested reader can refer to more technologically oriented texts for further information [56].

4.4.2. Surfactants

For emulsion adhesives, the presence and nature of surfactants is also an important parameter. Their action is two-fold: by diffusing to the surface of the film, they may change its properties and its ability to form a good contact, and by plasticizing the polymer they can change its bulk properties. However, while some studies have appeared on their effect on peel properties [57,58], not much is known of their influence on tack.

5. Influence of experimental parameters

5.1. Velocity, temperature, time and pressure of contact

In a probe tack test, several experimental parameters can be varied independently, such as the pressure, p, exerted by the probe on the surface of the adhesive during the compressive stage, the duration of the contact stage called contact time, t_c, or the separation rate of the probe, V_{deb}. Since a change in experimental parameters can sometimes lead to a change in fracture mechanism, the evolution of the adhesion energy with these parameters is in general complex, and deserves to be described a little further. When the debonding rate, V_{deb}, is varied, the characteristic strain rates applied to the adhesive layer are changed accordingly, which, in turn, modifies its viscoelastic response. The response of the material, which depends on its spectrum of relaxation times, is also a function of the temperature of the sample. The contact time, t_c, and pressure, p, can control the size of the real contact area, but also the degree of relaxation of the adhesive when the debonding starts. Effectively, this means that the initial condition of the debonding part of the test will depend on the applied pressure and contact time in the compressive part of the test. Depending on the bulk properties of the material, or on the features of the probe–adhesive interface, such as roughness, this difference in initial condition can have a negligible effect (typically for soft PSA on smooth surfaces) or a large one (for stiffer PSA on rough surfaces).

5.1.1. Time–temperature superposition principle

The time–temperature (t–T) superposition principle is based on the idea that when a polymer is deformed, a change in the characteristic deformation rate is equivalent to a change in temperature. In the context of tack tests, one can assume that the characteristic deformation rate is the V_{deb}/h_o, so that an increase in V_{deb} would be equivalent for example to a decrease in temperature. This t–T superposition, which is very widely used in linear rheology [32], where small deformations are applied, can apparently also be used for the large deformations typical of tack experiments [34]. However, a necessary condition for this t–T equivalence to apply is that no change in fracture mechanism should be induced by a change in the initial condition of the test as defined in Section 5.1. If this change in initial condition causes a qualitative change in the mechanisms involved in the debonding stage at the microscopic level, the rate dependence of the adhesion energy has no reason to follow the t–T principle. As an example, a series of tack experiments [8] at a contact time of 60 s obey the t–T superposition very well, while the same series of experiments at a contact time of 1 s does not as shown in Fig. 16. In this case, the short contact time did not allow the full

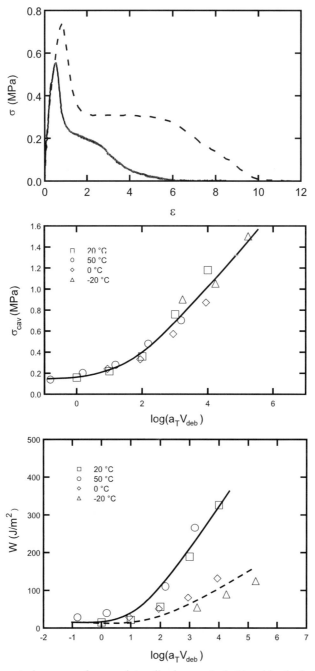

Fig. 16. (a) Stress–strain curves of an acrylate adhesive on steel at two identical values of reduced debonding velocity and for a contact time of 1 s. —, $V_{deb} = 100\ \mu$m/s, $T = 0°$C; - - -, $V_{deb} = 1000$ μm/s, $T = 20°$C. (b) Maximum stress σ_{max} vs. reduced debonding rate; and (c) adhesion energy, W, vs. reduced debonding rate for the same adhesive. Note that the low temperature data for W do not follow t–T superposition. Data from [8].

relaxation of the adhesive at the lower temperature, causing therefore a change in the initial condition. Experimentally, this change had a moderate effect on the cavitation process, but triggered a significant change in the later stage fibril formation process as shown in Fig. 16. Consequently, a master curve for σ_{max} could be built, but not so for W as shown in Fig. 16b,c. This type of effect is more dominant when the surface is rough and the time of contact does not allow the adhesive to fully relax.

Another interesting example where $t-T$ superposition fails is that of a change of temperature leading to a thermodynamic phase transition in the material (melting of a crystalline phase) and therefore to a change of structure [46].

In both examples, an experimental parameter has caused a change in the *initial* condition of the debonding test which, in turn, modified the microscopic separation mechanisms.

5.1.2. Effect of changing the temperature

As explained earlier, the temperature of the test is important since it modifies the relaxation times of the polymer and therefore the rheological behavior of the material: an increase of temperature near the T_g of the polymer decreases its elastic modulus, thus leading to a better contact area and a larger adhesion energy. On the other hand, when temperature is too high, the viscosity of the system becomes very low, decreasing the adhesion energy again. Therefore a maximum in tack as a function of temperature is usually observed at approximately $T_g + 50°C$ as described earlier and shown in Fig. 17. Although a proper formulation can extend the useful temperature range of a PSA or modify the temperature difference between the T_g and the maximum tack, it is extremely difficult to obtain a material which retains a high tackiness over a temperature range in excess of 50°C.

A more subtle effect of temperature is also displayed in Fig. 17 where experiments [34] show a rapid increase of the adhesion energy with temperature, related to a change in the rupture mechanism from interfacial fracture by crack propagation to failure by fibrillation and debonding of the fibrils. The interpretation of the authors is that the temperature change is responsible for a modification of the adhesive-probe contact area, which, in turn, results into different types of rupture mechanisms in the two temperature ranges.

5.1.3. Effect of changing the debonding velocity: cohesive to adhesive transition

The strain rate of an adhesive sample in a probe-tack test is related to the debonding velocity of the probe, however not in a straightforward way. Indeed, the separation stage induces very heterogeneous shear and elongation flows in the sample [31]. A common approximation, bypassing all these considerations, is, however, to consider that, at the beginning of the separation process at least, the

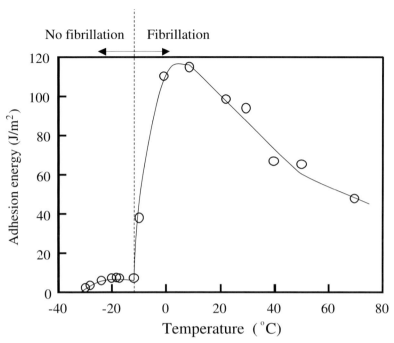

No fibrillation | Fibrillation

Fig. 17. Adhesion energy, W, as a function of temperature for a PEHA adhesive on steel. The final separation is always interfacial, but a clear transition between non-fibrillar and fibrillar debonding is observed at $T = -10°C$. The glass transition of the adhesive is around $-55°C$. Data from [34].

sample is homogeneously deformed at a frequency close to V_{deb}/h_o, where h_o is the sample thickness.

For a monodisperse linear polymer with well-defined relaxation times, its response can be well predicted by a Deborah number [17]. In the regime of cohesive failure (low De), the adhesion energy W is controlled by the value of ε_{max} which continuously increases with De. In this regime σ_{max} does not vary much with De. On the other hand, in the regime of adhesive failure (high De), ε_{max} is always low and W is mainly controlled by σ_{max} which increases with De.

In this case, a sharp drop of tack energy occurs at the transition from cohesive to adhesive failure, i.e. for $De = \tau V_{deb}/h$ between 100 and 1000. Such a transition is completely analogous to that observed by others in peel tests [59,60] and corresponds to a change in the *initiation* mechanism of failure as discussed in Section 3.2.4.

A similar change in mechanism with increasing velocity occurs for the debonding of a glass spherical indenter from a high-molecular weight PDMS. In this case, the debonding occurs in a situation of low confinement (value of a/h is low) with no cavitation. As shown in Fig. 18 [61], the adhesion energy is increasing with

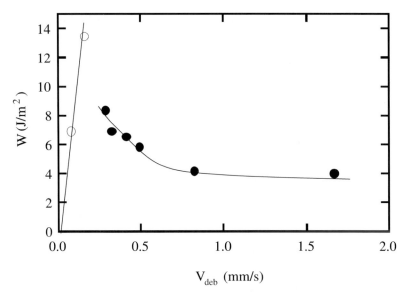

Fig. 18. Adhesion energy, W, as a function of V_{deb} for a spherical glass indenter (JKR geometry) on a high molecular weight PDMS polymer melt. $\circ$, Cohesive fibrillar fracture; $\bullet$, adhesive fracture. Data from [61].

velocity in the cohesive regime, where the separation occurs by bulk yielding and through eventual rupture of a single large column of polymer. If V_{deb} is further increased, adhesion energy drops abruptly then slowly decreases with V_{deb} and reaches an asymptotic behavior. In this regime the separation occurs through the propagation of an external radial crack.

A similar behavior has been observed for acrylic systems when a flat probe was used (high value of a/h). As shown in Fig. 19, the adhesion energy slowly increased in the fibrillation regime, and dropped when the fibril formation was suppressed [39].

These three experiments present two similar features: an increase of adhesion energy with V in the fibrillar regime, whether there is 'one' or many fibrils, and the drop in adhesion energy when going from a fibrillar to a non-fibrillar adhesive separation process, which corresponds to a change in initial fracture mechanism from bulk yield to crack propagation. This implies that when the deformation rate is high enough to prevent significant relaxation processes in the polymer, fibril formation can be suppressed and the tack energy drops.

It should be noted that while the transition from fibrillar to non-fibrillar fracture appears to be very general, it is not necessarily concomitant with a change from cohesive to adhesive failure. It was reported to be concomitant for linear polymers [17], but not for crosslinked or highly branched polymers [8,39].

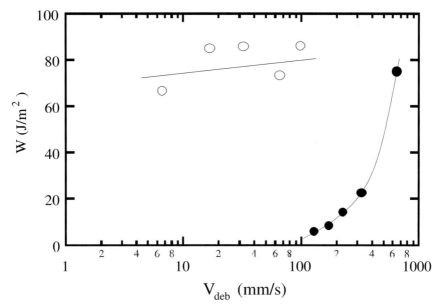

Fig. 19. Adhesion energy, W, as a function of V_{deb} for a PEHA-based adhesive on steel at $T = 20°C$ and 1 s. contact time. Transition from fibrillar (o) to non-fibrillar (•) fracture. Data from [39].

5.1.4. Role of contact time and pressure

The adhesion energy versus contact time is plotted in the upper panel of Fig. 20 for the case of an acrylic adhesive on steel [39]. The variation in W observed with increasing contact times was initially attributed to the increase of the real contact area. However, more recent and more detailed results have shown that the increase in the adhesion energy was also due to a larger value of ε_{max} as can be seen in the lower panel of Fig. 20 [39]. One can envision, therefore, two separate effects of the contact time. For very short contact times, the real area of contact may really be affected, giving, therefore, a lower value of σ_{max}. However, for intermediate contact times, it is the fibrillation process which is mainly affected and its effect can be seen more on W than on σ_{max}.

Presumably it is not the real contact area which is important in this case, but rather the degree of relaxation of the adhesive during the contact time. If the relaxation times of the adhesive are of the same order of magnitude than the time of contact, one expects large differences in the initial stress state of the adhesive layer depending on the contact time. These differences could then lead to a different response of the adhesive during the debonding stage as shown in the lower panel of Fig. 20. One should note that, in principle, one expects W to become independent of t_c for long contact times. Although this effect has been

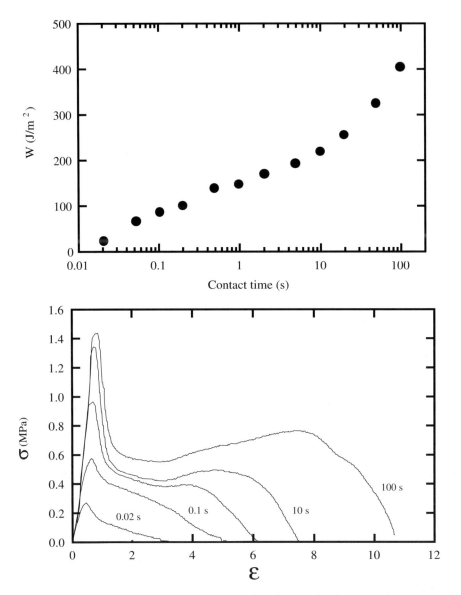

Fig. 20. Upper panel: W as a function of contact time for a PnBA adhesive on steel. Lower panel: corresponding stress–strain curves for selected contact times. Data from [39].

reported [34], experiments do not always show a plateau within the limits of the experimentally accessible range of t_c of a few hundred seconds.

Additionally, in some cases, the degree of roughness of the probe may lead to trapped air and inhomogeneous stress fields which are also bound to vary with the time of contact [15]. Experimentally, in the regime where relaxation times of the adhesive are important, adhesion is always lower and the time of contact always has a more marked effect for rough surfaces than for smooth ones [39,62,63].

In the same way and for the same reasons, a plateau in adhesion energy is observed for large contact pressures [34], but few systematic studies have been published for viscoelastic systems where relaxation processes during the time of contact can play a major role.

5.2. Effect of geometry

Most of the results which are discussed in this review were obtained with flat probe tack tests on thin adhesive films (typically, 20–100 μm). While this geometry has several advantages for the analysis of fundamental properties of PSA, it is by no means the only one that can be used and one should be careful to understand clearly what features of a tack curve are due to the material and which ones are due to the specific experimental geometry.

As discussed in Section 3.2.4, the experimental geometry and particularly the degree of confinement of the adhesive layer can have an important effect on the initial failure mechanisms which are observed.

Luckily for users, most PSA have properties of critical energy release rate, $\mathcal{G}_c$, and tensile modulus, E, in a range which makes them relatively insensitive to the degree of confinement. Therefore, tests done with spherical or flat indenters give relatively similar results.

However, for very soft (liquid-like) and very hard (solid-like) PSA, testing tack with the spherical probe, which typically applies a much lesser degree of confinement to adhesive film, may lead to significantly different failure mechanisms [31]. Three further comments should be added concerning the role of the geometry:

- The role of the geometry will be dominant only in the first stages of the debonding process since once a fibrillar structure is formed, the stress–strain curve is essentially representative of parallel tensile tests and ε_{max} at least, should be rather insensitive to the initial geometry.
- The stiffness of the experimental apparatus will also play a role in the deformation processes of the adhesive film during debonding. In probe tests, the stiffness of the apparatus is usually much larger than the stiffness of the film. However in a peel test this is no longer true and it is well known that significantly different results can be obtained with different backings. This will also hold in probe tests if a tack test is performed on a PSA on its backing. If the backing is soft and elastic it will act as an additional reservoir of elastic energy while if it is viscoelastic, it will strongly alter the rate and contact time dependence of the tack test of the adhesive. Therefore probe tests of adhesives should, whenever possible, be conducted without a backing.
- Finally, the notion of confinement also applies to peel tests; in this case, the confinement level may be given by the ratio between the width of the tape and

the thickness of the film. Since this ratio is usually large, results on peel tests should be correlated with flat probe tests, but experimental confirmation of this statement is presently not available in the literature.

6. Surface effects

PSA are normally designed to stick to most surfaces and to be therefore rather insensitive to the nature of the surface of the adherent. This is why, in the previous sections, we have considered mainly the rheological properties of the adhesive rather than the interfacial chemistry. However, based on the theoretical arguments set forth in Section 3, there are two major causes of poor adhesion of a PSA to a surface: low $\mathcal{G}_c$ and poor contact.

The first case is related to the arguments given in Section 6.1: if the interfacial $\mathcal{G}_c$ (static, but also dynamic) is too low, once cavities are nucleated at the interface, they can easily propagate, coalesce and debonding occurs without any fibril formation. The second case is discussed in Section 6.2 and occurs when the elastic modulus of the PSA is too high (Dahlquist's criterion is not met) or when the surface is too rough for the adhesive to conform to it during the short contact time.

Before we proceed, a word of caution is necessary here in terms of our use of the parameter $\mathcal{G}_c$. In fracture mechanics, $\mathcal{G}_c$ has the meaning of an energy per unit area necessary for a crack to advance. For elastic elastomers, $\mathcal{G}_c$ is a unique function of the velocity of the advancing crack and can be determined with a fracture mechanics test such as the JKR test [24].

Unfortunately, the independent determination of $\mathcal{G}_c$ by the same method is very difficult for viscoelastic materials [64,65] since it will depend on the history-dependent degree of relaxation of the adhesive. However as a phenomenological parameter associated with the amount of energy dissipated by the advance of a crack front (per unit area), it can be very useful and simplify the description of the mechanisms.

6.1. Adhesion on low $\mathcal{G}_c$ surfaces

Several studies have been undertaken to investigate the adhesion of model PSA on low energy surfaces or more generally on low $\mathcal{G}_c$ surface [7,66]. Examples of such surfaces are silicone rubbers, commonly used for release coatings, polyethylene, polypropylene and polycarbamates. These early results were somewhat contradictory and did not provide any explanation of the observed dependence on surface tension. Generally, tack decreased markedly when the surface energy of the adherend decreased well below that of the adhesive. More recently, peel experiments showed that for soft adhesives, it is not necessarily the surface tension of the adherend which is the controlling parameter, but rather the resistance to shear of the

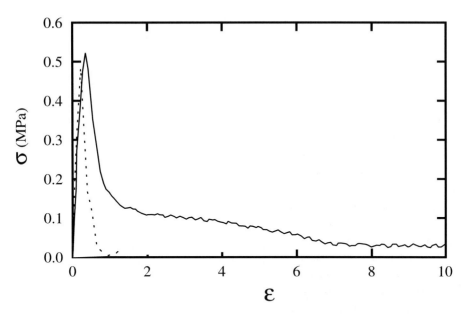

Fig. 21. Stress–strain curves of a soft adhesive (high and intermediate $\mathcal{G}_c/E$) on steel (solid line) and for a low $\mathcal{G}_c$ surface–adherent pair (dashed line). $V_{\text{deb}} = 100$ μm/s. Data from [40].

interface [67]. The cause for this effect, at the molecular level, remains, however, a controversial issue so we will focus here on a more macroscopic description of the tack experiment of a PSA on a low $\mathcal{G}_c$ surface. This description should be kept in mind when interpreting experimental data obtained on such surfaces.

The essential difference between a low energy surface and a high energy surface can be visualized through the comparative analysis of probe tack tests [40]. For a given experimental geometry, the key parameter controlling the behavior of the adhesive is the ratio of the energy release rate over the modulus $\mathcal{G}_c/E$. Three different cases can be observed as a function of $\mathcal{G}_c/E$:

- For high $\mathcal{G}_c/E$, the initial mechanism of failure is cavitation and fibrillation. The maximum extension of the fibrils is not controlled by the surface, but by the rheological behavior of the adhesive in elongation. This is the standard case for a PSA on steel or glass.
- For intermediate values of $\mathcal{G}_c/E$, the initial failure mechanism can still be cavitation, but the maximum extension of the fibrils is controlled by the surface. An example of this situation is given in Fig. 21: the measured σ_{max} is identical for both surfaces, but ε_{max} is very different. A more detailed analysis of the debonding mechanisms reveals that the initiation of failure occurs at about the same level of stress, but the lateral propagation of the existing defects is much faster for the low $\mathcal{G}_c$ situation. If this lateral propagation is fast enough, it prevents any growth of cavities in the bulk of the adhesive and therefore the

formation of the elongated foam of bridging fibrils. What happens is rather a coalescence of adjoining cracks and global debonding of the adhesive film from the probe at relatively low levels of deformation.

- Finally, for very low values of $\mathcal{G}_c/E$, the initiation mechanism is no longer cavitation, but internal crack propagation. This can occur at lower levels of stress so that both σ_{max} and ε_{max} are significantly decreased. An example of that situation is given in Fig. 22. This very different initiation mechanism is also clearly apparent from the video images of the debonding process shown in Fig. 22b.

This description is simple and yet very general. It can explain why transitions from interfacial separation to fibrillation are observed by changing V_{deb}, the surface roughness or surface chemistry or the contact time. In each one of these cases, the change in the experimental parameter had an effect both on E and on $\mathcal{G}_c$. However, the magnitude of this effect is generally very different and sometimes in opposite directions resulting in a very large change in $\mathcal{G}_c/E$.

6.2. Effect of surface roughness

The effect of surface roughness is in many ways much more complicated, and it is impossible at present to give trends generally valid for all PSA–surface pairs. It has been discussed theoretically first for perfectly elastic systems [68] and then for more viscoelastic systems [14]. Experiments have shown consistently that for PSA (unlike for other types of adhesives) the presence of a high level of surface roughness was detrimental for adhesion.

This, it was argued, was due to the limitation of the real surface of contact due to the asperities [13,14]. As discussed briefly in Section 3.1, when an elastic layer is brought into contact with a rough hard surface, it tries to comply to the topography of the surface, but if the balance between the amplitude of the roughness and the elastic modulus does not comply to Eq. 1, contact may be incomplete.

This first simplistic picture is, however, inconsistent with the experimental observation that even for low modulus PSA, an effect of surface roughness is observed. This effect is, however, magnified when the PSA has a relatively high elastic modulus.

A more realistic view may be that the presence of a surface roughness creates an inhomogeneous strain field around the surface and creates therefore pockets of residual tensile stress which will become preferential nucleation sites for cavities. More systematic experiments [69] have shown that the amplitude of the surface roughness had a direct effect on the level of stress at which the cavities appeared, as shown in Fig. 23.

On the other hand, experiments with triblock-based adhesives have shown that if nucleation is not affected (for example at high temperature), the propagation

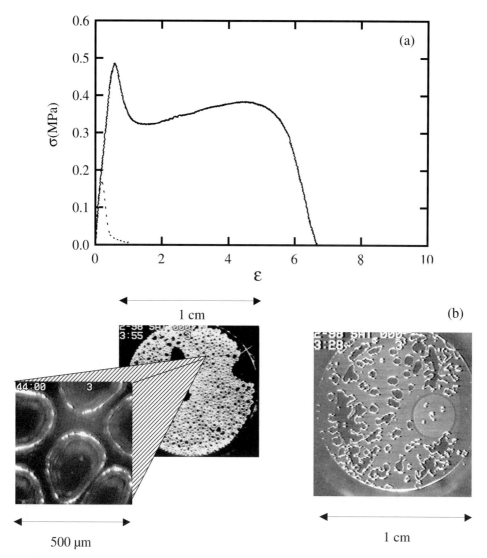

Fig. 22. (a) Stress–strain curves of the same hard elastic adhesive on steel (solid line) and on a low $\mathcal{G}_c$ surface–adherent pair (dashed line). (b) Video captures of the debonding process for both surfaces. $V_{\text{deb}} = 1$ μm/s. Note the relatively small cavities observed for the steel surface and the large irregularly shaped interfacial cracks observed for the low energy surface. Data from [40].

of the cracks can be greatly affected by the roughness [63]. In this regime, as shown in Fig. 24, σ_{max} is hardly affected but the fibril formation can be completely suppressed on smooth surfaces, by a rapid lateral propagation of nucleating cavities in an analogous way as what is observed for the intermediate $\mathcal{G}_c/E$ case discussed in the preceding section.

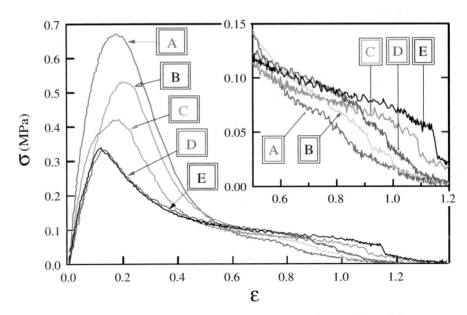

Fig. 23. Stress–strain curves of an acrylic PSA on rough surfaces with a different average amplitude R_a of the roughness but with the same wavelength. A, $R_a = 11$ nm; B, $R_a = 23$ nm; C, $R_a = 46$ nm; D, $R_a = 114$ nm; E, $R_a = 148$ nm. $V_{deb} = 30$ μm/s; $T = 20°C$. Data from [69].

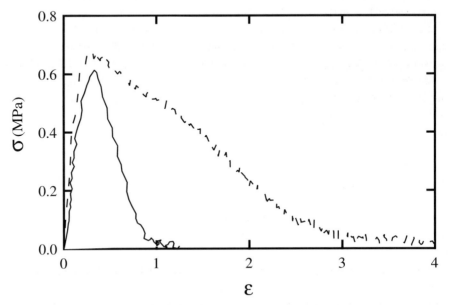

Fig. 24. Stress–strain curves for an SIS +50 wt% resin adhesive on a rough (dashed line) and on a smooth (solid line) steel surface [71]. $T = 60°C$; $V_{deb} = 100$ μm/s.

7. Conclusions

In this chapter, we have attempted to review, both the main material properties which are required for a PSA to be tacky and how experimental parameters affect the practical evaluation of tackiness. Although it must be clear from the previous sections that the bonding and debonding of a PSA from a surface is a complicated process, some important parameters have been identified:

- The elastic modulus, E', obtained from small strain oscillatory rheological measurements in the linear viscoelastic regime.
- The maximum extension of the adhesive in a tensile test, ε_{max}, or its elongational rheological properties.
- The spectrum of relaxation times of the adhesive.

E' should be in a window of 0.01–0.1 MPa. Above that level, proper bonding and fibril formation are reduced and below that level, viscoelastic dissipation during the debonding process will be too low. ε_{max} controls the maximum extension of the fibrils and therefore plays a major role in the measured debonding energy. A relatively small ε_{max} is typically desirable for removable PSA, while a larger value is often characteristic of semi-permanent ones. A reasonable idea of the value of ε_{max} can, in principle, be obtained by a characterization of the adhesive in elongational deformation.

Finally, a broad spectrum of relaxation times is necessary to ensure both quick bonding (short relaxation times) and reasonable fibril stability (long relaxation times) as well as to impart a broad temperature window of use.

In addition to these important material parameters of the PSA, the surface can also play a role in controlling tackiness in certain cases. Rough surfaces can prevent proper bonding and, by forming defects, initiate failure during debonding, both effects reducing tack. Low energy surfaces can also influence tack by preventing fibril formation and the relevant parameter to predict whether this will occur or not is the ratio of the critical energy release rate, $\mathcal{G}_c$, over the elastic modulus, E.

8. Further reading

Hammond, F.H., Tack. In: D. Satas (Ed.), *Handbook of Pressure Sensitive Adhesive Technology*, Vol. 1. Van Nostrand Reinhold, New York, 1989, pp. 38–60.

Zosel, A., Fracture energy and tack of pressure sensitive adhesives. *Adv. Press. Sens. Adhes. Technol.*, **1**, 92–127 (1992).

Creton, C. In: Meijer, H.E.H. (Ed.), *Materials Science of Pressure-Sensitive Adhesives*, Vol. 18, 1st edn. VCH, Weinheim, 1997, pp. 707–741.

Gay, C. and Leibler, L., *Phys. Today*, 47–52 (1999).

References

1. Chang, E.P., *J. Adhes.*, **34**, 189–200 (1991).
2. Chang, E.P., *J. Adhes.*, **60**, 233–248 (1997).
3. Tse, M.F. and Jacob, L., *J. Adhes.*, **56**, 79–95 (1996).
4. Zosel, A., *J. Adhes.*, **30**, 135–149 (1989).
5. Urahama, Y., *J. Adhes.*, **31**, 47–58 (1989).
6. Hammond, F.H., *ASTM Spec. Tech. Publ.*, **360**, 123–134 (1964).
7. Zosel, A., *Colloid Polym. Sci.*, **263**, 541–553 (1985).
8. Lakrout, H., Sergot, P. and Creton, C., *J. Adhes.*, **69**, 307–359 (1999).
9. Tordjeman, P., Papon, E. and Villenave, J.-J., *J. Polym. Sci. B Polym. Phys.*, **38**, 1201–1208 (2000).
10. Chuang, H.K., Chiu, C. and Paniagua, R., *Adhes. Age*, september, 18–23 (1997).
11. Crosby, A.J. and Shull, K.R., *J. Polym. Sci. B Polym. Phys.*, **37**, 3455–3472 (1999).
12. Dahlquist, C.A. In: Patrick, R.L. (Ed.), *Pressure-Sensitive Adhesives*, Vol. 2. Dekker, New York, 1969, pp. 219–260.
13. Creton, C. and Leibler, L., *J. Polym. Sci. B Polym. Phys.*, **34**, 545–554 (1996).
14. Hui, C.Y., Lin, Y.Y. and Baney, J.M., *J. Polym. Sci. B Polym. Phys.*, **38**, 1485–1495 (2000).
15. Gay, C. and Leibler, L., *Phys. Rev. Lett.*, **82**, 936–939 (1999).
16. Gay, C. and Leibler, L., *Phys. Today*, 47–52 (1999).
17. Lakrout, H., Creton, C., Ahn, D. and Shull, K.R., *Macromolecules*, **34**, 7448–7458 (2001).
18. Gent, A.N. and Lindley, P.B., *Proc. R. Soc. Lond. A*, **249**, 195–205 (1958).
19. Williams, M.L. and Schapery, R.A., *Int. J. Fract. Mech.*, **1**, 64–71 (1965).
20. Chikina, I. and Gay, C., *Phys. Rev. Lett.*, **85**, 4546–4549 (2000).
21. Lin, Y.Y., Hui, C.Y. and Conway, H.D., *J. Polym. Sci. B Polym. Phys.*, **38**, 2769–2784 (2000).
22. Creton, C. and Lakrout, H., *J. Polym. Sci. B Polym. Phys.*, **38**, 965–979 (2000).
23. Gent, A.N. and Schultz, J., *J. Adhes.*, **3**, 281–294 (1972).
24. Maugis, D. and Barquins, M., *J. Phys. D: Appl. Phys.*, **11**, 1989–2023 (1978).
25. Barquins, M. and Maugis, D., *J. Adhes.*, **13**, 53–65 (1981).
26. Creton, C. In: Meijer, H.E.H. (Ed.), *Materials Science of Pressure-Sensitive Adhesives*, Vol. 18, 1st edn. VCH, Weinheim, 1997, pp. 707–741.
27. Good, R.J. and Gupta, R.K., *J. Adhes.*, **26**, 13–36 (1988).
28. Baljon, A.R.C. and Robbins, M.O., *Science*, **271**, 482–483 (1996).
29. McKinley, G.H. and Hassager, O., *J. Rheol.*, **43**, 1195–1212 (1999).
30. Gersappe, D. and Robbins, M.O., *Europhys. Lett.*, **48**, 150–155 (1999).
31. Crosby, A.J., Shull, K.R., Lakrout, H. and Creton, C., *J. Appl. Phys.*, **88**, 2956–2966 (2000).
32. Ferry, J.D., *Viscoelastic Properties of Polymers*, 3rd edn., Vol. 1. Wiley, New York, 1980.
33. Doi, M. and Edwards, S.F., *The Theory of Polymer Dynamics*, Oxford, 1986.
34. Zosel, A., *Adv. Press. Sens. Adhes. Technol.*, **1**, 92–127 (1992).
35. Ahn, D. and Shull, K.R., *Langmuir*, **14**, 3637–3645 (1998).
36. Aubrey, D.W. and Ginosatis, S., *J. Adhes.*, **12**, 189–198 (1981).
37. Chan, H. and Howard, G.J., *J. Adhes.*, **9**, 279–301 (1978).
38. Krenceski, M.A. and Johnson, J.F., *Polym. Eng. Sci.*, **29**, 36–43 (1989).
39. Zosel, A., *Int. J. Adhes. Adhes.*, **18**, 265–271 (1998).
40. Creton, C., Hooker, J.C. and Shull, K.R., *Langmuir*, **17**, 4948–4954 (2001).
41. Zosel, A., *J. Adhes.*, **34**, 201–209 (1991).
42. Webster, I., *Int. J. Adhes. Adhes.*, **19**, 29–34 (1999).

43. Flanigan, C.M., Crosby, A.J. and Shull, K.R., *Macromolecules*, **32**, 7251–7262 (1999).
44. Komatsuzaki, S., *Tappi Hot Melt Symposium*, 1999, pp. 171–178.
45. Spence, B.A. and Higgins, J.W., *J. Adhes. Seal. Counc.*, 237–255 (1997).
46. de Crevoisier, G., Fabre, P., Corpart, J.-M. and Leibler, L., *Science*, **285**, 1246–1249 (1999).
47. Aymonier-Marçais, A., Papon, E., Villenave, J.J., Tordjeman, P., Pirri, R. and Gérard, P., *Proceedings of Euradh*, Lyon, 2000, pp. 170–175.
48. Portigliatti, M., Hervet, H. and Léger, L., *C.R. Acad. Sci. Paris, IV*, **1**, 1187–1196 (2000).
49. Aubrey, D.W. and Sherriff, M., *J. Polym. Sci. Polym. Chem. Ed.*, **16**, 2631–2643 (1978).
50. Aubrey, D.W. and Sherriff, M., *J. Polym. Sci. Polym. Chem. Ed.*, **18**, 2597–2608 (1980).
51. Class, J.B. and Chu, S.G., *J. Appl. Polym. Sci.*, **30**, 805–814 (1985).
52. Class, J.B. and Chu, S.G., *J. Appl. Polym. Sci.*, **30**, 825–842 (1985).
53. Class, J.B. and Chu, S.G., *J. Appl. Polym. Sci.*, **30**, 815–824 (1985).
54. Kim, H.J. and Mizumachi, H., *J. Adhes.*, **49**, 113–132 (1995).
55. Tancrede, J.M., *J. Adhes. Seal. Counc.*, (**Spring**) 13–27 (1998).
56. Satas, D. In: Satas, D. (Ed.), *Handbook of Pressure Sensitive Adhesive Technology*, 2nd edn., Vol. 1, Van Nostrand Reinhold, New York, 1989, p. 940.
57. Zosel, A. and Schuler, B., *J. Adhes.*, **70**, 179–195 (1999).
58. Delgado, J., Kinning, D.J., Lu, Y.Y. and Tran, T.V., *Proceedings of 19th Annual Conference of the Adhesion Society*, 1996, pp. 342–345.
59. Gent, A.N. and Petrich, R.P., *Proc. R. Soc. Lond. A*, **310**, 433–448 (1969).
60. Benyahia, L., Verdier, C. and Piau, J.M., *J. Adhes.*, **62**, 45–73 (1997).
61. Ondarçuhu, T., *J. Phys. II*, **7**, 1893–1916 (1997).
62. Zosel, A., *J. Adhes. Sci. Tech.*, **11**, 1447–1457 (1997).
63. Hooker, J.C., Creton, C., Tordjeman, P. and Shull, K.R., *Proceedings of 22nd Annual Meeting of the Adhesion Society*, Panama City, 1999, pp. 415–417.
64. Lin, Y.Y., Hui, C.Y. and Baney, J.M., *J. Phys. D: Appl. Phys.*, **32**, 2250–2260 (1999).
65. Baney, J.M., *J. Appl. Phys.*, **86**, 4232–4241 (1999).
66. Toyama, M., Ito, T., Nukatsuka, H. and Ikeda, M., *J. Appl. Polym. Sci.*, **17**, 3495–3502 (1973).
67. Zhang Newby, B.M., Chaudhury, M.K. and Brown, H.R., *Science*, **269**, 1407–1409 (1995).
68. Fuller, K.N.G. and Tabor, D., *Proc. R. Soc. Lond. A*, **A345**, 327–342 (1975).
69. Chiche, A., Pareige, P. and Creton, C., *C.R. Acad. Sci. Paris, IV*, **1**, 1197–1204 (2000).
70. Lakrout, H., PhD. Thesis, Université Paris, 7, 1998.
71. Hooker, J.C., Creton, C., Unpublished results.

Chapter 15

Contact mechanics

KENNETH R. SHULL [*]

*Department of Materials Science and Engineering, Northwestern University,
Evanston, IL 60201, USA*

1. Introduction

A substantial body of literature is devoted to the mechanics of contact between
elastic solids. A book on this topic, written by Johnson and appearing in 1985,
gives a very useful review of this subject [1]. The purpose of this chapter is to
provide an up-to-date description of contact mechanics methods in the context
of adhesion science, focusing on the underlying mathematical description of the
most relevant contact problems. More detailed descriptions of the applications
of these methods are discussed in other chapters in this volume. Some of the
mathematical expressions used here are very well established, and date back to the
work of Hertz in the late 1800s [2]. Much of the formalism described here is quite
recent, however, and is not included in previous review texts. More recent work
includes the extension to the contact of thin adhesive layers, and the development
of models that quantitatively account for viscoelastic effects. We begin in the
following section by introducing some of the fundamental expressions of contact
mechanics for axially symmetric geometries. The effects of adhesive forces on
these geometries are then discussed in Section 3. Many of the concepts introduced
in these sections can be applied to non-axisymmetric geometries as well. The
contacting cylinders described in Section 4 are given as one example of such an
alternate geometry. The contact of linear viscoelastic materials is described in
Section 5, before concluding with a brief summary.

2. Fundamental contact expressions

The common assumption of contact mechanics is that the two surfaces in contact are frictionless, so that shear stresses cannot be sustained at the interface.

---

[*] E-mail: k-shull@northwestern.edu

This 'frictionless' assumption is often appropriate for very stiff materials where adhesive forces are relatively unimportant, but it is often not the case for softer materials such as elastomers, where adhesive forces play a very important role. In these cases, a 'full-friction' boundary condition, where sliding of the two surfaces is not allowed, is often more appropriate. In many important cases (contact of a very thick, incompressible elastic layer, for example) there is little or no practical difference for these two boundary conditions. Nevertheless, in the discussion that follows, we are careful to indicate that boundary condition (frictionless or full-friction) that formally applies in each case. In all cases we assume that the contacting materials are isotropic and homogeneous, each being characterized by two independent elastic constants.

2.1. Elastic half spaces and axisymmetric geometries

Sneddon has developed analytic expressions for the relationship between load and displacement for adhesionless, axisymmetric, rigid punches of an arbitrary profile in contact with an elastic half space, where the film thickness h is much larger than the contact radius a [3]. The simplest situation is for that of a flat, rigid, cylindrical punch, illustrated schematically in Fig. 1. Here the contact radius remains equal to the punch radius as the indenter is pushed into the half space. The stress and displacement fields in the half space gradually decay to zero at depths into the elastic layer that greatly exceed the contact radius. For small strains the relationship between the compressive load P and the displacement δ are linearly related through the compliance C, defined as the ratio of the displacement to the load. For a rigid, frictionless indenter in contact with an elastic half space, C is equal to C_0, given by [1,3]:

$$C_0 = \frac{1}{2E^*a} \tag{1}$$

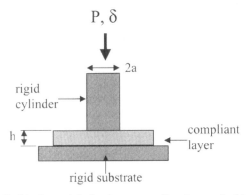

Fig. 1. Flat, cylindrical punch indenting a compliant layer of arbitrary thickness.

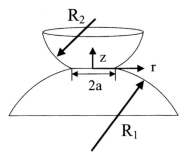

Fig. 2. Two curved surfaces in contact.

Here E^* is the effective modulus of the elastic half space, which includes its Young modulus, E, and its Poisson ratio, ν

$$E^* = \frac{E}{1 - \nu^2} \tag{2}$$

This combination of E and ν appears frequently in plane-strain fracture problems of the sort discussed in subsequent sections, and is introduced here for this reason.

Eqs. 1 and 2 can be generalized to the case where two curved elastic half spaces are in contact with one another, as illustrated in Fig. 2 [1]. In this case the displacement fields in the two half spaces are added to one another. For the hypothetical case where the contact area between the two materials remains fixed, the overall compliance is still given by Eq. 1, with an effective modulus that involves the elastic constants, E_1, E_2, ν_1 and ν_2, of both half spaces:

$$\frac{1}{E^*} = \frac{1 - \nu_1^2}{E_1^*} + \frac{1 - \nu_2^2}{E_2^*} \tag{3}$$

While this expression describes the compliance of the system at a fixed contact area, it does not provide any information about the development of this contact area as the two surfaces are brought into contact with one another. If the two half spaces are spheres with radii of curvature R_1 and R_2, the displacement and loads are given by the respective Hertzian values, δ_H and P_H, which can both be expressed in terms of the contact radius, a [1]:

$$P_H = \frac{4E^* a^3}{3R} \tag{4}$$

$$\delta_H = \frac{a^2}{R} \tag{5}$$

Here R is a function of R_1 and R_2, the radii of curvature of the individual half spaces:

$$\frac{1}{R} = \frac{1}{R_1} + \frac{1}{R_2} \tag{6}$$

Our notation is to refer to these non-adhesive values of the load and displacement as P' and δ', respectively. Actual values of the load and displacement will generally differ from these values when the contacting surfaces adhere to one another. These differences are quantified in subsequent sections. Barquins has given values of P' for a variety of other frictionless, axisymmetric half spaces, including rounded and flat-ended cones, etc. [4]. The reader is referred to this reference for details.

Frictional contributions to the interfacial stresses vanish in two important situations. For a rigid frictionless punch against an incompressible material ($E_1^* \gg E_2^*$, $\nu_2 = 0.5$), the radial displacement at the interface vanishes. If the elastic constants of both materials are identical ($E_1^* = E_2^*$, $\nu_1 = \nu_2$), the radial displacements for the frictionless case are continuous across the interface, and are not affected by the addition of interfacial friction. In both cases all of the expressions given above are unaffected by the nature of the frictional boundary condition [1]. Finally, it is important to realize that P', δ' and C are not independent of one another, but are related by the following expression:

$$C = \frac{\partial \delta'}{\partial P'} \tag{7}$$

This expression is a useful check on approximate relationships between C, δ' and P', given, for example, in the following section. Eq. 7 is a general expression that is valid for both the frictionless and full-friction boundary conditions.

2.2. Modifications due to finite size effects

In the previous examples we have assumed that the thickness of the sample is much larger than the contact radius. Modifications to these expressions are necessary when the compliant material is reduced in thickness so that it cannot be appropriately represented as an elastic half space. Important experimental examples of this situation are illustrated in Fig. 3. The important parameter is the thickness of the compliant layer, h, relative to the contact radius, a. In the examples that we consider, an elastic material with a thickness of h is confined between a rigid substrate and a rigid indenter. The following approximate modifications to P', δ' and C have been obtained [5]:

$$P' = P_{\mathrm{H}} f_P(a/h), \qquad f_P(a/h) = 1 + \beta(a/h)^3 \tag{8}$$

$$\delta' = \delta_{\mathrm{H}} f_\delta(a/h), \qquad f_\delta(a/h) = 0.4 + 0.6\exp(-1.8a/h) \tag{9}$$

$$C = C_0 f_C(a/h),$$

$$\frac{1}{f_C(a/h)} = 1 + \{0.75/[(a/h) + (a/h)^3] + 2.8(1 - 2\nu)/(a/h)\}^{-1} \tag{10}$$

These correction factors are plotted in Fig. 4 for an incompressible layer ($\nu = 0.5$).

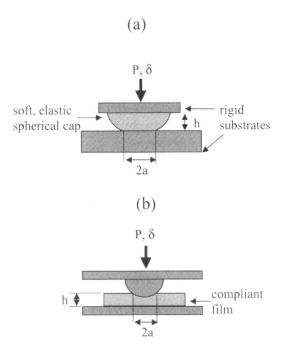

Fig. 3. Curved geometries where confinement effects are important.

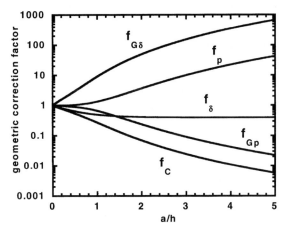

Fig. 4. Finite size correction factors.

Note that for $a/h = 0$, P' reduces to P_H, δ' reduces to δ_H and C reduces to C_0. All three expressions assume a full-friction boundary condition between the thin layer and the supporting substrate, meaning that shear stresses at this interface are not allowed to relax as the materials are pressed into contact with one another. For the compliance expression (Eq. 10) a full-friction boundary condition is also assumed for the interface between the compliant layer and the indenter itself. The expression for δ' is unaffected by the nature of the indenter/layer interface, and corresponds to a frictionless or full-friction boundary condition. Finally, the coefficient β in Eq. 8 is equal to 0.25 for a full-friction boundary condition at the indenter interface, and is equal to 0.15 for a frictionless boundary condition at this interface. These expressions are valid when ν, Poisson's ratio for the elastic material, is close to 0.5, although a set of compliance expressions that are valid for all values of ν have recently been developed [6].

3. Inclusion of adhesive forces

The expressions above do not account for adhesive interactions between the contact surfaces, i.e., the contact stresses are purely compressive. Adhesive interactions allow tensile forces to be supported, so that the actual values of the load and displacement for a given contact area are lower than the respective values of P' and δ' given by the above expressions. The effects of these adhesive forces on the contact problem are discussed in this section.

3.1. Linear elastic fracture mechanics approach

Our treatment of adhesive interactions is based on an energy balance, and is very similar to the derivation presented by Johnson et al. in their classic paper [7]. These authors assumed a geometry of contacting spheres with small contact areas ($a/h = 0$). Here we use a more generalized version in order to readily account for geometrical effects [8].

Our starting point is the definition of the energy release rate, $\mathcal{G}$. This quantity has the units of a surface energy (energy/area), and describes the amount of energy that is available to decrease the contact area, A, by a unit amount:

$$\mathcal{G} \equiv \frac{\partial}{\partial A}(U_E + U_M) \tag{11}$$

where U_E is the elastic energy of the system and U_M is the mechanical potential energy associated with the applied load. Because the elastic energy is a state function that is independent of the detailed deformation history, any hypothetical deformation history can be used to develop an expression for U_E in terms of the current values of P, δ and a. It is useful here to consider a hypothetical deformation history that takes place in two stages. In the first stage adhesive interactions are 'turned off', and the displacement is increased until the contact area is equal to the contact area of interest. Note that displacement and the load are related to one another by the elastic properties of the system, so that we can think of either the load or the displacement as the independent variable. Letting

U_1 represent the elastic energy associated with this portion of the deformation process, we have:

$$U_1 = \int_0^{\delta'} P'\mathrm{d}\delta \tag{12}$$

where P' and δ' define the evolution of the load and displacement because adhesive interactions are ignored in this first stage in the deformation process.

In the second stage of our hypothetical deformation process, the contact radius is fixed, and the displacement (or load) is decreased until it is equal to the value of interest. The energy U_2 associated with this portion of the deformation process is:

$$U_2 = \int_{P'}^{P} P\,\mathrm{d}\delta \bigg|_a \tag{13}$$

This quantity can be rewritten if we use the definition of compliance: $C = \partial\delta/\partial P|_a$. Using the chain rule, a substitution can be made and the integral for U_2 can be solved:

$$U_2 = \int_{P'}^{P} P\frac{\partial\delta}{\partial P}\mathrm{d}P \bigg|_a = C\int_{P'}^{P} P\,\mathrm{d}P \bigg|_a = C\left(\frac{P^2}{2} - \frac{P'^2}{2}\right) \tag{14}$$

The overall elastic energy is then obtained as the sum of U_1 and U_2:

$$U_\mathrm{E} = U_1 + U_2 = \int_0^{\delta'} P'\mathrm{d}\delta - \frac{P'^2 C}{2} + \frac{P^2 C}{2} \tag{15}$$

The mechanical potential energy is determined by the distance over which the load has been moved, or $U_\mathrm{M} = -P\delta$. This simple expression can be rewritten by rewriting the compliance expression in the following form:

$$C = \frac{\delta' - \delta}{P' - P} \tag{16}$$

from which we obtain:

$$U_\mathrm{M} = -P[\delta' - C(P' - P)] \tag{17}$$

Substitution of Eqs. 15 and 17 into Eq. 11 for $\mathcal{G}$ gives:

$$\mathcal{G} = \frac{\partial}{\partial A}\left(\int_0^{\delta'} P'\mathrm{d}\delta\right) - \frac{(P'-P)^2}{2}\frac{\partial C}{\partial A} - C(P'-P)\frac{\partial P'}{\partial A} - P\frac{\partial\delta'}{\partial A} \tag{18}$$

Use of the following expressions:

$$\frac{\partial}{\partial A}\left[\int_0^{\delta'} P'd\delta\right] = P'\frac{\partial\delta'}{\partial A} \tag{19}$$

and

$$\frac{\partial\delta'}{\partial A} = \frac{\partial\delta'}{\partial P'}\frac{\partial P'}{\partial A} = C\frac{\partial P'}{\partial A} \tag{20}$$

gives the final solution for $\mathcal{G}$:

$$\mathcal{G} = -\frac{(P'-P)^2}{2}\frac{\partial C}{\partial A} = -\frac{(P'-P)^2}{4\pi a}\frac{\partial C}{\partial a} \tag{21}$$

It is sometimes useful to use Eq. 16 to rewrite this expression in terms of the displacement:

$$\mathcal{G} = -\frac{(\delta'-\delta)^2}{2C^2}\frac{\partial C}{\partial A} = -\frac{(\delta'-\delta)^2}{4\pi aC^2}\frac{\partial C}{\partial a} \tag{22}$$

The versions of Eqs. 21 and 22 involving the contact radius, a, have been obtained by assuming axial symmetry, with $A = \pi a^2$.

3.1.1. The JKR limit

For contact between two spheres, in the case where the contact radius is small ($a/h \to 0$), we can set P' to P_H and C to C_0 to obtain the following expression:

$$\mathcal{G} = \frac{(4E^*a^3/3R - P)^2}{8\pi E^*a^3} \tag{23}$$

which can be rearranged to give the following relationship for a in terms of P:

$$a^3 = \frac{3R}{4E^*}\left\{P + 3\pi\mathcal{G}R + [6\pi\mathcal{G}RP + (3\pi\mathcal{G}R)^2]^{1/2}\right\} \tag{24}$$

This equation is the well-known expression derived originally by Johnson et al. [7]. Formally, these authors considered the equilibrium case, where $\mathcal{G}$ is equal to the thermodynamic work of adhesion. Our approach is more closely analogous to the derivation of Maugis and Barquins, and is based on *mechanical* equilibrium between the adhesive forces and the bulk deformation of the elastic material. With this assumption, Eq. 23 can be used to obtain a meaningful value for $\mathcal{G}$ for every point along a loading or unloading curve. Adhesion hysteresis is almost always observed, with the value for $\mathcal{G}$ obtained for unloading exceeding the value obtained for loading. For elastomers, $\mathcal{G}$ is often found to be a function of the crack velocity, given by the rate of change of the contact radius [9]. A variety of approaches for predicting the relationship between $\mathcal{G}$ and v have been proposed, as discussed below in the section on viscoelastic contact mechanics. If this relationship between $\mathcal{G}$ and v is known, Eq. 23, or the modified versions given in the following section, can be used to predict the relationship between the load and displacement for an actual experiment [8,10].

3.1.2. Finite size corrections to $\mathcal{G}$

The expressions for the energy release rate can be generalized to the adhesion of relatively thin layers by using the approximate form of the compliance described in Section 2.2. The 'JKR' limit of the previous section corresponds to the case where $a/h = 0$. Eq. 10 can be combined with Eq. 21 or Eq. 22 to give expressions for the energy release rate that are not dependent on the assumption that $a/h = 0$. Simple expressions are not readily obtained for compressible systems. For incompressible materials with $v = 0.5$, however, the following relatively simple expressions are obtained for $\mathcal{G}$ [5]:

$$\mathcal{G} = \frac{(P'-P)^2}{8\pi E^* a^3} f_{Gp}(a/h), \quad f_{Gp}(a/h) = \frac{0.56 + 1.5(a/h) + 3(a/h)^3}{(0.75 + (a/h) + (a/h)^3)^2} \tag{25}$$

$$\mathcal{G} = \frac{E^*(\delta' - \delta)^2}{2\pi a} f_{G\delta}(a/h), \quad f_{G\delta}(a/h) = 1 + 2.67\left(\frac{a}{h}\right) + 5.33\left(\frac{a}{h}\right)^3 \tag{26}$$

with P' and δ' given by Eqs. 8 and 9, respectively. These correction factors are plotted as a function of a/h in Fig. 4. Note that both geometric correction factors, f_{GP} and $f_{G\delta}$, reduce to unity at $a/h = 0$. Also, both expressions correspond to the full-friction boundary condition, since this is the boundary condition for the compliance expression from which they were derived.

3.2. Expressions for the stress distribution

The compliance based treatment of contact mechanics is useful because it relies only on directly measurable quantities, these being the load, displacement and contact radius. It does not rely on information relating to the detailed stress distribution within the material. Often, however, it is useful to have more information about this stress distribution. Analytic theories of non-adhesive contact generally assume that the contact is frictionless, so that the interface is not able to support shear stresses. The Hertzian form of the radial distribution of the normal stress, or contact pressure, for frictionless contact is given by the following expression [1]:

$$\sigma'_{zz}(r) = -\frac{3P'}{2\pi a^2}\left\{1 - (r/a)^2\right\}^{1/2} \tag{27}$$

where r is radial distance from the axis of symmetry and we have used the primed

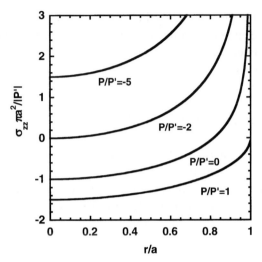

Fig. 5. JKR stress distributions for different values of P/P'.

notation to indicate that the stress and overall load correspond to the non-adhesive case. The negative sign accounts for the fact that positive stresses are considered to be tensile, whereas positive loads are compressive according to the normal contact mechanics convention.

In the adhesive case tensile loads can be supported, so that the actual load P is generally less than P'. An additional stress field arises that is proportional to the difference, $P' - P$. For frictionless contact with $a/h = 0$, this additional normal stress, referred to here as σ_{zz}^{adh}, is given by the following expression [3]:

$$\sigma_{zz}^{adh}(r) = \frac{P' - P}{2\pi a^2}\left\{1 - (r/a)^2\right\}^{-1/2} \tag{28}$$

The overall stress distribution is obtained by adding σ_{zz}' and σ_{zz}^{adh}. For example, the 'JKR' stress distribution, σ_{zz}^{jkr}, is obtained by adding these contributions and using the Hertzian expression for P':

$$\sigma_{zz}^{jkr}(r) = -\frac{3P'}{2\pi a^2}\left\{1 - (r/a)^2\right\}^{1/2} + \frac{P' - P}{2\pi a^2}\left\{1 - (r/a)^2\right\}^{-1/2} \tag{29}$$

with $P' = P_H$. This stress distribution is plotted in Fig. 5 for different values of the normalized load, P/P'. For a rigid punch adhering to an elastic half space ($a/h = 0$), the adhesive component of the stress field is independent of the shape of the indenter. The non-adhesive load, P', and its associated stress field, σ_{zz}', does depend on the shape of the indenter. Because the adhesive stress field is dominant at the periphery of the contact zone ($r \approx a$), much of our focus below is on this contribution.

The expressions given above are valid only for frictionless contacts in the limit where a/h is very small. Additional information about the stress distribution within the contact zone in the region close to the crack tip can be obtained from the stress intensity factor. The stress field for small, positive values of the quantity $a - r$ is given by the following expression [11]:

$$\sigma_{zz}(r) = \frac{K_{\mathrm{I}}}{\{2\pi (a - r)\}^{1/2}}, \qquad \sigma_{rz}(r) = \frac{K_{\mathrm{II}}}{\{2\pi (a - r)\}^{1/2}} \tag{30}$$

where K_{I} and K_{II} are the respective mode I and mode II stress intensity factors. Note that $K_{\mathrm{II}} = 0$ for frictionless contact, and that $K_{\mathrm{II}} = 0$ in general for a rigid indenter in contact with a thick, incompressible material ($a/h = 0$, $v = 0.5$), and in the case where the contacting materials have identical elastic properties. For the full friction case, more highly confined systems ($a/h > 1$) will be characterized by a finite value of K_{II}, corresponding to a non-zero phase angle, Ψ, defined as follows [11]:

$$\Psi = \tan^{-1}(K_{\mathrm{II}}/K_{\mathrm{I}}) \tag{31}$$

The stress intensity for a rigid punch on an elastic half space is given by the following expression:

$$K_{\mathrm{I}} = \lim_{r \to a} \{2\pi (a - r)\}^{1/2} \sigma_{zz}^{\mathrm{adh}}(r) = \frac{P' - P}{2a^{3/2}\pi^{1/2}} \tag{32}$$

For plane strain conditions (corresponding to the contact problems of interest here), $\mathscr{G}$, K_{I} and K_{II} are related to one another by the following relationships:

$$\mathscr{G} = \frac{1}{2E^*}\{K_{\mathrm{I}}^2 + K_{\mathrm{II}}^2\},$$
$$K_{\mathrm{I}} = (2E^*\mathscr{G})^{1/2}\cos\Psi, \qquad K_{\mathrm{II}} = (2E^*\mathscr{G})^{1/2}\sin\Psi \tag{33}$$

For frictionless contact, $K_{\mathrm{II}} = 0$, and combination of Eqs. 32 and 33, with $P' = P_{\mathrm{H}}$, gives Eq. 23 for the energy release rate.

For an unconfined system ($a/h = 0$) under tensile loading, the stress distribution is dominated by the singularity at the edge of contact. The magnitude of this stress singularity, as quantified by K_{I} and K_{II}, decreases with increasing values of a/h. For the full-friction boundary condition, values for K_{I} and K_{II} can be obtained by combining Eqs. 25 and 33 to give the following:

$$K_{\mathrm{I}} = \frac{P' - P}{2a^{3/2}\pi^{1/2}} f_{Kp}\cos\Psi, \qquad K_{\mathrm{II}} = \frac{P' - P}{2a^{3/2}\pi^{1/2}} f_{Kp}\sin\Psi \tag{34}$$

where f_{Kp} is the finite size correction factor for the stress intensity factor:

$$f_{Kp} = f_{Gp}^{1/2} = \frac{(0.56 + 1.5(a/h) + 3(a/h)^3)^{1/2}}{0.75 + (a/h) + (a/h)^3} \tag{35}$$

Expressions for the phase angle, Ψ, are not yet available, although these can be determined by finite element methods. For example, Lin et al. have used a finite

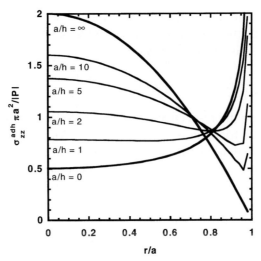

Fig. 6. Adhesive stress distributions for $\nu = 0.5$, for different values of a/h.

element method to determine Ψ for the flat punch geometry of Fig. 1 [6]. For $\nu = 0.49$, they obtain $\Psi = 29°$ for $a/h = 10$ and $\Psi = 32°$ for $a/h = 20$. As mentioned previously, the phase angle is zero for $a/h = 0$ and $\nu = 0.5$.

As a/h increases towards infinity, the magnitude of the stress singularity at the edge of the contact zone continues to decrease. The stress in this case is dominated by hydrostatic stresses that reach a maximum at the center of the contact zone and decay to zero at the edges. For a highly confined incompressible system ($a/h = \infty$ and $\nu = 0.5$), the adhesive contribution to the stress can be approximated by the following expression [12]:

$$\sigma_{zz}^{\text{adh}}(r) = \frac{2\left(P' - P\right)}{\pi a^2}\left(1 - (r/a)^2\right) \tag{36}$$

Fig. 6 shows the distribution of adhesive stresses for varying values of a/h. The curves for $a/h = 0$ and $a/h = \infty$ correspond to Eqs. 28 and 36, respectively. The curves for intermediate values of a/h were obtained by finite element methods, and correspond to a full-friction boundary condition [13]. The change in the shape of the stress distribution with increasing values of a/h is responsible for the development of elastic shape instabilities that appear at relatively large strains [13].

3.3. Cohesive zone models and stress in the crack tip region

While the stress intensity factor describes the adhesive stress field near the edge of the contact area, it obviously breaks down at $r = a$, where an infinite stress is predicted. In reality, the stress remains finite because of attractive forces within

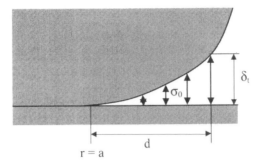

Fig. 7. Schematic illustration of the cohesive zone.

a cohesive zone immediately outside the area of contact. Cohesive zone models, which self-consistently relate the stress distribution to the shape of the contacting solids immediately outside the contact area, are required in order to understand the stress distribution in the immediate vicinity of the contact edge. The simplest model is the Dugdale model illustrated in Fig. 7, where a constant tensile stress, σ_0, is assumed to exist over a region of width d outside the contact radius. In this case the adhesive stresses are given by the following equation [14]:

$$\sigma_{zz}^{\text{adh}}(r) = \frac{2\sigma_0}{\pi} \arctan \left(\frac{((r+d)/a)^2 - 1}{1 - (r/a)^2} \right)^{1/2} \qquad r < a$$

$$\sigma_{zz}^{\text{adh}}(r) = \sigma_0 \qquad\qquad\qquad\qquad a \leq r \leq a+d$$

$$\sigma_{zz}^{\text{adh}}(r) = 0 \qquad\qquad\qquad\qquad r > a+d \qquad (37)$$

In the derivation of Eq. 37, no assumption is made about the relative size of the cohesive zone. The underlying assumption of the fracture mechanics treatment employed throughout this chapter is that the cohesive zone is small in comparison to the overall dimensions of the contact zone, i.e. $d \ll a$. The opposite case, where $d \gg a$, corresponds to the 'DMT' theory of Derjaguin, Muller and Toporov [15]. The 'DMT' condition of large cohesive zone size is almost never met in practice, with the exception of very stiff, nano-scale contacts, or in cases where the cohesive zone corresponds to a liquid meniscus

If the cohesive zone is assumed to be small ($d \ll a$), the adhesive stress can be written in the following form:

$$\sigma_{zz}(r) = \frac{2\sigma_0}{\pi} \arctan \left(\frac{d}{(a-r)} \right)^{1/2} \qquad (38)$$

Away from the immediate vicinity of the crack tip, the stress distribution has the form given by Eq. 30, with the following stress intensity factor:

$$K_{\text{I}} = 4\sigma_0 (d/2\pi)^{1/2} \qquad (39)$$

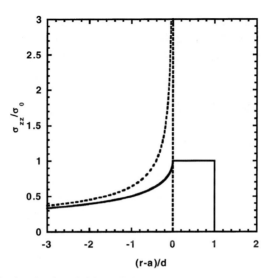

Fig. 8. Stress distribution in the vicinity of the edge of contact for the Dugdale/Barenblatt cohesive zone model (solid line, Eq. 38), compared to the asymptotic form given by the stress intensity factor (dashed line, Eqs. 30 and 39).

The predictions of Eqs. 38 and of 30 (with K_{I} given by Eq. 39) are compared in Fig. 8. The cohesive forces eliminate the divergence in the stress field, and modify the distribution of stresses over a distance that is comparable with the cohesive zone width, d. At distances from the crack tip that significantly exceed d, the stress field is accurately represented by the square route singularity associated with the stress intensity factor.

The energy associated with the cohesive zone in the Dugdale model is simply the product of the cohesive stress, σ_0, and the distance, δ_t, over which this stress operates. For a mode one crack ($K_{\mathrm{II}} = 0$) we can therefore write two expressions for the energy release rate:

$$\mathcal{G} = \sigma_0 \delta_t \tag{40}$$

and

$$\mathcal{G} = \frac{K_{\mathrm{I}}^2}{2E^*} = \frac{4\sigma_0^2 d}{\pi E^*} \tag{41}$$

Equating these two expressions for $\mathcal{G}$ gives the following relationship between d and δ_t:

$$d = \frac{\pi E^* \delta_t}{4\sigma_0} \tag{42}$$

For a given cohesive zone model, specified by the characteristic stress σ_0 and the range of the interaction δ_t, the width of the cohesive zone is proportional to the modulus of the elastic material. In a viscoelastic material the effective modulus, and hence the width of the cohesive zone, depends on the crack velocity. The consequences of this are discussed more thoroughly in Section 5.4.

4. Contacting cylinders and rolling contact mechanics

While the contact geometries considered in this chapter are restricted primarily to axisymmetric geometries, a variety of other useful geometries can be considered. Our discussion of alternate geometries is limited here to contacting cylinders, illustrated schematically in Fig. 9. In this case a long half-cylinder lies parallel to a rigid surface. The analysis is very similar to the analysis given in the previous sections, but we normalize P and P' by the length ℓ of the cylinder to obtain the corresponding quantities P_ℓ and P'_ℓ. Also, because currently available expressions for the finite size corrections are quite complicated [16], we assume frictionless contact with $a/h = 0$ at all times. One useful feature of the cylindrical geometry is that it can be 'rolled' across the surface by applying a torque τ to the cylinder. For now we ignore the effects of the applied torque, and consider only the effects of an applied, compressive load. The non-adhesive load and displacement for a frictionless interface are given by the following expressions [1]:

$$P'_\ell = 4\pi E^* b^2 / 16R \qquad (43)$$

and

$$\delta' = \frac{b^2}{4R} \{2\ln(4R/b) - 1\} \qquad (44)$$

The compliance, C_ℓ for a cylinder of unit length, is given by the following expression:

$$C_\ell = \frac{\partial \delta'}{\partial P'_\ell} = \frac{\partial \delta'/\partial b}{\partial P'_\ell/\partial b} = \frac{2}{\pi E*} \{\ln(4R/b) + 1/2\} \qquad (45)$$

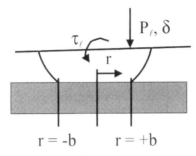

Fig. 9. Cylinder in contact with a flat rigid surface, showing the applied normal load, normal displacement and torque.

The energy release rate is obtained from Eq. 21. With $P = \ell P_\ell$, $P' = \ell P'_\ell$, $C = C_\ell/\ell$ and $A = 2b\ell$ we obtain the following equation:

$$\mathcal{G} = -\frac{\left(P'_\ell - P_\ell\right)^2}{4}\frac{\partial C_\ell}{\partial b} = \frac{\left(P'_\ell - P_\ell\right)^2}{2\pi E^* b} \tag{46}$$

This expression for the energy release rate is valid for symmetric loading, where the same stress field is transferred to the contact lines at each side of the cylinder. The situation is much more complicated in the presence of an applied torque, as described by Barquins [17] and as shown schematically in Fig. 9. In this figure, where the cylinder is being rolled in the counterclockwise direction, the left contact edge is a receding crack and the right contact edge is an advancing crack. The advancing and receding cracks are characterized by the respective energy release rates, $\mathcal{G}_a$ and $\mathcal{G}_r$, which are no longer equal to one another when a torque is applied to the cylinder. Expressions for these energy release rates can be obtained by starting with the following expression for the distribution of contact stresses under the cylinder:

$$\sigma_{zz}(r) = -\frac{2P'_\ell\left(1 - (r/b)^2\right)^{1/2}}{b\pi} + \frac{P'_\ell - P_\ell}{\pi b\left\{1 - (r/b)^2\right\}^{1/2}}$$
$$+ \frac{2\tau_\ell P'_\ell r}{\pi b^3\left(P'_\ell - P_\ell\right)\left\{1 - (r/b)^2\right\}^{1/2}} \tag{47}$$

The first term is the Hertzian compressive stress associated with the non-adhesive load, P'_ℓ, and the second term represents the tensile component arising from the adhesive load $P'_\ell - P_\ell$. These two contributions to the stress are symmetric about the middle of the contact zone ($r = 0$), and are the cylindrical counterpart to the JKR stress distribution given by Eq. 29. The third term is due to the applied torque, and is antisymmetric about $r = 0$. This term increases the magnitude of the stress singularity at the advancing crack ($r = b$), and decreases the magnitude of the stress singularity at the receding crack ($r = -b$). Values of K_{Ia} and K_{Ir}, the respective mode I stress intensity factors at the advancing and receding contact lines, can be obtained from the following expressions:

$$K_{\mathrm{Ia}} = \lim_{r \to +b}\{2\pi(b-r)\}^{1/2}\sigma_{zz}(r), \qquad K_{\mathrm{Ir}} = \lim_{r \to -b}\{2\pi(b-r)\}^{1/2}\sigma_{zz}(r) \tag{48}$$

which when combined with Eq. 47 gives:

$$K_{\mathrm{Ia}} = \frac{P'_\ell - P_\ell}{(\pi b)^{1/2}} + \frac{2\tau_\ell P'_\ell}{\pi^{1/2}b^{3/2}\left(P'_\ell - P_\ell\right)} \tag{49}$$

and

$$K_{\mathrm{Ir}} = \frac{P'_\ell - P_\ell}{(\pi b)^{1/2}} - \frac{2\tau_\ell P'_\ell}{\pi^{1/2}b^{3/2}\left(P'_\ell - P_\ell\right)} \tag{50}$$

The corresponding expressions for the energy release rates, obtained from Eq. 33 with $K_{\text{II}} = 0$ are:

$$\mathcal{G}_{a} = \frac{1}{2E^{*}} \left\{ \frac{P'_{\ell} - P_{\ell}}{(\pi b)^{1/2}} + \frac{2\tau_{\ell} P'_{\ell}}{\pi^{1/2} b^{3/2} \left(P'_{\ell} - P_{\ell} \right)} \right\}^{2} \tag{51}$$

and

$$\mathcal{G}_{r} = \frac{1}{2E^{*}} \left\{ \frac{P'_{\ell} - P_{\ell}}{(\pi b)^{1/2}} - \frac{2\tau_{\ell} P'_{\ell}}{\pi^{1/2} b^{3/2} \left(P'_{\ell} - P_{\ell} \right)} \right\}^{2} \tag{52}$$

Note that Eq. 46 is recovered in each case when no torque is applied to the cylinder ($\tau_{\ell} = 0$).

An advantage of the cylindrical geometry is that the crack velocity can be more directly controlled than it can be for the axially symmetric geometries. Also, the fact that the non adhesive load increases with the square of the contact length, and not with the cube as with the spherical geometries, can give greater sensitivity in some cases. The introduction of torque can be a useful addition in that it enables one to simultaneously obtain information pertaining to both advancing and receding cracks, provided that the torque and load are both measured. In this case one is left with two equations for the unknown quantities $\mathcal{G}_{a}$ and $\mathcal{G}_{r}$ that must be solved simultaneously. Various approximations can often be employed to simplify the analysis of the data, as described recently by She et al. [18,19].

5. Accounting for linear viscoelastic effects

The results presented in the previous sections assume that the contacting materials have well-defined elastic constants. In fact, most materials have at least some viscoelastic character, and it is important to understand how these effects should be taken into account. Viscoelastic effects enter into our analysis in two ways. First, it is possible that the overall elastic response of the system, described by the effective elastic constant, E^{*}, is time-dependent. In the case where adhesion is present, the stress near the crack tip will be defined by stress intensity factors, K_{I} and K_{II} that are themselves time-dependent. A unique energy release rate cannot be defined in this case. We refer to this macroscopic manifestation of viscoelastic behavior as 'large-scale viscoelasticity'. In this case one needs a procedure for determining the stress intensity factor that describes the current state of stress in the vicinity of the contact perimeter. Appropriate expressions for K_{I} are an essential result of treatments of large scale viscoelasticity, and these expressions are provided in Section 5.1.

Once the stress intensity factor is known, a single critical value does not generally describe the adhesive behavior. Instead, a relationship exists between the stress intensity factor and the resultant crack velocity, v. This manifestation of viscoelasticity is observed even when the relationships between load, displacement and contact radius are adequately described by a single value of the elastic constant, E^*. In this case the bulk of the material is behaving in a purely elastic manner, and the dissipative processes responsible for the velocity dependence of K_{I} are confined to a region that is small relative to the overall sample size. A well-defined stress intensity factor can still be defined, but the nature of the cohesive zone depends on the crack velocity. This situation is referred to as small-scale viscoelasticity, and is discussed in section 5.2.

In our discussion of large-scale and small-scale viscoelasticity, we consider a linear viscoelastic material with a stress relaxation modulus that can be written in terms of an initial modulus, E_0^*, and a relaxation function $\phi(t)$:

$$E^*(t) \equiv E_0^* \phi(t) \tag{53}$$

A related creep compliance function, $C^*(t)$, can also be defined:

$$C^*(t) \equiv \frac{\psi(t)}{E_0^*} \tag{54}$$

Note that $\phi(0) = \psi(0) = 1$, and that these two functions are related to one another through the following expression:

$$s^2 \mathcal{L}(\phi)\mathcal{L}(\psi) = 1 \tag{55}$$

where $\mathcal{L}$ is the Laplace transform and s is the Laplace transform variable. The full time dependence of either ϕ or ψ is sufficient to completely characterize the behavior of a linearly viscoelastic material.

5.1. Large-scale viscoelasticity

The review of large-scale viscoelasticity given here is divided into four parts, determined by the presence or absence of adhesion, and by the direction of motion of the contact line. The stress intensity factor is zero for non-adhesive contact, whereas it has some positive value for adhesive contact. Advancing contact refers to the case where the contact radius increases with time, whereas receding contact refers to the case where the contact radius is decreasing with time. The simplest case of advancing non-adhesive contact is treated in Section 5.1.1, followed by a discussion of receding non-adhesive contact in Section 5.1.2. The effects of adhesion are then introduced, with adhesive advancing contact being discussed in Section 5.1.3 and receding adhesive contact being discussed in Section 5.1.4. In all cases we consider only the axisymmetric geometries of Section 3, although extension of these concepts to contacting cylinders is relatively straightforward. Also, we refer to the stress intensity factor as K_{I}, although as mentioned above, the phase angle will be relatively small but finite for highly confined systems characterized by high values of a/h.

5.1.1. Non-adhesive advancing contact

The problem of non-adhesive contact on curved viscoelastic half-spaces ($a/h = 0$) has been formulated by Ting [20,21]. The treatment here is equivalent, although it has been generalized in order to account for finite size effects and the eventual inclusion of adhesive interactions. Because the elastic modulus does not appear in the relationship between contact radius and the non-adhesive displacement, δ', Eq. 9 still holds even for a viscoelastic material, and we can write the following:

$$\delta'(t) = \delta'_{el}(t) \equiv \frac{a^2(t)}{R} f_\delta(t) \tag{56}$$

where the time dependence of f_δ, and of the other geometric correction factors used below, enters through the time dependence of the contact radius, a. The effects of stress relaxation can be accounted for by first defining an elastic load P'_{el}, which depends only on the instantaneous modulus and the current value of the contact radius. The expression for this non-adhesive load is identical in form to the expression for the non-adhesive load given previously (Eq. 8):

$$P'_{el}(t) = \frac{4 f_p(t) E_0^* a^3(t)}{3R} \tag{57}$$

Incremental changes in the contact radius produce incremental changes in the elastic load. These contributions to the load then relax with a time dependence that is described by the relaxation function $\phi(t)$. The overall load is obtained as the superposition of the loads imposed at different loading times, giving the overall non-adhesive load as the following convolution integral:

$$P'(t) = \int_0^t \phi(t - t') \frac{\partial}{\partial t'} \left(P'_{el}(t') \right) dt' \tag{58}$$

5.1.2. Non-adhesive receding contact

The general case of advancing and receding contact is shown in Fig. 10, where representative values of the contact radius are plotted as a function of time. At the time t_m, the contact radius reaches its maximum value. For some time t larger than t_m, the contact radius will be equal to the value it had at a corresponding time $t_1(t)$ during the advancing portion of the experiment. In the absence of adhesion, Ting showed that the load is independent of the contact history for the intermediate times between $t_1(t)$ and t, and can be written in the following form:

$$P'(t) = \int_0^{t_1(t)} \phi(t - t') \frac{\partial}{\partial t'} \left(P'_{el} \right) dt' \tag{59}$$

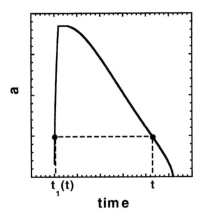

Fig. 10. Schematic representation of the contact radius as a function of time for the case where
the contact radius increases to a maximum value and then decreases. The contact radii at times
t (from the receding portion of the experiment) and $t_1(t)$ (from the advancing portion of the
experiment) are equal to one another.

This equation is valid because the contact stresses are equal to zero outside the
current contact area. Prior forces outside the current contact area do not affect the
present value of the load. These forces do affect the displacement, however, which
is given by the following expression:

$$\delta'(t) = \delta'_{el}(t) - \int_{t_m}^{t} \psi\left(t - t'\right) \frac{\partial \Delta'(t')}{\partial t'}\, dt' \tag{60}$$

with

$$\Delta'(t) = \int_{t_1(t)}^{t} \phi\left(t - t'\right) \frac{\partial \delta'_{el}}{\partial t'}\, dt' \tag{61}$$

5.1.3. Advancing adhesive contact

In the presence of adhesive interactions between a viscoelastic material and a rigid
surface, the stress intensity factor has a positive value, and the actual load is no
longer equal to P'. For $a/h = 0$, the relationships between P, P' and K_I have been
considered by Schapery for advancing contact [22], and more recently by Hui et
al. for both advancing contact [23] and receding contact [24]. The description
given here is based largely on the work of Hui et al. and readers are referred to
these references for details.

For a viscoelastic material, we need to introduce the relaxation function to
obtain the relationship between P', P and K_I. The procedure is similar to the
procedure used to develop Eq. 58 for P'. For a mode I crack ($\psi = 0$) in a linearly

elastic material Eq. 34 can be rearranged to give $P' - P$ as a function of K_I. For a linearly viscoelastic material, contributions to $P' - P$ decay with time as described by the relaxation function, and the current value of $P' - P$ is obtained from the following convolution integral:

$$P'(t) - P(t) = 2\sqrt{\pi} \int_0^t \phi(t - t') \frac{\partial}{\partial t'} \left(\frac{K_I a^{3/2}}{f_{Kp}} (t') \right) dt' \tag{62}$$

Because $\psi(t)$ and $\phi(t)$ are related to one another by Eq. 55, Eq. 62 can be inverted to give the following expression for $K_I(t)$ [24]:

$$K_I(t) = \frac{f_{Kp}(t)}{2\sqrt{\pi} \left(a(t)^{3/2} \right)} \int_0^t \psi(t - t') \frac{\partial}{\partial t'} \left(P'(t) - P(t) \right) dt' \tag{63}$$

The expression for the displacement can be understood by analogy to the perfectly elastic case, where Eqs. 1, 10 and 16 can be combined to give:

$$\delta = \delta'_{el} - \frac{f_c}{2E^*a} \left(P' - P \right) \tag{64}$$

For a viscoelastic system δ'_{el} is replaced by $\delta'(t)$ as given by Eq. 60, and a convolution integral involving the creep function ψ is used to describe the influence of $P'(t) - P(t)$ on the evolution of the displacement:

$$\delta(t) = \delta'(t) - \frac{f_c(t)}{2E_0^* a(t)} \int_0^t \psi(t - t') \frac{\partial}{\partial t'} \left(P'(t') - P(t') \right) dt' \tag{65}$$

5.1.4. Receding adhesive contact

Receding adhesive contact, where the contact radius is decreasing with time, is the most important and also the most complicated situation. With the inclusion of the finite size correction factor f_{Kp} the expression given by Lin et al. [24] for the stress intensity factor for the debonding phase of the experiment can be written in the following form:

$$K_I(t) = \phi(t - t_1(t)) K_I(t_1(t))$$

$$+ \frac{f_{kp}}{2\sqrt{\pi} (a(t))^{3/2}} \int_{t_m}^t \psi(t - t') \frac{\partial}{\partial t'} \left(P'_{eff}(t') - P(t') \right) dt' \tag{66}$$

with

$$P'_{eff}(t) = \int_0^{t_1(t)} \phi(t - t') \frac{\partial}{\partial t'} \left(P_{el}(t') \right) dt' \tag{67}$$

and

$$P_{\mathrm{el}}(t) = \int\limits_0^t \psi(t-t')\frac{\partial}{\partial t'}\left(P(t')\right)dt' \tag{68}$$

The first term in Eq. 66 describes the relaxation of the stress field established during the advancing phase of the experiment, when the contact radius is increasing. For systems exhibiting a substantial adhesion hysteresis, the second term will be much larger than the first term. This second term describes the additional stress fields formed in the vicinity of the crack tip during the debonding phase of the experiment.

After accounting for the geometric correction factor, the following expression for the displacement is obtained from the results of Lin et al. [24]:

$$\delta(t) = \delta'_{\mathrm{eff}}(t) - \frac{f_{\mathrm{c}}(t)}{2E_0^* a(t)} \int\limits_{t_{\mathrm{m}}}^t \psi(t-t')\frac{\partial}{\partial t'}\left(P'_{\mathrm{eff}}(t') - P(t')\right)dt' \tag{69}$$

with

$$\delta'_{\mathrm{eff}}(t) = \delta\left(t_1(t)\right) - \int\limits_{t_{\mathrm{m}}}^t \psi(t-t')\frac{\partial}{\partial t'}\left(A(t')\right)dt' \tag{70}$$

and

$$A(t) = \int\limits_{t_1(t)}^t \phi(t-t')\frac{\partial}{\partial t'}\left(\delta(t_1(t'))\right)dt' \tag{71}$$

5.1.5. General comments

Any axisymmetric test where a rigid, spherical indenter is brought into contact with an adhesive layer of thickness h and then removed is completely described by the equations given in Sections 5.1.3 and 5.1.4. In the absence of adhesion, $K_1 = 0$, and the results of Sections 5.1.1 and 5.1.2 are recovered, with $P = P'$ and $\delta = \delta'$ at all times, and $P = P'_{\mathrm{eff}}$, $\delta' = \delta'_{\mathrm{eff}}$ and $\Delta = \Delta'$ for receding contact. Eqs. 65 and 69 give the displacement in terms of the applied load and the contact area. These relationships between P, δ and a are completely independent of the stress intensity factor, and are determined only by the viscoelastic properties of the adhesive layer. In principle, these viscoelastic properties can be obtained from an experiment where P, δ and a are all independently measured. In practice it is often more useful to measure the viscoelastic properties independently, and use Eqs. 65 and 69 to verify that these properties accurately describe the measured values for the displacement. If agreement between the measured and predicted values of the displacement is obtained, then one can be confident that accurate values for the stress concentration factor are obtained from Eqs. 63 and 66. The magnitude of this stress concentration factor determines how fast the contact area either advances or recedes, as described in the following section on small-scale viscoelasticity. Recently, Giri et al. have used these methods in the analysis of data obtained from the indentation of viscoelastic latex films [25,26]. Hui and Lin have also used this approach in their analysis of the previously published data of Falsafi and Tirrell [27] for viscoelastic advancing contact.

5.2. Small-scale viscoelasticity

The fracture mechanics formalism outlined above provides a method for experimentally determining the stress intensity factor, which will control the rate at which the contact area changes. In essence, K_I is a crack driving force that determines the rate at which adhesive failure occurs. In an axisymmetric geometry the crack velocity, $v = -da/dt$, is a function of K_I. In the absence of large-scale viscoelasticity, the bulk energy supplied to the crack tip region is uniquely related to the stress intensity factor. In this case the energy-based descriptions based on $\mathcal{G}$ and the stress-based approaches based on K_I are completely equivalent. Large-scale viscoelasticity is a complicating factor, because in this case a unique energy release rate can no longer be determined. Because the energy dissipation in the crack tip region is determined by the local stress state, which is in turn specified by K_I, the stress intensity factor is a more appropriate measure of the crack driving force than the energy release rate for viscoelastic materials [24]. With this approach, the effects of large-scale and small-scale viscoelasticity can be considered in the same sample [28].

For simplicity, our discussion here is focused on situations where the effective modulus decays from E_0^* at low times to E_∞^* at large times. If the relaxation to a state characterized by E_∞^* occurs quickly in comparison to the time scale of the experiment, ϕ and ψ in the expressions for K_I can be approximated as E_∞^*/E_0^* and E_0^*/E_∞^*, respectively. The results from Section 3.2 are recovered, with $E^* = E_\infty^*$. In this case a well-defined energy release rate exists, and can be used to describe the driving force for crack propagation. The classic work of Maugis and Barquins is an excellent example [9]. These authors studied the adhesion between a rigid indenter and a thick, nearly incompressible polyurethane elastomer ($a/h \approx 0$, $v \approx 0.5$, $K_{II} \approx 0$). The relationship between $\mathcal{G}$ and v obtained by Maugis and Barquins can be written in the following form:

$$\mathcal{G} = \mathcal{G}_0 \Phi(v), \qquad \Phi(v) = \left\{ 1 + \left(v/v^* \right)^n \right\} \tag{72}$$

with $\mathcal{G}_0 \approx 0.08 \text{ J/m}^2$, $n = 0.6$ and $v^* = 22 \text{ nm/s}$. Eq. 33 can be used to obtain an equivalent expression relating the crack velocity to the stress concentration factor:

$$K_I = K_{Ic}\Phi_k(v), \qquad K_{Ic} = \left(2E_\infty^* \mathcal{G}_0\right)^{1/2},$$

$$\Phi_k(v) = \Phi_k^{1/2}(v) \approx \left\{1 + (v/v^*)^{n/2}\right\} \tag{73}$$

The zero velocity limit, $\mathcal{G}_0$, represents the threshold value required for crack motion, and is equal to the thermodynamic work of adhesion under true equilibrium conditions. In developing a description of the effects of small-scale viscoelasticity, the aim is to develop a viscoelastic cohesive zone model that gives reasonable predictions for $\mathcal{G}_0$, v^* and n.

A common starting point for viscoelastic cohesive zone models is the Dugdale model, mentioned above in Section 3.3. This model applies in the zero velocity limit, with $\mathcal{G}_0 = \sigma_0 \delta_t$, where σ_0 is the stress applied across the adhesive zone and δ_t is the range of the adhesive forces responsible for this stress. Viscoelastic versions of this cohesive model have a long history, beginning with the work of Schapery [29], further described by Greenwood and Johnson [30], and more recently by de Gennes [31]. Johnson has recently given a summary of results that forms the basis for much of what is presented here [28]. Because the starting point is the stress σ_0 applied across the cohesive zone, it is most convenient to use the creep compliance function $\psi(t)$ to define the viscoelastic response of the material in the cohesive zone. A very simple model for the creep function is the following single exponential form, which can be used to characterize the transition from a glassy state with a modulus of E_0^*, to a rubbery state with a modulus of E_∞^*:

$$\psi(t) = k - (k-1)\exp(-t/\tau) \tag{74}$$

with $k = E_0^*/E_\infty^*$. Note that the creep function defined by Eq. 74 increases from 1 to k as the time increases from 0 to ∞. Use of a single relaxation time to characterize the transition between the rubber and glassy regimes is a substantial approximation, but the procedure outlined below can be generally applied to more complicated forms of the compliance function.

In the vicinity of the crack tip the stresses and strains are amplified. For a moving crack tip, the strain rate is therefore amplified as well, so that the effective modulus of the material within the cohesive zone is increased. According to Eq. 42, the width d of the cohesive zone will be an increasing function of the crack speed. The zero velocity value for d, which we refer to as d_0, is obtained by using E_∞^*, the fully relaxed modulus, in Eq. 42:

$$d(v=0) = d_0 = \frac{\pi E_\infty^* \delta_t}{4\sigma_0} \tag{75}$$

For very high velocities the unrelaxed modulus must be used to describe the stiffness of the material in the vicinity of the crack tip, giving the following for the width of the cohesive zone:

$$d(v = \infty) = \frac{\pi E_0^* \delta_t}{4\sigma_0} = \frac{k d_0}{\psi(0)} = k d_0 \tag{76}$$

In general we can define a characteristic time, t^*, so that d can be written in the following form:

$$d(v) = \frac{k d_0}{\psi(t^*)} \tag{77}$$

The characteristic time t^* is related to the time, d/v, that it takes for the stress field to pass through the Dugdale zone. The actual value of t^* is a third of this time [28], giving the following expression for t^*:

$$t^* = \frac{d(v)}{3v} \tag{78}$$

Eqs. 77 and 78 can be combined to give the following expression that must be solved for t^*:

$$t^*/\tau = \frac{k}{\psi(t^*)} \frac{v^*}{v} \tag{79}$$

with,

$$v^* = d_0/3\tau \tag{80}$$

The multiplication factor appearing in Eq. 69 is given by the following ratio of cohesive zone widths:

$$\frac{\mathcal{G}}{\mathcal{G}_0} = \Phi(v) = \frac{d(v)}{d_0} = \frac{k}{\psi(t^*)} \tag{81}$$

Eq. 79 can be solved numerically, using any form of the compliance function, to give t^* as a function of v. These values for t^* can then be used in Eq. 81 to obtain $\Phi(v)$. The results for the simple compliance expression of Eq. 74, given originally by Greenwood and Johnson [30], are plotted in Fig. 11 for $k = 10$, 100 and 1000. The solid lines represent the following empirical fit to these curves:

$$\frac{\mathcal{G}_0}{\mathcal{G}} = \frac{1}{\Phi(v)} = \left[\frac{1}{1 + (v/v^*)^{0.5}} + \frac{1}{k} \right] \tag{82}$$

The model gives $n = 0.5$, which is in reasonably good agreement with the data obtained by Maugis and Barquins. Of course choice of a more realistic form of the compliance function will affect the results. The important thing from the point of this chapter is that a theoretical formalism exists to relate experimentally accessible quantities such as v^* and n to the viscoelastic properties of the material and the detailed nature of the interfacial forces.

The essence of the approach to viscoelastic contact is that it is reasonable to separate the effects of 'large-scale' and 'small-scale' viscoelasticity. Large-scale

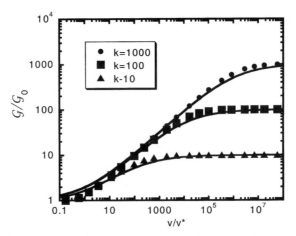

Fig. 11. Viscoelastic enhancement factor as a function of the normalized crack velocity. Symbols represent the numerical solution of Eqs. 74, 79 and 81 and the solid lines represent the corresponding empirical relationships given by Eq. 82.

viscoelastic effects occur throughout a macroscopic sample, and are dominated by the longer relaxation times. These effects determine how the overall applied load and displacement result in local stress state near the contact edge that is characterized by the stress intensity factor. The response of the system to this applied stress intensity factor depends on the local, 'small-scale' viscoelastic response, dominated by the faster relaxation times. These are confined to a region that is presumed to be small enough so that they do not noticeably affect the overall loads and displacements that are applied to the sample. This distinction between 'large-scale' and 'small-scale' viscoelastic effects is somewhat arbitrary, and breaks down altogether if the cohesive zone size d is comparable to the overall sample dimensions. Also, non-linear effects, associated for example with bulk yielding of a sample, can be very important for sufficiently large cohesive zone stresses. Application of these ideas to experimental systems is currently an active area of research.

6. Summary

The purpose of this chapter has been to give a description of some of the most useful contact mechanics expressions as they relate to studies of adhesion. The primary assumption regarding the properties of the materials themselves is that a linear constitutive model is obeyed throughout the strained region, with the possible exception of a relatively small cohesive zone at the contact edge. Many of the results obtained for simple linear elastic behavior are analytic. Linear viscoelasticity can be handled as well, although in this case numerical approaches are often required. Non-linear effects, including bulk yielding of the sample at locations that are well-removed from the cohesive zone, have not been considered in this chapter. In terms of the experimental geometry, our primary focus has been on axisymmetric geometries, with the cylindrical geometry of Section 4 included as one of many alternative geometries that can also be considered. Contact mechanics methods are continually being adapted to other geometries, including membranes, contacting fibers, etc. One can view the specific examples described in this chapter as case studies in a field that is continually being applied in new areas.

Acknowledgements

I have benefited greatly from correspondence with several colleagues in the preparation of this chapter. These interactions have been particularly important with regard to the section on large-scale viscoelasticity. This section could not have been completed without the help of Dr. C.Y. Hui and Dr. Y.Y. Lin. Helpful comments regarding the entire manuscript from Dr. A.J. Crosby are also acknowledged, as are comments from Dr. W. Unertl, who also supplied preprints of his own unpublished work on viscoelastic contact mechanics.

References

1. Johnson, K.L., *Contact Mechanics*. Cambridge University Press, Cambridge, 1985.
2. Hertz, H. and Reine, J., *Angew. Math.*, **92**, 156 (1882).
3. Sneddon, I.N., *Int. J. Eng. Sci.*, **3**, 47 (1965).
4. Barquins, M., *Wear*, **158**, 87 (1992).
5. Shull, K.R., Ahn, D., Chen, W.-L., Flanigan, C.M. and Crosby, A.J., *Macromol. Chem. Phys.*, **199**, 489 (1998).
6. Lin, Y.Y., Hui, C.Y. and Conway, H.D., *J. Polym. Sci. Polym. Phys.*, **38**, 2769 (2000).
7. Johnson, K.L., Kendall, K. and Roberts, A.D., *Proc. R. Soc. London, A*, **324**, 301 (1971).
8. Crosby, A.J. and Shull, K.R., *J. Polym. Sci., Polym. Phys.*, **37**, 3455 (1999).
9. Maugis, D. and Barquins, M., *J. Phys. D*, **11**, 1978 (1989).
10. Barquins, M. and Maugis, D., *J. Adhes.*, **13**, 53 (1981).
11. Lawn, B., *Fracture of Brittle Solids*, 2nd ed. Cambridge University Press, Cambridge, 1993.
12. Gent, A.N., *Rubber Chem. Technol.*, **67**, 549 (1994).
13. Shull, K.R., Flanigan, C.M. and Crosby, A.J., *Phys. Rev. Lett.*, **84**, 3057 (2000).
14. Maugis, D., *J. Colloid Interface Sci.*, **150**, 243 (1992).
15. Derjaguin, B.V., Muller, V.M. and Toporov, Y.P., *J. Colloid Interface Sci.*, **53**, 314 (1975).
16. Gwo, T.-J. and Lardner, T.J., *J. Tribol.*, **112**, 460 (1990).
17. Barquins, M., *J. Nat. Rubber Res.*, **5**, 199 (1990).
18. She, H.Q., Malotky, D. and Chaudhury, M.K., *Langmuir*, **14**, 3090 (1998).
19. She, H.Q. and Chaudhury, M.K., *Langmuir*, **16**, 622 (2000).

20. Ting, T.C.T., *J. Appl. Mech.*, **33**, 845 (1966).
21. Ting, T.C.T., *J. Appl. Mech.*, **35**, 248 (1968).
22. Schapery, R.A., *Int. J. Fract.*, **39**, 163 (1989).
23. Hui, C.-Y., Baney, J.M. and Kramer, E.J., *Langmuir*, **14**, 6570 (1998).
24. Lin, Y.Y., Hui, C.Y. and Baney, J.M., *J. Phys. D*, **32**, 2250 (1999).
25. Giri, M., Bousfield, D.B. and Unertl, W.N., *Langmuir*, **17**, 2973 (2001).
26. Giri, M., Bousfield, D.B. and Unertl, W., *Tribol. Lett.*, **9**, 33 (2000).
27. Falsafi, A., Deprez, P., Bates, F.S. and Tirrell, M., *J. Rheol.*, **41**, 1349 (1997).
28. Johnson, K.L., In: Tsukruk, V.V. and Wahl, K.J. (Eds.), *Microstructure and Microtriboloby of Polymer Surfaces*. American Chemical Society, Washington, DC, 2000, p. 24.
29. Schapery, R.A., *Int. J. Fract.*, **11**, 141 (1975).
30. Greenwood, J.A. and Johnson, K.L., *Philos. Mag. A.*, **43**, 697 (1981).
31. de Gennes, P.G., *Langmuir*, **12**, 4497 (1996).

Chapter 16

Measurement methods for fiber–matrix adhesion in composite materials

LAWRENCE T. DRZAL [a],[*] and PEDRO J. HERRERA-FRANCO [b]

[a] *Departments of Chemical Engineering and Materials Science and Mechanics, Composite Materials and Structures Center, Michigan State University, East Lansing, MI 48824-1326, USA*
[b] *Centro de Investigación Científica de Yucatán, A.C., Calle 43 # 130, Col Chuburná de Hidalgo, C.P. 97200, Mérida, Yucatán, México*

1. Fiber–matrix adhesion measurement methods

It is well known that the level of adhesion between fibers and matrix affect the ultimate mechanical properties of the composite, not only in the off-axis but also in the direction parallel to the fibers. Experimental evidence for this dependence is available in investigations of the adhesion between carbon fibers and an epoxy matrix in which the fibers were systematically surface-treated to provide a wide range of fiber–matrix adhesion [1,2]. These results have shown that as a result of differences in the level of adhesion, fundamental differences exist in not only the fiber–matrix interfacial shear strength but in the mechanical properties and in the interfacial failure modes for each of the fiber–matrix combinations as well [3–6]. Properties such as the elastic modulus and Poisson's ratio were relatively insensitive to the surface treatment. However, the inelastic properties both in the fiber and transverse to the fiber, were significantly different and dependent on the level of fiber–matrix adhesion. While the fiber-dominated strength properties (such as the longitudinal tensile, compressive and flexural strength) showed only moderate sensitivity, the off-axis strength properties (such as the transverse tensile and flexural strength properties, in-plane and interlaminar shear strength) were shown to be highly sensitive to the fiber–matrix adhesion level. Furthermore, Modes I and II fracture toughness also changed significantly with the varying degrees of fiber–matrix adhesion.

One of the major findings in this experimental study has been the correlation between the single-fiber failure mode observed during the fiber fragmentation

---
[*] Corresponding author. E-mail: drzal@egr.msu.edu

tests and the mechanical behavior of the composite materials under various loading conditions. The critical parameters affecting the mechanical properties of a given fiber–matrix combination were identified to be the level of adhesion between fiber and matrix and the interphase morphology.

In polymer matrix composites, there appears to be the optimum level of fiber–matrix adhesion which provides the best mechanical properties. Several models which relate the structure and properties of composites to fiber–matrix interfacial behavior have been proposed based either on mechanical principles with some assumptions made about the level of fiber–matrix adhesion in the composite or have taken a surface chemistry approach in which the fiber–matrix interphase was assumed to be the only factor of importance in controlling the final properties of the composite. Neither effort has had much success.

A growing body of experimental evidence points to the existence of a region different in structure and composition near the fiber–matrix interface, i.e. an 'interphase'. These results have led to an understanding of the inter-relationships between fiber, interface, and matrix, giving birth to the concept of the interphase, i.e. a three-dimensional region existing between the bulk fiber and bulk matrix [1]. This interphase includes the two-dimensional region of contact between the fiber and matrix (the interface) but also incorporates the region of some finite thickness extending on both sides of the interface in both the fiber and matrix. The 'interphase' concept also allows for the inclusion of both interfacial as well as material mechanisms. For example, it has been shown that the fiber and matrix surface energy as well as the chemical bonding of the polymer on the fiber surface contribute to adhesion. Likewise, the material properties of the polymer near the fiber surface control the stress transfer and failure mode between the fiber and matrix. Furthermore, chemical and thermal shrinkage arises in specimens during cure and cool-down as well as from the differences in the mechanical properties of the constituents. These residual stresses that develop in the interphase can greatly affect the fiber–matrix adhesion [7]. The complexity of the interphase can best be visualized from the schematic shown in Fig. 1. The desirability to develop structure–processing–property relationships incorporating the fiber–matrix interphase increases the need to characterize the fiber–matrix adhesion and the interphase [8,9]. Therefore, it would be desirable to have a technique which would allow measurements of fiber–matrix adhesion level in a high fiber-volume fraction composite that has been subjected to the same processing or environmental exposures encountered either during manufacturing and fabrication or while in service. Then, processing effects, moisture and solvent sorption, thermal exposures and fatigue, could be properly evaluated for their effect on composite properties.

There have been several techniques developed to measure fiber–matrix adhesion levels and the effect of the interphase on the effective properties of composites. These methods can broadly be classified into three separate categories:

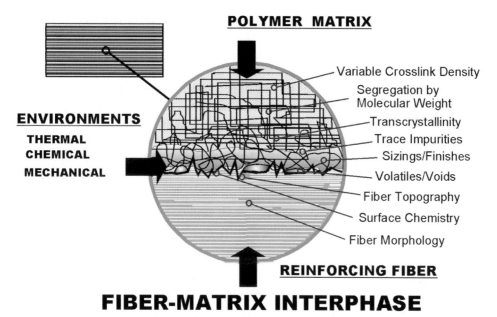

POLYMER MATRIX

Variable Crosslink Density

Segregation by
Molecular Weight

Transcrystallinity

Trace Impurities

Sizings/Finishes

Volatiles/Voids

Fiber Topography

Surface Chemistry

Fiber Morphology

ENVIRONMENTS

THERMAL
CHEMICAL
MECHANICAL

REINFORCING FIBER

FIBER-MATRIX INTERPHASE

Fig. 1. Characteristics of the fiber–matrix interphase in a composite material.

direct methods, indirect methods and composite lamina methods. The *direct* methods include the fiber pull-out method, the single-fiber fragmentation method, the embedded-fiber compression method and the microindentation method. The *indirect* methods for fiber–matrix adhesion testing include: the variable curvature method, the slice compression test, the ball compression test; the use of dynamical mechanical analysis and voltage-contrast X-ray spectroscopy. The *composite lamina* methods include: the 90° transverse flexural and tensile tests, three- and four-point shear, ±45° and edge delamination tests, the short-beam shear test method and the Mode I and Mode II fracture tests.

It should be pointed out that while the indirect methods provide a qualitative method of ranking the adhesion between fiber and matrix and the composite lamina test methods actually measure fiber–matrix interface sensitive composite properties, the direct methods not only provide a measure of fiber–matrix adhesion but can also provide information about fiber–matrix failure mode and a method to measure the energy involved in fracture of the fiber–matrix interface. This last parameter is important in relating fiber–matrix adhesion to composite toughness.

1.1. Direct methods

The direct methods of characterizing the fiber–matrix adhesion and the interphase, have relied on the use of single-fiber–matrix test methods for measuring adhesion and failure modes. The first technique proposed was the fiber pull-out method

[9], which was developed in the early stages of composite research when the fibers were much larger and easier to handle than they are today. There have been variations in the experimental details pertaining to the fabrication of the test coupon and to the execution of this test mainly in the matrix portion, but overall, the procedures to fabricate samples, the experimental protocols, and data analysis remain the same. In the pull-out version, the fiber is pulled out of the matrix which can be a block of resin, a disc, or a droplet [10]. The use of very small droplets reduces the difficulties in preparing thin discs of resin and can reduce the variability in exit geometry [10]. These advantages have made this test very popular in the last decade. In this test, the load and displacements are monitored continuously and upon fiber pull-out, the load registered at complete debonding of the fiber from the matrix is converted into an apparent interfacial shear strength. The advantage of this method is that it allows testing of brittle and/or opaque matrices.

Another popular method, is the embedded single-fiber fragmentation test. Here a single fiber is totally encapsulated in the polymeric matrix shaped into a tensile dogbone-shaped coupon, which in turn is loaded in tension. An interfacial shear stress transfer mechanism is relied upon to transfer tensile forces to the encapsulated fiber through the interphase from the polymeric matrix [11,12]. The fiber tensile strength σ_f is exceeded and the fiber fractures inside the matrix tensile coupon. This process is repeated, producing shorter and shorter fragments until the remaining fragment lengths are no longer sufficient in size to produce further fracture through this stress transfer mechanism. A simple shear-lag analysis is applied to analyze the experimental data based on the length of the resulting fiber fragments, the fiber diameter and the fiber tensile strength in order to calculate the interfacial shear strength.

Another method proposed over three decades ago by Outwater and Murphy [13] uses a single fiber aligned axially in a rectangular prism of matrix. A small hole is drilled in the center of the specimen through the fiber. The prism is placed under a compressive load and the propagation of an interfacial crack is followed with increasing load. The Mode II fracture toughness of the interface can be calculated from these data based on ε_r, the strain in the resin, E_f, the tensile modulus of the fiber, τ, the frictional shear stress, x the length of the interfacial crack and a is the fiber diameter.

An in-situ microindentation measurement technique has also been proposed for measuring the fiber interfacial shear strength [14]. It involves the preparation of a polished cut surface of a composite in which the fibers are oriented perpendicular to the surface. A small hemispherical indenter is placed on an individual fiber and the force and displacements are monitored to the point at which the fiber detaches from the matrix.

1.2. Indirect methods

The indirect methods for fiber–matrix adhesion-level measurement are shown in Fig. 2. These include: the variable curvature method; the slice compression test, the ball compression test, the fiber-bundle pull-out test; the use of dynamical–mechanical thermal analysis; and voltage-contrast X-ray photoelectron spectroscopy (VCXPS).

Narkis [15] cleverly proposed the use of a single-fiber specimen in which the fiber is embedded along the center line in the neutral plane of a uniform cross-section beam. The beam is placed in nonuniform bending according to an elliptical bending geometry with the aid of a template. This causes the shear stress to build up from one end of the fiber according to the gradient of curvature of the specimen. Careful observation of the fiber in the specimen allows location of the point at which the fiber fails as a result of a maximum shear stress criterion. The stress along the fiber is calculated as a function of the matrix tensile modulus, the beam width, the first moment of transformed cross-sectional area, and constants from the equation of the ellipse. Some of the advantages of this technique are that a single fiber or fiber tow can be used, the results do not depend on fiber strength and sample preparation is relatively easy. Some of the disadvantages are that the debond front is not so easy to detect, and the results are sensitive to the location of the single-fiber layer within the cross-section of the coupon.

The slice compression test has been applied to polymer–matrix composites even though it was developed to probe the interface in ceramic matrix composites [16]. A thin slice sample of unidirectional composite is produced with the cut surface perpendicular to the fiber axis. The surfaces are cut and polished to be parallel to each other and perpendicular to the fibers. The thin slice is loaded in compression in the fiber axis direction with two plates. One of the plates is made of a very hard material such as silicon nitride and the other of a soft material, e.g. pure aluminum which can deform as the fibers are compressed into it. The thickness of the slice must be controlled to allow the fibers to debond without failing in compression as well as allowing them to slide inside through the matrix. The depth of the fiber indentation into the plate can be related to the interfacial shear strength [17].

Carman et al. [18] developed a test called the meso-indentation test which used a hard spherical ball indenter to apply a compressive force to a surface of the composite perpendicular to the fiber axis. The indenter was much larger than the diameter of a single fiber; therefore, when the ball was forced into the end of the composite, it made a permanent depression in the material. From the size of the depression and the force–deflection curve, they calculated a mean hardness pressure as a function of strain in the coupon. Qualitative differences have been reported in tests conducted on carbon fiber–epoxy composites where the fiber–matrix adhesion had been varied systematically.

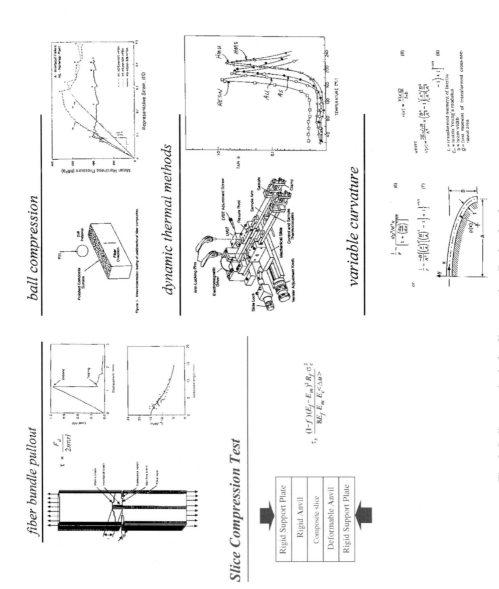

Fig. 2. Indirect methods for fiber–matrix adhesion-level measurement.

The fiber-bundle pull-out method [19] is similar to the single-fiber pull-out method except that instead of using a single fiber, a bundle of fibers is used. A coupon is fabricated in which a bundle of fibers or a lamina of unidirectional fibers is cast in a block of matrix. Transverse notches are cut into the coupon near the end of the fiber bundle. The coupon is loaded in tension with the load applied parallel to the fiber axes. The load versus displacement curve can be monitored and the debonding point detected. In a similar manner to the way data are reduced for the single-fiber pull-out test, the interfacial shear strength between the bundle of fibers and matrix can be calculated.

Ko et al. [20] examined a carbon fiber–epoxy system in which the interfacial properties have been varied by the use of dynamic mechanical analysis. They report a change in the tan δ peak attributable to changes in the fiber–matrix adhesion. Chua [21,22] also measured a shift in the loss factor for glass–polyester systems that corresponded to changes in the condition of the fiber–matrix interphase. Perret [23] measured both the loss factor and the change in the shear modulus with increasing displacement and detected a change in composite properties with a change in the fiber–matrix adhesion. Yuhas et al. [24] have used ultrasonic wave attenuation to establish correlations with short beam shear data. This method was useful for poorly bonded systems but was not sensitive to well bonded interfaces. Wu used localized heating coupled with acoustic emission events to detect interfacial debonding [25].

Laser Raman spectroscopy (LRS) can be applied to the fiber–matrix interface in order to determine the actual stresses that exist at the interface. LRS is a visible light spectroscopy that relies on the inelastic scattering of visible light photons from a surface. Certain chemical groups in a material or on a surface can scatter incident radiation at characteristic frequencies. Tuinstra and Koening [26] showed that certain characteristic frequencies in the Raman active bands of graphite and other fibers are sensitive to the level of applied stress or strain. There is a measurable shift in the characteristic frequency which is proportional to the applied strain. A small 1-μm spot generated by a laser beam can be scanned along a fiber surface and provide the Raman information which can be converted to the local stresses in the fiber. A transparent matrix incorporating fibers having a Raman active band (e.g. aramid, high-modulus graphite) can be analyzed with this method [27].

A recent method for a determining information about fiber–matrix adhesion is a technique identified as voltage-contrast X-ray photoelectron spectroscopy (VCXPS) [28]. This method relies on the VCXPS characterization of the fracture surface of high-volume fraction fiber composites. A unidirectional coupon is fractured in an opening mode to produce a fracture surface. This fracture surface containing fibers and polymer is placed inside of an XPS spectrometer for analysis. X-ray photons are directed at the surface causing the emission of photoelectrons. These electrons are collected and analyzed for quantity and energy which contains useful information about the atomic composition of the surface as well as the molecular environment of the atoms on the surface. During the process of photoelectron emission nonconducting (insulating) samples will acquire a charge and cause peaks to shift from their neutral position. This happens in nonconductive materials like polymers, but does not happen in conducting materials like carbon fibers. As a result, the carbon peak begins to split into two peaks as charge builds up on the surface. One carbon peak due to the conductive carbon fiber stays at the neutral position while the other portion due to the polymer shifts depending on the magnitude of the charge on the surface. The height and width of the peaks and the shift in energy is related to the content of the conducting carbon fiber and nonconductive polymer remaining on the fracture surface. As a result the ratio of the two carbon peaks is a qualitative indicator of the degree of adhesion. For example, if the ratio of the nonconductive carbon peak to the conductive carbon peak is large, the fracture surface contains a large amount of nonconductive polymer and very little conductive carbon fiber. This can be interpreted as being due to good adhesion between the fiber in the matrix causing failure to occur in the weaker polymer matrix between fibers. On the other hand, if the ratio of the nonconductive carbon peak to the conductive carbon peak is small, many bare carbon fibers are exposed on the fracture surface, indicating poor adhesion between the fiber and the matrix. In cases where the same carbon fibers are used with various polymeric matrices, a semi-quantitative relationship between this parameter and fiber–matrix adhesion has been developed.

1.3. Composite lamina tests

Composite laminate tests are often used to measure fiber–matrix adhesion. The obvious tests to be conducted are those in which the fiber–matrix interface dominates the results, such as shear properties. Numerous techniques have been developed for measuring shear properties in fiber-reinforced composite laminates. The most commonly used test methods for in-plane shear characterization are the [±45]$_S$ tension test [29] and the Iosipescu test [30]. To determine the interlaminar shear strength, the short beam shear test [31] is more frequently used. In all these cases, standard protocols exist for preparing the samples, conducting the tests, reporting the data and analyzing the results. These include American Society for Testing and Materials (ASTM) and Automotive Composites Consortium (ACC) standards (Automotive Structural Composite Materials, 1994). A careful experimental study has been published relating differences in fiber–matrix adhesion to these tests [32].

1.4. Summary

Overall the use of any of the direct, indirect or composite lamina tests in the hands of a skilled experimenter can provide a consistent way of ranking fiber–matrix adhesion regardless of the method chosen. However, one should be aware that there are various issues related to the use of these tests that limit their applicability. One issue is the identification of the appropriate parameter for characterizing the fiber–matrix interface. All of the direct and indirect tests have been developed with the goal of measuring the fiber–matrix interfacial shear strength. However, several of these tests are really fracture tests and are more properly used if the interfacial fracture energy is calculated. On the other hand interfacial fracture energy is rarely used to evaluate or measure fiber–matrix adhesion or to design composite materials. Another factor that must be considered is the preparation of the samples. The single-fiber tests are very sensitive to careful preparation of samples and the careful selection of fibers for testing within those samples. Testing conditions are likewise very important. While normally one would conduct any of these tests at reasonably slow strain rates, in microtesting, the strain rates used are only nominally slow. These strain rates become extraordinarily high when taking into account the small dimensions of the distances over which these tests are conducted. There is also evidence that in dealing with viscoelastic polymer composites, creep effects can be important and must be considered. Finally, the data analysis methods associated with these techniques rely on the assumption of a value for the modulus of the matrix near the fiber surface for reduction of the test results into a usable parameter whether it is strength or energy. The literature contains numerous references indicating that the structure of the polymer near the fiber surface can be quite different from the bulk polymer. Indeed the modulus in some cases can be quite a bit lower or higher than the bulk matrix depending on the system investigated [33]. At the present time there is no accurate method for measuring the interface modulus that may exist in dimensions of a few tens to a few hundreds of nanometers from the fiber surface. Until such a quantitative measurement is available, it will not be possible to accurately relate interfacial tests, whether single fiber or microscopic, to composite properties.

The remaining part of this chapter will review the three most common direct methods for measuring fiber–matrix adhesion, focusing on the sample preparation and fabrication, the experimental protocols and the underlying theoretical analyses upon which evaluation of these methods are based. In addition, finite-element nonlinear analyses and photoelastic analyses will be used to identify differences in the state of stress that is induced in each specimen model of the three different techniques. In order to provide an objective comparison between the three different techniques to measure the interfacial shear strength for the prospective user, data and a carbon fiber–epoxy resin system will be used as a baseline system throughout this chapter. However, these methods and procedures can be applied for adhesion measurements to any fiber–matrix combination.

2. Pull-out and the microbond technique

2.1. The pull-out technique

The pull-out experiment, which is believed to possess some of the characteristics of fiber pull-out in composites, consists of a fiber or filament embedded in a matrix block or thin disk normal to the surface of the polymer. A steadily increasing force is applied to the free end of the fiber in order to pull it out of the matrix [9]. The load and displacement are monitored as the fiber is pulled axially until either pull-out occurs or the fiber fractures. The strength of the fiber–matrix interface can be calculated to a first approximation by balancing the tensile stress (σ_f) on the fiber and shear stress (τ) acting on the fiber–matrix interface, obtaining a simple relationship of the form $\tau = (\sigma_f/2)(d/L)$, where it is assumed that the shear stress is uniformly distributed along the embedded length, and d, the fiber diameter to be constant along the embedded length [9].

Several theoretical models [34–46] have been proposed to determine the state of stress developed at the fiber–matrix interface. Greszczuk [34] considered the case of an elastic matrix in which the shear stress distribution was no longer uniform and the load transferred between the fiber and matrix did not change uniformly along the fiber. He showed that distribution of stresses and forces depend on the properties of the elastic matrix. Lawrence [35] further developed Greszczuk's theory by including the effect of friction. Takaku and Arridge [36] considered the effect of the embedded length on the debonding stress and the pull-out stress and also the effect of Poisson's contraction on the variation of pull-out stresses. Gray [37] reviewed and applied the previously mentioned theories [34,35] to calculate the maximum shear stress when the fiber is pulled out from the elastic matrix. He concluded that the mixture of adhesional bonding and friction resistance that occurs in a pull-out test specimen depends on the length of the embedded fiber. The contribution from adhesion increases with embedded fiber length, whereas the frictional resistance to pull-out due to friction decreases. Laws [38] was able to calculate the load–displacement curve of a pull-out test based on Lawrence's theory [35], the crack spacing and strength of an aligned short fiber composite, and the effect of the interfacial and frictional bonds on pull-out.

Banbaji [39,40] presented a theoretical model that considered the effect of normal transverse stresses on the pull-out force. He first analyzed the case where the normal stress is constant, and then the case in which the stresses depend on the way the tensile force changes during an actual test. He applied the results to a polypropylene–cement system.

Another variation of the pull-out technique was reported by Hampe [41]. The main difference in his approach was in the geometry of the matrix material used to fabricate the test specimens. He used a small amount of polymer in the form of a hemisphere formed on the surface of a metal plate which in turn was mounted on an electronic balance capable of measurements with an accuracy of ± 0.1 mN. A motor-driven support was used to apply an increasing load pulling on the free end of the fiber out of the matrix hemisphere using speeds from 5 μm min^{-1} to 5 mm min^{-1}. The displacement was also monitored continuously with an accuracy of ± 0.3 mN. All data acquisition and processing is performed using a personal computer and considerable precision in the measurements was obtained from this approach.

2.2. Theoretical considerations

Theoretical considerations performed by Chua and Piggott [42–45] about the pull-out process are described in this section. They assumed that both fiber and matrix behave elastically and that stress transfer occurs at the fiber–matrix interface without yielding or slippage (perfect bonding) based on previous work by Greszczuk and Lawrence [34,35]. It should be mentioned that they did not consider any bonding across the end of the fiber. They recognized that the pull-out process is governed by five different variables: interfacial pressure (p), friction coefficient (μ) along the debonded length, work of fracture of the interface (G_i), the embedded fiber length (L) and the free fiber length (l_f). They developed a relationship for the tensile stress within the fiber (σ_f) at any point along the embedded length:

$$\sigma_f = \sigma_{fe} \frac{\sinh \dfrac{n(L-x)}{r}}{\sinh(ns)} \tag{1}$$

where $s = L/r$, L is the length of the embedded fiber, r is the radius of the fiber, σ_{fe} is the average tensile stress on the stress at the polymer surface. The geometric terms are defined in Fig. 3 and n is:

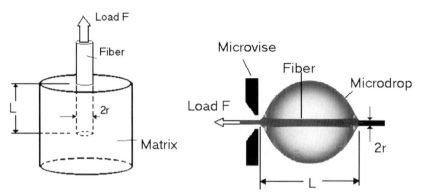

Fig. 3. Schematic representation of the fiber pull-out and microbond techniques showing geometrical parameters.

$$n^2 = \frac{E_m}{E_f(l + v_m) \ln \dfrac{R}{r}} \tag{2}$$

where E_f and E_m are Young's moduli of the fiber and matrix, respectively, v_m is the matrix Poisson's ratio, and R is the radius of the polymer specimen. The shear stress at the interface is calculated from the equilibrium of forces exerted on a differential fiber element, of length dx, to give the well known equation:

$$\tau_i = \frac{r}{2} \frac{d\sigma_f}{dx} \tag{3}$$

which, can be expressed in terms of the tensile stress within the fiber as:

$$\tau_i = n\sigma_{fe} \frac{\cosh \dfrac{n(L-x)}{r}}{2 \sinh(ns)} \tag{4}$$

Again, it should be pointed out that no bonding is considered across the end of the fiber. During the pull-out process there are three possible routes to failure during a pull-out test [46].

(1) Failure may occur when the maximum shear stress reaches the interface strength (τ_{iu}), which has a maximum absolute value at $x = 0$, that is, at the surface where the fiber leaves the block of polymer. The debonding force is then $F_d = \pi r^2 \sigma_{fe}$. From Eq. 4 is obtained:

$$F_d = \frac{2\pi r^2 \tau_{iu} \tanh(ns)}{n} \tag{5}$$

(2) Yielding at the interface might also occur if its yield strength τ_{iu} is reached, in which case a constant shear stress distribution can be assumed along the embedded fiber length, as long as work hardening effects are negligible and thus:

$$F_d = 2\pi r L \tau_{iy} \tag{6}$$

(3) From experimental results, it has been observed that the fiber–matrix interface might fail catastrophically. This failure mode can be attributed to the stress concentration at the fiber and matrix junction, and at this point, failure initiates and rapidly propagates along the interface. Another approach for fiber–matrix interface failure is based on energy methods. Failure may occur if the interface fractures with work of fracture (G_i) per unit area of interface. The source of the required fracture surface energy is the strain energy stored within the specimen components [42–44,47].

Chua and Piggott [44,45] considered only the extensional strain energy (U_L) stored in the embedded fiber length and the shear strain energy (U_m) in the matrix immediately surrounding the fiber which is given in the following equation where

n is given by Eq. 2 and $s = L/r$:

$$U_L + U_m = \frac{\pi r^3 \sigma_{fe}^2 \coth(ns)}{2n E_f} \tag{7}$$

Equating the total strain energy to $2\pi r L G_c$, where G_c is the unit fracture energy at the interface, the debonding load P_d is found to be:

$$P_d = 2\pi r \sqrt{E_f G_c r (ns) \tanh(ns)} \tag{8}$$

After debonding, friction at the interface has to be overcome [10,37,39,43] in order for pull-out to proceed. Friction at the interface is due to the normal compressive stresses that are caused by the pressure p_o acting on the fiber from the matrix, where p_o is the pressure exerted due to Poisson's contraction of the matrix at the moment the fiber emerges from the polymer. Such stresses arise from resin shrinkage resulting from the curing of the specimen and from dissimilar coefficients of thermal expansion of the matrix and the reinforcing fiber [40,44,48].

After failure of adhesion, the interfacial shear stress $\tau_i = \mu p$, increases with increasing pull-out distance [46], but the normal stress from the matrix acting on the fiber will produce a reduction in cross-sectional area due to Poisson's effect resulting in a reduction of interfacial normal stresses.

In a typical force–displacement plot (Fig. 4) describing the pull-out process, the first peak is attributed to debonding and frictional resistance to slipping, whereas subsequent lower peaks are attributed to friction and stick–slip mechanism, giving rise to a serrated portion of the curve [15,37,38,41]. Because of relaxation at

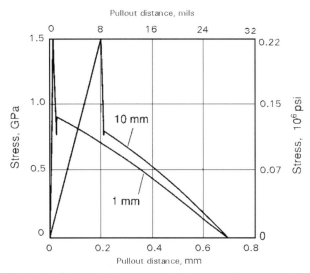

Fig. 4. Typical force–displacement plot from a pull-out experiment.

both the free and embedded lengths of the fiber, the slope of a curve (τ_i) against pull-out distance gives only an approximate value of the interfacial shear stress (μp_o) where p_o is the pressure exerted by the matrix shrinkage at the moment the fiber emerges from the polymer. The experimental value for shear stress (τ_{exp}) obtained from the slope of the pull-out curve is related to the true value of $\tau_i = \mu p_o$ by Eq. 9:

$$\tau_i = \frac{\tau_{exp}}{1 + \dfrac{2\tau_{exp}(1+2L)}{E_f r}} \tag{9}$$

In addition to shrinkage stresses resulting from curing as external pressure p_e is applied, the shear stress increases as:

$$\tau_i = \mu(p_e + p_o) \tag{10}$$

The value of μ can be determined from experimental results by plotting τ_i against p_e and evaluating the slope of the curve. The intersection of the curve with the vertical axis, should give μp_o. If no external pressure is applied, μ can be estimated from the curve of the pull-out force F_A as a function of the pull-out distance:

$$F_A = \frac{\pi d^2 E_f (1+\nu_m)}{\nu_f E_m \left(1 - \dfrac{e^{\eta(y/L)}}{e^{\eta}}\right)} \tag{11}$$

where ν_f is the fiber Poisson's ratio, $\eta = 2\mu\nu E_m L / E_f r(1+\nu_m)$, and y is the pull-out distance. The effect of the external pressure applied to the specimen is more noticeable on the pull-out force at short embedded lengths [40,49].

Most researchers have used the interfacial shear strength as a criterion for fracture, that is, when the interfacial shear stress exceeds the shear strength of the interface, then a crack may propagate. The debonding process can also be treated using the fracture energy as the failure criterion, that is, debonding occurs when the work done by the applied load minus the energy stored in the system is equal to or larger than the adhesive fracture energy, denoted G_a, i.e. the amount of energy to separate a unit area of interface [50–52]. A debond will propagate from the embedded length end when the applied force is [53]:

$$F_i = 2\pi (R^2 r_f - r_f^3)^{1/2}(E_m G_a)^{1/2} + 2\pi r_f c \tau_f \tag{12}$$

where c is the length of debond and τ_f is the frictional stress in the debond region. Propagation of the debond from the loaded end of the linearly elastic fiber embedded in an inextensible matrix takes place when the applied force is given by:

$$F_o^2 = 4\pi^2 r_f^3 F_f G_a \tag{13}$$

If friction is considered, the previous equation is modified as:

$$F_f = 4\pi r_f^{3/2}(E_f G_a)^{1/2} + 2\pi r_f c \tau_f \tag{14}$$

According to this model [54], the growth of an interfacial crack is stable after the initiation of a debond at the loaded fiber end. It was also found that the debonding force increased linearly with crack length due to friction in the debond region. The force of debonding increased after reaching a critical length and then leveled off. Then, no further increase in force was necessary to continue the debond process.

2.3. The microbond technique

In an analysis of the microdrop or microbond method, Penn and Lee [55] considered the existence of an initial microcrack of length a at the fiber–matrix interface. They also considered the effect of the strain energy contributed by the free fiber length to the crack propagation process, and using an energy balance they derived an expression for the debonding force P_d of the microdrop:

$$P_d = \frac{2\pi r \sqrt{r G_c E_f}}{1 + \mathrm{csch}^2(ns)} \tag{15}$$

The microbond technique was analyzed by Nairn [56] and more recently by Scheer and Nairn [57]. In the first paper, a variational mechanics analysis of the stresses in the microbond specimen was completed. In the second paper, a new, more complete shear-lag analysis was presented. The proposed shear-lag result for analyzing the microbond test results is given by the equation in terms of the force of debonding:

$$F_d = \pi r_1^2 \left(\sqrt{\frac{2G_{ic}}{r_1 C_{33s}}} - \frac{D_{3s}\Delta T}{C_{33s}} \right) \tag{16}$$

In this equation, G_{ic} is the interfacial toughness, ΔT is the difference between the stress-free temperature and the specimen temperature, and

$$C_{33} = \frac{1}{2}\left(\frac{1}{E_A} + \frac{V_1}{V_2 E_m} \right) - \frac{V_2 A_3^2}{V_1 A_0} \tag{17}$$

$$D_{3s} = \frac{1}{2}(\alpha_A - \alpha_m) \tag{18}$$

$$A_0 = \frac{V_2(1-\nu_T)}{V_1 E_T} + \frac{1-\nu_m}{E_m} + \frac{1+\nu_m}{V_1 E_m} \tag{19}$$

$$A_3 = \left(\frac{\nu_A}{E_A} + \frac{V_1 \nu_m}{V_2 E_m} \right) \tag{20}$$

In these expressions, E_A and E_T are the axial and transverse moduli of the fiber, G_A is the axial shear modulus of the fiber, ν_A and ν_T are the axial and transverse Poisson's ratios of the fiber, α_A and α_T are the axial and transverse thermal expansion coefficients of the fiber, and E_m, G_m, ν_m and α_m are the tensile modulus, shear modulus, Poisson's ratio, and thermal expansion coefficient of the matrix.

2.4. Experimental apparatus and procedure

The following experimental procedure can be used to measure the interfacial shear strength by means of the microbond technique for a thermosetting resin matrix.

(1) First, the ends of 100 mm (4 in.) long single-carbon IM-6 fibers are taped to parallel sides of a specially built frame using double-stick tape.

(2) After mixing and degassing the resin, the microdrops are applied to the fibers using a syringe and needle. A small drop of the thermoset resin is made to flow to the needle tip and it is allowed to come in contact with a fiber. After retraction of the needle tip, some of the resin remains forming a microdrop around the fiber. The microdrop size ranges from 80 to 200 μm (3 to 8 mils) in diameter.

(3) The microdrops for this resin are allowed to cure at room temperature for 24 h, and are then post-cured for 2 h at 75°C (165°F) and for 2 h at 125°C (255°F). (This procedure would be adjusted depending on the matrix chemistry.) After curing, the fiber is affixed to an aluminum plate and kept in a desiccator to wait testing.

(4) The small aluminum plate is attached to a 50 or 250 g (2 or 9 oz) load cell. The droplet is gripped with micrometer blades, which are brought together until they nearly touch the fiber. The micrometer blade, mounted on a translation stage, is made to move away from the load cell and become parallel to the longitudinal axis of the fiber, at a speed of 0.11 mm min^{-1} (4 mils min^{-1}). The position and speed of the translation stage is controlled remotely using a motorized actuator. The translation of the stage causes the microdrop to be sheared off of the fiber surface. The force required to debond the microdrop is recorded using a pen plotter.

Fig. 5 shows a photograph of the droplet pull-off test apparatus. The entire apparatus is mounted on an optical microscope stage for measurement of the embedded length and fiber diameter. The procedure outlined here is similar to those described by other investigators [58–63].

Thermoplastic matrices may also be used with the microdrop method [58,61] A method to form thermoplastic matrix material microdrops in various fiber–thermoplastic systems has been reported by Gaur et al. [58]. They measured the interfacial shear strength of carbon and aramid fibers embedded in four thermoplastic resins: polyetheretherketone (PEEK), polyphenylene sulfide (PPS),

1 LOAD CELL
2 BLADE MICROMETER
3 X-Y TRANSLATION STAGE
4 ACTUATOR
5 BASE

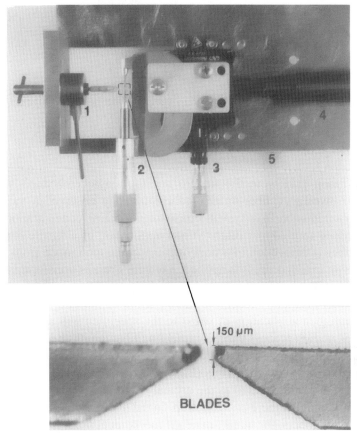

Fig. 5. Apparatus used to perform the microbond technique showing details of the microvise used to hold the microdrop while a load is applied on the fiber.

polycarbonate (PC), and polybutylene terephthalate (PBT). The procedure to produce microdrops is the following. A small, thin piece of film (about 2–30 mm, or 0.08 to 1.2 in.) is used. A longitudinal cut is made on the film piece, along nearly its entire length, to form two strips joined at one end for a distance of 50–100 μm (2 to 4 mils). The strips are suspended on the horizontal fibers already affixed to a holding frame, and the thermoplastic is melted on the fibers. Upon melting, nearly uniform-sized droplets are obtained. Their lengths are controlled by the film thickness. This procedure is shown in Fig. 6.

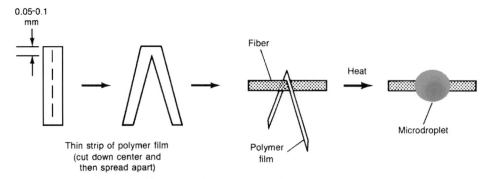

Fig. 6. Procedure to form microdrops on a fiber from thermoplastic polymers [59].

3. Single-fiber fragmentation technique

Kelly and Tyson [11] were the first to use the single embedded-fiber tensile specimen to investigate fiber–matrix adhesion. They observed a multiple-fiber fracturing phenomenon in a system consisting of a low concentration of brittle fibers embedded in a copper matrix, upon application of a tensile force. In this method, a fiber is embedded in a dogbone-shaped coupon made from the matrix material. As an external stress is applied, a tensile stress is transferred to the fiber through an interfacial shear. When the external stress is increased, the tensile strain in the fiber will eventually reach the failure strain of the fiber and the fiber will fracture. Continued application of stress to the specimen will result in repetition of this fragmentation process until all remaining fiber lengths become so short that the shear stress transfer along their lengths can no longer build up enough tensile stresses to cause any further failures (Fig. 7). This maximum final fragmentation length of the fiber is referred to as the critical length, l_c. When this critical length is reached it is said that there is 'saturation' in the fiber fragmentation process. For successful experiments, the matrix material mechanical properties should be such that its strain to failure is at least three times higher than that of the fiber. The shear stress at the interface is assumed to be constant along the short fiber critical length (also assumed to have a constant diameter). An average shear strength (τ) can be determined from a simple balance of force which results in:

$$\tau = \frac{\sigma_f d}{2 l_c} \tag{21}$$

where d is the fiber diameter for a circular fiber cross-section. Since the fiber–matrix interface is placed under shear, the calculated value of τ is often used as an estimator of the fiber–matrix interface shear strength [12].

SINGLE-FIBER-FRAGMENTATION TEST

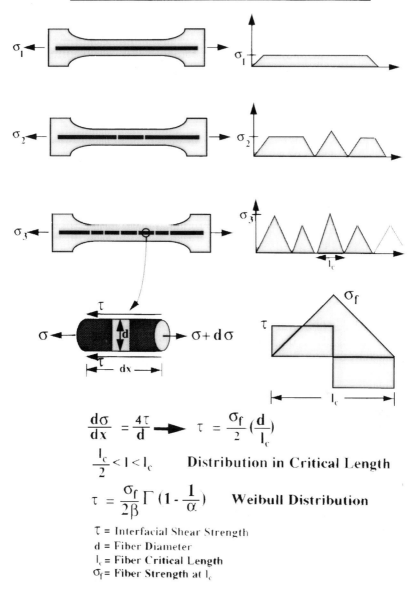

$$\frac{d\sigma}{dx} = \frac{4\tau}{d} \longrightarrow \tau = \frac{\sigma_f}{2}\left(\frac{d}{l_c}\right)$$

$$\frac{l_c}{2} < l < l_c \qquad \textbf{Distribution in Critical Length}$$

$$\tau = \frac{\sigma_f}{2\beta}\Gamma\left(1-\frac{1}{\alpha}\right) \qquad \textbf{Weibull Distribution}$$

τ = Interfacial Shear Strength
d = Fiber Diameter
l_c = Fiber Critical Length
σ_f = Fiber Strength at l_c

Fig. 7. Schematic representation of the single-fiber fragmentation process.

3.1. Characterization of the critical fiber fragment length

Most high-performance engineering fibers have strengths that vary considerably along their length because of flaws introduced through handling or manufacturing or because of intrinsic anomalies of the material [66,67]. These randomly spaced flaws or point defects introduce a slight variation in strength, which, depending on its value may or may not prevent the fiber from fracturing once the average σ_f is reached. When the built-up stress in the fiber approaches the true value of σ_f, the fiber will instantaneously and repeatedly break into shorter and shorter pieces until the remaining fragments are not long enough for the linear stress to build up from either end to exceed the fiber strength anywhere. This final length is usually referred to as the critical length l_c. Any fragment with a length slightly exceeding l_c will break in two, yielding, at the conclusion of the experiment, a random distribution of fragment lengths between $l_c/2$ and l_c.

Drzal et al. [12,68,69] recognized the random nature of this problem and used a two-parameter Weibull distribution to characterize the distribution of fiber fragment length. Then, using the arithmetic mean fragment length, that is, the original unbroken length divided by the number of fragments, and an average value for σ_f at the critical length (l_c) extrapolated from simple tension tests, they obtained expressions for the mean interfacial shear strength:

$$\bar{\tau} = \frac{\sigma_f}{2\beta^2} \Gamma\left(1 - \frac{1}{\alpha}\right) \tag{22}$$

$$\mathrm{var}(\tau) = \frac{\sigma^2}{4\beta^2}\left[\Gamma\left(1 - \frac{2}{\alpha}\right) - \Gamma^2\left(1 - \frac{1}{\alpha}\right)\right] \tag{23}$$

where β and α are the maximum likelihood estimates of the scale and shape parameters, respectively, and Γ is the gamma function.

Bascom and Jensen [67], used an approach similar to that of Drzal and coworkers. Wimolkiatisak et al. [70] found that the fragmentation length data fitted both the Gaussian and Weibull distributions equally well. Fraser et al. [71] developed a computer model to simulate the stochastic fracture process and, together with the shear-lag analysis, described the shear transmission across the interface. Netravali et al. [72], used a Monte Carlo simulation of a Poisson–Weibull model for the fiber strength and flaw occurrence to calculate an effective interfacial shear strength τ_i using the relationship:

$$\tau_i = K \frac{\sigma_f d}{2 l} \tag{24}$$

where K is a correction factor to be determined from the Weibull–Poisson model, l is the mean fiber fragment length. They proposed a value of $K = 0.889$ for brittle fibers.

Since the measured fiber fragment lengths are distributed between ($l_c/2$) and l_c, an average fragmentation length can be approximated as $l = 0.75l_c$, and the given correction factors K, yield errors of 7% and 19%, respectively, with regard to the average value of 0.75.

A procedure based on a newly developed statistical theory for estimating the effect of different sizings on the interfacial strength was present by Hui et al. [73]. The interfacial shear strength was given by:

$$\tau = \left(\frac{d}{2l_{\mathrm{f}}^{-\infty}\chi_\infty}\right)^{(1+\rho)/\rho} \left(\frac{\sigma_{\mathrm{o}}^{\rho} l_{\mathrm{o}} 2}{d}\right)^{1/\rho} \tag{25}$$

where $l_{\mathrm{f}}^{-\infty}$ is the average fragment length at saturation and can be obtained from the plateau value of the experimental data, χ_∞ is obtained from the equation:

$$l_{\mathrm{f}}^{-\infty} = \delta_{\mathrm{c}}\chi_\infty(\rho) \tag{26}$$

where ρ and $\sigma_{\mathrm{o}}^{\rho} l_{\mathrm{o}}$ are Weibull strength parameters determined from experimental data. Hui et al. [73] also found that the Weibull parameters are very sensitive to the residual stresses due to the thermal coefficient mismatch between fiber and matrix.

The experimental determination of the strength of individual fibers at very short lengths is very difficult, and most analyses extrapolate mean strength and strength distribution data obtained from longer gage lengths. Asloum et al. [74] studied the dependence of the strength of high-strength carbon fibers on gage length by means of the Weibull model. They showed that the mathematical form of the estimator chosen and the sample size, when larger than about 20, do not influence the results of the analysis. Also, they found that neither the three-parameter nor the two-parameter Weibull distribution is appropriate for describing the critical length dependence on gage length of the fiber during testing. Furthermore, it was shown that a linear logarithmic dependence of strength on gage length is the most accurate, and simple method for extrapolating the fiber tensile strength at short lengths.

Other researchers have reported on the influence of mechanical properties of the matrix and fiber on the critical aspect ratio and, consequently, on the stress transfer in single-fiber composites. The effect of adhesion as affected by the surface treatment on the fiber, the ratio of elastic moduli of fiber and matrix, and temperature on the critical aspect ratio was analyzed experimentally by Asloum et al. [75]. Rao and Drzal [76] and Drzal [77] studied the dependence of the interfacial shear strength on the bulk material matrix properties using model compounds based on epoxy/amine chemistry. AS4 carbon fibers were used as the subject for these measurements with both a difunctional epoxy (DGEBA) system as well as a tetrafunctional epoxy (MY720) system. In order to produce matrices with a range of matrix properties from brittle, elastic to ductile, plastic, amine curing agents were carefully selected. The fiber–matrix interfacial chemistry was kept constant throughout this study by always using the same amount of curing agent. They found that for the difunctional as well as the tetrafunctional epoxy system, the interfacial shear strength (measured using the single-fiber fragmentation technique) decreases nonlinearly with decreasing modulus of the matrix. Linear elastic analysis yields a nearly linear relationship, for both systems, between interfacial shear stress and the product of strain to final break and the square root of the matrix shear modulus. A linear relationship is also found between the difference in test temperature and glass transition temperature of the cured matrix and the interfacial shear strength. Additionally, the failure mode is seen to remain interfacial as the ductility of the matrix changes. Termonia [78] used a finite-difference approach to show that the critical length for efficient stress transfer to the fiber is a function of the ratio between elastic moduli of the fiber and matrix. In his model he also considered the dependence of the critical length on the adhesion by including an adhesion factor. A decrease in the adhesion is seen to increase the critical length, particularly when the adhesion factor becomes less than 30%.

Folkes and Wong [79], in their study of adhesion between fiber and matrix of thermoplastic composites, noticed that the formation of transcrystalline morphology around glass fibers in polypropylene has an effect on the critical fiber length, probably through the change in local interphase modulus.

In a recent study, Galiotis et al. [80] used Raman spectroscopy to determine strain profiles along the fiber fragment length, on surface-treated and non-treated carbon fiber. They reported that for the treated fiber, debonding at the crack tip initiates at the fiber fracture strain. The maximum interfacial shear stress per increment of load is obtained at certain distance from the crack tip which is equal in size to the combined debonding and matrix yielding zones. The maximum interfacial shear strength profiles for the non-treated fiber indicated that the load transfer between fiber and matrix is obtained through friction only.

Verpoest et al. [81] presented a micromechanical analysis to show that for increasing applied strains, the fiber aspect ratio can reach values which are lower than those predicted by Kelly's shear-lag analysis. They also proposed that the single-fiber fragmentation test could be used to estimate the different components of the interface shear strength, that is, the bond strength, the friction strength and the matrix yield strength.

3.2. Stress analysis

The distribution of stress around discontinuous fibers in composites has been studied by a number of researchers. Theoretical analyses have been performed by Cox [82] and Rosen [83]. In these models only fiber axial stress distribution and the fiber–matrix interfacial shear stress distribution are determined. Amirbayat and Hearle [84] studied the effect of different levels of adhesion on the stress distribution, that is, no bond, no adhesion, perfect bond, and the intermediate case of limited friction. They also considered the inhibition of slippage by frictional forces resulting from interfacial pressure due to Poisson's lateral contractions of the matrix but did not consider the shrinkage of the matrix during curing.

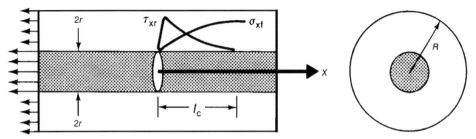

Fig. 8. Micromechanical model of the single-fiber fragmentation test [88].

Theocaris [85] proposed a model that incorporates an interphase which he named a mesophase, which constitutes a boundary layer between the main phases of the composite. From a physical basis, a continuous and smooth transition of the properties from one phase to the other is assumed. Because the mechanical properties of this region also contribute to the composite properties, the determination of the local mechanical modulus is important. Dynamic mechanical analysis is used to identify the mesophase properties, primarily the glass transition temperature (T_g), through changes in the loss modulus peak.

Lhotellier and Brinson [86] developed a mathematical model that includes the mechanical properties of the interphase, the stress concentration near fiber breaks, and the elasto-plastic behavior of both the matrix and the interphase.

Whitney and Drzal [87] presented an analytical model to predict the stresses in a system consisting of a broken fiber surrounded by an unbounded matrix. The model (Fig. 8) is based on the superposition of the solutions to two axisymmetric problems, an exact far-field solution and an approximate transient solution. The approximate solution is based on the knowledge of the basic stress distribution near the end of the broken fiber, represented by a decaying exponential function multiplied by a polynomial. Equilibrium equations and the boundary conditions of classical theory of elasticity are exactly satisfied throughout the fiber and matrix, while compatibility of displacements is only partially satisfied. The far-field solution away from the broken fiber end satisfies all the equations of elasticity. The model also includes the effects of expansional, hygrothermal strains and considers orthotropic fibers of the transversely isotropic class.

The axial normal stress σ_x in the fiber:

$$\sigma_x = [1 - (4.75\bar{x} + 1)e^{-4.75\bar{x}}]C_1\varepsilon_o \tag{27}$$

where $\bar{x} = x/l_c$, ε_o is the far-field strain, C_1 is a constant dependent on material properties, expansional strains, and the far-field axial strain. It can be noticed that σ_x is independent of the fiber radius. The critical length l_c is defined such that the axial stress recovers 95% of its far-field value, that is:

$$\sigma_x(l_c) = 0.95C_1\varepsilon_0 \tag{28}$$

The interfacial shear stress is given by the expression:

$$\tau_{xR}(\bar{x}, r) = -4.75\mu C_1 \varepsilon_0 \bar{x} e^{-4.75\bar{x}} \tag{29}$$

where

$$\mu = \left(\frac{G_m}{E_{1f} - 4\nu_{12f}G_m} \right)^{1/2} \tag{30}$$

E_{1f} denotes the axial elastic modulus of the fiber, where as ν_{12f} is the longitudinal Poisson's ratio of the fiber, determined by measuring the radial contraction under an axial tensile load in the fiber axis direction and G_m denotes the matrix shear modulus. It should be noted that the negative sign in the expression for the shear stress is introduced to be consistent with the definition of an interfacial shear stress in classical theory of elasticity. The radial stress at the interface is given by:

$$\sigma_r(\bar{x}, r) = [C_2 + \mu^2 C_1(4.75\bar{x} - 1)e^{-4.75\bar{x}}]\varepsilon_0 \tag{31}$$

Constants C_2 and C_1 are dependent on material properties, expansional strains and the far-field strain. Numerical results are normalized by σ_o, which represents the far-field fiber stress in the absence of expansional strains. In particular:

$$\sigma_0 = C_3 \varepsilon_0 \tag{32}$$

The constants C_1, C_2, and C_3 are given by:

$$C_1 = E_{1f}\left(1 - \frac{\bar{\varepsilon}_{1f}}{\varepsilon_0}\right) + \frac{4K_f G_m \nu_{12f}}{K_f + G_m}\left[\nu_{12f} - \nu_m + \frac{(1+\nu_m)\bar{\varepsilon}_m - \bar{\varepsilon}_{2f} - \nu_{12f}\bar{\varepsilon}_{1f}}{\varepsilon_0}\right] \tag{33}$$

$$C_2 = \frac{2K_f G_m}{K_f + G_m}\left[\nu_{12f} - \nu_m + \frac{(1+\nu_m)\bar{\varepsilon}_m - \bar{\varepsilon}_{2f} - \nu_{12f}\bar{\varepsilon}_{1f}}{\varepsilon_0}\right] \tag{34}$$

$$C_3 = E_{1f} + \frac{4K_f G_m \nu_{12f}(\nu_{12f} - \nu_m)}{K_f + G_m} \tag{35}$$

and

$$K_f = \frac{E_m}{2\left(2 - \dfrac{E_{2f}}{2G_{2f}} - \dfrac{2\nu_{12f}^2 E_{2f}}{E_{1f}}\right)} \tag{36}$$

where E_{2f} is the radial elastic modulus of the fiber and K_f is the plane-strain bulk modulus of the fiber. A numerical example is presented for a single-fiber composite of AS4 fiber–Epon 828 epoxy matrix. The specimens were cured at 75°C (167°F) and post-cured at 125°C (257°F). The difference between room temperature, 21°C (69.8°F) and the post-cure temperature yields $\Delta T = 104$°C (219.2°F), which is the worst case for thermal residual stresses. Because it is most likely that some residual stresses will be relieved during cool-down from the

Table 1

Fiber-matrix material properties [87]

Property	Epoxy (Epon 828)	AS-4
E_1 GPa (Msi)	3.8 (0.55)	241 (35)
E_2 GPa (Msi)	3.8 (0.55)	21 (3)
ν_{12}	0.35	0.25
G_{23} GPa (Msi)	1.4 (0.20)	8.3 (1.2)
α_1 $10^{-6}/°C$ ($10^{-6}/°F$)	68 (32)	−0.11 (−0.5)

post-cure temperature, in this example the value $\Delta T = 75°C$ (167°F) was chosen. The material properties are given in Table 1.

Figs. 9–11 show plots of σ_{xf}/σ_o, τ_{rx}/σ_o, and σ_r/σ_o, where the far-field stress was used to normalize out the numerical results. The axial fiber stress and the interfacial shear stress are relatively insensitive to thermal strains, but the radial strain is quite sensitive to thermal strains.

It should also be noted that the peak value of interfacial shear predicted by this model occurs at a small distance from the broken end of the fiber. This was also noticed by other researchers [88–91].

Because of the fragmentation of a fiber embedded in a block of resin, a penny-shaped crack may result at the point of fiber fracture radially outward from

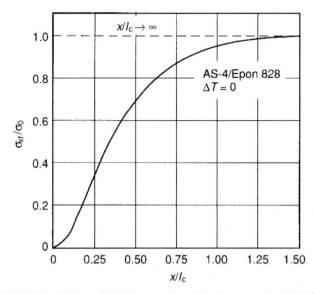

Fig. 9. Distribution of axial stress along fiber fragment length [88].

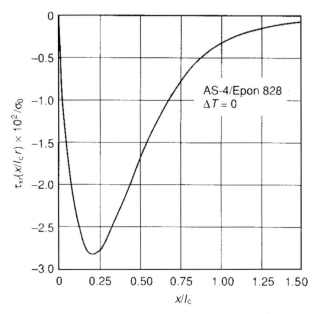

Fig. 10. Distribution of shear stress along fiber fragment length [88].

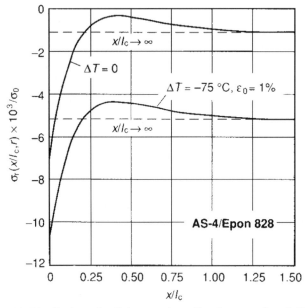

Fig. 11. Distribution of radial stress along fiber fragment length [88].

the fiber axis. Depending on the amount of energy released upon fiber failure, this microcrack may or may not propagate into the matrix and/or the interphase, depending on the level of adhesion between fiber and matrix. Then, the resulting stress distribution will depend on the mechanical properties of the fiber, the matrix and the interphase and the extent of the damaged zone around the fiber break. As pointed out by Ko et al. [20], when the interfacial shear strength or the matrix shear strength are low, the open mode deformation of the penny-shaped crack in the matrix could induce interfacial or near-interfacial shear failure arresting radial crack growth and propagating the crack growth in a direction along the interphase. After this, any further interfacial damage should be caused by the shear stress component rather than the radial stress component.

The micromechanics of stress transfer for the fragmentation test is further complicated by the different failure mechanisms that could be present and the mechanical properties of the fiber, matrix and interphase and the extent of damage in the neighborhood of the fiber break. Depending on the level of interaction between fiber and matrix, upon fiber failure, several events can take place either individually or simultaneously. If the bond is 'weak' an interfacial crack will grow and the fiber ends will slip. The stress transfer will be due to friction and/or a mechanical interlocking mechanism. If the bond is 'strong', matrix cracking will occur and the initiated crack could either propagate radially as a penny-shaped crack, at 45° as a concoidal crack, or a combination of both. The matrix nonlinear behavior such as yielding or strain hardening may further complicate matters [92].

Another three-dimensional axisymmetric stress distribution for the stress around fiber breaks was obtained by Nairn [93] using variational mechanics. In this study, breaks interaction was also included and it was assumed that both fiber and matrix were linearly elastic and a perfect adhesion at the fiber–matrix interface. To account for the stress singularity at the matrix crack tip of the fiber break, the matrix plastic-model was also included. Imperfect adhesion to mimic a failed fiber–matrix interface was added to this model to study the mechanism of interfacial failure, that is, the stress conditions that cause the extent of interfacial failure or its increase. It was suggested that due to the complexity of the multi-axial stress state, a simple maximum stress failure criterion was unrealistic and an energy release rate analysis was necessary to calculate the total energy release rate associated with the growth of interfacial damage.

In a more recent paper, Ho and Drzal [94] used a three-phase nonlinear finite-element analysis to investigate the stress transfer phenomenon in the single-fiber fragmentation test. The effect of fiber properties, interphase properties and thickness on the stress distribution in the vicinity of the fiber break was evaluated. Also, the stress fields for various fiber–matrix interface debonding conditions and the effect of frictional stress transfer were investigated. In this model, the fiber was assumed to be a linear elastic transversely isotropic material, the interphase as an elastic material and the matrix as a nonlinear material. It was found that the stress transfer length, that is, the section where the shear stress is not zero, is longer when the true nonlinear behavior of the matrix is considered than that from the linear elastic analysis. It was concluded that the linear elastic analysis overestimated the stress transfer efficiency and predicts a shorter stress transfer region. It was also found that four regions of stress transfer are distinguished in this model. The first region is the constant shear stress region resulting from the friction force between the two debonded bodies. The second region is a shear plateau in front of the crack tip as a result of the plastic behavior of the matrix. The third region is the shear decaying zone. The fourth region corresponds to a zero shear stress zone where the stress transfer does not occur. Tripathi et al. [95], used an axisymmetric finite-element model to study the effect of matrix properties (elastic modulus, yield and/or draw strengths and yield strain) on the interfacial shear stress in a short embedded fiber and the value of the interfacial shear strength obtained from the fragmentation test. They concluded that the stress transfer is strongly influenced by the plastic behavior of the matrix. Also, the interfacial shear stress at the interface never exceeds the shear yield strength of the matrix; therefore, it is not possible to measure the interfacial shear strength unless the interphase is stiffer than the matrix.

Nairn and Liu [96] used a Bessel–Fourier series stress function and added polynomial terms to provide a nearly exact solution to the stress transfer from the matrix to a fragmented fiber through an imperfect interphase. This solution satisfies equilibrium and compatibility every place and satisfies exactly most boundary conditions with the exception of the fiber axial stress. They also proposed the use of an interphase parameter, D_s, and provide a physical interpretation as:

$$D_s = \frac{r_1 G_i}{t_i} \tag{37}$$

where r_1 is the fiber radius, G_i and t_i are the interphase shear stiffness and thickness, respectively. D_s has units of modulus and is related to the effective shear stiffness of the interphase. This parameter can also be interpreted as the ability of the interphase to transfer stress from the matrix to the fiber. Because this problem can be viewed as a penny-shaped crack there should be a stress singularity at the crack tip, at least in the matrix axial stress. Such singularity is not captured in the present analysis because of the consideration of the fiber stress being zero at the fracture surface in the average instead of uniformly zero. Nevertheless, the solution presented converges to a singularity. As the number of Bessel–Fourier terms increases, the axial stress in the matrix at the crack tip increases. For a perfect interface, $D_s = \infty$, and the stress transfers back into the fiber in approximately 30 fiber diameters. As D_s decreases, the stress transfer gets slower and as it approaches zero, the interfacial shear stress also approaches zero, because there should be complete debonding at the fiber–matrix interface. From experimental results, values for D_s of the order of 500 agreed well with the stress distribution at the undamaged interface and for the exclusion or damaged zone, a value of 5 was in good agreement with the experimental data.

3.3. Experimental apparatus and procedure

Thermoset test coupons for the single-fiber fragmentation test can be fabricated by a casting method with the aid of a silicone room temperature vulcanizing (RTV) 664 eight-cavity mold. ASTM 24 mm (2.5 in.) tensile dogbone specimen cavities with a 3.175 mm wide by 1.587 mm thick by 25.40 mm ($1/8 \times 1/16 \times 1.0$ in.) gage section are molded into a $76.20 \times 203.20 \times 12.70$ mm ($3 \times 8 \times 0.5$ in.) silicone piece. Sprue slots are molded in the center of each dogbone to a depth of 0.7938 mm (1/32 in.) and through the end of the silicone piece.

Single fibers of approximately 150 mm (6.0 in.) in length are selected by hand from a fiber bundle. Single filaments are carefully separated from the fiber tow without touching the fibers, except at the ends. Once selected, a filament is mounted in the mold and held in place with a small amount of rubber cement at the end of the sprue. The rubber cement is not in contact with the cavity that contains the grip sections or the gage length section in the mold. The rubber cement is allowed to dry, and the resin is added with the aid of a disposable pipette. The long, narrow tip should be removed so that the resin can readily enter and exit the pipette chamber. Air bubbles are avoided by first degassing the silicone mold and the resin in a vacuum chamber before filling the mold cavities. The assembly is transferred to an oven where the curing cycle is completed. After cool down to room temperature, the mold can be curled away from the specimens parallel to the fiber to prevent fiber damage. The test samples can then be stored in a desiccator until ready for analysis. Prior to testing, the coupons should be inspected for defects, and any defective one should be discarded. The single-fiber coupons used in this study were also prepared following the same curing procedure as for the microdrop preparation.

The specimens are tested in uniaxial tension, using a micro-straining machine capable of applying enough load to the tensile coupon (Fig. 12) that is fitted to the microscope stage so that the $x-y$ stage controls manipulate the jig position. This allows the operator to assess the fiber fracture process along the entire gage length of the coupon. A transmitted light polarizing microscope should be configured such that there is one polarizer below and one above the test coupon. Since the embedded fiber is located in the center of the polymeric coupon and therefore is difficult to observe at high magnification with standard microscope objectives, the microscope should be equipped with a long working distance 20× or greater objective lens. The fiber diameter is measured using a calibrated filar eyepiece, or more accurately, using a Cue Micro 300, Digital Video Caliper (Olympus Corporation, Lake Success, New York).

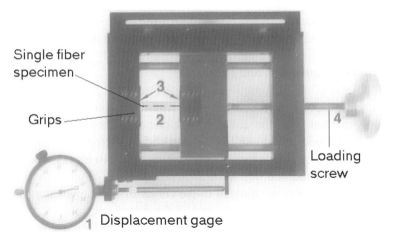

Fig. 12. Single-fiber fragmentation technique apparatus.

3.3.1. Measurement of the critical fiber fragment length

The fiber fragment lengths have to be measured for dogbone test specimen. Although a large statistical sampling of the interface occurs with the fragmentation test (for some matrix combinations, in a gage length of 20 mm (0.8 in.), the number of breaks is of the order of 50 to 100), this process could be tedious and time consuming, because each test coupon must be analyzed individually.

Traditionally, the fiber failure positions have been measured optically using a microscope equipped with a calibrated filar eyepiece, leaving room for errors due to the inherent limitations of light microscopes. The optical method also requires that the matrix be transparent in order to experimentally interpret the failure mode of the fiber–matrix interface.

An acoustic emission technique (AE) has been developed recently for the measurement of fiber fragment length distributions [97,98]. This technique is based on the fact that the speed of wave propagation is a function of the specimen material itself. It may also be influenced, however, by geometrical parameters, dominant frequency of the emitted ultrasonic signal, and, more important, the deformation of the material. The experimental configuration of the AE apparatus is shown in Fig. 13. A simple algorithm incorporating the average wave speed in the epoxy (or any other matrix), the distance between the two receiving transducers, the offset distance between one specific receiving transducer and the fixed grip, time intervals and the corresponding strains, was used to obtain the location of the fiber breaks, the fiber fragment lengths, and fiber aspect ratios. The applied stress at each fiber fracture can also be determined and from this evaluate the strength of the fiber at short gage lengths [98].

A comparison of aspect ratios measured by optical and acoustic emission meth-

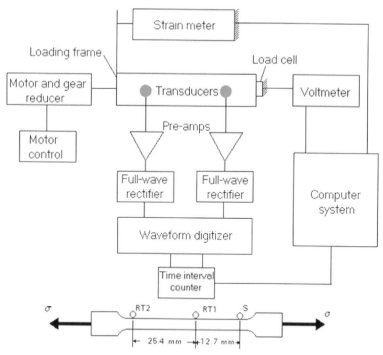

Fig. 13. Experimental configuration used in the acoustic emission technique to measure fiber fragment length [97].

ods for glass fibers in two different epoxy blends was made. Some discrepancies were obtained between the acoustic emission and optical techniques for low aspect ratios in the brittle epoxy blend. This may have been because one of the AE probes used had a diameter of 1 mm (0.04 in.). Better agreement was obtained for the interfacial shear strength for the flexible resin blend using both acoustical and optical techniques. One advantage of the acoustic measurement technique is that it does not require a transparent matrix and therefore can be used in matrices such as metal and many polymeric matrices.

An alternative technique to measure the fiber fragment length has been developed by Waterbury and Drzal [99]. It uses a special software package called 'Fiber track', together with a computer-interfaced translation stage. While the coupon is translated at a constant velocity, the operator presses a 'mouse' button as each break passes a set of cross hairs. The time intervals between breaks are stored in RAM memory and converted in displaced distances by entering the overall distance traveled. These distances are written to a magnetic disk and combined with fiber strength and diameter data to produce Weibull distributions and gamma function calculations to find the interfacial shear strength.

This technique thereby makes use of the human ability to discriminate events and to identify breakpoints visually while freeing the operator from much of the drudgery associated with manual operation of the test. The entire process of test coupon mounting, loading, fiber fragment length measurement, data storage, and interfacial shear calculation requires approximately 6 min per fiber, which represents a considerable improvement over the current state of the art.

3.3.2. Photoelastic evaluation of the interphase

Some common polymeric matrices can be considered optically isotropic in an undeformed state. However, when subjected to stresses, whether due to externally applied loads or thermally induced stresses from differential shrinkage during sample cool-down from cure temperature, the material becomes optically anisotropic (birefringent). If the resin is sufficiently transparent, it can be studied with polarized light [1,2,8,69,100]. Thus, it would be beneficial, for a better understanding of the fiber–matrix interactions, to study the stress birefringence adjacent to the fibers, before, during, and after application of load in a single embedded-fiber test.

An Olympus BHA transmitted light microscope with its polarizer and analyzer set at extinction, and long working distance objectives was used to observe the different stress patterns resulting from interphase changes.

Fig. 14 shows a series of photoelastic stress patterns generated for a series of intermediate modulus carbon fibers with varying levels of surface treatment (e.g. IM6-U, untreated; IM6-100%, full regular surface treatment, etc.) in an epoxy matrix at increasing levels of strain. Normal chromatic cycles were observed in the epoxy at low strains but they disappeared at higher load levels. Isochromatic fringes were not observed under high strain conditions present at the fiber fragment ends; instead a light bulbous region was observed. At low strains after a fiber break, extensive resin birefringence can be seen around the fiber ends. With increasing strain, this birefringence activity rapidly extends down the fragment away from the break. It is observed that in this process the stresses progress along the fragment an incremental distance with continuously increasing load. This suggests that forces are transferred from matrix to fiber by a stick–slip mechanism, that is, the stresses build up, then they appear to release and move ahead an incremental amount repetitively. For this particular fiber combination, each fiber fragment fails interfacially at low levels of strain. With increasing load, the fiber fragments interact with the matrix only through weak frictional forces along debonded areas, and through adhesive forces at a short central portion of the fiber fragment, resulting in a very low interfacial shear strength.

IM6-U fibers possess a weak structural layer on their surface that cannot support high shear loads. The photoelastic patterns then correspond to failure of the fiber through fracture of the outer layers as well as along the interface.

IM6-100% and IM6-600% fibers behave, under polarized light in a completely different manner from the untreated fibers. Fig. 15 shows photoelastic stress

IM6-U/EPON 828

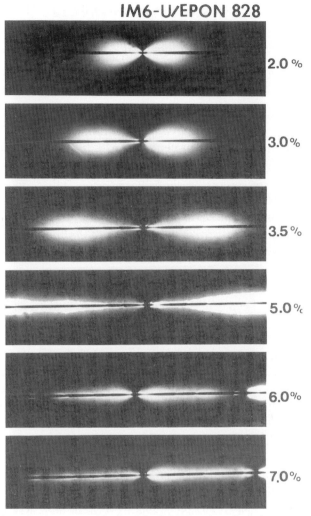

2.0 %

3.0 %

3.5 %

5.0 %

6.0 %

7.0 %

Fig. 14. Photoelastic patterns obtained for IM6-U carbon fibers in an epoxy matrix.

patterns corresponding to IM6-100% fibers. At the fiber break, the stress builds up at the ends of the fiber. However, at higher strain levels, a narrow very intense region of photoelastic activity remains around the fiber, while the initial bulbous region moves away from the fiber ends toward the center of the fiber fragment in shorter increments than those observed for IM6-U fibers. The interfacial shear strength that was measured from this fiber–matrix system was almost two times greater than that obtained for the IM6-U fiber, indicating a higher degree of fiber–matrix interaction.

In the case of IM6-600% fibers, the initial bulbous region moves away from the fiber ends towards the center of the fiber, but there is no sign of force transfer

IM6-100/EPON 828

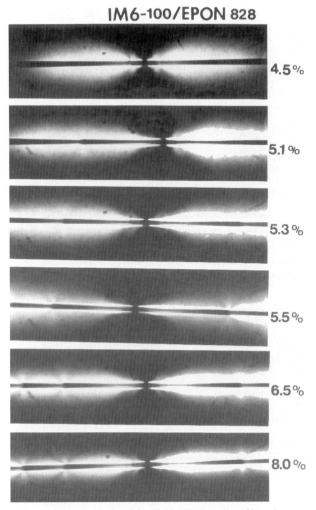

Fig. 15. Photoelastic patterns obtained for IM6-100% carbon fibers in an epoxy matrix.

by a stick–slip mechanism (see Fig. 16). Fig. 17 shows IM6 fiber fragments at high strain values and after unloading. It is interesting to observe that there exists a considerable difference between the failure mode characteristics of each fiber–matrix level of adhesion. In the case of the IM6-U fibers, the crack propagated very rapidly at low strain levels. IM6-100% fibers failed in small increments, and to achieve the same amount of damage as the for IM6-U fibers, higher loads were applied. Also, it is observed, that even when loads are applied at a constant rate, the failure path does not progress at the same rate as the load, but only when the load has attained certain values. Drzal and Rich [69] also noticed in their work with Type A fibers (Hercules, Inc.) and epoxy matrices

IM6-600/EPON 828

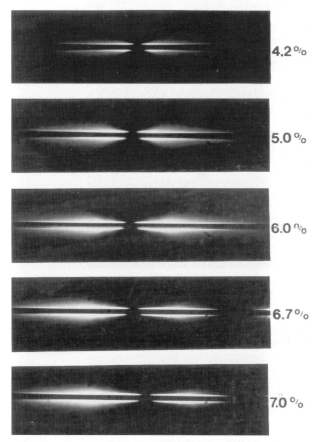

4.2%

5.0%

6.0%

6.7%

7.0%

Fig. 16. Photoelastic patterns obtained for IM6-600% carbon fibers in an epoxy matrix.

that there exists a definite relationship between the level of fiber–matrix adhesion and the fracture behavior of composites. Subsequent examination after unloading of each fiber–matrix interphase and comparison of the photoelastic photographs suggested that the narrow intense area that remains around the fiber is a region where an interfacial crack has passed, whereas the bulbous region at the tip of the photoelastically active zone appears to be the plastically deformed region due to the moving crack tip.

It is thus evident that photoelastic observations, before, during, and after application of loads to a single-fiber composite coupon could help to elucidate the fiber–matrix interactions, as well as to judge the effect of surface chemistry and morphology on the properties of the fiber–matrix interface.

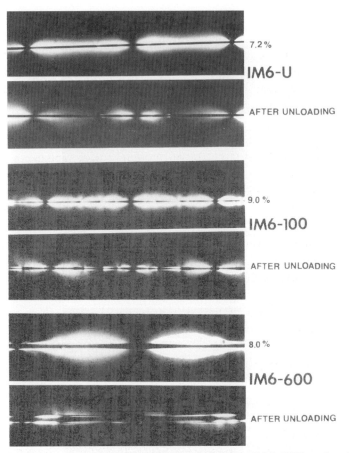

Fig. 17. Photoelastic patterns obtained for IM6-U, IM6-100%, IM6-600% carbon fibers corresponding to a whole fiber fragment at some value of strain and after unloading, to show crack propagation mode and permanent deformations in surrounding matrix.

4. Microdebonding–microindentation technique

Mandell and coworkers [101–104] first proposed an alternative technique to measure the interfacial shear strength. Single fibers perpendicular to a cut and polished surface of a regular high fiber-volume fraction composite are compressively loaded to produce debonding and/or fiber slippage [102]. In contrast to conventional methods, which use a model system to provide information on fiber–matrix adhesion, this microindentation technique is an in-situ interface test for real composites and has the advantage of reflecting actual processing conditions. It can allow determination of the interface strength due to fatigue or environmental exposure or possibly monitor interface properties of parts in service [104].

4.1. Stress analysis

The microindentation test is run on individual selected fibers on a polished cross-section. An individual fiber in a composite is surrounded by neighboring fibers located at various distances, distributed in a variety of arrangements, which range from a hexagonal array to random and dispersed configurations. The diameter of the tested fiber and the distance to the nearest-neighbor fiber are recorded for each test on a micrograph, and a simplified axisymmetric finite-element model (FEM) is used. This model includes the fiber, surrounding matrix, and average composite properties beyond the matrix [104]. It was shown that the maximum shear stress along the interface is insensitive to probe stiffness as long as the contact area does not approach the interface. Fig. 18 shows results for a case in which the fiber and matrix are considered to have the same mechanical properties. The finite-element results are compared with those obtained from an analytical solution of the Hertz contact problem of a point load on a half-space with imaginary boundaries. Good agreement was found between the FEM and the analytical solution in spite of slight differences in loading conditions. Figs. 19 and 20 show results for Nicalon (SiC) fibers in an aluminosilicate glass matrix and for HMU carbon fibers in a borosilicate glass matrix system, respectively. The stress distribution for the Nicalon fibers is very similar to that for the isotropic homogeneous case because of the low E_f/E_m ratio. It can also be noticed that the anisotropy of the carbon fibers affects the stress distribution by spreading out the shear stress transfer along the length of the fiber.

Fig. 21 shows interfacial shear stresses for a carbon fiber–epoxy system. The low modulus of the epoxy matrix produces a similar effect on the stress distribution, as in the case of isotropic fibers. Calculations of interfacial shear strength assume either a maximum interfacial shear stress criterion or a maximum radial tensile stress criterion, ignoring other stress components and residual stresses in both cases. The interfacial shear strength (τ_i) is calculated from:

$$\tau_i = \sigma_{fd} \left(\frac{\tau_{max}}{\sigma_f} \right)_{FEM} \tag{38}$$

where σ_{fd} is the average compressive stress applied to the fiber end at debonding and $(\tau_{max}/\sigma_f)_{FEM}$ is the ratio of the maximum shear stress to applied stress. Fig. 22 shows normalized interfacial shear stress (τ_{max}/σ_f) as a function of G_m/E_f (matrix shear modulus/fiber axial elastic modulus) for a variety of materials. Calculations of the interfacial tensile strength σ_i^r for tensile radial stress at the surface is given by:

$$\sigma_i^r = \sigma_{fd} \left(\frac{\sigma_{max}^r}{\sigma_f} \right)_{FEM} \tag{39}$$

Here $(\sigma_{max}^r/\sigma_f)_{FEM}$ is the ratio of the maximum radial tensile stress at the surface

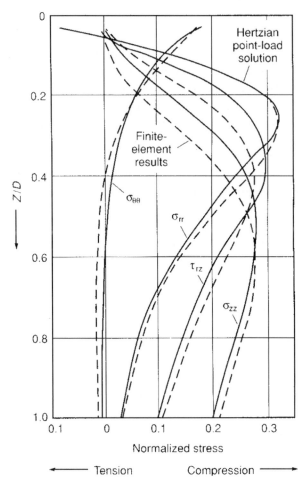

Fig. 18. Interfacial stress components along imaginary interface normal to free surface in a homogeneous isotropic material: comparison of (—) Hertzian point-load solution with (---) finite-element results. D is fiber diameter [104].

to the applied fiber pressure resulting from the finite-element analysis. In all cases, the FEM results are calculated for the ratio of matrix layer thickness to fiber diameter of 0.40. It is evident that further refinements are required to treat the free-surface problem adequately and the accuracy of stress in this area is still uncertain. Also, the effect of thermal and elastic residual stresses should be included because they should be very significant near the free surface.

To avoid any cyclic loading on the specimen during visual detection of fiber debonding, Netravali et al. [105] developed a test procedure to monitor the load and depth of indentation continuously as a characteristic change in the load–depth curve. They used a slightly modified microindentation technique to determine the

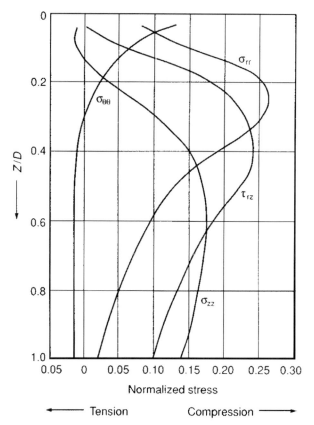

Fig. 19. Interfacial stress distribution for Nicalon (SiC) fibers in an aluminosilicate glass matrix obtained using the microindentation technique [104].

interfacial shear strength between E-glass fibers and an epoxy matrix. In addition, acoustic emissions generated by various events are monitored and the rate of loading and specimen geometry can be adjusted to simulate different situations that can occur in actual service. Thin samples of approximately 4 to 10 fiber diameters thick were used, and the assumption of constant shear stress along the fibers was made. The shear stress was calculated from the relation:

$$\tau_a = \frac{F}{\pi d t} \tag{40}$$

where F is the load value, d is the diameter of the fiber, and t is the thickness of the specimen. Differential shrinkage between the epoxy and the fiber after elevated temperature curing generates a hydrostatic pressure, P, which is given by:

$$P = \alpha_m E_m \Delta T (1 + \nu_m) \tag{41}$$

where α_m is the linear coefficient of thermal expansion of the matrix material, E_m

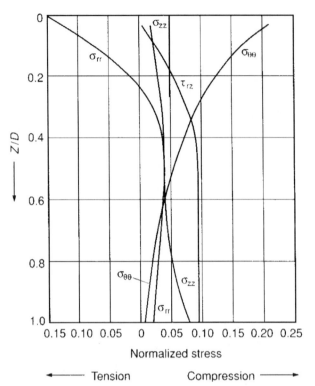

Fig. 20. Interfacial stress distribution for HMU carbon fibers in borosilicate glass matrix obtained using the microindentation technique [104].

is the matrix Young's modulus, ΔT is the difference between the curing and room temperatures, and ν_m is Poisson's ratio of the matrix. It should be noticed that the linear coefficient of thermal expansion of the fiber is considered to be negligible and also that its elastic modulus is several times greater than that of the matrix. As a consequence, lateral expansion of the fiber at the point of application of the force was also considered negligible.

Because of its thickness, the specimen bends upon application of the indentation force, resulting in radial compression at the top and tension at the bottom. The resulting stress distribution is calculated from the elastic analysis of the theory of plates according to:

$$\sigma_r = 6 \left(\frac{d}{t} \right) \left(\frac{F}{2\pi dt} \right) \left(\frac{1}{2} + \frac{\rho^2}{d^2 - \rho^2} \ln \frac{\rho}{d} \right) \tag{42}$$

where σ_r is the radial stress at the top surface, d is the fiber diameter, t is the specimen thickness and ρ is the inside radius of the brass annulus on which the specimen is rigidly mounted.

Marshall et al. [106–108] used a nanohardness apparatus to determine interfa-

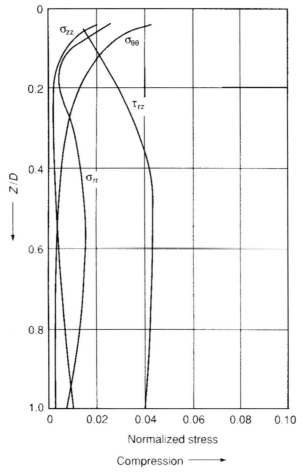

Fig. 21. Interfacial stress distribution for a carbon fiber–epoxy matrix system [104].

cial mechanical properties in fiber-reinforced ceramic composite materials. They calculated the interfacial sliding friction using the relationship:

$$\tau_F = \frac{\sigma_f^2}{4E_f}\frac{r}{L} \tag{43}$$

where σ_f is the pressure applied to the fiber, r is the radius of the fiber, L is the distance the fiber has been displaced into the matrix, and E_f is the elastic modulus of the fiber.

The microdebonding indentation system was developed by the Dow Chemical Company [109], and overcomes some deficiencies of the method described by Mandell and co-workers. This fully automated instrument is designed for use outside the research environment. It is based on a Zeiss optical microscope. A

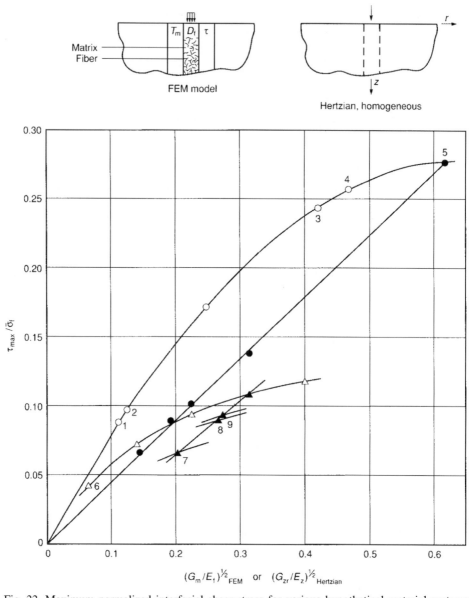

Fig. 22. Maximum normalized interfacial shear stress for various hypothetical material systems and trends from Hertzian solutions for transversely isotropic properties reference to an imaginary interface position. (○) FEM model, isotropic fibers; (●) Hertzian with $G_{zr} = 14$ MPa, E_z varying; (△) FEM with $E_z = 276$ GPa, matrix varying; (▲) FEM with $G_m = 28$ GPa, carbon fiber E_z varying, (1) S-Glass-epoxy, (2) E-Glass-epoxy, (3) NIcalon-1723 glass, (4) Nicalon-BMAS, (5) Hertzian isotropic, (6) Carbon-epoxy, (7) P100-aluminum, (8) P55-aluminum, (9) HMU-borosilicate [104].

diamond-tipped stylus mounted on the objective holder of the microscope is used to compress single fibers into the specimen. A load cell attached to the sample holder senses initiation of fiber debonding. Other components include a precision-controlled motorized stage with three degrees of freedom (linear motion in three orthogonal axes), a television camera and monitor, and an IBM PC-AT-compatible computer. The specimens are prepared following standard metallographic techniques, assuring that the fibers of the composite are always perpendicular to the surface of the specimen holder. The force required to debond the fiber is input to an algorithm that calculates the interfacial shear strength. In the commercial system the test data are reduced using a closed-form algorithm derived from Mandell's finite-element analysis using the method of least squares, resulting in an expression that is a function of σ (the axial stress in the fiber at debond), G_m (the shear modulus of the matrix), E_f (the axial tensile modulus of the fiber), T_m (the distance from the tested fiber to the nearest adjacent fiber), and d (the diameter of the selected fiber).

Ho and Drzal [110] used a nonlinear finite-element model to perform a complete parametric analysis for the microindentation test for composite interfacial shear strength measurements. Effects of material parameters such as matrix and interphase properties and interphase thickness on the interfacial shear strength and load displacement data were evaluated. It was found that for a carbon fiber–epoxy composite, the stress perturbation resulting from the indentation load diminishes at approximately 36 fiber diameters below the free surface. Also, the stress field for samples with different aspect ratios and supporting materials are identical when the fiber aspect ratio and supporting materials were identified, the fiber aspect ratio is larger than the critical value. It was also found that the stress field of the indented fibers are significantly affected by the local fiber-volume distribution, which in turn is responsible for the large data scatter of the test results. As can be seen in Fig. 23, when the composite fiber-volume fraction changes from 0.1 to 0.5, the interfacial shear stress varies approximately 35%. Therefore, if the interfacial shear stress is invariant, fibers associated to laminates with high fiber-volume fraction will fail under lower applied load due to the higher interfacial shear stress imposed by the larger constraints. It is also noticed that the empirical equation developed at Dow [111], agrees well with the nonlinear finite-element model when the fiber-volume fraction is between 0.3 and 0.5. The classical shear-lag and modified shear-lag models compare best to the finite-element model and the ITS (indentation testing system) empirical equation only for very low fiber volume fractions ($V_f = 0.1$ approximately).

The interfacial stress distributions along the interface in the axial direction near the fiber ends for a model with $V_f = 0.36$ are shown in Fig. 24. It is noticed that the interfacial axial (σ_{ZZ}), radial (σ_{rr}), hoop ($\sigma_{\theta\theta}$), and shear stress (τ_{rz}) stress distributions for models with different fiber-volume fractions are of similar shapes but the shear stress is higher for higher-volume fractions. The radial stress is

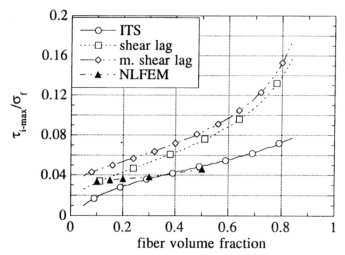

Fig. 23. Normalized maximum interfacial shear stress obtained from a nonlinear finite element analysis (NLFEM), empirical (ITS) equations, shear-lag and modified shear-lag equations as a function of fiber volume fraction [110].

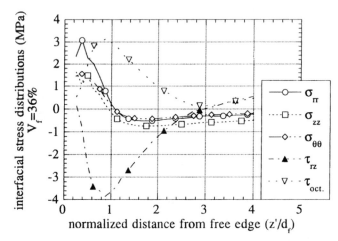

Fig. 24. Normalized interfacial stress distributions along longitudinal z-direction for $V_f = 36\%$ [110].

compressive with its maximum value at a distance equal to half a fiber diameter below the free surface. It can then be seen that Mode I fracture is unlikely to occur. All the stresses except the hoop stresses have the maximum value at a distance below the free surface. The maximum value of the shear and octahedral stresses also occur at a point located below the free surface. If the failure criterion used is that of a maximum stress, then interface shear or debonding is expected to occur

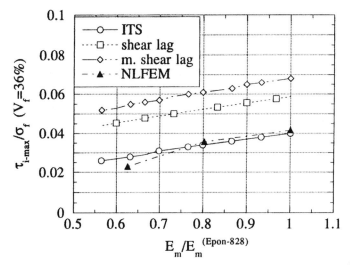

Fig. 25. Interfacial residual stresses along z-direction due to elastic–plastic behavior ($V_f = 36\%$) [110].

below the free surface and rapidly propagate to the surface. As soon as this failure is observed, the test is stopped.

Because of the nonlinear behavior of the matrix, repeated cycles of loading, necessary for optical inspection of the debonding process introduces residual stresses. The residual radial, hoop and axial interfacial stresses are positive, while the residual shear stress is negative at regions below the sample free surface (Fig. 25). The negative value of the interfacial shear stress will result in a higher apparent interface shear strength value. Similar trends are observed when the matrix shows a viscoelastic behavior; however, the magnitudes of the residual stresses are so low that they should have no noticeable effect on the interface shear strength values.

4.2. Experimental apparatus and procedure

The microdebonding indentation system described here [109], is a fully automated instrument designed to be used both as a research tool and for quality control purposes [111,112]. The system is constructed on a Mitutoyo Finescope optical microscope. A diamond-tipped stylus mounted on a collar fitted to the objective on the microscope is used to compress selected single fibers into the specimen. The probe has a 90° cone with a user-selected tip radius. Initiation of fiber debonding is sensed by a weighing mechanism from a Sartorius model L610. Other components include a precision-controlled motorized stage with three degrees of freedom (linear motion in three orthogonal axes). Klinger linear motion stages for translations in the x and y axes replace the usual stage. The fine focusing control of the microscope is controlled by a Klinger stepping motor with ± 0.04 μm resolution. The stage controllers and balance read-out are interfaced to a microcomputer. The specimen surface is monitored using a video camera mounted on the microscope. The specimen is prepared following standard metallographic techniques, assuring that the fibers of the composite are perpendicular to the surface of the specimen holder. The force (f_g) required to debond the fiber is input to an algorithm that calculates the interfacial shear strength as a function of fiber diameter (d_f), shear modulus of the matrix (G_m), axial tensile modulus of the fiber (E_f), and the distance from the tested fiber to the nearest adjacent fiber (T_m). The fibers to be indented must be in the neighborhood of other fibers but no closer than 2 μm and no further than half a fiber diameter. Once a fiber is selected, its coordinates, diameter and distance to its nearest neighbor are recorded. Fiber–matrix properties are also entered for analysis. The program directs the controller to move the fiber end selected through an offset to place the fiber below the indenter tip. The specimen is moved to within 4 μm of the indenter, at this point the stage is slowed to the rate and step size selected by the user. Once the indenter contacts a fiber, a real time plot of load versus displacement is obtained. Fiber debonding is visible as a dark shadow around the fiber. When the shadow appears around at least one sixth of the fiber it is said to be debonded. Upon debonding, an interfacial shear strength is calculated and an average value is obtained from tests on several individual fibers.

The specimen preparation procedure is as follows.

(1) Square chips of composite laminate approximately 1.25 cm per side are cut using a water-cooled diamond saw. The squares are trimmed with a scalpel to remove burrs from the cutting process.

(2) Tape is placed over one end of a cylindrical section of phenolic pipe 2.54 cm in diameter and the squares are held in place using spring holders. The squares are oriented with the fiber axis parallel to the pipe axis. The pipe section is filled with a low exotherm epoxy resin (9 parts diglycidyl ether of bisphenol A (Shell Epon 828) to 1 part diethylene triamine (Aldrich Chemical). The filled samples are left to cure at room temperature overnight.

(3) Specimens are polished using a Struers Abramin counter-rotating polisher. The polisher is set at a force of 50 N and a speed of 150 rpm and lubricated with water. Specimens are polished using paper with grits 1000, 2400, and 4000 for 4, 5, and 6 min, respectively. A final relief polishing step is performed using a Vibromet I polisher with 0.05 micron gamma alumina suspended in water. This final step is carried overnight and then the samples are thoroughly rinsed with distilled water, air dried, and placed in a desiccator until needed for testing.

5. Advantages and limitations of the methods

Aside from the theoretical analyses which have been discussed earlier and compare the assumptions and limitations of each method, it is also very useful to compare experimental results from each of these fiber–matrix adhesion methods with the same fiber–matrix system. A baseline system of continuous high-performance, intermediate-modulus polyacrylonitrile-based graphite-designated MAGNAMITE® TYPE IM6 with systematic variations in their surface chemistry has been identified for this purpose. The IM6-U fibers are 'as received', that is, as removed from the heat treatment ovens without any further processing. The IM6-100% and IM6-600% are surface-treated with a commercial electrochemical oxidation step to 100% and 600% of their commercial levels, which optimizes the adhesion to epoxy matrices. A di-functional epoxy, diglycidyl ether of bisphenol-A (DGEBA) (Epon 828, Shell Chemical Company) with 14.5 wt% of *m*-phenylenediamine (*m*PDA) curing agent (Aldrich Chemical) was selected as the baseline matrix. Single-fiber microbond and fragmentation specimens as well as ~50% volume fraction unidirectional composites were fabricated according to accepted sample preparation procedures.

5.1. Comparison with experimental results

In order to provide an objective comparison between the three single-fiber techniques values of measured interfacial shear strength for IM6-U, IM6-100% and IM6-600% carbon fiber–Epon 828 *m*PDA epoxy resin systems are summarized in Table 2.

Values of interfacial shear strength measured with the single-fiber fragmentation technique was used as a reference to compare the results from the single-fiber techniques.

The single-fiber fragmentation technique appears to be very sensitive to interphase conditions as reflected by the aspect ratios (critical fiber fragment length

Table 2

Summary of interfacial shear strength values

System	Method			
	Single fiber fragmentation technique		Microindentation technique	Microbond technique
	Ksi/MPa	(l_c/d)	Ksi/MPa	Ksi/MPa
IM6-U Epon 828 *m*PDA	3180/21.9	99	3870/26.6	2160/14.9
IM6-100% Epon 828 *m*PDA	5740/39.6	67	5400/37.2	2750/19.0
IM6-600% Epon 828 *m*PDA	6870/47.4	52	6200/42.7	2180/15.0

divided by fiber diameter). The microindentation technique seems to agree well with the results obtained from the single-fiber fragmentation method. An interfacial shear strength equal to $\tau_{av} = 3870 \pm 750$ psi (26.6 ± 5.2 MPa) was obtained for IM6-U fibers using the ITS. This value is slightly higher than the one obtained from the single-fiber fragmentation technique. For IM6-100%, $\tau_{av} = 5400 \pm 1540$ psi (37.2 ± 11.0 MPa) and for IM6-600%, $\tau_{av} = 6200 \pm 1640$ psi (42.7 ± 11.3 MPa) were obtained also using the ITS. The last two values of interfacial shear strength are slightly lower than the values obtained from the single-fiber fragmentation technique; however, higher standard deviations were obtained for the IM6-100% and IM6-600% fibers than for IM6-U fibers. It was observed from the photoelastic characterization of the interphase that there is a marked difference in failure mode due to differences in fiber surface chemistry. When probing the interphase with the ITS, a failure criterion has to be defined in terms of load drop during testing. This means that, as soon as the force applied to the fiber tip drops a specified amount, the test will stop and the interfacial shear strength calculated. This criterion could be more appropriate for fibers which exhibit a low adhesion level like the IM6-U fibers where an interfacial crack propagates rapidly at low load levels. When the adhesion level between fiber and matrix is increased, the failure criterion should be changed accordingly since load transfer at the interface is governed by a stick–slip mechanism, and the interfacial crack propagates in shorter intervals as in the case of IM6-100% fibers or through plastic deformation of the matrix as for IM6-600% fibers. In some cases, fiber splitting or crushing was observed for fibers that exhibited high adhesion to the matrix. Fiber damage was observed more frequently with IM6-100% and IM6-600%, whereas damage to both matrix and fiber occurred more frequently on IM6-U fibers.

Comparing the single-fiber fragmentation and the microbond techniques, it can be seen that the second yields lower results. Also, it is observed that for IM6-100% and IM6-600% fibers, the interfacial shear strength measured with the microbond technique is considerably lower than the value obtained from the single-fiber fragmentation and the microindentation techniques. Several reasons can be mentioned to explain this difference. First, because of high surface-to-volume ratios and the evaporation of curing agent during microdroplet formation, bulk properties of the microdroplets can be significantly different than the bulk matrix produced in the single-fiber technique matrix. It has been indicated by Rao et al. [76] that significant diffusion of the curing agent out of the droplet can occur due to the high ratio of vapor pressure of the curing agent to the surface tension of the drop. This effect is particularly noticeable for the 100% and 600% surface treatments where the level of adhesion is higher and the critical length to achieve debonding is shorter than the critical length of untreated IM6-U fibers. In order to achieve full cure, a modified curing cycle needs to be used, which will also yield slightly lower bulk mechanical properties of the epoxy resin because of the lower stoichiometric content of the curing agent. Other studies have not reported this problem in their work with carbon fiber–epoxy matrix systems, and aramid–epoxy matrix systems [7,90], but different curing agents were used in these studies. Also, it has been shown [36–41,113] that the mechanical events that occur in the pull-off technique are different to those occurring in the single-fiber technique. In the microbond technique, the total fracture energy is contributed by the strain energy stored in the free and embedded fiber length, as well as in the matrix immediately surrounding the fiber. In the single-fiber fragmentation method, the total energy is contributed by the embedded fiber and the surrounding matrix. The free fiber length should be made as short as possible during the experiment in order to obtain accurate results. The small size of the droplet may also influence the interfacial shear stress due to local variation in the adhesion properties along the fiber [76].

6. Summary

6.1. Microbond–pull-out technique

The major advantages of the pull-out and microbond techniques can be summarized as follows: (1) the value of force at the moment of debonding can be measured; (2) these techniques can be used for almost any fiber–matrix combination.

On the other hand, there are serious inherent limitations to the pull-out technique. (1) Because the debonding force is a function of the embedded length, when using very fine reinforcing fibers with diameters ranging from 5 to 50 μm (0.2 to 2 mils), the maximum embedded length is of the range of 0.05 to 1.0 mm (0.002 to 0.04 in.). Longer embedded lengths cause fiber fracture. It is extremely difficult to keep the length to such short values and both the preparation of square fiber ends and the handling of the specimens is very difficult. (2) The meniscus that is formed on the fiber by the resin makes the measurement of the embedded length very inaccurate. (3) For microdrop specimens, the small size makes the failure process difficult to observe. (4) Most important, the state of stress in the droplet can vary both with size and with variations in the location of points of contact between the blades and the microdrop. (5) Furthermore, it has also been shown by Rao et al. [50] that the mechanical properties of the microdrop vary with size because of variations of concentration of the curing agent. (6) For a given fiber–matrix combination, a relatively large scatter in the test data is obtained from the microdrop or the pull-out tests. Such wide distributions of shear strengths have been attributed mainly to testing parameters such as position of the microdrop in the loading fixture, droplet gripping, faulty measurement of fiber diameters, and so on [10]. In addition, variations in the chemical, physical, or morphological nature of the fiber along its length will affect the results of interfacial shear strength measurements, which only consider very small sections [10,56,59].

6.2. Single-fiber fragmentation technique

The advantages can be summarized as follows: (1) this technique has the advantage of yielding a large amount of information for statistical sampling; (2) as mentioned before, in the case of transparent matrices, the failure process can be observed under polarized light; and (3) this technique replicates the events in-situ in the composite. The shortcomings can be summarized as follows: (1) the matrix must have a strain limit at least three times greater than the fiber; (2) the matrix should have sufficient toughness to avoid fiber fracture induced failure; (3) the fiber strength should be known at the critical length; and (4) despite of the sophisticated statistical techniques used to characterize the fiber fragment length distribution, the shear strength is calculated using an oversimplified representation of the mechanical events occurring at the interphase.

6.3. Microindentation–microdebonding technique

The advantages are: (1) allows in-situ measurement of force debonding; (2) allows probing of the interface in the 'real' environment; (3) it yields multiple data points; and (4) data collection is fast and automated. The disadvantages are: (1) the failure mode or locus of failure can not be observed; (2) there exists the possibility of inducing artifacts by the surface preparation procedure; (3) the assumptions made to calculate the interfacial shear stress may not be valid; (4) crushing of fibers is observed very frequently, limiting the variety of fibers to be tested.

It is clear that each of these methods for measuring fiber–matrix adhesion in composite materials requires some assumptions to be made about the material properties in the interphase and also requires that the system studied conform to the boundary conditions established for the analysis of the results. None of these techniques offers a complete and unambiguous method for measuring the interfacial shear strength between fiber and matrix. However, each method has proven to be sensitive to slight changes in adhesion in a given fiber–matrix system. For proper interpretation of test results, the model methods can provide a clear idea of the changes in the level of adhesion and failure mode between fiber and matrix resulting from surface chemical, fiber coating, or composite processing changes.

References

1. Drzal, L.T., Rich, M.J. and Lloyd, P.F., Adhesion of graphite fibers to epoxy matrices, I. The role of fiber surface treatment. *J. Adhes.*, **16**, 1–30 (1982).
2. Drzal, L.T., Rich, M.J., Koenig, M. and Lloyd, P.F., Adhesion of graphite fibers to epoxy matrices, II. The effect of fiber finish. *J. Adhes.*, **16**, 133–152 (1983).

3. Madhukar, M.S. and Drzal, L.T., Fiber–matrix adhesion and its effect on composite mechanical properties, I. Inplane and interlaminar shear behavior of graphite/epoxy composites. *J. Compos. Mater.*, **25**, 932–957 (1991).
4. Madhukar, M.S. and Drzal, L.T., Fiber–matrix adhesion and its effect on composite mechanical properties, II. Longitudinal (0°) and transverse (90°) tensile and flexure behavior of graphite/epoxy composites. *J. Compos. Mater.*, **25**, 958–991 (1991).
5. Madhukar, M.S. and Drzal, L.T., Fiber–matrix adhesion and its effect on composite mechanical properties, III. Longitudinal (0°) compressive properties of graphite/epoxy composites. *J. Compos. Mater.*, **26**, 310–333 (1992).
6. Madhukar, M.S. and Drzal, L.T., Fiber–matrix adhesion and its effect on composite mechanical properties, IV. Mode I and Mode II fracture toughness of graphite/epoxy composites. *J. Compos. Mater.*, **26**, 936–968 (1992).
7. Gorbatkina, Y., *Adhesive Strength of Fibre–Polymer Systems*. Ellis Norwood, New York, NY, 1992.
8. Drzal, L.T., Composite interphase characterization. *SAMPE J.*, **19**, 7–13 (1983).
9. Broutman, L.J., Measurement of the fiber–polymer matrix interfacial strength. In: *Interfaces in Composites*, American Society for Testing and Materials, Philadelphia, PA, 1969, STP 452, pp. 27–41.
10. Miller, B., Muri, P. and Rebenfeld, L., A microbond method for determination of the shear strength of a fiber/resin interface. *Compos. Sci. Technol.*, **28**, 17–32 (1987).
11. Kelly, A. and Tyson, W.R., Tensile properties of fiber-reinforced metals: copper/tungsten and copper/molybdenum. *J. Mech. Phys. Solids*, **13**, 329–350 (1965).
12. Drzal, L.T., Rich, M.J., Camping, J.D. and Park, W.J., Interfacial shear strength and failure mechanisms in graphite fiber composites. In: *35th Annual Technical Conference*, Reinforced Plastics/Composites Institute, The Society of the Plastics Industry, 1980, Paper 20-C.
13. Outwater, J.O. and Murphy, M.C., The influences of environment and glass finishes on the fracture energy of glass–epoxy joints. In: *Proceedings of the 24th Annual Technical Conference*, The Society of the Plastics Industry, 1969, Paper 16-D.
14. Mandell, J.F., Chen, J.-H. and McGarry, F.J., *A microdebonding test for in-situ fiber–matrix bond and moisture effects*. Department of Materials Science and Engineering, Massachusetts Institute of Technology, Cambridge, MA, Feb. 1980, Research Report R80-1.
15. Brandon, D.G. and Fuller Jr., E.R., *Ceram. Eng. Soc. Proc.*, **10**, 871 (1989).
16. Shafry, N., Brandon, D.G. and Terasaki, M., *Euro-Ceramics*, **3**, 453–457 (1989).
17. Narkis, M., Chen, E.J.H. and Pipes, R.B., Review of methods for characterization of interfacial fiber–matrix interactions. *Polym. Compos.*, **9**, 245–251 (1988).
18. Carman, G.P., Lesko, J.J., Reifsnider, K.L. and Dillard, D.A., *J. Compos. Mater.*, **27**, 303–329 (1993).
19. Gopal, P., Dharani, L.R., Subramanian, N. and Blum, F.D., *J. Mater. Sci.*, **29**, 1185–1190 (1994).
20. Ko, Y.S., Forsman, W.C. and Dziemianowicz, T.S., Carbon fiber-reinforced composites: effect of fiber surface on polymer properties. *Polym. Eng. Sci.*, **22**, 805–814 (1982).
21. Chua, P.S. and Piggott, M.R., The glass fibre–polymer interface. IV. Controlled shrinkage polymers. *Compos. Sci. Technol.*, **22**, 245–258 (1985).
22. Chua, P.S., Characterization of the interfacial adhesion using tan delta. *SAMPE Q.*, **18**, 10–15 (1987).
23. Perret, P., Gerard, J.F. and Chabert, B., A new method to study the fiber–matrix interface

in unidirectional composites: application for carbon fiber–epoxy composites. *Polym. Test.*, **7**, 405–418 (1987).

24. Yuhas, D.E., Dolgin, B.P., Vorres, C.L., Nguyen, H. and Schriver, A., Ultrasonic methods for characterization of interfacial adhesion in spectra composites. In: Ishida, H. (Ed.), *Interfaces in Polymer, Ceramic and Metal Matrix Composites*. Elsevier, Amsterdam, 1988, pp. 595–609.

25. Wu, W.L., *Thermal Technique for Determining the Interface and/or Interplay Strength in Polymeric Composites*. National Institute of Standards and Technology, Polymers Division Preprint 1989.

26. Tuinstra, T. and Koening, J.L., *J. Compos. Mater.*, **4**, 492–499 (1970).

27. Kim, J.-K. and Mai, Y.-W. (Eds.), In: *Engineered Interfaces in Fiber Reinforced Composites*. Elsevier, London, 1998, pp. 21–24.

28. Miller, J.D., Harris, W.C. and Zajac, G.W., *Surf. Interface Anal.*, **20**, 977–983 (1993).

29. Petit, P.H., A simplified method of determining the inplane shear stress–strain response of unidirectional composites. American Society for Testing and Materials, Philadelphia, PA, 1969, *STP 460*, p. 63.

30. Walrath, D.E. and Adams, D.F., *Analysis of the Stress State in an Iosipescu Shear Test Specimen*. Composite Materials Research Group, Department of Mechanical Engineering, University of Wyoming, Laramie, June 1983, Department Report UWME-DR-301-102-1.

31. Whitney, J.M., Daniel, I.M. and Pipes, R.B., *Experimental Mechanics of Fiber Reinforced Composite Materials*. Society for Experimental Stress Analysis, 1982, Monograph 4. Brookfield Center, Connecticut, Prentice Hall, Englewood Cliffs, 1982.

32. Drzal, L.T. and Madhukar, M.S., *J. Mater. Sci.*, **28**, 569–610 (1993).

33. Dirand, X., Hilaire, E., Lafontaine, E., Mortaigne, B. and Nardin, M., *Composites*, **25**, 645–652 (1994).

34. Greszczuk, L.B., Theoretical studies of the mechanics of the fiber–matrix interface in composites. In: *Interfaces in Composites*. American Society for Testing and Materials, Philadelphia, PA, 1969, STP 452, pp. 42–58.

35. Lawrence, P., Some theoretical considerations of fibre pull-out from an elastic matrix. *J. Mater. Sci.*, **7**, 1–6 (1972).

36. Takaku, A. and Arridge, R.G.C., The effect of interfacial radial and shear stress on fibre pull-out in composite materials. *J. Phys. D.*, **6**, 2038–2047 (1973).

37. Gray, R.J., Analysis of the effect of embedded fibre length on the fibre debonding and pull-out from an elastic matrix. *J. Mater. Sci.*, **19**, 861–870 (1984).

38. Laws, V., Micromechanical aspects of the fibre–cement bond. *Composites*, April, 145–151 (1982).

39. Banbaji, J., On a more generalized theory of the pull-out test from an elastic matrix. Part I. Theoretical considerations. *Compos. Sci. Technol.*, **32**, 183–193 (1988).

40. Banbaji, J., On a more generalized theory of the pull-out test from an elastic matrix. Part II. Application to polypropylene–cement system. *Compos. Sci. Technol.*, **32**, 195–207 (1988).

41. Hampe, A., *Grundlegende Untersuchungen an Faserverstarkten Polymeren*. BAM, Federal Institute for Materials Research and Testing, Berlin, Federal Republic of Germany, August 13, 1987.

42. Chua, P.S. and Piggott, M.R., The glass fibre–polymer interface. I. Theoretical considerations for single fibre pull-out tests. *Compos. Sci. Technol.*, **22**, 33–42 (1985).

43. Chua, P.S. and Piggott, M.R., The glass fibre–polymer interface. II. Work of fracture and shear stresses. *Compos. Sci. Technol.*, **22**, 107–119 (1985).

44. Chua, P.S. and Piggott, M.R., The glass fibre–polymer interface. III. Pressure and coefficient of friction. *Compos. Sci. Technol.*, **22**, 185–196 (1985).

45. Chua, P.S. and Piggott, M.R., The glass fibre–polymer interface. IV. Controlled shrinkage polymers. *Compos. Sci. Technol.*, **22**, 245–258 (1985).

46. Bowling, J. and Groves, G.W., The debonding and pull-out of ductile wires from a brittle matrix. *J. Mater. Sci.*, **14**, 431–442 (1979).

47. Kelly, A., Interface effects and the work of fracture of a fibrous composite. *Proc. R. Soc. London A*, **319**, 95–111 (1970).

48. Piggott, M.R., Debonding and friction at fibre–polymer interfaces. I. Criteria for failure and sliding. *Compos. Sci. Technol.*, **30**, 295–306 (1987).

49. Piggott, M.R., Sanadi, A., Chua, P.S., Andison, D., Mechanical interactions in the interfacial region of fibre reinforced thermosets. In: *Composite Interfaces*. Proc. 1st Int. Conf. Composite Interface, May 1986, pp. 109–121.

50. Rao, V., Herrera-Franco, P., Ozzello, A.D. and Drzal, L.T., A direct comparison of the fragmentation test and the microbond pull-out test for determining the interfacial shear strength. *J. Adhes.*, **34**, 65–77 (1991).

51. Kim, J. K., Baillie, C. and Mai, Y. W., *J. Mater. Sci.*, **27**, 3143–3154 (1991).

52. Zhou, M., Kim, J.-K. and Mai, Y.M., *J. Mater. Sci.*, **27**, 3155–3166 (1992).

53. Hsueh, C.-H., *Mater. Sci. Eng.*, **A154**, 125–132 (1992).

54. Wang, C.-J., *Mater. Sci.*, **32**, 483–490 (1997).

55. Penn, L.S. and Lee, S.M., Interpretation of experimental results in the single pull-out filament test. *J. Compos. Technol. Res.*, **11**, 23–30 (1989).

56. Nairn, J.A., *Mech. Mater.*, **13**, 131 (1992).

57. Scheer, R.J. and Nairn, J.A., A comparison of several fracture mechanics methods for measuring the interfacial toughness with microbond tests. *J. Adhes.*, (1994).

58. Gaur, U., Besio, G. and Miller, B., Measuring fiber/matrix adhesion in thermoplastic composites. *Plast. Eng.*, Oct., 43–45 (1989).

59. Penn, L.S. and Lee, S.M., Interpretation of the force trace for Kevlar/epoxy single filament pull-out tests. *Fibre Sci. Technol.*, **17**, 91–97 (1985).

60. Gaur, U. and Miller, B., Microbond method for determination of shear strength of a fiber/resin interface: evaluation of experimental parameters. *Compos. Sci. Technol.*, **34**, 35–51 (1989).

61. McAlea, K.P. and Besio, G.J., Adhesion between polybutylene terephthalate and E-glass measured with a microdebond technique. *Polym. Compos.*, **9**, 285–290 (1988).

62. Penn, L.S., Tesoro, G.C. and Zhou, H.X., Some effects of surface-controlled reactions of Kevlar 29 on the interface in epoxy composites. *Polym. Compos.*, **9**, 184–191 (1988).

63. Ozzello, A., Grummon, D.S., Drzal, L.T., Kalantar, J., Loh, I.-H. and Moody, R.A., Interfacial shear strength of ion beam modified UHMW-PE fibers in epoxy matrix composites. *Proc. Mater. Res. Soc. Symp.*, **153**, 217–222 (1989).

64. Latour Jr., R.A., Black, J. and Miller, B., Fatigue behavior characterization of the fibre–matrix interface. *J. Mater. Sci.*, **24**, 3616–3620 (1989).

65. McAlea, K.P. and Besio, G.J., Adhesion kinetics of polybutylene to E-glass fibers. *J. Mater. Sci. Lett.*, **7**, 141–143 (1988).

66. Henstenburg, R.B. and Phoenix, S.L., Interfacial shear strength studies using the single-filament-composite test. Part II. A probability model and Monte Carlo simulation, *Polym. Compos.*, **10** (1989).

67. Bascom, W.D. and Jensen, R.M., Stress transfer in single fiber resin tensile tests. *J. Adhes.*, **19**, 219–239 (1986).

68. Drzal, L.T., Effect of graphite fiber–epoxy adhesion on composite fracture behavior. In:

Proceedings of the 2nd U.S.–Japan ASTM Conference on Composite Materials. American Society for Testing and Materials, Philadelphia, PA, 1985.

69. Drzal, L.T., Rich, M.J. and Koenig, M., Adhesion of graphite fibers to epoxy matrices. III. The effect of hygrothermal exposure. *J. Adhes.*, **18**, 49–72 (1985).

70. Wimolkiatisak, A.S. and Bell, J.P., Interfacial shear strength and failure modes of interphase-modified graphite–epoxy composites. *Polym. Compos.*, **10**, 162–172 (1989).

71. Fraser, W.A., Ancker, F.H., Dibenedetto, A.T. and Elbirli, B., Evaluation of surface treatments for fibers in composite materials. *Polym. Compos.*, **4**, 238–248 (1983).

72. Netravali, A.N., Henstenburg, R.B., Phoenix, S.L. and Schwartz, P., Interfacial shear strength studies using the single-filament composite test. I. Experiments on graphite fibers in epoxy. *Polym. Compos.*, **10**, 226–241 (1989).

73. Hui, C.Y., Shia, D. and Berglund, L.A., Estimation of interfacial shear strength: an application of a new statistical theory for the single fiber composite test. *Compos. Sci. Technol.*, **59**, 2037–2046 (1999).

74. Asloum, El M., Donnet, J.B., Guilpain, G., Nardin, M. and Schultz, J., On the determination of the tensile strength of carbon fibers at short lengths. *J. Mater. Sci.*, **24**, 3504–3510 (1989).

75. Asloum, El M., Nardin, M. and Schultz, J., Stress transfer in single-fiber composites: effect of adhesion, elastic modulus of fibre and matrix and polymer chain mobility. *J. Mater. Sci.*, **24**, 1835–1844 (1989).

76. Rao, V. and Drzal, L.T., The dependence of interfacial shear strength on matrix and interphase properties. *Polym. Compos.*, **12**, 48–56 (1991).

77. Drzal, L.T., Intrinsic material limitations in using interphase modifications to alter fiber–matrix adhesion in composites. *Mat. Res. Soc. Symp.*, Proceedings, Materials Research Society, Vol. 170, 1990.

78. Termonia, Y., Theoretical study of the stress transfer in single fibre composites. *J. Mater. Sci.*, **22**, 504–508 (1987).

79. Folkes, M.J. and Wong, W.K., Determination of interfacial shear strength in fibre-reinforced thermoplastic composites. *Polymer*, **28**, 1309–1314 (1987).

80. Galiotis, C., Melanitis, N., Tetlow, P.L. and Davies, C.K.L., Interfacial shear stress mapping in model composites. In: *Proceedings of the 5th Technical Conference*. American Society for Composites, June 12–14, 1990.

81. Verpoest, I., Desaeger, M. and Keunings, R., Critical review of direct micromechanical test methods for interfacial strength measurements in composites. In: Ishida, H. (Ed.), *Controlled Interphases in Composite Materials*. Proc. 3rd Int. Conf. Composite Interfaces, Elsevier, NY, 1990, pp. 653–666.

82. Cox, H.L., The elasticity and strength of paper and other fibrous materials. *Br. J. Appl. Phys.*, **3**, 72–79 (1952).

83. Rosen, B.W., Mechanics of composite strengthening. In: *Fiber Composite Materials*. Chapter 3. American Society for Metals, 1964, pp. 37–75.

84. Amirbayat, J. and Hearle, J.W.S., Properties of unit composites as determined by the properties of the interface. Part I. Mechanism of matrix–fibre load transfer. *Fiber Sci. Technol.*, **2**, 123–141 (1969).

85. Theocaris, P.S., *The Mesophase Concept in Composites*. Springer, New York, NY, 1987.

86. Lhotellier, F.C. and Brinson, H.F., Matrix–fiber stress transfer in composite materials: elasto-plastic model with an interphase layer. *Compos. Struct.*, **10**, 281–301 (1988).

87. Whitney, J.M. and Drzal, L.T., Axisymmetric stress distribution around an isolated fiber fragment. In: *Toughened Composites*. American Society for Testing and Materials, Philadelphia, PA, 1987, STP 937, pp. 179–196.

88. Ostrowski, J., Will, G.T. and Piggott, M.R., Poisson's stresses in fibre composites. I. Analysis. *J. Strain Anal.*, **19**, 43–49 (1984).

89. Carrara, A.S. and McGarry, F.J., Matrix and interface stresses in a discontinuous fiber composite model. *J. Compos. Mater.*, **2**, 222–243 (1968).

90. Soh, A.K., On the determination of interfacial stresses in a composite material. *Strain*, **21**, 163–172 (1985).

91. Pu, L. and Sadowski, M.A., Strain gradient effects on force transfer between embedded microflakes. *J. Compos. Mater.*, **2**, 138–151 (1968).

92. Pisanova, E.V., Zhandarov, S.F. and Dovgyalo, V.A., Interfacial adhesion and failure modes in single filament thermoplastic composites. *Polym. Compos.*, **15**, 147–155 (1994).

93. Nairn, J.A., A variational mechanics analysis of the stresses around breaks in embedded fibers. *Mech. Mater.*, **13**, 131–154 (1992).

94. Ho, H. and Drzal, L.T., *Compos. Eng.*, **5**, 1231–1244 (1995).

95. Tripathi, D., Chen, F. and Jones, F.R., The effect of matrix plasticity on the stress fields in a single filament composite and the value of interfacial shear strength obtained from the fragmentation test. *Proc. R. Soc. London A*, **452**, 621–653 (1996).

96. Nairn, J.A. and Liu, Y.C., Stress transfer into a fragmented anisotropic fiber through an imperfect interface. *Int. J. Solids Struct.*, **34**, 1255–1281 (1996).

97. Netravali, A.N., Topoleski, L.T.T., Sachse, W.H. and Phoenix, S.L., An acoustic emission technique for measuring fiber fragment length distributions in the single-fiber-composite test. *Compos. Sci. Technol.*, **35**, 13–29 (1989).

98. Sachse, W., Netravali, A.N. and Baker, A.R., An enhanced, acoustic emission-based, single-fiber-composite test. *J. Nondestruct. Eval.*, **11**(3/4) (1992).

99. Waterbury, M.C. and Drzal, L.T., On the determination of fiber strengths by in situ fiber strength testing. *J. Compos. Technol. Res.*, **13**, 22–28 (1991).

100. Ashbee, K.H.G. and Ashbee, E., Photoelastic study of epoxy resin/graphite fiber load transfer. *J. Compos. Mater.*, **22**, 602–615 (1988).

101. Mandell, J.F., Chen, J.H. and McGarry, F.J., In: *Proceedings of the 35th Annual Conference on Reinforced Plastics*. Composites Institute, Society of the Plastics Industry, 1980, Paper 26D.

102. Grande, D.H., *Microdebonding Test for Measuring Shear Strength of the Fiber/Matrix Interface in Composite Materials*. M.S. Thesis, Massachusetts Institute of Technology, Cambridge, MA, June 1983.

103. Mandell, J.F., Grande, D.H., Tsiang, T.H. and McGarry, F.J., *A modified microdebonding test for direct in-situ fiber/matrix bond strength determination in fiber composites*. Massachusetts Institute of Technology, Cambridge, MA, Dec. 1984, Research Report R84-3.

104. Grande, D.H., Mandell, J.F. and Hong, K.C.C., Fibre–matrix bond strength studies of glass, ceramic and metal matrix composites. *J. Mater. Sci.*, **23**, 311–328 (1988).

105. Netravali, A.N., Stone, D., Ruoff, S. and Topoleski, T.T.T., Continuous microindentor push through technique for measuring interfacial shear strength of fiber composites. *Compos. Sci. Technol.*, **34**, 289–303 (1989).

106. Marshall, D.B., *An indentation method for measuring matrix–fiber frictional stresses in ceramic composites*. American Ceramics Society, Dec. 1984, pp. c-259–c-260.

107. Marshall, D.B. and Evans, A.G., Failure mechanisms in ceramic fiber/ceramic–matrix composites. *J. Am. Ceram. Soc.*, **68**, 225–231 (1985).

108. Marshall, D.B. and Oliver, W.C., Measurement of interfacial mechanical properties in fiber reinforced ceramic composites. *J. Am. Ceram. Soc.*, **70**, 542–548 (1987).

109. Tse, M.K., U.S. Patent 4,662,228, 1987.

110. Ho, H. and Drzal, L.T., Composites Part A, 27A, 1996, pp. 961–971.
111. Caldwell, D.L., Babbington, D.A. and Johnson, C.F., Interfacial bond strength deter-mination in manufactured composites. In: Jones, F.R. (Ed.), *Interfacial Phenomena in Composite Materials*. Butterworths, London, 1989, pp. 44–52.
112. Caldwell, D.L. and Cortez, F.M., *Mod. Plast.*, September, 132 (1988).
113. Désarmot, G. and Favre, J.P., Advances in pull-out testing and data analysis. *Compos. Sci. Technol.*, **42**, 151–187 (1991).

Chapter 17

The durability of adhesive joints

A.J. KINLOCH [*]

Department of Mechanical Engineering, Imperial College of Science, Technology and Medicine, Exhibition Rd., London SW7 2BX, UK

1. Introduction

One of the most important requirements of an adhesive joint is the ability to retain a significant proportion of its load-bearing capability for long periods under the wide variety of environmental conditions which are likely to be encountered during its service-life. Unfortunately, for the reasons outlined later, one of the most hostile environments for joints involving high surface-energy substrates, e.g. metals, glasses and ceramics, is water and this is, of course, one of the most commonly encountered. Indeed, empirical laboratory investigations established [1–6] many years ago that water, either in the liquid or the vapour form, is the most hostile environment that is commonly encountered; and in natural outdoor climatic trials it is invariably the presence of moisture which is responsible for environmental attack rather than, say, oxygen or ozone.

Thus, in many applications, which may involve some of the most critical uses of adhesives technology, the bonded joints are exposed to an environment which also happens to be one of the most potentially damaging. Indeed, the aspect of the durability of adhesive joints to aqueous environments is undoubtedly one of the most important challenges that the adhesives community faces. In particular, there are two main challenges to develop: (a) 'adhesive systems' (i.e. combinations of an adhesive/primer/surface pretreatment/substrate type, all of which may interact to affect the joint durability) which possess excellent long-term durability, and which are 'environmentally friendly' and cost effective; and (b) test methods and models to accurately rank and predict the service-life from short-term experiments, and thereby convince the potential user that an adequate durability will be realised. Clearly, these two aims are strongly inter-linked.

———————
[*] E-mail: a.kinloch@ic.ac.uk

The present chapter will, therefore, discuss the aspects of environmental attack with special reference to the effect of moisture, especially when elevated temperatures and/or applied stresses are also involved. Further, it should be noted that the discussions will largely, although not exclusively, centre upon those environments/adhesive systems where the interfacial regions are attacked. This is not to say that the bulk adhesive itself may not be attacked by the ingressing liquid, and if this does occur then the joint may well be appreciably weakened. However, the general polymeric materials literature, including that specialising in adhesives chemistry, should enable the adhesive user to select an adhesive type which will be sufficiently durable from the 'bulk' viewpoint to withstand the service environment. (Also, of course, if the adhesive is merely plasticised by the ingressing moisture, then the mechanical properties of the joint may be essentially recovered if the water is desorbed upon the joint being dried.) The more assiduous problem, and the far more difficult to design against, is that of the environment attacking the interfacial regions; and when this occurs the mechanical properties of the joint are invariably *not* recoverable when the water is desorbed from the adhesive upon the joint being dried. Thus, it is this aspect of the environment attacking the interfacial regions on which the present discussions will concentrate.

The above comments are seen to be reinforced by observations on the failure path in joints before and after environmental attack. The locus of joint failure of adhesive joints when initially prepared is usually by cohesive fracture in the adhesive layer, or possibly in the substrate materials. However, a classic symptom of environmental attack is that, after such attack, the joints exhibit some degree of apparently interfacial failure between the substrate (or primer) and the substrate. The extent of such apparently interfacial attack increases with time of exposure to the hostile environment. In many instances environmental attack is not accompanied by gross corrosion and the substrates appear clean and in a pristine condition, whilst in other instances the substrates may be heavily corroded. However, as will be shown later, first appearances may be deceptive. For example, to determine whether the failure path is truly at the interface, or whether it is in the oxide layer, or in a boundary layer of the adhesive or primer (if present), requires the use of modern surface analytical methods: one cannot rely simply upon a visual assessment. Also, the presence of corrosion on the failed surfaces does not necessarily imply that it was a key aspect in the mechanism of environmental attack. In many instances, corrosion only occurs once the intrinsic adhesion forces at the adhesive/substrate interface, or the oxide layer itself, have failed due the ingressing liquid: the substrate surface is now exposed and a liquid electrolyte is present so that post-failure corrosion of the substrate may now result.

There are many references [5,6] that illustrate the deleterious effect that water may have upon adhesively bonded joints where the interfacial regions have suffered attack and so led to significant weakening of the joint. For example, Fig. 1 imparts [7] an appreciation of the extent of the problem, although worst cases may

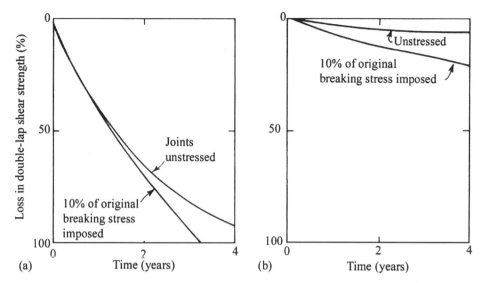

Fig. 1. Effect of outdoor weathering on the strength of epoxy-polyamide/aluminium alloy (chromic-acid etched pretreatment) structural joints [7]. (a) Hot/wet tropical site. (b) Hot/dry desert site.

be found in the literature! It is evident that the hot/wet tropical environment was by far the more hostile and that the presence of an applied load increased the rate of loss of strength. Indeed, the joints exposed under this loading at the tropical site fell apart shortly after three years had elapsed.

The present chapter will firstly discuss the various mechanisms of attack which may explain the loss of durability, such as that seen in Fig. 1. During these discussions the various methods which may be employed to extend the service-life will also be illustrated. Again the important role of the interfacial regions, i.e. the 'interphase', of the bonded joint will be evident, since most of these methods are based upon some form of pretreatment for the substrates prior to bonding. Secondly, the kinetic factors behind the mechanisms will be reviewed, and thirdly methods for ranking and predicting the service-life from short-term experiments will be considered.

The important role that continuum fracture mechanics [6,8] may play in (a) identifying the mechanisms of environmental attack, (b) ranking the durability of different adhesives, and (c) quantitatively predicting the lifetime of bonded joints and components will be emphasised. In particular, the combination of cyclic-fatigue loading and the presence of an aggressive environment is shown to be able to give a quantitative assessment of the environmental resistance of a adhesive system within a matter of weeks, as opposed to the more typical accelerated-ageing tests which involve exposing the joint, unstressed, in water for many, many months. Also, in unstressed tests, the water temperature is often selected to be relatively high, well above any likely service-temperature, in order to try to produce a large accelerated-ageing factor. This frequently leads to unrepresentative failure mechanisms being observed in these tests, and very misleading results therefore being obtained. Such elevated test temperatures are not needed in the aforementioned fracture-mechanics tests.

2. Mechanisms of environmental attack

2.1. Introduction

The interfacial regions have been highlighted as the regions where moisture may attack adhesive joints and the various mechanisms of environmental attack which have been postulated to explain the loss of durability are considered below. It should be emphasised that they should not necessarily be viewed as mutually exclusive mechanisms. Certainly there is ample evidence that no single mechanism can explain all the different examples which have been reported; and which one is operative in any situation does depend upon the exact details of the adhesive system and the service environment. Also, it is noteworthy that the exact details of the mechanism of environmental attack may be dependent upon the timescale, temperature and stress level which the bonded joint experiences, and these aspects are also discussed below.

2.2. Interface stability

2.2.1. Thermodynamic considerations

Typically when adhesive joints undergo fracture in a relatively dry environment they do not exhibit failure via a true interfacial failure between the adhesive and substrate. Instead, although visually interfacial failure may have appeared to have occurred, detailed examination of the fracture surfaces reveals the presence of a thin layer of adhesive retained on the substrate. This observation is particularly relevant to the structural adhesives where the adhesive is relatively strong and tough. However, in the presence of an aggressive environment, then true interfacial failure may indeed occur, and may often result at a very low applied load. The obvious question is why an interface, which can withstand comparatively high stresses when initially prepared, should be so unstable in the presence of liquids, such as water.

Now the intrinsic stability of any adhesive/substrate interface in the presence of a liquid environment may be assessed from the thermodynamic arguments advanced by Gledhill and Kinloch [9]. The thermodynamic work of adhesion is defined as the energy required to separate unit area of two phases forming an interface. If only secondary forces (e.g. van der Waals forces) are acting across the interface which is believed to be the main mechanism of adhesion of most adhesives, then the work of adhesion, W_A, in an inert medium may be expressed by:

$$W_A = \gamma_a + \gamma_s - \gamma_{as} \tag{1}$$

where γ_a and γ_s are the surface free energies of the adhesive and substrate, respectively, and γ_{as} is the interfacial free energy. In the presence of a liquid (denoted by the suffix 'l') this expression must be modified and the work of adhesion W_{Al} is now given by:

$$W_{Al} = \gamma_{al} + \gamma_{sl} - \gamma_{as} \tag{2}$$

where γ_{al} and γ_{sl} are the interfacial free energies between the adhesive/liquid and substrate/liquid interfaces, respectively.

For an adhesive/substrate interface the work of adhesion, W_A, in an inert atmosphere, for example dry air, usually has a positive value indicating thermodynamic stability of the interface. However, in the presence of a liquid the thermodynamic work of adhesion, W_{Al}, may well have a negative value indicating that the interface is now unstable and will dissociate. Thus, calculation of the terms W_A and W_{Al} may enable the durability of the interface to be predicted [5,6,9].

An example of the above mechanism of attack comes from recent work [10] on the durability of a typical aerospace rubber-toughened, epoxy-film adhesive bonding aluminium-alloy substrates, where the surface pretreatment which was employed for the aluminium-alloy prior to bonding was a simple grit-blasting and solvent degreasing (i.e. GBD) treatment. (This simple pretreatment is not undertaken in practice in the aerospace industry, but was used in these studies for comparative purposes. Surface pretreatments which are typically employed by the aerospace community are discussed below.) The test methodology employed was based upon using cyclic-fatigue fracture-mechanics tests which were conducted in both a 'dry' environment of $23 \pm 1°C$ and 55% relative humidity and a 'wet' environment of immersion in distilled water at $28 \pm 1°C$ using the tapered-double-cantilever-beam specimen [11,12] (see Fig. 2).

The tapered-double-cantilever-beam (TDCB) specimen [11,12] shown in Fig. 2 is a standard linear-elastic fracture-mechanics (LEFM) test specimen and the value of the adhesive fracture energy, G_c, may be deduced from:

$$G_c = \frac{P_c^2}{2B} \frac{dC}{da} \tag{3}$$

where P_c is the load for crack propagation, B is the specimen width, C is the compliance of the specimen (i.e. deflection, δ/load, P) and a is the crack length. The value of dC/da may be accurately determined via direct experimental measurements or via an analytical beam theory analysis, corrected for specimen

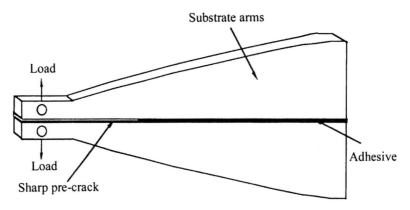

Fig. 2. Schematic of the tapered-double-cantilever-beam specimen.

geometry and crack-tip rotation effects [11,12]. In the case of cyclic-fatigue loading, the maximum strain-energy release-rate, G_{max}, applied during a fatigue cycle is given by:

$$G_{max} = \frac{P_{max}^2}{2B} \frac{dC}{da} \qquad (4)$$

where P_{max} is the maximum value of the applied load during a fatigue cycle.

The results for the relationship between the rate of fatigue crack growth per cycle, da/dN, and the maximum strain-energy release-rate, G_{max}, applied during the fatigue cycle are shown in Fig. 3, where the surface pretreatment which was employed for the aluminium-alloy prior to bonding was a simple grit-blasting and solvent-degreasing treatment. The poor durability of this adhesive system is immediately obvious from these tests. For example, the threshold value, G_{th}, of the maximum strain-energy release-rate applied in a fatigue cycle (i.e. the value of G_{max} below which no fatigue crack growth will occur) is reduced from 125 J/m^2 in the 'dry' environment to only 25 J/m^2 in the 'wet' environment. The locus of failure of the joints which had been exposed to the 'wet' environment was visually assessed as being completely interfacial, exactly along the adhesive/aluminium-oxide interface. This assessment was confirmed initially from the electron micros-copy studies, and then from using X-ray photoelectron spectroscopy (XPS). For an epoxy-adhesive/aluminium-oxide interface, as formed in the case of an oxide layer generated via grit-blasting and assuming that only secondary intermolecular forces are acting across the interface, then the thermodynamic work of adhe-sion, W_A, in a dry environment has a large positive value of about 230 mJ/m^2. This indicates that the interface will be thermodynamically stable. However, in the presence of ingressing water molecules, the thermodynamic work of adhe-sion, W_{Al}, now becomes negative in value, i.e. approximately -140 mJ/m^2. This change from a positive to a negative value of the thermodynamic work of adhesion

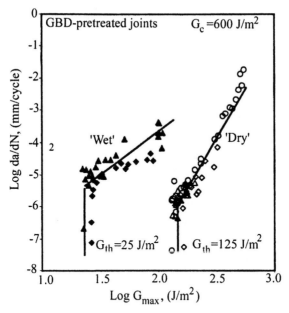

Fig. 3. Logarithmic crack growth rate per cycle, da/dN, versus logarithmic G_{max} for aluminium-alloy substrates (grit-blasted and degreased (GBD) pretreatment) bonded using an epoxy-film adhesive. Circles are for tests conducted in a 'dry' environment of 23°C and 55% relative humidity and dots for tests conducted in a 'wet' environment of distilled water at 28°C. Different styles of the symbols represent replicate tests [10].

provides a driving force for the displacement of the adhesive on the substrate by the, more polar, water molecules. Thus, the epoxy-adhesive/aluminium-oxide interface for these joints is unstable in the presence of an aqueous environment and this explains the change in the locus of failure to one of complete interfacial failure, accompanied by the dramatic decrease in the value of G_{th} observed in the 'wet' cyclic-fatigue tests.

Finally, it is noteworthy that if water (or indeed other highly polar liquids) is the environment of interest, then metallic and ceramic substrates are those which result in joints most likely to exhibit poor durability. This is a consequence, of course, of the relatively polar nature of their surfaces and their high surface free energies. Thus, ingressing water molecules are preferentially attracted to the surfaces of these substrates and will displace the physisorbed molecules of the adhesive. These comments are also reflected [5,6] in the values of W_A and W_{Al} for joints based upon carbon-fibre-reinforced plastic (CFRP) substrates typically being of the order of 90 mJ/m^2 and 30 mJ/m^2, respectively. The positive values of both of these terms indicate that the durability of adhesively bonded CFRP joints should not represent a major problem. This is indeed found to be the case, from the aspect of the stability of the interface. (Although problems may arise if (a) the adhesive or CFRP matrix suffer a loss of cohesive properties, due to plasticisation by water for example, or (b) the initially formed interface is inherently weak, due to, for example, release agent not being effectively removed prior to bonding, or excessive water, oil, etc., not being removed from the CFRP prior to an adhesively bonded repair being undertaken.)

2.2.2. Interfacial primary forces

An approximate indication of the contribution to interface stability from the additional establishment of primary chemical bonds acting across the interface may be deduced by taking the chemical bond energy to be of the order of 250 kJ/mol and assuming a coverage of 0.25 nm^2 per adsorbed site. This yields an intrinsic work of adhesion of +1650 mJ/m^2 and from energetic considerations it would be unlikely that water would readily displace such a chemisorbed layer, although it should be noted that more information is required before definitive calculations and predictions can be undertaken.

One technique for establishing interfacial primary bonds has been the use of organometallic silane-based primers. Some years ago, Gettings and Kinloch [13] investigated a range of silane-based primers deposited onto mild-steel substrates. This resulted in joints possessing different durabilities. However, the only silane primer which was found to be effective was γ-glycidoxypropyltrimethoxysilane, and this was the only primer for which there was any evidence, from static secondary ion mass spectroscopy (SIMS) and XPS, that the primer had polymerised to give a polysiloxane *and* had established primary chemical bonds, i.e. –Fe–O–Si– bonds, across the interface. Undoubtedly, both of these aspects are important if the presence of the primer is to increase the durability of the joint. If the former does not occur then the silane layer, which is invariably thicker than a monolayer, will act as a low molecular-weight, weak boundary layer. Also, if no primary chemical bonds are established, then the thermodynamic arguments predict that the joint durability would not be significantly increased. The type of primary bonds could not be precisely identified from the data. However, they are probably mainly covalent in nature but with some degree of ionic character. Similar studies have been conduced on stainless steel [14]. More recently, Abel et al. [15] and Adams et al. [16] have examined the interaction of γ-glycidoxypropyltrimethoxysilane with aluminium-alloy substrates. By using the latest techniques of high-mass resolution time-of-flight with secondary ion mass spectroscopy (ToF–SIMS) these groups have identified the presence of $-Al-O-Si^+$ fragments, which provides evidence for the formation of covalent –Al–O–Si– bonds across the interface. Again, this particular silane primer was found to give an increase in joint durability [16].

2.3. Role of the oxide

2.3.1. Introduction

The stability of the oxide layer on metallic substrates, in situ in the adhesive joint, may be considered from two different aspects: (a) electrochemical corrosion of the substrate, and (b) subtle changes occurring in the nature and properties of the oxide layer on the substrate.

2.3.2. Electrochemical corrosion of the substrate

2.3.2.1. Introduction. Electrochemical corrosion of the substrate may lead to a loss of joint strength and greatly reduce the service-life. However, one has to be wary of misinterpreting any initial observations of the failed joints: often signs of gross corrosion are seen on the fracture surfaces of joints which have suffered environmental attack, but such corrosion may have frequently resulted after interfacial failure had occurred due to one of the mechanisms given in Section 2.2. Thus, in such cases, corrosion of the substrates is a *post-failure* event, rather than a primary cause of environmental failure.

Nevertheless, there are many instances where electrochemical corrosion mechanisms may play a primary role in affecting the service performance of bonded joints. It should be noted that such mechanisms of attack involve both the presence of (a) anodic sites, where reaction with the metallic substrate occurs and electrons are generated, and (b) cathodic sites, where the electrons are consumed. The major reaction leads to the generation of hydroxyl ions, and the liquid present at these sites will become strongly basic and so possess a relatively high pH. Thus, typically an aqueous (electrolyte) layer needs to be present, since, without such an aqueous film, no electrical current can flow from the anodic to the cathodic sites. These aspects are illustrated, for example, by the schematic electrochemical corrosion mechanism for an organic coating on a steel substrate shown in Fig. 4, which is discussed in detail in Section 2.3.2.2.

Now the observed loss of joint strength may arise from a variety of electro-chemical-corrosion mechanisms:

(a) The oxides which are the corrosion products may have no cohesive strength or even dissolve [17–19]. Thus, failure of the joint may occur via fracture through the corrosion products under the bonded area. This mechanism of attack has been observed in the industrially important case of the adhesive-bonding of zinc-coated electrogalvanised steel substrates [19]. Also, such oxide dissolution may leave voids under the adhesive layer. Such voids may then act as sites for water accumulation and further attack. Alternatively, weakening of the substrate due to corrosion may lead to the joint failing via the substrate fracturing in a region far removed from the bonded area.

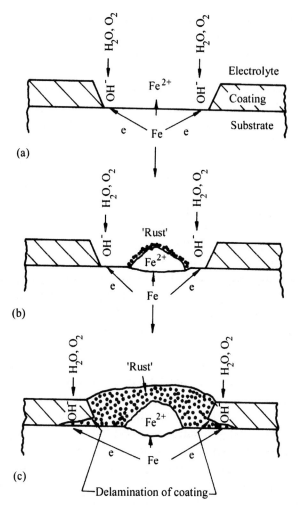

Fig. 4. General schematic mechanism of corrosion as a primary mechanism of environmental attack upon an organic coating/steel interface [33].

(b) Interfacial failure may occur. since the hydroxyl ions, which are generated as a result of the electrochemical corrosion process, may cause the very rapid hydrolysis of the intrinsic molecular bonds at the adhesive/substrate interface [20]. This mechanism is often termed 'cathodic delamination' or 'cathodic disbondment'. This mechanism of attack on the interface involves both the presence of cations, such as strongly basic sodium or potassium cations, and hydroxyl ions to give a strongly (local) basic solution.

(c) The hydroxyl ions generated during the electrochemical process may attack the polymeric coating or adhesive. This mechanism is, also, often termed 'cathodic delamination' or 'cathodic disbondment', although the locus of failure is now actually cohesive through the polymeric adhesive layer, albeit possibly very close (i.e. of the order of a few nanometres) to the interface [21,22]. However, although it is a cohesive failure through the polymer, the mechanism of attack clearly involves the nature and stability of the oxide layer on the underlying substrate, since the oxide is involved in the electrochemical corrosion process. This mechanism of electrochemical attack on the polymeric adhesive layer usually occurs at a somewhat higher pH than that needed for electrochemical attack on the adhesive/substrate interface, i.e. the mechanism described above.

(d) As always, of course, in a given adhesive system more than one of the above mechanisms of environmental attack may occur simultaneously [23].

Now, with the advent of surface-specific analytical techniques (such as XPS, ToF–SIMS, etc.) it has been demonstrated that the presence of ions from an electrolyte can give an indication of the prior electrochemical history of an electrode surface [24]. Thus, cations from the electrolyte solution (e.g. Na^+, Mg^{2+}) migrate to cathodic sites, whilst anions (e.g. Cl^-, SO_4^{2-}) migrate to anodic sites. This simple observation has been extremely useful in adhesion studies, since it has allowed the relatively subtle electrochemical mechanisms that have led to the failure of coatings [25,26] and adhesive joints [19,27–31] to be deduced. Examples to illustrate these aspects are given below.

2.3.2.2. Steel substrates. In studies of low-carbon steels coated with a polymeric layer it is well-established [32,33] that the exposed ('rusting') metal is anodic to the adjacent coated metal, at which the cathodic reduction of water and oxygen occurs to yield hydroxyl ions. A general simplified scheme [33] for the corrosion of steel under an organic adhesive or coating is shown in Fig. 4. Water and oxygen, as well as some liquid electrolyte, are required to establish an electrochemical cell at the adhesive/substrate interface. As a result of small differences between oxide structure, or contaminant level, the steel develops local anodic and cathodic areas, i.e. small differences between electrochemical potentials are established. At the anodic sites iron dissolves:

$$Fe \rightarrow Fe^{2+} + 2e \tag{5}$$

and at the cathodic sites the major reaction is:

$$4e + 2H_2O + O_2 \rightarrow 4OH^- \tag{6}$$

i.e. oxygen is reduced, although other reactions are possible under other specific conditions. If the anodic and cathodic sites are in electrical contact with each other a corrosion current will flow. Thus, the liquid collected at the interface will become strongly alkaline (i.e. possess a relatively high pH) and the Fe^{2+} ions will form $Fe(OH)_2$ by hydrolysis and by reaction with the cathodically generated hydroxyl ions; eventually corrosion products of trivalent iron are formed, e.g. FeOOH, Fe_2O_3 or Fe_3O_4 though a complex series of reactions. The coating may then fail via any, or a combination, of the mechanisms of environmental attack stated in Section 2.3.2.1.

Recently, Dickie et al. [19] studied the failure of joints, which consisted of zinc-coated electrogalvanised (EG) steel bonded using an epoxy-paste automotive adhesive, when exposed to cyclic-fatigue loading in the presence of water at 28°C. The exact sequence of events leading up to the environmentally assisted fatigue failure process was envisaged in the following manner. Firstly, the joint edges (i.e. exposed steel) were cathodically protected by the anodic dissolution of the exposed zinc coating. Secondly, at the crack tip and environs, anodic activity occurred at the adhesive/metal (i.e. zinc oxide) interface, leading to the creation of a voluminous corrosion product with little cohesive strength. (Ready access of the exposure medium to the crack tip was ensured by the cyclic nature of the loading.) Thirdly, as a result of the fatigue loading in the aqueous environment, the crack propagated within the deposit of the corrosion product ('white rust') but close to the adhesive/zinc oxide interface. Thus, subsequent surface analyses of both the 'metal side' and 'adhesive side' of the failed joint gave X-ray photoelectron spectra characteristic of the 'white rust' degradation product. The 'white rust' corrosion product was extremely fine in scale (perhaps of colloidal dimensions), and this aspect was confirmed by the atomic force microscopy (AFM) studies. Indeed, from using both AFM and scanning electron microscopy (SEM), it was readily confirmed that on both the 'adhesive side' and the 'metal side' of the failed EG steel joint there was a surface which had the distinct appearance of a metallic oxide. For example, Fig. 5 shows the images from AFM for the 'adhesive side' and the 'metal side' of a failed joint. The surface topography appears to be identical in both cases, and furthermore is identical to that of the EG steel surface prior to bonding. The needle-like columns on the surface was identified from the XPS studies as zinc oxide.

2.3.2.3. Aluminium-alloy substrates. In the case of aluminium alloys, the aerospace grades of clad aluminium alloys may present a particular problem, and the reasons for this have been considered by Reil [17]. With clad aluminium alloys the electrode potential of the clad layer, which is a thin layer of pure aluminium, is generally higher than the base alloy. This choice is deliberate in that the clad material is selected to be anodic with respect to the base alloy so that in a corrosive environment the cladding will be consumed via dissolution, thus protecting the base alloy. This idea is very effective in protecting the structure from surface corrosion such as pitting. On the clad alloy, pitting is less likely to occur, compared to the unclad alloy, due to the nature of the material, and where pits do form and penetrate the clad surface, its anodic nature will cause the pit to grow laterally once the base alloy is reached, instead of penetrating into the base alloy. However, whilst this mechanism of corrosion inhibition may be effective for exposed aluminium-alloy structures, if one considers the mechanisms whereby

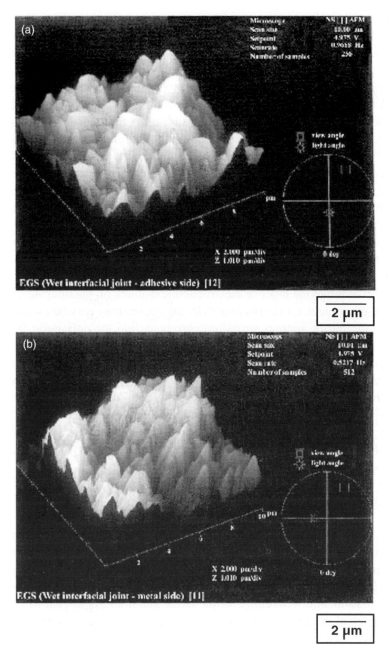

Fig. 5. Atomic force microscopy (AFM) images of the fracture surfaces of an electrogalvanised (EG) steel joint bonded using an epoxy-paste adhesive [19]. (a) 'Adhesive' side. (b) 'Metal' side.

clad aluminium alloy achieves its corrosion resistance, then the presence of the sacrificial clad layer is actually undesirable in the context of adhesive bonding; i.e. once the clad layer has disappeared the joint strength will not be very high!

2.3.2.4. Role of the chemistry of the adhesive/primer. The role of the stability of the polymeric layer (whether a coating, an adhesive or a primer) has already been highlighted with respect to the possibility of the hydroxyl ions, generated as a result of the electrochemical-corrosion mechanism, possibly attacking the polymer and causing a cohesive failure through the polymer layer, albeit very close to the interface. However, the nature of the polymeric layer may also lead to certain constituents being preferentially adsorbed onto the oxide layer, for example the curing agent for a thermosetting primer or adhesive. This may render the polymeric layer more susceptible to attack by the hydroxyl ions, due to a decrease in the local crosslink density of the polymer. Also, any unreacted curing agent may be dissolved in water (trapped, for example, in interfacial voids) and dramatically change the local pH, and so increase the rate of electrochemical dissolution of the oxide layer or electrochemical attack on the interface.

On the other hand, on the positive side, the opportunity may be taken to include corrosion inhibitors, such as strontium chromate, into the polymeric adhesive, or primer, layer. Such inhibitors will slowly leach out and are especially selected to retard the rate of the electrochemical-corrosion mechanism. They typically achieve this by increasing the polarisation of the anodic sites by reaction with the ions of the corroding metallic substrate to produce (a) thin passive films, or (b) salt layers of limited solubility which coat the anodic sites [34].

Thus, again, we return to the idea of needing to consider the 'adhesive system' when (a) formulating and designing for good joint durability, and (b) predicting the service-life of bonded components and structures.

2.3.3. Oxide stability

Another aspect of oxide stability, and often a more assiduous problem, is the observation that very subtle changes may occur in the nature and stability of the oxide, in situ in the adhesive joint, which mechanically weaken the oxide and lead to premature failure through the oxide layer. This is thought to arise from a subtle hydration, and weakening, of the oxide and does not appear to involve any electrochemical-corrosion mechanism. (Although such corrosion may well occur, post-failure, once fracture through the oxide has resulted in fracture of the joint.)

This mechanism has been clearly identified in the case of the durability of aluminium-alloy joints. One example [10] is that of aluminium-alloy joints bonded using an aerospace epoxy-film adhesive where the aluminium-alloy substrate was subjected to a phosphoric-acid anodising (PAA) surface treatment, but where the primer (which is normally used in such an adhesive system) was omitted. Under

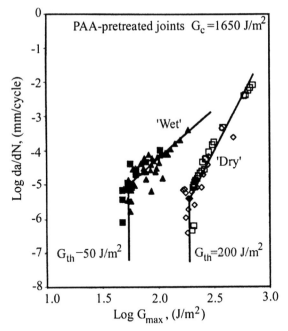

Fig. 6. Logarithmic crack growth rate per cycle, da/dN, versus logarithmic G_{max} for aluminium-alloy substrates (phosphoric-acid anodised (PAA) pretreatment) bonded using an epoxy-film adhesive. Cicles are for tests conducted in a 'dry' environment of 23°C and 55% relative humidity and dots for tests conducted in a 'wet' environment of distilled water at 28°C. Different styles of the symbols represent replicate tests [10].

both monotonic and 'dry' cyclic-fatigue loading the PAA-pretreated joints failed via cohesive crack growth through the adhesive layer, accompanied by a relatively high value of the adhesive fracture energy, G_c, and fatigue threshold value, G_{th}, respectively. However, in the 'wet' cyclic-fatigue tests, the PAA-pretreated joints visually failed by crack growth along the adhesive/oxide interface, accompanied by a relatively high low value of G_{th} (see Fig. 6). The results of the XPS studies from the failure surfaces of the 'wet' cyclic-fatigue tests are shown in Fig. 7, together with the reference spectra of the PAA-pretreated aluminium alloy and adhesive for comparison purposes. Of particular note in these spectra is the aluminium 2p peak at approximately 74 eV in Fig. 7b–d. As may be readily seen, the 'oxide' side of the failed joint (see Fig. 7d) has a very similar spectrum to that of the PAA-oxide surface prior to bonding (see Fig. 7b). This is not unexpected, of course. Secondly, however, the 'adhesive' side of the failed surface (see Fig. 7c) also has a spectrum very similar to that of the PAA oxide surface prior to bonding. Indeed, the spectrum of the 'adhesive' side is not at all similar to that of the reference, 'control' adhesive spectrum (see Fig. 7a). Thus, in the PAA-pretreated joints, fatigue crack growth in the 'wet' environment has occurred within the oxide

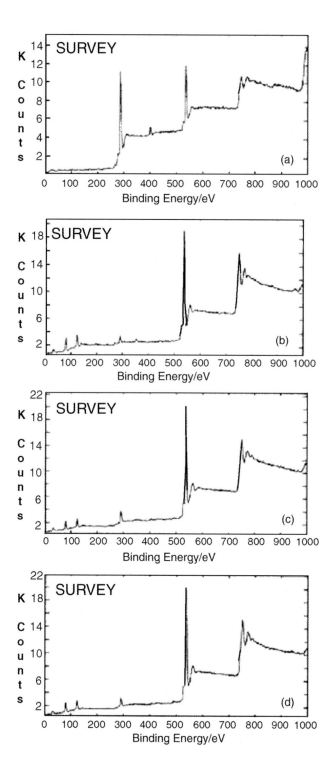

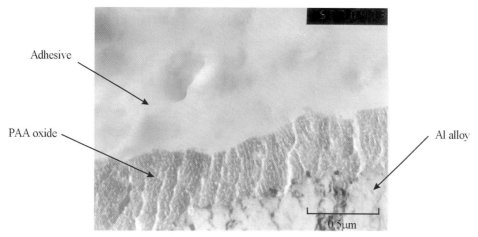

Fig. 8. Transmission electron micrograph of a cross-section of the interphase region of a PAA-pretreated joint from a 'dry' cyclic-fatigue test [10].

layer itself, and not along the adhesive/oxide interface as supposed from a simple visual examination.

To investigate this conclusion further, transmission electron micrographs of cross-sections of the interphase region, cut normal to the fracture plane of the joint as described above, were examined. A cross-section from a joint fatigue tested in the 'dry' environment is shown in Fig. 8. As may be seen, the thickness of the PAA-generated oxide is about 400 to 600 nm and the oxide was found, using micro-area electron-diffraction techniques, to be amorphous in nature. Both of these results are in accord with previous work [6,35,36]. Previous studies [6,35–38] have also shown that the oxide (a) has a hexagonal-type porous microstructure, with pores a few tens of nanometres in width which run almost the full thickness of the oxide, and (b) contains a small concentration of phosphates, present as PO_4^{3-} ions, in its surface regions which may inhibit water attack on the PAA-generated oxide. However, from Fig. 8, it is not really possible to observe the porous microstructure of the PAA-generated oxide, nor to ascertain whether the adhesive has penetrated the oxide to form a 'micro-composite' interphase between the adhesive and the underlying aluminium alloy. Therefore, energy-filtered transmission electron microscopy (EFTEM) was employed to determine the extent of penetration of the adhesive into the pores of the oxide. The resulting micrograph is shown in Fig. 9. In this figure the porous microstructure may be

Fig. 7. X-ray photoelectron spectra with respect to the PAA-pretreated joints [10]. (a) 'Control' adhesive. (b) 'As-prepared' PAA-generated oxide. (c) 'Adhesive' side of failure surface from TDCB 'wet' fatigue test. (d) 'Oxide' side of failure surface from TDCB 'wet' fatigue test.

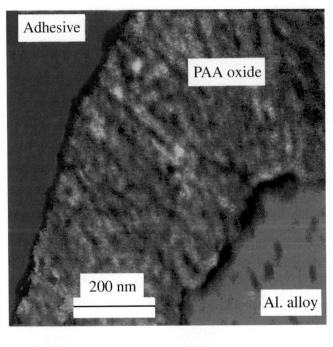

Fig. 9. Micrograph from energy-filtered transmission electron microscopy of a cross-section of the interphase region of a PAA-pretreated joint [10].

readily discerned and clearly there is no strong carbon signal from within the oxide layer. Thus, no penetration, or only very limited penetration, of the oxide by the adhesive has occurred.

From the above results, three reasons are suggested for the locus of failure being in the oxide layer for the PAA-treated joints when subjected to 'wet' cyclic-fatigue loading, and why this is accompanied by a correspondingly low value of G_{th}. Firstly, it has been reported [39] that the thermodynamics for water to displace an adhesive from a phosphated-oxide layer are far less favourable than for a GBD-pretreated oxide. Secondly, there are unfilled pores, i.e. voids, in the oxide layer of the PAA-pretreated joints, which will make the oxide layer intrinsically weak. Thirdly, and most importantly, such unfilled pores will allow the relatively rapid ingress of water into the interphase and permit a relatively high concentration of water molecules to develop in the pores. Thus, in the PAA-pretreated joints both (a) the kinetics of water penetration and (b) the presence of sites where relatively high concentrations of water molecules may accumulate are favourable to environmental attack on the interphase being initiated by the ingressing water.

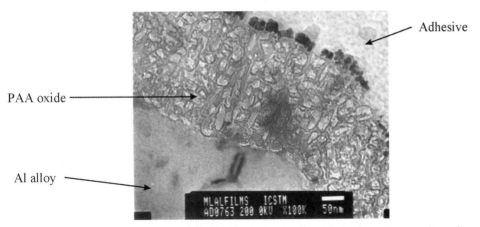

Fig. 10. Transmission electron micrograph of a cross-section of the interphase region of a PAA-pretreated joint from a 'wet' cyclic-fatigue test [10].

Indeed, the effects of attack by water on the PAA-generated oxide in these joints may be observed from the transmission electron micrograph of a cross-section from a PAA-pretreated joint shown in Fig. 10. Compared to Fig. 8, the oxide layer in Fig. 10 after 'wet' cyclic-fatigue testing has changed in appearance and possesses a somewhat coarser microstructure. Also, and most noticeably, a region, adjacent to the adhesive layer, of the oxide of about 20 nm in thickness is denser than the rest of the oxide, and this region is virtually continuous across the oxide layer. Using micro-area electron diffraction, this denser region was found to be still amorphous in nature and, using X-ray dispersive analysis, only the elements aluminium and oxygen were found to be present. An interesting aspect from this latter observation is the fact that the elements sodium and chlorine were not present, and their absence was also confirmed from the XPS studies of the failure surfaces of the joints. This interest arises since the presence of sodium and chlorine are typical indicators of a corrosion-driven mechanism, as noted above. Thus, it seems unlikely that any electrochemical-corrosion mechanism is involved in the weakening of the oxide layer in these studies on aluminium-alloy joints. Hence, from these results, it is suggested that these regions, about 20 nm in thickness, observed in Fig. 10 represent regions where hydration of the original oxide layer has occurred, which arises from water accumulating in the unfilled oxide pores. This conclusion is in agreement with the results from a recent study [40] which employed electrochemical impedance spectroscopy to detect hydration, albeit on a larger scale than in the present work, on an aluminium substrate under an unconstrained epoxy coating. Hydrated forms of aluminium oxide are known [36] to be relatively weak and, thus, the formation of an hydrated oxide layer, between the underlying substrate and the adhesive layer, would explain the observed failure path being through this weakened-oxide layer, accompanied by a relatively low value of G_{th} of $50 \, \text{J/m}^2$. It should be noted that these proposed hydration products which are formed in situ in the joint are comparatively subtle in nature. For example, from comparing Figs. 8 and 10, it may be seen that there is no evidence of any significant thickening of the oxide layer due to exposure of the joint to the 'wet' environment, as has been observed [36,38] on bare (i.e. unbonded) oxides and has been hypothetically proposed [36] to arise in situ in an adhesive joint upon exposure to an aqueous environment. This observation also raises a major question mark over the development of non-destructive tests for detecting such a mechanism of environmental attack. Since, whilst relatively thick oxide growth under an adhesive layer may be readily detected [41], using ultrasonics for example, the subtle type of oxide degradation shown in Fig. 10 would be impossible to readily detect using current non-destructive test techniques.

In the above study, the reason why the oxide is susceptible to such attack by ingressing moisture appears to be due to the inability of the high-viscosity adhesive to penetrate into the relatively deep porous microstructure of the PAA-generated oxide. This leads to unfilled pores which act as sites for water molecules to aggregate, which then attack and weaken the oxide. Indeed, if a low-viscosity primer is applied prior to bonding (i.e. joints pretreated with phosphoric-acid anodised and primed (PAAP)), as is usually undertaken in practice, then complete penetration of the oxide layer now occurs, as may be seen from Fig. 11. Thus, water cannot now accumulate in unfilled pores and no hydration of the oxide is observed. Further, a 'micro-composite' interphase is formed between the underlying aluminium-alloy substrate and the adhesive layer. This results in a greatly increased surface area for interfacial bonding, compared to a planar interface. Also, the formation of the 'micro-composite' interphase will tend to reduce the local stress-concentrations, since this region will possess an intermediate modulus between that of the relatively low-modulus polymeric-adhesive and high-modulus aluminium-alloy substrate. Hence, an interphase with a graded stiffness, with respect to the various layers, has now been created. From the aspect of possible surface chemistry effects, the primer does contain phenolic- and silane-based additives which may increase the adhesion to the oxide surface (via the formation of relatively strong chemical bonds or acid–base interactions [15,16,42]) and corrosion inhibitors (which will retard the hydration of the oxide layer, see above). Unfortunately, it is impossible to really separate out which of these various factors are the most important. However, it is clear that, for the PAAP-pretreated joints, these various factors lead to a value of G_{th} being ascertained from the 'wet' cyclic-fatigue tests which is very similar to that obtained from the 'dry' fatigue tests, and so the joints possess excellent durability; since, this is the type of adhesive system which is the basis for many aerospace adhesively bonded structures, this is clearly good news!

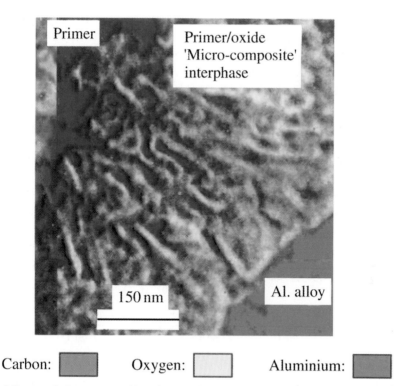

Carbon: ▨   Oxygen: ▨   Aluminium: ▨

Fig. 11. Micrograph from energy-filtered transmission electron microscopy of a cross-section of the interphase region of a phosphoric-acid anodised and primed (PAAP) pretreated joint [10].

3. Kinetics of environmental attack

3.1. Introduction

The kinetics of environmental attack will be governed by the rate-determining step in the overall mechanism of failure and very little information exists on such details. Indeed, the lack of such information has been a severe handicap in developing life prediction models, where clearly a knowledge of the kinetics is a crucial aspect which is needed in many types of time-dependent models.

3.2. Role of diffusion of water

Gledhill and co-workers [9,43] have studied the kinetics of environmental failure for epoxy/steel joints where the mechanism of attack had been found to be via the displacement of adhesive on the oxide surface by the ingressing moisture, as described in Section 2.2.1. The rate of dissociation of the interface is shown as a function of the water immersion temperature in the form of an Arrhenius

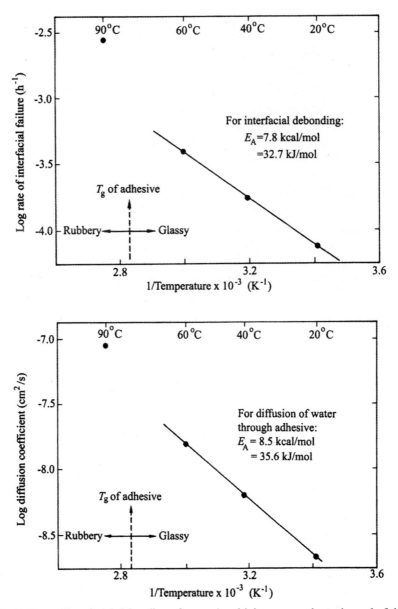

Fig. 12. (a) Rate of interfacial debonding of epoxy/steel joints versus the reciprocal of the water immersion temperature [9,43]. (b) Diffusion coefficient for water through the adhesive layer versus the reciprocal of the water immersion temperature [9,43].

plot in Fig. 12a and several interesting features are evident. Firstly, for the three lowest temperatures a linear correlation exists which yields an activation energy for the displacement of adhesive by water of 32 kJ/mol. Secondly, the rate of interfacial debonding at an immersion temperature of 90°C is far greater than would be expected from this activation energy. It was suggested that the main cause for this is that the glass transition temperature, T_g, of the adhesive is 85°C, and at temperatures above this, segmental motions of the polymer chains greatly increase. This permits diffusing species to penetrate far more readily than would be predicted from experiments conducted below T_g. The independently measured coefficients for water diffusion through the bulk adhesive are also shown as a function of temperature in the form of an Arrhenius plot in Fig. 12b. These data yield an activation energy of 36 kJ/mol, with the rate of diffusion at 90°C being greater than expected from the lower temperature results. These data and values of activation energies clearly suggest that the rate of interface debonding in these joints is controlled by the availability of water at the interface, which in turn is governed by the diffusion of water through the adhesive.

The above studies showed that, for this particular adhesive system, the rate of diffusion of water through the adhesive to the interface was the rate controlling step. Now, if the diffusion of water through the adhesive is Fickian in nature, then the concentration profile of water as a function of time into the joint may be calculated [6,43,44], and such information may then be readily used in life-prediction models, as discussed below. It should be noted that the values of the diffusion coefficient given in Fig. 12b, which are very typical for structural epoxy adhesives, lead to the conclusion that, at ambient temperatures, it will take at least a year or more for the adhesive layer in a joint, say about 20 mm × 10 mm in size (as often used in single-overlap shear joints), to reach its uniform, equilibrium concentration of water, although of course, depending upon the details of the 'adhesive system', even complete failure of the joint due to environmental attack may have occurred well before this time is reached.

However, other workers [45–47] have shown that the situation is not always so simple, and that for some adhesive systems the diffusion of water along the adhesive/substrate *interface* is far quicker than that through the bulk adhesive. Indeed, using an elegant Fourier-transform infrared multiple internal-reflection (FTIR–MIR) technique, Linossier et al. [47] not only demonstrated this aspect but also observed that the rate of interfacial diffusion of water was a function of the surface pretreatment used for the substrate prior to bonding.

High-frequency dielectric measurements have also been undertaken [48,49] in studies on water ingress into various adhesive systems. However, whilst yielding much valuable information and offering great potential for studying water ingress into adhesive joints, the spatial resolution is not yet available to differentiate between water diffusion through the adhesive versus that along the adhesive/substrate interface.

Finally, a further possible complication is that the diffusion and solubility coefficients of water into the adhesive, or along the interface, may be somewhat dependent upon the level and nature of the stress concentrations which act upon the adhesive layer. Also, water diffusion into the adhesive layer may well result in swelling of the layer, which in turn will change the level and nature of the stress concentrations which are acting. Further, these factors will be influenced by the ingressing water affecting the elastic and viscoelastic properties of the adhesive, as well as decreasing its glass transition temperature, T_g. Hence, as may be appreciated, the ingress of water into an adhesive joint is a very complex phenomenon and is very difficult to model accurately, with experimental validation of any model being virtually impossible.

3.3. Other kinetic factors

Notwithstanding, in other adhesive systems the situation is even more complex, since the rate of diffusion of water through the adhesive, or along the interface, is clearly not the rate controlling step. For example, when an effective silane-based primer was used in the aforementioned epoxy/steel joints, the rate of loss of joint strength was far lower than for the joints where no silane primer was used and the kinetics of attack were no longer governed by the diffusion of water through the adhesive. This is especially significant since the water would be expected to diffuse faster through the polysiloxane primer by several orders of magnitude, although the solubility coefficient of water in the polysiloxane primer would be far lower than in a typical structural adhesive. Another example of a more complex rate-controlling step is in the case of the mechanism of joint attack via hydration and weakening of the aluminium oxide layer (see Section 2.3.3). Here, Davis et al. [50] have proposed that the rate-determining step is the slow dissolution of the $AlPO_4$ layer on the outermost surface of the PAA oxide. This then exposes Al_2O_3 which rapidly hydrates to the mechanically weak boehmite form of aluminium oxide, through which joint failure now occurs.

It is noteworthy, that frequently a general correlation may be found [44] between the loss of joint strength and the uptake of water into the adhesive layer, at least until the adhesive layer is saturated. This arises from both factors initially increasing with increasing time. However, it is important not to be misled by this apparently simple and appealing correlation. It must be recognised that (a) the exact relation will depend, of course, upon other factors such as choice of pretreatment, and (b) there is no general relationship between the diffusion and the solubility coefficients of water in the adhesive layer and the loss of joint strength, although obviously any feature which will decrease the rate of water ingress, all other factors being equal, will tend to increase the service-life. Hence, the use of sealants around joint edges (which will essentially lengthen the diffusion path), designing long access paths for moisture, etc. will all be of help in ensuring durable joints. However, it must be generally accepted that water will ingress into the joint eventually, and therefore the adhesive system must be selected to possess sufficient stability in the presence of moisture.

Finally, in the case of the mechanisms which involve electrochemical corrosion of the substrate (see Section 2.3.2), it is of interest to note that environmental attack may be initiated from a great number of distinct points, rather than a single advancing front. The reasons for this observation are due to such factors as (a) regions of low ionic conductivity in the polymeric adhesive layer, (b) sites on the substrate which are relatively anodically active, for example due to local changes in the chemical composition of the substrate, (c) local voids at, or near, the interface, and (d) the relative diffusion rates for water and oxygen through the adhesive layer, etc. These factors again emphasise the importance of the kinetic aspects of environmental attack, and emphasise our current lack of understanding of such aspects.

3.4. Accelerated-ageing test methods

From the above discussions, it may be concluded that to obtain significant water ingress into a typical adhesive joint at ambient temperature, and to observe any appreciable degree of environmental attack, may take many months, if not years, of exposing the joint to the aqueous environment. Accelerated-ageing tests are therefore a very essential tool in order to accelerate the kinetics of environmental attack, and so obtain durability data in a realistic time-scale for the development and selection of adhesive systems. As noted above, the lack of current knowledge on the details of the kinetics of the mechanisms of environmental attack certainly hinders the development of such tests.

A common form of accelerated-ageing tests typically involves exposing bonded joints (e.g. single-overlap shear joints) in water, or the environment of interest (e.g. a corrosive salt-spray), at a relatively high temperature, for example, maybe 6 months exposure in boiling water, or at least water at, say, 70°C and periodically removing some of the joints from the test environment and then measuring any loss of strength of these joints. A major problem which may be encountered with such an approach, and a reason why such accelerated tests may be very misleading, is succinctly summed up by the question: "When did boiling an egg ever produce a chicken?"

Thus, it is important to ensure that accelerated-ageing test methodologies are selected which do give the same outcome (i.e. the same mechanisms of ageing) as would be seen in real life. The aim of the tests being, of course, to accelerate the mechanisms of environmental attack; not to produce mechanisms different to those seen in real life. (This problem may be particularly accentuated when the ageing temperature is above the curing temperature or the glass transition temperature, T_g, of the adhesive.) This dichotomy of trying to ensure a reasonable degree of accelerated ageing from the tests, at the same time as ensuring that the mechanism of attack is the same as that observed at the far lower service-temperature, is the 'Achilles heel' of accelerated-ageing tests. This is especially the case for those tests which simply rely upon exposing unstressed basic test geometries, such as the lap-shear and butt-joint tests, to water and then testing them after given intervals of time, since the only acceleration factor involved is that of the temperature of the aqueous environment. For example, if a relatively low water temperature, more typical of the service environment, is employed, then any adhesive system which possesses even a reasonably adequate durability may not show any statistical decrease in strength for many, many months, if not for several years, by which time the data may be of little relevance.

The above factors were behind the development of the cyclic-fatigue fracture-mechanics test described above, coupled with the fact that the data generated may also be used to undertake quantitative predictions of the service-life of adhesive joints, as discussed later. A major factor in these tests is that the cyclic loading maintains a sharp crack in the fracture-mechanics test specimen. (The stress concentration associated with this sharp crack is 'felt' by the nearby interfacial regions and thus gives rise to a stress concentration on such nearby regions, which accelerates the environmental attack mechanism.) A main reason for the choice of cyclic-fatigue loading is that, under such loading, the crack does not blunt, as is usually observed when a *constant* applied load is employed. This is important, since when the crack blunts, the local stress concentration diminishes around the crack tip, and this causes any crack (whether located in the adhesive layer or in the interfacial regions) to cease growing and the test to effectively end [51]. Thus, since a stress concentration is maintained on the interphase regions adjacent to the propagating crack tip, a relatively high water temperature is now *not* needed to accelerate the environmental attack mechanisms in such tests. Also, in the cyclic-fatigue tests, the levels of the applied load required at the threshold values of G_{th} are relatively low. Hence, such tests do not lead to unrealistic mechanisms of environmental attack being recorded, such as a cohesive failure of the adhesive (a) due to high-temperature hydrolytic degradation or (b) creep of the adhesive.

The well-known Boeing wedge-test [6] (see Fig. 13) is also a form, of course, of an adhesively bonded double-cantilever-beam test subjected to an applied load. Thus, it also may suffer from the crack blunting effect referred to above, which effectively ends the durability test and so may give very misleading interpretations concerning the durability-ranking of different adhesive systems. Further, since in the Boeing wedge-test the load is applied via a fixed displacement (i.e. the inserted wedge), the applied load may relax due to creep of the adhesive; such relaxation of the adhesive's macromolecules will be accelerated by moisture uptake in the adhesive. Therefore, adhesives which are prone to creep may often appear to possess superior durability when compared to those where no, or little, creep occurs, since, as the applied load diminishes, the stress concentration around the crack tip (which acts on the nearby interfacial regions) will also decrease. Another problem with the Boeing wedge-test is that the length of the initially inserted crack influences the applied stresses at the crack tip (and the applied-strain-energy

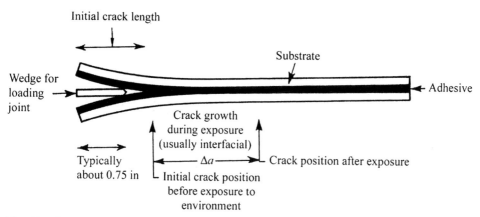

Fig. 13. The Boeing wedge-test for assessing joint durability. Typical exposure to the hot/wet environment is 1 h, or longer, at 50° or 60°C, and 100% relative humidity. The environmental crack growth, Δa, is then determined, which is a measure of joint durability.

release rate available to drive a crack) [11]. Hence, the length of the initially inserted crack must also be closely controlled, which is not always easy to do when the initial crack is typically inserted via a sharp hammer-blow to the back-edge of the wedge. The Boeing wedge-test was originally devised to act as a quality-control test for ensuring that the surface pretreatment of the substrates had been satisfactorily undertaken, and in this role it has been successful. Nevertheless, the Boeing wedge-test may be used [6,52] to undertake basic investigations and, when considerable attention is paid to several critical aspects of the test procedures noted above, it may be a useful qualitative test method. However, for it to be a *quantitative* test method, also requires that it be designed so that it follows the requirements of being a linear-elastic fracture-mechanics (i.e. a LEFM) test and this aspect is rarely checked, albeit rigorously assessed.

However, other approaches have also successfully been adopted. For example, Dickie and Ward [53] have studied single-lap joints exposed to a high humidity at moderately elevated temperatures but maintained a constant stress on the joints. Also, periodically the joints were removed from the high-humidity environment and exposed to a salt solution for a short time period. Using this accelerated-ageing test they were able to rank the durability performance of various adhesive systems in a comparatively short timescale. Further, they reported that not only were the kinetics of mechanisms of environmental attack accelerated, but also the exact details of the mechanisms were affected by the levels of the applied load. For example, for joints which consisted of bonded galvanised steel substrates, the effect of relatively high applied loads was to prevent the formation of an effective barrier of corrosion products, i.e. passivation of the substrate surface was prevented. This allowed the electrochemical corrosion process to proceed unimpeded, and hence at a faster rate than for similar, but unstressed, joints. Thus, indeed, one has to be cautious about the levels of applied load in any accelerated-ageing test. For example, as noted above, if the stress levels are too high, then the failure mechanism may well be via a creep fracture which occurs cohesively through the adhesive layer, which may not at all be the mechanism of attack of interest in the actual bonded component.

A variant on the above single-lap shear tests is where one, or more, holes are drilled through the centre of the adhesively bonded overlap in order to try to accelerate further the ingress of water, and hence the kinetics of attack [54,55]. However, such holes are not located near the end regions of the lap joint, where the maximum stress concentrations will occur and failure will initiate when the joint is loaded. Whether the drilling of such holes has any significant effect on the kinetics of attack is a much debated topic, but it certainly appears that any resulting increase in the kinetics of attack may in some cases be relatively small [56,57].

Another approach which has been adopted to developing accelerated-ageing tests is to use the adhesive as a coating for the substrate [58–61]. Here, of course, the advantage is that the environment has ready access to the interface via the 'open face' of the adhesive coating. Thus, complete saturation of the coating (i.e. the adhesive layer under study) may be attained in a few days, as opposed to the many months needed if the adhesive was bonding two substrates and the environment could only gain access via the edges of the bonded joint. However, one problem which arises is how to test these adhesive-coated substrate specimens. Some authors [58–60] have used a peel-test method and adopted a fracture-mechanics approach [62,63] to analyse the data obtained. Hence, they have ascertained values of the adhesive fracture energy, G_c, as a function of the time allowed for environmental attack to occur. On the other hand, Chang et al. [61] have again used a fracture-mechanics approach but have devised a novel specimen where the adhesive coating was notched to cause the adhesive coating to debond in the region of the interface. However, it remains to be demonstrated whether these open-face specimens do provide a sound basis for an accelerated-ageing test which correctly ranks different adhesive systems with respect to the durability recorded when they are used in their service environment. For example, in such specimens the adhesive coating may readily deform. Hence, a corrosion product on the substrate surface under the coating can readily grow and thicken, and generate relatively large tensile stresses at the advancing debonding front. This may be compared to the case when, in a joint, the adhesive layer is constrained between two substrates and no such gross corrosion mechanism can readily develop — and hence no such tensile stresses are generated [41]. Thus, in the adhesive joint, a very different mechanism of environmental attack may be observed compared to that in the analogous 'open-face' test specimen.

Nevertheless, in all such ranking tests, even those that do provide an accurate ranking order, the question which now arises is:

"How does the order of ranking so obtained (a) relate to the actual lifetime that will be seen in bonded components and structures subjected to changing environmental conditions, and (b) relate to the cost effectiveness of the adhesive system chosen, i.e. has a very conservative 'overkill' been made in the choice of the adhesive system needed to attain a given durability, to the extent that adhesive bonding is now not an economically viable option for the joining process?"

There are only basically two approaches to answering such important questions. One approach to is rely upon a very experienced team of adhesive specialists which can answer these questions based upon the rankings obtained from accelerated-ageing tests, further component tests on the 'highest-ranked' adhesive systems and, most importantly, lengthy experience on how these various tests relate to the in-service performance of a wide range of adhesive systems. This approach is obviously difficult to adopt with confidence in a new, emerging technology. (From the author's own experience, any companies which have such valuable experts should undertake cataloguing their 'folklore' knowledge before they retire!) The second approach is to devise a methodology whereby the ranking tests can be further used to give quantitative predictions of the service-life of the adhesive system when used in 'real' applications, as discussed below.

4. Life prediction models

4.1. Introduction

Clearly it would be of immense benefit to all those involved with adhesives technology if the long-term service-life of bonded joints and components could be reliably predicted from short-term models. However, to predict accurately the service-life from short-term models requires a knowledge of the mechanisms and kinetics of attack.

4.2. A 'strength of materials' approach

When the loss of strength of a joint is due to plasticisation of the adhesive (and/or the substrates, if polymeric in nature) *and* the locus of joint failure is cohesive in the adhesive, or the substrates, then the loss of strength as a function of time in a given environment may be predicted, in principle, via a classic 'strength of materials' approach. In such a case, the change of the elastic and failure properties of the materials forming the joint may be measured as a function of time in the environment (being extrapolated via a theoretical or empirical approach, as necessary) and these data may then be combined with a finite-element analysis (FEA) model of the joint. Hence, when coupled with a suitable failure criterion (such as a critical plastic shear strain or a critical strain acting at a critical distance), the strength of the joint as a function of time in the service environment may then be predicted. This approach has been adopted [64] with good results for CFRP substrates bonded using an epoxy adhesive when exposed to an accelerated test environment of water at 50°C. (The locus of joint failure was always found to be cohesive through the adhesive layer and a failure criterion of a critical stress acting over a critical distance was employed, although it was recognised that this was basically an empirical criterion.)

However, the major problems with such an approach are (a) that the bulk mechanical properties of the adhesive are difficult to measure accurately unless representative bulk specimens (having attained their equilibrium water content, which may take several years) can be prepared, and even then the measured failure properties are very prone to being inaccurate and irreproducible, (b) that FEA models struggle to cope with singularities in the stresses and/or strains around important geometric features such as adhesive fillets, and (c) most importantly, that this approach breaks down when the failure of the joint is interfacial — for example, what is the failure criterion when the locus of failure is at, or very near, the interface? For all these reasons, the present author firmly believes that a fracture-mechanics approach is by far the best approach to model the service-life. This is because, in comparison to the 'strength of materials' approach: (a) the fracture-mechanics parameters may be readily and accurately measured directly from using joints which are tested in the environment of interest; (b) singularities in the finite-element analyses are easily overcome; and (c) the fracture-mechanics parameters may be measured equally readily for any locus of failure.

4.3. A fracture-mechanics approach

4.3.1. Introduction

The initial approach was to consider relatively simple epoxy/steel joints exposed to an aqueous environment where the mechanism was identified as being due to the rupture of interfacial secondary forces, as predicted from a consideration of the thermodynamics of the system, and the kinetics were found to be governed by the diffusion of water through the adhesive, to the interface. Thus, in this adhesive system, one might hope to describe a model to predict quantitatively the service-life, and Kinloch and co-workers [6,43,65] have proposed the following model.

The first stage is the accumulation of a critical concentration of water in the interfacial regions which must be exceeded for environmental attack to occur, and the rate of attaining this critical concentration was governed by the rate of diffusion of water through the adhesive. This requirement of a critical concentration of water for significant attack to initiate appears to be frequently observed from experimental studies of attack by moisture upon many different types of adhesive joints [6,58,65–70], although its value would clearly be expected to be dependent upon the particular adhesive system being studied. Nevertheless, it is not immediately evident why such a critical concentration should exist. It may arise from the absorbed water forming clusters within the adhesive, and the careful studies conducted by Comyn and colleagues [44,71] on the diffusion of water in adhesives shows that clustering of water molecules may occur. Only the molecularly dispersed water molecules would contribute to the process of environmental attack. Alternatively, when the mechanism of environmental attack is via hydration of the oxide, then the critical water concentration may correspond to the critical level needed to form a particular hydrate, which is a well-documented phenomenon [72].

The second stage involves a loss of integrity of the interfacial regions. In the particular case of the epoxy/mild-steel joints, this arises from the rupture of interfacial secondary bonds. In other adhesive systems, however, any of the mechanisms discussed above could be envisaged.

The third stage concerns the ultimate failure of the adhesive joint. However, for the joint to fracture, or lose an appreciable amount of its original strength upon subsequent testing, it is usually not necessary for the weakening of the interfacial regions to have proceeded completely through the joint. From basic fracture-mechanics considerations only a relatively small environmental crack is required to have developed before a substantially decreased failure time, under a constant load test, or a diminished joint fracture stress is observed. Indeed, with many joint geometries subjected to an imposed load and moisture, catastrophic failure will occur when the environmental crack, which is growing by the mechanisms outlined above, attains a critical length and the applied-strain-energy release rate now attains the value of the adhesive fracture energy, G_c. The value of G_c may be determined via independent tests under monotonic loading using, for example, the tapered-double-cantilever-beam specimen (see Fig. 2) or the double-cantilever-beam specimen [11,12].

The above methodology successfully predicted the loss of joint strength as a function of the time that the epoxy/steel joints were exposed to water at various temperatures [6,43,65]. However, several problems do occur in applying this model in practice. For example, the critical concentration of water necessary for the environmental attack mechanism to initiate has to be known, or has to be fixed empirically by an initial calculation using known strength-loss data. Also, the detailed kinetics of the mechanism have to be known.

4.3.2. Predicting the cyclic-fatigue life

One important area where durability models have been successfully developed is in the prediction of the joint behaviour under the combined effects of an aggressive environment and cyclic-fatigue loads. In this approach the fracture-mechanics data under cyclic-fatigue loading are measured in the environment of interest. As noted above, a major feature of such tests is that the cyclic loading maintains a sharp crack in the fracture-mechanics test specimen, which provides a good 'acceleration factor'. Therefore, the crack does not blunt, as it usually does under a constant applied load, which thereby causes the crack to cease growing and the test to effectively end. Also, in the fatigue tests, the levels of the applied load required at the threshold values of G_{th} are relatively low and do not therefore lead to unrealistic mechanisms of environmental attack being induced.

Another basic major advantage is that the cyclic-fatigue fracture-mechanics data may be gathered in a relatively short time-period, but may be applied to other designs of bonded joints and components, whose lifetime may then be predicted over a far longer time-span. Obviously, the fracture-mechanics tests need to be conducted under similar test conditions and environments as the joints, or components, whose service-life is to be predicted. This is important since the fracture-mechanics test specimens do need to exhibit a similar mechanism and locus of failure (e.g. cohesively through the adhesive layer, or interfacially between the adhesive and substrate, or through the oxide layer on the metallic substrate, etc.) as observed in the joints, or components, whose lifetime is to be ranked and predicted.

The overall approach in developing the life-prediction model is [73,74] as follows.

(a) To derive a mathematical relationship for the rate of fatigue crack growth per cycle, da/dN, and the maximum strain-energy release-rate, G_{max}, applied during the fatigue cycle, i.e. for the experimental results such as shown in Figs. 3 and 6. The complete relationship between these parameters is typically of a sigmoidal form and may be expressed by a modified form of the Paris law, namely:

$$\frac{da}{dN} = DG_{max}^n \left[\frac{1 - (G_{th}/G_{max})^{n_1}}{1 - (G_{max}/G_c)^{n_2}} \right] \tag{7}$$

where G_{th} and G_c are the values of the cyclic-fatigue threshold strain-energy release-rate and the constant displacement-rate adhesive fracture energy, respectively. The empirical constants, D, n, n_1 and n_2 may be obtained by fitting the above expression to the experimental data given in Figs. 3 and 6, for example.

(b) Next, the variation of the maximum strain-energy release-rate, G_{max}, with the length, a, of the growing fatigue crack needs to be deduced as a function of the maximum applied load in the fatigue cycle for the joint, or component, of interest. This theoretical expression is typically derived by using analytical or FEA models of the joint or the bonded component. For example, for a single-lap joint loaded in tension which was studied, the relationship between the maximum load, T_{max} (N/m), per unit width applied in a fatigue cycle and the resulting value of G_{max} (J/m^2) for (hypothetical) cracks of length a (mm) in the lap joint was found from

FEA studies [74] to be:

$$\frac{G_{max}}{T_{max}^2} = 22a^4 + 5 \times 10^{-9} \tag{8}$$

(c) Then, the experimental relationship (Eq. 7), determined from step (a) above may be re-arranged to give:

$$N_f = \int_{a_o}^{a_f} \frac{1}{DG_{max}^n} \frac{1 - (G_{max}/G_c)^{n_2}}{1 - (G_{th}/G_{max})^{n_1}} \, da \tag{9}$$

where a_o and a_f are the initial (i.e. inherent) and final flaw sizes, respectively, and may be estimated from Griffith's equation (Eq. 9). This equation may now be combined with the theoretical expression from step (b) above (i.e. Eq. 8) in the case of the lap joint which was studied in the above example. Thus, by this substitution, we may eliminate the term G_{max} to give a theoretical expression for the number, N_f, of cycles for failure as a function of the maximum load in a fatigue cycle applied to the joint, component or structure.

(d) Finally, this substituted expression may be integrated (between the limits of the initial, Griffith, flaw size and the crack length at final failure) to give the number, N_f, of cycles needed to cause failure as a function of the maximum applied load in the fatigue cycle. Hence, in the given example, the predicted number of cycles to failure for the single-lap joints may be deduced as a function of the cyclically applied load.

The cyclic-fatigue lifetimes for the single-lap joints predicted [74] using the above approach are compared with the experimental results in Fig. 14. In this figure, the maximum load, T_{max}, per unit width applied in the fatigue cycle, is shown as a function of the number of cycles, N_f, to failure. (It should be noted that the lap joints did not contain any artificially induced initial cracks.) The overall agreement between the theoretical model and the experimental results is relatively good, bearing in mind that the fatigue life of the single-lap joints has been predicted from first principles with no empirical 'fitting factors' being employed. The modelling studies give a threshold value of T_{max} of approximately 75 kN/m, which is equivalent to about 25% of the initial failure load, or stress, of the lap joints. This predicted value of 75 kN/m may be compared with the measured value of 90 kN/m, which is equivalent to 30% of the initial fracture strength of the lap joints. However, as may be seen, whilst the agreement from the models around the threshold portion of the T_{max} versus N_f plots is good, that the agreement is clearly poorer as one moves to higher values of T_{max}, i.e. to lower values of N_f. A possible reason for this may be the inherent scatter in the experimental fracture-mechanics data in the linear region of the graphs such as that shown in Figs. 3 and 6. Nevertheless, it should be noted that, in comparison with metallic materials, the linear region in the graphs of logarithm da/dN versus logarithm G_{max} (see Figs. 3

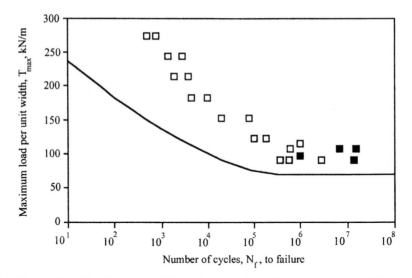

Fig. 14. The number, N_f, of cycles to failure for single-lap joints as a function of the maximum load, T_{max}, per unit width applied in a fatigue cycle. The lap joints were cycled at 5 Hz in water at 28°C. The points represent the experimental data (filled points indicate where the test was stopped prior to failure) whilst the lines are the theoretically predicted lifetimes [74].

and 6, for example) for polymeric adhesives is often relatively steep. This implies that, for adhesive joints, the rate of fatigue crack growth may rapidly increase for relatively small increases in the applied-strain-energy release rate, G_{max}, and hence for relatively small increases in T_{max}. Thus, it may be argued that predicting a lower-limit (threshold) load (below which cyclic-fatigue crack growth will not be observed) is the appropriate design philosophy in the case of adhesively bonded joints. Further, it was found that the predictions were not significantly dependent upon the value of the inherent flaw size, a_o, which was employed. Therefore, as noted above, the present models are clearly capable of achieving very good predictions in this respect and, indeed, the above methodology has recently [74] been extended to predict the fatigue performance of bonded components.

4.3.3. Future developments

The most productive methodology at the time of writing for predicting the service-life would appear to be based upon determining the cyclic-fatigue threshold value of G_{th} in the environment of interest, and employing this value to predict a lower-limit (threshold) load, i.e. a load below which crack growth will not be observed. As demonstrated above, this can be achieved by combining the experimentally measured value of G_{th} with results from FEA modelling. Such FEA modelling can be readily used to deduce the variation of the strain-energy release rate, G, with the length, a, of a growing crack, as a function of the loads applied to the joint, or to the component or structure, of interest. (These comments are especially appropriate with the development of 'cohesive zone models', also known as 'interface elements', which offer an excellent route for combining fracture-mechanics parameters with FEA modelling studies.) Further, it is suggested that one may conservatively use this value of G_{th} from cyclic-fatigue loading for other time-dependent loading conditions. This reasoning is based upon the fact that cyclic-fatigue conditions (in the environment of interest) will invariably lead to the lowest value of G_{th}, due to the very damaging nature of cyclic-fatigue test conditions.

5. Concluding remarks

The present chapter has emphasised the types of loading conditions and environments which may be experienced by a bonded component and which may greatly reduce the service-life of adhesive joints, unless consideration is given to all aspects of the material selection, choice of surface pretreatment and the design of the bonded component; i.e. unless an 'adhesive system' approach is adopted.

The chapter has clearly demonstrated that, when due care and attention is applied in considering the adhesive system, adhesive joints are capable of surviving under some of the most arduous environments for long periods of time. In discussing the above, it has been only natural to concentrate on some of the most demanding applications for adhesives, such as for example the aerospace industry. However, it should be emphasised that very few applications are quite as demanding, either in terms of the applied loadings, the nature of the environments which may be encountered, or in the very long service-lives which are required. For less demanding applications, the critical balance which the adhesives technologist and design engineer must make is that between attaining the necessary service-life but at an economically viable cost. Thus, there are two major current challenges.

Firstly, to continue to develop test methods and durability models which permit the service-life of an adhesive system to be accurately ranked, *and* quantitatively predicted, from relatively short-term tests. The use of various fracture-mechanics techniques have been detailed in the present chapter and they appear to be gaining increasing recognition as the preferred methodology, especially with the development of 'cohesive zone models' (also known as 'interface elements'), which offer an excellent route for coupling fracture-mechanics parameters with FEA modelling studies.

Secondly, to move away from the very complex multi-stage surface pretreatments which are typically needed to give the most durable joints, but are expensive and present health and safety problems. Hence, there is a need to develop surface treatments for metallic substrates, such as the organosilane primers, which will give joints possessing a long service-life in the most demanding of environments, but which are more suitable for general engineering applications. Also, in many industries, such as the automotive industry, the manufacturer will typically not undertake any surface pretreatment prior to adhesive bonding. Since it seems clear that some form of surface pretreatment will continue to be necessary for good durability, this implies that (a) any effective 'primer' may have to be added directly to the adhesive formulation (and diffuse to the interface and react prior to the adhesive hardening), and/or (b) the materials supplier will need to surface pretreat the substrate prior to supplying it to the final assembly manufacturer. This latter route is the one which has been developed in the car industry for bonding aluminium alloy. Here, the aluminium alloy is surface pretreated (in order to subsequently give good joint durability) and coated with a protective press lubricant (which is compatible with, and displaced by, the adhesive during the later bonding operation) by the materials supplier. Such routes would appear to be the obvious ones to pursue for a wide range of industrial applications where adhesives technology offers many significant advantages compared to other more traditional means of joining, and where good durability is a key requirement.

References

1. Kerr, C., MacDonald, N.C. and Orman, S., *J. Appl. Chem.*, **17**, 62 (1967).
2. Orman, S. and Kerr, C., In: Alner, D.J. (Ed.), *Aspects of Adhesion, 6*. University of London Press, London, 1971, p. 64.
3. Sharpe, L.H., *Appl. Polym. Symp.*, **3**, 353 (1969).
4. Albericci, P., In: Kinloch, A.J. (Ed.), *Durability of Structural Adhesives*. Applied Science Publishers, London, 1983, p. 317.
5. Kinloch, A.J. (Ed.), *Durability of Structural Adhesives*. Applied Science Publishers, London, 1983.
6. Kinloch, A.J., *Adhesion and Adhesives: Science and Technology*. Chapman and Hall, London, 1987.
7. Cotter, J.L., In: Wake, W.C. (Ed.), *Developments in Adhesives, 1*. Applied Science Publishers, London, 1977, p. 1.
8. Kinloch, A.J., *Proc. Inst. Mech. Eng.*, **211**, Part G, 307 (1997).
9. Gledhill, R.A. and Kinloch, A.J., *J. Adhes.*, **6**, 315 (1974).
10. Kinloch, A.J., Little, M.S.G. and Watts, J.F., *Acta Mater.*, **48**, 4543 (2000).
11. Blackman, B.R.K. and Kinloch, A.J., *Structural Adhesives: Test Protocol*. European Structural Integrity Society, Technical Committee TC4, 1999 (http://www.me.ic.ac.uk/materials/AACgroup/index.html).
12. Blackman, B.R.K. and Kinloch, A.J., In: Moore, D.R., Pavan, A. and Williams, J.G. (Eds.), *Fracture Mechanics Testing Methods for Polymers, Adhesives and Composites*. Elsevier, Amsterdam, 2001, p. 225.
13. Gettings, M. and Kinloch, A.J., *J. Mater. Sci.*, **12**, 2511 (1977).
14. Gettings, M. and Kinloch, A.J., *Surf. Interface Anal.*, **6**, 40 (1979).
15. Abel, M.-L., Digby, R.P., Fletcher, I.W. and Watts, J.F., *Surf. Interface Anal.*, **29**, 115 (2000).

16. Adams, A.N., Kinloch, A.J., Digby, R.P. and Shaw, S.J., in: Anderson, G.L. (Ed.), *Proceedings of the 23rd Annual Meeting of the Adhesion Society*. The Adhesion Society, Virginia, 2000, p. 219.
17. Reil, F.J., *SAMPE J.*, **7(4)**, 16 (1971).
18. Ritter, J.J., *J. Coat. Technol.*, **54(695)**, 51 (1982).
19. Dickie, R.A., Haack, L.P., Jethwa, J.K., Kinloch, A.J. and Watts, J.F., *J. Adhes.*, **66**, 1 (1998).
20. Koehler, E.L., *Corrosion*, **40(1)**, 5 (1984).
21. Hammond, J.S., Holubka, J.W., DeVries, J.E. and Dickie, R.A., *Corros. Sci.*, **21**, 239 (1981).
22. Leidheiser Jr., H., *J. Adhes. Sci. Technol.*, **1**, 79 (1987).
23. Fitzpatrick, M.F., Ling, J.S.G. and Watts, J.F., *Surf. Interface Anal.*, **29**, 131 (2000).
24. Castle, J.E. and Epler, D.C., *Surf. Sci.*, **53**, 286 (1975).
25. Hammond, J.S., Holubka, J.W. and Dickie, R.A., *J. Coat. Technol.*, **51**, 45 (1979).
26. Watts, J.F. and Castle, J.E., *J. Mater. Sci.*, **18**, 2987 (1983).
27. deVries, J.E., Holubka, J.W. and Dickie, R.A., *J. Adhes. Sci. Technol.*, **3**, 189 (1989).
28. deVries, J.E., Haack, L.P., Holubka, J.W. and Dickie, R.A., *J. Adhes. Sci. Technol.*, **3**, 203 (1989).
29. Dickie, R.A., Holubka, J.W. and DeVries, J.E., *J. Adhes. Sci. Technol.*, **4**, 57 (1990).
30. Davis, S.J. and Watts, J.F., *J. Mater. Chem.*, **6**, 479 (1996).
31. Fitzpatrick, M.F. and Watts, J.F., *Surf. Interface Anal.*, **27**, 705 (1999).
32. Leidheiser Jr., H. and Wang, W., *J. Coat. Technol.*, **53**, 77 (1981).
33. van Ooij, W.J., In: Brewis, D.M. and Briggs, D. (Eds.), *Industrial Adhesion Problems*. Orbital Press, Oxford, 1985, p. 87.
34. Trethewey, K.R. and Chamberlain, J., *Corrosion*. Longman Scientific and Technical, Harlow, 1988, p. 228.
35. Bishopp, J.A. and Thompson, G.E., *Surf. Interface Anal.*, **20**, 485 (1993).
36. Venables, J.D., *J. Mater. Sci.*, **19**, 2431 (1984).
37. Kinloch, A.J. and Smart, N.R., *J. Adhes.*, **12**, 23 (1981).
38. Davies, R.J. and Kinloch, A.J., In: Allen, K.W. (Ed.), *Adhesion, 13*. Elsevier, Amsterdam, 1989, p. 8.
39. Carre, A. and Schultz, J., *J. Adhes.*, **15**, 151 (1983).
40. Davis, G.D., Whisnant, P.L. and Venables, J.D., *J. Adhes. Sci. Technol.*, **9**, 433 (1995).
41. Vine, K., Cawley, P. and Kinloch, A.J., *J. Adhes.*, **77**, 125 (2001).
42. Zaporozhskaya, E.A., Ginsberg, L.V. and Donstov, A.A., *Vysokomol. Soedin. Ser. A*, **22**, 1222 (1980).
43. Gledhill, R.A., Kinloch, A.J. and Shaw, S.J., *J. Adhes.*, **11**, 3 (1980).
44. Comyn, J., In: Kinloch, A.J. (Ed.), *Durability of Structural Adhesives*. Applied Science Publishers, London, 1983, p. 85.
45. Nguyen, T., Byrd, E., Bentz, D. and Lin, C., *Prog. Org. Coat.*, **27**, 181 (1996).
46. Zanni-Deffarges, M.P. and Shanahan, M.E.R., *Int. J. Adhes. Adhes.*, **15**, 137 (1995).
47. Linossier, I., Gaillard, F., Romand, M. and Nguyen, T., *J. Adhes.*, **70**, 221 (1999).
48. Halliday, S.T., Banks, W.M. and Pethrick, R.A., *Proc. Inst. Mech. Eng., Part L*, **213**, 27 (1999).
49. Affrossman, S., Banks, W.M., Hayward, D. and Pethrick, R.A., *Proc. Inst. Mech. Eng., Part C*, **214**, 87 (2000).
50. Davis, G.D., Sun, T.S., Ahearn, J.S. and Venables, J.D., *J. Mater. Sci.*, **17**, 1807 (1982).
51. Gledhill, R.A., Kinloch, A.J. and Shaw, S.J., *J. Mater. Sci.*, **14**, 1769 (1979).
52. Cognard, J., *J. Adhes.*, **20**, 1 (1986).

53. Dickie, R.A. and Ward, S.M., In: van Ooij, W.J. and Anderson, H.R. (Eds.), *Mittal Festschrift*. VSP, Zeist, 1979, p. 1.
54. Shannon, R.W. and Thrall, E.W., *J. Appl. Polym. Sci., Appl. Polym. Symp.*, **32**, 131 (1977).
55. Arrowsmith, D.J. and Maddison, A., *Int. J. Adhes. Adhes.*, **7**, 15 (1987).
56. Olusanya, A., Tully, K., Mera, R. and Broughton, W.R., *Report CMMT(A)98*. National Physical Laboratory, U.K., 1998.
57. Broughton, W.R. and Hinopoulos, G., *Report CMMT(A)206*. National Physical Laboratory, U.K., 1999.
58. Jackson, R.S., Kinloch, A.J., Gardham, L.M. and Bowditch, M.R., In: Ward, T.C. (Ed.), *Proceedings of the 19th Annual Meeting of the Adhesion Society*. The Adhesion Society, Virginia, 1996, p. 147.
59. Moidu, A.K., Sinclair, A.N. and Spelt, J.K., *J. Adhes.*, **65**, 239 (1998).
60. Tu, Y. and Spelt, J.K., *J. Adhes.*, **72**, 359 (2000).
61. Chang, T., Sproat, E.A., Lai, Y.-H., Shepard, N.E. and Dillard, D.A., *J. Adhes.*, **60**, 153 (1997).
62. Kinloch, A.J., Lau, C.C. and Williams, J.G., *Int. J. Fract.*, **66**, 45 (1994) (http://www.me.ic.ac.uk/materials/AACgroup/index.html).
63. Moidu, A.K., Sinclair, A.N. and Spelt, J., *Test. Eval.*, **23**, 241 (1995).
64. John, S.J., Kinloch, A.J. and Matthews, F.L., *Composites*, **22**, 121 (1991).
65. Kinloch, A.J., *J. Adhes.*, **10**, 193 (1979).
66. Brewis, D.M., Comyn, J. and Tegg, J.L., *Int. J. Adhes. Adhes.*, **1**, 35 (1980).
67. Brewis, D.M., Comyn, J., Cope, B.C. and Moloney, A.C., *Polymer*, **21**, 1477 (1980).
68. DeLollis, N.J., *Natl. SAMPE Symp.*, **22**, 673 (1977).
69. Brewis, D.M., Comyn, J., Cope, B.C. and Moloney, A.C., *Polym. Eng. Sci.*, **21**, 797 (1981).
70. Brewis, D.M., Comyn, J., Raval, A.K. and Kinloch, A.J., *Int. J. Adhes. Adhes.*, **10**, 247 (1990).
71. Brewis, D.M., Comyn, J. and Tegg, J.L., *Polymer*, **21**, 134 (1980).
72. Glasstone, S. and Lewis, D., *Elements of Physical Chemistry*. Macmillan, London, 1964, p. 382.
73. Curley, A.J., Jethwa, J.K., Kinloch, A.J. and Taylor, A.C., *J. Adhes.*, **66**, 39 (1998).
74. Curley, A.J., Hadavinia, H., Kinloch, A.J. and Taylor, A.C., *Int. J. Fract.*, **103**, 41 (2000).

Chapter 18

Ultrasonic inspection of adhesive bonds

JOSEPH L. ROSE *

Engineering Science and Mechanics Department, The Pennsylvania State University, 411E Earth and Engineering Sciences Building, University Park, PA 16802, USA

1. Introduction

The subject of adhesive bonding inspection with ultrasonics has been a subject of deep concern for decades (Table 1). Many excellent review articles have been written. See, for example, Segal and Rose [1], Rose and Nestleroth [2], Hagemaier [3], Light and Kwun [4], Rose [5], Hagemaier [6], Cawley [7], De Sterke [8], Munns and Georgiou [9], and Arnold [10].

If the adhesive bonding manufacturing process is carried out correctly, the bonds are strong and can perform well for long periods of time. On the other hand, such subtle errors in the bonding process as poor surface preparation or incorrect curing can lead to disastrous results in adhesive and cohesive strength, respectively. Also, in service environmental conditions and fatigue can lead to serious defects and hence adhesive bond failure.

Let us consider some of the most common problems in an adhesively bonded joint. Bonding problems in a three-layer step-lap joint, for example, are illustrated in Fig. 1. Ultrasonic techniques to find voids, weak cohesive properties, and delaminations are available and are quite reliable. The weak interface or kissing bond detection situation, on the other hand, has a limited number of solutions useful only in special situations. No generalized technique is yet available to depict a weak interface or kissing bond reliably, although a number of promising techniques have emerged.

Basic ultrasonic experimental techniques to help understand the ultrasonic inspection techniques for adhesive bond inspection can be found in Krautkramer [11] and more detail on theoretical aspects of wave propagation in Rose [12].

A brief review of some basic principles in ultrasonics required to appreciate the ultrasonic adhesive bond inspection process is outlined next. Of paramount

---

* Tel.: +1-814-863-8026; Fax: +1-814-863-8164; E-mail: jlresm@engr.psu.edu

Table 1

Popular ultrasonic inspection choices for adhesive bond evaluation

Technique	Problem area						
	Dimensional checks	Voids	Cohesive weakness	Delaminations	Corrosion attack	Weak interface	Primary limitations
Normal longitudinal	X	X	X	X	X	Maybe with very high frequency	Thin parts
Normal shear						Often possible	Contact couplants
Resonance			X				Interface properties
Oblique incidence					X	Often possible	Test setup
Guided waves			X	X	X	Often possible	Complex technology
Acoustic microscopy			X	X	X	Often possible	Thick parts

importance is the concept of reflection factor illustrated in Fig. 2. As an ultrasonic wave impinges onto an interface between two materials, a portion of the energy is transmitted and a portion is reflected. The idealized reflection factor formula is presented in Fig. 2. In order to go beyond the normal incidence ultrasonic technique, it is necessary to consider the refraction principle illustrated in Fig. 3. As an ultrasonic wave impinges onto an interface between two materials at some oblique incident angle, a portion of the energy is reflected and some is refracted. The refraction process is governed by Snell's Law. Note that at the interface, there is a mode conversion process that occurs, and actually two waves reflect and also two waves refract into the second material, a longitudinal wave and a shear wave. The two basic modes of wave propagation, longitudinal and shear, can be differentiated by looking at the particle velocity vector. For longitudinal waves the particle velocity vector is in the same direction as the wave vector, oscillating back and forth in establishing either tensile or compressive wave propagation. On the other hand, for shear wave propagation, the particle velocity vector is at 90 degrees to the direction of the wave vector. The shear wave propagation can be considered to be either vertical or horizontal. Actually, all sorts of wave propagation, whether in unbounded or bounded media can be considered as a superposition of longitudinal and shear wave propagation in the media. For more details on normal incidence, oblique incidence and basic principles of wave propagation, see [12].

The most common Ultrasonic Bond Inspection Procedures are illustrated in Fig. 4. Each has its special benefits for solving specific adhesive bonding inspec-

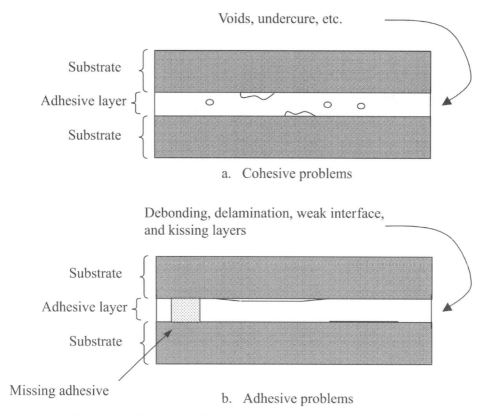

Fig. 1. (a, b) Common bonding problems in a three-layer step-lap joint.

tion problems. In some cases, all of the techniques presented in Fig. 4 might be useful for solving a particular adhesive bond inspection problem. On the other hand, depending on the nature of the adhesive bond, perhaps only one or two of the techniques outlined in Fig. 4 might be useful. A normal beam contact pulse-echo technique is illustrated in Fig. 4a. Notice that a thin oil film is required as a couplant in order to efficiently transfer ultrasonic energy from the piezoelectric transducer into the substrate for subsequent impingement onto the adhesive layer. Note that the piezoelectric effect, responsible for the basic operation of the transducer, is a principle associated with the conversion of electrical energy to mechanical vibrational energy. The reverse piezoelectric effect, now in turn, used in a reception mode, converts mechanical energy back into electrical energy. It is also possible to use a shear wave normal beam transducer similar to that presented in Fig. 4a as long as an adhesive or honey is used as a couplant, since there is not any transfer of shear wave energy across water and very little across an oil film. A normal beam longitudinal wave immersion pulse-echo technique is illustrated in Fig. 4b. There are many benefits of using this technique. As an example, a fluid

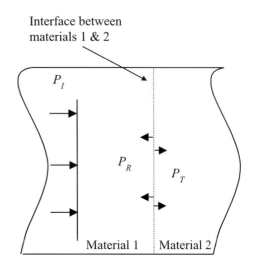

P_I – Incident ultrasonic
shear wave

P_R – Reflected ultrasonic
shear wave

P_T – Transmitted ultrasonic
shear wave

$$R_{12} = \frac{P_R}{P_I} = \frac{W_2 - W_1}{W_2 + W_1}$$

$$W = \rho c$$

W	Acoustic impedance
ρ	Density
c	Wave velocity

Fig. 2. Reflection factor, R_{12}, principle for a wave impinging upon an interface between two different materials.

medium allows focusing of an ultrasonic transducer to take place. Consistency in the pressure between the ultrasonic transducer and the substrate to which the ultrasonic energy is introduced is also possible compared to the contact technique. Variations on this technique can also be considered. In Fig. 4c, a normal beam contact through-transmission technique is presented. In this case one transducer is used as a sender and the second as a receiver. They are scanned together over the test part in trying to examine characteristics of the adhesive bond layer. This technique can also be used in an immersion mode as illustrated in Fig. 4d. Another variation on the immersion technique can be considered by using a bubbler technique whereby a small water film exists between the transducers, as the two transducers are scanned over the part in question. In the aircraft industry, a normal beam squirter technique in through-transmission is quite popular in a fashion similar to that presented in Fig. 4e. There is a housing surrounding the transducer whereby a water supply forces a water column onto the test part. The squirter system then allows all of the ultrasonic energy to travel along the water column allowing ultrasonic waves to be introduced into the adhesive bond test structure. As the ultrasonic energy travels through the structure, it then again encounters another water column where it meets with the receiving piezoelectric transducer. The water flow, by way of a pump, can maintain constant coupling conditions in the scanning system as the two transducers and the squirter system move over

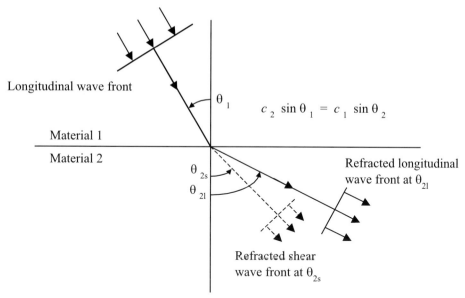

Longitudinal wave front

Material 1

Material 2

θ_1

$$c_2 \sin \theta_1 = c_1 \sin \theta_2$$

Refracted longitudinal wave front at θ_{21}

θ_{2s}

θ_{21}

Refracted shear wave front at θ_{2s}

Fig. 3. Refraction principle for ultrasonic waves impinging upon an interface between two materials at oblique incidence.

the entire test part. Another technique that makes use of just one transducer, but is actually a double through-transmission procedure, is presented in Fig. 4f. In this case, the ultrasonic transducer travels through a fluid medium into the test structure, across the fluid medium onto a reflector plate; the ultrasonic energy then comes back across the fluid medium into the adhesive bond layer across the fluid medium to the receiving transducer. Certainly many other variations on the schemes presented in Fig. 4c could be presented that could lead to a successful adhesive bond inspection program.

Data acquired from all of the techniques could be presented in a C-scan format which is a plot of amplitude versus two-dimensional position of the adhesive bonding area of concern. The ultrasonic C-scan test principle is illustrated in Fig. 5. In the example shown, a focused transducer is used that allows ultrasonic energy to propagate into a small area of the adhesive bond. Focusing is accomplished by way of a curved transducer, or some other technique that allows a time delay profile to occur across the face of the transducer, hence modifying the interference phenomenon in such a way that the beam can be focused to an area of concern inside the test material. In Fig. 5, note that an electronic gate or window could be set to examine signals directly from the adhesive bond layer, or perhaps even later in time domain allowing ultrasonic energy to travel completely through to a reflector plate and back to a receiving transducer. Obviously, in the case shown for a complete delamination, you can see that an echo comes back from the adhesive interface leading to a situation with no backwall echo at all. This is only one test

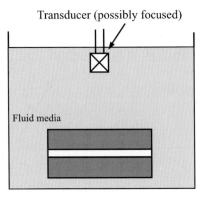

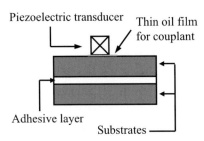

a. Normal longitudinal contact in
   the pulse-echo mode

b. Longitudinal normal incidence
   in the immersion pulse-echo mode

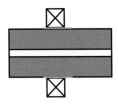

c. Normal longitudinal contact in
   the through transmission mode

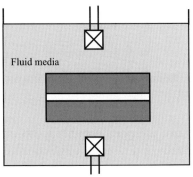

d. Longitudinal normal incidence in the
   immersion through transmission mode

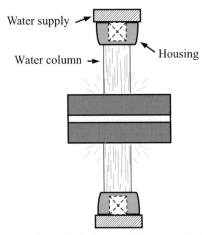

e. Longitudinal normal incidence in the
   squirter through transmission mode

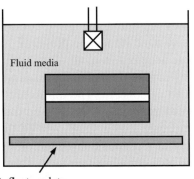

f. Longitudinal normal incidence in the
   immersion double through transmission
   mode

Fig. 4. (a–f) Common normal beam longitudinal wave inspection techniques for adhesive bond inspection.

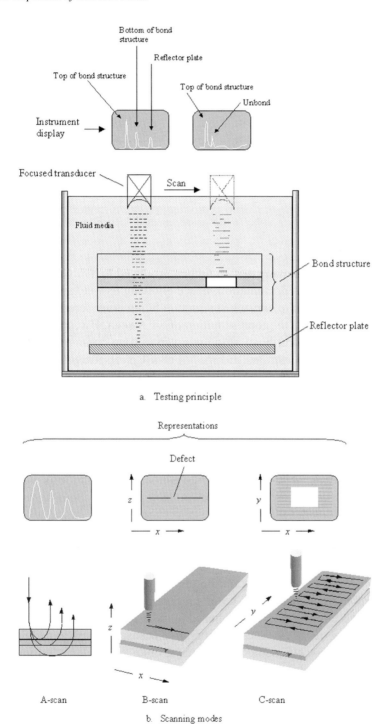

Fig. 5. Ultrasonic C-scan principle (a) compared with A-scan and B-scan (b).

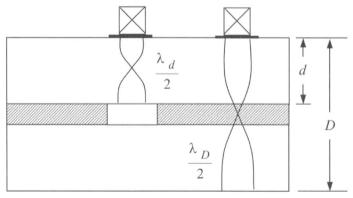

Fig. 6. Resonance method.

algorithm possibility, of many that could be introduced in hardware or software to provide a total C-scan representation, hence allowing us to examine amplitude at a particular depth over the entire test structure being considered. For comparison purposes, Fig. 5b shows the representation of an A-scan, B-scan, and the popular C-scan. An A-scan is simply an amplitude versus time domain signal display. A rectified waveform is illustrated in Fig. 5. For a B-scan image, a cross-sectional image of a structure is actually produced, whereby all of the echos coming back from that cross-section are recorded as illustrated in Fig. 5b of amplitude depicted as a function of x and z. In the C-scan situation, amplitude is depicted again, but this time as a function of x and y. The amplitudes can be presented in a gray scale or color format to provide us with some idea of the variations that might occur in the bonded structure.

Another variation of the normal beam ultrasonic technique is illustrated in Fig. 6 for a so-called resonance method. In this case, continuous wave ultrasonic energy travels from the ultrasonic transducer through the entire adhesive bond structure. When encountering a delamination, notice that the wave resonance set up is of a different wavelength than one where ultrasonic energy can nicely traverse the entire adhesive bond structure. There are many resonance techniques and instrumentation possibilities on the market that work very well for cohesive bonding problems and in some cases are even acceptable for adhesive defect situations. Standing waves are established and the frequency observed is equal to the wave velocity over twice the thickness, in this case either twice d or twice D.

Oblique incidence ultrasonic test techniques like those illustrated in Fig. 7 are also quite useful. In fact, it has been found in earlier studies that the normal beam shear wave incidence technique is often more sensitive to subtle interfacial weakness problems than a normal beam longitudinal incident wave. It is difficult, however, to produce shear wave incidence in a normal beam mode because of the somewhat permanent adherence to the substrate that is required. As a result, it was

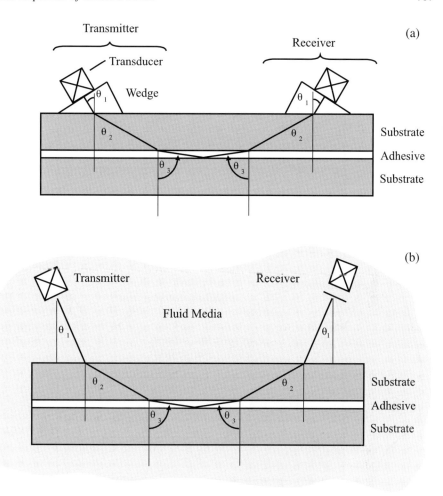

Fig. 7. (a,b) Common oblique incidence adhesive bond inspection procedures.

found that by using the oblique incidence technique and the refraction principle that it became possible to have ultrasonic shear wave energy impinge onto the interface of the adhesive layer and hence provide excellent sensitivity to the subtle defects in that adhesive interface. The procedures illustrated in Fig. 7 show possibilities of contact and immersion. In Fig. 7a, a transducer wedge is required, where a longitudinal wave can be sent from the transducer along the wedge at some particular angle, whereby refraction of shear wave energy can take place into the substrate with subsequent impingement onto the adhesive layer. Keep in mind though, it is still sometimes useful to consider the longitudinal refraction wave also as an impingement onto the adhesive layer. Both possibilities can be used depending on the problem at hand. We see therefore a refraction from the wedge to the substrate and refraction again from the substrate to the adhesive layer whereby

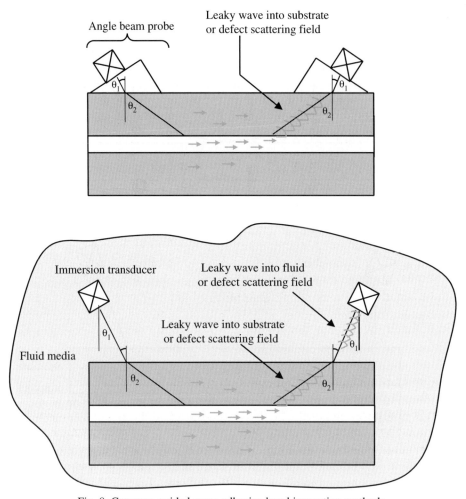

Fig. 8. Common guided wave adhesive bond inspection methods.

reflection occurs at the lower interface followed by subsequent refractions again into the second receiving transducer. These two transducers can be moved together over the entire structure in producing a C-scan type image output to be used for defect analysis in the adhesively bonded structure. The concepts are similar for that of an immersion probe illustrated in Fig. 7b. Quite often, the angle theta as incident from the wedge or from the transducer into the fluid medium will have to be adjusted in order to obtain the best possible sensitivity to the characteristics of the adhesive layer, particularly for the case of a weak interface or a kissing bond. This same oblique incidence concept can be considered for Lamb wave excitation and horizontal shear wave excitation in producing guided waves in the structure. This is the subject presented in Fig. 8 where common guided wave adhesive bond inspection procedures are illustrated for a typical three-layer structure, consisting of substrate, adhesive bond layer and substrate. Again, entry into the adhesive bond layer is by way of refraction whereby the refracted angle into the adhesive bond layer is considered at 90 degrees. If we have the correct selection of angles and frequency, a nicely behaved wave packet is formed that allows ultrasonic energy to travel along the adhesive bond layer, where leakage of energy into the upper substrate or actual defect scattering itself, can cause ultrasonic energy to travel to the receiving transducer #2. Guided waves can be produced in either a contact or immersion mode as illustrated in Fig. 8. To understand guided wave techniques for testing an adhesive bond, it is necessary to understand the concepts of dispersion curves and analysis where the possible phase velocity versus frequency values that might occur for the particular adhesive bond structure can be displayed. For more information on this subject, the reader is referred to Rose [12].

Proper gate settings are also required for oblique incidence or for guided wave inspection. In both cases, specific angles are usually required to acquire good results based on appropriate mode choice.

A review of various aspects of the normal beam technique, either pulse-echo or through-transmission is presented next along with sections on oblique incidence, guided waves and acoustic microscopy.

2. Normal beam ultrasonics

Normal beam ultrasonic inspection like that illustrated in Fig. 1a is obviously the most simple to use and can also provide a great deal of information. Gross defects are immediately found. As early as 1973, ultrasonic normal beam procedures were studied for predicting adhesive bond strength; see Rose and Meyer [13]. In Rose and Meyer, a 10-MHz narrow band transducer was used to study the bond line where both interface echoes are superimposed to form a single bond-line echo. It was shown [13] that the bond-line echo amplitude was inversely proportional to the bond failure load. Sample C-scan results were also obtained by gating the bond-line echo. In coming up with a figure of merit for the bond, amplitude values were also weighted with respect to position in considering a shear stress distribution across the bond. Ultrasonic spectroscopy was also considered where different amplitude versus frequency response curves were obtained for different thicknesses. Since then, it was found that this approach is limited to certain bond system and defect situations. Rose et al. [14] again used normal beam ultrasonics but selected various features of the bond-line signal to produce a 'feature map' which looked like a C-scan, but was of wave velocity, attenuation at a particular frequency, or some other feature. Higher-frequency probes were used to resolve the echoes from each interface. Frequencies around 30 MHz could often locate weak interface zones in a bonded structure; see

Nestleroth et al. [15]. Segal et al. [16] considered some established novel signal processing schemes to assist in adhesive bond inspection. Sinclair et al. [17] and Filimonov [18] employed acoustic resonance methods for dynamic elastic modulus measurements in adhesively bonded structures. Yost and Cantrell [19], Achenbach and Parikh [20] and Nagy et al. [21] considered a nonlinear response of bonded structures to estimate material characteristics. In Achenbach and Parikh [20], failure was preceded by nonlinear behavior of thin boundary layers at the interfaces. Billson and Hutchins [22] considered lasers and EMATS in bond investigations. It was shown that this non-contact technique was reasonable when compared to that obtained by conventional piezoelectric transducers. Ince et al. [23] also characterized bonds with laser-generated ultrasound and through-transmission measurements.

Light and Harvey [24] presented a paper that analyzed the pulse-echo ultrasonic squirter technique for the first layer delamination in a composite structure. Parikh and Achenbach [25] established a framework for studying the nonlinear viscoelastic behaviors of adhesive layers in an attempt at advancing the state of the art in a nondestructive evaluation of adhesively bonded structures.

Fraisse et al. [26] looked at ultrasonic measurements for the adhesive layer. Thickness was negligible compared to the ultrasonic wavelength. Some excellent results were reported. Hsu and Patton [27] considered a water-coupled focused beam broadband pulse technique referred to as the 'dripless bubbler' technique. In this case, ultrasound at a low-frequency of 1 MHz was used where the interfacial echoes were unresolved. The technique presented here was similar to earlier work presented by [13]. A variety of practical adhesive bonding test situations associated with tilting, rivets, thickness variations, and paint were considered, and excellent results were obtained. Pangraz and Arnold [28] considered a microscopic description of the interface binding force. As a result, ultrasonic data were taken in a normal beam through-transmission technique. Good results were obtained. They considered the binding force of glass polymer composites with different humidity considerations, as an example, for a polymer bond. Baltzersen et al. [29] considered inspection of coupler joints in glass-reinforced polymer-piping systems. The attenuation was noted to increase approximately linearly with frequency pointing to the benefits of low-frequency transducers for ultrasonic inspection of the adhesive bond situations. A variety of other ultrasonic testing techniques are also reported. Holland et al. [30] employed an ultrasonic 5.5-MHz linear array medical imaging system to examine disbonded regions in bonded aluminum plates. Since then, of course, array technology is finding many applications in nondestructive evaluation. Interesting experiments and results are reported in the paper. Ultrasonic spectroscopy techniques for a layer between two similar substrates were studied by Lavrentyev et al. [31]. Interference of the ultrasonic signals between the front and back interface layer resulted in minima in the amplitude versus frequency spectrum of the signals that would allow us to extract thickness information as well as quality characteristics of the adhesive bond layer. Light et al. [32] introduced a dry coupled ultrasonic technique for evaluating bond-line quality in order to overcome the adversity associated with the liquid couplant usually required in the contact inspection technique to reliably transfer energy from the piezoelectric transducer into the test part. Promise for dry couplant was demonstrated in the paper. A novel approach of detecting nonlinearities due to bond deterioration was studied by Tang et al. [33]. Some promise for solving the complex adhesive bond degradation problem was shown in the paper. Some additional interesting reading on normal beam ultrasonic test techniques for adhesive bonding structures as well as some other interesting subtleties of ultrasonic analysis are reported in references [34–45].

3. Oblique incidence techniques

A common ultrasonic oblique incidence technique is shown in Fig. 7. Besides the benefit of a different look into the adhesive bond layer by way of refraction, there are other benefits of the oblique incidence technique. Worth noting, of course, is the fact that it can now have longitudinal and shear waves impinging onto the interface between the substrate and adhesive layer. This allows us to achieve excellent sensitivity for a variety of different problems that might be found in the adhesive bond layer. Motivation for some of the work associated with the oblique incidence technique was obtained by looking at normal beam shear wave techniques onto the adhesive layer. Quite often excellent sensitivity was obtained for weak interface situations. Because of the difficulty of the technique, though, in coupling the transducer to the test part, it was decided by many investigators to consider the refraction process of an oblique incidence where ultrasonic shear wave would directly impinge onto the interface in question. Some work reported by Pilarski and Rose [46] outlines the oblique incidence inspection technique that can lead to this special sensitivity for the weak interface situation. The weak interface in the paper was modeled as a spring having a variety of different boundary conditions at the interface going from a welded or perfect bond onto a smooth boundary condition associated with the kissing bond. Intermediate conditions were also studied. The theoretic work presented led to some very exciting experiments that are reported in Pilarski and Rose [46] and in more detail in Pilarski and Rose [47]. C-scan presentations or F-map techniques were illustrated in the paper which shows the oblique incidence reflection factor shear wave results to be quite sensitive to a weak interface. Additional experiments are reported in Rose et al. [48]. A whole host of test specimens was studied with a variety of different interfacial defects. Excellent experimental results were obtained. More discussion of the problem associated with the kissing bond at the interface and the ultrasonic detection possibilities is reported by Nagy [49]. Some interesting oblique incidence experiments that employ a goniometer to provide an assessment of an interface condition in an adhesive bond are reported by Lavrentyev and Rokhlin [50]. Cawley et al. [51] and Cawley and Pialucha [52] report the use of ultrasonic reflection coefficients for both normal and oblique incidence for the detection and characterization of various oxide layers and other problems at the adhesive adherent interface. Good results were obtained.

4. On guided wave inspection techniques

A simplified presentation of the guided wave inspection technique is illustrated in Fig. 8. Good experimental techniques for guided wave propagation in an adhesively bonded structure rely heavily on theoretical aspects of guided waves, commonly referred to as dispersion analysis. When looking at a particular adhesively bonded structure as an example, a three-layer structure of a substrate, adhesive layer, substrate, a boundary value problem should be solved whereby a governing wave equation is examined along with appropriate boundary conditions a utilization of the theory of elasticity and an assumed harmonic equation. A transcendental equation is then obtained in most cases calling for a numerical solution that leads to the dispersion curves of phase velocity versus frequency. In some cases, a group velocity versus frequency and attenuation versus frequency will also be defined depending on the problem statement. Details on the generation of these guided wave theoretical procedures can be found in Rose [12]. The dispersion curve establishes the guidelines for data acquisition and analysis, primarily in selecting the appropriate angle of incidence, which relates to phase velocity by way of Snell's Law, and of course frequency. Selection of a phase velocity and appropriate frequency for all of the points that might be selected on a dispersion curve in a plot of phase velocity versus frequency must be carefully done. Some points will behave better than others with respect to being able to define a sensitive ultrasonic guided wave test for finding certain defect situations. From point to point on the dispersion curve, the actual wave structure across the adhesive bond changes, with certain points providing excellent sensitivity to the interface conditions and others more sensitive to the cohesive situation inside the layer. Other points might simply inspect the substrate materials themselves. The approach is complex, the mathematical details are considerable and the reader is again referred to literature on ultrasonic waves in solid media, as an example, in Rose [12].

5. Sample dispersion curve and wave structure

A sample dispersion curve is illustrated in Fig. 9 for a three-layer adhesively bonded structure of aluminum–epoxy–aluminum. Points on the curve show pos-

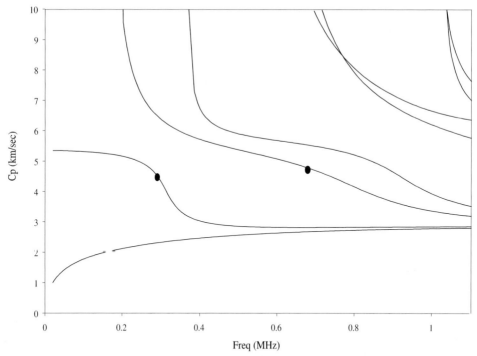

Fig. 9. Sample dispersion curves for a three-layer structure (aluminum 2.54 mm–epoxy 0.254 mm–aluminum 2.54 mm).

sible inspection points where each point has a different penetration power and sensitivity to certain kinds of defects in either the substrate, epoxy layer, or actual interface between the epoxy and aluminum substrate. Sample wave structure values to help determine sensitivity regions are illustrated in Fig. 10. Note that for some points, that is for a specific phase velocity and frequency value, nicely behaved wave structures can be found that are useful for carrying out a specific inspection. Again, details on the generation and interpretation of these dispersion curves can be found in Rose [12].

Many choices for carrying out an inspection with guided waves can be obtained from the dispersion curves. Many different combinations of phase velocity and frequency might work equally well for finding specific cohesive or adhesive type defects. As an example, wave structures for two specific points are illustrated in Fig. 10. In Fig. 10a, for example, the in-plane displacement component, which could be quite sensitive to certain kinds of defects along the interface or in the adhesive layer, is quite dominant along the interface and in the adhesive layer. In Fig. 10b, however, from the power distribution, it is expected to have best results for substrate inspection only if needed. This point would not be selected for adhesive bond inspection. Theoretical results are useful primarily for establishing

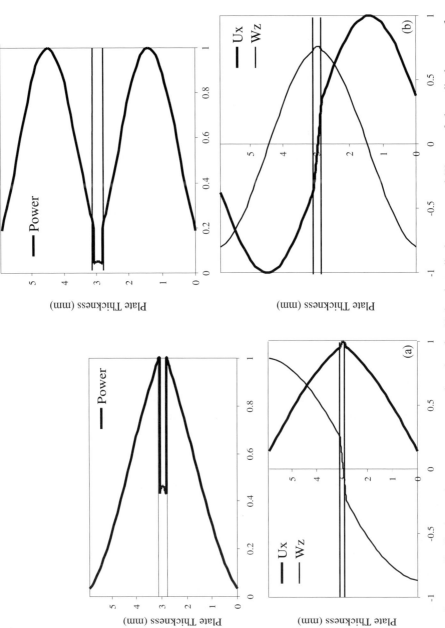

Fig. 10. Sample power distribution and wave structure results taken for U_x in plane displacement, and W_z, out of plane displacement from specific points on a dispersion curve: (a) $f = 0.293$ MHz, $C_p = 4.48$ km/s; (b) $f = 0.664$ MHz, $C_p = 4.85$ km/s.

trends and demonstrating the wave structure variations from one point to another on the dispersion curve. As a result, tuning processes are useful, that is to vary phase velocity and frequency in an attempt to find all defects in the adhesive bond layer.

Rose and Ditri [53] examined both pulse-echo and through-transmission Lamb wave inspection techniques for adhesive bond inspection. Excellent sensitivity was achieved for a variety of different adhesive bond preparation strategies. Cawley et al. [54] also looked at Lamb wave inspection possibilities for adhesive bonds where emphasis was placed on the true guided waves that actually occur in the adhesive layer. Zeros of the reflection coefficient that might occur and the reflection coefficient amplitude from the adhesive adherent interface were discussed along with the relationship with guided wave possibilities in an adhesive layer. Correlations with the zeros of the reflection coefficient and a detailed examination of the reflection coefficient amplitude provided insight into the guided wave technique for adhesive bond examination. Nagy and Adler [55] study guided wave inspection potential in adhesive joints. Fahr et al. [56] considered an acousto-ultrasonics and pattern recognition technique for their analysis of various adhesive bond situations. The acousto-ultrasonic technique was one that places a normal beam probe transducer on one section of a test specimen and a normal beam receiver at some other position on the test structure. What actually occurs between the two transducers is a guided wave ultrasonic transmission from one point to another. This commonly referred to signature technique is extremely useful for a number of different inspection problems as long as the structure is similar and the ultrasonic transducer size and frequency bandwidth characteristics for a particular pulser are just about identical. We can then obtain a signature of a situation, which can be correlated with damage type, or high quality of the adhesive bonded structure being considered. Lowe and Cawley [57] explores the use of plate waves for adhesive bond inspection. Rose et al. [58] present a paper on Lamb waves for aircraft bond inspection with special considerations of mode, frequency, and probe geometries that are required to successfully launch and receive the guided waves. A number of sample problems in the aircraft industry are discussed. Lih et al. [59] studied the ultrasonic evaluation thermal degradation in adhesive bonds. A leaky Lamb wave phenomenon based on angular insonification is used to determine the adhesive bond properties. Comparison was made with the reference dispersion curves in order to depict slight changes in the development of the dispersion curves that could relate to damage mechanisms in the adhesive bond. Lowe and Cawley [60] report some measurements on diffusion bonding joints, which can be directly applied to adhesive bonding situations. Some interesting results are reported that look at two transducers in oblique incidence pitch catch. In examining the frequencies and angles at which minima can occur in the reflected signals in establishing relationships with bond quality. Rose et al. [61] introduce a somewhat sophisticated guided wave approach to practical

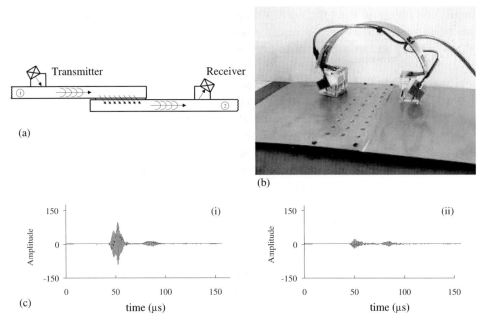

(a)

(b)

(c)

Fig. 11. A lap-splice inspection sample problem. (a) Ultrasonc through-transmission approach for lap splice joint inspection. (b) Double spring 'hopping probe' used for the inspection of a lap splice joint. (c) Signal of a lap splice joint inspection showing good and poor results $f = 1.5$ MHz, $\theta_i = 31°$: (i) is a good bonded region and (ii) a poorly bonded region.

aging aircraft inspection problems of delaminations and corrosion detection in a lap-splice joint and in a tear-strap structure. A double spring hopping probe is used to acquire data over the curved aircraft structure. See Fig. 11 for a brief description of a technique that is being considered for lap-splice joint inspection in the field today. A resonance tuning concept is presented that shows how variations in phase velocity and frequency can be used to obtain maximum guided wave penetration power across a step-lap joint or tear strap in coming up with figures of merit for the structure with respect to bond integrity. A tear-strap inspection procedure is illustrated in Fig. 12 followed by a skin to honeycomb bond in Fig. 13. These figures provide some basic insight into the guided wave approach to adhesive bond inspection. Although fairly simple in concept, great care must be given to mode and frequency selection to allow appropriate sensitivity to the problem being investigated. Rose et al. [62] consider guided waves for composite patch repair of aging aircraft. By sending ultrasonic energy from one side of the patch to another it becomes possible to examine the ultrasonic leakage of energy into the patch if prepared properly. A straightforward technique allows us to evaluate the quality of the bond in the repair process of a section of the aircraft. The composite patch problem was also examined by Scala and Doyle [63]. The utilization of ultrasonic leaky interface waves between the overlays

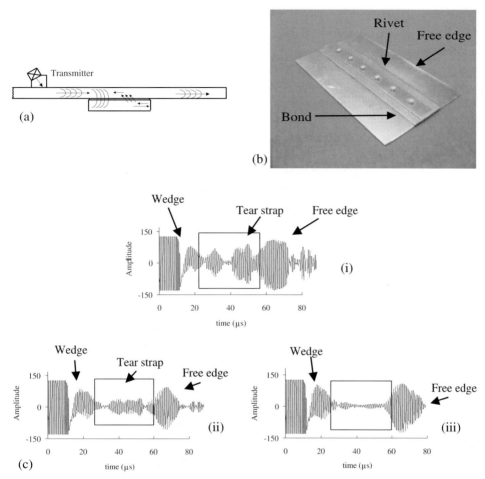

Fig. 12. A tear-strap inspection sample problem showing in the proper time domain gated region the decrease in amplitude and signature change for the poorly bonded tear strap and the loss of signal amplitude for the debonded tear strap. (a) Ultrasonic pulse-echo approach for tear strap inspection. (b) Underside of a tear strap. (c) Ultrasonic signals for the sample tear strap inspection (pulse-echo) $f = 1.1$ MHz, $\theta_i = 45°$: (i) good adhesion, (ii) poor adhesion and (iii) bad adhesion (debonded).

in the substrate aluminum alloy materials is also explored. A laser ultrasonic excitation technique was also considered in the study. Singher et al. [64] measured a sandwich bond strength with a spring model at the interface in establishing guidelines for looking at the generation and detection of guided waves across a step-lap joint. An optical reception technique was considered that produced some very nice results for this adhesive bond inspection examination problem. Mustafa et al. [65] used Lamb waves for imaging, quite nicely, disbonds and adhesive joints with Lamb waves. Chimenti [66] in a review article on guided

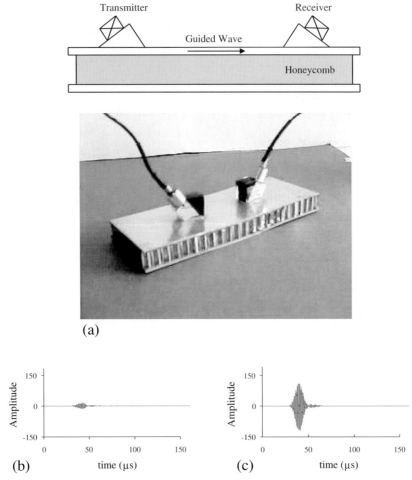

(a)

(b)   time (μs)   (c)   time (μs)

Fig. 13. Ultrasonic guided wave inspection concept for skin to core adhesive bond evaluation. (a) An example of the guided wave inspection setup. (b) Signal for good bond. (c) Signal for poor bond.

waves and plates presented some very interesting ideas and analysis for using guided waves in materials characterization and analysis with one possibility being, of course, adhesive bond inspection. Rose et al. [67] examine ultrasonic guided wave inspection of a Boeing 747 tear-strap structure. Qu and Liu [68] looked at the effect of residual stress on guided waves and layered media with some very interesting results. Rose et al. [69] considered guided wave techniques for an examination of titanium to titanium diffusion bonding. The techniques presented in that paper can be applied to adhesive bond inspection. One really interesting thing that occurred in that paper was the introduction of two modes simultaneously whereby one mode was strongly affected by variations in the bond line, and one mode almost independent of what had happened in the bond line. This provided for us a very interesting feature of an amplitude ratio of the two modes that could correlate nicely with adhesive bond quality. Amplitude itself, as a feature, is not reliable, but an amplitude ratio is quite reliable. This interesting result produces something that hopefully could be extended in the future to guided wave analysis for reliable inspection of ultrasonically adhesive bonded structures. Vine et al. [70] examined the correlation of ultrasonic measurements with toughness changes during the environmental degradation of adhesive joints. Excellent results were obtained on a number of different situations with a consideration of the detrimental effect of moisture on adhesive bond strength. Xiu et al. [71] also present an interesting article on the ultrasonic characterization of thin plate bonding. Zhu et al. [72] present some guided wave techniques for hidden corrosion detection associated with the development of corrosion emanating from poorly bonded interface conditions for a variety of different adhesive bonded structures. It is shown in the paper how certain modes might disappear depending on corrosion levels and location inside a structure. Todd and Challis [73] consider Lamb waves along with artificial neural networks. The artificial neural network technique being used today goes beyond some of the techniques introduced decades ago on pattern recognition analysis. Efficient routines of establishing relationships between features and adhesive bond quality become available. Rose et al. [74] and Rose and Soley [75] review a whole host of additional problems associated with bonded components and hidden corrosion detection on naval aircraft. Special algorithms are introduced that draw upon the theoretical aspects of guided wave analyses and special transducer designs of either comb type configurations or angle beam entries to come up with procedures and algorithms for characterizing various situations in aging aircraft. Sample problems reported include lap-splice inspection, tear-strap inspection, honeycomb skin to core delamination detection, and multi-layer structure analysis, and looking at defects in a second or third layer.

6. Acoustic microscopy techniques

The basic approach to ultrasonic microscopy is illustrated in Fig. 14. Both normal beam impingements can be considered to a focal depth as illustrated as well as surface waves across the test piece. It might be pointed out that the frequency ranges are much higher than those considered in guided wave analysis, going way beyond 50 MHz at times on up to the GHz region. It becomes possible to examine subtle defects in adhesive bond layer in a laboratory environment. Sklar et al. [76] used quantitative acoustic microscopy to study coated aluminum at frequencies on up to 1 GHz. Zeller et al. [77] look at adhesive adherent interlayer measurements by acoustic microscopy with promising results. A comparison of a variety of different techniques including pulsed infrared laser shearography with acoustic

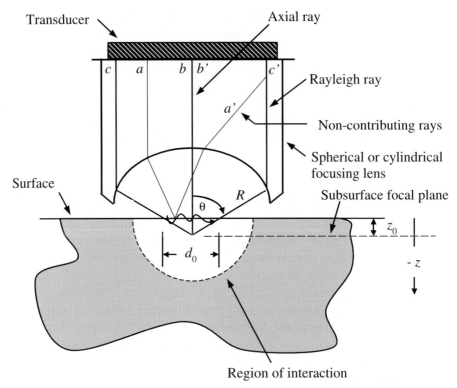

Fig. 14. Schematic of a typical acoustic scanning microscope lens.

microscopy for debonds in a structure is presented by Light and Schaefer [78]. Zeller et al. [79] characterized epoxy-coated oxide films with acoustic microscopy with technology certainly being applied to subtle adhesively bonded structures. Zinin et al. [80] determined density and elastic constants of an oxide film by acoustic microscopy with techniques suitable for adhesive bonding evaluation. It might be pointed out that all of the acoustic microscopy techniques are excellent for laboratory use but still have some way to go with respect to practical field utilization for adhesive bond inspection problems.

References

1. Segal, E. and Rose, J.L., Nondestructive testing techniques for adhesive bond joints. In: Sharpe, R.S. (Ed.), *Research Techniques in Nondestructive Testing*, Vol. IV. Academic Press, London, 1980.
2. Rose, J.L. and Nestleroth, J.B., Advanced ultrasonic techniques for adhesive bond and composite material inspection. In Search of Excellence in a Rapidly Changing World. The Adhesive and Sealant Council, 1984.
3. Hagemaier, D.J., Nondestructive inspection (of A ℓ alloy adhesive bonds). *Adhesive Bond-*

ing of Aluminum Alloys. Marcel Dekker, New York, NY, 1985, pp. 337–423.

4. Light, G.M. and Kwun, H., Nondestructive evaluation of adhesive bond quality. State-of-the-art review, NTIAC-89-1, June 1989.

5. Rose, J.L., Ultrasonic nondestructive evaluation technology for adhesive bond and composite material inspection. In: *Adhesive Bonding*, Ch. 15. Plenum Press, New York, NY, 1991, pp. 425–448.

6. Hagemaier, D.J., Adhesive-bonded joints. In: *Nondestructive Evaluation and Quality Control*. ASM Handbook, Vol. 17, 1992, pp. 610–640.

7. Cawley, P., NDT of adhesive bonds. In: *Flight-Vehicle Materials, Structure, and Dynamics — Assessment and Future Directions, Vol. 4. Tribological Materials and NDE*. ASME, 1993, pp. 349–363.

8. De Sterke, A., Inspection of adhesively bonded joints in glass-reinforced plastic pipe systems: Search for NDT techniques. In: *Proceedings — Conference on the Inspection of Structural Composites*, London, 9–10 June, 1994, Paper 14.

9. Munns, I.J. and Georgiou, G.A., Non-destructive testing methods for adhesively bonded joint inspection — A review. *Insight*, **37**, 941–952 (1995).

10. Arnold, W., Nondestructive determination of the strength of adhesive joints. In: *Trends in NDE Science and Technology*. Proc. 14th World Conf. on NDT, 1, 1997, pp. 93–98.

11. Krautkramer, J., *Ultrasonic Testing of Materials*. Springer, New York, NY, 4th ed., 1990.

12. Rose, J.L., *Ultrasonic Waves in Solid Media*. Cambridge University Press, 1999.

13. Rose, J.L. and Meyer, P.A., Ultrasonic procedures for predicting adhesive bond strength. *Mater. Eval.*, **31**, 109–114 (1973).

14. Rose, J.L., Nestleroth, J.B. and Balasubramaniam, K., Utility of feature mapping in ultrasonic non-destructive evaluation. *Ultrasonics*, **26**, 124–131 (1988).

15. Nestleroth, J.B., Rose, J.L., Lecuru, D. and Budillon, E., An ultrasonic F-scan inspection technique for the detection of surface preparation variances in adhesively bonded structures. *Rev. Prog. QNDE*, **6B**, 1787–1795 (1986).

16. Segal, E., Dickstein, P., Segal, Y., Kenig, S. and Dodiuk, H., Novel method of processing pulse echo data in adhesive bond inspection. *J. Nondestruct. Eval.*, **9**, 1–17 (1990).

17. Sinclair, A.N., Dickstein, P.A., Spelt, J.K., Segal, E. and Segal, Y., Acoustic resonance methods for measuring dynamic elastic modulus of adhesive bonds. In: *Dynamic Elastic Modulus Measurements in Materials*. ASTM STP 1045, 1990, pp. 162–179.

18. Filimonov, S.A., Resonance method as a means for inspecting the quality of adhesive joints. *Sov. J. Nondestruct. Test.*, **26**, 853–860 (1991).

19. Yost, W.T. and Cantrell, J.H., Material characterization using acoustic nonlinearity parameters and harmonic generation: engineering materials. *Rev. Prog. QNDE*, **13B**, 1669–1676 (1990).

20. Achenbach, J.D. and Parikh, O.K., Ultrasonic analysis of nonlinear response and strength of adhesive bonds. *J. Adhes. Sci. Technol.*, **5**, 601–618 (1991).

21. Nagy, P.B., McGowan, P. and Adler, L., Acoustic nonlinearities in adhesive joints. *Rev. Prog. QNDE*, **9**, 1685–1692 (1991).

22. Billson, D.R., Hutchins, D.A., Laser-EMAT ultrasonic measurements of bonded metals. *Nondestr. Test. Eval.*, **10(1)**, 43–53 (1991).

23. Ince, R., Thompson, G.E. and Dewhurst, R.J., Characterisation of adhesive bonds from inspection by laser-generated ultrasound. *J. Adhes.*, **43**, 135–159 (1993).

24. Light, G.M. and Harvey, D.P., Development of a pulse-echo ultrasonic squirter for detection of first layer delamination in composite structures. *1988 Review of Progress in Quantitative NDE*, July 31–August 5, 1988.

25. Parikh, O.K. and Achenbach, J.D., Analysis of nonlinearly viscoelastic behavior of adhesive bonds. *J. NDE*, **11**, 221–226 (1992).

26. Fraisse, P., Schmit, F. and Zarembowitch, A., Ultrasonic inspection of very thin adhesive layers. *J. Appl. Phys.*, **72**, 3264–3271 (1993).

27. Hsu, D.K. and Patton, T.C., Development of ultrasonic inspection for adhesive bonds in aging aircraft. *Mater. Eval.*, **51**, 1390–1397 (1993).

28. Pangraz, S. and Arnold, W., Quantitative determination of nonlinear binding forces by ultrasonic technique. *Rev. Prog. QNDE*, **13B**, 1995–2001 (1994).

29. Baltzersen, O., Bang, J., Moursund, B. and Melve, B., Ultrasonic inspection of adhesive bonded coupler joints in GRP piping systems. *J. Reinforced Plast. Compos.*, **14**, 362–377 (1995).

30. Holland, M.R., Johnston, P.H., Handley, S.M. and Miller, J.G., Detection of disbonded regions in bonded aluminum plates using an ultrasonic 7.5 MHz linear array medical imaging system. *Rev. Prog. Quant. Nondestruct. Eval.*, **14B**, 1513–1520 (1995).

31. Lavrentyev, A.I., Huang, W., Chu, Y.C. and Rokhlin, S.I., Ultrasonic spectroscopy of a layer between dissimilar substrates. *Rev. Prog. Quant. Nondestruct. Eval.*, **14B**, 1561–1568 (1995).

32. Light, G.M., Goodlin, D.L. and Bloom, E.A., Nondestructive evaluation of the integrity of adhesively bonded structures. In: *10th Int. Conf. Composite Materials*, Vancouver, BC, August 8–11, 1995.

33. Tang, Z., Cheng, A. and Achenbach, J.D., Ultrasonic evaluation of adhesive bond degradation by detection of the onset of nonlinear behavior. *J. Adhes. Sci. Technol.*, **13**, 837–854 (1999).

34. Dewen, P.N. and Cawley, P., Cohesive property determination in bonded joint with composite adherends. *Measurement*, **11**, 361–379 (1993).

35. Dewen, P.N. and Cawley, P., Ultrasonic determination of the cohesive properties of bonded joints by measurement of reflection coefficient and bondline transit time. *J. Adhes.*, **40**, 207–227 (1993).

36. Fraisse, P., Schmit, F. and Zarembowitch, A., Ultrasonic inspection of very thin adhesive layers. *J. Appl. Phys.*, **72**, 3264–3271 (1993).

37. Hsu, D.K., Patton, T.C., Aglan, H.A. and Shroff, S., Fatigue-induced disbonds in adhesive lap splices of aluminum and their ultrasonic detection. Proceedings of SPIE — The International Society for Optical Engineering Nondestructive Inspection of Aging Aircraft, San Diego, CA, 1993.

38. Pialucha, T.P. and Cawley, P., The detection of thin embedded layers using normal incidence ultrasound. *Ultrasonics*, **32**, 431–440 (1994).

39. Barnard, D.J. and Hsu, D.K., NDI of aircraft fuselage structures using the dripless bubbler ultrasonic scanner. In: *Nondestructive Evaluation of Aging Aircraft, Airports, and Aerospace Hardware*, SPIE 2945, Scottsdale, AZ, 3–5 December, 1996.

40. Light, G.M. and Bloom, E.A., Use of dry-coupled ultrasonic inspection techniques for evaluation of bondline quality. In: *Nondestructive Evaluation of Materials and Composites*, SPIE Proceedings Vol. 2944, Scottsdale, AZ, Dec. 3–5, 1996.

41. Qu, J., Ultrasonic nondestructive characterization of adhesive bonds. Georgia Institite of Technology, Annual Report #NAS 1.26:206473, NASA/CR-97-206473, NAG1-1810, 1997.

42. Berndt, T.P. and Green, R.E. Jr., Feasibility study of a nonlinear ultrasonic technique to evaluate adhesive bonds. In: Green, R.E. Jr. (Ed.), *Nondestructive Characterization of Materials, VIII*. Plenum Press, New York, NY, 1998.

43. Liu, G., Qu, J., Jacobs, L. and Li, J., Characterizing the curing of adhesive joints by a nonlinear ultrasonic technique. *Rev. Prog. Quant. NDE*, **18B**, 2191–2199 (1999).

44. Qu, J., Ultrasonic nondestructive characterization of adhesive bonds. Georgia Institute of Technology, Final Report #NAG1-1810, 1999.

45. Lavrentyev, A.I. and Rokhlin, S.I., Models for ultrasonic characterization of environmental interfacial degradation in adhesive joints. *Rev. Prog. Quant. Nondestruct. Eval.*, **13B**, 1531–1538 (1994).

46. Pilarski, A. and Rose, J.L., A transverse-wave ultrasonic oblique-incidence technique for interfacial weakness detection in adhesive bonds. *J. Appl. Phys.*, **63**, 300–307 (1988).

47. Pilarski, A. and Rose, J.L., Ultrasonic oblique incidence for improved sensitivity in interface weakness determination. *NDT Int.*, **21**, 241–246 (1988).

48. Rose, J.L., Dale, J. and Ngoc, T.D.C., Ultrasonic oblique incidence experiments for interface weakness. *Br. J. Non Destr. Test.*, **32**, 449–452 (1999).

49. Nagy, P.B., Ultrasonic detection of kissing bonds at adhesive interfaces. *J. Adhes.*, **5**, 619–630 (1991).

50. Lavrentyev, A.I. and Rokhlin, S.I., Ultrasonic evaluation of environmental degradation of adhesive joints. *Rev. Prog. Quant. Nondestruct. Eval.*, **13B**, 1539–1546 (1994).

51. Cawley, P., Pialucha, T.P. and Zeller, B.D., The characterisation of oxide layers in adhesive joints using ultrasonic reflection measurements. *Proc. R. Soc. London, Ser. A*, **452**, 1903–1926 (1996).

52. Cawley, P. and Pialucha, T., The detection of a weak adhesive/adherend interface in bonded joints. In: *Ultrasonic International 93*, Vienna, 6–8 July, Butterworths-Heinemann, London, 1993.

53. Rose, J.L. and Ditri, J.J., Pulse-echo and through transmission Lamb wave techniques for adhesive bond inspection. *Br. J. Non Destr. Test.*, **34**, 591–594 (1992).

54. Cawley, P., Pialucha, T.P. and Lowe, M.J.S., A comparison of different methods for the detection of a weak adhesive/adherend interface in bonded joints. *Rev. Prog. Quant. NDE*, **12**, 1531–1538 (1993).

55. Nagy, P.B. and Adler, L., Nondestructive evaluation of adhesive joints by guided waves. *J. Appl. Phys.*, **66**, 4658–4663 (1989).

56. Fahr, A., Youssef, Y. and Tanary, S., Adhesive bond evaluation using acousto-ultrasonics and pattern recognition analysis. *J. Acoust. Emission*, **12**, 39–44 (1994).

57. Lowe, M.J.S. and Cawley, P., The applicability of plate wave techniques for the inspection of adhesive and diffusion bonded joints. *J. NDE*, **13**, 185–199 (1994).

58. Rose, J.L., Ditri, J.J. and Pilarski, A., Lamb waves for aircraft bond inspection. *Journal for Italian Society for NDT* (Il Giornale delle Prove Non Distruttive Monitoraggio Diagnostica), **15**, 70–76 (1994).

59. Lih, S.S., Mal, A.K. and Bar-Cohen, Y., Ultrasonic evaluation of thermal degradation in adhesive bonds. *Rev. Prog. Quant. Nondestruct. Eval.*, **14B**, 1481–1488 (1995).

60. Lowe, M. and Cawley, P., Comparison of reflection coefficient minima with dispersion curves for ultrasonic waves in embedded layers. *Rev. Prog. Quant. Nondestruct. Eval.*, **14B**, 1505–1512 (1995).

61. Rose, J.L., Rajana, K.M. and Hansch, M.K.T., Ultrasonic guided waves for NDE of adhesively bonded structures. *J. Adhes.*, **50**, 71–82 (1995).

62. Rose, J.L., Rajana, K.M. and Barshinger, J.N., Guided waves for composite patch repair of aging aircraft. Presented at Symposium on Composite Aircraft Repair, Vancouver, BC, 1995.

63. Scala, C.M. and Doyle, P.A., Ultrasonic leaky interface waves for composite-metal adhesive bond characterization. *J. Nondestruct. Eval.*, **14**, 49–59 (1995).

64. Singher, L., Segal, Y., Segal, E. and Shamir, J., Measurement of a sandwich bond strength. *Rev. Prog. Quant. Nondestruct. Eval.*, **14B**, 1481–1488 (1995).

65. Mustafa, V., Chahbaz, A., Hay, D.R., Brassard, M., Dubois, S., Imaging of disbond in adhesive joints with Lamb waves. In: *Nondestructive Evaluation of Materials and Composites*. SPIE 2944, Scottsdale, AZ, 3–5 December, 1996.

66. Chimenti, D.E., Guided waves in plates and their use in materials characterization. *Appl. Mech. Rev.*, **50**, 247–284 (1997).

67. Rose, J.L., Barshinger, J. and Meyer, P., Ultrasonic guided wave inspection of a 747 tear strap structure. ASNT Fall Conference and Quality Testing Show, 1997.

68. Qu, J. and Liu, G., Effect of residual stress on guided waves in layered media. *Rev. Prog. Quant. NDE*, **16B**, 1635–1642 (1998).

69. Rose, J.L., Zhu, W. and Zaidi, M., Ultrasonic NDT of titanium diffusion bonding with guided waves. *Mater. Eval.*, **56**, 535–539 (1998).

70. Vine, K., Cawley, P. and Kinloch, A.J., The correlation of ultrasonic measurements with toughness changes during the environmental degradation of adhesive joints. *Rev. Prog. Quant. NDE*, **18**, 1525–1532 (1999).

71. Xie, Q., Lavrentyev, I. and Rokhlin, S.I., Ultrasonic characterization of thin plate bonding. *Rev. Prog. QNDE*, **17**, 1355–1362 (1998).

72. Zhu, W., Rose, J.L., Barshinger, J.N. and Agarwala, V.S., Ultrasonic guided wave NDT in hidden corrosion detection. *Res. Nondestruct. Eval.*, **10**, 205–225 (1998).

73. Todd, C.P.D. and Challis, R.E., Quantitative classification of adhesive bondline dimensions using Lamb waves and artificial neural networks. *IEEE Trans. Ultrasonics, Ferroelectrics, and Frequency Control*, **46**, 167–181 (1999).

74. Rose, J.L., Soley, L.E., Hay, T. and Agarwala, V.S., Ultrasonic guided waves for hidden corrosion detection in Naval aircraft. CORROSION NACExpo 2000, 55th Annual Conference and Exposition, March 26–31, 2000, Orlando, FL.

75. Rose, J.L. and Soley, L., Ultrasonic guided waves for the detection of anomalies in aircraft components. Mater. Eval., 50, 2000, 1080–1086.

76. Sklar, Z., Briggs, G.A.D., Cawley, P. and Kinloch, A.J., Quantitative acoustic microscopy of anodised and coated aluminium at frequencies up to 1 GHz. *J. Mater. Sci.*, **30**, 3752–3760 (1995).

77. Zeller, B.D., Kinloch, A.J., Cawley, P., Zinin, P., Briggs, G.A.D., Thompson, G.E. and Zhou, X., Adhesive-adherend interlayer measurement by acoustic microscopy. *Rev. Prog. Quant. NDE*, **16**, 1237–1244 (1997).

78. Light, G.M. and Schaefer, L., Comparison of acoustic microscopy, pulsed infrared thermography (PIRT), and laser shearography for the detection of debonds in honeycomb structures. 43rd SAMPE Symposium and Exhibition, Anaheim, CA, May 31–June 4, 1998.

79. Zeller, B.D., Kinloch, A.J., Cawley, P., Zinin, P., Lefeuvre, O., Briggs, G.A.D., Thompson, G.E. and Zhou, X., Characterisation of epoxy coated oxide films with acoustic microscopy. *Rev. Prog. Quant. NDE*, **17**, 1261–1268 (1998).

80. Zinin, P., Lefeuvre, O., Briggs, G.A.D., Zeller, B.D., Cawley, P., Kinloch, A.J., Zhou, X. and Thompson, G.E., Determination of density and elastic constants of a thin PAA oxide film by acoustic microscopy. *J. Acoust. Soc. Am.*, **106**, 2560–2567 (1999).

Chapter 19

The design of adhesively bonded joints

L.J. HART-SMITH [*]

Phantom Works, The Boeing Company, Huntington Beach, CA, USA

1. Introduction

Designing successful adhesively bonded joints is straightforward, provided that one pays close attention to a small number of critical issues, none of which is difficult to comprehend and none of which is any more difficult to comply with than to ignore. The first is that the joint must *never* be designed to be weaker than the surrounding structure, unless one is deliberately planning a weak-link fuse with no damage tolerance. The second is that the bond stress distribution must *never be uniform* if a *durable* bond is sought. The third is that bonding is totally unreliable in the absence of *appropriate surface preparation*. (It must be remembered that all of the joints that disbonded in service were built in accordance with specifications. A very real problem is that not all *approved* surface preparations can be relied upon to keep the glue stuck.) And the fourth, which is often overlooked, is that in weight critical structures, it is the weight of the entire structure that must be minimized, *NOT* the weight of the splices.

The international aerospace community can look back proudly to many entirely successful applications of adhesively bonded secondary and primary structures, some of which have served for more than 50 years without failure and others of which are scheduled to remain in service for yet another 30 years. Unfortunately, what is remembered the most are the relatively few failures, all caused by inappropriate surface preparations. These have had a disproportionate negative influence on the subsequent applications of adhesive bonding. Yet the resolution of these problems was not at all difficult technically, and the cures were easy to implement — and they *have* been implemented, albeit not universally. There

---

[*] Corresponding author. E-mail: john.hart-smith@boeing.com

has been a noticeable lack of dedication to revising repair manuals to update the process specifications, particularly on out-of-production aircraft, with the result that far too many components are 'repaired' using the very same processes that led to their failure in the first place (see [1]). There seems to be two reasons for this. The discredited procedures produce products that do not fall apart immediately, so there has been a denial by some of the existence of any such problems. Also, no *after-the-fact* inspection has been able to differentiate between bonded structures that will not fall apart in service and those that will, given enough time and exposure to moisture. Successful bonding has always relied upon diligent in-process control and has always been remarkably tolerant of all but the grossest deviations of good practice.

The surface–preparation issue is just as serious for fibrous composite structures (see [2]) as for ones made from metal, and is addressed in other chapters of these volumes. Cleanliness alone is not sufficient; the adhesive must want to *adhere* to the substrate, so the surface must be activated appropriately. No glue will bond to a totally *inert* surface (see [3]). It suffices to say, here, that all the mechanics-based design procedures are rendered worthless if the glue does not stay stuck!

That said, there remains one other chronic misunderstanding about the behavior of adhesives in bonded structures. This is the oversimplified 'analysis' method, still found in most handbooks and in most of the few university courses that actually teach engineering design, whereby it is assumed that the strength of a bonded joint can be equated to the product of the area over which the adhesive extends and some fictitious *uniform* shear- (or peel-) stress allowable. This method actually used to work for the wooden airframes made circa World War I because, with a slope less than 1-in-16, the glues of the day were always stronger than the wood. A lot has changed since then. Today, bonded structures are made from materials far stronger than wood and there is a need for more realistic and less simplistic analysis tools. Some short-overlap test coupons might still obey such a rule, although there is a discussion later explaining why even this is generally not so, but doubling the bonded area in a reasonably well-designed bonded joint will *not* double the joint strength. In most cases, doing so would have no effect at all, as explained in ref. [4].

Given that the adhesive in a bonded joint must never be allowed to become the weakest potential failure mode, it is finally necessary to introduce the issue of the multiplicity of potential failure modes *in a structural joint*, and the importance of assessing *all* of them each time. With an understanding that adhesive shear and peel, for example, are governed by different mechanisms, it becomes possible to see the benefits of not accepting the lowest possible strength, but of minor redesign (such as tapering the ends of the overlaps) to suppress one or more failure modes and to attain the highest one possible. The standard riveted-joint design practice of seeking the geometry for a joint that becomes simultaneously critical in rivet shear, skin bearing, tension-through-the-hole, shear-out, and rivet bending has always represented a *sub-optimum* solution. [1] In a *truly optimized* design, only one or two variables will become critical; *all* of the others will remain sub-critical. And the validation of an optimized bonded joint design is that the joint itself will *not* fail; the structure *outside* the joint should fail first. This is not an issue of conservatism *versus* added weight. There are no penalties associated with good bonded joint designs, when the correct evaluation criteria are employed. Joints (splices) typically involve only about 2–5% of the structural weight. The true measure of joint efficiency is how highly can the remaining 95% or more be safely stressed, *not* how many pounds of load can be transferred through how many pounds of splice.

2. The PABST adhesively bonded wide-body fuselage

During the late 1970s and early 1980s, the then Douglas Aircraft Company (now part of Boeing) designed and tested a wide-body aircraft fuselage under contract to the US Air Force Wright Laboratories, Dayton, in which adhesive bonding was employed widely in the design (see [5]). The program, referred to as the Primary Adhesively Bonded Structure Technology (PABST) program, also undertook extensive durability testing to confirm the reliability of the phosphoric-acid surface treatment, in conjunction with a primer (BR-127), including both phenolic and epoxy resins. The use of phenolic resin in the primer is significant, because the renowned Redux bonded structures made by de Havilland in England and Fokker in Holland, which have stayed bonded during as much as 50 years in service, used a vinyl–phenolic adhesive, sometimes over a merely grit-blasted surface (see [6]). A good indication of the effectiveness of both the PABST bonded joints and the process used to design and analyze them is revealed in Fig. 1, showing a 1-inch-deep bonded stiffener ripped apart by skin wrinkles in a test panel with the adhesive remaining totally intact without even any crazing in the adhesive fillets.

The portion of this work to be discussed here is the longitudinal splices in the skin. Every second splice was designed as an entirely bonded double-strap splice, with no fail-safe rivets. (A small number of rivets located judiciously in low-stress areas could have served as both valuable tooling aids to position components and as a means for electrical grounding, but the decision was made to demonstrate that they were *structurally* unnecessary.) The intervening riveted splices at the manufacturing breaks employed selective bonded doublers at the most critical

---

[1] In scientific terms, this is a *constrained* optimum. It is an approach that has been just as harmful to business practices, in which all the wrong costs are minimized, as it has been to joints in aircraft structures for the past 70 years.

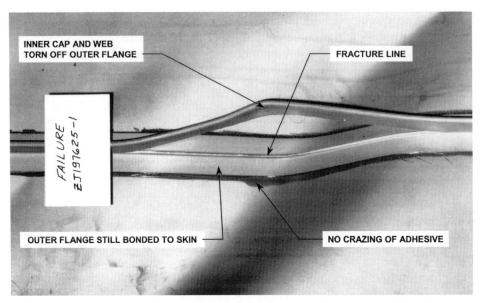

INNER CAP AND WEB
TORN OFF OUTER FLANGE   FRACTURE LINE

FAILURE ZJ197625-1

OUTER FLANGE STILL BONDED TO SKIN   NO CRAZING OF ADHESIVE

Fig. 1. Failure of metallic stiffener without damage to adhesive in bonded panel.

rivet rows, without adding to any eccentricity in load path. These simple joints outperformed by far the more complex riveted splices used in production. Had any of these splices failed, the fuselage would have exploded and the building in which the test was performed would have been destroyed. That did not happen. The message from that program, in the present context, was how easy it was to design these critical bonded joints. The actual design process was reduced a simple table look-up for the appropriate skin gauge (thickness) which defined both the overlap of the joint and the thickness of the splice plates. Only one table was needed; it is reproduced here as Fig. 2.

This table was approximated by the simple empirical formula that the overlap, at each end of the butt joint, was approximately 30 times[2] the thickness of the central skin. The splice plates were made one gauge thicker than half the skin thickness, because fatigue testing had revealed a preponderance of failures in the middle of the splices, where the skins butted together, rather than in the nominally

---

[2] This practice continues to this day, for bonded patches over cracks, or other damage, in metallic structures. The preliminary design overlap for one-sided patches is 60 times the skin thickness, on each side of the crack. This ratio is not universal, of course; it was developed for aluminum adherends bonded together by toughened epoxy adhesives. But the process by which the overlaps were established, which is described later, *is* universal and can be repeated for other materials. It should be noted, however, that nearly isotropic carbon–epoxy laminates have about the same Young's modulus as aluminum alloys and that the same factors can be applied directly.

CENTRAL SHEET THICKNESS t_i(IN.)	0.040	0.050	0.063	0.071	0.080	0.090	0.100	0.125
SPLICE SHEET THICKNESS t_o (IN.)	0.025	0.032	0.040	0.040	0.050	0.050	0.063	0.071
RECOMMENDED OVERLAP[1] ℓ (IN.)	1.21	1.42	1.68	1.84	2.01	2.20	2.39	2.84
STRENGTH OF 2024-T3 ALUMINUM (LB/IN.)	2600	3250	4095	4615	5200	5850	6500	8125
POTENTIAL ULTIMATE BOND STRENGTH (LB/IN.)[2,3]	7699	8562	9628	10,504	10,888	11,865	12,151	13,910

[1]BASED ON 160°F DRY OR 140°F/100-PERCENT RH PROPERTIES NEEDING LONGEST OVERLAP.

VALUES APPLY FOR TENSILE OR COMPRESSIVE IN-PLANE LOADING. FOR IN-PLANE SHEAR LOADING, SLIGHTLY DIFFERENT LENGTHS APPLY.

[2]BASED ON –50°F PROPERTIES GIVING LOWEST JOINT STRENGTH AND ASSUMING TAPER OF OUTER SPLICE STRAPS THICKER THAN 0.050 IN. STRENGTH VALUES CORRECTED FOR ADHEREND STIFFNESS IMBALANCE.

[3]FOR NOMINAL ADHESIVE THICKNESS $\eta = 0.005$ IN. FOR OTHER THICKNESSES, MODIFY STRENGTHS IN RATIO $\sqrt{\eta/0.005}$.

Fig. 2. Design overlaps for adhesively bonded splices in PABST fuselage.

equally stressed skin at the ends of the bonded overlap when each splice plate was made precisely half as thick as the skin. The need for close to stiffness balance is explained later; it is a matter of maximizing the potential shear load that could be carried by the adhesive layer if only the adherends were strong enough. The scientific derivation of these overlaps is also explained.

It can be seen from the entries in Fig. 2 that the potential adhesive shear strength for the thinnest (0.040 inch) skin is roughly three times the adherend strength outside the joint. But, for the thickest skin shown (0.125 inch), this ratio has fallen to less than 2-to-1. The reason for this is that the adherend strength is directly proportional to its thickness, while the adhesive bond strength is proportional only to the square root of the thickness of the adherends (see [7]). At somewhere between 3/16 and 1/4 of an inch this ratio of strengths falls below unity, which is why the table is terminated where it is. (Multi-step joints are needed for thicker skins, as is discussed later.)

Except for the need to employ peel-stress relief, as discussed later, the process of designing the bonded splices in the PABST fuselage really was as simple as looking up entries for the appropriate skin thickness in Fig. 2. The astute reader will observe that no mention has yet been made of the service temperature. This is because the calculations on which Fig. 2 is based were repeated for the highest and lowest service temperatures, and for room temperature as well. The design overlap is actually set by the maximum temperature, where the adhesive is softest, as is explained later, and the joint strength by the minimum temperature, where the bond strength is least because the adhesive is more brittle then. The tensile strength of aluminum alloy skins is insensitive to operating temperatures between −67 and +160°F.

3. Relief from induced peel stresses

The simple table in Fig. 2 actually did need a little supplemental information to complete the design. This is provided in Fig. 3, which shows that, beyond a certain splice-plate thickness, the outer edges of the splice straps needed to be tapered down to thin ends to prevent failure of the adhesive bonds as the result of peel stresses that peak at the outer ends of the overlap because the adhesive applies shear stresses to one side of the splice plates with no balancing stresses on the other side. The splice plates would bend off were it not for the peel stresses that provide a restoring moment. The generation of these peel stresses is explained later.

Were it not for this peel-stress relief, the bonded splices would be prevented from developing their full potential shear strength. Some 'scholars' would recommend not incurring the quantifiable cost of this tapering and have advocated instead the development of far more complicated analyses in which the adhesive shear and peel stresses are interacted and failure at a much lower applied load is predicted. This may be a realistic way to analyze a poorly designed bonded joint, but it is hardly the best way to encourage the use of more adhesively bonded

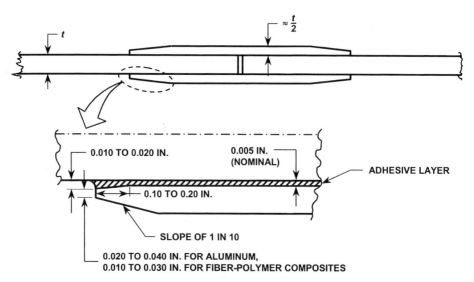

Fig. 3. Design of peel-stress relief for adhesively bonded splice straps and stiffeners in PABST bonded fuselage.

structures. [3] The maximum tip thickness is 0.030 ± 0.010 inch for aluminum structures, for which the adhesive layer sets the peel-strength limit, but is reduced to 0.020 ± 0.010 inch for composite structures made from laminating resins, which are typically weaker in interlaminar tension than are adhesives in peel.

Only the external tapering shown in Fig. 3 was used on the splice plates for the PABST fuselage. The local thickening of the adhesive layer, on the other side of the plate, was incorporated only in the flanges of the extruded stiffeners, where the benefits could be achieved for no added cost. The same feature was employed on the stringers for the wing of the SAAB-340 Commuter Aircraft. There is a strict limit on this added glue thickness, because heat-cured adhesives tend to flow out of the cavity under capillary action, leaving a non-structural void and nullifying the peel stress relief. However, there is no such limit for high-viscosity room-temperature-curing paste adhesives. Engineers at NASA Goddard took advantage of this in the design of steel fittings (splice members) for carbon–epoxy tubes for space applications, as shown in Fig. 4. (Doing so required that air bubbles in the mixed adhesive paste needed to removed by vacuum before the adhesive was applied to the adherends.) The peel stresses induced by 0.10-inch thick steel plates bonded to carbon epoxy substrates are considerable. Application of reverse tapering, at a 1-in-10 slope, of the splice plates was found by test to increase the joint strengths by a factor of 10! This success reinforces the author's recommendations not to blindly accept joint failure at the weakest possible of failure modes, but to deliberately seek to enhance the strength using techniques that are simple to understand and equally simple to apply.

The mechanism for peel stress relief using external tapering, as in Fig. 3, is that the eccentricity in load path is thereby reduced locally and the flexibility of the tip of the splice plate increased, permitting more deflection and less resistance. In the case of the internal tapering shown in Fig. 4, the benefit is derived primarily from the added gauge length (of the adhesive layer) over which the peel stresses act, so that the flexibility is achieved by transversely stretching a low-modulus polymer instead of a high-modulus metal alloy.

---

[3] A Boeing Seattle colleague, Jon Gosse, has introduced the author to a strain-invariant procedure for analyzing combinations of shear and peel stresses in adhesive layers (as well as in the matrix of fibrous composite laminates) (see ref. [8]). Analyses made with these new techniques have confirmed how much stronger bonded structures can be if their geometry is modified to preclude the weakest of the potential failure mechanisms. In this context, the ability to assess all stress components simultaneously is a very useful *design* tool. What it must *not* be used for is to justify the *acceptance* of inferior designs merely because it is now possible to analyze their strength.

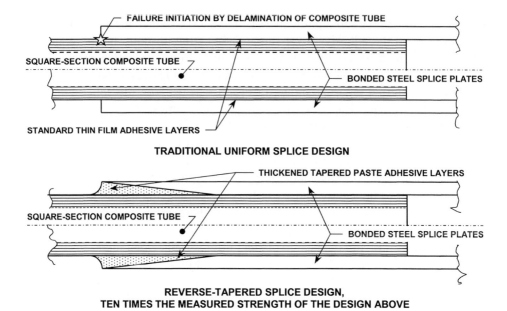

Fig. 4. NASA Goddard design of peel-stress relief for steel fittings bonded to composite tubes.

4. The origin of induced peel stresses

The only circumstances under which adhesive peel stresses are significant is when they prevent the adhesive from developing its intrinsic shear strength. It is for this reason that they are discussed before the shear stresses through which the loads are actually transferred.

Fig. 5 explains how the eccentricity in load path, between the bottom face of the upper splice plate and its centroid, creates a moment as shear loads are transferred that can be balanced only by normal (peel) stresses developed in the adhesive. Note that this effect exists even if there is no primary eccentricity in load path of the type found in single-lap and single-strap bonded joints, for which the corresponding phenomena are explained later. In both cases, peel stresses peak at the edge of the overlap and decay to negligibility away from any discontinuities, oscillating as they do to satisfy the requirement that there is no net vertical force on either splice plate.

Fig. 5 has been prepared for laminated composite adherends. In this case, the peel stresses will develop interlaminar tensile stresses in both the splice plate and the skin. These will cause failure in the manner shown unless the splice plates are tapered sufficiently to reduce the peel stresses to insignificance. In the case of metallic structures, it would be the adhesive layer that would fail under this mechanism. Obviously, these peel stresses will be more severe for thicker adherends and will be negligible for extremely thin members, which is why the

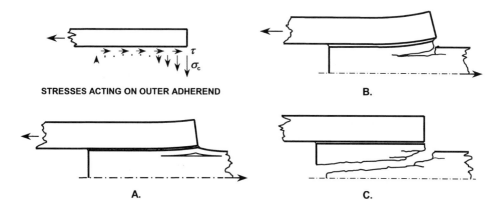

STRESSES ACTING ON OUTER ADHEREND

A, B, AND C INDICATE FAILURE SEQUENCE

Fig. 5. Generation of induced peel stresses in double-lap and double-strap bonded joints.

limits shown in Fig. 2 were developed. The shear and peel stresses are governed by different power laws in relation to adherend thickness, as explained in other chapters in these volumes. They are almost never simultaneously equally critical.

5. The origin of, and need for, *nonuniform* adhesive shear stresses

Despite the importance of designing out potential induced peel stresses in maximizing the strength of adhesively bonded joints, by far the *most* important factor is that the adhesive shear stress distribution is *naturally* highly non-uniform and that such joints would have extremely limited durability if this were *not* so. The variability comes about as the result of enforcing compatibility of deformations, as is explained in Fig. 6.

Fig. 6 indicates how the adherend stresses drop to zero at one or other end of the bonded overlap and that, as a consequence of this, there are differential movements between the adherends, across the bond line, that result in adhesive shear stresses, and strains, that peak at the ends and are reduced throughout the elastic trough in the interior. If the load is high enough, the adhesive will go 'plastic'[4] in the load transfer zones at the ends. These zones are shown by

---

[4] The adhesive does not *yield* in the classic sense of ductile metals. Instead, a series of fractures (hackles) at 45° to the bond surface develop as the adhesive is strained beyond its elastic limit, reducing the once-continuous adhesive layer to a series of discrete ligaments that are bent under what remains a shear load at the macro level. When the load is removed, the ligaments recover elastically, almost back to the original configuration, with virtually no offset, However, the cracks remain, the adhesive is permanently 'softened', and the density of the hackles increases as the

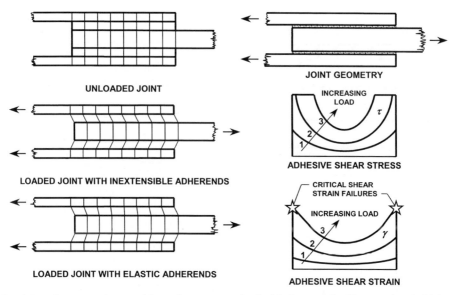

Fig. 6. Development of non-uniform shear stresses in double-lap and double-strap bonded joints.

closed-form analysis, in ref. [9], to have a constant width, independent of the total overlap. Most of the load is transferred there, with very little in the middle of the joint. However, the low-stressed interior is absolutely vital to the *durability* of adhesively bonded joint. The adhesive will creep, *locally*, at the ends of the overlap, at quite low sustained loads, because of the severity of the exponent in the typical adhesive shear stress and strain distributions. This was demonstrated experimentally during the PABST program (see [5]). However, it was found to be harmless, provided that the *remainder* of the adhesive, and the adherends, remained elastic. Then, when the load was removed, the adherends, and adhesive, at the ends of the overlap, were restored to their original positions. In other words, while creep *occurred*, it did not *accumulate*. The lightly stressed elastic trough acted as an anchor, or memory, whenever the total load that could be applied to the adhesive was limited by the strength of the adherends. Conversely, when the same sustained loads were applied to short-overlap test coupons, at typically a 0.5-inch overlap of 0.375–0.5-inch-thick plate lap instead of the 2 inches or so for real structural joints, the *entire* bonded overlap was subjected to creep — and there was no recovery mechanism available. Under these circumstances, in which the load capacity of the adherends could overpower the adhesive, complete failure

---

adhesive is strained closer to its ultimate failure point. While scientifically 'unrepresentative' of the actual mechanics of the adhesive failure process, the elastic–perfectly plastic model is an extremely useful mathematical technique with which to characterize, at the macro level, the nonlinear behavior of even brittle adhesives.

of the adhesive occurred in relatively few cycles. The structurally configured bonded joints, on the other hand, survived 4 years of 3-shifts-a-day 7-days-a-week in a hot/wet humidity chamber under cyclic loads almost sufficient to yield the aluminum. The minimum stress level was set at 1/10th of the maximum for these designs, and it is *this* requirement (for the hot/wet environment) that actually *sized* the overlaps — *not* any critical feature at the highly stressed ends of the overlap. The exact fraction 1/10 was an educated best guess, but it worked — and thereby explained why adhesive bonds worked so well in service even when previously misunderstood test coupons had predicted that they would not.

It is now clear that a scarf joint, designed to achieve as close as possible to a uniform adhesive shear stress, should be limited in application to lightly loaded situations like the repair of thin-skinned composite structures. Such joints would creep intolerably as a joint between strong, thick members. Scarf joints work admirably when the adherends are made from wood, or other weak materials, but they are far from optimum for modern high strength materials. The problem is that the slope must be so low that *all* of the adhesive is lightly stressed, not just some of it. Apart from that, even the smallest of finite tip thicknesses causes such a significant stress concentration in the adhesive that it is more reliable to analyze such joints as a stepped-lap joint with very many small steps and an accurate thickness for the thinnest steps, down to a single 0.005-inch thick layer of composite material.

6. Adverse effect of stiffness imbalance between adherends

The adhesive in Fig. 6 is equally critical, in shear, at both ends of the joint[5] because the two splice plates combined have the *same* extensional stiffness as the central skin. In the event that one adherend were stiffer than the other was, this maximum transfer of load would be diminished by a reduction in shear stress at the end of the overlap from which the stiffer adherend extended. This is explained in Fig. 7, which shows how the differential displacement across the bond layer is reduced by the stiffer adherend.

One should *always* strive for stiffness balance between the members being joined together, whether by adhesive bonding or mechanical fastening. (The designs in Fig. 2 deviated slightly from this goal but only because there was yet another failure mode, in the adherends rather than the adhesive, which, left

---

[5] There are no tensile normal stresses where the skin ends, so peel is less of a problem there than at the other end, but whatever problem there is should be taken care of by appropriate peel stress relief. The tapering does not affect the equality of load transfer at each end of stiffness-balanced joints; it merely redistributes it and reduces *both* the peak peel and shear stresses in the process, as explained in ref. [10].

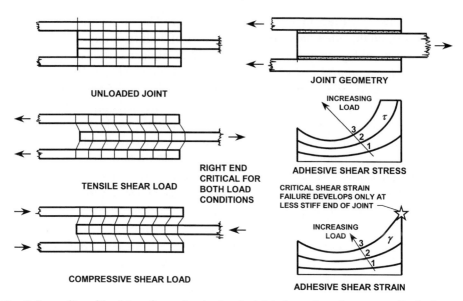

UNLOADED JOINT

JOINT GEOMETRY

TENSILE SHEAR LOAD

RIGHT END
CRITICAL FOR
BOTH LOAD
CONDITIONS

ADHESIVE SHEAR STRESS

CRITICAL SHEAR STRAIN
FAILURE DEVELOPS ONLY AT
LESS STIFF END OF JOINT

COMPRESSIVE SHEAR LOAD

ADHESIVE SHEAR STRAIN

Fig. 7. Inequality of load transfer, and reduction in joint shear strength, as a result of adherend stiffness imbalance.

unattended, would have limited the joint performance far more than a minimum stiffness imbalance that could be tolerated because the adhesive was not critical.)

Fig. 8 is a plot of the effect of adherend stiffness imbalance on the strength of long-overlap double-lap joints (no out-of-plane bending), assuming that no other variables influence the behavior. It is seen that the effect, which is characterized mathematically by a simple formula in refs. [9,7], is quite significant. In the absence of thermal mismatch, as between composite and metallic adherends, this equation for the relative adhesive shear strengths of long-overlap joints is simply

Adhesive shear strength of unbalanced joint
―――――――――――――――――――――――――――――――――
Adhesive shear strength of balanced joint

$$= \text{the lesser of} \quad \sqrt{\frac{1 + \left(\dfrac{E_1 t_1}{E_2 t_2}\right)}{2}} \quad \text{and} \quad \sqrt{\frac{1 + \left(\dfrac{E_2 t_2}{E_1 t_1}\right)}{2}}, \qquad (1)$$

in which Et is the extensional stiffness of each adherend, of thickness, t, and Young's modulus, E. (Different ratios would apply for the other failure modes.) The reduction in strength is the result of decreasing the load transferred at what becomes the less critical end. Fig. 8 is plotted in such a way as to indicate the loss of strength from splice plates that are too thin as well as too thick. (The adherends in Fig. 8 are identified as inner or outer, rather than by indices 1 and 2. Note also that these thicknesses refer to each adhesive layer. In earlier publications, the author and others have sometimes followed different conventions with the result

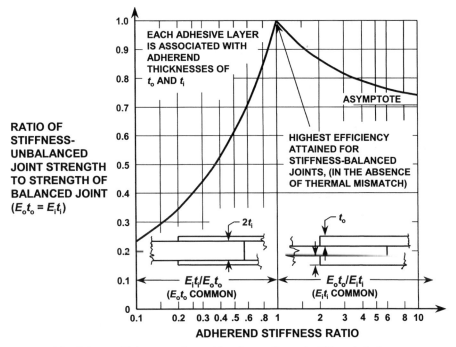

Fig. 8. Loss of joint strength resulting from adherend stiffness imbalance.

that, while each document is consistent, individual equations exhibit potentially confusing differences between documents because of the consideration of one or two layers of adhesive. From now on, the author will standardize on the consideration of one adhesive layer at a time, no matter whether there actually are one or two layers of adhesive in the joint. This should remove any ambiguities in the future.) The message from Fig. 8 is nevertheless clear; balanced joints are the strongest. [6]

7. Detrimental consequences of adherend *thermal* mismatch

It is always possible to design around potential adherend stiffness imbalances, by correcting the design to eliminate them, but there is no such cure available for the reductions in strength that arise when thermally dissimilar materials are

---

[6] The note in Fig. 8 about the absence of thermal mismatch is to draw attention to possible residual thermal effects that can change joint strengths with respect to tensile and compressive shear loads. In such cases, if there is a strong bias between the external loads, stronger joints for loads of one sign can be achieved by deliberate adherend stiffness imbalance, but only at the expense of reducing the strength of the joint to resist loads of the opposite sign even more.

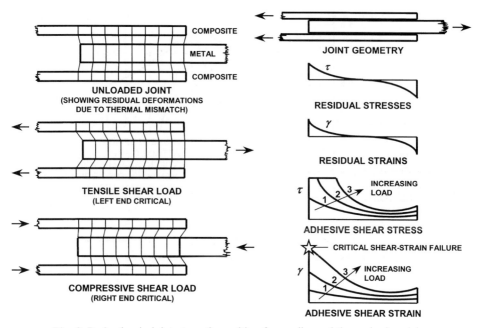

Fig. 9. Reduction in joint strength resulting from adherend thermal mismatch.

bonded together. This happens, from time to time, particularly when composite laminates are bonded to stepped titanium end plates in which bolted joints are located. This design is common for the tails of many current military fighter and attack aircraft and is used in an even more demanding situation at the wing root of the F/A-18 Hornet aircraft. The load intensity transmitted through that particular bond is almost 30,000 lbs./in., indicating that adhesively bonded structures need not be confined to their usual applications in lightly loaded structures.

The mechanism whereby adherend thermal mismatch causes a reduction in bonded joint strength is explained in Fig. 9, for a carbon–epoxy to titanium bonded joint typically cured at 250°F or 350°F and operated at as low a temperature as −67°F.

The titanium tries to shrink after bonding, but the composite laminate resists this because of its much lower coefficient of thermal expansion. The net result is to preload the adhesive in opposite directions at each end of the overlap. Consequently, when mechanical loads are applied, the residual thermal stresses in the adhesive will increase the tensile strength and decrease the compressive strength of the joint. It is, of course, possible to deliberately incorporate a determinable amount of adherend stiffness imbalance to take advantage of this phenomenon whenever the design tension and compression loads differ, but an ideal result will exist for only one temperature. This, of course, should be for the most critical temperature/load combination but it would also be necessary to also check that taking advantage of this approach did not aggravate the joint inefficiencies for the lesser loads at the other end of the operational envelope.

Some researchers have indicated that they believe that these residual thermal stresses will creep out of the structure, just as others believe that they will disappear from the resin in typical fiber/polymer composites. There are even test data on short-overlap bonded test coupons purporting to show that there is no such effect. The former is probably true, although the latter is definitely not, but there are also experiments to show that thermally unbalanced laminates between metal and composites or between composite of different lay-ups bow like bi-metallic strips, and that the stress-free temperature at which they flatten out is only slightly below the cure temperature of the adhesive. The effect is very real and it is easy to calculate, for long strips of such thermally unbalanced materials, just how much creep would be needed to relieve these thermal stresses. Only the end zones of the adhesive are appreciably stressed by this phenomenon, which is independent of length once a short transitional length has been exceeded. But eliminating these stresses by creep would require relative displacements that *are* proportional to the length of the bonded overlap.

8. Effect of overlap length on strength of bonded joints (with no eccentricities in load path)

Perhaps the most misunderstood characteristic of adhesively bonded joints is that increasing the overlap does *not* increase the joint strength of the bond, once quite a short transitional overlap has been exceeded. This is true for both uniformly thick adherends in simple joints and for stepped-lap joints, for which it is necessary to increase the number of steps if more joint strength is needed. (It is even true for very long single-lap joints with a primary eccentricity in load path causing out-of-plane bending, although there is a dependency of all three joint strengths (adherend bending, adhesive shear and adhesive peel) on bonded overlap for practical designs.) Fig. 10 explains the influence of bonded overlap on joint strength, showing how the initial increase in strength, in proportion to the overlap, for short-overlap test coupons is followed by a plateau that cannot be raised unless some property *other* than the overlap is altered. The load is actually transferred through narrow load-transfer zones at one or both ends of the joint, with a deep elastic trough in the interior of the joint. Whether the overlap is 5 inches or 5 miles, the joint strength will be the same. And so will the peak adhesive stresses and strains.

Given this characteristic, it is appropriate to explain how to choose the *most suitable* overlap, which is something very different from a mathematical optimum solution. The process is actually very straightforward. Minimizing total weight (not just that of the splice) will usually demand that all *unnecessary* overlap be

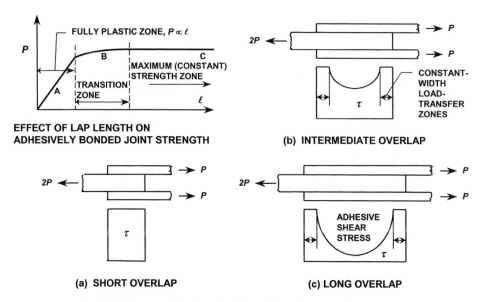

Fig. 10. Effect, or lack thereof, of bonded overlap on joint strength.

removed. This is easy to agree on; it is the definition of what is necessary that is critical. It was established during the PABST program, cited earlier, that the *minimum* adhesive shear stress needed to be low enough to prevent the occurrence of adhesive creep away from the ends of any bonded overlap, be it in a joint or as a large-area doubler. The procedure developed then, which is described in Fig. 2, was not only validated at the time, but has since become the basis for establishing minimum overlaps for bonded patches over cracked metal structures (see [11,12]). The length of the elastic trough, between the load transfer zones, should be $6/\lambda$, to ensure that the minimum stress will be no more than 10% of the maximum. Here, λ is the exponent of the exponential shear–stress curve in Fig. 11, given by

$$\lambda = \sqrt{\frac{G}{\eta}\left(\frac{1}{E_1 t_1} + \frac{1}{E_2 t_2}\right)}, \tag{2}$$

in which G is the adhesive shear modulus, η is the adhesive layer thickness, and Et is the extensional stiffness of each adherend, with the subscripts 1 and 2 discriminating between them. The characteristic length $1/\lambda$ has a physical meaning, too. It is the distance over which the total elastic strength of the joint could be transferred by adhesive stressed uniformly to the maximum adhesive shear strength τ_p. (This characteristic length has been referred to by the symbol $1/\beta$ in much of the literature on bonded crack-patching.)

There is no purpose in having any greater overlap, other than say 0.25 inch for assembly tolerances, so the process of establishing the most appropriate design

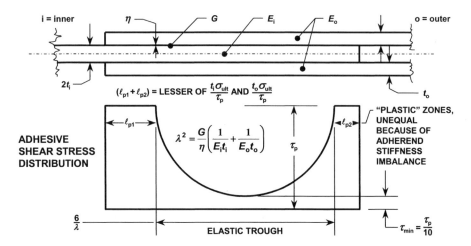

- PLA3TIC ZONES LONC ENOUGH TO TRANSFER *ENTIRE* ADHEREND STRENGTH
- ELASTIC TROUGH WIDE ENOUGH TO PREVENT CREEP AT ITS MIDDLE
- DESIGN OVERLAP = $\ell_{p1} + \ell_{p2} + (5/\lambda)$
- CHECK ON ADEQUATE STRENGTH BY NOT EXCEEDING MAXIMUM ADHESIVE SHEAR STRAIN

Fig. 11. Explanation of the establishment of the design overlap for bonded joints.

overlap is straightforward. First, 'plastic' load transfer zones are established at each end through which the *entire* strength of the (weaker) adherend can be transferred. This is independent of the nominal local stress intensity identified by structural analysis, other than that this nominal load will probably size the thickness, and strengths, of the adherends. This practice is to ensure that the adhesive layer cannot be transformed into a weak-link fuse as the result of redistribution of loads caused by even small damage or defects in the vicinity of the joint. This step has two incredible benefits. It means that bonded joints can be completely designed, or pre-designed, before the stress analysis is complete and, more importantly, that any belated stressing *cannot* possibly cause any change in design overlap. This procedure also makes it possible to design reliable repairs even when internal loads reports are unavailable. It does not matter whether the load transfer is identical at each end of the joint; the *combination* will always be the same, at an increment of overlap equal to

$$\Sigma \ell_p = \text{the lesser of } \frac{\sigma_{1\,\text{ult}} \times t_1}{\tau_p} \quad \text{or} \quad \frac{\sigma_{2\,\text{ult}} \times t_2}{\tau_p}. \tag{3}$$

To this must be added the elastic trough, of length

$$\ell_e = \frac{6}{\lambda}, \tag{4}$$

so that the total design overlap is

$$\text{Design Overlap} = \ell_p + \ell_e + \text{tolerance}. \tag{5}$$

Strictly, this formula is slightly conservative. If one were to reduce the load transferred through the 'plastic' end zones by the increment transferred through the elastic trough, which is easily shown to be (τ_p/λ), the design overlap would be reduced slightly to

$$\text{Design Overlap} = \ell_p + \frac{4}{\tau_p} + \text{tolerance}, \tag{6}$$

although the added length $(2/\lambda)$ may have served well as a surrogate from the usually omitted positioning tolerance.

Eqs. 2–6 have been formulated for one-sided bonded joints. For double-lap or double-strap joints, there needs to be some self-evident modifications. The easiest way to approach this is to say that t_2 refers to *one* of the two splice straps and that t_1 refers to *half* the inner adherend thickness, so that one is dealing with the load transferred through *one* bond layer.

Because these equations have not addressed adhesive *strains*, there is one further check to be made. This is for the potential shear strength of the adhesive bond, to ensure that the maximum possible adhesive shear strain $\gamma_{\max} = \gamma_e + \gamma_p$ is not exceeded, in which the elastic adhesive shear strain γ_e is related to the other adhesive properties by the standard stress–strain relation

$$\gamma_e = \frac{\tau_p}{G}. \tag{7}$$

The formula governing this bond strength limit P, per adhesive layer, for long-overlap joints, has four possible values whenever both adherend stiffness imbalance and thermal mismatch are present, with different tensile and compressive strengths. For tensile loads,

$$P_{\text{tens}} = \tau_{\text{avg}}\ell$$

$$= \text{the lesser of } \quad \frac{\tau_p}{\lambda}\left(\frac{1 + E_1 t_1/E_2 t_2}{2}\right)\sqrt{1 + 2\frac{\gamma_p}{\gamma_e}} \pm E_1 t_1 (\alpha_2 - \alpha_1)\Delta T$$

$$\text{and} \quad \frac{\tau_p}{\lambda}\left(\frac{1 + E_2 t_2/E_1 t_1}{2}\right)\sqrt{1 + 2\frac{\gamma_p}{\gamma_e}} \pm E_2 t_2 (\alpha_1 - \alpha_2)\Delta T. \tag{8}$$

For compressive loads, the strength is given by the least of the following four possibilities.

$$P_{\text{comp}} = \tau_{\text{avg}}\ell$$

$$= \text{the lesser of } \quad \frac{\tau_p}{\lambda}\left(\frac{1 + E_1 t_1/E_2 t_2}{2}\right)\sqrt{1 + 2\frac{\gamma_p}{\gamma_e}} \mp E_1 t_1 (\alpha_2 - \alpha_1)\Delta T$$

$$\text{and} \quad \frac{\tau_p}{\lambda}\left(\frac{1 + E_2 t_2/E_1 t_1}{2}\right)\sqrt{1 + 2\frac{\gamma_p}{\gamma_e}} \mp E_2 t_2 (\alpha_1 - \alpha_2)\Delta T. \tag{9}$$

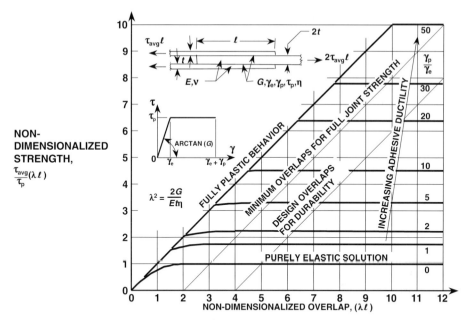

Fig. 12. Shear strength of balanced double-lap joints.

These formulae are notable for the absence of any involvement of the overlap ℓ. For short-overlap test coupons, of course, the corresponding formula

$$P = \tau_p \ell \tag{10}$$

does involve the overlap but imposes no corresponding limit on the adhesive strains, which is why creep can occur easily and the 'joint' can fail prematurely. Note that Eqs. 8–10 are formulated for *single*-lap joints (neglecting out-of-plane bending) and that the modifications described above would be needed for *double*-lap joints. The effect of the cut-offs created by Eqs. 8 and 9 is to impose an upper limit on the adherend thicknesses t_1 and t_2, beyond which it would be necessary to use stepped-lap joint designs to increase the adhesive joint strength to keep up with that of thicker adherends. This is why the entries in Fig. 2 stop at a central adherend thickness of 0.125 inch.

Fig. 12 depicts the special case of Eqs. 8 and 9 in which there is neither adherend stiffness imbalance nor thermal mismatch. It therefore indicates the strong effect of the toughening of typical structural adhesives, via the elastic–plastic model. Eq. 6 is plotted in Fig. 12, being expressed in nondimensionalized form by

$$(\lambda \ell) = \frac{\tau_{avg}}{\tau_p}(\lambda \ell) + 4 = \sqrt{1 + 2\frac{\gamma_p}{\gamma_e}} + 4. \tag{11}$$

The reasons why the overlap defined by Eq. 6 is to be preferred over any alternative estimates is that no greater length would result in any greater strength, while sufficiently shorter lengths will lead to bond failures by creep rupture. All of the above calculations are repeated for a range of temperatures, usually three, encompassing the service environments. It has been found that the longest (governing) overlaps are usually established for the hottest/wettest environment, while the limiting joint strengths, which impose the upper limit on adherend thicknesses, are established at the lowest service temperature.

9. Effect of operating environment on joint bond strength

Since the strength of structural bonded joints, as opposed to test coupons, is established by the *integral* of the stress distribution, rather than by any single directly measurable adhesive property like shear strength or modulus, it is appropriate to explain why, other than for the effects on the most appropriate bonded overlap discussed above, the service environment has surprisingly little effect on the strength of real bonded joints. The need for this explanation is the far greater variations in the lap-shear strengths commonly reported as if they were somehow related to the strengths of real bonded joints, an issue that is discussed later. The stress–strain curves measured on thick-adherend test coupons for a typical modified (toughened) epoxy adhesive are shown in Fig. 13. It is quite apparent that there are substantial differences between these curves, with the adhesive being far more brittle at low temperatures than at room temperature, and far softer at elevated temperatures.

Nevertheless, it is equally evident that the *areas* under these curves are *not*

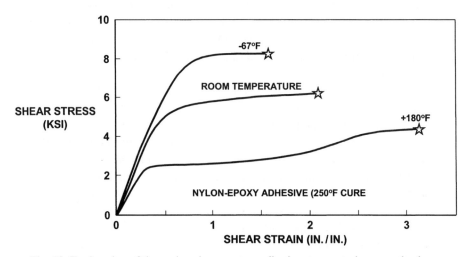

Fig. 13. Explanation of thermal environment on adhesive stress–strain curves in shear.

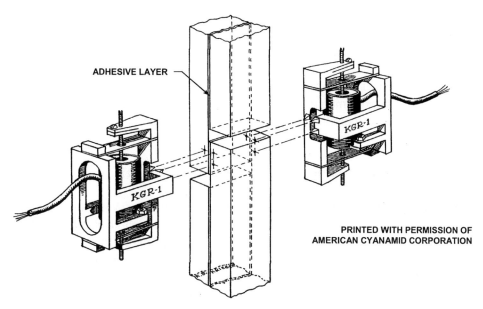

ADHESIVE LAYER

KGR-1

KGR-1

PRINTED WITH PERMISSION OF
AMERICAN CYANAMID CORPORATION

Fig. 14. Thick-adherend test coupon and Kreiger KGR-1 extensometer.

greatly different. This is why the environment has far less effect on the strength of well-designed bonded joints than is commonly understood. It was established in ref. [9] that, for long-overlap joints, the adhesive shear strength *of the joint* was limited by the square root of the adhesive strain-energy in shear and not by any individual property.

Fig. 14 shows the instrument with which these curves are obtained. The aluminum-to-aluminum specimen is quite thick, 0.375 inch or sometimes 0.5 inch thick, to create a close-to-uniform state of shear stress in the adhesive. Even so, precise analyses show significant nonuniformities. Nevertheless, the shear stresses are far more uniform than they are in typical adhesively bonded joints between uniformly thin adherends.

Ref. [13] contains further information about this instrument and test coupon. It should be noted that the amplification of the relative displacement signal is extremely great and that it has been found necessary to correct the initial modulus of the curve to compensate for the small distortions of the aluminum adherends themselves. This is accomplished by subjecting an equivalent one-piece notched coupon, with no adhesive layer, to the same loads as the bonded coupons, with the instruments mounted in the same locations. Although there were earlier such curves measured on napkin-ring test coupons, the coupons for that set-up were difficult to fabricate because the radial width of the bond layer was so narrow and the adhesive tended run out. In addition, it was difficult to ensure uniformity of bond thickness around the perimeter. It is for these reasons that the Krieger approach was the first to be applied widely in generating such curves. It is worthy of note that the former McDonnell Aircraft Company, now part of Boeing, in St. Louis, uses the complete stress–strain curve measured in the manner described above as part of its incoming acceptance criteria for the rolls of adhesive sent by *all* manufacturers. These stress–strain curves are ideally suited to the widely used elastic–plastic approximation of adhesive nonlinear behavior used as the basis of bonded joint strength prediction and design.

10. The elastic–plastic adhesive shear mathematical model

Actual adhesive shear stress–strain curves like those in Fig. 13 are simplified for analysis purposes. The most widely used model is the linearly elastic, perfectly plastic model developed by the author originally under contract to NASA Langley in the early 1970s. This model is described in Fig. 15 and is the basis of the A4E. series of computer codes, of which the A4EI code for stepped-lap joints and doublers covering variable adhesive properties and adhesive porosity and voids is perhaps the best known. This particular code was developed under contract to the USAF at the Wright Laboratories in Dayton (see [14]).

There are certain 'tricks of the trade' in using this model. They are easy to follow and implement, but it is vital that they be employed. The first is that an elastic–plastic model fitted to the ultimate failure point will seriously under-

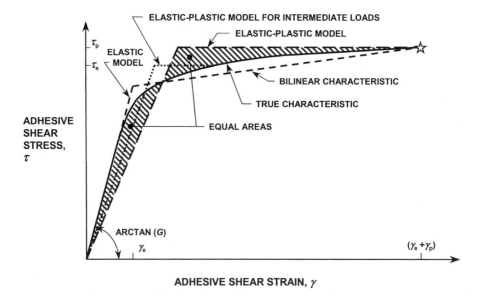

Fig. 15. Elastic–plastic and bi-linear adhesive shear models.

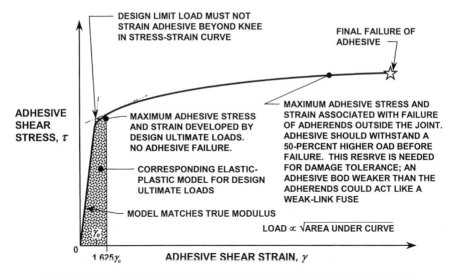

Fig. 16. Adhesive shear design model based on restricting design limit loads below the adhesive elastic capability.

estimate the initial shear modulus for the small loads that occur most of the time and over-estimate the elastic joint strength. It has been customary to create both a linearly elastic model, matching the real initial modulus and an ultimate strength model and carrying out both analyses with appropriate adhesive models. This could have been avoided if the more-complex bi-elastic two-straight-line model in Fig. 15 had been coded, but the equations are more complicated and the answers would be no more accurate, even though a single model could then closely represent all load levels. A more recent approach, based on a better understanding of the actual mechanical behavior of the adhesive is defined in Fig. 16. Here, the design limit load is set not to exceed the knee in the stress–strain curve.[7] The reason for this is that higher shear strains are associated with progressively greater

---

[7] This approach may seem to be conservative by some standards. Indeed, the practice for designing bonded composite repairs in Australia is to set a strain limit *twice* as high a this for frequently occurring fatigue loads, with the ultimate load not to exceed 80% of the ultimate adhesive shear strain. Because of the square root term in Eq. 11, this latter limit is close to tantamount to a further 1.5 factor between design ultimate and the adhesive ultimate capacity, for each environment. They follow this practice because, otherwise, the opportunities to apply such patches would be too limited. They validated this approach by tests, with no failures in service other than those caused by a combination of compressive applied loads and intense residual thermal stresses. On the other hand, there is less need to exceed the lower limit when designing original joints without constraints from existing structure. For example, the bonded fuselage splices for the PABST fuselage were

permanent damage of the adhesive layer. Then, based on the knowledge that the joint strength is proportional to the square root of the area under the stress–strain curve, the *equivalent* design ultimate load, higher by a standard factor of 1.5, is associated with an ultimate shear strain slightly less than twice that for limit load. The remainder of the shear deformation is reserved for one-time situations like maintaining a residual strength after local damage and for load redistribution around manufacturing defects. Under these circumstances, the simple elastic–plastic model will always suffice and only one model need be used for all load levels.

When first developed, most carbon–epoxy composite laminates were made from unidirectional tape. It was rapidly learned that it was best to have 0° fibers adjacent to the adhesive bond, parallel to the load direction, to develop the full strength of the adhesive. It immediately became apparent that it was very unwise to try and transmit load through a 90° ply adjacent to the bond. Stepped-lap joints, discussed later, were designed so as to avoid this known weakness. What was not anticipated at the time, but has become apparent since, is that the introduction of woven fabrics has created a situation whereby it is impossible to avoid typically 50% of the fibers adjacent to the bond being in the *wrong* direction for transmitting a shear load. This has introduced a further mode of failure not yet covered by the model whereby it is the composite laminate, which fails because of the shear transfer and not the adhesive. The A4EI code *should* be (and some day *will* be) modified to permit consideration of this additional mode of failure, as indicated in Fig. 17. In the interim, some users of the code have modified the adhesive properties to simulate interlaminar failures in the laminate by changing the adhesive properties that are input to the analysis. This technique works well for laminates made from cloth layers, but it needs to be remembered that this is unnecessary and unreasonably conservative for laminates made exclusively from tape, except for the mistake of a 90° ply adjacent to one adherend or other.

It should be noted that the preceding representations are merely *mathematical* models enabling the analyses to extend beyond the earlier elastic solutions. The actual adhesive bonds do *not* yield in the classical ductile metal sense. What actually happens is that the adhesive fails under the tensile component of the applied combination of shear and peel stresses, as explained by Gosse [8]. The failure mode for the peel-dominated case is a simple single fracture surface parallel to the adherends which, once started, will not arrest. Under dominant in-plane shear loads, however, the 'failure' mode is a series of hackles inclined at

---

not even strained beyond the knee in the stress–strain curve for even 1.33 times cabin pressure proof-pressure loads. The real message about both of these limits is that there *needs* to be such a limit for bonded joints to not wear out in service, because deviation from linear behavior of the adhesive *is* associated with irreversible damage, even if *some* of it can be tolerated.

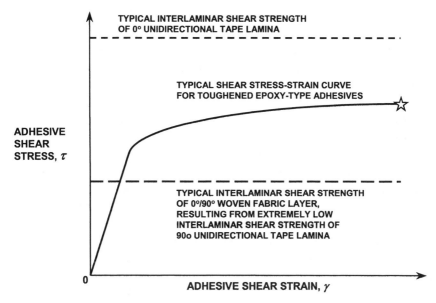

Fig. 17. Addition of adherend interlaminar failures to adhesive shear design model.

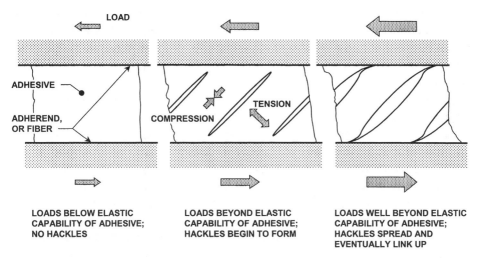

Fig. 18. Closely spaced hackles developed in thin adhesive bonds loaded to failure in shear.

roughly 45° to the adherend surfaces. These short cracks are effectively arrested as they approach the adherend surfaces, as shown in Fig. 18.

The density and individual length of the hackles increases with the application of further loads, but without such increases, the effect of the hackles is to reduce the once continuous adhesive layer to a series of discrete ligaments that are bent elastically. At the micro level, adhesive 'failure' occurs in the absence of significant nonlinear behavior. A joint in which the adhesive is loaded beyond the knee in the stress–strain curve will, on unloading, return nearly to the *original* condition, with none of the permanent offset associated with unloading ductile metal structures loaded beyond their yield point. The hackles cannot be restored, but will not spread rapidly until a higher load level is applied. [8] Bonded joints can still retain virtually infinite fatigue lives with the adhesive strained slightly beyond the end of its elastic capability, because the rest of the adhesive remains undamaged and, therefore, no creep can accumulate in the regions containing the cracks. This is why it makes sense to restrict the design limit loads to not exceed the adhesive elastic capability. Otherwise, the adhesive would be damaged repeatedly and we do not know how severe the consequent reduction in service life would be. Given that the use of stepped-lap joints (described later) instead of uniformly thick adherends is a straightforward technique with which to circumvent potential strength limits, this rational design philosophy is not all that restrictive in regard to the applications of adhesive bonding.

11. Effect of adherend material on adhesive 'lap-shear strength'

Even today there are many who still believe that it is possible to design (size) adhesively bonded joints via the overly simplistic formula that the joint strength is the product of the bonded area and some fictitious uniform bond 'shear allowable' measured on lap-shear coupons. To dispel this myth that joint strength would be doubled if the bonded area were, it is necessary only to show, by test, that the so-called 'allowable strength' varies with the thickness of the adherends and with the metals the adherends are made from. If the concept of a universal uniform allowable had any merit, it should at least be constant for common 1-inch wide half-inch overlaps. That this is not so is recorded in Fig. 19, in which the strengths of bonded single-lap joints made with successively stiffer adherends are recorded as a function of joint overlap. The stiffer and stronger the adherends are, the stronger the adhesive appears to become! Fig. 19 shows tests performed decades ago at the Picatinny Arsenal (see [15]). The results ought to have undermined the myth of a universal adhesive shear strength, regardless of joint geometry, but they did not.

During the PABST program, the supplier of one candidate adhesive tried to create the impression that their product was superior to that of their competitors by submitting durability tests performed on aluminum adherends twice as thick as

---

[8] Under *reversed* loads sufficiently intense to create hackles in *both* directions, this very long life is lost since a saw-tooth fracture surface is created. Restricting the peak strain in the adhesive is even more important then; it is not just a matter of locally reduced stiffness, it is a matter of strengths locally reduced to zero.

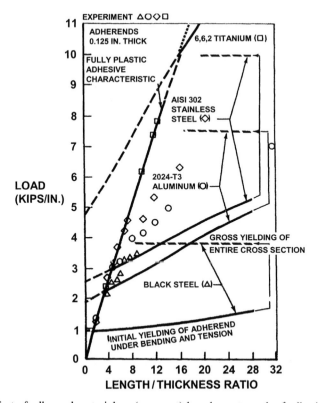

Fig. 19. Effect of adherend material on (apparent) lap-shear strength of adhesive bonds.

anyone else's (0.125 inch instead of 0.063 inch). Some such results are shown in Fig. 20, from a different series of tests, but the scheme backfired. There was no way to convert their test data to what they would have been if the adherends had been half as thick, so they were eliminated from consideration.

If the lap-shear strength were any indicator of the strength of real adhesively bonded joints, adhesives would surely be used far more than they are. What other method of transmitting load could, without *any* change in the fastening material, conveniently become stronger whenever the members being joined together were stronger? The simple truth is that the lap-shear strength is dangerously misleading for design purposes, no matter how useful it is as one half of a quality-control procedure for the production of bonded joints, as is explained in ref. [16].

12. Ultra-simplified analysis techniques for bonded structures

All standard bonded joints and doublers have one characteristic in common. This is that load transfer between the members is confined to narrow zones adjacent

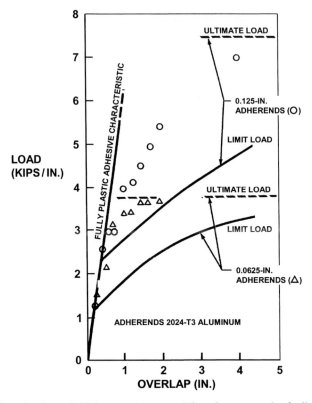

Fig. 20. Effect of adherend thickness on (apparent) lap-shear strength of adhesive bonds.

to the ends of the various members, with virtually no load transferred anywhere else, unless a very long gentle taper in adherend thickness is involved. Even with fully elastic adhesive behavior the load transfer is equivalent to a uniform shear stress τ_p developed over a short characteristic length $1/\lambda$, where τ_p is the peak adhesive stress defined in Fig. 15 and λ is the exponent of the shear stress distribution defined in Eq. 2. *Everywhere* else, it is reasonable to assume that the adhesive stress is zero. Such a model really simplifies the various compatibility-of-deformations analyses that define the widths of nonlinear load-transfer zones. It also means that, within quite a short distance from the edge of any adherend, the adhesive is unaware of the presence or absence of any remote load-transfer zones. The adhesive at the edge of a bonded doubler, for instance, is just as highly stressed as at the edge of a bonded joint made between the same adherends. This is explained in Fig. 21.

This same characteristic also explains why the local tapering of the ends of outer adherends to achieve relief from induced peel stresses is insensitive to the precise amount of tapering applied, provided that it is sufficient. In a stiffness-

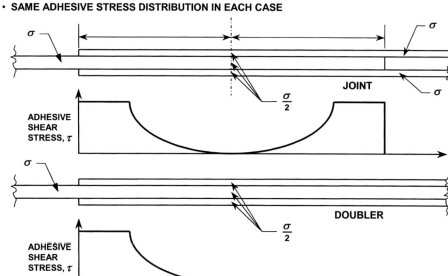

- **SAME ADHESIVE STRESS DISTRIBUTION IN EACH CASE**

- **SAME MAXIMUM ADHESIVE SHEAR STRAINS FOR SAME ADHERENDS AT SAME LOAD**

Fig. 21. Identity between adhesive shear stresses in bonded joints and bonded doublers.

balanced joint, for example, the integrals of the shear stresses at each end of a long-overlap joint *must* be equal, even if the precise distributions differ because only one end is tapered. This is explained in Fig. 22; the *untapered* internal end of the joint is always equally critical, at the same applied load level, no matter how much *less* critical the tapered end may be.

This simplified analysis method does not precisely satisfy compatibility of deformations. Nevertheless, its simplicity leads to close approximations in joints containing tapered adherends, for which few exact closed-form solutions exist (see [17]).

13. Stepped-lap joints to apply bonding to thicker structures

The nominally uniformly thick adherends discussed above are associated with a limited total shear strength of the adhesive layer that varies with the adherend thickness according to the simple relationship in Eq. 8. This potential strength is proportional to the square root of the adherend thicknesses and will eventually be overpowered by the adherend strengths that are directly proportional to their thicknesses. It is explained in a later section, on damage tolerance and two-dimensional load redistribution around local damage, that it is almost never acceptable to have the bond layer as the weak link. The simplest design modification to overcome

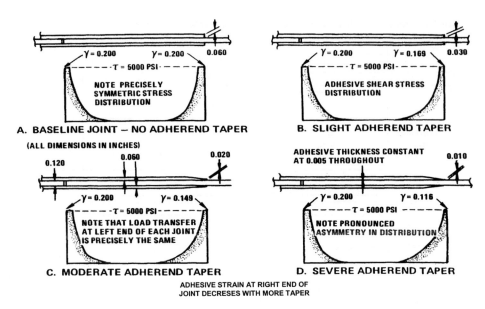

Fig. 22. Insensitivity of joint shear strength to degree of tapering to relieve peel-stresses.

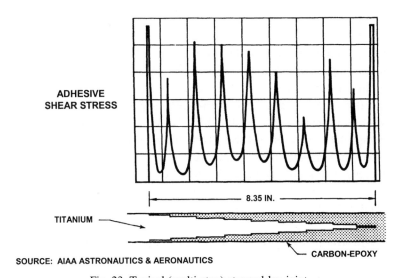

Fig. 23. Typical (multi-step) stepped-lap joint.

this constraint is the stepped-lap joint, an example of which is shown in Fig. 23. It is this kind of joint that secures the wing skins of the F/A-18 Hornet aircraft to the root fittings attaching them to the fuselage. Despite the high cost of manufacture, because each ply of composite material must be ended precisely at the end of the appropriate step to avoid internal skin wrinkles, this structurally efficient joint is

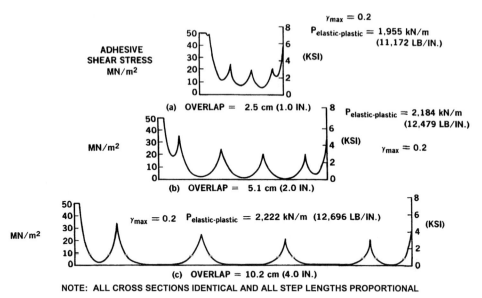

ADHESIVE SHEAR STRESS MN/m²

$\gamma_{max} = 0.2$

$P_{elastic-plastic} = 1,955$ kN/m (11,172 LB/IN.)

(KSI)

(a) OVERLAP = 2.5 cm (1.0 IN.)

MN/m²

$P_{elastic-plastic} = 2,184$ kN/m (12,479 LB/IN.)

(KSI)

$\gamma_{max} \approx 0.2$

(b) OVERLAP = 5.1 cm (2.0 IN.)

MN/m²

$\gamma_{max} = 0.2$ $P_{elastic-plastic} = 2,222$ kN/m (12,696 LB/IN.)

(KSI)

(c) OVERLAP = 10.2 cm (4.0 IN.)

NOTE: ALL CROSS SECTIONS IDENTICAL AND ALL STEP LENGTHS PROPORTIONAL

Fig. 24. Almost total independence of shear strength for stepped-lap joint on step lengths.

so compact that the seemingly less expensive alternative of a multi-row bolted joint would actually cost far more because of an associated increase in size of the much larger titanium forging needed. In addition, the bonded joint is far lighter and thinner, using less material than a joint full of holes and their associated stress concentrations would be.

Each step of this joint is governed by exactly the same differential equations as apply for bonded joints without steps, and the joint inherits the same characteristics, with an increment of load transferred at each end of each step, with little in between. In addition, they share the independence between joint strength and total overlap. This is shown in Fig. 24, in which a series of otherwise identical joints was analyzed to show how insensitive the strength was to overlap. The prime design variable for stepped-lap bonded joints is the *number of steps*, and *NOT* the total bond area.

Analysis with the A4EI computer code, however, has predicted that the strength of a stepped-lap joint would continue to increase with the number of steps, right up to the practical limit of one 0.005-inch thick ply per step.

Practical experience has produced a number of critical but simple design rules to avoid premature failures in stepped-lap bonded joints. These are illustrated in the following example of optimizing detail joint proportions by use of the A4EG computer code, an earlier version of the A4EI code in which the adhesive properties were held constant for the entire joint. Fig. 25 shows an original design, already reflecting considerable prior knowledge of the subject, and an improved configuration based on the insight gained by this level of analysis. (The almost

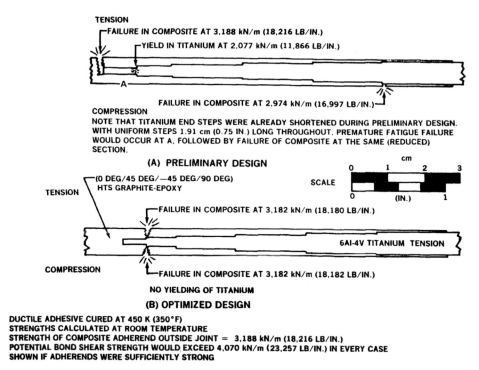

TENSION
FAILURE IN COMPOSITE AT 3,188 kN/m (18,216 LB/IN.)
YIELD IN TITANIUM AT 2,077 kN/m (11,866 LB/IN.)

Fig. 25. Original and improved designs of stepped-lap bonded joint.

perfect match between the second tensile and compressive joint strengths was a fluke and not a realistic expectation for optimization.)

Even though the first (thinnest) step at the end of the titanium plate was already far shorter than the other steps, the initial analysis indicated that it was likely that this step would be fatigued off at its root in service where the static analysis had indicated yielding. This has been observed in test coupons. Similarly, the most critical location for the carbon–epoxy composite material was predicted to be the corresponding location at the other end of the joint. These two steps were shortened in the next design and the critical locations moved elsewhere, accompanied by an increase in projected joint strength. The analysis includes both tensile and compressive mechanical loads because of the different contributions from residual thermal stresses. The titanium in the bonded overlap tries to shrink during the cool-down after cure. This induces adhesive shear strains of opposite signs at each end of the overlap. These compound with external tension at the left (composite) end of the joint and with external compression at the right (titanium) end of the joint, as explained earlier. But, in addition to that, the high shear stresses at the left end are associated with high residual tensile stresses in the thinnest titanium step as compatibility is enforced even *before* the mechanical loads are applied. This translates into a maximum permissible ℓ/t ratio for the end

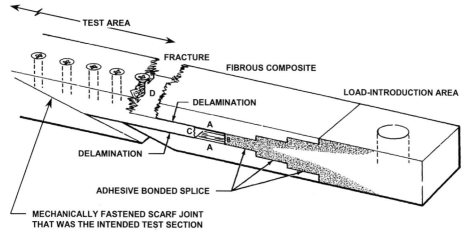

TEST AREA

FRACTURE

FIBROUS COMPOSITE

LOAD-INTRODUCTION AREA

DELAMINATION

D

A

C

B

A

DELAMINATION

ADHESIVE BONDED SPLICE

MECHANICALLY FASTENED SCARF JOINT
THAT WAS THE INTENDED TEST SECTION

INITIAL FAILURE AT "A" BECAUSE THICKNESS "B" IS EXCESSIVE AND
LOAD IN FIBERS "C" CANNOT BE UNLOADED THROUGH RESIN MATRIX

FINAL FAILURE AT "D" IS BY NET-SECTION TENSION ON THE TOP FACE
AND SHEAROUT (NOT SHOWN) ON THE LOWER SURFACE

Fig. 26. Delamination of boron–epoxy composite laminate caused by poor step-plate detail.

step. A typical dimension is 0.030 inch thick for an overlap of only 0.375 inch. The thicker the end step is made, the greater are the strength-decreasing wrinkles (or fibers terminated outside the joint), so increasing the titanium tip thickness is not a viable option. The most critical conditions, which size the overlap on the end step, are associated with the coldest operating environment, where the adhesive is strongest and stiffest, rather than with the hot/wet conditions that size the overlap for uniformly thick adherends. For a peak adhesive shear stress of 7000 psi in double shear and a titanium yield strength of about 135 ksi, the ℓ/t ratio for the end step should be about 10-to-1, which is consistent with the dimensions cited above.

Fig. 26 shows what happened when far less attention was paid to detail than is suggested above, presumably because the 'joint' was actually only a load-introduction tab at the end of a test coupon and not the actual test area that was believed to be far weaker.

The core of the laminate in Fig. 26 was an all-0° block of boron–epoxy some 0.1 inch thick. (This represents bad design practice even for laminates remote from splices. There should have been interspersed cross plies and angle pies.) The only load path available for transferring load out of the highly stressed filaments butting up against the excessively thick end step on the aluminum step-plate fitting is via a *single* layer of resin on each side, which is even less capable of accomplishing this task than a layer of adhesive on each side of the end step would be. Avoiding the unrestrained delaminations caused by this failure mechanism *requires* that the end step thickness be limited not to exceed the capabilities of the adhesive and laminating resin to transfer loads between the load carrying members, which are the fibers and the metal and *not* layers of mathematically homogenized composite material.

It is not necessary that *every* step in a stepped-lap bonded joint contain some adhesive so lightly stressed that it cannot creep. Indeed, as indicated above, this would be virtually impossible for the outermost steps. However, it *is* necessary that some of the internal steps be long enough to ensure the existence of some such very lightly stressed adhesive. The other design constraint, of minimizing the highest stress and strain in the adhesive at the ends of the overlap, is best addressed by an increase in the number of steps, taking care to avoid placing 90° fiber layers (or sides of woven fabrics) adjacent to steps in the titanium plate.

Stepped-lap joints are inherently free from the harmful effects of induced peel stresses at the ends of the bonded overlap because the end steps must inevitably be thin. The significance of this benefit can be assessed from a recent comparison between measured bonded joint strengths and the predictions of various theories [18]. An unrecognized misapplication of the A4EI code [9] to bonded joints between uniformly thick adherends has inadvertently quantified the loss of strength that can be caused by induced peel stresses when no attempt is made to alleviate them. The conclusion reached in ref. [18] was that the A4EI code over-estimated the joint strengths by as much as a factor 2. This discrepancy should properly be ascribed to the loss of strength caused by the peel stresses that had not been designed out of the uniformly thick test coupons and serves as a convincing indicator of the importance of designing to eliminate induced peel stresses.

Just as stepped-lap joints become necessary once the adherends have exceeded some determinable thickness, it follows equally that they are unnecessary for bonding thin adherends together. The primary underlying reason for this is that bonded joints cannot be scaled. Full-strength layers of adhesive are produced only in the range of about 0.005 to 0.010 inch thickness for heat-cured film adhesives. A secondary reason is that peaks in the shear and peel stresses are associated with different nondimensionalized elasto-geometric parameters. These phenomena are not as widely known as they ought to be. They are, in fact, the unrecognized explanation of the failure of attempts to reduce the cost of developmental test

---

[9] The appropriate code for this exercise is the little known A4EM code, for double-lap joints, which is a dimensionalized version of the nondimensionalized A4EB code developed for NASA Langley in ref. [9]. The better known A4EI code covers net-section failures in the adherends and shear failures in the adhesive and omitted consideration of induced peel stresses because they should inherently be insignificant in any reasonably well designed stepped-lap joint. The A4EM code covers the same variables, without any steps, but also predicts failures caused by adhesive peel or interlaminar tension because peel-stress relief in double-lap joints requires good design practice and is not inherent.

programs by validating concepts with sub-scale components. Everything *else* about a bonded joint *can* be scaled, but it is to no avail if the adhesive layer cannot also be scaled.

14. Effects of flaws, porosity, and variations in bond-line thickness

The bonded joints between typical thin structural elements have an extensive capability to tolerate the load redistribution that is caused by local flaws and porosity with no loss whatever in strength or durability. This derives from the same minimum overlaps needed to provide resistance to creep rupture that were discussed earlier. Figs. 26–29, taken from ref. [19], address this issue in the context of the longitudinal skin splices in the PABST forward fuselage, where the thickness was 0.050 inch of 2024-T3 aluminum alloy. Fig. 27 shows the adhesive stress distribution at room temperature for a load of 1000 lbs./in., which corresponds with a $1.3 \times P$ proof pressure load. Significantly, this load does not even exceed the elastic capability of the adhesive for this environment.

Fig. 28, also based on calculations with the A4EI computer code, shows the effect of a 0.5-inch-wide flaw 0.25 inch away from one edge of the 2.0-inch overlap. The peak shear stress, at the edge of the overlap, was not increased. The load that had would have been transferred through the flawed area is predicted to be transferred through narrow zones immediately adjacent to each side of the flaw.

Flaws like those in Fig. 28 should be recorded but left alone because it is not exposed to any source of water *unless* it is 'repaired' by drilling holes and

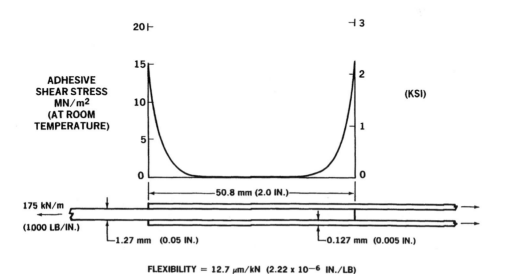

Fig. 27. Adhesive shear stresses in defect-free bonded fuselage double-lap splice.

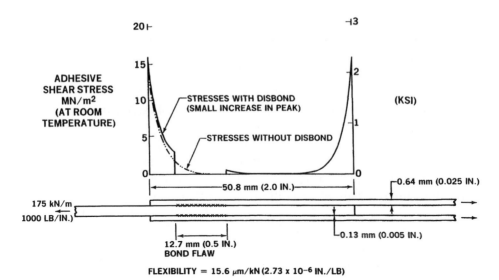

Fig. 28. Load redistribution due to a flaw in an adhesively bonded joint.

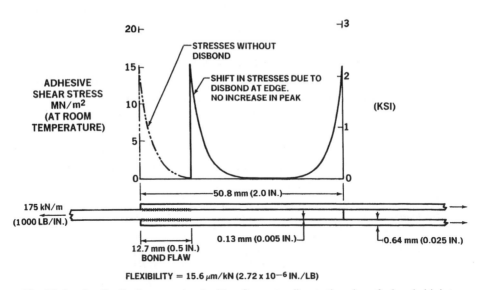

Fig. 29. Load redistribution associated with a flaw extending to the edge of a bonded joint.

injecting resin to fill the cavity so that the disbond can no longer be detected. Even if the lost load transfer in Fig. 28 could be restored, the joint strength could not be increased. Worse, trying to 'restore' the defect will inevitably break the environmental protection afforded by the anodized surface and primer. Such repairs should be looked upon as tantamount to sabotage; they cannot possibly do any good and will most likely do a lot of harm.

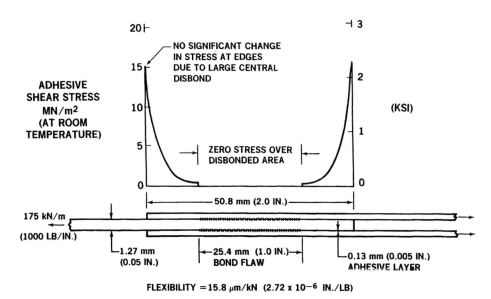

Fig. 30. Absence of effect of disbonds in the lightly loaded elastic trough in the adhesive.

Fig. 29 shows a flaw of the same size as in Fig. 28, but exposed to an edge. While there is no loss of short-term strength, because the load-transfer zone is simply moved inwards without changing the peak stress and strain, it *does need sealing* to protect it from what is known as the freeze/thaw cycle. Water that can migrate from the edge of the flaw to the interior would expand when it freezes at high altitude. This swelling would progressively increase the size of the damage, leading eventually to failure. One cannot rely on restoring the bond strength in the damaged area, so repairs by resin injection are inappropriate. What is needed is a tough (rubber-based) sealant to ensure that no moisture enters the joint.

If the bond defect were caused by a trapped air bubble in the middle of the overlap, as in Fig. 30, there would be no load transfer to redistribute because no load would have been transferred there, even if the bond were defect-free. In this case, the requirement would be that such defects were separated by sufficient undamaged bonding to still provide the necessary anchor to protect the joint against creep. Also, if the disbond were too large, one would need to check that the skin over the flaw would not buckle under in-plane compression loads.

For the PABST program, far more generous acceptance limits were established than are customary even today. Flaws up to 0.25 inch in diameter would be left unrepaired if they had occurred within 0.5 inches of an edge of the bonded members and voids up to 0.5 inch in diameter would be left alone in the interior. The manufacturing techniques were good enough that there were few flaws to assess but, actually, far larger flaws could have been tolerated structurally.

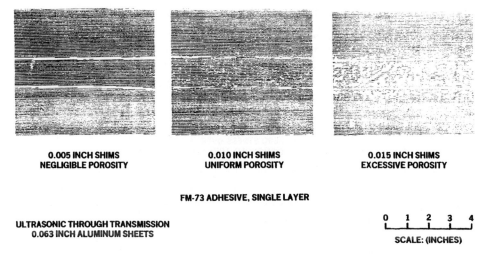

0.005 INCH SHIMS   **0.010 INCH SHIMS**   **0.015 INCH SHIMS**
NEGLIGIBLE POROSITY   **UNIFORM POROSITY**   **EXCESSIVE POROSITY**

FM-73 ADHESIVE, SINGLE LAYER

ULTRASONIC THROUGH TRANSMISSION
0.063 INCH ALUMINUM SHEETS

0   1   2   3   4

SCALE: (INCHES)

Fig. 31. Relationship between adhesive layer thicknesses and the formation of porosity.

However, to keep the costs low, it is necessary to *restrict* the size of tolerable flaws. The reason for this is that a bond flaw, once detected, must be evaluated, dispositioned and possibly repaired. The costs of doing this for each and every flaw is far greater than fixing the source of the flaws for once and for all. Unfortunately, with the current business pre-occupation with deferring costs, or transferring them to other departments, rather than truly minimizing their total, it is very difficult to implement long-term cost savings. However, it is worth recording that the Hagerstown plant of Fairchild Industries (bought by Rohr and now part of B.F. Goodrich) once did correct the surface geometry of the bonding tools for the wings of what was then the SF-340 commuter aircraft and found that the six consecutive defect-free panels they then made were the least expensive of all (see [20]).

Porosity is another class of detectable defects that has invited a more active response than have the far more serious global processing errors that cannot be detected after the fact. Ironically, if more attention had been paid to the origin of porosity, it would have been apparent that repairs would be futile. In a series of tests conducted in conjunction with the PABST program (see [21]), test panels were made in which the thickness of the cured adhesive layers was controlled by shims of varying thickness between the panels which were pushed together by platens instead of vacuum bags, to keep them flat. Fig. 31 shows ultrasonic scans of such panels, revealing that porosity would not occur for layers of normal thickness (0.005–0.010 inch). Porosity occurred only for *greater* thicknesses of about 0.010 to 0.015 inch. For still greater bond layer thickness, capillary action would cause the individual pores of porosity to coalesce into large totally disbonded areas.

What the relationship between bond thickness and the occurrence or absence of porosity means is that, other than for porosity caused by pre-bond moisture that brings many further problems in its wake, areas of porosity will be surrounded by thinner and therefore much *stiffer* areas of bond. In other words, the porous areas are *less* likely to fail than the surrounding defect-free bonds. Even if the thicker porous areas contained no voids, they would still contribute very little to the transfer of load because of the reduction in stiffness caused by the unplanned added thickness. The correct solution to this problem is to ensure that the bonding tool contours are correct and to *improve venting* during cure, so that there will be no further problems, and not to tolerate the porosity and develop repairs that will, at best, hide it from subsequent detection, and create a situation whereby such defects, once accepted, would establish a new lower acceptance level — with even more detectable defects to add to the costs. It should be noted that SAAB includes a grid of structurally insignificant vent holes in large-area doublers, to prevent air bubbles or volatiles being trapped by pinching off the edges of the overlap. Doing so not only improves the quality and reduces the cost of the bonds, it saves even more money by controlling total thicknesses so that bonded panels fit correctly to adjacent structures.

15. Two-dimensional load redistribution around flaws and damage

The reason why it is so important that bonded joints should never be designed to be weaker than the surrounding structure is that bonded joints with gross local defects can share some of the characteristics of through cracks in stressed skin. Just as the remote skin stress must be restricted more and more for longer cracks to not fast fracture, the nominal load transmitted through a bonded joint could need to be decreased for larger flaw sizes — unless the joint had been designed so that no defect could spread, no matter how large it already was. This is the characteristic that *can* be designed into bonded joints that is even more effective than discrete crack stoppers in stiffened structures. The governing phenomena are described in Fig. 32. If the adherends bonded together were stronger than the bond, the two-dimensional redistribution of the load around the initial damage or flaw would overload the adjacent as-yet-undamaged bond, spreading the weakness at an ever-increasing rate.

This process has nothing to do with the *nominal* applied loads, in the sense that there would always be a defect large enough to spread, no matter how low the skin stress was. The only way to prevent this potential failure from occurring is to design the bonded joint to be *stronger* than the surrounding structure. Any secondary failure would then have to be in the form of fatigue cracks induced in the skin at each end of the ineffective bond, for metallic structures, or as delaminations at the same location for fibrous composite adherends. Such

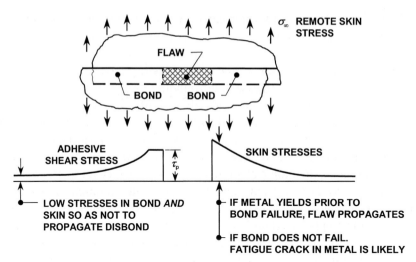

Fig. 32. Analogy between local bond flaws and cracks in metallic skins.

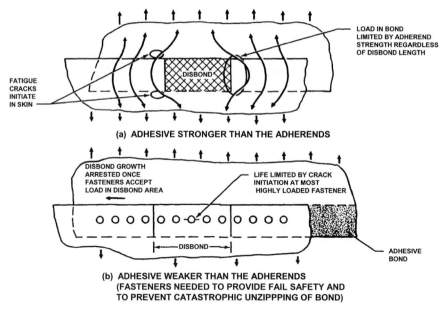

Fig. 33. Description of automatic disbond arrest whenever the bond is stronger than the surrounding adherends.

secondary failures are easier to detect than disbonds and would grow less rapidly. This restraint of further damage to the bond layer is described in Fig. 33.

The importance of not designing weak-link fuses into bonded structure cannot be overemphasized. There is an added benefit, of course with this design philosophy, the design of the bonded joints can proceed with confidence without waiting for any internal loads analysis.

16. Single-lap and single-strap (flush) joints with primary eccentricities in load path

The bonded joints discussed so far remain flat, at the global level, as the load is transferred from one adherend to another. This flatness usually requires a two-sided bond to achieve symmetry in the load paths. Individual face sheets bonded to honeycomb core are effectively so well stabilized as to free them from bending, too. However, the thinnest of structures require only a one-sided bond to transfer all of the load that is needed. In some cases, as with transverse (circumferential) splices around fuselage structures and with spanwise splices in wings, aerodynamic drag reduction requires that external bonded straps be avoided. It is sometimes more practical to make one-sided bonded joints, even though the inherent crack arrest capability described above is thereby usually lost through peel stresses induced at the ends of the overlaps. Whatever the justification for one-sided single-lap and single-strap bonded joints, there are so many of them that it is necessary to have a rational procedure to design them and that requires a physical understanding of their unique characteristics that are not shared by double-lap and double-strap joints, for example. One-sided stepped-lap joints are particularly efficient, provided that there is no thermal mismatch between adherends, because any eccentricity in load path is related to the depths of individual steps and not to the *overall* thickness of the members being joined.

The same general principles apply to all forms of one-sided bonded joints, simple and complex. The primary weakness, or limitation on strength, is now in the adherends, due to the combination of applied membrane stresses and bending stresses resulting from the eccentricity in load path. Local reductions in membrane stresses usually involve further increases in eccentricity in the load path, unless one is willing to employ stepped-lap designs to minimize them. So the primary design goal, and the only option available for uniformly thick adherends is to decrease the bending moments by making the overlap long enough to limit the effects of the eccentricity. The eccentricity for uniformly thick adherends between identical adherends is equal to the thickness of either member, so the satisfactory resolution of this design challenge will be in the form of an ℓ/t ratio. For the PABST program, the value adopted was 80-to-1 for aluminum skins bonded by toughened epoxy adhesives, although the adhesive would have little impact on this number. [10] This family of designs, regardless of applied stress level, may appear

---

[10] It may seem to some that this is equivalent to the 30-to-1 approximation characterizing the double-lap joints, but there is actually no relation. The sizes were set by adhesive characteristics for double-lap joints but by adherend bending for single-lap joints.

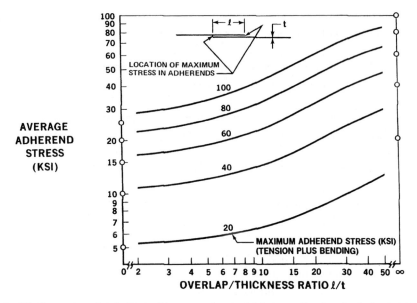

Fig. 34. Strong beneficial effect of large overlap-to-thickness ratios for single-lap joints.

to entail excessively heavy splices. However, this is definitely *not* the case. The problem is with the 'performance' metrics. The objective *should* be, but usually is not, to minimize the weight of the *total structure*. Far too often, a 'figure of merit' is established as the ratio of pounds of load transferred through pounds of splice, totally ignoring the fact that, for one-sided splices, the safe operating stress in all the *rest* of the structure is determined by the joint details and is *not* an independent issue. The splices typically weigh some 2–5% of the entire structural weight, so minimizing the weight of a 2% increment while increasing the weight of the other 98% is not really a smart thing to do. Fig. 34 shows, for aluminum adherends, how the combined bending and stretching stresses compare with the membrane stresses alone as a function of ℓ/t ratio, and how great the benefits are from using long-overlaps for one-sided joints.

The application of *tensile* loads to single-lap joints will *decrease* the eccentricity in load path below the nominal value. However, *compressive* loads will *increase* the eccentricity and accelerate failure by instability. One should therefore avoid transferring compressive loads through unstabilized one-sided joints. This is obvious, but one also needs to recognize the need *not* to use *linear* structural analyses when the physics of the situation *demands* the use of *nonlinear* analyses, even if it is not customary to do so. The classic single-lap bonded joint analyses by Goland and Reissner [22] identified the need for nonlinear analyses for this class of problems as long ago as 1944 but, even today, some large finite-element analyses are modified to 'simplify' them by *eliminating* eccentricities in load paths.

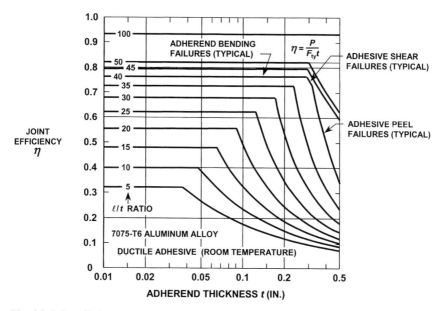

Fig. 35. Joint efficiency chart for bonded single-lap joints between identical adherends.

It is also possible for one-sided bonded joints to fail in the adhesive, by shear or by peel. However, unlike the situation with double-sided bonded joints, these strengths *are* influenced by overlap length, as first explained by Goland and Reissner [22], with both peak stresses reduced appreciably by longer overlaps, as explained elsewhere in these volumes and in refs. [23,24][11]. All three potential failure mechanisms are characterized in Fig. 35, showing how each prevails for certain combinations of design variables, and how each prevents the joint from achieving a strength in excess of that of the structure outside the joint. The joint efficiency defined in Fig. 34 is the ratio of the load at which the joint would fail and the membrane strength of the adherends outside the joint. Because of the eccentricity in load path it is not possible for this ratio to ever exceed unity.

When the adherends are identical, they deflect out of plane to minimize the total strain energy in the structure, including the joint. This means that the amount of deflection will be such as to align the deformed adherends as closely as possible with the line of the remotely applied loads, reducing the original static bending moment to barely one quarter of the nominal value if the overlap is long enough.

---

[11] Many authors, including this one, have published papers on the imprecision of the classic work by Goland and Reissner [22]. However, in ref. [24], the present author showed that their results are essentially numerically accurate, in spite of this, by correcting not only Goland and Reissner's use of imprecise boundary conditions at the ends of the overlap but also correcting the mistake he had made in his earlier attempt to correct this in ref. [23].

The analyses needed for this class of joints *MUST* be geometrically nonlinear; linear analyses are totally inappropriate, as explained in ref. [23]. For single-lap joints between identical adherends, the out-of-plane bending deflections will always reduce the actual bending moment at the ends of the bonded overlap *below* what a linear analysis would predict. However, when the adherends are different, the bending deformations are concentrated at the end of the joint from which the thinner adherend extends, and the peak bending moment can be far greater than the initial static eccentricity would indicate. In addition, just as with double-lap joints, adherend stiffness imbalance in single-lap joints greatly diminishes the adhesive joint strength by reducing the adhesive load transfer at the end from which the thicker adherend extends. It is far better to locally reduce the thickness of the thicker adherend to match that of the thinner adherend rather than merely overlapping them to join them together. This is equally true for riveted and bolted single-lap joints.

The simple design modification shown in Fig. 3 to severely reduce the harmful effects of induced peel stresses can be applied equally effectively to single-lap joints. It is even more important to do so than for double-lap joints because the peel stresses are inherently higher in one-sided bonded joints.

17. Bonded joints with directly applied peel loads

Despite the superior ability of adhesive to transfer load by shear rather than by peel, there are some structural configurations of bonded joints in which the primary applied load is one of normal tension, or peel. In a few instances rigid blocks are joined together with a close-to-uniform stress distribution. However in most cases, such as when a stiffener is bonded to a skin and both the skin and stiffener flange are thin, the peel-stress distribution is anything but uniform, peaking at the edge of the flanges or, in a badly designed joint with flanges that are too narrow, under the web of the stiffener. Fig. 36 identifies some of the major considerations pertaining to this class of bonded joints.

The actual peel-stress distribution is oscillatory, which is why the peak peel stress is orders of magnitude higher than the average peel stress. There are also certain characteristics unique to pressurized fuselages that were learned during the PABST program. For a circular fuselage, the skin can expand only a certain amount under cabin pressure. This restricts the maximum possible relative motion between skin and stringers and between skin and the frame outer flange. Without this constraint, these bonded joints would be far more likely to fail. It is important to include this feature in any tests for these strengths; experience showed that, otherwise, the joints would fail prematurely. Ref. [5] contains far more detailed discussion of these issues.

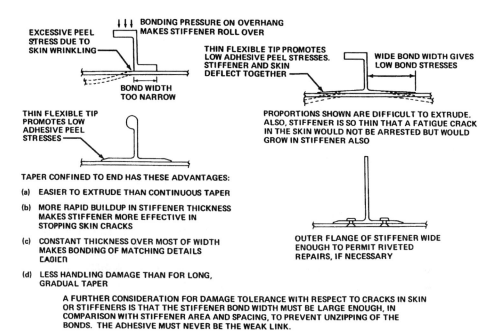

EXCESSIVE PEEL STRESS DUE TO SKIN WRINKLING

↓ ↓ ↓ BONDING PRESSURE ON OVERHANG MAKES STIFFENER ROLL OVER

THIN FLEXIBLE TIP PROMOTES LOW ADHESIVE PEEL STRESSES. STIFFENER AND SKIN DEFLECT TOGETHER

WIDE BOND WIDTH GIVES LOW BOND STRESSES

BOND WIDTH TOO NARROW

THIN FLEXIBLE TIP PROMOTES LOW ADHESIVE PEEL STRESSES

PROPORTIONS SHOWN ARE DIFFICULT TO EXTRUDE. ALSO, STIFFENER IS SO THIN THAT A FATIGUE CRACK IN THE SKIN WOULD NOT BE ARRESTED BUT WOULD GROW IN STIFFENER ALSO

TAPER CONFINED TO END HAS THESE ADVANTAGES:

(a) EASIER TO EXTRUDE THAN CONTINUOUS TAPER

(b) MORE RAPID BUILDUP IN STIFFENER THICKNESS MAKES STIFFENER MORE EFFECTIVE IN STOPPING SKIN CRACKS

(c) CONSTANT THICKNESS OVER MOST OF WIDTH MAKES BONDING OF MATCHING DETAILS EASIER

(d) LESS HANDLING DAMAGE THAN FOR LONG, GRADUAL TAPER

OUTER FLANGE OF STIFFENER WIDE ENOUGH TO PERMIT RIVETED REPAIRS, IF NECESSARY

A FURTHER CONSIDERATION FOR DAMAGE TOLERANCE WITH RESPECT TO CRACKS IN SKIN OR STIFFENERS IS THAT THE STIFFENER BOND WIDTH MUST BE LARGE ENOUGH, IN COMPARISON WITH STIFFENER AREA AND SPACING, TO PREVENT UNZIPPING OF THE BONDS. THE ADHESIVE MUST NEVER BE THE WEAK LINK.

Fig. 36. Design considerations for bonded joints subjected to primary peel loads.

18. Step-by-step design procedures

Having explained the phenomena governing the behavior of adhesively bonded joints, it is appropriate to conclude this chapter with an outline of the steps involved in the typical joint-design process.

The first step is the establishment of the *design* loads *for the joints*, which will usually differ from those in the adjacent structure. Once the nominal local structural loads have been established, it is necessary to decide on the appropriate thickness(es) of material. These are a function of the design operating stresses for the basic materials, and are influenced by considerations of fatigue and damage tolerance as well as the more customary static ultimate strength condition. Once the thickness(es) of the adherends has been established, the *joint design* loads are given as the product of each adherend thickness and its *ultimate* static strength(s) — in tension, compression, and shear. From this point on, what have been traditionally misunderstood as structural design loads can be disregarded. If they are not, the bonded joint will have no damage tolerance and can act as a weak-link fuse. Obviously, the joint-design load cannot exceed the weaker of the adherends and can be further limited by a common practice of reducing the skin thickness away from padded-up areas at the joints. However, it should be noted that this technique is usually confined to mechanically fastened splices. An issue of profound importance is that this design process leaves the joint designs

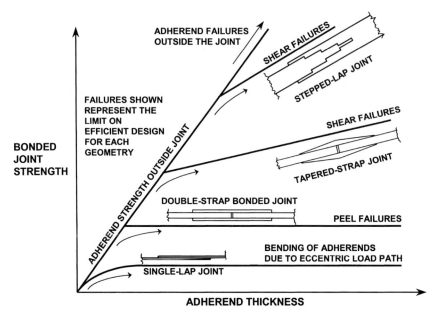

Fig. 37. Selection of bonded joint configuration as a function of adherend thickness.

unaltered when, as is customary, it is discovered that the original structural design loads were too low, at some time *after* the drawings had been released, making it extremely difficult to change the design overlap, which might require larger skin sizes, or possibly the thickness of the members. No revision of the joint analyses is necessary or even possible when this rational procedure is followed.

The second point to be established is the joint configuration — double-lap, single-lap, stepped-lap, or scarfed, as indicated in Fig. 37. The thicker the adherends, the more complicated the joint design needs to be and, conversely, the thinner the adherends, the simpler the joint configuration can be.

Let us start with the simplest joint to analyze, the double-lap joint that is free from primary out-of-plane bending effects. There are only two design variables to be established — the overlap, and whether or not the outer adherends (splice members) need to be tapered down in thickness at their ends. The bonded overlap is established in accordance with Eq. 6, as the sum of the plastic load-transfer zones, ℓ_p, through which the entire strength of the weaker adherend can be transferred, the elastic trough $5/\lambda$ to ensure a sufficiently low minimum adhesive shear stress to prevent the accumulation of creep, and some positioning tolerance for manufacturing, that experience shows may need to be 0.25 inch for large panels, regardless of the typical drawing tolerance of ±0.03 inch unless the relative location of all adjacent panels is controlled by precision-located holes needed for jigless assembly. No greater overlap can transfer any more load. (One must be very wary of the inappropriate, but common, design process whereby

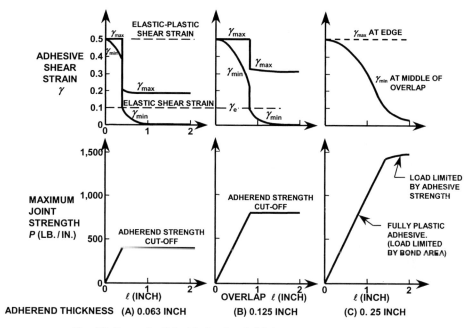

Fig. 38. Strength of double-lap bonded joints at room temperature.

a bonded joint strength is 'established' as the product of the total bond area and some fictitious uniform bond shear stress 'allowable'. This 'method' falsely creates the illusion that bonded joints can be made stronger by increasing their area. This is simply not so!) If the adherends are so thin that the load-transfer zone at the more critical end is computed to have a length less than $1/\lambda$, the adhesive will not be strained beyond its elastic capability. The reduced peak adhesive shear stress will then be $\tau_p \times$ the nondimensionalized load-transfer length (λd). The total design overlap can then be set conservatively at ($6/\lambda$), to save the complication of establishing the more precise length in excess of ($5/\lambda$). The design overlap needs to be established throughout a range of temperatures sufficient to encompass the operating environments. The highest service temperature customarily establishes the design overlap because it is then that the adhesive is weakest (lowest thick-adherend shear stress τ_p), maximizing the width of the plastic load-transfer zones, and softest, maximizing the width of the elastic trough through the reduced shear modulus G. The lowest service temperature usually establishes the limiting joint strength because that is when the adhesive is stiffest and the load-transfer zones the narrowest. This is evident in Figs. 38–40, in which double-lap shear strengths, and maximum and minimum adhesive shear strains, are plotted as a function of total overlap for a range of operating temperatures.

These three figures have been drawn to scale for aluminum adherends bonded together by typical 250°F-curing toughened epoxy adhesive. It is assumed that

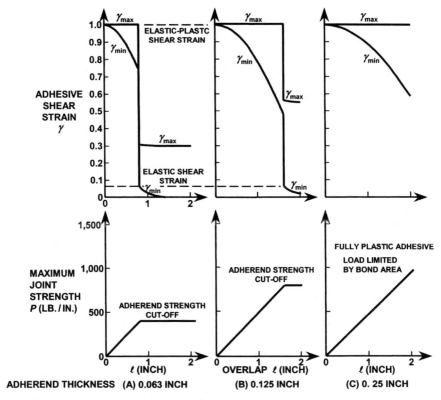

Fig. 39. Strength of double-lap bonded joints at maximum service temperature.

the appropriate amount of tapering has been applied at the ends of the outer adherends to prevent premature peel-induced failures. These three figures shown that 0.25-inch-thick central adherends cannot be bonded successfully by two 0.125-inch-thick splice plates, no matter what the service temperature, because there is absolutely no reserve strength to redistribute the loads around any local defect, no matter how small it may be. The peak adhesive shear strains are seen to be higher for 0.125-inch-thick central adherends than for 0.063-inch-thick central adherends, as one should expect, since the applied load can be doubled. The length of the plastic load-transfer zones is doubled, for the same reason. The most important message from these three figures is that, below the critical adherend thickness, there is an abrupt precipice at some overlap, beyond which, once exceeded, the adherend is simply not strong enough to load the adhesive to failure. This is why bonded joints between thin adherends last forever if the interfaces are environmentally stable. In this case, the critical central adherend thickness is 0.125 inch because, in that case, the precipice just disappears at the minimum operating temperature. The logical design process is to identify this

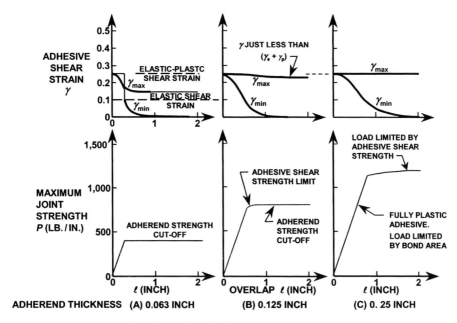

Fig. 40. Strength of double-lap bonded joints at minimum service temperature.

limiting central adherend thickness and to compute the overlap needed to transfer the adherend ultimate strength (not any nominal design load, as explained above), using Eq. 6 and to add an appropriate manufacturing/assembly tolerance. No greater overlap can ever add to the joint strength unless a more complex bonded joint geometry is adopted, as would be necessary for adherends thicker than the upper limit established by this method.

It is also necessary to protect against peel induced failures in the adhesive or, more likely, in composite adherends, by tapering the ends of the adherends as shown in Fig. 3, down to a tip thickness of 0.030 ± 0.01 inch for aluminum splice plates or 0.020 ± 0.010 inch for composite adherends and/or splice plates.

The critical design case for single-lap or single-strap bonded joints will not be found in the adhesive, unless no steps are taken to protect against induced peel stresses by tapering the ends of the bonded overlaps. Assuming that this has been done properly, the design overlap is established by consideration of the combination of membrane and bending stresses in the adherends. In the case of the PABST adhesively bonded aluminum fuselage cited earlier, it was found that the appropriate nondimensionalized design overlap was $\ell = 80t$, in conjunction with tapering of the adherend tips in accordance with Fig. 3. It must be noted that such a joint, in theory, has no tolerance for load redistribution around locally defective bonds. However, in the tests on this actual fuselage, it was found that the necessary hole size to cause sufficient load redistribution to create such a condition was so much larger than the metallic skins could tolerate that we never did succeed in making the bond critical. [12]

The successful design of stepped-lap bonded joints for thicker adherends depends, primarily on having sufficient steps and that the end steps not be too long. This is really a specialist task and the reader is referred to ref. [19] for discussions of the steps involved. Notwithstanding this, all of the *principles* involved have been identified in Fig. 25, here.

The literature developed by the English and European (particularly the Dutch) companies that made extensive use of adhesively bonded structures contain empirical design charts based on average bond stresses that varied with the parameter $\ell/\sqrt{t}$. The real nondimensionalized mathematical parameter should have been $\ell^2/(t\eta)$, but the adherend stiffness and *all* adhesive variables, including the thickness that was maintained by the process of manufacturing, were held constant for these procedures. These design tools were based on contemporary scientific understanding, but cannot be directly converted into design procedures for bonding fibrous composite components together the way the model presented here can. On the other hand, it should be noted that, more than 50 years ago, our colleagues on the other side of the Atlantic Ocean knew *not* to try and design in terms of fictitious *universal* adhesive shear strengths in the manner that so many students are taught today.

19. Concluding remarks

The design details presented in this chapter are intended to make it clear how easy it is to design adhesively bonded joints that are stronger than the surrounding structures, no matter how thick or highly loaded the structure may be. All this can be accomplished without ever introducing *unnecessary* complexity. It is important to understand that three potential failure modes are possible and that it is necessary to consciously design to prevent premature failures by the weakest mode, which is usually one of induced peel stresses. It is hoped that the explanation of how remarkably tolerant bonded joints are to local defects, provided that the basic joint has been designed to be stronger than the surrounding structure, will lead to a more rational policy on repairs to bonded structures damaged in service. Local,

---

[12] Actually, the full-size full-length bonded splices on the PABST test barrel were all of double lap-design. Every alternate splice was a riveted lap splice with adhesively bonded doublers to reduce the fastener bearing stresses below those needed to start fatigue cracks, so full-scale single-lap bonded joints were never tested on anything larger than a full-size in-plane-shear test panel. Nevertheless, all of even the bonded single-lap test coupons designed to develop the required fatigue life in the adherends never failed. Only those coupons with artificially short overlaps to *enforce* a premature failure in the bond ever did so.

rather than global, repairs to failures caused by improper surface preparation, are an exercise in futility, no matter what analysis/design process is employed. Far too many of today's procedures, that go beyond sealing the structure to prevent the ingress of moisture, are tantamount to sabotage in the sense that they do nothing to increase the residual strength of the structure and are likely only to ensure a decrease in the remaining life by breaking the environmental protection. It is hoped that the present explanations of what happens, and why, will lead to better design procedures in future.

Far more bonded aircraft structures will be made in the future than have been in the past when it is finally conceded that after-the-fact non-destructive inspections (NDI) to assess the strength [13] of adhesive bonds are as *unnecessary* as they have proved to be *impossible* to implement (see [26]). All successful bonds have been made reliably by controlling the *process*, without relying on inspecting each individual part. Conversely, every part that failed in service due to environmental attack failed because the processes were either *not* controlled or *were* controlled diligently to the *wrong* values, no matter how manner inspections were made subsequently to prove that the bonds had not *yet* fallen apart (see [1]). Without suitable surface preparations, no reliance can be placed on any of the analyses described in this article, or elsewhere. It is hoped that the methods of analysis described here will so enhance the understanding of the damage tolerance of properly designed bonded joints and of the necessarily non-uniform distribution of adhesive stresses that the *desire* to rely on end-item inspections in place of in-process control will eventually diminish and be forgotten, so that it can no longer hamper the application of this valuable technology.

References

1. Hart-Smith, L.J. and Rashi-Dian, D., On the futility of performing local repairs on adhesively bonded metallic structures failing globally from environmental attack, McDonnell Douglas Paper MDC 94K0100. Presented at 16th European SAMPE Conference, Salzburg,

---

[13] Ultrasonic NDI *can* be useful in identifying gaps between the adherends, which can identify misfitting details, discrepancies in tooling surfaces that need to be corrected, or a need to switch from Outer-Mold-Line (OML) bond tools to envelope bagging (see [20]) to improve fit, but it has *never* been able to establish strength or durability the way the wedge-crack test coupon (see [5]) has, and there is no indication that it ever can. Nevertheless, there *is* a way of verifying that bonded structures in service are *not* about to fall apart with a test that can be applied to actual structures years after they were built. It is described in ref. [25] and involves the bonding of small tabs to the structures at their time of manufacture and verifying that the tabs will not separate when a drop of water is applied while the tab is loaded in peel. The same tab would also indicate that the entire structure *is* in jeopardy long before there would be any other indication if the surface had *not* been prepared properly.

Austria, May 30–June 1, 1995. In: Lang, R.W. and Erath, M.A. (Eds.), *Proceedings of HIGH TECH IN SALZBURG — Creativity in Advanced Materials and Process Engineering*, 1995, pp. 357–368.

2. Hart-Smith, L.J., Redmond, G. and Davis, M.J., The curse of the nylon peel ply, McDonnell Douglas Paper MDC 95K0072. In: *Proceedings of the 41st International SAMPE Symposium and Exhibition*, Anaheim, CA, March 25–28, 1996, pp. 303–317.

3. Mahoney, C.L., Fundamental Factors Influencing the Performance of Structural Adhesives, *Internal Report*, Dexter Adhesives and Structural Materials Division, The Dexter Corporation, circa 1990.

4. Hart-Smith, L.J., Adhesive bonding of aircraft primary structures, Douglas Paper 6979. Presented to *SAE Aerospace Congress and Exhibition*, Los Angeles, CA, October 13–16, 1980; *SAE Trans. 801209*. Reprinted in: De Frayne, L. (Ed.), *High Performance Adhesive Bonding*, Society of Manufacturing Engineers, Dearborn, MI, 1983, pp. 99–113.

5. Thrall E.W., Jr., and Shannon, R.W. (Eds.), *Adhesive Bonding of Aluminum Alloys*. Marcel Dekker, New York, 1985.

6. *Bonded Aircraft Structures*, Bonded Structures Ltd., Duxford, Cambridge, England, 1957, Conference Proceedings.

7. Hart-Smith, L.J., Adhesive-bonded joints for composites — phenomenological considerations, Douglas Paper 6707. In: *Proceedings of Technology Conferences Associates Conference on Advanced Composites Technology*, El Segundo, California, March 14–16, 1978, pp. 163–180; reprinted as Designing adhesive bonds, in *Adhesives Age* **21**, 32–37 (1978).

8. Gosse, J.H., Strain invariant failure criteria for fiber reinforced polymeric composite materials. In: *Proceedings of the 13th ICCM Conference*, Beijing, China (2001), paper ID 1687.

9. Hart-Smith, L.J., Adhesive-bonded double-lap joints. NASA Langley Contract Report NASA CR-112235, January, 1973.

10. Hart-Smith, L.J., Joining of Composites, Douglas Paper 7820, published as Joints. In: C.A. Dostal (Ed.), *Engineered Materials Handbook, Vol. 1, Composites*, ASM International, Metals Park, OH, November 1987, pp. 479–495. Also published as: Joining, mechanical fastening. In: S.M. Lee (Ed.), *International Encyclopedia of Composites, Vol. 2*. VCH, New York, 1990, pp. 438–460. Second revised edition published in 2001.

11. Hart-Smith, L.J. and Davis, M.J., An object lesson in false economies — the consequences of *not* updating repair procedures for older adhesively bonded panels, with M.J. Davis, Royal Australian Air Force, McDonnell Douglas Paper MDC 95K0074. In: *Proceedings of the 41st International SAMPE Symposium and Exhibition*, Anaheim, CA, March 25–28, 1996, pp. 279–290.

12. Hart-Smith, L.J., *Recent Expansions in the Capabilities of Rose's Closed-Form Analyses for Bonded Crack Patching*, Boeing Paper MDC 00K0108, presented to 9th Australian International Aerospace Congress, Canberra, Australia, March 5–8, 2001.

13. Krieger R.B., Jr., Stress analysis concepts for adhesive bonding of aircraft primary structure. In: Johnson, W.S. (Ed.),*Adhesively Bonded Joints: Testing, Analysis and Design, ASTM STP 981*. American Society for Testing and Materials, Philadelphia, PA, 1988, pp. 264–275.

14. Hart-Smith, L.J., Design methodology for bonded-bolted composite joints, USAF Contract Report AFWAL-TR-81-3154, February 1982. Summarized in Bonded-Bolted Composite Joints, Douglas Paper 7398. Presented to *AIAA/ASME/ASCE/AHS 25th Structures, Structural Dynamics and Materials Conference*, Palm Springs, CA, May 14–16, 1984. Published in: *J. Aircraft*, **22**, 993–1000 (1985).

15. Hart-Smith, L.J., Differences between adhesive behavior in test coupons and structural joints, Douglas Paper 7066. Presented to *ASTM Adhesives Committee D-14 Meeting*, Phoenix, AZ, March 11–13, 1981.

16. Hart-Smith, L.J., The bonded lap-shear test coupon — useful for quality assurance, but dangerously misleading for design data, McDonnell Douglas Paper MDC 92K0922. In: *Proceedings of the 38th International SAMPE Symposium and Exhibition*, Anaheim, CA, May 10–13, 1993, pp. 239–246.

17. Hart-Smith, L.J., Simple approximate analyses of adhesive bonds in joints, doublers, and bonded patches, Boeing Paper, to be published under the auspices of the CRAS program.

18. Tsai, H.C., Alper, J. and Barrett, D., Failure analysis of composite bonded joints, AIAA-2000-1428.

19. Hart-Smith, L.J., Further developments in the design and analysis of adhesive-bonded structural joints, Douglas Paper 6922. Presented to *ASTM Symposium on Joining of Composite Materials*, Minneapolis, MN, April 16, 1980. In: K.T. Kedward (Ed.), *ASTM STP 749*, 1981, pp. 3–31.

20. Hart-Smith, L.J. and Strindberg, G., Developments in adhesively bonding the wings of the SAAB 340 and 2000 aircraft, McDonnell Douglas Paper MDC 94K0098. Presented to *2nd PICAST and 6th Australian Aeronautical Conference*, Melbourne, Australia, March 20–23, 1995. Abridged version in: Proceedings, Vol. 2, pp. 545–550. Full paper published in: *Proceedings of the Instution of Mechanical Engineers, Part G, Journal of Aerospace Engineering*, Vol. 211, 1997, pp. 133–156.

21. Hart-Smith, L.J., Adhesive layer thickness and porosity criteria for bonded joints, USAF Contract Report AFWAL-TR-82-4172, December 1982. See also: Effects of flaws and porosity on strength of adhesive-bonded joints, Douglas Paper 7388. In: *Proceedings of the 29th National SAMPE Symposium and Exhibition*, Reno, NV, April 3–5, 1984, pp. 840–852.

22. Goland, M. and Reissner, E., The stresses in cemented joints. *J. Appl. Mech.* **29** (*Trans. ASME*) **11**, A17–A27 (1944).

23. Hart-Smith, L.J., Adhesive-bonded single-lap joints, NASA Langley Contract Report NASA CR-112236, January, 1973.

24. Hart-Smith, L.J., The Goland and Reissner bonded lap joint analysis revisited yet again — but this time essentially validated, Boeing Paper MDC 00K0036, to be published under the auspices of the CRAS program.

25. Hart-Smith, L.J., Reliable nondestructive inspection of adhesively bonded metallic structures without using any instruments, McDonnell Douglas Paper MDC 94K0091. In: *Proceedings of the 40th International SAMPE Symposium and Exhibition*, Anaheim, CA, May 8–11, 1995, pp. 1124–1133.

26. Hart-Smith, L.J., How to get the best value for each dollar spent inspecting composite and bonded aircraft structures, McDonnell Douglas Paper MDC 92K0121. In: *Proceedings of the 38th International SAMPE Symposium and Exhibition*, Anaheim, CA, May 10–13, 1993, pp. 226–238.

Author Index

Subject Index